Repeated Games and Reputations

Repeated Games and Reputations

Long-Run Relationships

George J. Mailath and Larry Samuelson

UNIVERSITY PRESS
2006

OXFORD
UNIVERSITY PRESS

Oxford University Press, Inc., publishes works that further
Oxford University's objective of excellence
in research, scholarship, and education.

Oxford   New York
Auckland   Cape Town   Dar es Salaam   Hong Kong   Karachi
Kuala Lumpur   Madrid   Melbourne   Mexico City   Nairobi
New Delhi   Shanghai   Taipei   Toronto

With offices in
Argentina   Austria   Brazil   Chile   Czech Republic   France   Greece
Guatemala   Hungary   Italy   Japan   Poland   Portugal   Singapore
South Korea   Switzerland   Thailand   Turkey   Ukraine   Vietnam

Copyright © 2006 by Oxford University Press

Published by Oxford University Press, Inc.
198 Madison Avenue, New York, New York 10016

www.oup.com

Oxford is a registered trademark of Oxford University Press

All rights reserved. No part of this publication may be reproduced,
stored in a retrieval system, or transmitted, in any form or by any means,
electronic, mechanical, photocopying, recording, or otherwise,
without the prior permission of Oxford University Press.

Library of Congress Cataloging-in-Publication Data
Mailath, George Joseph.
Repeated games and reputations : long-run relationships / George J. Mailath, Larry Samuelson.
p. cm.
Includes bibliographical references and index.
ISBN-13 978-0-19-530079-6
ISBN 0-19-530079-3
1. Game theory.   2. Economics, Mathematical.   I. Samuelson, Larry, 1953–   II. Title.
HB144.M32 2006
519.3—dc22   2005049518

1 3 5 7 9 8 6 4 2

Printed in the United States of America
on acid-free paper

To Loretta

GJM

To Mark, Brian, and Samantha

LS

Acknowledgments

We thank the many colleagues who have encouraged us throughout this project. In particular, we thank Martin Cripps, Jeffrey Ely, Eduardo Faingold, Drew Fudenberg, Qingmin Liu, Stephen Morris, Georg Nöldeke, Ichiro Obara, Wojciech Olszewski, Andrew Postlewaite, Patrick Rey, Andrzej Skrzypacz, and Ennio Stacchetti for comments on various chapters; Roberto Pinheiro for exceptionally close proofreading; Bo Chen, Emin Dokumaci, Ratul Lahkar, and José Rodrigues-Neto for research assistance; and a large collection of anonymous referees for detailed and helpful reports. We thank Terry Vaughn and Lisa Stallings at Oxford University Press for their help throughout this project.

We also thank our coauthors, who have played an important role in shaping our thinking for so many years. Their direct and indirect contributions to this work are significant.

We thank the University of Copenhagen and University College London for their hospitality in the final stages of the book.

We thank the National Science Foundation for supporting our research that is described in various chapters.

Contents

1 **Introduction** 1
    1.1 Intertemporal Incentives 1
    1.2 The Prisoners' Dilemma 3
    1.3 Oligopoly 4
    1.4 The Prisoner's Dilemma under Imperfect Monitoring 5
    1.5 The Product-Choice Game 7
    1.6 Discussion 8
    1.7 A Reader's Guide 10
    1.8 The Scope of the Book 10

Part I Games with Perfect Monitoring

2 **The Basic Structure of Repeated Games with Perfect Monitoring** 15
    2.1 The Canonical Repeated Game 15
        2.1.1 *The Stage Game* 15
        2.1.2 *Public Correlation* 17
        2.1.3 *The Repeated Game* 19
        2.1.4 *Subgame-Perfect Equilibrium of the Repeated Game* 22
    2.2 The One-Shot Deviation Principle 24
    2.3 Automaton Representations of Strategy Profiles 29
    2.4 Credible Continuation Promises 32
    2.5 Generating Equilibria 37
        2.5.1 *Constructing Equilibria: Self-Generation* 37
        2.5.2 *Example: Mutual Effort* 40
        2.5.3 *Example: The Folk Theorem* 41
        2.5.4 *Example: Constructing Equilibria for Low δ* 44
        2.5.5 *Example: Failure of Monotonicity* 46
        2.5.6 *Example: Public Correlation* 49
    2.6 Constructing Equilibria: Simple Strategies and Penal Codes 51
        2.6.1 *Simple Strategies and Penal Codes* 51
        2.6.2 *Example: Oligopoly* 54
    2.7 Long-Lived and Short-Lived Players 61
        2.7.1 *Minmax Payoffs* 63
        2.7.2 *Constraints on Payoffs* 66

3 The Folk Theorem with Perfect Monitoring — 69

- 3.1 Examples — 70
- 3.2 Interpreting the Folk Theorem — 72
  - 3.2.1 *Implications* — 72
  - 3.2.2 *Patient Players* — 73
  - 3.2.3 *Patience and Incentives* — 75
  - 3.2.4 *Observable Mixtures* — 76
- 3.3 The Pure-Action Folk Theorem for Two Players — 76
- 3.4 The Folk Theorem with More than Two Players — 80
  - 3.4.1 *A Counterexample* — 80
  - 3.4.2 *Player-Specific Punishments* — 82
- 3.5 Non-Equivalent Utilities — 87
- 3.6 Long-Lived and Short-Lived Players — 91
- 3.7 Convexifying the Equilibrium Payoff Set Without Public Correlation — 96
- 3.8 Mixed-Action Individual Rationality — 101

4 How Long Is Forever? — 105

- 4.1 Is the Horizon Ever Infinite? — 105
- 4.2 Uncertain Horizons — 106
- 4.3 Declining Discount Factors — 107
- 4.4 Finitely Repeated Games — 112
- 4.5 Approximate Equilibria — 118
- 4.6 Renegotiation — 120
  - 4.6.1 *Finitely Repeated Games* — 122
  - 4.6.2 *Infinitely Repeated Games* — 134

5 Variations on the Game — 145

- 5.1 Random Matching — 145
  - 5.1.1 *Public Histories* — 146
  - 5.1.2 *Personal Histories* — 147
- 5.2 Relationships in Context — 152
  - 5.2.1 *A Frictionless Market* — 153
  - 5.2.2 *Future Benefits* — 154
  - 5.2.3 *Adverse Selection* — 155
  - 5.2.4 *Starting Small* — 158
- 5.3 Multimarket Interactions — 161
- 5.4 Repeated Extensive Forms — 162
  - 5.4.1 *Repeated Extensive-Form Games Have More Subgames* — 163
  - 5.4.2 *Player-Specific Punishments in Repeated Extensive-Form Games* — 165
  - 5.4.3 *Extensive-Form Games and Imperfect Monitoring* — 167
  - 5.4.4 *Extensive-Form Games and Weak Individual Rationality* — 168
  - 5.4.5 *Asynchronous Moves* — 169
  - 5.4.6 *Simple Strategies* — 172

	5.5	Dynamic Games: Introduction	174
		5.5.1 *The Game*	175
		5.5.2 *Markov Equilibrium*	177
		5.5.3 *Examples*	178
	5.6	Dynamic Games: Foundations	186
		5.6.1 *Consistent Partitions*	187
		5.6.2 *Coherent Consistency*	188
		5.6.3 *Markov Equilibrium*	190
	5.7	Dynamic Games: Equilibrium	192
		5.7.1 *The Structure of Equilibria*	192
		5.7.2 *A Folk Theorem*	195
6	**Applications**		**201**
	6.1	Price Wars	201
		6.1.1 *Independent Price Shocks*	201
		6.1.2 *Correlated Price Shocks*	203
	6.2	Time Consistency	204
		6.2.1 *The Stage Game*	204
		6.2.2 *Equilibrium, Commitment, and Time Consistency*	206
		6.2.3 *The Infinitely Repeated Game*	207
	6.3	Risk Sharing	208
		6.3.1 *The Economy*	209
		6.3.2 *Full Insurance Allocations*	210
		6.3.3 *Partial Insurance*	212
		6.3.4 *Consumption Dynamics*	213
		6.3.5 *Intertemporal Consumption Sensitivity*	219

Part II Games with (Imperfect) Public Monitoring

7 The Basic Structure of Repeated Games with Imperfect Public Monitoring 225

7.1	The Canonical Repeated Game	225
	7.1.1 *The Stage Game*	225
	7.1.2 *The Repeated Game*	226
	7.1.3 *Recovering a Recursive Structure: Public Strategies and Perfect Public Equilibria*	228
7.2	A Repeated Prisoners' Dilemma Example	232
	7.2.1 *Punishments Happen*	233
	7.2.2 *Forgiving Strategies*	235
	7.2.3 *Strongly Symmetric Behavior Implies Inefficiency*	239
7.3	Decomposability and Self-Generation	241
7.4	The Impact of Increased Precision	249
7.5	The Bang-Bang Result	251

	7.6	An Example with Short-Lived Players	255
		7.6.1 *Perfect Monitoring*	256
		7.6.2 *Imperfect Public Monitoring of the Long-Lived Player*	260
	7.7	The Repeated Prisoners' Dilemma Redux	264
		7.7.1 *Symmetric Inefficiency Revisited*	264
		7.7.2 *Enforcing a Mixed-Action Profile*	267
	7.8	Anonymous Players	269

8 Bounding Perfect Public Equilibrium Payoffs — 273

	8.1	Decomposing on Half-Spaces	273
	8.2	The Inefficiency of Strongly Symmetric Equilibria	278
	8.3	Short-Lived Players	280
		8.3.1 *The Upper Bound on Payoffs*	280
		8.3.2 *Binding Moral Hazard*	281
	8.4	The Prisoners' Dilemma	282
		8.4.1 *Bounds on Efficiency: Pure Actions*	282
		8.4.2 *Bounds on Efficiency: Mixed Actions*	284
		8.4.3 *A Characterization with Two Signals*	287
		8.4.4 *Efficiency with Three Signals*	289
		8.4.5 *Efficient Asymmetry*	291

9 The Folk Theorem with Imperfect Public Monitoring — 293

	9.1	Characterizing the Limit Set of PPE Payoffs	293
	9.2	The Rank Conditions and a Public Monitoring Folk Theorem	298
	9.3	Perfect Monitoring Characterizations	303
		9.3.1 *The Folk Theorem with Long-Lived Players*	303
		9.3.2 *Long-Lived and Short-Lived Players*	303
	9.4	Enforceability and Identifiability	305
	9.5	Games with a Product Structure	309
	9.6	Repeated Extensive-Form Games	311
	9.7	Games of Symmetric Incomplete Information	316
		9.7.1 *Equilibrium*	318
		9.7.2 *A Folk Theorem*	320
	9.8	Short Period Length	326

10 Private Strategies in Games with Imperfect Public Monitoring — 329

	10.1	Sequential Equilibrium	329
	10.2	A Reduced-Form Example	331
		10.2.1 *Pure Strategies*	331
		10.2.2 *Public Correlation*	332
		10.2.3 *Mixed Public Strategies*	332
		10.2.4 *Private Strategies*	333
	10.3	Two-Period Examples	334
		10.3.1 *Equilibrium Punishments Need Not Be Equilibria*	334
		10.3.2 *Payoffs by Correlation*	337
		10.3.3 *Inconsistent Beliefs*	338

10.4 An Infinitely Repeated Prisoner's Dilemma	340
10.4.1 *Public Transitions*	340
10.4.2 *An Infinitely Repeated Prisoners' Dilemma: Indifference*	343
11 Applications	**347**
11.1 Oligopoly with Imperfect Monitoring	347
11.1.1 *The Game*	347
11.1.2 *Optimal Collusion*	348
11.1.3 *Which News Is Bad News?*	350
11.1.4 *Imperfect Collusion*	352
11.2 Repeated Adverse Selection	354
11.2.1 *General Structure*	354
11.2.2 *An Oligopoly with Private Costs: The Game*	355
11.2.3 *A Uniform-Price Equilibrium*	356
11.2.4 *A Stationary-Outcome Separating Equilibrium*	357
11.2.5 *Efficiency*	359
11.2.6 *Nonstationary-Outcome Equilibria*	360
11.3 Risk Sharing	365
11.4 Principal-Agent Problems	370
11.4.1 *Hidden Actions*	370
11.4.2 *Incomplete Contracts: The Stage Game*	371
11.4.3 *Incomplete Contracts: The Repeated Game*	372
11.4.4 *Risk Aversion: The Stage Game*	374
11.4.5 *Risk Aversion: Review Strategies in the Repeated Game*	375

Part III Games with Private Monitoring

12 Private Monitoring	**385**
12.1 A Two-Period Example	385
12.1.1 *Almost Public Monitoring*	387
12.1.2 *Conditionally Independent Monitoring*	389
12.1.3 *Intertemporal Incentives from Second-Period Randomization*	392
12.2 Private Monitoring Games: Basic Structure	394
12.3 Almost Public Monitoring: Robustness in the Infinitely Repeated Prisoner's Dilemma	397
12.3.1 *The Forgiving Profile*	398
12.3.2 *Grim Trigger*	400
12.4 Independent Monitoring: A Belief-Based Equilibrium for the Infinitely Repeated Prisoner's Dilemma	404
12.5 A Belief-Free Example	410
13 Almost Public Monitoring Games	**415**
13.1 When Is Monitoring Almost Public?	415
13.2 Nearby Games with Almost Public Monitoring	418

13.2.1 *Payoffs*	418
13.2.2 *Continuation Values*	419
13.2.3 *Best Responses*	421
13.2.4 *Equilibrium*	421
13.3 Public Profiles with Bounded Recall	423
13.4 Failure of Coordination under Unbounded Recall	425
13.4.1 *Examples*	425
13.4.2 *Incentives to Deviate*	427
13.4.3 *Separating Profiles*	428
13.4.4 *Rich Monitoring*	432
13.4.5 *Coordination Failure*	434
13.5 Patient Players	434
13.5.1 *Patient Strictness*	435
13.5.2 *Equilibria in Nearby Games*	437
13.6 A Folk Theorem	441
14 Belief-Free Equilibria in Private Monitoring Games	**445**
14.1 Definition and Examples	445
14.1.1 *Repeated Prisoners' Dilemma with Perfect Monitoring*	447
14.1.2 *Repeated Prisoners' Dilemma with Private Monitoring*	451
14.2 Strong Self-Generation	453

Part IV Reputations

15 Reputations with Short-Lived Players	**459**
15.1 The Adverse Selection Approach to Reputations	459
15.2 Commitment Types	463
15.3 Perfect Monitoring Games	466
15.3.1 *Building a Reputation*	470
15.3.2 *The Reputation Bound*	474
15.3.3 *An Example: Time Consistency*	477
15.4 Imperfect Monitoring Games	478
15.4.1 *Stackelberg Payoffs*	480
15.4.2 *The Reputation Bound*	484
15.4.3 *Small Players with Idiosyncratic Signals*	492
15.5 Temporary Reputations	493
15.5.1 *Asymptotic Beliefs*	494
15.5.2 *Uniformly Disappearing Reputations*	496
15.5.3 *Asymptotic Equilibrium Play*	497
15.6 Temporary Reputations: The Proof of Proposition 15.5.1	500
15.6.1 *Player 2's Posterior Beliefs*	500
15.6.2 *Player 2's Beliefs about Her Future Behavior*	502

	15.6.3 *Player 1's Beliefs about Player 2's Future Behavior*	503
	15.6.4 *Proof of Proposition 15.5.1*	509

16 Reputations with Long-Lived Players — 511

- 16.1 The Basic Issue — 511
- 16.2 Perfect Monitoring and Minmax-Action Reputations — 515
  - 16.2.1 *Minmax-Action Types and Conflicting Interests* — 515
  - 16.2.2 *Examples* — 518
  - 16.2.3 *Two-Sided Incomplete Information* — 520
- 16.3 Weaker Reputations for Any Action — 521
- 16.4 Imperfect Public Monitoring — 524
- 16.5 Commitment Types Who Punish — 531
- 16.6 Equal Discount Factors — 533
  - 16.6.1 *Example 1: Common Interests* — 534
  - 16.6.2 *Example 2: Conflicting Interests* — 537
  - 16.6.3 *Example 3: Strictly Dominant Action Games* — 540
  - 16.6.4 *Example 4: Strictly Conflicting Interests* — 541
  - 16.6.5 *Bounded Recall* — 544
  - 16.6.6 *Reputations and Bargaining* — 546
- 16.7 Temporary Reputations — 547

17 Finitely Repeated Games — 549

- 17.1 The Chain Store Game — 550
- 17.2 The Prisoners' Dilemma — 554
- 17.3 The Product-Choice Game — 560
  - 17.3.1 *The Last Period* — 562
  - 17.3.2 *The First Period, Player 1* — 562
  - 17.3.3 *The First Period, Player 2* — 565

18 Modeling Reputations — 567

- 18.1 An Alternative Model of Reputations — 568
  - 18.1.1 *Modeling Reputations* — 568
  - 18.1.2 *The Market* — 570
  - 18.1.3 *Reputation with Replacements* — 573
  - 18.1.4 *How Different Is It?* — 576
- 18.2 The Role of Replacements — 576
- 18.3 Good Types and Bad Types — 580
  - 18.3.1 *Bad Types* — 580
  - 18.3.2 *Good Types* — 581
- 18.4 Reputations with Common Consumers — 584
  - 18.4.1 *Belief-Free Equilibria with Idiosyncratic Consumers* — 585
  - 18.4.2 *Common Consumers* — 586
  - 18.4.3 *Reputations* — 587
  - 18.4.4 *Replacements* — 588
  - 18.4.5 *Continuity at the Boundary and Markov Equilibria* — 590
  - 18.4.6 *Competitive Markets* — 594

18.5	Discrete Choices	596
18.6	Lost Consumers	599
	18.6.1 *The Purchase Game*	599
	18.6.2 *Bad Reputations: The Stage Game*	600
	18.6.3 *The Repeated Game*	601
	18.6.4 *Incomplete Information*	603
	18.6.5 *Good Firms*	607
	18.6.6 *Captive Consumers*	608
18.7	Markets for Reputations	610
	18.7.1 *Reputations Have Value*	610
	18.7.2 *Buying Reputations*	613

Bibliography 619

Symbols 629

Index 631

Repeated Games and Reputations

1 Introduction

1.1 Intertemporal Incentives

In Puccini's opera *Gianni Schicchi*, the deceased Buoso Donati has left his estate to a monastery, much to the consternation of his family.[1] Before anyone outside the family learns of the death, Donati's relatives engage the services of the actor Gianni Schicchi, who is to impersonate Buoso Donati as living but near death, to write a new will leaving the fortune to the family, and then die. Anxious that Schicchi do nothing to risk exposing the plot, the family explains that there are severe penalties for tampering with a will and that any misstep puts Schicchi at risk. All goes well until Schicchi (acting as Buoso Donati) writes the new will, at which point he instructs that the entire estate be left to the great actor Gianni Schicchi. The dumbstruck relatives watch in horror, afraid to object lest their plot be exposed and they pay the penalties with which they had threatened Schicchi.

Ron Luciano, who worked in professional baseball as an umpire, occasionally did not feel well enough to umpire. In his memoir, Luciano writes,[2]

> Over a period of time I learned to trust certain catchers so much that I actually let them umpire for me on bad days. The bad days usually followed the good nights.... On those days there wasn't much I could do but take two aspirins and call as little as possible. If someone I trusted was catching... I'd tell them, "Look, it's a bad day. You'd better take it for me. If it's a strike, hold your glove in place for an extra second. If it's a ball, throw it right back. And please, don't yell."... No one I worked with ever took advantage of the situation.

In each case, the prospect for opportunistic behavior arises. Gianni Schicchi sees a chance to grab a fortune and does so. Any of Luciano's catchers could have tipped the game in their favor by making the appropriate calls, secure in the knowledge that Luciano would not expose them for doing his job, but none did so. What is the

---

1. Our description of *Gianni Schicchi* is taken from Hamermesh (2004, p. 164) who uses it to illustrate the incentives that arise in isolated interactions.
2. The use of this passage (originally from Luciano and Fisher 1982, p. 166) as an illustration of the importance of repeated interactions is due to Axelrod (1984, p. 178), who quotes and discusses it.

	2		
	A	B	C
A	5, 5	0, 0	12, 0
B	0, 0	2, 2	0, 0
C	0, 12	0, 0	10, 10

(Row labels belong to player 1.)

Figure 1.1.1 A modified coordination game. Pure-strategy Nash equilibria include *AA* and *BB* but not *CC*.

difference between the two situations? Schicchi anticipates no further dealings with the family of Buoso Donati. In the language of game theory, theirs is a one-shot game. Luciano's catchers know there is a good chance they will again play games umpired by Luciano and that opportunistic behavior may have adverse future consequences, even if currently unexposed. Theirs is a repeated game.

These two stories illustrate the basic principle that motivates interest in repeated games: *Repeated interactions give rise to incentives that differ fundamentally from those of isolated interactions.* As a simple illustration, consider the game in figure 1.1.1. This game has two strict Nash equilibria, *AA* and *BB*. When this game is played once, players can do no better than to play *AA* for a payoff of 5. If the game is played twice, with payoffs summed over the two periods, there is an equilibrium with a higher average payoff. The key is to use first-period play to coordinate equilibrium play in the second period. The players choose *CC* in the first period and *AA* in the second. Any other first-period outcome leads to *BB* in the second period. Should one player attempt to exploit the other by playing *A* in the first period, he gains 2 in the first period but loses 3 in the second. The deviation is unprofitable, and so we have an equilibrium with a total payoff of 15 to each player.

We see these differences between repeated and isolated interactions throughout our daily lives. Suppose, on taking your car for a routine oil change, you are told that an engine problem has been discovered and requires an immediate and costly repair. Would your confidence that this diagnosis is accurate depend on whether you regularly do business with the service provider or whether you are passing through on vacation? Would you be more willing to buy a watch in a jewelry store than on the street corner? Would you be more or less inclined to monitor the quality of work done by a provider who is going out of business after doing your job?

Repeated games are the primary tool for understanding such situations. This preliminary chapter presents four examples illustrating the issues that arise in repeated games.

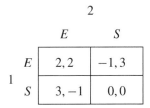

Figure 1.2.1 The prisoners' dilemma.

1.2 The Prisoners' Dilemma

The prisoners' dilemma is perhaps the best known and most studied (and most abused) of games.[3] We will frequently use the example given in figure 1.2.1.

We interpret the prisoners' dilemma as a partnership game in which each player can either exert effort (E) or shirk (S).[4] Shirking is strictly dominant, while higher payoffs for both players are achieved if both exert effort. In an isolated interaction, there is no escape from this "dilemma." We must either change the game so that it is not a prisoners' dilemma or must accept the inefficient outcome. Any argument in defense of effort must ultimately be an argument that either the numbers in the matrix *do not* represent the players preferences, or that some other aspect of the game is not really descriptive of the actual interaction.

Things change considerably if the game is repeated. Suppose the game is played in periods $0, 1, 2, \ldots$. The players make their choices simultaneously in each period, and then observe that period's outcome before proceeding to the next. Let a_i^t and a_j^t be the actions (exert effort or shirk) chosen by players i and j in period t and let u_i be player i's utility function, as given in figure 1.2.1. Player i maximizes the *normalized discounted sum of payoffs*

$$(1-\delta) \sum_{t=0}^{\infty} \delta^t u_i(a_i^t, a_j^t),$$

where $\delta \in [0, 1)$ is the common discount factor. Suppose that the strategy of each player is to exert effort in the first period and to continue to do so in every subsequent period as long as both players have previously exerted effort, while shirking in all other circumstances. Suppose finally that period τ has been reached, and no one has

---

3. Binmore (1994, chapter 3) discusses the many (unsuccessful) attempts to extract cooperation from the prisoners' dilemma.
4. Traditionally, the dominant action is called *defect* and the dominated action *cooperate*. In our view, the words *cooperate* and *defect* are too useful to restrict their usage to actions in a single (even if important) game.

yet shirked. What should player i do? One possibility is to continue to exert effort, and in fact to do so forever. Given the strategy of the other player, this yields a (normalized, discounted) payoff of 2. The only other candidate for an optimal strategy is to shirk in period τ (if one is ever going to shirk, one might as well do it now), after which one can do no better than to shirk in all subsequent periods (since player j will do so), for a payoff of $(1-\delta)[3+\sum_{t=\tau+1}^{\infty} \delta^{t-\tau} 0] = 3(1-\delta)$. Continued effort is optimal if $2 \geq 3(1-\delta)$, or

$$\delta \geq \tfrac{1}{3}.$$

By making future play depend on current actions, we can thus alter the incentives that shape these current actions, in this case allowing equilibria in which the players exert effort. For these new incentives to have an effect on behavior, the players must be sufficiently patient, and hence future payoffs sufficiently important.

This is not the only equilibrium of the infinitely repeated prisoners' dilemma. It is also an equilibrium for each player to relentlessly shirk, regardless of past play. Indeed, for any payoffs that are feasible in the stage game and strictly individually rational (i.e., give each player a positive payoff), there is an equilibrium of the repeated game producing those payoffs, if the players are sufficiently patient. This is an example of the *folk theorem* for repeated games.

1.3 Oligopoly

We can recast this result in an economic context. Consider the Cournot oligopoly model. There are n firms, denoted by $i = 1, \ldots, n$, who costlessly produce an identical product. The firms simultaneously choose quantities of output, with firm i's output denoted by $a_i \in \mathbb{R}_+$ and with the market price then given by $1 - \sum_{i=1}^{n} a_i$. Given these choices, the profits of firm j are given by $u_j(a_1, \ldots, a_n) = a_j(1 - \sum_{i=1}^{n} a_i)$.

For any number of firms n, this game has a unique Nash equilibrium, which is symmetric and calls for each firm i to produce output a^N and earn profits $u_i(a^N, \ldots, a^N)$, where[5]

$$a^N = \frac{1}{n+1},$$

and $$u_i(a^N, \ldots, a^N) = \left(\frac{1}{n+1}\right)^2.$$

When $n = 1$, this is a monopoly market. As the number of firms n grows arbitrarily large, the equilibrium outcome approaches that of a competitive market, with zero price and total quantity equal to one, while the consumer surplus increases and the welfare loss prompted by imperfect competition decreases.

---

5. Firm j's first-order condition for profit maximization is $1 - 2a_j - \sum_{i \neq j} a_i = 0$ or $a_j = 1 - A$, where A is the total quantity produced in the market. It is then immediate that the equilibrium must be symmetric, giving a first-order condition of $1 - (n+1)a^N = 0$.

The analysis of imperfectly competitive markets has advanced far beyond this simple model. However, the intuition remains that less concentrated markets are more competitive and yield higher consumer welfare, providing the organizing theme for many discussions of merger and antitrust policy.

It may instead be reasonable to view the interaction between a handful of firms as a repeated game.[6] The welfare effects of market concentration and the forces behind these effects are now much less clear. Much like the case of the prisoners' dilemma, consider strategies in which each firm produces $1/2n$ as long as every firm has done so in every previous period, and otherwise produces output $1/(n+1)$. The former output allows the firms to jointly reproduce the monopoly outcome in this market, splitting the profits equally, whereas the latter is the Nash equilibrium of the stage game. As long as the discount factor is sufficiently high, these strategies are a subgame-perfect equilibrium.[7] Total monopoly profits are $1/4$ and the profits of each firm are given by $1/4n$ in each period of this equilibrium. The increased payoffs available to a firm who cheats on the implicit agreement to produce the monopoly output are overwhelmed by the future losses involved in switching to the stage-game equilibrium. If these "collusive" strategies are the equilibrium realized in the repeated game, then reductions in the number of firms may have no effect on consumer welfare at all. No longer can we regard less concentrated markets as more competitive.

1.4 The Prisoners' Dilemma under Imperfect Monitoring

Our first two examples have been games of *perfect monitoring*, in the sense that the players can observe each others' actions. Consider again the prisoners' dilemma of section 1.2, but now suppose a player can observe the outcome of the joint venture but cannot observe whether his partner has exerted effort. In addition, the outcome is either a success or a failure and is a random function of the partners' actions. A success appears with probability p if both partners exert effort, with probability q if one exerts effort and one shirks, and probability r if both shirk, where $p > q > r > 0$.

This is now a game of *imperfect public monitoring*. It is clear that the strategies presented in section 1.2 for sustaining effort as an equilibrium outcome in the repeated game—exert effort in the absence of any shirking, and shirk otherwise—will no longer

---

6. To evaluate whether the stage game or the repeated game is a more likely candidate for usefully examining a market, one might reasonably begin with questions about the qualitative nature of behavior in that market. When firm i sets its current quantity or price, does it consider the effect that this quantity and price might have on the future behavior of its rivals? For example, does an airline wonder whether a fare cut will prompt similar cuts on the part of its competitors? Does an auto manufacturer ask whether rebates and financial incentives will prompt similar initiatives on the part of other auto manufacturers? If so, then a repeated game is the obvious tool for modelling the interaction.

7. After some algebra, the condition is that

$$\frac{1}{16}\left(\frac{n+1}{n}\right)^2 \leq \frac{1}{1-\delta}\left(\frac{1}{4n} - \delta\frac{1}{(n+1)^2}\right).$$

work, because players cannot tell when someone has shirked. However, all is not lost. Suppose the players begin by exerting effort and do so as long as the venture is successful, switching to permanent shirking as soon as a failure is observed. For sufficiently patient players, this is an equilibrium (section 7.2.1 derives the necessary condition $\delta \geq 1/[3p - 2q]$). The equilibrium embodies a rather bleak future, in the sense that a failure will eventually occur and the players will shirk thereafter, but supports at least some effort.

The difficulty here is that the "punishment" supporting the incentives to exert effort, consisting of permanent shirking after the first failure, is often more severe than necessary. This was no problem in the perfect-monitoring case, where the punishment was safely off the equilibrium path and hence need never be carried out. Here, the imperfect monitoring ensures that punishments will occur. The players would thus prefer the punishments be as lenient as possible, consistent with creating the appropriate incentives for exerting effort. Chapter 7 explains how equilibria can be constructed with less severe punishments.

Imperfect monitoring fundamentally changes the nature of the equilibrium. If nontrivial intertemporal incentives are to be created, then over the course of equilibrium play the players will find themselves in a punishment phase infinitely often. This happens despite the fact that the players know, when the punishment is triggered by a failure, that both have in fact followed the equilibrium prescription of exerting effort. Then why do they carry through the punishment? Given that the other players are entering the punishment phase, it is a best response to do likewise. But why would equilibria arise that routinely punish players for offenses not committed? Because the expected payoffs in such equilibria can be higher than those produced by simply playing a Nash equilibrium of the stage game.

Given the inevitability of punishments, one might suspect that the set of feasible outcomes in games of imperfect monitoring is rather limited. In particular, it appears as if the inevitability of some periods in which players shirk makes efficient outcomes impossible. However, chapter 9 establishes conditions under which we again have a folk theorem result. The key is to work with asymmetric punishments, sliding along the frontier of efficient payoffs so as to reward some players as others are penalized.[8] There may thus be a premium on asymmetric strategies, despite the lack of any asymmetry in the game.

The players in this example at least have the advantage that the information they receive is *public*. Either both observe a success or both a failure. This ensures that they can coordinate their future behavior, as a function of current outcomes, so as to create the appropriate current incentives. Chapters 12–14 consider the case of *private monitoring*, in which the players potentially receive different private information about what has transpired. It now appears as if the ability to coordinate future behavior in response to current events has evaporated completely, and with it the ability to support any outcome other than persistent shirking. Perhaps surprisingly, there is still considerable latitude for equilibria featuring effort.

---

8. This requires that the imperfect monitoring give players (noisy) indications not only that a deviation from equilibrium play has occurred but also who might have been the deviator. The two-signal example in this section fails this condition.

1.5 ■ The Product-Choice Game   7

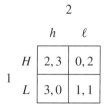

Figure 1.5.1 The product-choice game.

1.5 The Product-Choice Game

Consider the game shown in figure 1.5.1. Think of player 1 as a firm who can exert either high effort (H) or low effort (L) in the production of its output. Player 2 is a consumer who can buy either a high-priced product, h, or a low-priced product, ℓ. For example, we might think of player 1 as a restaurant whose menu features both elegant dinners and hamburgers, or as a surgeon who treats respiratory problems with either heart surgery or folk remedies.

Player 2 prefers the high-priced product if the firm has exerted high effort, but prefers the low-priced product if the firm has not. One might prefer a fine dinner or heart surgery from a chef or doctor who exerts high effort, while preferring fast food or an ineffective but unobtrusive treatment from a shirker. The firm prefers that consumers purchase the high-priced product and is willing to commit to high effort to induce that choice by the consumer. In a simultaneous move game, however, the firm cannot observably choose effort before the consumer chooses the product. Because high effort is costly, the firm prefers low effort, no matter the choice of the consumer.

The stage game has a unique Nash equilibrium, in which the firm exerts low effort and the consumer purchases the low-priced product. Suppose the game is played infinitely often, with perfect monitoring. In doing so, we will often interpret player 2 not as a single player but as a succession of short-lived players, each of whom plays the game only once. We assume that each new short-lived player can observe signals about the firm's previous choices.[9] As long as the firm is sufficiently patient, there is an equilibrium in the repeated game in which the firm exerts high effort and consumers purchase the high-priced product. The firm is deterred from taking the immediate payoff boost accompanying low effort by the prospect of future consumers then purchasing the low-priced product.[10] Again, however, there are other equilibria, including one in which low effort is exerted and the low priced product purchased in every period.

---

9. If this is not the case, we have a large collection of effectively unrelated single-shot games that happen to have a common player on one side, rather than a repeated game.
10. Purchasing the high-priced product is a best response for the consumer to high effort, so that no incentive issues arise concerning player 2's behavior. When consumers are short-lived, there is no prospect for using future play to alter current incentives.

Now suppose that consumers are not entirely certain of the characteristics of the firm. They may attach high probability to the firm's being "normal," meaning that it has the payoffs just given, but they also entertain some (possibly very small) probability that they face a firm who fortuitously has a technology or some other characteristic that ensures high effort. Refer to the latter as the "commitment" type of firm. This is now a game of incomplete information, with the consumers uncertain of the firm's type. As long as the firm is sufficiently patient, then in any Nash equilibrium of the repeated game, the firm's payoff must be arbitrarily close to 2. This result holds no matter how unlikely consumers think the commitment type, though increasing patience is required from the normal firm as the commitment type becomes less likely.

To see the intuition behind this result, suppose that we have a candidate equilibrium in which the normal firm receives a payoff less than $2 - \varepsilon$. Then the normal and commitment types must be making different choices over the course of the repeated game, because an equilibrium in which they behave identically would induce consumers to choose h and would yield a payoff of 2. Now, one option open to the normal firm is to mimic the behavior of the commitment type. If the normal firm does so over a sufficiently long period of time, then the short-run players (who expect the normal type of firm to behave differently) will become convinced that the firm is actually the commitment type and will play their best response of h. Once this happens, the normal firm thereafter earns a payoff of 2. Of course, it may take a while for this to happen, and the firm may have to endure much lower payoffs in the meantime, but these initial payoffs are not very important if the firm is patient. If the firm is patient enough, it thus has a strategy available that ensures a payoff arbitrarily close to 2. Our initial hypothesis, that the firm's equilibrium payoff fell short of $2 - \varepsilon$, must then have been incorrect. Any equilibrium must give the (sufficiently patient) normal type of firm a payoff above $2 - \varepsilon$.

The common interpretation of this argument is that the normal firm can acquire and maintain a *reputation* for behaving like the commitment type. This reputation-building possibility, which excludes many of the equilibrium outcomes of the complete-information game, may appear to be quite special. Why should consumers' uncertainty about the firm take precisely the form we have assumed? What happens if there is a single buyer who reappears in each period, so that the buyer also faces intertemporal incentive considerations and may also consider building a reputation? However, the result generalizes far beyond the special structure of this example (chapter 15).

1.6 Discussion

The unifying theme of work in repeated games is that links between current and future behavior can create incentives that would not be apparent if one examined a current interaction in isolation. We routinely rely on the importance of such links in our daily lives.

Markets create boundaries within which a vast variety of behavior is possible. These markets can function effectively only if there is a shared understanding of what constitutes appropriate behavior. Trade in many markets involves goods and services

1.6 ■ Discussion

whose characteristics are sufficiently difficult to verify as to make legal enforcement a hopelessly clumsy tool. This is an especially important consideration in markets involving expert providers, with the markets for medical care and a host of maintenance and repair services being the most prominent examples, in which the party providing the service is also best positioned to assess the service. More generally, legal sanctions cannot explain why people routinely refrain from opportunistic behavior, such as attempting to renegotiate prices, taking advantage of sunk costs, or cheating on a transaction. But if such behavior reigned unchecked, our markets would collapse.

The common force organizing market transactions is the prospect of future interactions. We decline opportunities to renege on deals or turn them to our advantage because we expect to have future dealings with the other party. The development of trading practices that transferred enough information to create effective intertemporal incentives was a turning point in the development of our modern market economy (e.g., Greif 1997, 2005; Greif, Milgrom, and Weingast 1994).

Despite economists' emphasis on markets, many of the most critical activities in our lives take place outside of markets. We readily cede power and authority to some people, either formally through a political process or through informal acquiesence. We have social norms governing when one is allowed to take advantage of another and when one should refrain from doing so. We have conventions for how families are formed, including who is likely to mate with whom, and how they are organized, including who has an obligation to support whom and who can expect resources from whom. Our society relies on institutions to perform some functions, whereas other quite similar functions are performed outside of institutions—the helpless elderly are routinely institutionalized, but not infants.

A unifying view of these observations is that they reflect equilibria of the repeated game that we implicitly play with one another.[11] It is then no surprise, given the tendency for repeated games to have multiple equilibria, that we see great variation across the world in how societies and cultures are organized. This same multiplicity opens the door to the possibility that we might think about designing our society to function more effectively. The theory of repeated games provides the tools for this task.

The best known results in the theory of repeated games, the folk theorems, focus attention on the multiplicity of equilibria in such games, a source of great consternation for some. We consider multiple equilibria a virtue—how else can one hope to explain the richness of behavior that we observe around us?

It is also important to note that the folk theorem characterizes the payoffs available to arbitrarily patient players. Much of our interest and much of the work in this book concerns cases in which players are patient enough for intertemporal incentives to have some effect, but not arbitrarily patient. In addition, we are concerned with the properties of equilibrium behavior as well as payoffs.

---

11. Ellickson (1991) provides a discussion of how neighbors habitually rely on informal intertemporal arrangements to mediate their interactions rather than relying exclusively on current incentives, even when the latter are readily available. Binmore (1994, 1998) views an understanding of the incentives created by repeated interactions as being sufficiently common as to appropriately replace previous notions, such as the categorical imperative, as the foundation for theories of social justice.

1.7 A Reader's Guide

Chapter 2 is the obvious point of departure, introducing the basic tools for working with repeated games, including the "dynamic programming" approach to repeated games.

The reader then faces a choice. One can proceed through the chapters in part I of the book, treating games of perfect monitoring. Chapter 3 uses constructive arguments to prove the folk theorem. Chapter 4 examines a number of issues, such as what we should make of an infinite horizon, that arise in interpreting repeated games, while chapter 5 pushes the analysis beyond the confines of the canonical repeated game. Chapter 6 illustrates the techniques with a collection of economic applications.

Alternatively, one can jump directly to part II. Here, chapters 7 and 8 present more powerful (though initially seemingly more abstract) techniques that allow a unified treatment of games of perfect and imperfect monitoring. These allow us to work with the limiting case of perfectly patient players as well as cases in which players may be less patient and in which the sufficient conditions for a folk theorem fail. Chapter 9 presents the public monitoring folk theorem, and chapter 10 explores features that arise out of imperfections in the monitoring process. Chapter 11 again provides economic illustrations.

The reader now faces another choice. Part III (chapters 12–14) considers the case of private monitoring. Here, we expect a familiarity with the material in chapter 7. Alternatively, the reader can proceed to part IV, on reputations, whether arriving from part I or part II. Chapters 15 and 16 form the core here, presenting the classical reputation results for games with a single long-lived player and for games with multiple long-lived players, with the remaining chapters exploring extensions and alternative formulations.

For an epilogue, see Samuelson (2006, section 6).

1.8 The Scope of the Book

The analysis of long-run relationships is relevant for virtually every area of economics. The literature on intertemporal incentives is vast. If a treatment of the subject is to be kept manageable, some things must be excluded. We have not attempted a comprehensive survey or history of the literature.

Our canonical setting is one in which a fixed stage game is played in each of an infinite number of time periods by players who maximize the average discounted payoffs. We concentrate on cases in which players have the same discount factor (see remark 2.1.4) or on cases in which long-lived players with common discount factors are joined by myopic short-lived players.

We are sometimes interested in cases in which the players are quite patient, typically captured by examining the limit as their discount factors approach 1, because intertemporal incentives are most effective with patient players. An alternative is to work directly with arbitrarily patient players, using criteria such as the limit-of-the-means payoffs or the overtaking criteria to evaluate payoffs. We touch on this subject briefly (section 3.2.2), but otherwise restrict attention to discounted games.

1.8 ■ The Scope of the Book

In particular, we are also often interested in cases with nontrivial intertemporal incentives, but sufficient *im*patience to impose constraints on equilibrium behavior. In addition, the no-discounting case is more useful for studying payoffs than behavior. We are interested not only in players' payoffs but also in the strategies that deliver these payoffs.

We discuss finitely repeated games (section 4.4 and chapter 17) only enough to argue that there is in general no fundamental discontinuity when passing from finite to infinitely repeated games. The analyses of finite and infinite horizon games are governed by the same conceptual issues. We then concentrate on infinitely repeated games, where a more convenient body of recursive techniques generally allows a more powerful analysis.

We concentrate on cases in which the stage game is identical across periods. In a *dynamic game*, the stage game evolves over the course of the repeated interaction in response to actions taken by the players.[12] Dynamic games are also referred to as *stochastic games*, emphasizing the potential randomness in the evolution of the state. We offer an introduction to such issues in sections 5.5–5.7, suggesting Mertens (2002), Sorin (2002), and Vieille (2002) as introductions to the literature.

A literature in its own right has grown around the study of *differential games*, or perfect-monitoring dynamic games played in continuous time, much of it in engineering and mathematics. As a result of the continuous time setting, the techniques for working with such games resemble those of control theory, whereas those of repeated games more readily prompt analogies to dynamic programming. We say nothing about such games, suggesting Friedman (1994) and Clemhout and Wan (1994) for introductions to the topic.

The first three parts of this book consider games of complete information, where players share identical information about the structure of the game. In contrast, many economic applications are concerned with cases of incomplete information. A seller may not know the buyer's utility function. A buyer may not know whether a firm plans to continue in business or is on the verge of absconding. A potential entrant may not know whether the existing firm can wage a price war with impunity or stands to lose tremendously from doing so. In the final part of this book, we consider a special class of games of incomplete information whose study has been particularly fruitful, namely, reputation games. A more general treatment of games of incomplete information is given by Zamir (1992) and Forges (1992).

The key to the incentives that arise in repeated games is the ability to establish a link between current and future play. If this is to be done, players must be able to observe or monitor current play. Much of our work is organized around assumptions about players' abilities to monitor others' behavior. Imperfections in monitoring can impose constraints on equilibrium payoffs. A number of publications have shown that

---

12. For example, the oligopolists in section 1.3 might also have the opportunity to invest in cost-reducing research and development in each period. Each period then brings a new stage game, characterized by the cost levels relevant for the period, along with a new opportunity to affect future stage games as well as secure a payoff in the current game. Inventory levels for a firm, debt levels for a government, education levels for a worker, or weapons stocks for a country may all be sources of similar intertemporal evolution in a stage game. Somewhat further afield, a bargaining game is a dynamic game, with the stage game undergoing a rather dramatic transformation to becoming a trivial game in which nothing happens once an agreement is reached.

these constraints can be relaxed if players have the ability to communicate with one another (e.g., Compte 1998 and Kandori and Matsushima 1998). We do not consider such possibilities.

We say nothing here about the reputations of expert advisors. An expert may have preferences about the actions of a decision maker whom he advises, but the advice itself is cheap talk, in the sense that it affects the expert's payoff only through its effect on the decision maker's action. If this interaction occurs once, then we have a straightforward cheap talk game whose study was pioneered by Crawford and Sobel (1982). If the interaction is repeated, then the expert's recommendations have an effect not only on current actions but also possibly on how much influence the expert will have in the future. The expert may then prefer to hedge the current recommendation in an attempt to be more influential in the future (e.g., Morris 2001).

Finally, the concept of a reputation has appeared in a number of contexts, some of them quite different from those that appear in this book, in both the academic literature and popular use. Firms offering warranties are often said to be cultivating reputations for high quality, advertising campaigns are designed to create a reputation for trendiness, forecasters are said to have reputations for accuracy, or advisors for giving useful counsel. These concepts of reputation touch on ideas similar to those with which we work in a number of places. We have no doubt that the issues surrounding reputations are much richer than those we capture here. We regard this area as a particularly important one for further work.

Part I

Games with Perfect Monitoring

2 The Basic Structure of Repeated Games with Perfect Monitoring

2.1 The Canonical Repeated Game

2.1.1 The Stage Game

The construction of the repeated game begins with a *stage game*. There are n players, numbered $1, \ldots, n$.

We refer to choices in the stage game as *actions*, reserving *strategy* for behavior in the repeated game. The set of pure actions available to player i in the stage game is denoted A_i, with typical element a_i. The set of pure action *profiles* is given by $A \equiv \prod_i A_i$. We assume each A_i is a compact subset of the Euclidean space $\mathbb{R}^k$ for some k. Some of the results further assume each A_i is finite.

Stage game payoffs are given by a continuous function,

$$u : \prod_i A_i \to \mathbb{R}^n.$$

The set of mixed actions for player i is denoted by $\Delta(A_i)$, with typical element α_i, and the set of mixed profiles by $\prod_i \Delta(A_i)$. The payoff function is extended to mixed actions by taking expectations.

The set of stage-game payoffs generated by the pure action profiles in A is

$$\mathscr{F} \equiv \{v \in \mathbb{R}^n : \exists a \in A \ s.t. \ v = u(a)\}.$$

The set of *feasible* payoffs,

$$\mathscr{F}^\dagger \equiv \mathrm{co}\mathscr{F},$$

is the convex hull of the set of payoffs $\mathscr{F}$.[1] As we will see, for sufficiently patient players, intertemporal averaging allows us to obtain payoffs in $\mathscr{F}^\dagger \setminus \mathscr{F}$. A payoff $v \in \mathscr{F}^\dagger$ is *inefficient* if there exists another payoff $v' \in \mathscr{F}^\dagger$ with $v'_i > v_i$ for all i; the payoff v' strictly dominates v. A payoff is *efficient* (or *Pareto efficient*) if it is not inefficient. If, for $v, v' \in \mathscr{F}^\dagger$, $v'_i \geq v_i$ for all i with a strict inequality for some i, then v' *weakly dominates* v. A feasible payoff is *strongly efficient* if it is efficient and not weakly dominated by any other feasible payoff.

By Nash's (1951) existence theorem, if the stage game is finite, then it has a (possibly mixed) Nash equilibrium. In general, because payoffs are given by continuous functions on the compact set $\prod_i A_i$, it follows from Glicksberg's (1952) fixed point theorem that the infinite stage games we consider also have Nash equilibria. It

---

1. The convex hull of a set $\mathscr{A} \subset \mathbb{R}^n$, denoted $\mathrm{co}\mathscr{A}$, is the smallest convex set containing $\mathscr{A}$.

is common when working with infinite action stage games to additionally require that the action spaces be convex and u_i quasi-concave in a_i, so that pure-strategy Nash equilibria exist (Fudenberg and Tirole 1991, section 1.3.3).

For ease of reference, we list the maintained assumptions on the stage game.

Assumption 2.1.1
1. A_i *is either finite, or a compact and convex subset of the Euclidean space* $\mathbb{R}^k$ *for some k. We refer to compact and convex action spaces as* continuum action spaces.
2. *If A_i is a continuum action space, then $u : A \to \mathbb{R}^n$ is continuous, and u_i is quasiconcave in a_i.*

Remark 2.1.1 **Pure strategies given continuum action spaces** When action spaces are continua, to avoid some tedious measurability details, we only consider pure strategies. Because the basic analysis of finite action games (with pure or mixed actions) and continuum action games (with pure actions) is identical, we use α_i to both denote pure or mixed strategies in finite games, and pure strategies only in continuum action games.
◆

Much of the work in repeated games is concerned with characterizing the payoffs consistent with equilibrium behavior in the repeated game. This characterization in turn often begins by identifying the worst payoff consistent with individual optimization. Player i always has the option of playing a best response to the (mixed) actions chosen by the other players. In the case of pure strategies, the worst outcome in the stage game for player i, consistent with i behaving optimally, is then that the other players choose the profile $a_{-i} \in A_{-i} \equiv \prod_{j \neq i} A_j$ that minimizes the payoff i earns when i plays a best response to a_{-i}. This bound, player i's *(pure action) minmax payoff*, is given by

$$\underline{v}_i^p \equiv \min_{a_{-i} \in A_{-i}} \max_{a_i \in A_i} u_i(a_i, a_{-i}).$$

The compactness of A and continuity of u_i ensure $\underline{v}_i^p$ is well defined. A *(pure action) minmax profile* (which may not be unique) for player i is a profile $\hat{a}^i = (\hat{a}_i^i, \hat{a}_{-i}^i)$ with the properties that $\hat{a}_i^i$ is a stage-game best response for i to $\hat{a}_{-i}^i$ and $\underline{v}_i^p = u_i(\hat{a}_i^i, \hat{a}_{-i}^i)$. Hence, player i's minmax action profile gives i his minmax payoff and ensures that no alternative action on i's part can raise his payoff. In general, the other players will not be choosing best responses in profile $\hat{a}^i$, and hence $\hat{a}^i$ will not be a Nash equilibrium of the stage game.[2]

A payoff vector $v = (v_1, \ldots, v_n)$ is *weakly (pure action) individually rational* if $v_i \geq \underline{v}_i^p$ for all i, and is *strictly (pure action) individually rational* if $v_i > \underline{v}_i^p$ for all i. The set of feasible and strictly individually rational payoffs is given by

$$\mathscr{F}^{\dagger p} \equiv \{v \in \mathscr{F}^\dagger : v_i > \underline{v}_i^p, i = 1, \ldots, n\}.$$

The set of strictly individually rational payoffs generated by pure action profiles is given by

$$\mathscr{F}^p \equiv \{v \in \mathscr{F} : v_i > \underline{v}_i^p, i = 1, \ldots, n\}.$$

2. An exception is the prisoners' dilemma, where mutual shirking is both the unique Nash equilibrium of the stage game and the minmax action profile for both players. Many of the special properties of the prisoners' dilemma arise out of this coincidence.

2.1 ■ The Canonical Repeated Game

	H	T
H	1, −1	−1, 1
T	−1, 1	1, −1

Figure 2.1.1 Matching pennies.

Remark 2.1.2 **Mixed-action individual rationality** In finite games, lower payoffs can sometimes be enforced when we allow players to randomize. In particular, allowing the players other than player i to randomize yields the mixed-action minmax payoff,

$$\underline{v}_i \equiv \min_{\alpha_{-i} \in \Pi_{j \neq i} \Delta(A_j)} \max_{a_i \in A_i} u_i(a_i, \alpha_{-i}), \tag{2.1.1}$$

which can be lower than the pure action minmax payoff, $\underline{v}_i^p$.[3] A *(mixed) action minmax profile* for player i is a profile $\hat{\alpha}^i = (\hat{\alpha}_i^i, \hat{\alpha}_{-i}^i)$ with the properties that $\hat{\alpha}_i^i$ is a stage-game best response for i to $\hat{\alpha}_{-i}^i$ and $\underline{v}_i = u_i(\hat{\alpha}_i^i, \hat{\alpha}_{-i}^i)$.

In matching pennies (figure 2.1.1), for example, player 1's pure minmax payoff is 1, because for any of player 2's pure strategies, player 1 has a best response giving a payoff of 1. Pure minmax action profiles for player 1 are given by (H, H) and (T, T). In contrast, player 1's mixed minmax payoff is 0, implied by player 2's mixed action of $\frac{1}{2} \circ H + \frac{1}{2} \circ T$.[4]

We use the same term, *individual rationality*, to indicate both $v_i \geq \underline{v}_i^p$ and $v_i \geq \underline{v}_i$, with the context indicating the appropriate choice. We denote the set of feasible and strictly individually rational payoffs (relative to the mixed minmax utility, $\underline{v}_i$) by

$$\mathscr{F}^* \equiv \{v \in \mathscr{F}^\dagger : v_i > \underline{v}_i, i = 1, \ldots, n\}.$$

◆

2.1.2 Public Correlation

It is sometimes natural to allow players to use a public correlating device. Such a device captures a variety of public events that players might use to coordinate their actions. Perhaps the best known example is an agreement in the electrical equipment industry in the 1950s to condition bids in procurement auctions on the current phase of the moon.[5]

---

3. Allowing player i to mix will not change i's minmax payoff, because every action in the support of a mixed best reply is also a best reply for player i.
4. We denote the mixture that assigns probability $\alpha_i(a_i)$ to action a_i by $\sum_{a_i} \alpha_i(a_i) \circ a_i$.
5. A small body of literature has studied this case. See Carlton and Perloff (1992, pp. 213–216) for a brief introduction.

Definition 2.1.1 A public correlating device *is a probability space* $([0, 1], \mathscr{B}, \lambda)$, *where $\mathscr{B}$ is the Borel σ-algebra and λ is Lebesgue measure. In the stage game with public correlation, a realization* $\omega \in [0, 1]$ *of a public random variable is first drawn, which is observed by all players, and then each player i chooses an action* $a_i \in A_i$.

A stage-game action for player i in the stage game with public correlation is a (measurable) function $\mathsf{a}_i : [0, 1] \to \Delta(A_i)$. If $\mathsf{a}_i : [0, 1] \to A_i$, then a_i is a pure action. When actions depend nontrivially on the outcome of the public correlating device, we calculate player i's expected payoff in the obvious manner, by taking expectations over the outcome $\omega \in [0, 1]$. Any strategy profile $\mathsf{a} \equiv (\mathsf{a}_1, \ldots, \mathsf{a}_n)$ induces a joint distribution over $\prod_i \Delta(A_i)$. When evaluating the profitability of a deviation from a, because the realization of the correlating device is public, the calculation is *ex post*, that is, conditional on the realization of ω. If a is a Nash equilibrium of the stage game with public correlation, then every α in the support of a is a Nash equilibrium of the stage game without public correlation; in particular, most correlated equilibria (Aumann 1974) are not equilibria of the stage game *with* public correlation.

It is possible to replace the public correlating device with communication, by using *jointly controlled lotteries*, introduced by Aumann, Maschler, and Stearns (1968). For example, for two players, suppose they simultaneously announce a number from $[0, 1]$. Let ω equal their sum, if the sum is less than 1, and equal their sum minus 1 otherwise. It is easy to verify that if one player uniformly randomizes over his selection, then ω is uniformly distributed on $[0, 1]$ for any choice by the other player. Consequently, neither player can influence the probability distribution. We will not discuss communication in this book, so we use public correlating devices rather than jointly controlled lotteries.

Trivially, every payoff in $\mathscr{F}^\dagger$ can be achieved in pure actions using public correlation. On the other hand, not all payoffs in $\mathscr{F}^\dagger$ can be achieved in mixed actions without public correlation. For example, consider the game in figure 2.1.2. The set $\mathscr{F}$ is given by

$$\mathscr{F} = \{(2, 2), (5, 1), (1, 5), (0, 0)\}.$$

The set $\mathscr{F}^\dagger$ is the set of all convex combinations of these four payoff vectors. Some of the payoffs that are in $\mathscr{F}^\dagger$ but not $\mathscr{F}$ can be obtained via independent mixtures, ignoring the correlating device, over the sets $\{T, B\}$ and $\{L, R\}$. For example, $\mathscr{F}^\dagger$ contains $(20/9, 20/9)$, obtained by independent mixtures that place probability $2/3$ on T (or L) and $1/3$ on B (or R). A pure strategy that uses the public correlation device to place probability $4/9$ on (T, L), $2/9$ on each of (T, R) and (B, L), and $1/9$ on (B, R) achieves the same payoff. In addition, the public correlating device allows the players to achieve some payoffs in $\mathscr{F}^\dagger$ that cannot be obtained from independent

	L	R
T	2, 2	1, 5
B	5, 1	0, 0

Figure 2.1.2 The game of "chicken."

mixtures. For example, the players can attach probability $1/2$ to each of the outcomes (T, R) and (B, L), giving payoffs $(3, 3)$. No independent mixtures can achieve such payoffs, because any such mixtures must attach positive probability to payoffs $(2, 2)$ and $(0, 0)$, ensuring that the sum of the two players' average payoffs falls below 6.

2.1.3 The Repeated Game

In the repeated game,[6] the stage game is played in each of the periods $t \in \{0, 1, \ldots\}$. In formulating the notation for this game, we use subscripts to refer to players, typically identifying the element of a profile of actions, strategies, or payoffs corresponding to a particular player. Superscripts will either refer to periods or denote particular profiles of interest, with the use being clear from the context.

This chapter introduces repeated games of *perfect monitoring*. At the end of each period, all players observe the action profile chosen. In other words, the actions of every player are perfectly monitored by all other players. If a player is randomizing, only the realized choice is observed.

The set of period $t \geq 0$ histories is given by

$$\mathcal{H}^t \equiv A^t,$$

where we define the initial history to be the null set, $A^0 \equiv \{\varnothing\}$, and A^t to be the t-fold product of A. A history $h^t \in \mathcal{H}^t$ is thus a list of t action profiles, identifying the actions played in periods 0 through $t-1$. The addition of a period t action profile then yields a period $t+1$ history h^{t+1}, an element of the set $\mathcal{H}^{t+1} = \mathcal{H}^t \times A$. The set of all possible histories is

$$\mathcal{H} \equiv \bigcup_{t=0}^{\infty} \mathcal{H}^t.$$

A *pure strategy* for player i is a mapping from the set of all possible histories into the set of pure actions,[7]

$$\sigma_i : \mathcal{H} \to A_i.$$

A *mixed strategy* for player i is a mixture over the set of all pure strategies. Without loss of generality, we typically find it more convenient to work with behavior strategies rather than mixed strategies.[8] A *behavior strategy* for player i is a mapping

$$\sigma_i : \mathcal{H} \to \Delta(A_i).$$

Because a pure strategy is trivially a special case of a behavior strategy, we use the same notation σ_i for both pure and behavior strategies. Unless indicating otherwise,

---

6. The early literature often used the term *supergame* for the repeated game.
7. Because there is a natural bijection (one-to-one and onto mapping) between $\mathcal{H}$ and each player's collection of information sets, this is the standard notion of an extensive-form strategy.
8. Two strategies for a player i are *realization equivalent* if, fixing the strategies of the other players, the two strategies of player i induce the same distribution over outcomes. It is a standard result for finite extensive form games that every mixed strategy has a realization equivalent behavior strategy (Kuhn's theorem, see Ritzberger 2002, theorem 3.3, p. 127), and the same is true here. See Mertens, Sorin, and Zamir 1994, theorem 1.6, p. 66 for a proof (though the proof is conceptually identical to the finite case, the infinite horizon introduces some technical issues).

we then use the word *strategy* to denote a behavior strategy, which may happen to be pure. Recall from remark 2.1.1 that we consider only pure strategies for a player whose action space is a continuum (even though for notational simplicity we sometimes use α_i to denote the stage game action).

For any history $h^t \in \mathcal{H}$, we define the *continuation* game to be the infinitely repeated game that begins in period t, following history h^t. For any strategy profile σ, player i's *continuation strategy induced by* h^t, denoted $\sigma_i|_{h^t}$, is given by

$$\sigma_i|_{h^t}(h^\tau) = \sigma_i(h^t h^\tau), \quad \forall h^\tau \in \mathcal{H},$$

where $h^t h^\tau$ is the concatenation of the history h^t followed by the history h^τ. This is the behavior implied by the strategy σ_i in the continuation game that follows history h^t. We write $\sigma|_{h^t}$ for $(\sigma_1|_{h^t}, \ldots, \sigma_n|_{h^t})$. Because for each history h^t, $\sigma_i|_{h^t}$ is a strategy in the original repeated game, that is, $\sigma_i|_{h^t} : \mathcal{H} \to \Delta(A_i)$, the continuation game associated with each history is a *subgame* that is strategically identical to the original repeated game. Thus, repeated games have a recursive structure, and this plays an important role in their study.

An *outcome path* (or more simply, *outcome*) in the infinitely repeated game is an infinite sequence of action profiles $\mathbf{a} \equiv (a^0, a^1, a^2, \ldots) \in A^\infty$. Notice that an outcome is distinct from a history. Outcomes are infinite sequences of action profiles, whereas histories are finite-length sequences (whose length identifies the period for which the history is relevant). We denote the first t periods of an outcome $\mathbf{a}$ by $\mathbf{a}^t = (a^0, a^1, \ldots, a^{t-1})$. Thus, $\mathbf{a}^t$ is the history in $\mathcal{H}^t$ corresponding to the outcome $\mathbf{a}$.

The pure strategy profile $\sigma \equiv (\sigma_1, \ldots, \sigma_n)$ induces the outcome $\mathbf{a}(\sigma) \equiv (a^0(\sigma), a^1(\sigma), a^2(\sigma), \ldots)$ recursively as follows. In the first period, the action profile

$$a^0(\sigma) \equiv (\sigma_1(\varnothing), \ldots, \sigma_n(\varnothing))$$

is played. In the second period, the history $a^0(\sigma)$ implies that action profile

$$a^1(\sigma) \equiv (\sigma_1(a^0(\sigma)), \ldots, \sigma_n(a^0(\sigma)))$$

is played. In the third period, the history $(a^0(\sigma), a^1(\sigma))$ is observed, implying the action profile

$$a^2(\sigma) \equiv (\sigma_1(a^0(\sigma), a^1(\sigma)), \ldots, \sigma_n(a^0(\sigma), a^1(\sigma)))$$

is played, and so on.

Analogously, a behavior strategy profile σ induces a *path of play*. In the first period, $\sigma(\varnothing)$ is the initial mixed action profile $\alpha^0 \in \prod_i \Delta(A_i)$. In the second period, for each history a^0 in the support of α^0, $\sigma(a^0)$ is the mixed action profile $\alpha^1(a^0)$, and so on. For a pure strategy profile, the induced path of play and induced outcome are the same. If the profile has some mixing, however, then the profile induces a path of play that specifies, for each period t, a probability distribution over the histories $\mathbf{a}^t$. The underlying behavior strategy specifies a period t profile of mixed stage-game actions for each such history $\mathbf{a}^t$, in turn inducing a probability distribution $\alpha^{t+1}(\mathbf{a}^t)$ over period $t+1$ action profiles a^{t+1}, and hence a probability distribution over period $t+1$ histories $\mathbf{a}^{t+1}$.

2.1 ■ The Canonical Repeated Game

Suppose σ is a pure strategy profile. In period t, the induced pure action profile $a^t(\sigma)$ yields a flow payoff of $u_i(a^t(\sigma))$ to player i. An outcome $\mathbf{a}(\sigma)$ thus implies an infinite stream of stage-game payoffs for each player i, given by $(u_i(a^0(\sigma)), u_i(a^1(\sigma)), u_i(a^2(\sigma)), \ldots) \in \mathbb{R}^\infty$. Each player discounts these payoffs with the discount factor $\delta \in [0, 1)$, so that the *average discounted payoff* to player i from the infinite sequence of payoffs $(u_i^0, u_i^1, u_i^2, \ldots)$ is given by

$$(1-\delta) \sum_{t=0}^\infty \delta^t u_i^t.$$

The payoff from a pure strategy profile σ is then given by

$$U_i(\sigma) = (1-\delta) \sum_{t=0}^\infty \delta^t u_i(a^t(\sigma)). \tag{2.1.2}$$

As usual, the payoff to player i from a profile of mixed or behavior strategies σ is the expected value of the payoffs of the realized outcomes, also denoted $U_i(\sigma)$.

Observe that we normalize the payoffs in (2.1.2) (and throughout) by the factor $(1-\delta)$. This ensures that $U(\sigma) = (U_1(\sigma), \ldots, U_n(\sigma)) \in \mathscr{F}^\dagger$ for all repeated-game strategy profiles σ. We can then readily compare payoffs in the repeated game and the stage game, and compare repeated-game payoffs for different (common) discount factors.

Remark 2.1.3 **Public correlation notation** In the repeated game with public correlation, a t-period history is a list of t action profiles and t realizations of the public correlating device, $(\omega^0, a^0; \omega^1, a^1; \ldots; \omega^{t-1}, a^{t-1})$. In period t, as a measurable function of the period t history and the period t realization ω^t, a behavior strategy specifies $\alpha_i \in \Delta(A_i)$. As for games without public correlation, every t-period history induces a subgame that is strategically equivalent to the original game. In addition, there are subgames corresponding to the period t realizations ω^t.

Rather than explicitly describing the correlating device and the players' actions as a function of its realization, strategy profiles are sometimes described by simply specifying a correlated action in each period. Such a strategy profile in the repeated game with public monitoring specifies in each period t, as a function of history $h^{t-1} \in \mathscr{H}^{t-1}$, a correlated action profile, that is, a joint distribution over the action profiles $\alpha \in \prod_i \Delta A_i$. We also denote the reduction of the compound lottery induced by the public correlating device and subsequent individual randomization by α. The precise meaning will be clear from context. ◆

Remark 2.1.4 **Common discount factors** With the exception of the discussion of reputations in chapter 16, we assume that long-lived players share a common discount factor δ. This assumption is substantive. Consider the battle of the sexes in figure 2.1.3. The set of feasible payoffs $\mathscr{F}^\dagger$ is the convex hull of the set $\{(3, 1), (0, 0), (1, 3)\}$. For any common discount factor $\delta \in [0, 1)$, the set of feasible repeated-game payoffs is also the convex hull of $\{(3, 1), (0, 0), (1, 3)\}$. Suppose, however, players 1 and 2 have discount factors δ_1 and δ_2 with $\delta_1 > \delta_2$, so that player 1 is more patient than player 2. Then any repeated-game strategy that calls for (B, L) to be played

	L	R
T	0,0	3,1
B	1,3	0,0

Figure 2.1.3 A battle-of-the-sexes game.

in periods $0, \ldots, T-1$ and (T, R) to be played in subsequent periods yields a repeated game vector outside the convex hull of $\{(3, 1), (0, 0), (1, 3)\}$, being in particular above the line segment joining payoffs $(3, 1)$ and $(1, 3)$. This outcome averages over the payoffs $(3, 1)$ and $(1, 3)$, but places relatively high player 2 payoffs in early periods and relatively high player 1 payoffs in later periods, giving repeated-game payoffs to the two players of

$$\text{player 1:} \quad (1 - \delta_1^T) + 3\delta_1^T$$
$$\text{and player 2:} \quad 3(1 - \delta_2^T) + \delta_2^T.$$

Because $\delta_1 > \delta_2$, each player's convex combination is pushed in the direction of the outcome that is relatively lucrative for that player. This arrangement capitalizes on the differences in the two players' discount factors, with the impatient player 2 essentially borrowing payoffs from the more patient player 1 in early periods to be repaid in later periods, to expand the set of feasible repeated-game payoffs beyond those of the stage game. Lehrer and Pauzner (1999) examine repeated games with differing discount factors.

◆

2.1.4 Subgame-Perfect Equilibrium of the Repeated Game

As usual, a Nash equilibrium is a strategy profile in which each player is best responding to the strategies of the other players:

Definition 2.1.2 *The strategy profile σ is a* Nash equilibrium *of the repeated game if for all players i and strategies σ_i',*

$$U_i(\sigma) \geq U_i(\sigma_i', \sigma_{-i}).$$

We have the following formalization of the discussion in section 2.1.1 and remark 2.1.2 on minmax utilities:

Lemma 2.1.1 *If σ is a pure-strategy Nash equilibrium, then for all i, $U_i(\sigma) \geq \underline{v}_i^p$. If σ is a (possibly mixed) Nash equilibrium, then for all i, $U_i(\sigma) \geq \underline{v}_i$.*

Proof Consider a Nash equilibrium. Player i can always play the strategy that specifies a best reply to $\sigma_{-i}(h^t)$ after every history h^t. In each period, i's payoff is thus at least $\underline{v}_i^p$ if σ_{-i} is pure ($\underline{v}_i$ if σ_{-i} is mixed), and so i's payoff in the equilibrium must be at least $\underline{v}_i^p$ ($\underline{v}_i$, respectively).

■

We frequently make implicit use of the observation that a strategy of the repeated game with public correlation is a Nash equilibrium if and only if for almost all realizations of the public correlating device, the induced strategy profile is a Nash equilibrium.

In games with a nontrivial dynamic structure, Nash equilibrium is too permissive—there are Nash equilibrium outcomes that violate basic notions of optimality by specifying irrational behavior at out-of-equilibrium information sets. Similar considerations arise from the dynamic structure of a repeated game, even if actions are chosen simultaneously in the stage game. Consider a Nash equilibrium of an infinitely repeated game with perfect monitoring. Associated with each history that cannot occur in equilibrium is a subgame. The notion of a Nash equilibrium imposes no optimality conditions in these subgames, opening the door to violations of sequential rationality.

Subgame perfection strengthens Nash equilibrium by imposing the *sequential rationality requirement* that behavior be optimal in all circumstances, both those that arise in equilibrium (as required by Nash equilibrium) and those that arise out of equilibrium. In finite horizon games of perfect information, such sequential rationality is conveniently captured by requiring backward induction. We cannot appeal to backward induction in an infinitely repeated game, which has no last period. We instead appeal to the underlying definition of sequential rationality as requiring equilibrium behavior in every subgame, where we exploit the strategic equivalence of the repeated game and the continuation game induced by history h^t.

Definition 2.1.3 *A strategy profile σ is a* subgame-perfect equilibrium *of the repeated game if for all histories $h^t \in \mathcal{H}$, $\sigma|_{h^t}$ is a Nash equilibrium of the repeated game.*

The existence of subgame-perfect equilibria in a repeated game is immediate: Any profile of strategies that induces the *same* Nash equilibrium of the stage game after every history of the repeated game is a subgame-perfect equilibrium of the latter. For example, strategies that specify shirking after every history are a subgame-perfect equilibrium of the repeated prisoners' dilemma, as are strategies that specify low effort and the low-priced choice in every period (and after every history) of the product-choice game. If the stage game has more than one Nash equilibrium, strategies that assign any stage-game Nash equilibrium to each period t, *independently* of the history leading to period t, constitute a subgame-perfect equilibrium. Playing one's part of a Nash equilibrium is always a best response in the stage game, and hence, as long as future play is independent of current actions, doing so is a best response in each period of a repeated game, regardless of the history of play.

Although the notion of subgame perfection is intuitively appealing, it raises some potentially formidable technical difficulties. In principle, checking for subgame perfection involves checking whether an infinite number of strategy profiles are Nash equilibria—the set $\mathcal{H}$ of histories is countably infinite even if the stage-game action spaces are finite. Moreover, checking whether a profile σ is a Nash equilibrium involves checking that player i's strategy σ_i is no worse than an infinite number of potential deviations (because player i could deviate in any period, or indeed in any combination of periods). The following sections show that we can simplify this task immensely, first by limiting the number of alternative strategies that must be examined, then by organizing the subgames that must be checked for Nash equilibria into equivalence classes, and finally by identifying a simple constructive method for characterizing equilibrium payoffs.

2.2 The One-Shot Deviation Principle

This section describes a critical insight from dynamic programming that allows us to restrict attention to a simple class of deviations when checking for subgame perfection.

A *one-shot deviation* for player i from strategy σ_i is a strategy $\hat{\sigma}_i \neq \sigma_i$ with the property that there exists a unique history $\tilde{h}^t \in \mathcal{H}$ such that for all $h^\tau \neq \tilde{h}^t$,

$$\sigma_i(h^\tau) = \hat{\sigma}_i(h^\tau).$$

Under public correlation, the history $\tilde{h}^t$ includes the period t realization of the public correlating device. The strategy $\hat{\sigma}_i$ plays identically to strategy σ_i in every period other than t and plays identically in period t if the latter is reached with some history other than $\tilde{h}^t$. A one-shot deviation thus agrees with the original strategy everywhere except at one history where the one-shot deviation occurs. However, a one-shot deviation can have a substantial effect on the resulting outcome.

Example 2.2.1 Consider the grim trigger strategy profile in the infinitely repeated prisoners' dilemma of section 1.2. The equilibrium outcome when two players each choose grim trigger is that both players exert effort in every period. Now consider the one-shot deviation $\hat{\sigma}_1$ under which 1 plays as in grim trigger, with the exception of shirking in period 4 if there has been no previous shirking, that is, with the exception of shirking after the history (EE, EE, EE, EE). The deviating strategy shirks in *every* period after period 4, as does grim trigger, and hence we have an *outcome* that differs from the mutual play of grim trigger in infinitely many periods. However, once the deviation has occurred, it is a prescription of grim trigger that one shirk thereafter. The only deviation from the original *strategy* hence occurs after the history (EE, EE, EE, EE).
●

Definition 2.2.1 *Fix a profile of opponents' strategies σ_{-i}. A one-shot deviation $\hat{\sigma}_i$ from strategy σ_i is profitable if, at the history $\tilde{h}^t$ for which $\hat{\sigma}_i(\tilde{h}^t) \neq \sigma_i(\tilde{h}^t)$,*

$$U_i(\hat{\sigma}_i|_{\tilde{h}^t}, \sigma_{-i}|_{\tilde{h}^t}) > U_i(\sigma|_{\tilde{h}^t}).$$

Notice that profitability of $\hat{\sigma}_i$ is defined conditional on the history $\tilde{h}^t$ being reached, though $\tilde{h}^t$ may not be reached in equilibrium. Hence, a Nash equilibrium can have profitable one-shot deviations.

Example 2.2.2 Consider again the prisoners' dilemma. Suppose that strategies σ_1 and σ_2 both specify effort in the first period and effort as long as there has been no previous shirking, with any shirking prompting players to alternate between 10 periods of shirking and 1 of effort, regardless of any subsequent actions. For sufficiently

2.2 ■ The One-Shot Deviation Principle

large discount factors, these strategies constitute a Nash equilibrium, inducing an outcome of mutual effort in every period. However, there are profitable one-shot deviations. In particular, consider a history h^t featuring mutual effort in every period except $t-11$, at which point one player shirked, and periods $t-10, \ldots, t-1$, in which both players shirked. The equilibrium strategy calls for both players to exert effort in period t, and then continue alternating 10 periods of shirking with a period of effort. A profitable one-shot deviation for player 1 is to shirk after history h^t, otherwise adhering to the equilibrium strategy. There are other profitable one-shot deviations, as well as profitable deviations that alter play after more than just a single history. However, all of these deviations increase profits only after histories that do not occur along the equilibrium path, and hence none of them increases equilibrium profits or vitiates the fact that the proposed strategies are a Nash equilibrium.

●

Proposition 2.2.1 **The one-shot deviation principle** *A strategy profile σ is subgame perfect if and only if there are no profitable one-shot deviations.*

To confirm that a strategy profile σ is a subgame-perfect equilibrium, we thus need only consider alternative strategies that deviate from the action proposed by σ once and then return to the prescriptions of the equilibrium strategy. As our prisoners' dilemma example illustrates, this does not imply that the path of generated actions will differ from the equilibrium strategies in only one period. The deviation prompts a different history than does the equilibrium, and the equilibrium strategies may respond to this history by making different subsequent prescriptions.

The importance of the one-shot deviation principle lies in the implied reduction in the space of deviations that need to be considered. In particular, we do not have to worry about alternative strategies that might deviate from the equilibrium strategy in period t, and then again in period $t' > t$, and again in period $t'' > t'$, and so on. For example, we need not consider a strategy that deviates from grim trigger in the prisoners' dilemma by shirking in period 0, and then deviates from the equilibrium path (now featuring mutual shirking) in period 3, and perhaps again in period 6, and so on. Although this is obvious when examining such simple candidate equilibria in the prisoners' dilemma, it is less clear in general.

Proof We give the proof only for pure-strategy equilibria in the game without public correlation. The extensions to mixed strategies and public correlation, though conceptually identical, are notationally cumbersome.

If a profile is subgame perfect, then clearly there can be no profitable one-shot deviations.

Conversely, we suppose that a profile σ is not subgame perfect and show there must then be a profitable one-shot deviation. Because the profile is not subgame perfect, there exists a history $\tilde{h}^t$, player i, and a strategy $\tilde{\sigma}_i$, such that

$$U_i(\sigma_i|_{\tilde{h}^t}, \sigma_{-i}|_{\tilde{h}^t}) < U_i(\tilde{\sigma}_i, \sigma_{-i}|_{\tilde{h}^t}).$$

Let $\varepsilon = U_i(\tilde{\sigma}_i, \sigma_{-i}|_{\tilde{h}^t}) - U_i(\sigma_i|_{\tilde{h}^t}, \sigma_{-i}|_{\tilde{h}^t})$. Let $m = \min_{i,a} u_i(a)$ and $M = \max_{i,a} u_i(a)$. Let T be large enough that $\delta^T(M-m) < \varepsilon/2$. Then,

$$(1-\delta)\sum_{\tau=0}^{T-1} \delta^\tau u_i(a^\tau(\sigma_i|_{\tilde{h}^t}, \sigma_{-i}|_{\tilde{h}^t})) + (1-\delta)\sum_{\tau=T}^{\infty} \delta^\tau u_i(a^\tau(\sigma_i|_{\tilde{h}^t}, \sigma_{-i}|_{\tilde{h}^t}))$$

$$= (1-\delta)\sum_{\tau=0}^{T-1} \delta^\tau u_i(a^\tau(\tilde{\sigma}_i, \sigma_{-i}|_{\tilde{h}^t})) + (1-\delta)\sum_{\tau=T}^{\infty} \delta^\tau u_i(a^\tau(\tilde{\sigma}_i, \sigma_{-i}|_{\tilde{h}^t})) - \varepsilon,$$

so

$$(1-\delta)\sum_{\tau=0}^{T-1} \delta^\tau u_i(a^\tau(\sigma_i|_{\tilde{h}^t}, \sigma_{-i}|_{\tilde{h}^t})) < (1-\delta)\sum_{\tau=0}^{T-1} \delta^\tau u_i(a^\tau(\tilde{\sigma}_i, \sigma_{-i}|_{\tilde{h}^t})) - \frac{\varepsilon}{2},$$
(2.2.1)

because $\delta^T(M-m) < \varepsilon/2$ ensures that regardless of how the deviation in question affects play in period $t+T$ and beyond, these variations in play have an effect on player i's period t continuation payoff of strictly less than $\varepsilon/2$. This in turn implies that the strategy $\hat{\sigma}_i$, defined by

$$\hat{\sigma}_i(h^\tau) = \begin{cases} \tilde{\sigma}_i(h^\tau), & \text{if } \tau < T, \\ \sigma_i|_{\tilde{h}^t}(h^\tau), & \text{if } \tau \geq T, \end{cases}$$

$$= \begin{cases} \tilde{\sigma}_i(h^\tau), & \text{if } \tau < T, \\ \sigma_i(\tilde{h}^t h^\tau), & \text{if } \tau \geq T, \end{cases}$$

is a profitable deviation. In particular, strategy $\hat{\sigma}_i$ agrees with $\tilde{\sigma}_i$ over the first T periods, and hence captures the payoff gains of $\varepsilon/2$ promised by (2.2.1).

The strategy $\hat{\sigma}$ only differs from $\sigma_i|_{\tilde{h}^t}$ in the first T periods. We have thus shown that if an equilibrium is not subgame perfect, there must be a profitable T period deviation. The proof is now completed by arguing recursively on the value of T. Let $\hat{h}^{T-1} \equiv (\hat{a}^0, \ldots, \hat{a}^{T-2})$ denote the $T-1$ period history induced by $(\hat{\sigma}_i, \sigma_{-i}|_{\tilde{h}^t})$. There are two possibilities:

1. Suppose $U_i(\sigma_i|_{\tilde{h}^t \hat{h}^{T-1}}, \sigma_{-i}|_{\tilde{h}^t \hat{h}^{T-1}}) < U_i(\hat{\sigma}_i|_{\hat{h}^{T-1}}, \sigma_{-i}|_{\tilde{h}^t \hat{h}^{T-1}})$. In this case, we have a profitable one-shot deviation, after the history $\tilde{h}^t \hat{h}^{T-1}$ (note that $\hat{\sigma}_i|_{\hat{h}^{T-1}}$ agrees with σ_i in period T and every period after T).
2. Alternatively, suppose $U_i(\sigma_i|_{\tilde{h}^t \hat{h}^{T-1}}, \sigma_{-i}|_{\tilde{h}^t \hat{h}^{T-1}}) \geq U_i(\hat{\sigma}_i|_{\hat{h}^{T-1}}, \sigma_{-i}|_{\tilde{h}^t \hat{h}^{T-1}})$. In this case, we define a new strategy, $\bar{\sigma}_i$ as follows:

$$\bar{\sigma}_i(h^\tau) = \begin{cases} \hat{\sigma}_i(h^\tau), & \text{if } \tau < T-1, \\ \sigma_i|_{\tilde{h}^t}(h^\tau), & \text{if } \tau \geq T-1. \end{cases}$$

2.2 ■ The One-Shot Deviation Principle

Now,

$$U_i(\hat{\sigma}_i|_{\hat{h}^{T-2}}, \sigma_{-i}|_{\tilde{h}^t \hat{h}^{T-2}}) = (1-\delta)u_i(\hat{a}^{T-1}) + \delta U_i(\hat{\sigma}_i|_{\hat{h}^{T-1}}, \sigma_{-i}|_{\tilde{h}^t \hat{h}^{T-1}})$$
$$\leq (1-\delta)u_i(\hat{a}^{T-1}) + \delta U_i(\sigma_i|_{\tilde{h}^t \hat{h}^{T-1}}, \sigma_{-i}|_{\tilde{h}^t \hat{h}^{T-1}})$$
$$= U_i(\bar{\sigma}_i|_{\hat{h}^{T-2}}, \sigma_{-i}|_{\tilde{h}^t \hat{h}^{T-2}}),$$

which implies

$$U_i(\hat{\sigma}_i, \sigma_{-i}|_{\tilde{h}^t}) \leq U_i(\bar{\sigma}_i, \sigma_{-i}|_{\tilde{h}^t}),$$

and so $\bar{\sigma}_i$ is a profitable deviation at $\tilde{h}^t$ that only differs from $\sigma_i|_{\tilde{h}^t}$ in the first $T-1$ periods.

Proceeding in this way, we must find a profitable one-shot deviation. ■

A key step in the proof is the observation that because payoffs are discounted, any strategy that offers a higher payoff than an equilibrium strategy must do so within a finite number of periods. A backward induction argument then allows us to show that if there is a profitable deviation, there is a profitable one-shot deviation. Fudenberg and Tirole (1991, section 4.2) show the one-shot deviation principle holds for a more general class of games with perfect monitoring, those with payoffs that are *continuous at infinity* (a condition that essentially requires that actions in the far future have a negligible impact on current payoffs). In addition, the principle holds for sequential equilibria in any finite extensive form game (Osborne and Rubinstein 1994, exercise 227.1), as well as for perfect public equilibria of repeated games with public monitoring (proposition 7.1.1) and sequential equilibria of private-monitoring games with no observable deviations (proposition 12.2.2).

Suppose we have a Nash equilibrium σ that is not subgame perfect. Then, from proposition 2.2.1, there must be a profitable one-shot deviation from the strategy profile σ. However, because σ is a Nash equilibrium, no deviation can increase either player's equilibrium payoff. The profitable one-shot deviation must then occur after a history that is not reached in the course of the Nash equilibrium. Example 2.2.2 provided an illustration.

In light of this last observation, do we have a corresponding one-shot deviation principle for Nash equilibria? Is a strategy profile σ a Nash equilibrium if and only if there are no one-shot deviations whose differences from σ occur after histories that arise along the equilibrium path? The answer is no. It is immediate from the definition of Nash equilibrium that there can be no profitable one-shot deviations along the equilibrium path. However, their absence does *not* suffice for Nash equilibrium, as we now show.

Example 2.2.3 Consider the prisoners' dilemma, but with payoffs given in figure 2.2.1.[9] Consider the strategy profile in which both players play tit-for-tat, exerting effort in the first period and thereafter mimicking in each period the action chosen by the opponent

---

9. With the payoffs of figure 2.2.1, the incentives to shirk are independent of the action of the partner, and so the set of discount factors for which tit-for-tat is a Nash equilibrium coincides with the set for which there are no profitable one-shot deviations on histories that appear along the equilibrium path.

	E	S
E	3, 3	−1, 4
S	4, −1	1, 1

Figure 2.2.1 The prisoners' dilemma with incentives to shirk that depend on the opponent's action.

in the previous period. The induced outcome is mutual effort in every period, yielding an equilibrium payoff of 3. To ensure that there are no profitable one-shot deviations whose differences appear after equilibrium histories, we need only consider a strategy for player 1 that shirks in the first period and otherwise plays as does tit-for-tat. Such a strategy induces a cyclic outcome of the form $SE, ES, SE, ES, \ldots$, for a payoff of

$$(1-\delta)\left(4(1+\delta^2+\delta^4+\cdots)-1(\delta+\delta^3+\delta^5+\cdots)\right) = \frac{4-\delta}{1+\delta}.$$

There are then no profitable one-shot deviations whose differences from the equilibrium strategy appear after equilibrium histories if and only if

$$\delta \geq \tfrac{1}{4}.$$

However, when $\delta = 1/4$, the most attractive deviation from tit-for-tat in this game is perpetual shirking, which is *not* a one-shot deviation. For this deviation to be unprofitable, it must be that

$$3 \geq (1-\delta)4 + \delta = 4 - 3\delta,$$

and hence

$$\delta \geq \tfrac{1}{3}.$$

For $\delta \in [1/4, 1/3)$ tit-for-tat is thus not a Nash equilibrium, despite the absence of profitable one-shot deviations that differ from tit-for-tat only after equilibrium histories.

●

What goes wrong if we mimic the proof of proposition 2.2.1 in an effort to show that if there are no profitable one-shot deviations from equilibrium histories, then we have a Nash equilibrium? Proceeding again with the contrapositive, we would begin with a strategy profile that is not a Nash equilibrium. A profitable deviation may involve a deviation on the equilibrium path, as well as subsequent deviations off-the-equilibrium path. Beginning with a profitable deviation, and following the argument of the proof of proposition 2.2.1, we find a profitable one-shot deviation. The difficulty is that this one-shot deviation may occur off the equilibrium path. Although this is immaterial for subgame perfection, this difficulty scuttles the relationship between Nash equilibrium and profitable one-shot deviations along the equilibrium path.

2.3 Automaton Representations of Strategy Profiles

The one-shot deviation principle simplifies the set of alternative strategies we must check when evaluating subgame perfection. However, there still remains a potentially daunting number of histories to be evaluated. This evaluation can often be simplified by grouping histories into equivalence classes, where each member of an equivalence class induces an identical continuation strategy. We achieve this grouping by representing repeated-game strategies as automata, where the states of the automata represent equivalence classes of histories.

An *automaton* (or *machine*) $(\mathcal{W}, w^0, f, \tau)$ consists of a set of states $\mathcal{W}$, an initial state $w^0 \in \mathcal{W}$, an output (or decision) function $f: \mathcal{W} \to \prod_i \Delta(A_i)$ associating mixed action profiles with states, and a transition function, $\tau: \mathcal{W} \times A \to \mathcal{W}$. The transition function identifies the next state of the automaton, given its current state and the realized stage-game pure action profile.

If the function f specifies a pure output at state w, we write $f(w)$ for the resulting action profile. If a mixture is specified by f at w, $f^w(a)$ denotes the probability attached to profile a, so that $\sum_{a \in A} f^w(a) = 1$ (recall that we only consider mixtures over finite action spaces, see remark 2.1.1). We emphasize that even if two automata only differ in their initial state, they nonetheless are different automata.

Any automaton $(\mathcal{W}, w^0, f, \tau)$ with f specifying a pure action at every state induces an outcome $\{a^0, a^1, \ldots\}$ as follows:

$$a^0 = \sigma(\varnothing) = f(w^0),$$
$$a^1 = \sigma(a^0) = f(\tau(w^0, a^0)),$$
$$a^2 = \sigma(a^0, a^1) = f(\tau(\tau(w^0, a^0), a^1)),$$
$$\vdots$$

We extend this to identify the strategy induced by an automaton. First, extend the transition function from the domain $\mathcal{W} \times A$ to the domain $\mathcal{W} \times \mathcal{H} \backslash \{\varnothing\}$ by recursively defining

$$\tau(w, h^t) = \tau(\tau(w, h^{t-1}), a^{t-1}).$$

With this definition, we have the strategy σ described by $\sigma(\varnothing) = f(w^0)$ and

$$\sigma(h^t) = f(\tau(w^0, h^t)).$$

Similarly, an automaton for which f sometimes specifies mixed actions induces a path of play and a strategy.

Conversely, it is straightforward that any strategy profile can be represented by an automaton. Take the set of histories $\mathcal{H}$ as the set of states, the null history $\varnothing$ as the initial state, $f(h^t) = \sigma(h^t)$, and $\tau(h^t, a) = h^{t+1}$, where $h^{t+1} \equiv (h^t, a)$ is the concatenation of the history h^t with the action profile a.

This representation leaves us in the position of working with the full set of histories $\mathcal{H}$. However, strategy profiles can often be represented by automata with finite sets $\mathcal{W}$. The set $\mathcal{W}$ is then a partition on $\mathcal{H}$, grouping together those histories that prompt identical continuation strategies.

We say that a state $w' \in \mathscr{W}$ is *accessible* from another state $w \in \mathscr{W}$ if there exists a sequence of action profiles such that beginning at w, the automaton eventually reaches w'. More formally, there exists h^t such that $w' = \tau(w, h^t)$. Accessibility is not symmetric. Consequently, in an automaton $(\mathscr{W}, w^0, f, \tau)$, even if every state in $\mathscr{W}$ is accessible from the initial state w^0, this may not be true if some other state replaced w^0 as the initial state (see example 2.3.1).

Remark 2.3.1 **Individual automata** For most of parts I and II, it is sufficient to use a single automaton to represent strategy profiles. We can also represent a single strategy σ_i by an automaton $(\mathscr{W}_i, w_i^0, f_i, \tau_i)$. However, because every mixed strategy has a realization equivalent behavior strategy, we can always choose the automaton to have deterministic transitions, and so for any strategy profile represented by a collection of individual automata (one for each player), we can define a single "grand" automaton to represent the profile in the obvious way. The same is true for public strategies in public-monitoring games (the topic of much of part II), but is not true more generally (we will see examples in sections 5.1.2, 10.4.2, and 14.1.1).

Our use of automata (following Osborne and Rubinstein 1994) is also to be distinguished from a well-developed body of work, beginning with such publications as Neyman (1985), Rubinstein (1986), Abreu and Rubinstein (1988), and Kalai and Stanford (1988), that uses automata to *both* represent and impose restrictions on the complexity of strategies in repeated games. The technique in such studies is to represent each player's strategy as an automaton, replacing the repeated game with an automaton-choice game in which players choose the automata that will then implement their strategies. In particular, players in that work have preferences over the nature of the automaton, typically preferring automata with fewer states. In this book, players only have preferences over payoff streams and automata solely represent behavior.

◆

Remark 2.3.2 **Continuation profiles** When a strategy profile σ is described by the automaton $(\mathscr{W}, w^0, f, \tau)$, the continuation strategy profile after the history h^t, $\sigma|_{h^t}$, is described by the automaton obtained by using $\tau(w^0, h^t)$ as the initial state, that is, $(\mathscr{W}, \tau(w^0, h^t), f, \tau)$. If every state in $\mathscr{W}$ is accessible from w^0, then the collection of all continuation strategy profiles is described by the collection of automata $\{(\mathscr{W}, w, f, \tau) : w \in \mathscr{W}\}$.

◆

Example 2.3.1 We illustrate these ideas by presenting the automaton representation of a pair of players using grim trigger in the prisoners' dilemma (see figure 2.3.1). The set of states is $\mathscr{W} = \{w_{EE}, w_{SS}\}$, with output function $f(w_{EE}) = EE$ and $f(w_{SS}) = SS$. We thus have one state in which both players exert effort and one in which they both shirk. The initial state is w_{EE}. The transition function is given by

$$\tau(w, a) = \begin{cases} w_{EE}, & \text{if } w = w_{EE} \text{ and } a = EE, \\ w_{SS}, & \text{otherwise.} \end{cases} \quad (2.3.1)$$

Grim trigger is described by the automaton $(\mathscr{W}, w_{EE}, f, \tau)$, whereas the continuation strategy profile after any history in which EE is not played in at least one

2.3 ■ Automaton Representations

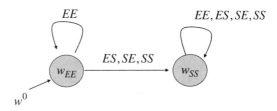

Figure 2.3.1 Automaton representation of grim trigger. Circles are states and arrows transitions, labeled by the profiles leading to the transitions. The subscript on a state indicates the action profile to be taken at that state.

period is described by $(\mathscr{W}, w_{SS}, f, \tau)$. Note that although every state in $\mathscr{W}$ is accessible from w_{EE}, the state w_{EE} is not accessible from w_{SS}.

●

The advantage of the automaton representation is that we need only verify the strategy profile induced by $(\mathscr{W}, w, f, \tau)$ is a Nash equilibrium, for each $w \in \mathscr{W}$, to confirm that the strategy profile induced by $(\mathscr{W}, w_0, f, \tau)$ is a subgame-perfect equilibrium. The following result is immediate from remark 2.3.2 (and its proof omitted).

Proposition 2.3.1 *The strategy profile with representing automaton $(\mathscr{W}, w^0, f, \tau)$ is a subgame-perfect equilibrium if and only if, for all $w \in \mathscr{W}$ accessible from w^0, the strategy profile induced by $(\mathscr{W}, w, f, \tau)$ is a Nash equilibrium of the repeated game.*

Each state of the automaton identifies an equivalence class of histories after which the strategies prescribe identical continuation play. The requirement that the strategy profile induced by each state of the automaton (i.e., by taking that state to be the initial state) corresponds to a Nash equilibrium is then equivalent to the requirement that we have Nash equilibrium continuation play after every history.

This result simplifies matters by transferring our concern from the set of histories $\mathscr{H}$ to the set of states of the automaton representation of a strategy. If $\mathscr{W}$ is simply the set of all histories $\mathscr{H}$, little has been gained. However, it is often the case that $\mathscr{W}$ is considerably smaller than the set of histories, with many histories associated with each state of $(\mathscr{W}, w, f, \tau)$, as example 2.3.1 shows is the case with grim trigger. Verifying that each state induces a Nash equilibrium is then much simpler than checking every history.

Remark 2.3.3 **Public correlation** The automaton representation of a pure strategy profile in the game with public correlation, $(\mathscr{W}, \mu^0, f, \tau)$ changes the description only in that the initial state is now randomly determined by a distribution $\mu^0 \in \Delta(\mathscr{W})$ and the transition function maps into probability distributions over states, that is, $\tau : \mathscr{W} \times A \to \Delta(\mathscr{W})$; $\tau_{w'}(w, a)$ is the probability that next period's state is w' when the current state is w and current pure action profile is a. A state $w' \in \mathscr{W}$ is *accessible* from $\mu \in \Delta(\mathscr{W})$ if there exists a sequence of action profiles such that beginning at some state in the support of μ, the automaton eventually reaches w' with positive probability. Proposition 2.3.1 holds as stated, with μ^0 replacing w^0, under public correlation.

◆

2.4 Credible Continuation Promises

This section uses the one-shot deviation principle to transform the task of checking that each state induces a Nash equilibrium in the repeated game to one of checking that each state induces a Nash equilibrium in an associated simultaneous move, or "one-shot" game.

Fix an automaton $(\mathcal{W}, w^0, f, \tau)$ where all $w \in \mathcal{W}$ are accessible from w^0, and let $V_i(w)$ be player i's average discounted value from play that begins in state w. That is, if play in the game follows the strategy profile induced by $(\mathcal{W}, w, f, \tau)$, then $V_i(w)$ is player i's average discounted payoff from the resulting outcome path. Although V_i can be calculated directly from the infinite sum, it is often easier to work with a recursive formulation, noting that at any state w, $V_i(w)$ is the average of current payoffs u_i (with weight $(1-\delta)$) and continuation payoffs (with weight δ). If the output function f is pure at w, then current payoffs are simply $u_i(f(w))$, whereas the continuation payoffs are $V_i(\tau(w, f(w)))$, because the current action profile $f(w)$ causes a transition from w to $\tau(w, f(w))$.

This extends to mixed-action profiles when A is finite. Payoffs are the expected value of flow payoffs under f, given by

$$\sum_a u_i(a) f^w(a),$$

and continuation payoffs are the expected value of the different states that are reached from the current realized action profile,

$$\sum_a V_i(\tau(w, a)) f^w(a).$$

Consequently, V_i satisfies the system of linear equations,

$$V_i(w) = (1-\delta) \sum_a u_i(a) f^w(a) + \delta \sum_a V_i(\tau(w, a)) f^w(a), \quad \forall w \in \mathcal{W}, \quad (2.4.1)$$

which has a unique solution in the space of bounded functions on $\mathcal{W}$.[10]

If an automaton with deterministic outputs is currently in state w, whether on the equilibrium path or as the result of a deviation, and if player i expects the other players to subsequently follow the strategy profile described by the automaton, then player i expects to receive a flow payoff of $u_i(a_i, f_{-i}(w))$ from playing a_i. The resulting action profile $(a_i, f_{-i}(w))$ then implies a transition to a new state $w' = \tau(w, (a_i, f_{-i}(w)))$. If *all* players follow the strategy profile in subsequent periods (a circumstance the one-shot deviation principle makes of interest), then player i expects a continuation value of $V_i(w')$. Accordingly, we interpret $V_i(w')$ as a *continuation promise* and view the profile as making such promises.

Intuitively, a subgame-perfect equilibrium strategy profile is one whose continuation promises are credible. Given the continuation promise $V_i(w')$ for player i at

---

10. The mapping described by the right side of (2.4.1) is a contraction on the space of bounded functions on $\mathcal{W}$, and so has a unique fixed point.

2.4 ■ Credible Continuation Promises

each state w', player i is willing to choose an action a_i' in the support of $f_i(w)$ if, for all $a_i \in A_i$,

$$(1-\delta)\sum_{a_{-i}} u_i(a_i', a_{-i})f^w(a_{-i}) + \delta \sum_{a_{-i}} V_i(\tau(w,(a_i', a_{-i})))f^w(a_{-i})$$
$$\geq (1-\delta)\sum_{a_{-i}} u_i(a_i, a_{-i})f^w(a_{-i}) + \delta \sum_{a_{-i}} V_i(\tau(w,(a_i, a_{-i})))f^w(a_{-i}),$$

or, equivalently, if for all $a_i \in A_i$,

$$V_i(w) \geq (1-\delta)\sum_{a_{-i}} u_i(a_i, a_{-i})f^w(a_{-i}) + \delta \sum_{a_{-i}} V_i(\tau(w,(a_i, a_{-i})))f^w(a_{-i}).$$

We say that continuation promises are *credible* if the corresponding inequality holds for each player and state.

Let $V(w) = (V_1(w), \ldots, V_n(w))$.

Proposition 2.4.1 *Suppose the strategy profile σ is described by the automaton $(\mathcal{W}, w^0, f, \tau)$. The strategy profile σ is a subgame-perfect equilibrium if and only if for all $w \in \mathcal{W}$ accessible from w^0, $f(w)$ is a Nash equilibrium of the normal form game described by the payoff function $g^w : A \to \mathbb{R}^n$, where*

$$g^w(a) = (1-\delta)u(a) + \delta V(\tau(w,a)). \tag{2.4.2}$$

Proof We give the argument for automata with deterministic output functions. The extension to mixtures is immediate but notationally cumbersome.

Let σ be the strategy profile induced by $(\mathcal{W}, w^0, f, \tau)$. We first show that if $f(w)$ is a Nash equilibrium in the normal-form game g^w, for every $w \in \mathcal{W}$, then there are no profitable one-shot deviations from σ. By proposition 2.2.1, this suffices to show that σ is subgame perfect. To do this, let $\hat{\sigma}_i$ be a one-shot deviation, $\hat{h}^t$ be the history for which $\sigma_i(\hat{h}^t) \neq \hat{\sigma}_i(\hat{h}^t)$, and $\hat{w}$ be the state reached by the history $\hat{h}^t$, that is, $\hat{w} = \tau(w^0, \hat{h}^t)$. Finally, let $a_i = \hat{\sigma}_i(\hat{h}^t)$. Then,

$$U_i(\sigma_i|_{\hat{h}^t}, \sigma_{-i}|_{\hat{h}^t}) = V_i(\hat{w}),$$

and, because $\hat{\sigma}_i$ is a one-shot deviation,

$$U_i(\hat{\sigma}_i|_{\hat{h}^t}, \sigma_{-i}|_{\hat{h}^t}) = (1-\delta)u_i\big(a_i, \sigma_{-i}|_{\hat{h}^t}(\varnothing)\big) + \delta V_i\big(\tau\big(\hat{w}, (a_i, \sigma_{-i}|_{\hat{h}^t}(\varnothing))\big)\big)$$
$$= (1-\delta)u_i(a_i, f_{-i}(\hat{w})) + \delta V_i(\tau(\hat{w},(a_i, f_{-i}(\hat{w})))).$$

Thus, if for all $\hat{w} \in \mathcal{W}$, $f(\hat{w})$ is a Nash equilibrium of the game $g^{\hat{w}}$, then no one-shot deviation is profitable.

Conversely, suppose there is some $\hat{w} \in \mathcal{W}$ accessible from w^0 and a_i such that

$$(1-\delta)u_i(a_i, f_{-i}(\hat{w})) + \delta V_i(\tau(\hat{w},(a_i, f_{-i}(\hat{w})))) > V_i(\hat{w}), \tag{2.4.3}$$

so $f(\hat{w})$ is not a Nash equilibrium of the game induced by $g^{\hat{w}}$. Again, by proposition 2.2.1, it suffices to show that there is a profitable one-shot deviation from σ.

	E	S
E	2, 2	−1, 3
S	3, −1	0, 0

Figure 2.4.1 The prisoners' dilemma from figure 1.2.1.

Because $\hat{w}$ is accessible from w^0, there is a history $\hat{h}^t$ such that $\hat{w} = \tau(w^0, \hat{h}^t)$. Let $\hat{\sigma}_i$ be the strategy defined by

$$\hat{\sigma}_i(h^\tau) = \begin{cases} a_i, & \text{if } h^\tau = \hat{h}^t, \\ \sigma_i(h^\tau), & \text{if } h^\tau \neq \hat{h}^t. \end{cases}$$

Then, $\hat{\sigma}_i$ is a one-shot deviation from σ_i, and it is profitable (from (2.4.3)). ∎

Public correlation does not introduce any complications, and so an essentially identical argument yields the following proposition (the notation is explained in remark 2.3.3).

Proposition 2.4.2 *In the game with public correlation, the strategy profile $(\mathcal{W}, \mu^0, f, \tau)$ is a subgame-perfect equilibrium if and only if for all $w \in \mathcal{W}$ accessible from μ^0, $f(w)$ is a Nash equilibrium of the normal form game described by the payoff function $g^w : A \to \mathbb{R}^n$, where*

$$g^w(a) = (1-\delta)u(a) + \delta \sum\nolimits_{w' \in \mathcal{W}} V(w')\tau_{w'}(w, a).$$

Example 2.4.1 Consider the infinitely repeated prisoners' dilemma with the stage game of figure 1.2.1, reproduced in figure 2.4.1. We argued in section 1.2 that grim trigger is a subgame-perfect equilibrium if $\delta \geq 1/3$, and in example 2.3.1 that the profile has an automaton representation (given in figure 2.3.1). It is straightforward to calculate $V(w_{EE}) = (2, 2)$ and $V(w_{SS}) = (0, 0)$. We can thus view the strategy profile as "promising" 2 if EE is played, and "promising" (or "threatening") 0 if not.

We now illustrate proposition 2.4.1. There are two one-shot games to be analyzed, one for each state. The game for $w = w_{EE}$ has payoffs

$$(1-\delta)u_i(a) + \delta V_i(\tau(w_{EE}, a)).$$

The associated payoff matrix is given in figure 2.4.2. If $\delta < 1/3$, SS is the only Nash equilibrium of the one-shot game, and so grim trigger cannot be a subgame-perfect equilibrium. On the other hand, if $\delta \geq 1/3$, both EE and SS are Nash equilibria. Hence, $f(w_{EE}) = EE$ is a Nash equilibrium, as required by proposition 2.4.1.[11]

---

11. There is no requirement that $f(w)$ be the only Nash equilibrium of the one-shot game.

2.4 ■ Credible Continuation Promises

	E	S
E	2, 2	$-(1-\delta), 3(1-\delta)$
S	$3(1-\delta), -(1-\delta)$	0, 0

Figure 2.4.2 The payoff matrix for $w = w_{EE}$.

	E	S
E	$2(1-\delta), 2(1-\delta)$	$-(1-\delta), 3(1-\delta)$
S	$3(1-\delta), -(1-\delta)$	0, 0

Figure 2.4.3 The payoff matrix for $w = w_{SS}$.

The payoff matrix of the one-shot game associated with w_{SS} is displayed in figure 2.4.3. For any value $\delta \in [0, 1)$, the only Nash equilibrium is SS, which is the action profile specified by f in the state w_{SS}. Putting these results together with proposition 2.4.1, grim trigger is a subgame-perfect equilibrium if and only if $\delta \geq 1/3$.

●

Just as there is no corresponding version of the one-shot deviation principle for Nash equilibrium, there is no result corresponding to proposition 2.4.1 for Nash equilibrium. We illustrate this in the final paragraph of example 2.4.2.

Example 2.4.2 We now consider tit-for-tat in the prisoners' dilemma of figure 2.2.1. An automaton representation of tit-for-tat is $\mathscr{W} = \{w_{EE}, w_{SS}, w_{ES}, w_{SE}\}$, $w^0 = w_{EE}$,

$$f(w_{a_1 a_2}) = a_1 a_2,$$

and

$$\tau(w_a, a'_1 a'_2) = w_{a'_2 a'_1}.$$

The induced outcome is that both players exert effort in every period, and the only state reached is w_{EE}.

The linear equations used to define the payoff function g^w for the associated one-shot games (see (2.4.2)), are

$$V_1(w_{EE}) = 3,$$
$$V_1(w_{SS}) = 1,$$
$$V_1(w_{ES}) = -(1-\delta) + \delta V_1(w_{SE}),$$

and

$$V_1(w_{SE}) = 4(1-\delta) + \delta V_1(w_{ES}).$$

Solving the last two equations gives

$$V_1(w_{ES}) = \frac{4\delta - 1}{1 + \delta}$$

and

$$V_1(w_{SE}) = \frac{4 - \delta}{1 + \delta}.$$

Now, *EE* is a Nash equilibrium of the game for $w = w_{EE}$ if

$$3 \geq (1 - \delta)4 + \delta V_1(w_{ES}) = (1 - \delta)4 + \frac{\delta(4\delta - 1)}{1 + \delta},$$

which is equivalent to

$$\delta \geq \tfrac{1}{4}.$$

Turning to the states w_{ES} and w_{SE}, *E* is a best response if

$$\frac{4\delta - 1}{1 + \delta} \geq (1 - \delta)1 + \delta 1 = 1,$$

giving

$$\delta \geq \tfrac{2}{3}.$$

Finally, *S* is a best response if

$$\frac{4 - \delta}{1 + \delta} \geq (1 - \delta)3 + \delta 3 = 3,$$

or

$$\delta \leq \tfrac{1}{4}.$$

Given the obvious inability to find a discount factor satisfying these various restrictions, we conclude that tit-for-tat is not subgame perfect.

The failure of a one-shot deviation principle for Nash equilibrium that we discussed earlier is also reflected here. For $\delta \geq 1/4$, the action profile *EE* is a Nash equilibrium of the game for state w_{EE}, the only state reached along the equilibrium path. This does *not* imply that tit-for-tat is a Nash equilibrium of the infinitely repeated game. Tit-for-tat is a Nash equilibrium only if a deviation to always defecting is unprofitable, or

$$3 \geq (1 - \delta)4 + \delta = 4 - 3\delta$$
$$\Rightarrow \delta \geq \tfrac{1}{3},$$

a more severe constraint than $\delta \geq 1/4$. We thus do not have a counterpart of proposition 2.4.1 for Nash equilibria (that would characterize Nash equilibria as corresponding to automata that generate Nash equilibria of the repeated game in every state reached in equilibrium).

2.5 Generating Equilibria

Many arguments in repeated games are based on constructive proofs. We are often interested in equilibria with certain properties, such as equilibria featuring the highest or lowest payoffs for some players, and the argument proceeds by exhibiting an equilibrium with the desired properties. When reasoning in this way, we are faced with the prospect of searching through a prohibitively immense set of possible equilibria. In the next two sections, we describe two complementary approaches for finding equilibria. This section, based on Abreu, Pearce, and Stacchetti (1990), introduces and illustrates the ideas of a self-generating set of equilibrium payoffs. Section 2.6, based on Abreu (1986, 1988), uses these results to introduce "simple" strategies and show that any subgame-perfect equilibrium can be obtained via such strategies.

For expositional clarity, we restrict attention to pure strategies in the next two sections. The notions of enforceability and pure-action decomposability, introduced in section 2.5.1, extend in an obvious way to mixed actions (and we do this in chapter 7), as do the notions of penal codes and optimal penal codes of section 2.6.

2.5.1 Constructing Equilibria: Self-Generation

We begin with a simple observation that has important implications. Denote the set of pure-strategy subgame-perfect equilibrium payoffs by $\mathscr{E}^p \subset \mathbb{R}^n$. For each $v \in \mathscr{E}^p$, σ^v denotes a pure-strategy subgame-perfect equilibrium yielding the payoff v. Suppose that for some action profile $a^* \in A$ there is a function, $\gamma : A \to \mathscr{E}^p$, with the property that, for all players i, and all $a_i \in A_i$,

$$(1-\delta)u_i(a^*) + \delta\gamma_i(a^*) \geq (1-\delta)u_i(a_i, a^*_{-i}) + \delta\gamma_i(a_i, a^*_{-i}).$$

Consider the strategy profile that specifies the action profile a^* in the initial period, and after any action profile $a \in A$ plays according to $\sigma^{\gamma(a)}$. Proposition 2.4.1 implies that this profile is a subgame-perfect equilibrium, with a value of $(1-\delta)u_i(a^*) + \delta\gamma_i(a^*) \in \mathscr{E}^p$.

Suppose now that instead of taking $\mathscr{E}^p$ as the range for γ, we take instead some arbitrary subset $\mathscr{W}$ of the set of feasible payoffs, $\mathscr{F}^\dagger$.

Definition 2.5.1 *A pure action profile a^* is enforceable on $\mathscr{W}$ if there exists some specification of (not necessarily credible) continuation promises $\gamma : A \to \mathscr{W}$ such that, for all players i, and all $a_i \in A_i$,*

$$(1-\delta)u_i(a^*) + \delta\gamma_i(a^*) \geq (1-\delta)u_i(a_i, a^*_{-i}) + \delta\gamma_i(a_i, a^*_{-i}).$$

In other words, when the other players play their part of an enforceable profile a^*, the continuation promises γ_i make (enforce) the choice of a_i^* optimal (incentive compatible) for i.

Definition 2.5.2 *A payoff $v \in \mathscr{F}^\dagger$ is pure-action decomposable on $\mathscr{W}$ if there exists a pure action profile a^* enforceable on $\mathscr{W}$ such that*

$$v_i = (1-\delta)u_i(a^*) + \delta\gamma_i(a^*),$$

where γ is a function enforcing a^*.

If a payoff vector is pure-action decomposable on $\mathscr{W}$, then it is "one-period credible" with respect to promises in the set $\mathscr{W}$. If those promises are themselves pure-action decomposable on $\mathscr{W}$, then the original payoff vector is "two-period credible." If a set of payoffs $\mathscr{W}$ is pure-action decomposable on itself, then each such payoff has "infinite-period credibility." One might expect such a payoff vector to be a pure-strategy subgame-perfect equilibrium payoff, and indeed this is correct.

Proposition 2.5.1 *Any set of payoffs $\mathscr{W} \subset \mathscr{F}^\dagger$ with the property that every payoff in $\mathscr{W}$ is pure-action decomposable on $\mathscr{W}$ is a set of pure-strategy subgame-perfect equilibrium payoffs.*

The proof views $\mathscr{W}$ as the set of states for an automaton. Pure-action decomposability allows us to associate an action profile and a transition function describing continuation values to each payoff profile in $\mathscr{W}$.

Proof For each payoff profile $v \in \mathscr{W}$, let $\tilde{a}(v)$ and $\gamma^v : A \to \mathscr{W}$ be the decomposing pure-action profile and its enforcing continuation promise. Consider the collection of automata $\{(\mathscr{W}, v, f, \tau) : v \in \mathscr{W}\}$, where the common set of states is given by $\mathscr{W}$, the common decision function by

$$f(v) = \tilde{a}(v)$$

for all $v \in \mathscr{W}$, and the common transition function by

$$\tau(v, a) = \gamma^v(a),$$

for all $a \in A$. These automata differ only in their initial state $v \in \mathscr{W}$.

We need to show that for each $v \in \mathscr{W}$, the automaton $(\mathscr{W}, v, f, \tau)$ describes a subgame-perfect equilibrium with payoff v. This will be an implication of proposition 2.4.1 and the pure-action decomposability of each $v \in \mathscr{W}$, once we have shown that

$$v_i = V_i(v),$$

where $V_i(v)$ is the value to player i of being in state v.

Because each $v \in \mathscr{W}$ is decomposable by a pure action profile, for any v we can define a sequence of payoff-action profile pairs $\{(v^k, a^k)\}_{k=0}^\infty$ as follows: $v^0 = v$, $a^0 = \tilde{a}(v^0)$, $v^k = \gamma^{v^{k-1}}(a^{k-1})$, and $a^k = \tilde{a}(v^k)$. We then have

$$v_i = (1 - \delta)u_i(a^0) + \delta v_i^1$$
$$= (1 - \delta)u_i(a^0) + \delta\{(1 - \delta)u_i(a^1) + \delta v_i^2\}$$
$$= (1 - \delta)\sum_{s=0}^{t-1} \delta^s u_i(a^s) + \delta^t v_i^t.$$

Since $v_i^t \in \mathscr{F}^\dagger$, $\{v_i^t\}$ is a bounded sequence, and taking $t \to \infty$ yields

$$v_i = (1 - \delta)\sum_{s=0}^\infty \delta^s u_i(a^s)$$

and so $v_i = V_i(v)$. ■

2.5 ■ Generating Equilibria

Definition 2.5.3 *A set $\mathscr{W}$ is* pure-action self-generating *if every payoff in $\mathscr{W}$ is pure-action decomposable on $\mathscr{W}$.*

An immediate implication of this result is the following important result.

Corollary 2.5.1 *The set $\mathscr{E}^p$ of pure-strategy subgame-perfect equilibrium payoffs is the largest pure-action self-generating set.*

In particular, it is clear that any pure-strategy subgame-perfect equilibrium payoff is pure-action decomposable on the set $\mathscr{E}^p$, because this is simply the statement that every history must give rise to continuation play that is itself a pure-strategy subgame-perfect equilibrium. The set $\mathscr{E}^P$ is thus pure-action self-generating. Proposition 2.5.1 implies that any other pure-action self-generating set is also a set of pure-strategy subgame-perfect equilibria, and hence must be a subset of $\mathscr{E}^p$.

Remark 2.5.1 **Fixed point characterization of self-generation** For any set $\mathscr{W} \subset \mathscr{F}^\dagger$, let $\mathscr{B}_a(\mathscr{W})$ be the set of payoffs $v \in \mathscr{F}^\dagger$ decomposed by $a \in A$ and continuations in $\mathscr{W}$. Corollary 2.5.1 can then be written as: The set $\mathscr{E}^p$ is the largest set $\mathscr{W}$ satisfying

$$\mathscr{W} = \cup_{a \in A} \mathscr{B}_a(\mathscr{W}).$$

When players have access to a public correlating device, a payoff v is decomposed by a distribution over action profiles α and continuation values. Hence the set of payoffs that can be decomposed on a set $\mathscr{W}$ using public correlation is

$$\operatorname{co}(\cup_{a \in A} \mathscr{B}_a(\operatorname{co}\mathscr{W})),$$

where $\operatorname{co}\mathscr{W}$ denotes the convex hull of $\mathscr{W}$. Hence, the set of equilibrium payoffs under public correlation is the largest convex set $\mathscr{W}$ satisfying

$$\mathscr{W} = \operatorname{co}(\cup_{a \in A} \mathscr{B}_a(\mathscr{W})).$$

◆

We then have the following useful result (a similar proof shows that the set of subgame-perfect equilibrium payoffs is compact).

Proposition 2.5.2 *The set $\mathscr{E}^p \subset \mathbb{R}^n$ of pure-strategy subgame-equilibrium payoffs is compact.*

Proof Because $\mathscr{E}^p$ is a bounded subset of a Euclidean space, it suffices (from corollary 2.5.1) to show that its closure $\overline{\mathscr{E}^p}$ is pure-action self-generating. Let v be a payoff profile in $\overline{\mathscr{E}^p}$, and suppose $\{v^{(\ell)}\}_{\ell=0}^\infty$ is a sequence of pure-strategy subgame-perfect equilibrium payoffs, converging to v, with each $v^{(\ell)}$ decomposable via the action profile $a^{(\ell)}$ and continuation promise $\gamma^{(\ell)}$.

If every A_i is finite, $\{(a^{(\ell)}, \gamma^{(\ell)})\}_\ell$ lies in the compact set $A \times (\overline{\mathscr{E}^p})^A$, and so there is a subsequence converging to $(a^\infty, \gamma^\infty)$, with $a^\infty \in A$ and $\gamma^\infty(a) \in \overline{\mathscr{E}^p}$ for all $a \in A$. Moreover, it is immediate that $(a^\infty, \gamma^\infty)$ decomposes v on $\overline{\mathscr{E}^p}$.

Suppose now that A_i is a continuum action space for some i. Although $(\overline{\mathscr{E}^p})^A$ is not sequentially compact, we can proceed as follows. For each ℓ and i, let $\underline{v}^{(\ell),i} \in \overline{\mathscr{E}^p}$ such that

$$\underline{v}_i^{(\ell),i} = \inf_{a_i' \neq a_i^{(\ell)}} \gamma_i^{(\ell)}(a_i', a_{-i}^{(\ell)})$$

and

$$\hat{\gamma}^{(\ell)}(a) = \begin{cases} \underline{v}^{(\ell),i}, & \text{if } a_i \neq a_i^{(\ell)} \text{ and } a_{-i} = a_{-i}^{(\ell)} \text{ for some } i, \\ \gamma^{(\ell)}(a^{(\ell)}), & \text{otherwise.} \end{cases}$$

Clearly, each $v^{(\ell)}$ is decomposed by $a^{(\ell)}$ and $\hat{\gamma}^{(\ell)}$. Let $\mathscr{A}^{(\ell)}$ denote the finite partition of A given by $\{A^{(\ell),k} : k = 0, \ldots, n\}$, where $A^{(\ell),i} = \{a \in A : a_i \neq a_i^{(\ell)}, a_{-i} = a_{-i}^{(\ell)}\}$ and $A^{(\ell),0} = A \setminus \cup_i A^{(\ell),i}$. Because $\hat{\gamma}^{(\ell)}$ is measurable with respect to $\mathscr{A}^{(\ell)}$, which has $n+1$ elements, we can treat $\hat{\gamma}^{(\ell)}$ as a function from the finite set $\{0, \ldots, n\}$ into $\overline{\mathscr{E}^p}$, and so there is a convergent subsequence, and the argument is completed as for finite A_i.

∎

2.5.2 Example: Mutual Effort

This and the following two subsections illustrate decomposability and self-generation. We work with the prisoners' dilemma shown in figure 2.5.1.

We first identify the set of discount factors for which there exists a subgame-perfect equilibrium in which both players exert effort in every period. In light of proposition 2.5.1, this is equivalent to identifying the discount factors for which there is a self-generating set of payoffs $\mathscr{W}$ containing $(2, 2)$. If such a set $\mathscr{W}$ is to exist, then the action profile EE is enforceable on $\mathscr{W}$, or,

$$(1-\delta)2 + \delta\gamma_1(EE) \geq (1-\delta)b + \delta\gamma_1(SE)$$

and

$$(1-\delta)2 + \delta\gamma_2(EE) \geq (1-\delta)b + \delta\gamma_2(ES),$$

for $\gamma(EE)$, $\gamma(SE)$ and $\gamma(ES)$ in $\mathscr{W} \subset \mathscr{F}^\dagger$. These inequalities are least restrictive when $\gamma_1(SE) = \gamma_2(ES) = 0$. Because the singleton set of payoffs $\{(0,0)\}$ is itself self-generating, we sacrifice no generality by assuming the self-generating set contains $(0, 0)$. We can then set $\gamma_1(SE) = \gamma_2(ES) = 0$. Similarly, the pair of inequalities is least restrictive when $\gamma_i(EE) = 2$ for $i = 1, 2$. Taking this to the case, the inequalities hold when

	E	S
E	2, 2	−c, b
S	b, −c	0, 0

Figure 2.5.1 Prisoners' dilemma, where $b > 2$, $c > 0$, and $b - c < 4$.

2.5 ■ Generating Equilibria

$$\delta \geq \frac{b-2}{b}. \qquad (2.5.1)$$

The inequality given by (2.5.1) is thus necessary for the existence of a subgame-perfect equilibrium giving payoff (2, 2). The inequality is sufficient as well, because it implies that the set $\{(0, 0), (2, 2)\}$ is self-generating. For the prisoners' dilemma of figure 2.4.1, we have the familiar result that $\delta \geq 1/3$.

2.5.3 Example: The Folk Theorem

We next identify the set of discount factors under which a pure-strategy folk theorem result holds, in the sense that the set of pure strategy subgame-perfect equilibrium payoffs $\mathscr{E}^P$ contains the set of feasible, strictly individually rational payoffs $\mathscr{F}^{\dagger p}$, which equals $\mathscr{F}^*$ for the prisoners' dilemma. Hence, we seek a pure-action self-generating set $\mathscr{W}$ containing $\mathscr{F}^*$. Because the set of pure-strategy subgame-perfect equilibrium payoffs is compact, if $\mathscr{F}^*$ is pure-action self-generating, then so is its closure $\overline{\mathscr{F}^*}$. The payoff profiles in $\overline{\mathscr{F}^*}$ are feasible and *weakly* individually rational, differing from those in $\mathscr{F}^*$ by including profiles in which one or both players receive their minmax payoff of 0. Because the set of pure-action subgame-perfect equilibria is the largest pure-action self-generating set, our candidate for the pure-action self-generating set $\mathscr{W}$ must be $\overline{\mathscr{F}^*}$.

If $\overline{\mathscr{F}^*}$ is pure-action self-generating, then every $v \in \overline{\mathscr{F}^*}$ is precisely the payoff of some pure strategy equilibrium. Note that this includes those v that are convex combinations of payoffs in $\mathscr{F}^p$ with *irrational* weights (reflecting the denseness of the rationals in the reals).

We begin by identifying the sets of payoffs that are pure-action decomposable using the four pure action profiles EE, ES, SE, and SS, and continuation payoffs in $\overline{\mathscr{F}^*}$.

Consider first EE. This action profile is enforced by γ on $\overline{\mathscr{F}^*}$ if

$$(1-\delta)2 + \delta\gamma_1(EE) \geq (1-\delta)b + \delta\gamma_1(SE)$$

and

$$(1-\delta)2 + \delta\gamma_2(EE) \geq (1-\delta)b + \delta\gamma_2(ES).$$

We are interested in the set of payoffs that can be decomposed using the action profile EE. The first step in finding this set is to identify the continuation payoffs $\gamma(EE)$ that are consistent with enforcing EE. Because $v_i \geq 0$ for all $v \in \overline{\mathscr{F}^*}$ and $(0, 0) \in \overline{\mathscr{F}^*}$, the largest set of values of $\gamma(EE)$ consistent with enforcing EE is found by setting $\gamma_1(SE) = \gamma_2(ES) = 0$ (which minimizes the right side). The set A_{EE} of continuation payoffs $\gamma(EE)$ consistent with enforcing EE is, then,

$$A_{EE} = \left\{ \gamma \in \overline{\mathscr{F}^*} : \gamma_i \geq \frac{(b-2)(1-\delta)}{\delta} \right\},$$

and the set of payoffs that are decomposable using EE and $\overline{\mathscr{F}^*}$ is

$$B_{EE} = \{\gamma \in \overline{\mathscr{F}^*} : \gamma = (1-\delta)(2, 2) + \delta\gamma(EE), \gamma(EE) \in A_{EE}\}.$$

This set is simply $\mathscr{B}_{EE}(\overline{\mathscr{F}^*})$ (see remark 2.5.1).

The incentive constraints for the action profile ES are

$$-(1-\delta)c + \delta\gamma_1(ES) \geq \delta\gamma_1(SS)$$

and

$$(1-\delta)b + \delta\gamma_2(ES) \geq (1-\delta)2 + \delta\gamma_2(EE).$$

We are again interested in the largest set of continuation values, in this case values $\gamma(ES)$, consistent with the incentive constraints. The second inequality can be ignored, because we can set $\gamma(EE) = \gamma(ES)$, implying $\gamma_2(EE) = \gamma_2(ES)$. As before, we minimize the right side of the first inequality by setting $\gamma_1(SS) = 0$ (which we can do by taking $\gamma(SS) = (0,0)$). Hence, the set of continuation payoffs $\gamma(ES)$ consistent with enforcing the profile ES is

$$A_{ES} = \left\{ \gamma \in \overline{\mathscr{F}^*} : \gamma_1 \geq \frac{c(1-\delta)}{\delta} \right\}$$

and the set of payoffs decomposable using ES and $\overline{\mathscr{F}^*}$ is

$$B_{ES} = \{\gamma \in \overline{\mathscr{F}^*} : \gamma = (1-\delta)(-c,b) + \delta\gamma(ES), \gamma(ES) \in A_{ES}\}.$$

A similar argument shows that the set of continuation payoffs consistent with enforcing the profile SE is

$$A_{SE} = \left\{ \gamma \in \overline{\mathscr{F}^*} : \gamma_2 \geq \frac{c(1-\delta)}{\delta} \right\},$$

and the set of payoffs decomposable using SE and $\overline{\mathscr{F}^*}$ is

$$B_{SE} = \{\gamma \in \overline{\mathscr{F}^*} : \gamma = (1-\delta)(b,-c) + \delta\gamma(SE), \gamma(SE) \in A_{SE}\}.$$

Finally, because SS is a Nash equilibrium of the stage game, and hence no appeal to continuation payoffs need be made when constructing current incentives to play SS, the set of continuation payoffs consistent with enforcing SS as an initial period action profile is the set

$$A_{SS} = \overline{\mathscr{F}^*},$$

and the set of payoffs decomposable using SS and $\overline{\mathscr{F}^*}$ is

$$B_{SS} = \{\gamma \in \overline{\mathscr{F}^*} : \gamma = \delta\gamma(SS), \gamma(SS) \in \overline{\mathscr{F}^*}\}.$$

A geometric representation of the sets $B_{a_1 a_2}$ is helpful. We can write them as:

$$B_{EE} = \{v \in (1-\delta)(2,2) + \delta\overline{\mathscr{F}^*} : v_i \geq (1-\delta)b, i = 1,2\},$$
$$B_{ES} = \{v \in (1-\delta)(-c,b) + \delta\overline{\mathscr{F}^*} : v_1 \geq 0\},$$
$$B_{SE} = \{v \in (1-\delta)(b,-c) + \delta\overline{\mathscr{F}^*} : v_2 \geq 0\},$$

and

$$B_{SS} = \delta\overline{\mathscr{F}^*}.$$

Each set is a subset of the convex combination of the flow payoffs implied by the relevant pure action profile and $\overline{\mathscr{F}^*}$, with the appropriate restriction implied by incentive compatibility. For example, the restriction in A_{EE} that $\gamma_i(EE) \geq (b-2)(1-\delta)/\delta$ implies

2.5 ■ Generating Equilibria

$$v_i = (1-\delta)2 + \delta \gamma_i(EE) \geq (1-\delta)2 + \delta\left(\frac{(b-2)(1-\delta)}{\delta}\right) = (1-\delta)b.$$

Similarly, the restriction in A_{ES} that $\gamma_1(ES) \geq c(1-\delta)/\delta$ implies

$$v_1 \geq (1-\delta)(-c) + \delta\frac{c(1-\delta)}{\delta} = 0.$$

Figure 2.5.2 illustrates these sets for a case in which they are nonempty and do not intersect. The set B_{EE} is nonempty when (2.5.1) holds (i.e., $\delta \geq (b-2)/b$) and includes all those payoff profiles in $\overline{\mathscr{F}^*}$ whose components both exceed $(1-\delta)b$. Hence, when nonempty, B_{EE} includes the upper right corner of the set $\overline{\mathscr{F}^*}$.

The set B_{SS} is necessarily nonempty and reproduces $\overline{\mathscr{F}^*}$ in miniature, anchored at the origin. The set B_{ES} is in the upper left, if it is nonempty, consisting of a shape with either three or four sides, one of which is the line $\gamma_2 = (1-\delta)b$ and one of which is the vertical axis, with the remaining two sides (or single side) being parallel to the efficient frontier (parallel to the efficient frontier between $(2,2)$ and the horizontal axis). Whether this shape has three or four sides depends on whether $(1-\delta)(-c,b) + \delta(2,2)$, the payoff generated by $(2,2)$, lies to the left (three sides) or right (four sides) of the vertical axis.

When do the sets B_{EE}, B_{ES}, B_{SE}, and B_{SS} have $\overline{\mathscr{F}^*}$ as their union? If they do, then $\overline{\mathscr{F}^*}$ is self-generating, and hence we can support any payoff in $\overline{\mathscr{F}^*}$ as a subgame-perfect equilibrium payoff. From figure 2.5.2 we can derive a pair of simple necessary and sufficient conditions for this to be the case. First, the point $(1-\delta)(-c,b) + \delta(2,2)$, the top right corner of B_{ES}, must lie to the right of the top-left corner of B_{EE}. This requires

$$(1-\delta)(-c) + \delta 2 \geq (1-\delta)b.$$

Applying symmetric reasoning to B_{SE}, this condition will suffice to ensure that any payoff in $\overline{\mathscr{F}^*}$ with at least one component above $(1-\delta)b$ is contained in either B_{EE},

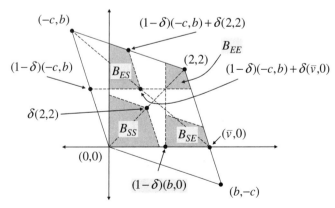

Figure 2.5.2 Illustration of the sets B_{EE}, B_{ES}, B_{SE}, and B_{SS}. This is drawn for $\delta = 1/2$ and the payoffs from figure 2.4.1. Because $b = c + 2$, the right edge of B_{ES} is a continuation of the right edge of B_{SS}; a similar comment applies to the top edges of B_{SS} and B_{SE}.

B_{ES}, or B_{SE}. To ensure that B_{SS} contains the rest, it suffices that the top right corner of B_{SS} lie above the bottom left boundary of B_{EE}, which is

$$\delta 2 \geq (1-\delta)b.$$

Solving these two inequalities, we have[12]

$$\delta \geq \max\left\{\frac{b+c}{b+c+2}, \frac{b}{b+2}\right\} = \frac{b+c}{b+c+2}.$$

For the prisoners' dilemma of figure 2.4.1, we have $\delta \geq 2/3$. The previous section showed that when $\delta \geq 1/3$, there exist equilibria supporting mutual effort in every period. When $\delta \geq 2/3$, there are also equilibria in which every other feasible weakly individually rational payoff profile is an equilibrium outcome.

2.5.4 Example: Constructing Equilibria for Low δ

We now return to the prisoners' dilemma in figure 2.4.1. We examine the set of subgame-perfect equilibria in which player 2 always plays E. Mailath, Obara, and Sekiguchi (2002) provide a detailed analysis. We proceed just far enough to illustrate how equilibria can be constructed using decomposability. Deviations from the candidate equilibrium outcome path trigger a switch to perpetual shirking, the most severe punishment available. It is immediate that if player 2's incentive constraint is satisfied, then so is player 1's, so we focus on player 2.

Let γ_2^t denote the normalized discounted value to player 2 of the continuation outcome path $\{(a_1^\tau, E)\}_{\tau=t}^\infty$ beginning in period t. Given the punishment of persistent defection, the condition that it be optimal for player 2 to continue with the equilibrium action of effort in period $t \geq 0$ is

$$(1-\delta)u_2(a_1^t, E) + \delta\gamma_2^{t+1} \geq (1-\delta)u_2(a_1^t, S) + \delta \times 0,$$

which holds if and only if

$$\gamma_2^{t+1} \geq \frac{1-\delta}{\delta}. \tag{2.5.2}$$

Thus, player 2's continuation value is always at least $(1-\delta)/\delta$ in any equilibrium in which 2 currently exerts effort.

Denote by $\mathscr{W}_2^{EE}$ the set of payoffs for player 2 that can be decomposed through a combination of mutual effort (EE) in the current period coupled with a payoff $\gamma_2 \in [(1-\delta)/\delta, 2]$. Then we have

$$v_2 \in \mathscr{W}_2^{EE} \iff \exists \gamma_2 \in [(1-\delta)/\delta, 2]$$
$$\text{s.t. } v_2 = (1-\delta)u_2(EE) + \delta\gamma_2 = (1-\delta)2 + \delta\gamma_2.$$

Hence, $\mathscr{W}_2^{EE} = [3 - 3\delta, 2]$.

---

12. Notice that if this condition is satisfied, we automatically have $\delta > (b-2)/b$, ensuring that B_{EE} is nonempty.

2.5 ■ Generating Equilibria

Similarly, denote by $\mathcal{W}_2^{SE}$ the set of payoffs for player 2 that can be decomposed using current play of *SE* and a continuation payoff $\gamma_2 \in [(1-\delta)/\delta, 2]$:

$$v_2 \in \mathcal{W}_2^{SE} \iff \exists \gamma_2 \in [(1-\delta)/\delta, 2]$$
$$\text{s.t. } v_2 = (1-\delta)u_2(SE) + \delta\gamma_2 = (1-\delta)(-1) + \delta\gamma_2.$$

This yields $\mathcal{W}_2^{SE} = [0, 3\delta - 1]$.

Collecting these results, we have

$$\mathcal{W}_2^{EE} = [3 - 3\delta, 2] \tag{2.5.3}$$

and

$$\mathcal{W}_2^{SE} = [0, \; 3\delta - 1]. \tag{2.5.4}$$

We now consider several possible values of the discount factor. First, suppose $\delta < 1/3$. Then it is immediate from (2.5.3)–(2.5.4) that $\mathcal{W}_2^{EE}$ and $\mathcal{W}_2^{SE}$ are both empty. Hence, there are no continuation payoffs available that can induce player 2 to exert effort in the current period, regardless of player 1's behavior. Applying an analogous argument to player 1, the only possible equilibrium payoff for this case ($\delta < 1/3$) is $(0, 0)$, obtained by perpetual shirking.

Suppose instead that $\delta \geq 2/3$. Then (2.5.3)–(2.5.4) imply that $\mathcal{W}_2^{EE} \cup \mathcal{W}_2^{SE} = [0, 2]$. In this case, proposition 2.5.1 implies that *every* payoff on the segment $\{(v_1, v_2): v_1 = \frac{8}{3} - \frac{v_2}{3}, v_2 \in [0, 2]\}$ (including irrational values) can be supported as an equilibrium payoff in the first period. This case is illustrated in figure 2.5.3. The line *SE* is given by the equation $\gamma_2 = (1-\delta)/\delta + v_2/\delta$ for $v_2 \in \mathcal{W}_2^{SE}$. The line *EE* is given by the equation $\gamma_2 = -2(1-\delta)/\delta + v_2/\delta$ for $v_2 \in \mathcal{W}_2^{EE}$. For any payoff in $[0, 2]$, either $v_2 \in \mathcal{W}_2^{EE}$ or $\mathcal{W}_2^{SE}$. If $v_2 \in \mathcal{W}_2^{EE}$, then the payoff v_2 can be achieved by coupling

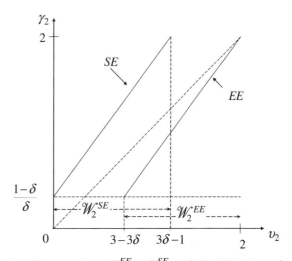

Figure 2.5.3 The case where $\mathcal{W}_2^{EE} \cup \mathcal{W}_2^{SE} = [0, 2]$; this is drawn for $\delta = 3/4$. The equation for the line labeled *SE* is $\gamma_2 = (1-\delta)/\delta + v_2/\delta$, and for the line *EE* is $\gamma_2 = -2(1-\delta)/\delta + v_2/\delta$.

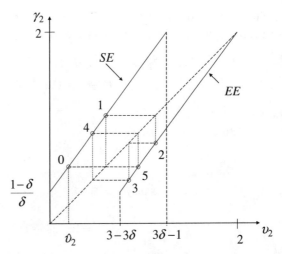

Figure 2.5.4 An illustration of the recursion (2.5.5) and (2.5.6) for an equilibrium with player 2 payoff $\hat{v}_2$, for $\delta = 3/4$. The outcome path is $(SE, (SE, EE, EE, SE, EE)^\infty)$, with the path after the initial period corresponding to the labeled cycle 12345.

current play of EE with a continuation payoff $\gamma_2 \in [(1-\delta)/\delta, 2]$ (yielding a point (v_2, γ_2) on the line labeled EE). If $v_2 \in \mathscr{W}_2^{SE}$, then v_2 can be achieved by coupling current play of SE with a continuation payoff $\gamma_2 \in [(1-\delta)/\delta, 2]$ (yielding a point (v_2, γ_2) on the line SE).

To construct an equilibrium with payoff to player 2 of $v_2 \in [0, 1]$, we proceed recursively as follows: Set $\gamma_2^0 = v_2$. Then,[13]

$$\gamma_2^{t+1} = \begin{cases} -2(1-\delta)/\delta + \gamma_2^t/\delta, & \text{if } \gamma_2^t \in \mathscr{W}_2^{EE}, \\ (1-\delta)/\delta + \gamma_2^t/\delta, & \text{if } \gamma_2^t \in [0, 3-3\delta). \end{cases} \quad (2.5.5)$$

An outcome path yielding payoff v_2 to player 2 is then given by

$$a^t = \begin{cases} EE, & \text{if } \gamma_2^t \in \mathscr{W}_2^{EE}, \\ SE, & \text{if } \gamma_2^t \in [0, 3-3\delta). \end{cases} \quad (2.5.6)$$

This recursion is illustrated in figure 2.5.4.

2.5.5 Example: Failure of Monotonicity

The analysis of the previous subsection suggests that the set of pure-strategy equilibrium payoffs is a monotonic function of the discount factor. When δ is less than $1/3$, the set of equilibrium payoffs is a singleton, containing only $(0, 0)$. At $\delta = 1/3$, we add the payoff $(8/3, 0)$ (from the equilibrium outcome path (SE, EE^∞)). For δ in the interval

---

13. There are potentially many equilibria with player 2 payoff v_2, arising from the possibility that a continuation falls into both $\mathscr{W}_2^{EE}$ and $\mathscr{W}_2^{SE}$. We construct the equilibrium corresponding to decomposing the continuation on $\mathscr{W}_2^{EE}$ whenever possible.

2.5 ■ Generating Equilibria

$[2/3, 1)$, we have all the payoffs in the set $\{(v_1, v_2) : v_1 = \frac{8}{3} - \frac{v_2}{3}, v_2 \in [0, 2]\}$. The next lemma, however, implies that the set of pure-strategy equilibrium payoffs is not monotonic.

Lemma 2.5.1 *For $\delta \in [1/3, 0.45)$, in every pure subgame-perfect outcome, if a player plays E in period t, then the opponent must play E in period $t + 1$. Consequently, the equilibrium payoff to player 1 (consistent with player 2 always exerting effort) is maximized by the outcome path (SE, EE^∞). Player 1's maximum payoff over the set of all subgame-perfect equilibria is decreasing in δ in this range.*

Proof Consider an outcome path in which a player (say, 2) is supposed to play E in period t and the other player is supposed to play S in period $t + 1$. As we saw from (2.5.2), we need player 2's period $t+1$ continuation value to be at least $(1 - \delta)/\delta$ to support such actions as part of equilibrium behavior. However, the continuation value is given by

$$(1 - \delta)u_2(Sa_2^{t+1}) + \delta\gamma_2^{t+2} \leq (1 - \delta) \times 0 + \delta\tfrac{8}{3} = \tfrac{8\delta}{3},$$

and $8\delta/3 \geq (1 - \delta)/\delta$ requires $\delta > 0.45$, a contradiction. Hence, if a player exerts effort, the opponent must exert effort in the next period.

Player 1's stage-game payoff is maximized by playing S while 2 plays E. Hence, player 1's equilibrium payoff is maximized, subject to 2 always exerting effort, by either the outcome path EE^∞ or (SE, EE^∞). If there is to be any profitable one-shot deviation from the latter, it must be profitable for player 2 to defect in the first period or player 1 to defect in the second period. A calculation shows that because $\delta \in [1/3, 0.45)$, neither deviation is profitable. Hence, the latter path is an equilibrium outcome path, supported by punishments of mutual perpetual shirking. Moreover, player 1's payoff from this equilibrium is a decreasing function of the discount factor, maximized at $\delta = 1/3$.

Finally, we need to show that, when $\delta \in [1/3, 0.45)$, the outcome path (SE, EE^∞) yields a higher payoff to player 1 than any other equilibrium outcome. Because $\delta < 0.8$, the outcome path (SE, EE^∞) yields a higher player 1 payoff than (ES, SE^∞), and so the outcome path **a** that maximizes 1's payoff must have SE in period 0, and by the first claim in the lemma, 1 must play E in period 1. If player 2 plays E as well, then the resulting outcome path is (SE, EE^∞). Suppose $\mathbf{a} \neq (SE, EE^\infty)$, so that player 2 plays S in period 1. But the outcome path (SE, EE^∞) yields a higher player 1 payoff than (SE, ES, SE^∞), a contradiction.
∎

Lemma 2.5.1 is an instance of an important general phenomenon. For the outcome path (SE, EE^∞) and $\delta \in (1/3, 0.45)$, player 2's incentive constraint holds strictly in every period. In other words, player 2's incentive constraint is still satisfied when 2's equilibrium continuation value is reduced by a small amount. This suggests that we should be able to increase player 1's total payoff by a corresponding small amount. However, there is a discreteness in incentives: The only way to increase player 1's total payoff is for him to play S in some future period, and, as lemma 2.5.1 reveals, for $\delta < 0.45$, this is inconsistent with player 2 playing E in the previous period. That is, the effect on payoffs of having player 1 choose S in some future period is sufficiently large to violate 2's incentive constraints. It is not possible to increase player 1's value by a small enough amount that player 2's incentive constraint is preserved.

There remains the possibility of mixing: Could we produce a smaller increase in player 1's value while preserving player 2's incentives by having player 1 choose S in some future period with a probability less than 1? Such a randomization will preserve player 2's incentives. However, it will not increase player 1's value, because in equilibrium, player 1 must be indifferent between E and S in any period in which he is supposed to randomize.

A public correlating device allows an escape from these constraints by allowing 1 to play S and E with positive probability in the same period, conditioning on the public signal, without indifference between S and E. In particular, player 1 can now be punished for not playing the action appropriate for the realization of the public signal, allowing incentives for mixing without indifference. Section 2.5.6 develops this possibility (see also the discussion just before proposition 7.3.4).

We conclude this subsection with some final comments on the structure of equilibrium. The outcome path (SE, EE^∞) is the equilibrium outcome path that maximizes 1's payoff, when 2 always chooses E, for $\delta \in [1/3, 1/\sqrt{3})$. The critical implication of $\delta < 1/\sqrt{3} \approx 0.577$ is $(1-\delta)/\delta > 3\delta - 1$ (see figure 2.5.5), so that player 2's payoff from (SE, SE, EE^∞) is negative, and applying (2.5.5) to any $\gamma_2 \in \mathscr{W}_2^{EE}$ with $\gamma_2 < 2$ eventually yields a value $\gamma_2^t \notin \mathscr{W}_2^{EE} \cup \mathscr{W}_2^{SE}$. Figure 2.5.5 illustrates the recursion (2.5.5)–(2.5.6) for $\delta = 1/\sqrt{3}$.

Finally, for $\delta \in [1/\sqrt{3}, 2/3)$, although $\mathscr{W}_2^{SE}$ is disjoint from $\mathscr{W}_2^{EE}$ (and so $[0, 2]$ is not self-generating), there are nontrivial self-generating sets and so subgame-perfect equilibria. The smallest value of δ with efficient equilibrium outcome paths in which 2 always chooses E and 1 chooses S more than once is $\delta = 1/\sqrt{3}$. Figure 2.5.5 illustrates this for the critical value of δ. The value $v' < 3\delta - 1$ in the figure is the value of the

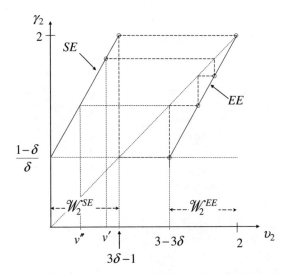

Figure 2.5.5 An illustration of the recursion for an equilibrium with player 2 payoff v', for $\delta = 1/\sqrt{3}$. The outcome path of this equilibrium is $(SE, EE, EE, EE, SE, EE^\infty)$. Any strictly positive player 2 payoff strictly less than v'' is not an equilibrium payoff.

2.5 ■ Generating Equilibria

outcome path $(SE, EE^3, SE, EE^\infty)$. For this value of δ, there are a countable number of efficient equilibrium payoffs, associated with outcome paths $(SE^x, EE^t, SE, EE^\infty)$, where $x \in \{0, 1\}$ and t is a nonnegative integer.

2.5.6 Example: Public Correlation

In this section, we illustrate the impact of allowing players to use a public correlating device. In particular, we will show that for the prisoners' dilemma of figure 2.5.1, the set of subgame-perfect equilibrium payoffs (with public correlation) is increasing in δ. We consider the case in which $c < b - 2$ (retaining, of course, the assumption $b - c < 4$). This moves us away from the $c = b - 2$ case of figure 2.4.1, on which we comment shortly.

With public correlation, it suffices for $\mathscr{E}^p(\delta) = \overline{\mathscr{F}^*}$ that each of the four sets in figure 2.5.2 be nonempty, and their union contains the extreme points $\{(0, 0), (\bar{v}, 0), (0, \bar{v}), (2, 2)\}$ of $\overline{\mathscr{F}^*}$. We have noted that $(2, 2) \in B_{EE}$ when $\delta \geq (b-2)/b$. From figure 2.5.2, it is clear that $(0, \bar{v}) \in B_{ES}$ when $\delta \geq c/(c+2)$. Hence, $\mathscr{E}^p(\delta) = \overline{\mathscr{F}^*}$ when

$$\delta \geq \max\left\{\tfrac{b-2}{b}, \tfrac{c}{c+2}\right\} = \tfrac{b-2}{b}.$$

(The equality is implied by our assumption that $c < b - 2$.) In this case, public correlation allows the players to achieve any payoff in $\overline{\mathscr{F}^*}$ (recall remark 2.5.1): First observe that each extreme point can be decomposed using one of the action profiles EE, ES, SE, or SS and continuations in $\overline{\mathscr{F}^*}$. Any payoff $v \in \overline{\mathscr{F}^*}$ can be written as $v = \alpha_1(0, 0) + \alpha_2(\bar{v}, 0) + \alpha_3(0, \bar{v}) + \alpha_4(2, 2)$, where α is a correlated action profile. The payoff v is decomposed by α and continuations in $\overline{\mathscr{F}^*}$, where the continuations depend on the realized action profile under α.

We now turn to values of $\delta < (b-2)/b$. Because B_{EE} is empty, EE cannot be enforced on $\overline{\mathscr{F}^*}$, even using public correlation. Hence, only the action profiles SS, SE, and ES can be taken in equilibrium. Recalling remark 2.5.1 again, the set $\mathscr{E}^p$ is the largest convex set $\mathscr{W}$ satisfying

$$\mathscr{W} = \text{co}(\mathscr{B}_{SS}(\mathscr{W}) \cup \mathscr{B}_{ES}(\mathscr{W}) \cup \mathscr{B}_{SE}(\mathscr{W})). \tag{2.5.7}$$

Because every payoff in $\mathscr{E}^p$ is decomposed by an action profile ES, SE, or SS and a payoff in $\mathscr{E}^p$, it is the discounted average value of the flow payoffs from ES, SE, and SS. Consequently, $\mathscr{E}^p$ must be a subset of the convex hull of $(-c, b)$, $(b, -c)$, and $(0, 0)$ (the triangle in figure 2.5.6). Let $\mathscr{W}$ denote the intersection of $\mathbb{R}_+^2$ and the convex hull of $(-c, b)$, $(b, -c)$, and $(0, 0)$, that is, $\mathscr{W} = \{(v : v_i \geq 0, v_1 + v_2 \leq b - c\}$.

The set $\mathscr{W}$ satisfies (2.5.7) if and only if the extreme points of $\mathscr{W}$ can be decomposed on $\mathscr{W}$. Because $(0, 0)$ can be trivially decomposed, symmetry allows us to focus on the decomposition of $(0, b - c)$.

Decomposing this point requires continuations $\gamma \in \mathscr{W}$ to satisfy

$$(0, b - c) = (1 - \delta)(-c, b) + \delta\gamma \tag{2.5.8}$$

and player 1's incentive constraint,

$$\gamma_1 \geq \frac{c(1 - \delta)}{\delta}. \tag{2.5.9}$$

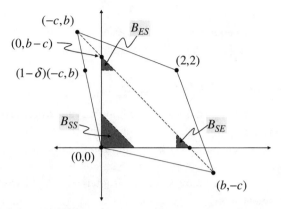

Figure 2.5.6 The sets $B_a = \mathscr{B}_a(\mathscr{W})$ for $a \in \{ES, SE, SS\}$ and $\mathscr{W} = \mathbb{R}_+^2 \cap \mathrm{co}(\{(-c, b), (b, -c), (0, 0)\})$, where $c < b - 2$ and $\delta < (b-2)/b$.

Equation (2.5.8) implies $\gamma_1 = c(1 - \delta)/\delta$, that is, (2.5.9) holds. The requirement that $\gamma \in \mathscr{W}$ is then equivalent to

$$0 \leq \gamma_2 \leq b - c - \gamma_1$$
$$= b - c - \frac{c(1-\delta)}{\delta}$$
$$= \frac{\delta b - c}{\delta},$$

that is, $\delta b \geq c$.

Finally, suppose $\delta b < c$, so that $\mathscr{W}$ does not satisfy (2.5.7). We now argue that the only set that can is $\{(0, 0)\}$, and so always SS is the only equilibrium. We derive a contradiction from the assumption that $v \neq (0, 0)$ can be decomposed by ES. Because $v = (1 - \delta)(-c, b) + \delta \gamma$, $\gamma_1 = [v_1 + (1 - \delta)c]/\delta \geq (1 - \delta)c/\delta$, and so $\gamma_2 \leq (b - c) - \gamma_1 \leq (\delta b - c)/\delta$. But $\delta b < c$ implies $\gamma_2 < 0$, which is impossible.

This completes the characterization of the prisoners' dilemma for this case. The set of equilibrium payoffs is:

$$\mathscr{E}^p = \begin{cases} \overline{\mathscr{F}^*}, & \text{if } \delta \geq \frac{b-2}{b}, \\ \mathbb{R}_+^2 \cap \mathrm{co}(\{(-c, b), (b, -c), (0, 0)\}), & \text{if } \delta \in [\frac{c}{b}, \frac{b-2}{b}), \\ \{(0, 0)\}, & \text{otherwise.} \end{cases}$$

The critical value of δ, $(b - 2)/b$, is less than $(b + c)/(b + c + 2)$, the critical value for a folk theorem without public correlation.

The set of subgame-perfect equilibrium payoffs is monotonic, in the sense that the set of equilibrium payoffs at least weakly expands as the discount factor increases. However, the relationship is neither strictly monotonic nor smooth. Instead, we take two discontinuous jumps in the set of payoffs that can be supported, one (at $\delta = c/b$) from the trivial equilibrium to a subset of the set of feasible, weakly individually rational payoffs, and one (at $\delta = (b - 2)/b$) to the set of all feasible and weakly individually rational payoffs. Stahl (1991) shows that the monotonicity property is preserved for other specifications of the parameters b and c consistent with the game being a prisoners' dilemma, though the details of the solutions differ. For example, when $c = b - 2$, as in figure 2.4.1, there is a single jump from the trivial equilibrium to being able to support the entire set of feasible, weakly individually rational payoffs.

2.6 Constructing Equilibria: Simple Strategies and Penal Codes

As in the previous section, we restrict attention to pure strategies (the analysis can be extended to mixed strategies at a significant cost of increased notation). By corollary 2.5.1, any action profile appearing on a pure-strategy subgame-perfect equilibrium path is decomposable on the set $\mathscr{E}^p$ of pure-strategy subgame-perfect equilibrium payoffs. By compactness of $\mathscr{E}^p$, there is a collection of pure-strategy subgame-perfect equilibrium profiles $\{\sigma^1, \ldots, \sigma^n\}$, with σ^i yielding the lowest possible pure-strategy subgame-perfect equilibrium payoff for player i. In this section, we will show that out-of-equilibrium behavior in any pure-strategy subgame-perfect equilibrium can be decomposed on the set $\{U(\sigma^1), \ldots, U(\sigma^n)\}$. This in turn leads to a simple recipe for constructing equilibria.

2.6.1 Simple Strategies and Penal Codes

We begin with the concept of a simple strategy profile:[14]

Definition 2.6.1 *Given $(n + 1)$ outcomes $\{\mathbf{a}(0), \mathbf{a}(1), \ldots, \mathbf{a}(n)\}$, the associated* simple strategy *profile $\sigma(\mathbf{a}(0), \mathbf{a}(1), \ldots, \mathbf{a}(n))$ is given by the automaton:*

$$\mathscr{W} = \{0, 1, \ldots, n\} \times \{0, 1, 2, \ldots\},$$
$$w^0 = (0, 0),$$
$$f(j, t) = a^t(j),$$

and

$$\tau((j, t), a) = \begin{cases} (i, 0), & \text{if } a_i \neq a_i^t(j) \text{ and } a_{-i} = a_{-i}^t(j), \\ (j, t+1), & \text{otherwise.} \end{cases}$$

A simple strategy consists of a prescribed outcome $\mathbf{a}(0)$ and a "punishment" outcome $\mathbf{a}(i)$ for each player i. Under the profile, play continues according to the outcome $\mathbf{a}(0)$. Players respond to any deviation by player i with a switch to the player i punishment outcome path $\mathbf{a}(i)$. If player i deviates from the path $\mathbf{a}(i)$, then $\mathbf{a}(i)$ starts again from the beginning. If some other player j deviates, then a switch is made to the player j punishment outcome $\mathbf{a}(j)$. A critical feature of simple strategy profiles is that the punishment for a deviation by player i is independent of when the deviation occurs and of the nature of the deviation. The profiles used to prove the folk theorem for perfect-monitoring repeated games (in sections 3.3 and 3.4) are simple.

14. Recall that $\mathbf{a}$ is an outcome path $(a^0, a^1, \ldots)$ with $a^t \in A$ an action profile.

We can use the one-shot deviation principle to identify necessary and sufficient conditions for a simple strategy profile to be a subgame-perfect equilibrium. Let

$$U_i^t(\mathbf{a}) = (1-\delta)\sum_{\tau=t}^{\infty} \delta^{\tau-t} u_i(a^\tau)$$

be the payoff to player i from the outcome path $(a^t, a^{t+1}, \ldots)$ (note that for any strategy profile σ, $U_i(\sigma) = U_i^0(\mathbf{a}(\sigma))$).

Lemma 2.6.1 *The simple strategy profile $\sigma(\mathbf{a}(0), \mathbf{a}(1), \ldots, \mathbf{a}(n))$ is a subgame-perfect equilibrium if and only if*

$$U_i^t(\mathbf{a}(j)) \geq \max_{a_i \in A_i}(1-\delta)u_i(a_i, a_{-i}^t(j)) + \delta U_i^0(\mathbf{a}(i)), \qquad (2.6.1)$$

for all $i = 1, \ldots, n$, $j = 0, 1, \ldots, n$, and $t = 0, 1, \ldots$

Proof The right side of (2.6.1) is the payoff to player i from deviating from outcome path $\mathbf{a}(j)$ in the tth period, and the left side is the payoff from continuing with the outcome. If this condition holds, then no player will find it profitable to deviate from any of the outcomes $(\mathbf{a}(0), \ldots, \mathbf{a}(n))$. Condition 2.6.1 thus suffices for subgame perfection.

Because a player might be called on (in a suitable out-of-equilibrium event) to play any period t of any outcome $\mathbf{a}(j)$ in a simple strategy profile, condition (2.6.1) is also necessary for subgame perfection.

∎

Remark 2.6.1 **Nash reversion and trigger strategies** A particularly *simple* simple strategy profile has, for all $i = 1, \ldots, n$, $\mathbf{a}(i)$ being a constant sequence of the same static Nash equilibrium. Such a simple strategy profile uses *Nash reversion* to provide incentives. We also refer to such Nash reversion profiles as *trigger strategy profiles* or *trigger profiles*.

A trigger profile is a *grim trigger profile* if the Nash equilibrium minmaxes the deviator. If the trigger profile uses the same Nash equilibrium to punish all deviators, grim trigger mutually minmaxes the players.

◆

A simple strategy profile specifies an equilibrium path $\mathbf{a}(0)$ and a *penal code* $\{\mathbf{a}(1), \ldots, \mathbf{a}(n)\}$ describing responses to deviations from equilibrium play. We are interested in optimal penal codes, embodying the most severe such punishments. Let

$$v_i^* = \min\{U_i(\sigma) : \sigma \in \mathscr{E}^p\}$$

be the smallest pure-strategy subgame-perfect equilibrium payoff for player i (which is well defined by the compactness of $\mathscr{E}^p$). Then:

Definition 2.6.2 Let $\{\mathbf{a}(i) : i = 1, \ldots, n\}$ be n outcome paths satisfying

$$U_i^0(\mathbf{a}(i)) = v_i^*, \quad i = 1, \ldots, n. \qquad (2.6.2)$$

2.6 ■ Constructing Equilibria

The collection of n simple strategy profiles $\{\sigma(i) : i = 1, \ldots, n\}$,

$$\sigma(i) = \sigma(\mathbf{a}(i), \mathbf{a}(1), \ldots, \mathbf{a}(n)),$$

is an optimal penal code *if*

$$\sigma(i) \in \mathscr{E}^p, \quad i = 1, \ldots, n.$$

Do optimal penal codes exist? Compactness of $\mathscr{E}^p$ yields the subgame-perfect outcome paths $\mathbf{a}(i)$ satisfying (2.6.2). The remaining question is whether the associated simple strategy profiles constitute equilibria.

The first statement of the following proposition shows that optimal penal codes exist. The second, reproducing the key result of Abreu (1988, theorem 5), is the punchline of the characterization of subgame-perfect equilibria: Simple strategies suffice to achieve any feasible subgame-perfect equilibrium payoff.

Proposition 2.6.1
1. *Let* $\{\mathbf{a}(i)\}_{i=1}^{n}$ *be n outcome paths of pure-strategy subgame-perfect equilibria* $\{\hat{\sigma}(i)\}_{i=1}^{n}$ *satisfying* $U_i(\hat{\sigma}(i)) = v_i^*$, $i = 1, \ldots, n$. *The simple strategy profile* $\sigma(i) = \sigma(\mathbf{a}(i), \mathbf{a}(1), \ldots, \mathbf{a}(n))$ *is a pure-strategy subgame-perfect equilibrium, for* $i = 1, \ldots, n$, *and hence* $\{\sigma(i)\}_{i=1}^{n}$ *is an optimal penal code.*
2. *The pure outcome path* $\mathbf{a}(0)$ *can be supported as an outcome of a pure-strategy subgame-perfect equilibrium if and only if there exist pure outcome paths* $\{\mathbf{a}(1), \ldots, \mathbf{a}(n)\}$ *such that the simple strategy profile* $\sigma(\mathbf{a}(0), \mathbf{a}(1), \ldots, \mathbf{a}(n))$ *is a subgame-perfect equilibrium.*

Hence, anything that can be accomplished with a subgame-perfect equilibrium in terms of payoffs can be accomplished with simple strategies. As a result, we need never consider complex hierarchies of punishments when constructing subgame-perfect equilibria, nor do we need to tailor punishments to the deviations that prompted them (beyond the identity of the deviator). It suffices to associate one punishment with each player, to be applied whenever needed.

Proof The "if" direction of statement 2 is immediate.

To prove statement 1 and the "only if" direction of statement 2, let $\mathbf{a}(0)$ be the outcome of a subgame-perfect equilibrium. Let $(\mathbf{a}(1), \ldots, \mathbf{a}(n))$ be outcomes of subgame-perfect equilibria $(\hat{\sigma}(1), \ldots, \hat{\sigma}(n))$, with $U_i(\hat{\sigma}_i) = v_i^*$. Now consider the simple strategy profile given by $\sigma(\mathbf{a}(0), \mathbf{a}(1), \ldots, \mathbf{a}(n))$. We claim that this strategy profile constitutes a subgame-perfect equilibrium. Considering arbitrary $\mathbf{a}(0)$, this argument establishes statement 2. For $\mathbf{a}(0) \in \{\mathbf{a}(1), \ldots, \mathbf{a}(n)\}$, it establishes statement 1.

From lemma 2.6.1, it suffices to fix a player i, an index $j \in \{0, 1, \ldots, n\}$, a time t, and action $a_i \in A_i$, and show

$$U_i^t(\mathbf{a}(j)) \geq (1-\delta)u_i(a_i, a_{-i}^t(j)) + \delta U_i^0(\mathbf{a}(i)). \tag{2.6.3}$$

Now, by construction, $\mathbf{a}(j)$ is the outcome of a subgame-perfect equilibrium—the outcome $\mathbf{a}(0)$ is by assumption produced by a subgame-perfect equilibrium, whereas each of $\mathbf{a}(1), \ldots, \mathbf{a}(n)$ is part of an optimal penal code. This ensures that for any t and $a_i \in A_i$,

$$U_i^t(\mathbf{a}(j)) \geq (1-\delta)u_i(a_i, a_{-i}(j)) + \delta U_i^d(\mathbf{a}(j), t, a_i), \qquad (2.6.4)$$

where $U_i^d(\mathbf{a}(j), t, a_i)$ is the continuation payoff received by player i in equilibrium $\sigma(j)$ after the deviation to a_i in period t. Because $\sigma(j)$ is a subgame-perfect equilibrium, the payoff $U_i^d(\mathbf{a}(j), t, a_i)$ must itself be a subgame-perfect equilibrium payoff. Hence,

$$U_i^d(\mathbf{a}(j), t, a_i) \geq U_i^0(\mathbf{a}(i)) = v_i^*,$$

which with (2.6.4), implies (2.6.3), giving the result. ∎

It is an immediate corollary that not only can we restrict attention to simple strategies but we can also take the penal codes involved in these strategies to be optimal.

Corollary 2.6.1 *Suppose $\mathbf{a}(0)$ is the outcome path of some subgame-perfect equilibrium. Then the simple strategy $\sigma(\mathbf{a}(0), \mathbf{a}(1), \ldots, \mathbf{a}(n))$, where each $\mathbf{a}(i)$ yields the lowest possible subgame-perfect equilibrium payoff v_i^* to player i, is a subgame-perfect equilibrium.*

2.6.2 Example: Oligopoly

We now present an example, based on Abreu (1986), of how simple strategies can be used in the characterization of equilibria. We do this in the context of a highly parameterized oligopoly problem, using the special structure of the latter to simplify calculations.

There are n firms, indexed by i. In the stage game, each firm i chooses a quantity $a_i \in \mathbb{R}_+$ (notice that in this example, action spaces are not compact). Given outputs, market price is given by $1 - \sum_{i=1}^n a_i$, when this number is nonnegative, and 0 otherwise. Each firm has a constant marginal and average cost of $c < 1$. The payoff of firm i from outputs $a_1, \ldots, a_n$ is then given by

$$u_i(a_1, \ldots, a_n) = a_i \left(\max\left\{ 1 - \sum_{j=1}^n a_j, 0 \right\} - c \right).$$

The stage game has a unique Nash equilibrium, denoted by $a_1^N, \ldots, a_n^N$, where

$$a_i^N = \frac{1-c}{n+1}, \quad i = 1, \ldots, n.$$

Firm payoffs in this Nash equilibrium are given by, for $i = 1, \ldots, n$,

$$u_i(a_1^N, \ldots, a_n^N) = \left(\frac{1-c}{n+1}\right)^2.$$

In contrast, the symmetric allocation that maximizes joint profits, denoted by $a_1^m, \ldots, a_n^m$, is

$$a_i^m = \frac{1-c}{2n}.$$

2.6 ■ Constructing Equilibria

Firm payoffs at this allocation are given by, for $i = 1, \ldots, n$,

$$u_i(a_1^m, \ldots, a_n^m) = \frac{1}{n}\left(\frac{1-c}{2}\right)^2.$$

As usual, this stage game is infinitely repeated with each firm characterized by the discount factor $\delta \in (0, 1)$. We restrict attention to *strongly symmetric equilibria* of the repeated game, that is, equilibria in which, after each history, the same quantity is chosen by every firm.

We can use the restriction to strongly symmetric equilibria to economize on notation. Let $q \in \mathbb{R}_+$ denote a quantity of output and $\mu : \mathbb{R}_+ \to \mathbb{R}$ be defined by

$$\mu(q) = q(\max\{1 - nq, 0\} - c). \tag{2.6.5}$$

Hence, $\mu(q)$ is the stage-game payoff obtained by each of the n firms when they each produce output level $q \in \mathbb{R}_+$. The function μ is essentially the function u, confined to the equal-output diagonal. The shift from $u(a)$ to $\mu(q)$ allows us to distinguish the (potentially asymmetric) action profile $a \in \mathbb{R}_+^n$ from the (commonly chosen) output level $q \in \mathbb{R}_+$. In keeping with the elimination of subscripts, we let $a^N \in \mathbb{R}_+$ and $a^m \in \mathbb{R}_+$ denote the output chosen by each firm in the stage-game Nash equilibrium and symmetric joint-profit maximizing profile, respectively.

Let $\mu^d(q)$ be the payoff to a single firm when every other firm produces output q and the firm in question maximizes its stage-game payoff. Hence (exploiting the symmetry to focus on 1's payoff),

$$\mu^d(q) = \max_{a_1 \in [0,1]} u_1(a_1, q, \ldots, q)$$
$$= \begin{cases} \frac{1}{4}(1 - (n-1)q - c)^2, & \text{if } 1 - (n-1)q - c \geq 0 \\ 0, & \text{otherwise.} \end{cases} \tag{2.6.6}$$

These functions exhibit the convenient structure of the oligopoly problem. The function (2.6.5) is concave, whereas (2.6.6) is convex in the relevant range, with the two being equal at the output corresponding to the unique Nash equilibrium. Figure 2.6.1 illustrates these functions. Notice that as q becomes arbitrarily large, the payoff $\mu(q)$ becomes arbitrarily small, as firms incur ever larger costs to produce so much output that they must give it away. This is an important feature of the example. The ability to impose arbitrarily large losses in a single period ensures that we can work with single-period punishments. If we had assumed an upper bound on quantities, punishments may require several periods. When turning to the folk theorem for more general games in the next chapter, the inability to impose large losses forces us to work with multiperiod punishments.

As is typically the case, our characterization of the set of subgame-perfect equilibrium payoffs begins with the lowest such payoff. A symmetric subgame-perfect equilibrium is an *optimal punishment* if it achieves the minimum payoff, for each player, possible under a (strongly symmetric) subgame-perfect equilibrium. We cannot assume the set of subgame-perfect equilibria is compact, because the action spaces are unbounded. Instead, we let v^* denote the infimum of the common payoff received in any subgame-perfect equilibrium and prove that the infimum can be achieved.

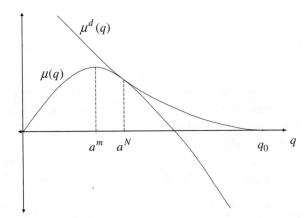

Figure 2.6.1 Illustration of the functions $\mu(q) = q(1 - nq - c)$, identifying the payoff of each firm when each chooses an output of q, and $\mu^d(q) = (1 - (n-1)q - c)^2/4$, identifying the largest profit available to one firm, given that the $n-1$ other firms produce output q.

To construct a simple equilibrium giving payoff v^*, we begin with the largest payoff that can be achieved in a (strongly symmetric) subgame-perfect equilibrium. Let $\bar{q}$ be the most collusive output level that can be enforced by v^*, that is,[15]

$$\bar{q} = \arg\max_{q \in [a^m, a^N]} \mu(q) = \arg\max_{q \in \left[\frac{1-c}{2n}, \frac{1-c}{n+1}\right]} q(1 - nq - c)$$

subject to the constraint

$$\mu(q) \geq (1-\delta)\mu^d(q) + \delta v^*. \tag{2.6.7}$$

The left side of the constraint is the payoff achieved from playing $\bar{q}$ in every period, and the right side is the payoff received from an optimal deviation from $\bar{q}$, followed by a switch to the continuation payoff v^*. Note that $\bar{q} \geq a^m$.

Because a^m is the joint-profit maximizing quantity, if $q = a^m$ satisfies (2.6.7), then the most collusive equilibrium output is $\bar{q} = a^m$. If (2.6.7) is not satisfied at $q = a^m$, then (noting that $\mu(q)$ is decreasing in $q > a^m$ and (2.6.7) holds strictly for $q = a^N$), the most collusive equilibrium output is the smallest $\bar{q} < a^N$ satisfying

$$\mu(\bar{q}) = (1-\delta)\mu^d(\bar{q}) + \delta v^*.$$

We now find a particularly simple optimal punishment. For any two quantities of output $(\bar{q}, \tilde{q})$, define a *carrot-and-stick* punishment $\sigma(\bar{q}, \tilde{q})$ to be strategies in which all firms play $\tilde{q}$ in the first period and thereafter play $\bar{q}$, with any deviation from these

---

15. We drop the nonnegativity constraint on prices in the following calculations because it is not binding. It is also clear from figure 2.6.1 that we lose nothing in excluding outputs in the interval $[0, a^m)$ from consideration. In particular, any payoff achieved by outputs in the interval $[0, a^m)$ can also be achieved by outputs larger than a^m, with the payoff from deviating from the former output being larger than from the latter. Any equilibrium featuring the former output thus remains an equilibrium when the latter is substituted. This restriction in turn implies that $\mu(q)$ and $\mu^d(q)$ are both decreasing functions, a property we use repeatedly.

2.6 ■ Constructing Equilibria

strategies causing this prescription to be repeated. Intuitively, $\tilde{q}$ is the "stick" and $\bar{q}$ the "carrot." The punishment specifies a single-period penalty followed by repeated play of the carrot. Deviations from the punishment simply cause it to begin again.

The basic result is that carrot-and-stick punishments are optimal. Note in particular that (2.6.8) implies $\tilde{q} > a^N$.

Proposition 2.6.2
1. There exists an output $\tilde{q}$ such that the carrot-and-stick punishment $\sigma(\bar{q}, \tilde{q})$ is an optimal punishment.
2. The optimal carrot-and-stick punishment satisfies

$$\mu^d(\tilde{q}) = (1-\delta)\mu(\tilde{q}) + \delta\mu(\bar{q}) = v^*, \tag{2.6.8}$$

$$\mu^d(\bar{q}) = \mu(\bar{q}) + \delta(\mu(\bar{q}) - \mu(\tilde{q})) \quad \text{if } \bar{q} > a^m, \tag{2.6.9}$$

and

$$\mu^d(\bar{q}) \leq \mu(\bar{q}) + \delta(\mu(\bar{q}) - \mu(\tilde{q})) \quad \text{if } \bar{q} = a^m. \tag{2.6.10}$$

3. Let v be a symmetric subgame-perfect equilibrium payoff with $\mu(q) = v$. Then v is the outcome of a subgame-perfect equilibrium consisting of simple strategies $\{\mathbf{a}^0, \sigma(\bar{q}, \tilde{q}), \ldots, \sigma(\bar{q}, \tilde{q})\}$, where $\mathbf{a}^0$ plays q in every period.

Because $\sigma(\bar{q}, \tilde{q})$ is an optimal punishment and $\bar{q}$ is the most collusive output, the best (strongly symmetric) subgame-perfect equilibrium payoff is given by an equilibrium that plays $\bar{q}$ in every period, with any deviation prompting a switch to the optimal punishment $\sigma(\bar{q}, \tilde{q})$. Any other symmetric equilibrium payoff can be achieved by playing an appropriate quantity q in every period, enforced by the optimal punishment $\sigma(\bar{q}, \tilde{q})$.

The implication of this result is that solving (the symmetric version of) this oligopoly problem is quite simple. If we are concerned with either the best or the worst subgame-perfect equilibrium, we need consider only two output levels, the carrot $\bar{q}$ and the stick $\tilde{q}$. Having found these two, every other subgame-perfect equilibrium payoff can be attained through consideration of only three outputs, one characterizing the equilibrium path and the other two involved in the optimal punishment. In addition, calculating $\bar{q}$ and $\tilde{q}$ is straightforward. We can first take $\bar{q} = a^m$, checking whether the most collusive output can be supported in a subgame-perfect equilibrium. To complete this check, we find the value of $\tilde{q}$ satisfying (2.6.8), given $\bar{q} = a^m$. At this point, we can conclude that the most collusive output a^m can indeed be supported in a subgame-perfect equilibrium and that we have identified $\bar{q} = a^m$ and $\tilde{q}$, if and only if the inequality (2.6.10) is also satisfied. If it is not, then the output a^m cannot be supported in equilibrium, and we solve the equalities (2.6.8)–(2.6.9) for $\bar{q}$ and $\tilde{q}$.

Proof *Statement 1.* Given v^*, the infimum of symmetric subgame perfect equilibrium payoffs, and hence $\bar{q}$, choose $\tilde{q}$ so that

$$(1-\delta)\mu(\tilde{q}) + \delta\mu(\bar{q}) = v^*.$$

We now argue that the carrot-and-stick punishment defined by $\sigma(\bar{q}, \tilde{q})$ is a subgame-perfect equilibrium. By construction, this punishment has value v^*. Because deviations from $\bar{q}$ would (also by construction) be unprofitable when punished by v^*, they are unprofitable when punished by $\sigma(\bar{q},\tilde{q})$. We need only argue that deviations from $\tilde{q}$ are unprofitable, or

$$(1-\delta)\mu^d(\tilde{q}) + \delta v^* \leq (1-\delta)\mu(\tilde{q}) + \delta\mu(\bar{q}) = v^*.$$

Suppose this inequality fails. Then, because v^* is the infimum of strongly symmetric subgame-perfect payoffs, there exists a strongly symmetric equilibrium σ^* such that

$$(1-\delta)\mu^d(\tilde{q}) + \delta v^* > (1-\delta)\mu(q^*) + U(\sigma^*|_{q^*}),$$

where q^* is the first-period output under σ^* and $U(\sigma^*|_{q^*})$ is the continuation value (from symmetry, to any firm) under σ^* after q^* has been chosen by all firms. Because σ^* is an equilibrium,

$$(1-\delta)\mu^d(\tilde{q}) + \delta v^* > (1-\delta)\mu^d(q^*) + \delta v^*.$$

Because $\mu^d(q)$ is decreasing in q, it thus suffices for the contradiction to show $\tilde{q} \geq q^*$. But σ^* can never prescribe an output lower than $\bar{q}$, because the latter is the most collusive output supported by the (most severe) punishment v^*. Hence, the carrot-and-stick punishment $\sigma(\bar{q},\tilde{q})$ features higher payoffs than σ^* in every period beyond the first. Because v^*, the value of the carrot-and-stick punishment, is no larger than the value of σ^*, it must be that the carrot-and-stick punishment generates a lower first-period payoff, and hence $\tilde{q} \geq q^*$. This ensures that $\sigma(\bar{q},\tilde{q})$ is an optimal punishment.

Statement 2. Suppose $\sigma(\bar{q},\tilde{q})$ is an optimal carrot-and-stick punishment. The requirement that firms not deviate from the stick output $\tilde{q}$ is given by

$$(1-\delta)\mu(\tilde{q}) + \delta\mu(\bar{q}) \geq (1-\delta)\mu^d(\tilde{q}) + \delta(1-\delta)\mu(\tilde{q}) + \delta^2\mu(\bar{q}).$$

Similarly, the requirement that agents not deviate from the carrot output $\bar{q}$ is

$$\mu(\bar{q}) \geq (1-\delta)\mu^d(\bar{q}) + (1-\delta)\delta\mu(\tilde{q}) + \delta^2\mu(\bar{q}).$$

Rearranging these two inequalities gives

$$\mu^d(\tilde{q}) \leq (1-\delta)\mu(\tilde{q}) + \delta\mu(\bar{q}) = v^* \qquad (2.6.11)$$

and

$$\mu^d(\bar{q}) \leq \mu(\bar{q}) + \delta(\mu(\bar{q}) - \mu(\tilde{q})), \qquad (2.6.12)$$

where the right side of (2.6.11) is the value of the punishment. If (2.6.11) holds strictly, we can increase $\tilde{q}$ and hence reduce $\mu(\tilde{q})$,[16] while preserving (2.6.12), because $\mu(\tilde{q})$ appears only on the right side of the latter (i.e., a more severe punishment cannot make deviations from the equilibrium path more attractive). This yields a lower punishment value, a contradiction. Hence, under an optimal carrot-and-stick punishment, (2.6.11) must hold with equality, which is (2.6.8).

It remains only to argue that if $\bar{q} > a^m$, then (2.6.12) holds with equality (which is (2.6.9)). Suppose not. Although unilateral adjustments in either $\bar{q}$ or $\tilde{q}$

---

16. Recall that we can restrict attention to $q \geq a^m$.

will violate (2.6.8), we can increase $\tilde{q}$ while simultaneously reducing $\bar{q}$ to preserve (2.6.8). From (2.6.8), the punishment value must fall (because μ^d is decreasing in q). Moreover, a small increase in $\tilde{q}$ in this manner will not violate (2.6.12), so we have identified a carrot-and-stick punishment with a lower value, a contradiction.

Statement 3. This follows immediately from the optimality of the carrot-and-stick punishment and proposition 2.6.1.

∎

Has anything been lost by our restriction to strongly symmetric equilibria? There are two possibilities. One is that, on calculating the optimal carrot-and-stick punishment, we find that its value is 0. The minmax values of the players are 0, implying that a 0 punishment is the most severe possible. A carrot-and-stick punishment whose value is 0 is thus the most severe possible, and nothing is lost by restricting attention to strongly symmetric punishments. However, if the optimal carrot-and-stick punishment has a positive expected value, then there are more severe, asymmetric punishments available.[17]

When will optimal symmetric carrot-and-stick punishments be optimal overall, that is, have a value of 0? Because 0 is the minmax value for the players, the subgame-perfect equilibrium given by the carrot-and-stick punishment $\sigma(\bar{q}, \tilde{q})$ can have a value of 0 only if

$$\mu^d(\tilde{q}) = 0,$$

that is, only if 0 is the largest payoff one can obtain by deviating from the stick phase of the equilibrium. (If $\mu^d(\tilde{q}) > 0$ with a punishment value of 0, a current deviation followed by perpetual payoffs of 0 is profitable, a contradiction.) Hence, the stick must involve sufficiently large production that the best response for any single firm is simply to shut down and produce nothing.

The condition $\mu^d(\tilde{q}) = 0$ requires that $\tilde{q}$ be sufficiently large, and hence $\mu(\tilde{q})$ sufficiently negative (indeed, as $\delta \to 1, \tilde{q} \to \infty$). To preserve the 0 equilibrium value, this negative initial payoff must be balanced by subsequent positive payoffs. For this to be possible, the discount factor must be sufficiently large. Let $(\bar{q}(\delta), \tilde{q}(\delta))$ be the optimal carrot-and-stick punishment quantities given discount factor δ, and $v^*(\delta)$ the corresponding punishment value. Then:

Lemma 2.6.2
1. *The functions $\bar{q}(\delta)$ and $v^*(\delta)$ are decreasing and $\tilde{q}(\delta)$ is increasing in δ.*
2. *There exists $\underline{\delta} < 1$, such that for all $\delta \in (\underline{\delta}, 1)$, the optimal strongly symmetric carrot-and-stick punishment gives a value of 0 and hence is optimal overall.*

Proof Suppose first that $\bar{q}(\delta) = a^m$. Then the output $\tilde{q}(\delta)$ and value $v^*(\delta)$ satisfy (from (2.6.8))

$$\mu^d(\tilde{q}(\delta)) - \mu(\tilde{q}(\delta)) = \delta(\mu(a^m) - \mu(\tilde{q}(\delta))) \quad (2.6.13)$$

and

$$\mu^d(\tilde{q}(\delta)) = v^*(\delta), \quad (2.6.14)$$

17. Abreu (1986) shows that if the optimal carrot-and-stick punishment gives a positive payoff, then it is the only strongly symmetric equilibrium yielding this payoff. If it yields a 0 payoff, there may be other symmetric equilibria yielding the same payoff.

while subgame perfection requires the incentive constraint

$$\mu^d(a^m) \leq \mu(a^m) + \delta(\mu(a^m) - \mu(\tilde{q}(\delta))).$$

The function $h(q) = \mu^d(q) - (1-\delta)\mu(q)$ is convex and differentiable everywhere. From (2.6.13),

$$h(a^N) + \int_{a^N}^{\tilde{q}(\delta)} h'(q)dq = \delta\mu(a^m),$$

and because $\tilde{q}(\delta) > a^N$, $h(a^N) = \delta\mu(a^N)$ implies $h'(\tilde{q}(\delta)) > 0$. It then follows that an increase in δ increases $\tilde{q}$ (to preserve (2.6.13)), thus decreasing v^* (from (2.6.14)) and preserving the incentive constraint. In addition, it is immediate that there exists $\delta < 1$ such that $\bar{q}(\delta) = a^m$.[18] As a result, we have established that there exists $\delta^* < 1$ such that $\bar{q}(\delta) = a^m$ if and only if $\delta \geq \delta^*$, and that the comparative statics in lemma 2.6.2(1) hold on $[\delta^*, 1)$.

Now consider $\delta < \delta^*$. Here, the carrot-and-stick punishment is characterized by

$$\mu^d(\tilde{q}(\delta)) - \mu(\tilde{q}(\delta)) = \delta\big(\mu(\bar{q}(\delta)) - \mu(\tilde{q}(\delta))\big) \qquad (2.6.15)$$

and

$$\mu^d(\bar{q}(\delta)) - \mu(\bar{q}(\delta)) = \delta\big(\mu(\bar{q}(\delta)) - \mu(\tilde{q}(\delta))\big), \qquad (2.6.16)$$

with the value of the punishment again given by $v^*(\delta) = \mu^d(\tilde{q}(\delta))$. This implies that the outputs $\bar{q} < a^N < \tilde{q}$ must be such that the payoff increment from deviating from the carrot must equal that of deviating from the stick.

Now fix δ and its corresponding optimal punishment quantities $\bar{q}(\delta), \tilde{q}(\delta)$ and payoff $v^*(\delta)$, and consider $\delta' > \delta$ (but $\delta' < \delta^*$). We then have

$$\mu^d(\tilde{q}(\delta)) < (1-\delta')\mu(\tilde{q}(\delta)) + \delta'\mu(\bar{q}(\delta)).$$

Because $d\mu^d(q)/dq > d\mu(q)/dq$ for $q > a^N$ (because the first function is convex and the second concave, and the two are tangent at a^N), there exists a value $q' > \tilde{q}(\delta) > a^N$ at which

$$\mu^d(q') = (1-\delta')\mu(q') + \delta'\mu(\bar{q}(\delta)).$$

The carrot-and-stick punishment $\sigma(\bar{q}(\delta), q')$ is thus a subgame-perfect equilibrium giving value $v = \mu^d(q') < v^*(\delta)$, ensuring that $v^*(\delta') < v^*(\delta)$. This in turn ensures that $\bar{q}(\delta') < \bar{q}(\delta)$, which, from (2.6.15)–(2.6.16), gives $\tilde{q}(\delta') > \tilde{q}(\delta)$. This completes the proof of statement 1.

For $v^* = (1-\delta)\mu(\tilde{q}) + \delta\mu(\bar{q})$ to equal 0, we need $\mu(\bar{q}) < 0$ and (as discussed just before the statement of the lemma) $\mu^d(\tilde{q}) = 0$. For sufficiently large $\delta \geq \delta^*$, $\bar{q}(\delta) = a^m$ and (because $\tilde{q}(\delta) > a^N$) $\delta\mu(a^m) - \mu^d(\tilde{q}(\delta)) > \delta\mu(a^m) - \mu(a^N) > 0$. From (2.6.13), $\mu(\tilde{q}(\delta)) \to -\infty$ as $\delta \to 1$ and so eventually $\tilde{q}(\delta) > (1-c)/(n-1)$.

∎

---

18. For δ sufficiently close to 1, permanent reversion to the stage-game Nash equilibrium suffices to support a^m as a subgame-perfect equilibrium outcome, and hence so must the optimal punishment.

2.7 Long-Lived and Short-Lived Players

We now consider games in which some of the players are short-lived. There are n long-lived players, numbered $1, \ldots, n$, and $N - n$ short-lived players numbered $n + 1, \ldots, N$.[19] As usual, A_i is the set of player i's pure stage-game actions, and $A \equiv \prod_{i=1}^{N} A_i$ the set of pure stage-game action profiles. A long-lived player i plays the game in every period and maximizes, as before, the average discounted payoff of the payoff sequence $(u_i^0, u_i^1, u_i^2, \ldots)$, given by

$$(1 - \delta) \sum_{t=0}^{\infty} \delta^t u_i(a^t).$$

Short-lived players are concerned only with their payoffs in the current period and hence are often referred to as myopic. One interpretation is that in each period, a new collection of short-lived players $n + 1, \ldots, N$ enters the game, is active for only that period, and then leaves the game. An example of such a scenario is one in which the long-lived players are firms and the short-lived players customers. Another example has a single long-lived player, representing a government agency or court, with the short-lived players being a succession of clients or disputants who appear before it. A common interpretation of the product-choice game is that player 2 is a sequence of customers, with a new customer in each period.

An alternative interpretation is that each so-called short-lived player represents a continuum of long-lived agents, such as a long-lived firm facing a competitive market in each period, or a government facing a large electorate in each period. Under this interpretation, members of the continuum are sometimes referred to as *small* players and the long-lived players as *large*. Each player's payoff depends on his own action, the actions of the large players, and the average of the small players' actions.[20] All players maximize the average discounted sum of payoffs. In addition to the plays of the long-lived players, histories of play are *assumed* to include only the distribution of play produced by the small players. Because each small player is a negligible part of the continuum, a change in the behavior of a member of the continuum does not affect the distribution of play, and so does not influence future behavior of any (small or large) player. For this reason, players whose individual behavior is unobserved are also called *anonymous*. As there is no link between current play of a small anonymous player and her future treatment, such a player can do no better than myopically optimize.

Remark 2.7.1 **The anonymity assumption** The assumption that small players are anonymous is critical for the conclusion that small players necessarily myopically optimize. Consider the repeated prisoners' dilemma of figure 2.4.1, with player 1 being a large

---

19. The literature also uses the terms *long-run* for long-lived, and *short-run* for short-lived.
20. Under this interpretation of the model described above, if $N = n + 1$, a small player's payoff depends only on her own action (in addition to the long-lived players' actions), not the average of the other small players' actions. Allowing the payoffs of each small player to also depend on the average of the other players' actions in this case does not introduce any new issues.

    We follow the literature in assuming that small players make choices so that the distribution over actions is well defined (i.e., the set of small players choosing any action is assumed measurable). Because we will be concerned with equilibria and with deviations from equilibria on the part of a single player, this will not pose a constraint.

player and player 2 a continuum of small players. Each of the small players receives the payoff given by the prisoners' dilemma, given her action and player 1's action. Player 1's payoff depends only on the aggregate play of the player 2's, which is equivalent to the payoff received against a single opponent playing a mixed strategy with probability of E given by the proportion of player 2's choosing E.

If, as described, the distribution of player 2's actions is public (but individual player 2 actions are private), then all player 2's myopically optimize, playing S in every period, and so the equilibrium is unique, and in this equilibrium, all players choose S in every period.

On the other hand, if all players observe the precise details of every player 2's choice, then if $\delta \geq 1/3$, an equilibrium exists in which effort is exerted after every history devoid of shirking, with any instance of shirking triggering a punishment phase of persistent shirking thereafter. It thus takes more than a continuum of players to ensure that the players are anonymous and behavior is myopic.

Consider now the same game but with a finite number of player 2s. Perfect monitoring requires that each player 2's action be publicly observed and so, for *any* finite number of player 2s and $\delta \geq 1/3$, always exerting effort is an equilibrium outcome. Consequently, there is a potentially troubling discontinuity as we increase the number of player 2s. Suppose $\delta \geq 1/3$. For a finite number (no matter how large) of player 2s, always exerting effort is an equilibrium outcome, whereas for a continuum of anonymous player 2s, it is not. Section 7.8 discusses one resolution: If actions are only imperfectly observed, then the behavior of a large but finite population of player 2s is similar to that of a continuum of anonymous player 2s.

◆

We present the formal development for short-lived players. Restricting attention to pure strategies for the short-lived players, the model can be reinterpreted as one with small anonymous players who play identical pure equilibrium strategies. The usual intertemporal considerations come into play in shaping the behavior of long-lived players. As might be expected, the special case of a single long-lived player is often of interest and is the typical example.

Again as usual, a period t history is an element of the set $\mathscr{H}^t \equiv A^t$. The set of all histories is given by $\mathscr{H} = \cup_{t=0}^{\infty} \mathscr{H}^t$, where $\mathscr{H}^0 = \{\varnothing\}$. A behavior strategy for player $i, i = 1, \ldots, n$, is a function $\sigma_i : \mathscr{H} \to \Delta(A_i)$.

The role of player i, for $i = n+1, \ldots, N$ is filled by a countable sequence of players, denoted $i_0, i_1, \ldots$. A behavior strategy for player i_t is a function $\sigma_i^t : \mathscr{H}^t \to \Delta(A_i)$. We then let $\sigma_i = (\sigma_i^0, \sigma_i^1, \sigma_i^2, \ldots)$ denote the sequence of such strategies. We will often simply refer to a short-lived player i, rather than explicitly referring to the sequence of player i's. Note that each player i_t observes the history $h^t \in \mathscr{H}^t$.

As usual, $\sigma|_{h^t}$ is the continuation strategy profile after the history $h^t \in \mathscr{H}$. Given a strategy profile σ, $U_i(\sigma)$ denotes the long-lived player i's payoff in the repeated game, that is, the average discounted value of $\{u_i(a^t(\sigma))\}$.

Definition *A strategy profile σ is a* Nash equilibrium *if,*
2.7.1
  1. for $i = 1, \ldots, n$, σ_i maximizes $U_i(\cdot, \sigma_{-i})$ over player i's repeated game strategies, and

2. for $i = n+1, \ldots, N$, all $t \geq 0$, and all $h^t \in \mathcal{H}^t$ that have positive probability under σ,
$$u_i(\sigma_i^t(h^t), \sigma_{-i}(h^t)) = \max_{a_i \in A_i} u_i(a_i, \sigma_{-i}(h^t)).$$

A strategy profile σ is a subgame-perfect equilibrium *if, for all histories $h^t \in \mathcal{H}$, $\sigma|_{h^t}$ is a Nash equilibrium of the repeated game.*

The notion of a one-shot deviation for a long-lived player is identical to that discussed in section 2.2 and applies in an obvious manner to short-lived players. Hence, we immediately have a one-shot deviation principle for games with long- and short-lived players (the proof is the same as that for proposition 2.2.1).

Proposition 2.7.1 **The one-shot deviation principle** *A strategy profile σ is subgame perfect if and only if there are no profitable one-shot deviations.*

2.7.1 Minmax Payoffs

Because the short-lived players play myopic best replies, given the actions of the long-lived players, the short-lived players must play a Nash equilibrium of the induced stage game. Let $B : \prod_{i=1}^{n} \Delta(A_i) \rightrightarrows \prod_{i=n+1}^{N} \Delta(A_i)$ be the correspondence that maps any mixed-action profile for the long-lived players to the corresponding set of static Nash equilibria for the short-lived players. The graph of B is denoted by $\mathbf{B} \subset \prod_{i=1}^{N} \Delta A_i$, so that $\mathbf{B}$ is the set of the profiles of mixed actions with the property that, for each player $i = n+1, \ldots, N$, α_i is a best response to α_{-i}. Note that under assumption 2.1.1, for all $\alpha_{LL} \in \prod_{i=1}^{n} \Delta(A_i)$, there exists $\alpha_{SL} \in \prod_{i=n+1}^{N} \Delta(A_i)$ such that $(\alpha_{LL}, \alpha_{SLL}) \in \mathbf{B}$. It will be useful for the notation to cover the case of no short-lived players, in which case, $\mathbf{B} = \prod_{i=1}^{n} \Delta(A_i)$. (If a player has a continuum action space, then that player does not randomize; see remark 2.1.1.)

For each of the long-lived players $i = 1, \ldots, n$, define

$$\underline{v}_i = \min_{\alpha \in \mathbf{B}} \max_{a_i \in A_i} u_i(a_i, \alpha_{-i}), \qquad (2.7.1)$$

and let $\hat{\alpha}^i \in \mathbf{B}$ be an action profile satisfying

$$\hat{\alpha}^i = \arg \min_{\alpha \in \mathbf{B}} \left\{ \max_{a_i \in A_i} u_i(a_i, \alpha_{-i}) \right\}.$$

The payoff $\underline{v}_i$, player i's *(mixed-action) minmax payoff with short-lived players*, is a lower bound on the payoff that player i can obtain in an equilibrium of the repeated game.[21] Notice that this payoff is calculated subject to the constraint that short-lived players choose best responses: The strategies $\hat{\alpha}^i_{-i}$ minmax player i, subject to the constraint that there exists some choice $\hat{\alpha}^i_i$ for player i that gives $(\hat{\alpha}^i_i, \hat{\alpha}^i_{-i}) \in \mathbf{B}$, that is, that makes the choices of the short-lived players a Nash equilibrium. Lower payoffs for player i might be possible if the short-lived players were not constrained to behave myopically (as when they are long-lived). In particular, reducing i's payoffs below $\underline{v}_i$

---

21. Restricting the long-lived players to pure actions gives player i's pure-action minmax payoff with short-lived players. We discuss pure-action minmaxing in detail when all players are long-lived and therefore restrict attention to mixed-action minmax when some players are short-lived.

	h	ℓ
H	2,3	0,2
L	3,0	1,1

	h	ℓ
H	2,3	1,2
L	3,0	0,1

Figure 2.7.1 The left game is the product-choice game of figure 1.5.1. Player 1's action L is a best reply to 2's choice of ℓ in the left game, but not in the right.

requires actions on the part of the short-lived players that cannot be best responses or part of a stage-game Nash equilibrium but that long-lived players potentially could be induced to play via the use of appropriate intertemporal incentives.

We use the same notation for minmax payoffs and strategies, $\underline{v}_i$ and $\hat{\alpha}^i$, as for the case in which all players were long-lived, because they are the appropriate generalization in the presence of short-lived players.

It is noteworthy that $\hat{\alpha}_i^i$ need not be a best response for player i to $\hat{\alpha}_{-i}^i$. When $\hat{\alpha}_i^i$ is not a best response, $u_i(\hat{\alpha}^i)$ is strictly smaller than $\underline{v}_i$. Hence, minmaxing player i may require i to incur a cost, in the sense of not playing a best response, to ensure that $\hat{\alpha}^i$ specifies stage-game best responses for the short-lived players.[22] If we are to punish player i through the play of $\hat{\alpha}^i$, player i will have to be subsequently compensated for any costs incurred, introducing a complication not found when all players are long-lived.[23]

Example 2.7.1 Consider the product-choice game of figure 1.5.1, where player 1 is a long-lived and player 2 a short-lived player, reproduced as the left game in figure 2.7.1. Every pure or mixed action for player 2 is a myopic best reply to some player 1 action. Hence, the constraint $\alpha \in \mathbf{B}$ on the minimization in (2.7.1) imposes no constraints on the action α_2 appearing in player 1's utility function. We have $\underline{v}_1 = 1$ and $\hat{\alpha}^1 = L\ell$, as would be the case with two long-lived players. In addition, L is a best response for player 1 to player 2's play of ℓ. There is then no discrepancy here between player 1's best response to the player 2 action that minmaxes player 1 and the player 1 action that makes such behavior a best response for player 2. It is also straightforward to verify, as a consequence, that the trigger-strategy profile in which play begins with Hh, and remains there until the first deviation, after which play is perpetual $L\ell$, is a subgame-perfect equilibrium for $\delta \geq 1/2$.

Now consider a modified version of this game, displayed on the right in figure 2.7.1. Once again, every pure or mixed action for player 2 is a myopic best

---

22. Formally, given any $\alpha \in \mathbf{B}$ that solves the minimization in (2.7.1), the subsequent maximization does not impose $a_i = \alpha_i$.

23. With long-lived players, we can typically assume that player 1 plays a stage-game best response when being minmaxed, using future play to create the incentives for the other players to do the minmaxing. Here the short-lived players must find minmaxing a stage-game best response, potentially requiring player 1 to not play a best response, being induced to do so by incentives created by future play.

2.7 ■ Long-Lived and Short-Lived Players

	h	ℓ	r
H	2, 3	1, 2	0, 0
L	3, 0	0, 1	−1, 0

Figure 2.7.2 A further modification to the product-choice game of figure 1.5.1.

reply to some player 1 action, implying no constraints on α_2 in (2.7.1). We have $\underline{v}_1 = 1$ with $\hat{\alpha}^1 = L\ell$ and $u_1(\hat{\alpha}^1) = 0$. In this case, however, $\hat{\alpha}^1$ does not feature a best response for player 1, who would rather choose H. If player 1 is to be minmaxed, 1 will have to choose L to create the appropriate incentives for player 2, and 1 will have to be compensated for doing so. Because $L\ell$ is not a Nash equilibrium of the stage game, there is no pure-strategy equilibrium in trigger strategies. However, $(Hh)^\infty$ can still be supported as a subgame-perfect equilibrium-outcome path for $\delta \geq 1/2$, using a carrot-and-stick approach: The profile specifies Hh on the outcome path and after any deviation one period of $L\ell$, after which play returns to Hh; in particular, if player 1 does not play L when required, $L\ell$ is specified in the subsequent period. The set of equilibrium payoffs for this example is discussed in detail in section 7.6.

Consider a further modification, displayed in figure 2.7.2. When player 2 is long-lived, then we have $\underline{v}_1 = 0$ and $\hat{\alpha}^1 = Hr$. When player 2 is short-lived, the restriction that player 2 always choose best responses makes action r irrelevant, because it is strictly dominated by ℓ. We then again have $\underline{v}_1 = 1$ and $\hat{\alpha}_1 = L\ell$.

●

This example illustrates two features of repeated games with short-lived players. First, converting some players from long-lived to short-lived players imposes restrictions on the payoffs that can be achieved for the remaining long-lived players. In the game of figure 2.7.2, player 1's minmax value increases from 0 to 1 when player 2 becomes a short-lived player. Payoffs in the interval (0, 1) are thus possible equilibrium payoffs for player 1 if player 2 is long-lived but not if player 2 is short-lived. As a result, the range of discount factors under which we can support relatively high payoffs for player 1 as equilibrium outcomes may be broader when player 2 is a long-lived rather than short-lived player. As we will see in the next section, the maximum payoff for long-lived players is typically reduced when some players are short-lived rather than long-lived.

Second, the presence of short-lived players imposes restrictions on the structure of the equilibrium. To minmax player 1, player 2 must play ℓ. To make ℓ a best response for player 2, player 1 must put at least probability 1/2 on L, effectively cooperating in the minmaxing. This contrasts with the equilibria that will be constructed in the proof of the folk theorem without short-lived players (proposition 3.4.1), where a player being minmaxed is allowed to play her best response to the minmax strategies, and the incentives for the punishing players to do so are created by future play.

2.7.2 Constraints on Payoffs

In restricting attention to action profiles drawn from the set **B**, short-lived players impose restrictions on the set of equilibrium payoffs that go beyond the specification of minmax payoffs.

Example 2.7.2 Return to the product-choice game (left game of figure 2.7.1). Minmax payoffs are 1 for both players, whether player 2 is a long-lived or short-lived player. When player 2 is long-lived, because $\overline{\mathscr{F}^*}$ contains $(8/3, 1)$, it follows from proposition 3.3.1 that there are equilibria (for sufficiently patient players) with payoffs for player 1 arbitrarily close to $8/3$.[24] This is impossible, however, when player 2 is short-lived. For player 1 to obtain a payoff close to $8/3$, there must be some periods in which the action profile Lh appears with probability at least $2/3$. But whether the result of pure independently mixed or correlated actions, this outcome cannot involve a best response for player 2. The largest equilibrium payoff for player 1, even when arbitrarily patient, is bounded away from $8/3$.

Consider, then, the problem of choosing $\alpha \in \mathbf{B}$ to maximize player 1's stage-game payoff. Player 2 is willing to play h as long as the probability that 1 plays H is at least $1/2$, and so 1's payoff is maximized by player 1 mixing equally between H and L and player 2 choosing the myopic best reply of h. The payoff obtained is often called the *(mixed-action) Stackelberg payoff*, because this is the payoff that player 1 can guarantee himself by publicly committing to his mixed action before player 2 moves.[25]

However, the mixed-action Stackelberg payoff cannot be achieved in any equilibrium. Indeed, there is no equilibrium of the repeated game in which player 1's payoff exceeds 2, the pure action Stackelberg payoff, independently of player 1's discount factor.

To see why this is the case, let $\bar{v}_1$ be the largest subgame-perfect equilibrium payoff available to player 1, and assume $\bar{v}_1 > 2$. Any period in which player 2 chooses ℓ or player 1 chooses H cannot give player 1 a payoff in excess of 2. If player 1 chooses L, the best response for 2 is ℓ, which again gives a payoff short of 2. Informally, there must be some period in which player 2 chooses h with positive probability and player 1 mixes between H and L. Consider the first such period, and consider the equilibrium that begins with this period. The payoff for player 1 in this equilibrium must be at least $\bar{v}_1$, because all previous periods

---

24. Using Nash reversion as a punishment and an algorithm similar to that in section 2.5.4 to construct an outcome path that switches appropriately between Hh and Lh, we can obtain precisely $(8/3, 1)$ as an equilibrium payoff.

25. More precisely, it is the supremum of such payoffs. When player 1 mixes equally between H and L, player 2 has two best replies, only one of which maximizes 1's payoff. However, by putting $\varepsilon > 0$ higher probability on H, 1 can guarantee that 2 has a unique best reply (see 15.2.2).

The *pure-action Stackelberg payoff* is the payoff that player 1 can guarantee by committing to a public action before player 2 moves (see 15.2.1). It is the subgame-perfect equilibrium payoff of the extensive form where 1 chooses $a_1 \in A_1$ first, and then 2, knowing 1's choice, chooses $a_2 \in A_2$.

2.7 ■ Long-Lived and Short-Lived Players

(if any) must give lower payoffs. However, because player 1 is mixing, he must be indifferent between H and L, and hence the payoff to player 1 when choosing H must be at least as large as $\bar{v}_1$. On the other hand, because 2 bounds the current period payoff from H and $\bar{v}_1$ bounds the continuation payoff by hypothesis, we have the following upper bound on the payoff from playing H, and hence an upper bound on $\bar{v}_1$,

$$\bar{v}_1 \leq (1-\delta)2 + \delta \bar{v}_1.$$

This inequality implies $\bar{v}_1 \leq 2$, contradicting our assumption that $\bar{v}_1 > 2$.

■

This result generalizes. Let i be a long-lived player in an arbitrary game, and define

$$\bar{v}_i = \sup_{\alpha \in \mathbf{B}} \min_{a_i \in \text{supp}(\alpha_i)} u_i(a_i, \alpha_{-i}), \tag{2.7.2}$$

where $\text{supp}(\alpha_i)$ denotes the support of the (possibly mixed) action α_i. Hence, we associate with any profile of stage-game actions, the minimum payoff that we can construct by adjusting player i's behavior within the support of his action. Essentially, we are identifying the least favorable outcome of i's mixture. We then choose, over the set of profiles in which short-lived players choose best responses, the action profile that maximizes this payoff. We have:

Proposition 2.7.2 *Every subgame-perfect equilibrium payoff for player i is less than or equal to $\bar{v}_i$.*

Proof The proof is a reformulation of the argument for the product-choice game. Let $v_i' > \bar{v}_i$ be the largest subgame-perfect equilibrium payoff available to player i and consider an equilibrium σ' giving this payoff. Let α^0 be the first period action profile. Then,

$$v_i' = (1-\delta)u_i(\alpha^0) + \delta E^{\sigma'}\{U(\sigma'|_{h^1})\}$$
$$\leq (1-\delta)\bar{v}_i + \delta v_i',$$

where the second inequality follows from the fact that α^0 must lie in $\mathbf{B}$, and player i must be indifferent between all of the pure actions in the support of α_i^0 if the latter is mixed. We can then rearrange this to give the contradiction, $v_i' \leq \bar{v}_i$.

■

As example 2.7.2 demonstrates, the action profile $\alpha^* = \arg\max_{\alpha \in \mathbf{B}} u_i(\alpha)$ may require player i to randomize to provide the appropriate incentives for the short-lived players. The upper bound $\bar{v}_i$ can then be strictly smaller than $u_i(\alpha^*)$. In equilibrium, i must be indifferent over all actions in the support of α_i^*. The payoff from the least lucrative action in this support, given by $\bar{v}_i$, is thus an upper bound on i's equilibrium payoffs in the repeated game.

3 The Folk Theorem with Perfect Monitoring

A folk theorem asserts that every feasible and strictly individually rational payoff is the payoff of some subgame-perfect equilibrium of the repeated game, when players are sufficiently patient.[1]

The ideas underlying the folk theorem are best understood by first considering the simpler *pure-action folk theorem* for infinitely repeated games of perfect monitoring: For every pure-action profile whose payoff strictly dominates the pure-action minmax, provided players are sufficiently patient, there is a subgame-perfect equilibrium of the repeated game in which that action profile is played in every period.

Because every feasible payoff can be achieved in a single period using a correlated action profile, identical arguments show that the *payoff folk theorem* holds in the following form:[2] Every feasible and strictly pure-action individually rational payoff is the payoff of some equilibrium with public correlation when players are sufficiently patient.

Although simplifying the argument (and strategy profiles) significantly, public correlation is not needed for a payoff folk theorem. We have already seen an example for the repeated prisoners' dilemma in section 2.5.3. As demonstrated there and in section 2.5.4, when players are sufficiently patient even payoffs that *cannot* be written as a rational convex combination of pure-action payoffs are *exactly* the average payoff of a pure-strategy subgame-perfect equilibrium without public correlation, giving a payoff folk theorem without public correlation. Of course, this requires a complicated nonstationary sequence of actions (and the required bound on the discount factor may be higher). As in section 2.5.3, the techniques of self-generation provide an indirect technique for constructing such sequences of actions. Those techniques are critical in proving the folk theorem for finite games with *imperfect* public monitoring and immediately yield (proposition 9.3.1) a perfect monitoring payoff folk theorem without public correlation. An alternative more direct approach is described in section 3.7.

The notion of individual rationality implicitly used in the last paragraph was relative to the pure-action minmax. The difficulty in proving the folk theorem, when individual rationality is defined relative to the mixed-action minmax, is that the imposition of punishments requires the punishing players to randomize (and so be indifferent over all the actions in the minmax profile). Early versions of the folk theorem provided

---

1. The term *folk* arose because its earliest versions were part of the informal tradition of the game theory community for some time before appearing in a publication.
2. The first folk theorems proved for our setting (subgame-perfect equilibria in discounted repeated games with perfect monitoring; Fudenberg and Maskin 1986) are of this form.

appropriate incentives for randomization by making the unconventional assumption that each player's choice of probability distribution over his action space was observable (this was referred to as the case of *observable mixtures*), so that indifference was not required. We never make such an assumption. The techniques of self-generation also allow a simple proof of the payoff folk theorem when individual rationality is defined relative to the mixed-action minmax (again, proposition 9.3.1). An alternative more direct approach is described in section 3.8.

3.1 Examples

We first illustrate the folk theorem for the examples presented in chapter 1.

Figure 3.1.1 shows the feasible payoff space for the prisoners' dilemma. The profile of pure and mixed minmax payoffs for this game is $(0, 0)$—because shirking ensures a payoff of at least 0, no player who chooses best responses can ever receive a payoff less than 0. The folk theorem then asserts (and we have already verified in section 2.5.3) that for any strictly positive profile in the shaded area of figure 3.1.1, there is a subgame-perfect equilibrium yielding that payoff profile for sufficiently high discount factors. This includes the payoffs $(2, 2)$ that arise from persistent mutual effort, as well as the payoffs very close to $(0, 0)$ (the payoff arising from relentless shirking), and a host of other payoffs, including asymmetric outcomes in which one player gets nearly 0 and the other nearly $8/3$.

The quantifier on the discount factor in this statement is worth noting. The folk theorem implies that for any candidate (feasible, strictly individually rational) payoff, there exists a sufficiently high discount factor to support that payoff as a subgame-perfect equilibrium. Equivalently, the set of subgame-perfect equilibrium payoffs converges to the set of feasible, strictly individually rational payoffs as the discount factor approaches unity. There may not be a single discount factor for which the entire

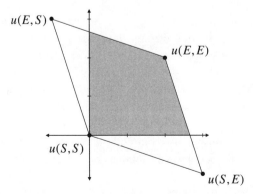

Figure 3.1.1 Feasible payoffs (the polygon and its interior) and folk theorem outcomes (the shaded area, minus its lower boundary) for the infinitely repeated prisoners' dilemma.

3.1 ■ Examples

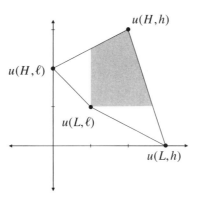

Figure 3.1.2 Feasible payoffs (the polygon and its interior) and folk theorem outcomes (the shaded area, minus its lower boundary) for the infinitely repeated product-choice game with two long-run players.

set appears as equilibrium payoffs, though, as we saw in section 2.5.3, this is the case for the prisoners' dilemma (see remark 3.3.1).

Chapter 1 introduced the product-choice game with the interpretation that the role of player 2 is filled by a sequence of short-lived players. However, let us consider the case in which the game is played by two long-lived players. Figure 3.1.2 illustrates the feasible payoffs. The pure and mixed minmax payoff is 1 for each player, with player 1 or 2 able to ensure a payoff of 1 by choosing L or ℓ, respectively. The folk theorem ensures that any payoff profile in the shaded area, except the lower boundary (to ensure strict individual rationality), can be achieved as a subgame-perfect payoff profile for sufficiently high discount factors.

Once again, the minmax payoff profile $(1, 1)$ corresponds to a stage-game Nash equilibrium and hence is a subgame-perfect equilibrium payoff profile, though this is not an implication of the folk theorem. The folk theorem implies that the highest payoff available to player 1 in a subgame-perfect equilibrium of the repeated game is (nearly) 8/3, corresponding to an outcome in which player 2 always buys the high-priced choice but player 1 chooses high effort with a probability only slightly exceeding 1/3. The highest available payoff to player 2 in a subgame-perfect equilibrium is 3, produced by an outcome in which player 2 buys the high-priced product and player 2 exerts high effort.

For our final example, consider the oligopoly game of section 2.6.2. In contrast to our earlier examples, the action spaces are uncountable; unlike the discussion in section 2.6.2, we make the spaces compact (as we do throughout our general development) by restricting actions to the set $[0, \bar{q}]$, where $\bar{q}$ is large. The set of payoffs implied by some pure action profile is illustrated in figure 3.1.3. The shaded region is $\mathscr{F}$, not its convex hull (in particular, note that all firms' profits must have the same sign, which is determined by the sign of the price-cost margin). The stage game has a unique Nash equilibrium, with payoffs $u_i(a^N) = [(1-c)/(n+1)]^2$, and minmax payoffs are $\underline{v}_i^p = \underline{v}_i = 0$ (player i chooses $a_i = 0$ in best response to the other players flooding the market, i.e., choosing $\bar{q}$). The folk theorem shows that in fact every feasible payoff

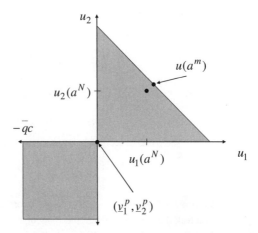

Figure 3.1.3 Payoffs implied by some pure-action profile in the two-firm oligopoly example of section 2.6.2. Stage-game Nash equilibrium payoffs (given by $u_i(a^N) = (1-c)^2/9$) and minmax payoffs ($\underline{v}_i^p = 0$) are indicated by two dots. The efficient frontier of the payoff set is given by $u_1 + u_2 = (1-c)^2/4$, where $(1-c)^2/4$ is the level of monopoly profits. The point $u(a^m)$ is equal division of monopoly profits, obtained by each firm producing half monopoly output.

in which all firms earn strictly positive profits can be achieved as a subgame perfect equilibrium outcome.

3.2 Interpreting the Folk Theorem

3.2.1 Implications

The folk theorem asserts that "anything can be an equilibrium." Once we allow repeated play and sufficient patience, we can exclude as candidates for equilibrium payoffs only those that are obviously uninteresting, being either infeasible or offering some player less than his minmax payoff. This result is sometimes viewed as an indictment of the repeated games literature, implying that game theory has no empirical content. The common way of expressing this is that "a theory that predicts everything predicts nothing."

Multiple equilibria are common in settings that range far beyond repeated games. Coordination games, bargaining problems, auctions, Arrow-Debreu economies, mechanism design problems, and signaling games (among many others) are notorious for multiple equilibria. Moreover, a model with a unique equilibrium would be quite useless for many purposes. Behavioral conventions differ across societies and cities, firms, and families. Only a theory with multiple equilibria can capture this richness. If there is a problem with repeated games, it cannot be that there are multiple equilibria but that there are "too many" multiple equilibria.

This indictment is unconvincing on three counts. First, a theory need not make precise predictions to be useful. Even when the folk theorem applies, the game-theoretic study of long-run relationships deepens our understanding of the incentives for opportunistic behavior and the institutional responses that might discourage such behavior. Repeated games help us understand why we might see asymmetric and nonstationary behavior in stationary symmetric settings, why some noisy signals may be more valuable than other seemingly just as noisy ones, why efficiency might hinge on the ability to punish some players while rewarding others, and why seemingly irrelevant details of an interaction may have tremendous effects on behavior. Without such an understanding, a useful model of behavior is beyond our grasp.

Second, we are often interested in cases in which the conditions for the folk theorem fail. The players may be insufficiently patient, the monitoring technology may be insufficiently informative, or some of the players may be short-lived (see section 2.7). There is much to be learned from studying the set of equilibrium payoffs in such cases. The techniques developed in chapters 7 and 8 allow us to characterize equilibrium payoffs when the folk theorem holds *and* when it fails.

Third, the folk theorem places bounds on payoffs but says nothing about behavior. The strategy profiles used in proving folk theorems are chosen for analytical ease, not for any putative positive content, and make no claims to be descriptions of what players are likely to do. Instead, the repeated game is a model of an underlying strategic interaction. Choosing an equilibrium is an integral part of the modeling process. Depending on the nature of the interaction, one might be interested in equilibria that are efficient or satisfy the stronger efficiency notion of renegotiation proofness (section 4.6), that make use of only certain types of (perhaps "payoff relevant") information (section 5.6), that are in some sense simple (remark 2.3.1), or that have some other properties. One of the great challenges facing repeated games is that the conceptual foundations of such equilibrium selection criteria are not well understood. Whether the folk theorem holds or fails, it is (only) the point of departure for the study of behavior.

3.2.2 Patient Players

There are two distinct approaches to capturing the idea that players are patient. Like much of the work inspired by economic applications, we consider players who discount payoff streams, with discount factors close to 1. Another approach is to consider preferences over the infinite stream of payoffs that directly capture patience. Indeed, the folk theorem imposing perfection was first proved for such preferences, by Aumann and Shapley (1976) for the *limit-of-means* preference $\succ^{LM}$ and by Rubinstein (1977, 1979a) for the limit-of-means and *overtaking* preferences $\succ^O$, where[3]

$$\{u_i^t\} \succ^{LM} \{\hat{u}_i^t\} \Leftrightarrow \liminf_{T \to \infty} \sum_{t=0}^{T} \frac{u_i^t - \hat{u}_i^t}{T} > 0$$

and

$$\{u_i^t\} \succ^O \{\hat{u}_i^t\} \Leftrightarrow \liminf_{T \to \infty} \sum_{t=0}^{T} u_i^t - \hat{u}_i^t > 0.$$

---

3. Because there is no guarantee that the infinite sum is well defined, the lim inf is used.

See Osborne and Rubinstein (1994) for an insightful discussion of these preferences and the associated folk theorems. Because a player who evaluates payoffs using limit-of-means is indifferent between two payoff streams that only differ in a finite number of periods, the perfection requirement is easy to satisfy in that case. Consider, for example, the following profile in the repeated oligopoly of section 2.6.2: The candidate outcome is $(a^m, \ldots, a^m)$ in every period; if a player deviates, then play $(\tilde{q}, \ldots, \tilde{q})$ for L periods (ignoring deviations during these L periods), and return to $(a^m, \ldots, a^m)$ (until the next deviation from $(a^m, \ldots, a^m)$), where L satisfies[4]

$$\mu^d(a^m) + L\mu(\tilde{q}) < (L+1)\mu(a^m).$$

This profile is a Nash equilibrium under limit-of-means and overtaking (as well as for discounting, if δ is sufficiently close to 1). Moreover, under limit-of-means, because the punishment is only imposed for a finite number of periods, it is costless to impose, and so the profile is perfect.

Overtaking is a more discriminating preference order than limit-of-means. In particular, a player who evaluates payoffs using overtaking strictly prefers the stream $u_i^0, u_i^1, u_i^2, \ldots$ to the stream $u_i^0 - 1, u_i^1, u_i^2, \ldots$ Consequently, though perfection is not a significant constraint with limit-of-means, it is with overtaking. In particular, the profile just described is not perfect with the overtaking criterion. A perfect equilibrium for overtaking requires sequences of increasingly severe punishments: Because minmaxing player i for L periods may require player j to suffer significantly in those L periods, the threatened punishment of j, if j does not punish i, may need to be significantly longer than the original punishment. And providing for that punishment may require even longer punishments.

Both overtaking and limit-of-means preferences suffer from two shortcomings: They are incomplete (not all payoff streams are comparable) and they do not respect mixtures, because lim inf is not a linear operator. Much of the more recent mathematical literature in repeated games has instead used *Banach limits*, which do respect mixtures and yield complete preferences.[5]

The tradition in economics has been to work with discounted payoffs, whereas the more mathematical literature is inclined to work with *limit* payoffs that directly capture patience. One motivation for the latter is the observation that many of the most powerful results in game theory—most notably the various folk theorems—are statements about patient players. If this is one's interest, then one might as well replace the technicalities involved in taking limits with the convenience of limit payoffs. If one

---

4. Recall that $\mu(q)$ is the stage-game payoff when all firms produce the same output q, and $\mu^d(q)$ is the payoff to a firm from myopically optimizing when all the other firms produce q.

5. Let ℓ_∞ denote the set of bounded sequences of real numbers (the set of possible payoff streams). A *Banach limit* is a linear function $\Lambda : \ell_\infty \to \mathbb{R}$ satisfying, for all $\mathbf{u} = \{u^t\} \in \ell_\infty$,

$$\liminf_{t \to \infty} u^t \leq \Lambda(\mathbf{u}) \leq \limsup_{t \to \infty} u^t$$

and displaying patience, in the sense that, if $\mathbf{u}$ and $\mathbf{v}$ are elements of ℓ_∞ with $v^t = u^{t+1}$ for $t = 0, \ldots,$ then

$$\Lambda(\mathbf{u}) = \Lambda(\mathbf{v}).$$

knows what happens with limit payoffs, then it seems that one knows what happens with arbitrarily patient players.

Many results in repeated games are indeed established for arbitrarily patient players, and patience often makes arguments easier. However, we are also often interested in what can be accomplished for fixed discount factors, discounting that may not be arbitrarily close to 1. In addition, although it is generally accepted that limit payoff results are a good guide to results for players with high discount factors, the details of this relationship are not completely known. We confine our attention to the case of discounting.

3.2.3 Patience and Incentives

The fundamental idea behind work in repeated games is the trade-off between current and future incentives. This balance appears most starkly in the criterion for an action to be enforceable (sections 2.5 and 7.3), which focuses attention on total payoffs as a convex combination of current and future payoffs. The folk theorems are variations on the theme that there is considerable latitude for the current behavior that might appear in an equilibrium if the future is sufficiently important.

The common approach to enhancing the impact of future incentives concentrates on making players more patient, increasing the weight on future payoffs. For example, player 1 can potentially be induced to choose high effort in the product-choice game if and only if the future consequences of this action are sufficiently important. This increased patience can be interpreted as either a change in preferences or as a shortening of the length of a period of play. This latter interpretation arises as follows: If players discount time continuously at rate r and a period is of length $\Delta > 0$, then the effective discount factor is $\delta = e^{-r\Delta}$. In this latter interpretation, Δ is a parameter describing the environment in which the repeated game is embedded, and the effective discount factor δ goes to 1 as the length of the period Δ goes to 0.[6] For perfect-monitoring repeated games, the preferred interpretation is one of taste, because the model does not distinguish between patience and short period length. However, as we discuss in section 9.8, this is not true with public monitoring.

An alternative approach to enhancing the importance of future incentives is to fix the discount factor and reduce the impact on current payoffs from different current actions. For example, we could reformulate the product-choice game as one in which player 1's payoff difference between high and low effort is given by a value $c > 0$ (taken to be 1 in our usual formulation) that reflects the cost of high effort. We could then examine the consequences for repeated-game payoffs of allowing c to shrink, for a fixed discount factor. As c gets small, so does the temptation to defect from an equilibrium prescription of high effort, making it likely that the incentives created by continuation payoffs will be powerful enough to induce high effort.

---

6. The latter interpretation is particularly popular in the bargaining literature (beginning with Rubinstein 1982). The fundamental insight in that literature is that the division of the surplus reflects the *relative* impatience of the two bargainers. At the same time, the absolute impatience of a bargainer arises from the delay imposed by the bargaining technology should a proposed agreement be rejected. On the strength of the intuition that in practice there is virtually nothing preventing parties from very rapidly making offers and counteroffers, interest has focused on the case in which bargainers become arbitrarily patient (while preserving relative rates of impatience).

These two approaches reflect different ways of organizing the collection of repeated games. The common approach fixes a stage game and then constructs a collection of repeated games corresponding to different discount factors. One then asks which of the games in this collection allow certain equilibrium outcomes, such as high effort in the product-choice game. The folk theorem tells us that patience is a virtue in this quest.

Alternatively, we could fix a discount factor and look at the collection of repeated games induced by a collection of stage games, such as the collection of product-choice games parameterized by the cost of high effort c. We could then again ask which of the games in this collection allow high effort. In this case, being relatively amenable to high effort (having a low value of c) will be a virtue.

Our development of repeated games in parts I–III follows the standard approach of focusing on the discount factor. We return to this distinction in chapter 15, where we find a first indication that focusing on families of stage games for a fixed discount factor may be of interest, and then in chapter 18, where this approach will play an important role.

3.2.4 Observable Mixtures

Many early treatments of the folk theorem assumed players' mixed strategies are observable and that public correlating devices are available. We are interested in cases with public correlation, as well as those without. However, we follow the now standard practice of assuming that mixed strategies are not observable.

The set of equilibrium payoffs for arbitrarily patient players is unaffected by the presence of a public correlating device or an assumption that mixed strategies are observable. However, public correlation may have implications for the strategies that produce these payoffs, potentially allowing an equilibrium payoff to be supported by the repeated play of a single correlated mixture rather than a possibly more complicated sequence of pure action profiles. Similarly, if players' mixtures themselves (rather than simply their realized actions) were observable, then we could design punishments using mixed minmax action profiles without having to compensate players for their lack of indifference over the actions in the support of these mixtures. In addition, we have seen in sections 2.5.3 and 2.5.6 that public correlation may allow us to attain a particular equilibrium payoff with a smaller discount factor. Observable mixtures can also have this effect.

3.3 The Pure-Action Folk Theorem for Two Players

The simplest nontrivial subgame-perfect equilibria in repeated games use Nash reversion as a threat. Any deviation from equilibrium play is met with permanent reversion to a static Nash equilibrium (automatically ensuring that the punishments are credible, see remark 2.6.1).[7] Because of their simplicity, such Nash-reversion equilibria are commonly studied, especially in applications. In the prisoners' dilemma or the product-choice game, the study of Nash-reversion equilibria could be the end of the

---

7. The seminal work of Friedman (1971) followed this approach.

3.3 ■ The Folk Theorem for Two Players

story. In each game, there is a Nash equilibrium of the stage game whose payoffs coincide with the minmax payoffs. Nash reversion is then the most severe punishment available, in the sense that any outcome that can be supported by any subgame-perfect equilibrium can be supported by Nash-reversion strategies. As the oligopoly game makes clear, however, not all games have this property. Nash reversion is not the most severe punishment in such games, and the set of Nash-reversion equilibria excludes some equilibrium outcomes.

Recall from lemma 2.1.1 that the payoff in any pure-strategy Nash equilibrium cannot be less than the pure-action minmax, $v_i^p = \min_{a_{-i}} \max_{a_i} u_i(a_i, a_{-i}) \equiv u_i(\hat{a}_i^i, \hat{a}_{-i}^i)$. The folk theorem for two players uses the threat of mutual minmaxing to deter deviations. Denote the *mutual minmax* profile $(\hat{a}_1^2, \hat{a}_2^1)$ by $\hat{a}$. It is worth noting that player i's payoff from the mutual minmax profile may be strictly less than v_i^p, that is, carrying out the punishment may be costly. Accordingly, the profile must provide sufficient incentives to carry out any punishments necessary, and so is reminiscent of the profile in example 2.7.1.

Proposition 3.3.1 **Two-player folk theorem with perfect monitoring** *Suppose $n = 2$.*

1. *For all strictly pure-action individually rational action profiles $\tilde{a}$, that is, $u_i(\tilde{a}) > v_i^p$ for all i, there is a $\underline{\delta} \in (0, 1)$ such that for every $\delta \in (\underline{\delta}, 1)$, there exists a subgame-perfect equilibrium of the infinitely repeated game with discount factor δ in which $\tilde{a}$ is played in every period.*
2. *For all $v \in \mathscr{F}^{\dagger p}$, there is a $\underline{\delta} \in (0, 1)$ such that for every $\delta \in (\underline{\delta}, 1)$, there exists a subgame-perfect equilibrium, with payoff v, of the infinitely repeated game with public correlation and discount factor δ.*

Remark 3.3.1 The order of quantifiers in the proposition is important, with the bound on the discount factor depending on the action profile. Only when A is *finite* and there is no public correlation is it necessarily true there is a single discount factor δ for which every strictly pure-action individually rational action profile is played in every period in some subgame-perfect equilibrium.

Similarly, there need be no single discount factor for which every payoff $v \in \mathscr{F}^{\dagger p}$ can be achieved as an equilibrium payoff in the game with public correlation. The folk theorem result in section 2.5.3 for the prisoners' dilemma is the stronger uniform result, because a single discount factor suffices for all payoffs in $\mathscr{F}^{\dagger p}$. The following two properties of the prisoners' dilemma suffice for the result: (1) The pure-action minmax strategy profile is a Nash equilibrium of the stage game; and (2) for any action profile a and player i satisfying $u_i(a) < v_i^p$, if a_i' is a best reply for i to a_{-i}, then $u_i(a_i', a_{-i}) \leq v_i^p$. Because the set of equilibrium payoffs is compact, this uniformity also implies that all weakly individually rational payoffs are equilibrium payoffs for large δ.

Figure 3.3.1 presents a game for which such a uniform result does not hold. Each player has a minmax value of 0. Consider the payoff $(1/2, 0)$, which maximizes player 1's payoff subject to the constraint that player 2 receive at least her minmax payoff. Any outcome path generating this payoff must consist exclusively of the action profiles *TL* and *TR*. There is no equilibrium, for any $\delta < 1$, yielding such an outcome path. If *TR* appears in the first period, then 2's equilibrium payoff is

	L	R
T	0, −1	1, 1
B	−1, −1	−1, 0

Figure 3.3.1 A game in which some weakly individually rational payoff profiles cannot be obtained as equilibrium payoffs, for any $\delta < 1$.

♦

at least $(1 - \delta) + \delta \times 0 > 0$, because 2 can be assured of at least her minmax payoff in the future. If TL appears in the first period, a deviation by player 2 to R again ensures a positive payoff.

◆

Proof *Statement 1.* Fix an action profile $\tilde{a}$ satisfying $u_i(\tilde{a}) > \underline{v}_i^p$ for all i, and (as usual) let $M = \max_{i, a \in A} u_i(a)$, so that M is the largest payoff available to any player i in the stage game.

The desired profile is the *simple strategy profile* (definition 2.6.1) given by $\sigma(\mathbf{a}(0), \mathbf{a}(1), \mathbf{a}(2))$, where $\mathbf{a}(0)$ is the constant outcome path in which $\tilde{a}$ is played in every period, and $\mathbf{a}(i)$, $i = 1, 2$, is the common *punishment* outcome path of L periods of mutual minmax $\hat{a}$ followed by a return to $\tilde{a}$ in every period (where L is to be determined).

The strategy profile is described by the automaton with states $\mathscr{W} = \{w(\ell) : \ell = 0, \ldots, L\}$, initial state $w^0 = w(0)$, output function

$$f(w(\ell)) = \begin{cases} \tilde{a}, & \text{if } \ell = 0, \\ \hat{a}, & \text{if } \ell = 1, \ldots, L, \end{cases}$$

and transition rule

$$\tau(w(\ell), a) = \begin{cases} w(0), & \text{if } \ell = 0 \text{ and } a = \tilde{a}, \text{ or } \ell = L \text{ and } a = \hat{a}, \\ w(\ell + 1), & \text{if } 0 < \ell < L \text{ and } a = \hat{a}, \\ w(1), & \text{otherwise.} \end{cases}$$

Because $u_i(\hat{a}) \leq \underline{v}_i^p < u_i(\tilde{a})$, there exists $L > 0$ such that

$$L \min_i (u_i(\tilde{a}) - u_i(\hat{a})) > M - \min_i u_i(\tilde{a}). \quad (3.3.1)$$

If L satisfies (3.3.1), then a sufficiently patient player i strictly prefers $L + 1$ periods of $u_i(\tilde{a})$ to deviating, receiving at most M in the current period and then enduring L periods of $u_i(\hat{a})$.

3.3 ■ The Folk Theorem for Two Players

We now argue that for sufficiently large δ,

$$u_i(\tilde{a}) \geq (1-\delta)M + \delta v_i^*, \tag{3.3.2}$$

where

$$v_i^* = (1-\delta^L)u_i(\hat{a}) + \delta^L u_i(\tilde{a}).$$

The left side of (3.3.2) is the repeated-game payoff from the equilibrium path of $\tilde{a}$ in every period. The right side of (3.3.2) is an upper bound on player i's payoff from deviating from the equilibrium. The deviation entails an immediate bonus, which is bounded above by M and then, beginning in the next period and hence discounted by δ, a "punishment phase" whose payoff is given by v_i^*. Note that there is a δ^* such that for all i and $\delta > \delta^*$, $v_i^* > \underline{v}_i^p$.

Substituting for v_i^* in (3.3.2) and rearranging gives

$$(1-\delta^{L+1})u_i(\tilde{a}) \geq (1-\delta)M + \delta(1-\delta^L)u_i(\hat{a}),$$

and dividing by $(1-\delta)$ yields

$$\sum_{t=0}^{L} \delta^t u_i(\tilde{a}) \geq M + \delta \sum_{t=0}^{L-1} \delta^t u_i(\hat{a}).$$

If (3.3.1) holds, there is a $\underline{\delta} \geq \delta^*$ such that the above inequality holds for all $\delta \in (\underline{\delta}, 1)$ and all i.

It remains to be verified that no player has a profitable one-shot deviation from the punishment phase when $\delta \in (\underline{\delta}, 1)$. This may be less apparent, as it may be costly to minmax the opponent. However, if any deviation from the punishment phase is profitable, it is profitable to do so in $w(1)$, because the payoff of the punishment phase is lowest in the first period and every deviation induces the same path of subsequent play. Following the profile in $w(1)$ gives a payoff of v_i^*, whereas a deviation yields a current payoff of at most $\underline{v}_i^p < v_i^*$ (because each player is being minmaxed) with a continuation of v_i^*, and hence strictly suboptimal.

Statement 2. Let α be the correlated profile achieving payoff v. The second statement is proved in the same manner as the first, once we have described the automaton. Recall (from remark 2.3.3) that the automaton representing a pure strategy profile in the game with public correlation is given by $(\mathcal{W}, \mu^0, f, \tau)$, where $\mu^0 \in \Delta(\mathcal{W})$ is the distribution over initial states and $\tau : \mathcal{W} \times A \to \Delta(\mathcal{W})$. The required automaton is a modification of the automaton from part 1 of the proof, where the state $w(0)$ is replaced by $\{w^a : a \in A\}$ (a collection of states, one for each pure-action profile) and the initial state is replaced by μ^0, a randomization according to α over $\{w^a : a \in A\}$. Finally, $f(w^a) = a$, and for states w^a, the transition rule is

$$\tau(w^a, a') = \begin{cases} \alpha, & \text{if } a' = a, \\ w(1), & \text{otherwise,} \end{cases}$$

where α is interpreted as the probability distribution over $\{w^a : a \in A\}$ assigning probability $\alpha(a)$ to w^a.

∎

In some games, it is a stage-game Nash equilibrium for each player to choose the action that minmaxes the other player. In this case, Nash-reversion punishments are the most severe possible, and there is no need to consider more intricate punishments. The prisoners' dilemma has this property, again making it a particularly easy game with which to work, though also making it rather special. In other games, mutual minmaxing behavior is not a stage-game Nash equilibrium, with it instead being quite costly to minmax another player. Mutual minmaxing is then a valuable tool, because it provides more severe punishments than Nash reversion, but now we must provide incentives for the players to carry out such a punishment. This gives rise to the temporary period of minmaxing used when constructing the equilibria needed to prove the folk theorem. The minmaxing behavior is sufficiently costly to deter deviations from the original equilibrium path, whereas the temporary nature of the punishment and the prospect of a return to the equilibrium path creates the incentives to adhere to the punishment should it ever begin.

3.4 The Folk Theorem with More than Two Players

3.4.1 A Counterexample

The proof of the two-player folk theorem involved strategies in which the punishment phase called for the two players to simultaneously minmax one another. This allowed the punishments needed to sustain equilibrium play to be simply constructed. When there are more than two players, there may exist no combination of strategies that simultaneously minmax all players, precluding an approach that generalizes the two-player proof.

We illustrate with an example taken from Fudenberg and Maskin (1986, example 3). The three-player stage game is given in figure 3.4.1. In this game, player 1 chooses rows, player 2 chooses columns, and player 3 chooses matrices. Each player's pure (and mixed) minmax payoff is 0. For example, if player 2 chooses the second column and player 3 the first matrix, then player 1 receives a payoff of at most 0. However, there is no combination of actions that simultaneously minmaxes all of the players. For any pure profile a with $u(a) = (0, 0, 0)$, there is a player with a deviation that yields that player a payoff of 1.

This ensures that the proof given for the two-player folk theorem does not immediately carry over to the case of more players but leaves the possibility that the result itself generalizes. However, this is not the case. To see this, let v^* be the minimum payoff attainable in any subgame-perfect equilibrium, for any player and any discount factor, in this three-player game. Given the symmetry of the game, if one player achieves a given payoff, then all do, obviating the need to consider player-specific minimum payoffs. We seek a lower bound on v^* and will show that this lower bound is at least $1/4$, for every discount factor.

3.4 ■ More than Two Players

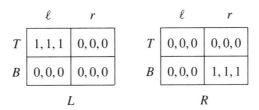

Figure 3.4.1 A three-player game in which players cannot be simultaneously minmaxed.

The first thing to note is that given any specification of stage-game actions, there is at least one player who, conditional on the actions of the other two players, can achieve a payoff of $1/4$ in the stage game. In particular, let $\alpha_i(1)$ be the probability attached to the first action (i.e., first row, column, or matrix) for player i. Then for any specification of stage game actions, there must be one player i for whom either $\alpha_j(1) \geq 1/2$ and $\alpha_k(1) \geq 1/2$ or $(1 - \alpha_j(1)) \geq 1/2$ and $(1 - \alpha_k(1)) \geq 1/2$. In the former case, playing the first action gives player i a payoff at least $1/4$, whereas in the second case playing the second action does so.

Now fix a discount factor δ and an equilibrium, and let player i be the one who, given the first-period actions of his competitors, can choose an action giving a payoff at least $1/4$ in the first period. Then doing so ensures player i a repeated-game payoff of at least

$$(1-\delta)\tfrac{1}{4} + \delta v^*.$$

If the putative equilibrium is indeed an equilibrium, then it must give an equilibrium payoff at least as high as the payoff $(1-\delta)\tfrac{1}{4} + \delta v^*$ that player i can obtain. Because this must be true of every equilibrium, the lower bound v^* on equilibrium payoffs must also satisfy this inequality, or

$$v^* \geq (1-\delta)\tfrac{1}{4} + \delta v^*,$$

which we solve for

$$v^* \geq \tfrac{1}{4}.$$

The folk theorem fails in this case because there are feasible strictly individually rational payoff vectors, namely, those giving some player a payoff in the interval $(0, 1/4)$, that cannot be obtained as the result of any subgame-perfect equilibrium of the repeated game, for any discount factor.

It may appear as if not much is lost by the inability to achieve payoffs less than $1/4$, because we rarely think of a player striving to achieve low payoffs. However, we could embed this game in a larger one, where it serves as a punishment designed to create incentives not to deviate from an equilibrium path. The inability to secure payoffs lower than $1/4$ may then limit the ability to achieve relatively lucrative equilibrium payoffs.

3.4.2 Player-Specific Punishments

The source of difficulty in the previous example is that the payoffs of the players are perfectly aligned, so that it is impossible to increase or decrease one player's payoff without doing so to all other players. In this section, we show that a folk theorem holds when we have the freedom to tailor rewards and punishments to players.

Definition 3.4.1 *A payoff $v \in \mathscr{F}^*$ allows player-specific punishment if there exists a collection $\{v^i\}_{i=1}^n$ of payoff profiles, $v^i \in \mathscr{F}^*$, such that for all i,*

$$v_i > v_i^i,$$

and, for all $j \neq i$,

$$v_i^j > v_i^i.$$

The collection $\{v^i\}_{i=1}^n$ constitutes a player-specific punishment *for v. If $\{v^i\}_{i=1}^n$ constitutes a player-specific punishment for v and $v^i \in \mathscr{F}$, then $\{v^i\}_{i=1}^n$ constitute* pure-action player-specific punishments *for v.*

Hence, we can define n payoff profiles, one profile v^i for each player i, with the property that player i fares worst (and fares worse than the candidate equilibrium payoff) under profile v^i. This is the type of construction that is impossible in the preceding example.

Recall that

$$\mathscr{F}^p \equiv \{v \in \mathscr{F} : v_i > \underline{v}_i^p\} = \{u(a) : u_i(a) > \underline{v}_i^p, \, a \in A\}$$

is the set of payoffs in $\mathscr{F}^{\dagger p}$ achievable in pure actions. The existence of pure-action player-specific punishments for a payoff $v \in \mathscr{F}^p$ or $\mathscr{F}^{\dagger p}$ turns out to be sufficient for v to be a subgame-perfect equilibrium payoff.

Proposition 3.4.1
1. *Suppose $v \in \mathscr{F}^p$ allows pure-action player-specific punishments in $\mathscr{F}^p$. There exists $\underline{\delta} < 1$ such that for all $\delta \in (\underline{\delta}, 1)$, there exists a subgame-perfect equilibrium with payoffs v.*
2. *Suppose $v \in \mathscr{F}^{\dagger p}$ allows player-specific punishments in $\mathscr{F}^{\dagger p}$. There exists $\underline{\delta} < 1$ such that for all $\delta \in (\underline{\delta}, 1)$, there exists a subgame-perfect equilibrium with payoffs v of the repeated game with public correlation.*

Before proving this proposition, we discuss its implications. Given proposition 3.4.1, obtaining a folk theorem reduces to obtaining conditions under which all feasible and strictly individually rational payoffs allow pure-action player-specific punishments. For continuum action spaces, $\mathscr{F}^p$ may well have a nonempty interior, as it does in our oligopoly example. If a payoff vector v' is in the interior of $\mathscr{F}^p$, we can vary one player's payoff without moving others' payoffs in lock-step. Beginning with a payoff v, we can then use variations on a payoff $v' < v$ to construct player specific punishments for v. This provides the sufficient condition that we combine with proposition 3.4.1 to obtain a folk theorem.

For finite games, there is no hope for $\mathscr{F}^p$ being convex (except the trivial case of a game with a single payoff profile) or having a nonempty interior. One can only assume directly that $\mathscr{F}^p$ allows player-specific punishment. However, the set $\mathscr{F}^{\dagger p}$ is necessarily convex. The full dimensionality of $\mathscr{F}^{\dagger p}$ then provides a sufficient condition for a folk theorem with public correlation.

3.4 ■ More than Two Players

Proposition 3.4.2 **Pure-minmax perfect monitoring folk theorem**

1. Suppose $\mathscr{F}^p$ is convex and has nonempty interior. Then, for every payoff v in $\{\tilde{v} \in \mathscr{F}^p : \exists v' \in \mathscr{F}^p, v'_i < \tilde{v}_i \forall i\}$, there exists $\underline{\delta} < 1$ such that for all $\delta \in (\underline{\delta}, 1)$, there exists a subgame-perfect equilibrium with payoffs v.

2. Suppose $\mathscr{F}^{\dagger p}$ has nonempty interior. Then, for every payoff v in $\{\tilde{v} \in \mathscr{F}^{\dagger p} : \exists v' \in \mathscr{F}^{\dagger p}, v'_i < \tilde{v}_i \ \forall i\}$, there exists $\underline{\delta} < 1$ such that for all $\delta \in (\underline{\delta}, 1)$, there exists a subgame perfect equilibrium, with payoffs v, of the repeated game with public correlation.

The set $\{\tilde{v} \in \mathscr{F}^p : \exists v' \in \mathscr{F}^p, v'_i < \tilde{v}_i \forall i\}$ equals $\mathscr{F}^p$ except for its lower boundary. When a set is convex, the condition that it have nonempty interior is equivalent to that of the set having *full dimension*.[8] The assumption that $\mathscr{F}^{\dagger p}$ have full dimension is Fudenberg and Maskin's (1986) full-dimensionality condition.[9]

Proof of Proposition 3.4.2 We only prove the first statement, because the second is proved mutatis mutandis. Fix $v \in \{\tilde{v} \in \mathscr{F}^p : \exists v' \in \mathscr{F}^p, v'_i < \tilde{v}_i \ \forall i\}$. Because $\mathscr{F}^p$ is convex and has nonempty interior, there exists $v' \in \text{int}.\mathscr{F}^p$ satisfying $v'_i < v_i$ for all i.[10] Fix $\varepsilon > 0$ sufficiently small that the $\sqrt{n}\varepsilon$-ball centered at v' is contained in $\text{int}.\mathscr{F}^p$. For each player i, let a^i denote the profile of stage-game actions that achieves the payoff profile

$$(v'_1 + \varepsilon, \ldots, v'_{i-1} + \varepsilon, v'_i, v'_{i+1} + \varepsilon, \ldots, v'_n + \varepsilon).$$

By construction, the profiles $a^1, \ldots, a^n$ generate player-specific punishments. Now apply proposition 3.4.1.

∎

Remark 3.4.1 **Dispensability of public correlation** The statement of this proposition and the preceding discussion may give the impression that public correlation plays an indispensable role in the folk theorem for finite games. Although it simplifies the constructive proof, it is unnecessary. Section 3.7 shows that the perfect monitoring folk theorem (proposition 3.4.2(2)) holds as stated without public correlation. Section 3.8 continues by showing that the proposition also holds (without public correlation) when $\mathscr{F}^{\dagger p}$ is replaced by $\mathscr{F}^*$, and hence pure minmax utilities replaced by (potentially lower) mixed minmax utilities.

◆

We now turn to the proof of proposition 3.4.1. The argument begins as in the two-player case, assuming that players choose a stage-game action profile $a \in \mathscr{A}$ giving payoff v. A deviation by player i causes the other players to minmax i for a sufficiently long period of time as to render the deviation suboptimal. Now, however, it is not obvious that the remaining players will find it in their interest to execute this punishment. In the two-player case, the argument that punishments would be optimally executed depended on the fact that the punishing players were themselves being minmaxed. This

---

8. The dimension of a convex set is the maximum number of linearly independent vectors in the set.
9. Though not usual, it is possible that $\mathscr{F}^{\dagger p}$ includes part of its lower boundary. For example, the payoff vector $(1/2, 1/2, 1)$ in the game of figure 3.4.2 is in $\mathscr{F}^{\dagger p}$ and lies on its lower boundary.
10. The interior of a set $\mathscr{A}$ is denoted $\text{int}.\mathscr{A}$.

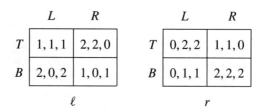

Figure 3.4.2 Impossibility of mutual minmax when player-specific punishments are possible.

imposed a bound on how much a player could gain by deviating from a punishment that allowed us to ensure that such deviations would be suboptimal. However, the assumption that v allows player-specific punishments does not ensure that mutual minmaxing is possible when there are more than two players. For example, consider the stage game in figure 3.4.2, where player 1 chooses rows, player 2 chooses columns, and player 3 matrices. The strictly individually rational payoff profile $(1, 1, 1)$, though not a Nash equilibrium payoff of the stage game, does allow player-specific punishments (in which v^i gives player i a payoff of 0 and a payoff of 2 to the other players). It is, however, impossible to simultaneously minmax all of the players. Instead, minmaxing player 1 requires strategies $(\cdot, L, r)$, minmaxing 2 requires $(B, \cdot, \ell)$, and minmaxing 3 requires $(T, R, \cdot)$.

As a result, we must find an alternative device to ensure that punishments will be optimally executed. The incentives to do so are created by providing the punishing players with a bonus on the completion of the punishment phase. The second condition for player-specific punishments ensures that it is possible to do so.

Proof of Proposition 3.4.1 Statement 1. Let a^0 be a profile of pure stage-game actions that achieves payoff v and $a^1, \ldots, a^n$ a collection of action profiles providing player-specific punishments. The desired profile is the *simple strategy profile* (definition 2.6.1), $\sigma(\mathbf{a}(0), \mathbf{a}(1), \ldots, \mathbf{a}(n))$, where $\mathbf{a}(0)$ is the constant outcome path in which a^0 is played in every period, and $\mathbf{a}(i), i = 1, \ldots, n$, is i's *punishment* outcome path of L periods of $\hat{a}^i$ followed by a^i in every period (where L is to be determined and $\hat{a}^i$ is the profile that minmaxes player i). Under this profile, the stage-game action profile a^0 is played in the first period and in every subsequent period as long as it has always been played. If agent i deviates from this behavior, then the profile calls for i to be minmaxed for L periods, followed by perpetual play of a^i. Should any player j (including i himself) deviate from either of these latter two prescriptions, then the equilibrium calls for the play of $\hat{a}^j$ for L periods, followed by perpetual play of a^j, and so on.[11]

An automaton for this profile has the set of states

$$\mathscr{W} = \{w(d) : 0 \leq d \leq n\} \cup \{w(i, t) : 1 \leq i \leq n,\ 0 \leq t \leq L - 1\},$$

---

11. In each case, simultaneous deviations by two or more players are ignored.

3.4 ■ More than Two Players

initial state $w^0 = w(0)$, output function $f(w(d)) = a^d$ and $f(w(i,t)) = \hat{a}^i$, and transition rule

$$\tau(w(d), a) = \begin{cases} w(j, 0), & \text{if } a_j \neq a_j^d,\ a_{-j} = a_{-j}^d, \\ w(d), & \text{otherwise,} \end{cases}$$

and

$$\tau(w(i,t), a) = \begin{cases} w(j, 0), & \text{if } a_j \neq \hat{a}_j^i,\ a_{-j} = \hat{a}_{-j}^i, \\ w(i, t+1), & \text{otherwise,} \end{cases}$$

where $w(i, L) \equiv w(i)$.

We apply proposition 2.4.1 to show that the profile is a subgame-perfect equilibrium. The values in the different states are:

$$V(w(d)) = u(a^d),$$

and

$$\begin{aligned} V(w(i,t)) &= (1 - \delta^{L-t})u(\hat{a}^i) + \delta^{L-t}V(w(i)) \\ &= (1 - \delta^{L-t})u(\hat{a}^i) + \delta^{L-t}u(a^i). \end{aligned}$$

We begin by verifying optimality in states $w = w(d)$. The one-shot game of proposition 2.4.1 has payoff function

$$g^w(a) = \begin{cases} (1-\delta)u(a) + \delta V(w(j,0)), & \text{if } a_j \neq a_j^d,\ a_{-j} = a_{-j}^d, \\ (1-\delta)u(a) + \delta u(a^d), & \text{otherwise.} \end{cases}$$

Setting $M \equiv \max_{i,a} u_i(a)$, the action profile a^d is a Nash equilibrium of g^w if, for all j

$$\begin{aligned} u_j(a^d) &\geq (1-\delta)M + \delta V_j(w(j,0)) \\ &= (1-\delta)M + \delta[(1-\delta^L)\underline{v}_j^p + \delta^L u_j(a^j)]. \end{aligned}$$

The left side of this inequality is the payoff from remaining in state w^d, whereas the right side is an upper bound on the value of a deviation followed by the accompanying punishment. This constraint can be rewritten as

$$(1-\delta)(M - u_j(a^d)) \leq \delta(1-\delta^L)(u_j(a^d) - \underline{v}_j^p) \\ + \delta^{L+1}(u_j(a^d) - u_j(a^j)). \quad (3.4.1)$$

As $\delta \to 1$, the left side converges to 0, and for $d \neq j$, the right side is strictly positive and bounded away from 0. Suppose $d = j$. Then the inequality can be rewritten as

$$M - u_j(a^j) \leq \frac{\delta(1-\delta^L)}{(1-\delta)}(u_j(a^j) - v_j^p).$$

Because

$$\lim_{\delta \to 1} \frac{\delta(1-\delta^L)}{(1-\delta)} = \lim_{\delta \to 1} \delta \sum_{t=0}^{L-1} \delta^t = L,$$

we can choose L sufficiently large that

$$M - u_j(a^j) < L(u_j(a^j) - v_j^p)$$

for all j. Consequently, for δ sufficiently close to 1, a^d is a Nash equilibrium of g^w for $w = w(d)$.

We now verify optimality in states $w = w(i, t)$. The one-shot game of proposition 2.4.1 in this case has payoff function

$$g^w(a) = \begin{cases} (1-\delta)u(a) + \delta V(w(j,0)), & \text{if } a_j \neq \hat{a}_j^i, \ a_{-j} = \hat{a}_{-j}^i, \\ (1-\delta)u(a) + \delta V(w(i,t+1)), & \text{otherwise.} \end{cases}$$

A sufficient condition for $\hat{a}^i$ to be a Nash equilibrium of g^w is that for all j,

$$V_j(w(i,t)) \geq (1-\delta)M + \delta V_j(w(j,0)),$$

that is,

$$(1-\delta^{L-t})u_j(\hat{a}^i) + \delta^{L-t}u_j(a^i) \geq (1-\delta)M + \delta[(1-\delta^L)v_j^p + \delta^L u_j(a^j)]. \quad (3.4.2)$$

Rearranging,

$$(1-\delta)u_j(\hat{a}^i) + \delta(1-\delta^{L-t-1})u_j(\hat{a}^i) + \delta^{L-t}(1-\delta^{t+1})u_j(a^i) + \delta^{L+1}u_j(a^i)$$
$$\geq (1-\delta)M + \delta(1-\delta^{L-t-1})v_j^p + \delta^{L-t}(1-\delta^{t+1})v_j^p + \delta^{L+1}u_j(a^j),$$

or

$$\delta^{L+1}\{u_j(a^i) - u_j(a^j)\} \geq (1-\delta)(M - u_j(\hat{a}^i)) + \delta(1-\delta^{L-t-1})(v_j^p - u_j(\hat{a}^i))$$
$$+ \delta^{L-t}(1-\delta^{t+1})(v_j^p - u_j(a^i)).$$

Because L is fixed, this last inequality is clearly satisfied for δ close to 1.

Statement 2. Let α^0 be the correlated profile that achieves payoff v, and $\alpha^1, \ldots, \alpha^n$ a collection of correlated action profiles providing player-specific punishments. The second statement is proved in the same manner as the first, with the correlated action profiles α^d replacing the pure action profiles a^d. We need only describe the automaton. Recall (from remark 2.3.3) that the automaton representing a pure strategy profile in the game with public correlation is given by $(\mathscr{W}, \mu^0, f, \tau)$, where $\mu^0 \in \Delta(\mathscr{W})$ is the distribution over initial states and $\tau : \mathscr{W} \times A \to \Delta(\mathscr{W})$. The set of states is

$$\mathscr{W} = (A \times \{0, 1, \ldots, n\}) \cup \{w(i, t) : 1 \leq i \leq n, \ 0 \leq t \leq L - 1\},$$

the distribution over initial states is given by $\mu^0((a, 0)) = \alpha^0(a)$, and the output function is $f((a, d)) = a$ and $f(w(i, t)) = \hat{a}^i$. For states $w(i, t)$, $t < L - 1$, the transition is deterministic and given by the transition rule

$$\tau(w(i, t), a) = \begin{cases} w(j, 0), & \text{if } a_j \neq \hat{a}^i_j, \ a_{-j} = \hat{a}^i_{-j}, \\ w(i, t+1), & \text{otherwise.} \end{cases}$$

For states $w(i, L - 1)$ the transition is stochastic, reflecting the correlated player-specific punishments α^i:

$$\tau(w(i, L - 1), a) = \begin{cases} w(j, 0), & \text{if } a_j \neq \hat{a}^i_j, \ a_{-j} = \hat{a}^i_{-j}, \\ (\alpha^i, i), & \text{otherwise,} \end{cases}$$

where (α^i, i) is interpreted as the probability distribution over $A \times \{i\}$ assigning probability $\alpha^i(a)$ to (a, i). Finally, for states (a, d), the transition is also stochastic, reflecting the correlated actions α^d:

$$\tau((a, d), a') = \begin{cases} w(j, 0), & \text{if } a'_j \neq a_j, \ a'_{-j} = a_{-j}, \\ (\alpha^d, d), & \text{otherwise,} \end{cases}$$

where (α^d, d) is interpreted as the probability distribution over $A \times \{d\}$ assigning probability $\alpha^d(a)$ to (a, d).

∎

3.5 Non-Equivalent Utilities

The game presented in section 3.4.1, in which the folk theorem does not hold, has the property that the three players always receive identical utilities. As a result, one can never reward or punish one player without similarly rewarding or punishing all players. The sufficient condition in proposition 3.4.2 is that the set $\mathscr{F}^p$ be convex and have a nonempty interior. In the interior of this set, adjusting one player's payoff has no implications for how the payoffs of the other players might vary.

How much stronger is the sufficient condition than the conditions necessary for the folk theorem? Abreu, Dutta, and Smith (1994) show that it suffices for the existence of player-specific punishments (and so the folk theorem) that no two players have identical preferences over action profiles in the stage game. Recognizing that the payoffs in a game are measured in terms of utilities and that affine transformations of utility functions preserve preferences, they offer the following formulation.

Definition 3.5.1 *The stage game satisfies* NEU (nonequivalent utilities) *if there are no two players i and j and constants c and $d > 0$ such that $u_i(a) = c + du_j(a)$ for all pure action profiles $a \in A$.*

Notice that a nontrivial symmetric game, in which players have the same utility function, need not violate the NEU condition. In a two-player symmetric game, $u_1(x, y) = u_2(y, x)$, whereas a violation of NEU requires $u_1(x, y) = u_2(x, y)$ for all x and y.

Proposition 3.5.1 *If the stage game satisfies NEU, then every $v \in \mathscr{F}^{\dagger p}$ allows player-specific punishments in $\mathscr{F}^{\dagger p}$. Moreover, every $v \in \mathscr{F}^*$ allows player-specific punishments in $\mathscr{F}^*$.*

Proof If $\mathscr{F}^{\dagger p}$ is nonempty, no player can be indifferent over all action profiles in A, and we will proceed under that assumption. For $j \neq i$, denote the projection of $\mathscr{F}^{\dagger}$ onto the ij-coordinate plane by $\mathscr{F}^{\dagger}_{ij}$, and its dimension by $\dim \mathscr{F}^{\dagger}_{ij}$. Clearly, $\dim \mathscr{F}^{\dagger}_{ij} \leq 2$. Because $\mathscr{F}^{\dagger}$ is convex and no player is indifferent over all action profiles, $\dim \mathscr{F}^{\dagger}_{ij} \geq 1$ for all $i \neq j$.

Claim 3.5.1. For all players i, either $\dim \mathscr{F}^{\dagger}_{ik} = 2$ for all $k \neq i$, or $\dim \mathscr{F}^{\dagger}_{ij} = 1$ for some $j \neq i$ and $\dim \mathscr{F}^{\dagger}_{ik} = 2$ for all $k \neq i, j$.

Proof. Suppose $\dim \mathscr{F}^{\dagger}_{ij} = \dim \mathscr{F}^{\dagger}_{ik} = 1$ for some $j \neq i$ and $k \neq i, j$. NEU implies that players i and j have payoffs that are perfectly negatively related, as do players i and k. But this implies that j and k have equivalent payoffs, violating NEU.

□

Claim 3.5.2. There exist n payoff profiles $\{\tilde{v}^1, \ldots, \tilde{v}^n\} \subset \mathscr{F}^{\dagger}$ with the property that for every pair of distinct players i and j,

$$\tilde{v}^i_i < \tilde{v}^j_i.$$

Note this is the second condition of player-specific punishments, and there is no assertion that these payoff profiles are strictly or weakly individually rational.

Proof. If $\dim \mathscr{F}^{\dagger}_{ij} = 1$ for some i and j, then i and j's payoffs are perfectly negatively related. Consequently, by the first claim, for each pair of players j and k, there exist some feasible payoff vectors v^{jk} and v^{kj} such that $v^{jk}_j > v^{kj}_j$ and $v^{kj}_k > v^{jk}_k$. Let $\{\theta_h : h = 1, \ldots, n(n-1)\}$ be a collection of weights satisfying $\theta_h > \theta_{h+1} > 0$ and $\sum_h \theta_h = 1$. For each player i, order the $n(n-1)$ payoff vectors v^{jk} in increasing size according to v^{jk}_i (break ties arbitrarily), and let $v^h(i)$ be the hth vector in the order. Define

$$\tilde{v}^i = \sum_h \theta_h v^h(i).$$

From convexity, $\tilde{v}^i \in \mathscr{F}^{\dagger}$. The construction of the vectors $v^h(i)$ ensures that for $i \neq j$, we have $\sum_{h \leq h'} v^h_i(i) \leq \sum_{h \leq h'} v^h_i(j)$ for all h' and strict inequality for at least one h'. Therefore,

3.5 ■ Non-Equivalent Utilities

$$0 > \sum_{h'=1}^{n(n-1)-1} (\theta_{h'} - \theta_{h'+1}) \sum_{h=1}^{h'} \left(v_i^h(i) - v_i^h(j)\right) + \theta_{n(n-1)} \sum_{h=1}^{n(n-1)} \left(v_i^h(i) - v_i^h(j)\right)$$

$$= \sum_{h'=1}^{n(n-1)-1} \theta_{h'} \left(v_i^{h'}(i) - v_i^{h'}(j)\right) + \sum_{h'=2}^{n(n-1)-1} \theta_{h'} \sum_{h=1}^{h'-1} \left(v_i^h(i) - v_i^h(j)\right)$$

$$- \sum_{h'=1}^{n(n-1)-1} \theta_{h'+1} \sum_{h=1}^{h'} \left(v_i^h(i) - v_i^h(j)\right) + \theta_{n(n-1)} \sum_{h=1}^{n(n-1)} \left(v_i^h(i) - v_i^h(j)\right)$$

$$= \sum_{h'=1}^{n(n-1)-1} \theta_{h'} \left(v_i^{h'}(i) - v_i^{h'}(j)\right) - \theta_{n(n-1)} \sum_{h=1}^{n(n-1)-1} \left(v_i^h(i) - v_i^h(j)\right)$$

$$+ \theta_{n(n-1)} \sum_{h=1}^{n(n-1)} \left(v_i^h(i) - v_i^h(j)\right)$$

$$= \sum_h \theta_h \left(v_i^h(i) - v_i^h(j)\right) = \tilde{v}_i^i - \tilde{v}_i^j,$$

and $\{\tilde{v}^i\}_i$ is the desired collection of payoffs.

□

It only remains to use the payoff vectors from this claim to construct player-specific punishments, for any $v \in \mathscr{F}^{\dagger p}$. Let $w^i \in \mathscr{F}^{\dagger}$ be a feasible payoff vector minimizing i's payoff (because $\mathscr{F}^{\dagger}$ is compact, the payoff w^i solves $\min\{v_i : v \in \mathscr{F}^{\dagger}\}$). Fix $v \in \mathscr{F}^{\dagger p}$ and consider the collection of payoff vectors $\{v^i\}_i$, where

$$v^i = \varepsilon(1-\eta)w^i + \eta\varepsilon\tilde{v}^i + (1-\varepsilon)v$$

(the constants $\eta > 0$ and $\varepsilon > 0$ are independent of i and to be determined). From convexity, $v^i \in \mathscr{F}^{\dagger}$. For any $\eta > 0$ and $\varepsilon > 0$, $v_i^i - v_i^j = \varepsilon(1-\eta)\left(w_i^i - w_i^j\right) + \eta\varepsilon\left(\tilde{v}_i^i - \tilde{v}_i^j\right) < 0$.[12] For sufficiently small ε, because $v_j > \underline{v}_j^p$, we have $v_j^i > \underline{v}_j^p$ and so $v^i \in \mathscr{F}^{\dagger p}$. Finally, for fixed ε, because $w_i^i < v_i$, for sufficiently small η, $v_i^i < v_i$, and $\{v^i\}_i$ is a player-specific punishment for v.

The second statement follows from the observation that for $v \in \mathscr{F}^*$, $v_j > \underline{v}_j$ and so $v^i \in \mathscr{F}^*$.

■

Using proposition 3.4.1, proposition 3.5.1 allows us to improve on proposition 3.4.2 in two ways. It weakens the full-dimensionality condition to NEU and includes any part of the lower boundary of $\mathscr{F}^p$ in the set of payoffs covered (see note 9 on page 83).

---

12. NEU is important here, since there is no guarantee that $w_i^i < w_i^j$. We could not use $u(\hat{a}^i)$ in place of w^i, because as we have argued, minmaxing player i may be very costly for j.

	L	R
T	0, −1	1, 0
B	0, −1	2, 1

Figure 3.5.1 A game with minimal attainable payoffs that are strictly individually rational.

Corollary 3.5.1 **Pure-minmax perfect monitoring folk theorem NEU**

1. *Suppose $\mathscr{F}^p$ is convex and the stage game satisfies NEU. Then, for every payoff v in $\mathscr{F}^p$, there exists $\underline{\delta} < 1$ such that for all $\delta \in (\underline{\delta}, 1)$, there exists a subgame-perfect equilibrium with payoffs v.*
2. *Suppose the stage game satisfies NEU. Then, for every payoff v in $\mathscr{F}^{\dagger p}$, there exists $\underline{\delta} < 1$ such that for all $\delta \in (\underline{\delta}, 1)$, there exists a subgame-perfect equilibrium, with payoffs v, of the repeated game with public correlation.*

Could an even weaker condition suffice? Abreu, Dutta, and Smith (1994) note that Benoit and Krishna (1985) offer a finite-horizon example that can be adapted to infinite-horizon games in which two out of three players have identical payoffs and the folk theorem fails. Hence, it appears unlikely that we can find a much weaker sufficient condition than NEU. Indeed, Abreu, Dutta, and Smith show that the NEU condition is very nearly necessary for the folk theorem. Define for each player i the payoff

$$f_i = \min\{v_i | v \in \mathscr{F}, v_j \geq \underline{v}_j, j = 1, \ldots, n\}.$$

Hence, f_i is i's *minimal attainable payoff*, the worst payoff that can be achieved for player i while restricting attention to payoff vectors that are (weakly) individually rational. If there is a stage-game action profile that gives every player their minmax payoff, as in the prisoners' dilemma, then $f_i = \underline{v}_i$ for all i. Even in the absence of such a profile, every player's minimal attainable payoff may be the minmax value, as long as for each player i there is some profile yielding the minmax payoff for i without forcing other players below their minmax value. However, it is possible that f_i is strictly higher than the minmax payoff. For example, consider the game in figure 3.5.1. Then the minmax value for each of players 1 and 2 is 0. However, $f_1 = 1$ (while $f_2 = 0$), because the lowest payoff available to player 1 that is contained in a payoff vector with only nonnegative elements is 1.[13]

Abreu, Dutta, and Smith (1994) then establish the following results. If for every mixed action profile α there is at most one player i whose best response to α gives a payoff no higher than f_i, then the NEU condition is necessary (as well as sufficient) for the folk theorem to hold. Notice that one circumstance in which there exists a strategy profile α to which *every* player i's best response to α gives a payoff no larger than f_i is

---

13. Minmaxing player 1 requires player 2 to endure a negative payoff.

3.6 Long-Lived and Short-Lived Players

This section presents a direct characterization of the set of subgame-perfect equilibrium payoffs of the game with one patient long-lived player and one or more short-lived players. We defer the general perfect-monitoring characterization result for games with long-lived and short-lived players to proposition 9.3.2.

From section 2.7.2, the presence of short-lived players imposes constraints on long-lived players' payoffs. Accordingly, there is no folk theorem in this case. Because there is only one long-lived player, the convexification we earlier obtained by public correlation is easily replaced by nonstationary sequences of outcomes.[15] As we will see in section 3.7, this is true in general but significantly complicated by the presence of multiple long-lived players. The argument in this section is also of interest because it presages critical features of the argument in section 3.8 dealing with mixed-action individual rationality.

For later reference, we first describe the set of feasible and strictly individually rational payoffs for n long-lived players. The payoff of a long-lived player i must be drawn from the interval $[\underline{v}_i, \bar{v}_i]$, where (recalling (2.7.1) and (2.7.2)),

$$\underline{v}_i = \min_{\alpha \in \mathbf{B}} \max_{a_i \in A_i} u_i(a_i, \alpha_{-i}) \tag{3.6.1}$$

and

$$\bar{v}_i = \max_{\alpha \in \mathbf{B}} \min_{a_i \in \mathrm{supp}(\alpha_i)} u_i(a_i, \alpha_{-i}). \tag{3.6.2}$$

Define

$$\mathscr{F} \equiv \{v \in \mathbb{R}^n : \exists \alpha \in \mathbf{B} \ s.t. \ v_i = u_i(\alpha), i = 1, \ldots, n\},$$
$$\mathscr{F}^\dagger \equiv \mathrm{co}(\mathscr{F}),$$

and

$$\mathscr{F}^* \equiv \{v \in \mathscr{F}^\dagger : v_i > \underline{v}_i, i = 1, \ldots, n\}.$$

We have thus defined the set of stage-game payoffs $\mathscr{F}$, its convex hull $\mathscr{F}^\dagger$, and the subset of these payoffs that are strictly individually rational, all for n long-lived players. These differ from their counterparts for a game with only long-lived players in the restriction to best-response behavior for the short-lived players. As has been our custom, we do not introduce new notation to denote these sets of payoffs, trusting the context to keep things clear.

---

14. Wen (1994) characterizes the set of equilibrium payoffs for games that fail NEU.
15. Unsurprisingly, public correlation allows a simpler argument; see Fudenberg, Kreps, and Maskin (1990). The argument presented here is also based on Fudenberg, Kreps, and Maskin (1990).

	L	C	R
T	1,3	0,0	6,2
B	0,0	2,3	6,2

Figure 3.6.1 The game for example 3.6.1.

We have defined $\mathscr{F}$ as the set of stage-game payoffs produced by profiles of mixed actions. When working with long-lived players, our first step was to define $\mathscr{F}$ as the set of pure-action stage-game payoffs. Why the difference? When working only with long-lived players, all of the extreme points of the set of stage-game payoffs produced by pure or mixed actions correspond to pure action profiles. It then makes no difference whether we begin by examining pure or mixed stage-game profiles, as the two sets yield the same convex hull $\mathscr{F}^\dagger$. When some players are short-lived, this equivalence no longer holds. The restriction to stage-game payoffs produced by profiles in the set **B** raises the possibility that some of the extreme points of the set of stage-game payoffs may be produced by mixtures. It is thus important to define $\mathscr{F}$ in terms of mixed action profiles. Example 3.6.1 illustrates.

Example 3.6.1 Consider the stage game in figure 3.6.1. The pure action profiles in the set **B** are (T, L) and (B, C). Denote a mixed action for player 1 by α_1^T, the probability attached to T. Then the set **B** includes

$$\{(\alpha_1^T, L) : \alpha_1^T \geq \tfrac{2}{3}\} \cup \{(\alpha_1^T, R) : \tfrac{1}{3} \leq \alpha_1^T \leq \tfrac{2}{3}\} \cup \{(\alpha_1^T, C) : \alpha_1^T \leq \tfrac{1}{3}\}.$$

Hence, $\underline{v}_1 = 1$ and $\bar{v}_1 = 6$. If we had defined $\mathscr{F}$ as the set of payoffs produced by pure action profiles in **B**, we would have $\mathscr{F} = \{1, 2\}$ and $\mathscr{F}^\dagger = \mathscr{F}^* = [1, 2]$, concluding that the set of equilibrium payoffs for player 1 in the repeated game is $[1, 2]$. Allowing $\mathscr{F}$ to include mixed action profiles gives $\mathscr{F} \supset [2/3, 1] \cup [4/3, 2] \cup \{6\}$. For any $v_1 \in [1, 6]$, there is then an equilibrium of the repeated game, for sufficiently patient player 1, with a player 1 payoff of v_1.[16]

●

Proposition 3.6.1 *Suppose there is one long-lived player, and $\underline{v}_1 < \bar{v}_1$. For every $v_1 \in (\underline{v}_1, \bar{v}_1]$, there exists $\underline{\delta}$ such that for all $\delta \in (\underline{\delta}, 1)$, there exists a subgame-perfect equilibrium of the repeated game with value v_1.*

Proof Let α' be a stage-game Nash equilibrium with payoff v_1' for player 1 (this equilibrium is pure if the stage game has continuum action spaces for all players). Note that $v_1' \in [\underline{v}_1, \bar{v}_1]$. Repeating the stage-game Nash equilibrium immediately gives a repeated-game payoff of v_1'. We separately consider payoffs below and

---

16. Proposition 3.6.1 ensures only that we can obtain payoffs in the interval $(1, 6]$. In this case, the lower endpoint corresponds to a stage-game Nash equilibrium and hence is easily obtained as an equilibrium payoff in the repeated game.

3.6 ■ Long-Lived and Short-Lived Players

above v_1'. Because we always require $\alpha \in \mathbf{B}$, the short-lived players' incentive constraints are always satisfied.

Case 1. $v_1 \in (v_1', \bar{v}_1]$.

Let $\bar{\alpha} \in \mathbf{B}$ be a stage-game action profile that solves the maximization problem in (3.6.2) (and hence $u_1(a_1, \bar{\alpha}_{-1}) \geq \bar{v}_1$ for all $a_1 \in \text{supp } \bar{\alpha}_1$).[17] Suppose the discount factor is sufficiently large that the following inequality holds:

$$(1 - \delta)(M - v_1') \leq \delta(v_1 - v_1'), \tag{3.6.3}$$

where M is the largest stage-game payoff.

The idea behind the equilibrium strategies is to reward player 1 with the relatively large payoff from $\bar{\alpha}$ when a discounted average of the payoffs he has so far received are "behind schedule," and to punish player 1 with the low payoff v_1' when it is "ahead of schedule." To do this, we recursively define a set of histories $\tilde{\mathcal{H}}^t$, a measure of the running total of player 1 expected payoffs $\zeta^t(h^t)$, and the equilibrium strategy profile σ. We begin with $\tilde{\mathcal{H}}^0 = \{\varnothing\}$ and $\zeta^0 = 0$. Given $\tilde{\mathcal{H}}^t$ and $\zeta^t(h^t)$, the strategy profile in period t is given by

$$\sigma(h^t) = \begin{cases} \alpha', & \text{if } \zeta^t(h^t) \geq (1 - \delta^{t+1})(v_1 - v_1') \text{ and } h^t \in \tilde{\mathcal{H}}^t, \text{ or if } h^t \notin \tilde{\mathcal{H}}^t, \\ \bar{\alpha}, & \text{if } \zeta^t(h^t) < (1 - \delta^{t+1})(v_1 - v_1') \text{ and } h^t \in \tilde{\mathcal{H}}^t. \end{cases}$$

The next period set of histories is

$$\tilde{\mathcal{H}}^{t+1} = \{h^{t+1} \in \mathcal{H}^{t+1} : h^t \in \tilde{\mathcal{H}}^t \text{ and } a^t \in \text{supp } \sigma(h^t)\},$$

and the updated running total is

$$\zeta^{t+1}(h^{t+1}) = \zeta^t(h^t) + (1 - \delta)\delta^t\left(u_1\left(a_1^t, \sigma_{-1}(h^t)\right) - v_1'\right).$$

Note that on $\tilde{\mathcal{H}}^t$, σ and ζ^t only depend on the history of player 1 actions. Hence, other than the specification of Nash reversion after an observable deviation by the short-lived players, the actions of the short-lived players do *not* affect ζ^t.

Because

$$\zeta^t(h^t) = (1 - \delta)\sum_{\tau=0}^{t-1} \delta^\tau u_1(a_1^\tau, \sigma_{-1}(h^\tau)) - (1 - \delta^t)v_1',$$

$\zeta^t(h^t)$ would equal $(1 - \delta^t)(v_1 - v_1')$ if player 1 expected to received a payoff of precisely v_1 in each of periods $0, \ldots, t - 1$ under σ_{-1} and his realized behavior. As a result, $\zeta^t(h^t) > (1 - \delta^t)(v_1 - v_1')$ implies that player 1 is ahead of schedule on h^t in terms of accumulating a "notional" repeated game payoff of v_1, whereas $\zeta^t(h^t) < (1 - \delta^t)(v_1 - v_1')$ indicates that player 1 is behind schedule.[18] It is

---

17. Note that when A_{-1} is finite, $\bar{\alpha}_{-1}$ may be mixed. The short-lived players play a Nash equilibrium of the stage game induced by 1's behavior. This stage game need not have a pure strategy Nash equilibrium (unless there is only one short-lived player).
18. The term $(1 - \delta^{t+1})$ cannot be replaced by $(1 - \delta^t)$ in the specification of σ. Under that specification, the first action profile is α' for all $v_1 \in (v_1', \bar{v}_1]$. It may then be impossible to achieve v_1 for v_1 close to $\bar{v}_1$.

important to remember in the following argument that for histories h^t in which the short-lived players do not follow σ_{-1}, player 1's actual accumulated payoff does *not* equal $(1 - \delta^t)v_1' + \zeta(h^t)$. On the other hand, for any strategy for player 1, $\tilde{\sigma}_1$, player 1's payoffs from the profile $(\tilde{\sigma}_1, \sigma_{-1})$ is

$$U_1(\tilde{\sigma}_1, \sigma_{-1}) = v_1' + E \lim_{t\to\infty} \zeta^t(h^t) \equiv v_1' + E\zeta^\infty(h), \qquad (3.6.4)$$

where expectations are taken over the outcomes $h \in A^\infty$.

We first argue that $\zeta^t(h^t) < v_1 - v_1'$ for all $h^t \in \mathcal{H}^t$. The proof is by induction. We have $\zeta^0 = 0 < v_1 - v_1'$. For the induction step, assume $\zeta^t < v_1 - v_1'$ and consider ζ^{t+1}. We break this into two cases. If $\zeta^t \geq (1 - \delta^{t+1})(v_1 - v_1')$ or $h^t \notin \tilde{\mathcal{H}}^t$, then $\zeta^{t+1} = \zeta^t + (1-\delta)\delta^t(u_1(a_1^t, \alpha_{-1}^t) - v_1') \leq \zeta^t < v_1 - v_1'$. If $\zeta^t < (1 - \delta^{t+1})(v_1 - v_1')$ and $h^t \in \tilde{\mathcal{H}}^t$, then $\zeta^{t+1} \leq \zeta^t + (1-\delta)\delta^t(M - v_1') < (1-\delta^{t+1})(v_1-v_1')+\delta^{t+1}(v_1-v_1') = v_1-v_1'$, where the final inequality uses (3.6.3).

We now argue that if player 1 does not observably deviate from σ_1 (i.e., for all $a_1^{t'} \in \text{supp } \sigma_1(h^{t'})$, $t' = 0, \ldots, t-1$) then $\zeta^t \geq (1-\delta^t)(v_1 - v_1')$. We again proceed via an induction argument, beginning with the observation that $\zeta^0 = 0 = (1-\delta^0)(v_1 - v_1')$, as required. Now suppose $\zeta^t \geq (1-\delta^t)(v_1 - v_1')$ and consider ζ^{t+1}. If $\zeta^t \geq (1-\delta^{t+1})(v_1 - v_1')$, then $\zeta^{t+1} = \zeta^t \geq (1-\delta^{t+1})(v_1 - v_1')$, the desired result. If $\zeta^t < (1-\delta^{t+1})(v_1 - v_1')$, then

$$\begin{aligned}\zeta^{t+1} &= \zeta^t + (1-\delta)\delta^t(u_1(a_1^t, \bar{\alpha}_2) - v_1') \\ &\geq \zeta^t + (1-\delta)\delta^t(\bar{v}_1 - v_1') \\ &\geq (1-\delta^t)(v_1 - v_1') + (1-\delta)\delta^t(v_1 - v_1') \\ &= (1-\delta^{t+1})(v_1 - v_1'),\end{aligned}$$

as required, where the first inequality notes that $\bar{\alpha}$ is played when $\zeta^t < (1-\delta^{t+1}) \times (v_1 - v_1')$ and applies (3.6.2), the next inequality uses the induction hypothesis that $\zeta^t \geq (1-\delta^t)(v_1 - v_1')$, and the final equality collects terms.

Hence, if in every period t, player 1 plays any action in the support of $\sigma_1(h^t)$, then $\lim_{t\to\infty} \zeta^t(h^t) = v_1 - v_1'$, and so from (3.6.4) player 1 is indifferent over all such strategies, and the payoff is v_1. Because play reverts to the stage-game Nash equilibrium α' after an observable deviation, to complete the argument that the profile is a subgame-perfect equilibrium, it suffices to argue that after a history $h^t \in \tilde{\mathcal{H}}^t$, player 1 does not have an incentive to choose an action $a_1 \notin \text{supp}\,\sigma_1(h^t)$. That is, it suffices to argue that the profile is Nash. But the expected value of $v_1' + \lim_{t\to\infty} \zeta(h^t)$ gives the period 0 value for any strategy of player 1 against σ_{-1}, and we have already argued that $\limsup_{t\to\infty} \zeta^t(h^t) \leq v_1 - v_1'$.

Case 2. $v_1 \in (\underline{v}_1, v_1')$.

This case is similar to case 1, though permanent Nash reversion can no longer be easily used to punish observable deviations because the payoff from the stage-game Nash equilibrium is more attractive than the target, v_1. Let $\hat{\alpha}$ solve (3.6.1) (and hence $u_1(a_1, \hat{\alpha}_{-1}) \leq \underline{v}_1$ for all a_1). Choose δ sufficiently large that

$$\delta(v_1 - v_1') < (1-\delta)(m - v_1'), \qquad (3.6.5)$$

where m is the smallest stage-game payoff.

3.6 ■ Long-Lived and Short-Lived Players

The strategies we define will only depend on the history of player 1 actions, and so the measure of the running total of player 1 expected payoffs ζ^t will also depend only on the history of player 1 actions. For any history $h^t \in \mathscr{H}^t$, $h_1^t \in A_1^t$ denotes the history of player 1 actions. We begin with $\tilde{\mathscr{H}}^0 = \{\varnothing\}$ and $\zeta^0 = 0$. The equilibrium strategy profile σ is defined as follows. Set $\zeta^0 = 0$, and given $\zeta^t(h_1^t)$, the strategy profile in period t is given by

$$\sigma(h^t) = \begin{cases} \alpha', & \text{if } \zeta^t(h_1^t) < (1-\delta^{t+1})(v_1 - v_1'), \\ \hat{\alpha}, & \text{if } \zeta^t(h_1^t) \geq (1-\delta^{t+1})(v_1 - v_1'). \end{cases}$$

Because period t behavior under σ depends only on the history of player 1 actions, we write $\sigma(h_1^t)$ rather than $\sigma(h^t)$. The updated running total is given by

$$\zeta^{t+1}(h_1^{t+1}) = \zeta^t(h_1^t) + (1-\delta)\delta^t \left(u_1(a_1^t, \sigma_{-1}(h_1^t)) - v_1' \right).$$

Again, only the actions of the long-lived player affect ζ^t.

We first argue that $\zeta^t(h_1^t) \leq (1-\delta^t)(v_1 - v_1') < 0$ for $t \geq 1$ for all $h_1^t \in A_1^t$. To do this, we first note that $\zeta^0 = 0$ and hence $\alpha_{-1}^0 = \hat{\alpha}_{-1}$, ensuring $\zeta^1 \leq (1-\delta)(v_1 - v_1') < (1-\delta)(v_1 - v_1')$. We then proceed via induction. Suppose $\zeta^t \leq (1-\delta^t)(v_1 - v_1')$. If $\zeta^t < (1-\delta^{t+1})(v_1 - v_1')$, then $\zeta^{t+1} \leq \zeta^t < (1-\delta^{t+1})(v_1 - v_1')$. Suppose instead that $\zeta^t \geq (1-\delta^{t+1})(v_1 - v_1')$. Then $\alpha_{-1}^t = \hat{\alpha}_{-1}$, and hence

$$\zeta^{t+1} \leq \zeta^t + (1-\delta)\delta^t(v_1 - v_1')$$
$$\leq (1-\delta^t)(v_1 - v_1') + (1-\delta)\delta^t(v_1 - v_1')$$
$$= (1-\delta^{t+1})(v_1 - v_1'),$$

where the second inequality uses the induction hypothesis.

We now argue that for any history h^t, if $\zeta^t(h_1^t) \geq (1-\delta^{t+1})(v_1 - v_1')$, then for all $a_1 \in A_1$,

$$\zeta^{t+1}(h_1^t a_1) \geq \zeta^t + (1-\delta)\delta^t(m - v_1')$$
$$> (1-\delta^{t+1})(v_1 - v_1') + \delta^{t+1}(v_1 - v_1')$$
$$= v_1 - v_1',$$

where the strict inequality uses (3.6.5).

Finally, for any history h^t, if $v_1 - v_1' < \zeta^t(h_1^t) < (1-\delta^{t+1})(v_1 - v_1')$ and player 1 chooses an action $a_1^t \in \text{supp } \sigma_1(h^t)$, then $\zeta^{t+1}(h_1^t a_1^t) = \zeta^t(h_1^t) > (v_1 - v_1')$.

Consider now a history h_1^t satisfying $\zeta^t(h_1^t) > v_1 - v_1'$ (note that ζ^0 is one such history). If player 1 chooses $a_1^{t'} \in \text{supp } \sigma_1(h^{t'})$, for all $t' \geq t$, then $\zeta^{t'}(h_1^{t'}) > v_1 - v_1'$ for all $t' \geq t$. Because $\zeta^{t'}(h_1^{t'}) \leq (1-\delta^{t'})(v_1 - v_1')$ for all $h_1^{t'}$, we then have that on any outcome $h \in A^\infty$ in which $\zeta^t(h_1^t) > v_1 - v_1'$ for some t and for all $t' \geq t$, player 1 does not observably deviate, $\zeta^\infty(h_1) \equiv \lim_{t \to \infty} \zeta^t(h_1^t) = v_1 - v_1'$.

We now apply the one-shot deviation principle (proposition 2.7.1) to complete the argument that the profile is an equilibrium. Fix an arbitrary history h^t and associated $\zeta^t(h_1^t)$. We first eliminate a straightforward case. If $\zeta^t(h_1^t) \leq v_1 - v_1'$, then $\sigma(h^t) = \alpha'$ and so $\zeta^{t+1}(h_1^t a_1) \leq \zeta^t(h_1^t) \leq v_1 - v_1'$ for all a_1. Consequently, at such a history, α' is specified in every subsequent period independent of history, and sequential rationality is trivially satisfied.

Suppose then $\zeta^t(h_1^t) > v_1 - v_1'$. Consider an action a_1 for which

$$\zeta^{t+1}(h_1^t a_1) > v_1 - v_1'. \tag{3.6.6}$$

Then,

$$[\zeta^t(h_1^t) + (1-\delta^t)v_1'] + \delta^t[(1-\delta)u_1(a_1, \hat{\alpha}_{-1}) + \delta \hat{U}_1(h_1^t a_1)] = v_1, \tag{3.6.7}$$

where $\hat{U}_1(h_1^t a_1)$ is the expected continuation from following σ after the history $h^t a_1$ (this follows from $v_1 - v_1' < \zeta^{t'}(h^{t'}) \leq (1-\delta^{t'})(v_1 - v_1')$, where $h^{t'}$ is a continuation of $h^t a_1$ and 1 follows σ_1 after $h^t a_1'$). This implies that the payoff from a one-shot deviation at h^t,

$$(1-\delta)u_1(a_1, \hat{\alpha}_{-1}) + \delta \hat{U}_1(h_1^t a_1),$$

is independent of a_1 satisfying (3.6.6), and player 1 is indifferent over all such a_1.

If $\zeta^t(h_1^t) \geq (1-\delta^{t+1})(v_1 - v_1')$, then $\sigma(h^t) = \hat{\alpha}$, and all a_1 satisfy (3.6.6). On the other hand, if $\zeta^t(h_1^t) < (1-\delta^{t+1})(v_1 - v_1')$, then $\sigma(h^t) = \alpha'$, and there is a possibility that for some action a_1, (3.6.6) fails. That is,

$$\zeta^t(h_1^t) + (1-\delta)\delta^t(u_1(a_1, \alpha_{-1}') - v_1') \leq v_1 - v_1'. \tag{3.6.8}$$

The history $(h_1^t a_1)$ is the straightforward case above, so that $\hat{U}_1(h_1^t a_1) = v_1'$ and the continuation payoff from a_1 is $(1-\delta)u_1(a_1, \alpha_{-1}') + \delta v_1'$. Equation (3.6.7) holds for $a_1' \in \operatorname{supp} \sigma_1(h^t)$, and so

$$\zeta^t(h_1^t) + (1-\delta^t)v_1' + \delta^t(1-\delta)v_1' + \delta^{t+1}\hat{U}_1' = v_1. \tag{3.6.9}$$

Using (3.6.9) to eliminate $\zeta^t(h_1^t)$ in (3.6.8) and rearranging yields

$$(1-\delta)\delta^t u_1(a_1, \alpha_{-1}') + \delta^{t+1}v_1' \leq (1-\delta)\delta^t v_1' + \delta^{t+1}\hat{U}_1',$$

and so the deviation to a_1 is unprofitable.

■

3.7 Convexifying the Equilibrium Payoff Set Without Public Correlation

Section 3.4 established a folk theorem using public correlation for payoffs in the set $\mathscr{F}^{\dagger p}$ rather than $\mathscr{F}^p$. This expansion of the payoff set is important because the former is more likely to satisfy the sufficient condition for player-specific punishments embedded in proposition 3.4.2, namely, that the set of payoffs has a nonempty interior. This is most obviously the case for finite games, where $\mathscr{F}^p$ necessarily has an empty interior.

3.7 ■ Convexifying the Equilibrium Payoff Set

In this section, we establish a folk theorem for payoffs in $\mathscr{F}^{\dagger p}$ as in proposition 3.4.2, but without public correlation.

We have already seen two examples. Section 3.6 illustrates how nonstationary sequences of outcomes can precisely duplicate the convexification of public correlation for the case of one long-lived player. Example 2.5.3 shows that when players are sufficiently patient, even payoffs that *cannot* be written as a rational convex combination of pure-action payoffs can be *exactly* the average payoff of a pure-strategy subgame perfect equilibrium, without public correlation. This ability to achieve exact payoffs will be important in the next section, when we consider mixed minmax payoffs.

It is an implication of the analysis in chapter 9 (in particular, proposition 9.3.1) that this is true in general (for finite games), and so the payoff folk theorem holds without public correlation. This section and the next (on punishing using mixed-action minmax) gives an independent treatment of this result. The results achieved using these techniques are slightly stronger than that of proposition 9.3.1 because they include the efficient frontier of $\mathscr{F}^*$ and cover continuum action spaces.

We first follow Sorin (1986) in arguing that for any payoff vector v in $\mathscr{F}^\dagger$, there is a sequence of pure actions whose average payoffs is precisely v. For ease of future reference, we state the result in terms of payoffs.

Lemma 3.7.1 *Suppose $v \in \mathbb{R}^n$ is a convex combination of $\{v(1), v(2), \ldots, v(\theta)\}$. Then, for all $\delta \in (1 - 1/\theta, 1)$, there exists a sequence $\{v^t\}_{t=0}^\infty$, $v^t \in \{v(1), v(2), \ldots, v(\theta)\}$, such that*

$$v = (1-\delta) \sum_{t=0}^\infty \delta^t v^t.$$

Proof By hypothesis, there exist θ nonnegative numbers $\lambda^1, \ldots, \lambda^\theta$ with $\sum_{k=1}^\theta \lambda^k = 1$ such that

$$v = \sum_{k=1}^\theta \lambda^k v(k).$$

Now we recursively construct a payoff path $\{v^t\}_{t=0}^\infty$ whose average payoff is v as follows. The first payoff profile is $v^0 = v(k)$, where k is an index for which λ^k is maximized. Now, suppose we have determined $(v^0, \ldots, v^{t-1})$, the first t periods of the payoff path. Let $I^\tau(k)$ be the indicator function for the τth element in the sequence:

$$I^\tau(k) = \begin{cases} 1 & \text{if } v^\tau = v(k), \\ 0 & \text{otherwise,} \end{cases}$$

and let $N^t(k)$ count the discounted occurrences of profile $v(k)$ in the first t periods,

$$N^t(k) = \sum_{\tau=0}^{t-1} (1-\delta)\delta^\tau I^\tau(k),$$

with $N^0(k) \equiv 0$. Set $v^t = v(k')$, where

$$k' = \arg\max_{k=1,\ldots,\theta} \{\lambda^k - N^t(k)\}, \tag{3.7.1}$$

breaking ties arbitrarily. Intuitively, $\max_{k=1,\ldots,\theta}\{\lambda^k - N^t(k)\}$ identifies the profile that is currently "most underrepresented" in the sequence, and the trick is to choose that profile to be the next element.

We now argue that as long as $\delta > 1 - 1/\theta$, the resulting sequence generates precisely the payoffs v. Because

$$\sum_{t=0}^{\infty}(1-\delta)\delta^t v^t = \sum_{k=1}^{\theta}\sum_{t=0}^{\infty}(1-\delta)\delta^t I^t(k)v(k) = \sum_{k=1}^{\theta}\lim_{t\to\infty} N^t(k)v(k),$$

we need only show that $\lim_{t\to\infty} N^t(k) = \lambda^k$ for each k. A sufficient condition for this equality is that for all k and all t, $N^t(k) \leq \lambda^k$ (because $\lim_{t\to\infty}\sum_{k=1}^{\theta} N^t(k) = 1 = \sum_{k=1}^{\theta}\lambda^k$).

We prove $N^t(k) \leq \lambda^k$ by induction. Let $k(t)$ denote the k for which $v^t = v(k)$. Consider $t = 0$. There must be at least one profile k for which $\lambda^k \geq 1/\theta$ and hence $\lambda^k > 1 - \delta$ (from the lower bound on δ), ensuring $N^1(k(0))(= 1-\delta) \leq \lambda^{k(0)}$. For all values $k \neq k(0)$, we trivially have $N^1(k) = 0 \leq \lambda^k$.

Suppose now $t > 0$ and $N^\tau(k) \leq \lambda^k$ for all $\tau < t$ and all k. We have $N^t(k(t-1)) = N^{t-1}(k(t-1)) + (1-\delta)\delta^{t-1}$ and $N^t(k) = N^{t-1}(k)$ for $k \neq k(t-1)$. From the definitions, we have

$$\sum_k \lambda^k - \sum_k N^{t-1}(k) = 1 - \sum_k N^{t-1}(k) = 1 - (1-\delta)\sum_{\tau=0}^{t-2}\delta^\tau = \delta^{t-1}.$$

Thus $\delta^{t-1}/\theta \leq \max_k\{\lambda^k - N^{t-1}(k)\}$. Hence (from (3.7.1))

$$N^t(k(t-1)) = N^{t-1}(k(t-1)) + (1-\delta)\delta^{t-1}$$
$$\leq N^{t-1}(k(t-1)) + (1-\delta)\theta[\lambda^{k(t-1)} - N^{t-1}(k(t-1))],$$

or

$$N^t(k(t-1)) \leq [1 - \theta(1-\delta)]N^{t-1}(k(t-1)) + \theta(1-\delta)\lambda^{k(t-1)}.$$

Because $0 < \theta(1-\delta) < 1$ and $N^{t-1}(k(t-1)) \leq \lambda^{k(t-1)}$, we have $N^t(k(t-1)) \leq \lambda^{k(t-1)}$. Hence $N^t(k) \leq \lambda^k$ for all k (because, for $k \neq k(t-1)$, the values $N^t(k)$ remain unchanged at $N^{t-1}(k) \leq \lambda^k$ [by the induction hypothesis]).
∎

We can thus construct deterministic sequences of action profiles that hit any target payoffs in $\mathscr{F}^\dagger$ exactly. However, dispensing with public correlation by inserting these sequences as continuation outcome paths wherever needed in the folk theorem strategies does not yet give us an equilibrium. The difficulty is that some of the continuation values generated by these sequences may fail to be even weakly individually rational. Hence, although the sequences may generate the required initial average payoffs, they may also generate irresistible temptations to deviate as they are played.

3.7 ■ Convexifying the Equilibrium Payoff Set

The key to resolving this difficulty is provided by the following, due to Fudenberg and Maskin (1991).

Lemma 3.7.2 *For all $\varepsilon > 0$, there exists $\underline{\delta} < 1$ such that for all $\delta \in (\underline{\delta}, 1)$ and all $v \in \mathscr{F}^\dagger$, there exists a sequence of pure action profiles whose discounted average payoffs are v and whose continuation payoffs at any time t are within ε of v.*

Proof Fix $\varepsilon > 0$ and $v \in \mathscr{F}^\dagger$. Set $\varepsilon' = \varepsilon/5$, and let $B_\varepsilon(v)$ be the open ball of radius ε centered at v, relative to $\mathscr{F}^\dagger$.[19] Finally, let $v^1, \ldots, v^\theta$ be payoff profiles with the properties that

1. $B_{\varepsilon'}(v)$ is a subset of the interior of the convex hull of $\{v^1, \ldots, v^\theta\}$;
2. $v^\ell \in B_{2\varepsilon'}(v)$, $\ell = 1, \ldots, \theta$; and
3. each v^ℓ is a rational convex combination of the same K pure-action payoff profiles $u(a(1)), \ldots, u(a(K))$, with some profiles possibly receiving zero weight.

The profile v^ℓ is a rational convex combination of $u(a(1)), \ldots, u(a(K))$ if there are nonnegative rationals, summing to unity, $\lambda^\ell(1), \ldots, \lambda^\ell(K)$ with $v^\ell = \sum_{k=1}^K \lambda^\ell(k) u(a(k))$. Because these are rationals, we can express each $\lambda^\ell(k)$ as the ratio of integers $r^\ell(k)/d$, where d does not depend on ℓ. Associate with each v^ℓ the cycle of d action profiles

$$(\underbrace{a(1), \ldots, a(1)}_{r^\ell(1) \text{ times}}, \underbrace{a(2), \ldots, a(2)}_{r^\ell(2) \text{ times}}, \ldots, \underbrace{a(K), \ldots, a(K)}_{r^\ell(K) \text{ times}}). \tag{3.7.2}$$

Let $v^\ell(\delta)$ be the average payoff of this sequence, that is,

$$v^\ell(\delta) = \frac{1-\delta}{1-\delta^d}\left[\frac{1-\delta^{r^\ell(1)}}{1-\delta}u(a(1)) + \delta^{r^\ell(1)}\frac{1-\delta^{r^\ell(2)}}{1-\delta}u(a(2))\right.$$
$$\left. + \cdots + \delta^{d-r^\ell(K)}\frac{1-\delta^{r^\ell(K)}}{1-\delta}u(a(K))\right].$$

Because $\lim_{\delta \to 1}(1-\delta^{r^\ell(k)})/(1-\delta^d) = r^\ell(k)/d$, $\lim_{\delta \to 1} v^\ell(\delta) = v^\ell$. Hence, because $B_{\varepsilon'}(v)$ is in the interior of the convex hull of the v^ℓ, for sufficiently large δ, $B_{\varepsilon'}(v)$ is a subset of the convex hull of $\{v^1(\delta), \ldots, v^\theta(\delta)\}$. Moreover, again for sufficiently large δ, $v^\ell(\delta) \in B_{3\varepsilon'}(v)$. Let $\underline{\delta}'$ be the implied lower bound on δ.

Let $\underline{\delta}'' > \underline{\delta}'$ satisfy $(\underline{\delta}'')^d > 1 - 1/\theta$. From lemma 3.7.1, if $\delta > \underline{\delta}''$, for each $v' \in B_{\varepsilon'}(v)$, there is a sequence $\{v^\tau\}_{\tau=0}^\infty$, with $v^\tau \in \{v^1(\delta), \ldots, v^\theta(\delta)\}$, whose average discounted value *using the discount factor* δ^d is exactly v'. We now construct another sequence by replacing each $v^\tau = v^\ell(\delta)$ in the original sequence with its corresponding cycle, given in (3.7.2). The average

---

19. Note that we are not assuming $\mathscr{F}^\dagger$ has nonempty interior (the full-dimension assumption).

discounted value of the resulting outcome path $\{a(k_t)\}$ using the discount factor δ is precisely v':

$$(1-\delta)\sum_{t=0}^{\infty}\delta^t u(a(k_t)) = (1-\delta^d)\sum_{\tau=0}^{\infty}\frac{(1-\delta)\delta^{d\tau}}{1-\delta^d}\left\{\frac{1-\delta^{r^{\ell(\tau)}(1)}}{1-\delta}u(a(1))\right.$$
$$+ \delta^{r^{\ell(\tau)}(1)}\frac{1-\delta^{r^{\ell(\tau)}(2)}}{1-\delta}u(a(2))$$
$$\left.+\cdots+\delta^{d-r^{\ell(\tau)}(K)}\frac{1-\delta^{r^{\ell(\tau)}(K)}}{1-\delta}u(a(K))\right\}$$
$$= (1-\delta^d)\sum_{\tau=0}^{\infty}\delta^{d\tau}v^{\tau} = v'.$$

At the beginning of each period $0, d, 2d, 3d, \ldots$, the continuation payoff must lie in $B_{3\varepsilon'}(v)$, because the continuation is a convex combination of $v^{\ell}(\delta)$'s, each of which lies in $B_{3\varepsilon'}(v)$. An upper bound on the distance that the continuation can be from v is then

$$(1-\delta^d)\max_{a,a'\in A}|u(a)-u(a')| + \delta^d 3\varepsilon'.$$

There exists $\underline{\delta} > \underline{\delta}''$ such that for all $\delta \in (\underline{\delta}, 1)$, $(1-\delta^d)\max|u(a)-u(a')| < \varepsilon'$, and so the continuations are always within $4\varepsilon'$ of v, and so within $5\varepsilon' = \varepsilon$ of v'.

We have thus shown that for fixed ε, for every $v \in \mathscr{F}^\dagger$ there is a $\underline{\delta}$ such that for all $\delta \in (\underline{\delta}, 1)$, for every $v' \in B_{\varepsilon'}(v)$, we can find a sequence of pure actions whose payoff is v' and whose continuation payoffs lie within ε of v'. Now cover $\mathscr{F}^\dagger$ with the collection of sets $\{B_{\varepsilon'}(v) : v \in \mathscr{F}^\dagger\}$. Taking a finite subcover and then taking the maximum of the $\underline{\delta}$ over this finite collection of sets yields the result.

∎

This allows us to establish:

Proposition 3.7.1 *Suppose $\mathscr{F}^{\dagger p}$ has nonempty interior. Then, for every payoff v in $\{\tilde{v} \in \mathscr{F}^{\dagger p} : \exists v' \in \mathscr{F}^{\dagger p}, v'_i < \tilde{v}_i \; \forall i\}$, there exists $\underline{\delta} < 1$ such that for all $\delta \in (\underline{\delta}, 1)$, there exists a subgame-perfect equilibrium with payoffs v.*

The idea of the proof is to proceed as in the proof of proposition 3.4.1, applied to $\mathscr{F}^\dagger$, choosing a discount factor large enough that each of the optimality conditions holds as a strict inequality by at least some $\varepsilon > 0$. For each payoff v^d, $d = 0, \ldots, n$, we then construct an infinite sequence of pure action profiles giving the same payoff and with continuation values within some ε' of this payoff. The strategies of proposition 3.4.1 can now be modified so that each state w^d is replaced by the infinite sequence of states implementing the payoff $u(a^d)$, with play continuing along the sequence in the absence of a deviation and otherwise triggering a punishment as prescribed in the original strategies. We must then argue that the slack in the optimality conditions (3.4.1)–(3.4.2) ensures that they continue to hold once the potential ε' error in continuation payoffs is introduced. Though conceptually straightforward, the details of this

3.8 Mixed-Action Individual Rationality

argument are tedious and so omitted. For finite games, this result is a special case of proposition 3.8.1 (which is proved in detail).

3.8 Mixed-Action Individual Rationality

Proposition 3.4.2 is a folk theorem for payoff profiles that strictly dominate players' pure-action minmax utilities. In this section, we extend this result for finite stage games to payoff profiles that strictly dominate players' mixed-action minmax utilities (recall that in many games, a player's mixed-action minmax utility is strictly less than his pure-action minmax utility). At the same time, using the results of the previous section, we dispense with public correlation.

The difficulty in using mixed minmax actions as punishments is that the punishment phase may produce quite low payoffs for the punishers. If the minmax actions are pure, the incentive problems produced by these small payoffs are straightforward. Deviations from the prescribed minmax profile can themselves be punished. The situation is more complicated when the minmax actions are mixed. There is no reason to expect player j to be myopically indifferent over the various actions in the support of the mixture required to minmax player i. Consequently, the continuations after the various actions in the support must be such that j is indifferent over these actions. In other words, strategies will need to adjust postpunishment payoffs to ensure that players are indifferent over any pure actions involved in a mixed-action minmax. This link between realized play during the punishment period and postpunishment play will make an automaton description of the strategies rather cumbersome.

We use the techniques outlined in the previous section to avoid the use of public correlation.[20] A more complicated argument shows that any payoff in $\mathscr{F}^*$ is an equilibrium payoff for patient players under NEU (see Abreu, Dutta, and Smith 1994). The following version of the full folk theorem suffices for most contexts.

Proposition 3.8.1 **Mixed-minmax perfect monitoring folk theorem** *Suppose A_i is finite for all i and $\mathscr{F}^*$ has nonempty interior. For any $v \in \{\tilde{v} \in \mathscr{F}^* : \exists v' \in \mathscr{F}^*, v'_i < \tilde{v}_i \,\forall i\}$, there exists $\underline{\delta} \in (0, 1)$ such that for every $\delta \in (\underline{\delta}, 1)$, there exists a subgame-perfect equilibrium of the infinitely repeated game (without public correlation) with discount factor δ giving payoffs v.*

Proof Fix $v \in \{\tilde{v} \in \mathscr{F}^* : \exists v' \in \mathscr{F}^*, v'_i < \tilde{v}_i \,\forall i\}$. Because $\mathscr{F}^*$ is convex and has a nonempty interior, there exists $v' \in \mathscr{F}^*$ and $\varepsilon > 0$ such that $B_{2n\varepsilon}(v') \subset \mathscr{F}^*$ and for every player i,

---

20. The presence of a public correlation device has no effect on player i's minmax utility. Given a minmax action profile with public correlation, select an outcome of the randomization for which player i's payoff is minimized, and then note that the accompanying action profile is a minmax profile giving player i the same payoff.

In some games with more than two players, player i's minmax payoff can be pushed lower than $\underline{v}_i$ if the other players have access to a device that allows them to play a correlated action profile, where this device is *not* available to player i. This is an immediate consequence of the observation that not all correlated distributions over action profiles can be obtained through independent randomization.

$$v_i - v_i' \geq 2\varepsilon. \tag{3.8.1}$$

Let $\kappa \equiv \min_i v_i' - \underline{v}_i$, and for each $\ell \in \mathbb{N} \equiv \{1, 2, \ldots\}$, let $L_\ell(\delta)$ be the integer satisfying

$$\frac{2}{\ell(1-\delta)\kappa} - 1 < L_\ell(\delta) \leq \frac{2}{\ell(1-\delta)\kappa}. \tag{3.8.2}$$

For each ℓ, there is $\delta^\dagger(\ell)$ such that $L_\ell(\delta) \geq 1$ for all $\delta \in (\delta^\dagger(\ell), 1)$. Moreover, an application of l'Hôpital's rule shows that

$$\lim_{\delta \to 1} \delta^{L_\ell(\delta)} = e^{-\frac{2}{\ell\kappa}},$$

and hence

$$\lim_{\ell \to \infty} \lim_{\delta \to 1} \delta^{L_\ell(\delta)} = 1.$$

Consequently, for all $\eta > 0$, there exists ℓ' and $\delta' : \mathbb{N} \to (0, 1)$, $\delta'(\ell) \geq \delta^\dagger(\ell)$ such that for all $\ell > \ell'$ and $\delta > \delta'(\ell)$, $1 - \eta < \delta^{L_\ell(\delta)} < 1$. We typically suppress the dependence of L on ℓ and δ in our notation.

Denote the profile minmaxing player j by $\hat{\alpha}_{-j}^j$, and suppose player j is to be minmaxed for L periods. Denote player i's realized average payoff over these L periods by

$$p_i^j \equiv \frac{1-\delta}{1-\delta^L} \sum_{\tau=1}^{L} \delta^{\tau-1} u_i(a(\tau)),$$

where $a(\tau)$ is the *realized* action profile in period τ. If the minmax actions are mixed, this average payoff is a random variable, depending on the particular sequence of actions realized in the course of minmaxing player j. Define

$$z_i^j \equiv \frac{1-\delta^L}{\delta^L} p_i^j.$$

Like p_i^j, z_i^j depends on the realized history of play. Because p_i^j is bounded by the smallest and largest stage-game payoff for player i, we can now choose ℓ_1 and $\delta_1 : \mathbb{N} \to (0, 1)$, $\delta_1(\ell) \geq \delta^\dagger(\ell)$ for all ℓ, such that for all $\ell > \ell_1$ and $\delta > \delta_1(\ell)$, $\delta^{L_\ell(\delta)}$ is sufficiently close to one that $|z_i^j| < \varepsilon/2$, for all i and j and all realizations of the sequence of minmaxing actions.

Deviations from minmax behavior outside the support of the minmax actions present no new difficulties because such deviations are detected. Deviations that only involve actions in the support, on the other hand, are not detected. Player i is only willing to randomize according to the mixture $\hat{\alpha}_i^j$ if he is indifferent over the pure actions in the support of the mixture. The technique is to specify continuation play as a function of z_i^j, following the L periods of punishment, to create the required indifference.

3.8 ■ Mixed-Action Individual Rationality

Define the payoff

$$v'(z^j) = (v'_1 + \varepsilon - z^j_1, \ldots, v'_{j-1} + \varepsilon - z^j_{j-1}, v'_j, v'_{j+1} + \varepsilon - z^j_{j+1}, \ldots, v'_n + \varepsilon - z^j_n).$$

Given $B_{2n\varepsilon}(v') \subset \mathscr{F}^*$ and our restriction that $|z^j_i| < \varepsilon/2$, we know that $B_\varepsilon(v'(z^j)) \subset \mathscr{F}^*$ and $v'_i(z^j) - v_i > 2n\varepsilon$.

From lemma 3.7.2, there exists $\delta_2 : \mathbb{N} \to (0, 1)$, $\delta_2(\ell) \geq \delta_1(\ell)$ for all ℓ, such that for any payoff $v'' \in \mathscr{F}^\dagger$, there is a sequence of pure action profiles, $\mathbf{a}(v'')$, whose average discounted payoff is v'' and whose continuation payoff at any point is within $1/\ell$ of v''.

We now describe the strategies in terms of phases:

Phase 0:   Play $\mathbf{a}(v)$,
Phase i, $i = 1, \ldots, n$:   Play L periods of $\hat{\alpha}^i$, followed by $\mathbf{a}(v'(z^i))$.

The equilibrium is specified by the following recursive rules for combining these phases. Play begins in phase 0 and remains there along the equilibrium path. Any unilateral deviation by player i causes play to switch to the beginning of phase i. Any unilateral deviation by player $j \neq k$ while in phase k causes play to switch to the beginning of phase j. A deviation from a mixed strategy is interpreted throughout as a choice of an action outside the support of the mixture. Mixtures appear only in the first L periods of phases $1, \ldots, n$, when one player is being minmaxed. Simultaneous deviations are ignored throughout.

We now verify that for some sufficiently large ℓ and $\underline{\delta}$, for all $\delta \in (\underline{\delta}, 1)$, these strategies (which depend on δ through $L_\ell(\delta)$) are an equilibrium.

Consider first phase 0. A sufficient equilibrium condition for player i to not find a deviation profitable is that for $i = 1, \ldots, n$

$$(1 - \delta)M + \delta(1 - \delta^L)\underline{v}_i + \delta^{L+1} v'_i \leq v_i - \frac{1}{\ell},$$

where the left side is an upper bound on the value of a deviation and the right side a lower bound on the equilibrium continuation value, with (from lemma 3.7.2) the $-1/\ell$ term reflecting the maximum shortfall from the payoff v_i of any continuation payoff in $\mathbf{a}(v)$. This inequality can be rearranged to give

$$(1 - \delta)(M - v_i) \leq \delta(1 - \delta^L)(v_i - \underline{v}_i) + \delta^{L+1}(v_i - v'_i) - \frac{1}{\ell}. \quad (3.8.3)$$

From (3.8.1), if $\ell > 1/\varepsilon$, $(v_i - v'_i) - 1/\ell > 2\varepsilon - 1/\ell > \varepsilon > 0$. Hence, there exists $\delta_3 : \mathbb{N} \to (0, 1)$, $\delta_3(\ell) \geq \delta_2(\ell)$ for all ℓ, such that, for all $\ell > 1/\varepsilon$ and all $\delta \geq \delta_3(\ell)$, $\delta^{L_\ell(\delta)}$ is sufficiently close to 1 that (3.8.3) is satisfied.

Now consider phase $j \geq 1$, and suppose at least L periods of this phase have already passed, so that play falls somewhere in the sequence $\mathbf{a}(v'(z^j))$. If $\ell > \ell_1$ and $\delta > \delta_2(\ell)$, a sufficient condition for player i to not find a deviation profitable is

$$(1 - \delta)M + \delta(1 - \delta^L)\underline{v}_i + \delta^{L+1} v'_i \leq v'_i - \frac{1}{\ell},$$

with the $-1/\ell$ term appearing on the right side because i's continuation value under $\mathbf{a}(v'(z^j))$ must be within $1/\ell$ of v'_i (if $i = j$) or within $1/\ell$ of $v'_i + \varepsilon - z^j_i$

(otherwise), and hence must be at least $v'_i - 1/\ell$. Using (3.8.2), a sufficient condition for this inequality is

$$(1-\delta)(M - v'_i) \leq \delta \left(1 - \delta^{\left(\frac{2}{\ell(1-\delta)\kappa} - 1\right)}\right)(v'_i - \underline{v}_i) - \frac{1}{\ell}. \tag{3.8.4}$$

We now note that as δ approaches 1, the right side converges to

$$\left(1 - e^{-\frac{2}{\ell\kappa}}\right)(v'_i - \underline{v}_i) - \frac{1}{\ell}. \tag{3.8.5}$$

Now, an application of l'Hôpital's rule, and using $\kappa \leq v'_i - \underline{v}_i$, shows that

$$\lim_{\ell \to \infty} \ell \left[\left(1 - e^{-\frac{2}{\ell\kappa}}\right)(v'_i - \underline{v}_i) - \frac{1}{\ell}\right] \geq 1.$$

Fix $\eta > 0$ small. There exists $\ell_4 \geq \max\{\ell_1, 1/\varepsilon\}$ such that for all $\ell > \ell_4$, the expression in (3.8.5) is no smaller than $(1-\eta)/\ell$. Consequently, there exists $\delta_4 : \mathbb{N} \to (0,1)$, $\delta_4(\ell) \geq \delta_3(\ell)$ for all ℓ, such that (3.8.4) holds for all $\ell > \ell_4$ and $\delta > \delta_4(\ell)$.

Now consider phase $j \geq 1$, and suppose we are in the first L periods, when player j is to be minmaxed. Player j has no incentive to deviate from the prescribed behavior and prompt the phase to start again, because j is earning j's minmax payoff and can do no better during each of the L periods, whereas subsequent play gives j a higher continuation payoff that does not depend on j's actions. Player i's *realized* payoff while minmaxing player j is given by

$$(1 - \delta^L)p_i^j + \delta^L \left[v'_i + \varepsilon - \frac{1-\delta^L}{\delta^L}p_i^j\right] = \delta^L(v'_i + \varepsilon),$$

which is independent of the actions i takes while minmaxing j. A similar characterization applies to any subset of the L periods during which j is to be minmaxed. Player i thus has no profitable deviation that remains within the support of $\hat{\alpha}_i^j$. Suppose player i deviates outside this support. If $\ell > \ell_1$ and $\delta > \delta_1(\ell)$, a deviation in period $\tau \in \{1, \ldots, L\}$ is unprofitable if

$$(1-\delta)M + \delta(1-\delta^L)\underline{v}_i + \delta^{L+1}v'_i \leq (1-\delta^{L-\tau+1})u_i(\hat{\alpha}^j) + \delta^{L-\tau+1}(v'_i + \varepsilon/2)$$

because $v'_i(z^j) = v'_i + \varepsilon - z_i^j > v'_i + \varepsilon/2$. As δ^L (and so $\delta^{L-\tau+1}$ for all τ) approaches 1, this inequality becomes $v'_i \leq v'_i + \varepsilon/2$. Consequently, there exists $\ell_5 \geq \ell_4$ and $\delta_5 : \mathbb{N} \to (0,1)$, $\delta_5(\ell) \geq \delta_4(\ell)$ for all ℓ, such that for all $\ell > \ell_5$ and $\delta > \delta_5(\ell)$, the above displayed inequality is satisfied.

We now fix a value $\ell > \ell_5$ and take $\underline{\delta} = \delta_5(\ell)$. For any $\delta \in (\underline{\delta}, 1)$, letting $L = L_\ell(\delta)$ as defined by (3.8.2) then gives a subgame-perfect equilibrium strategy profile with the desired payoffs.

∎

4 How Long Is Forever?

4.1 Is the Horizon Ever Infinite?

A common complaint about infinitely repeated games is that very few relationships have truly infinite horizons. Moreover, modeling a relationship as a seemingly more realistic finitely repeated game can make quite a difference. For example, in the finitely repeated prisoners' dilemma, the only Nash equilibrium outcome plays SS in every period.[1] Should we worry that important aspects of the results rest on unrealistic features of the model?

In response, we can do no better than the following from Osborne and Rubinstein (1994, p. 135, emphasis in original):

> In our view a model should attempt to capture the features of reality that the players *perceive* ... the fact that a situation has a horizon that is in some physical sense finite (or infinite) does not *necessarily* imply that the best model of the situation has a finite (or infinite) horizon.... If they play a game so frequently that the horizon approaches only very slowly then they may ignore the existence of the horizon entirely until its arrival is imminent, and until this point their strategic thinking may be better captured by a game with an infinite horizon.

The key consideration in evaluating a model is not whether it is a literal description of the strategic interaction of interest, nor whether it captures the behavior of perfectly rational players in the actual strategic interaction. Rather, the question is whether the model captures the behavior we observe in the situation of interest. For example, finite-horizon models predict that fiat money should be worthless. Because its value stems only from its ability to purchase consumption goods in the future, it will be worthless in the final period of the economy. A standard backward-induction argument then allows us to conclude that it will be worthless in the penultimate period, and the one before that, and so on. Nonetheless, most people are willing to accept money, even those who argue that there is no such thing as an interaction with an infinite horizon. The point

---

1. A standard backward induction argument shows that SS in every period is the only subgame-perfect outcome, because the stage game has a unique Nash equilibrium. The stronger conclusion that this is the only Nash equilibrium outcome follows from the observation that the stage-game Nash equilibrium yields each player his minmax utility. This latter property ensures that "incredible threats" (i.e., a non-Nash continuations) cannot yield payoffs below those of the stage-game Nash equilibrium, precluding the use of future play to create incentives for choosing any current actions other than the stage-game Nash equilibrium.

is that whether finite or infinite, the end of the horizon is sufficiently distant that it does not enter people's strategic calculations. Infinite-horizon models may then be an unrealistic literal description of the world but a useful model of how people behave in that world.

4.2 Uncertain Horizons

An alternative interpretation of infinitely repeated games begins by interpreting players' preferences for early payoffs as arising out of the possibility that the game may end before later payoffs can be collected. Suppose that in each period t, the game ends with probability $1 - \delta > 0$, and continues with probability $\delta > 0$. Players maximize the sum of realized payoffs (normalized by $1 - \delta$). Player i's expected payoff is then

$$(1 - \delta) \sum_{t=0}^{\infty} \delta^t u_i(a^t).$$

Because the probability that the game reaches period t is δ^t, and $\delta^t \to 0$ as $t \to \infty$, the game has a finite horizon with probability 1. However, once the players have reached any period t, they assign probability δ to play continuing, allowing intertemporal incentives to be maintained despite the certainty of a finite termination.

Viewing the horizon as uncertain in this way allows the model to capture some seemingly realistic features. One readily imagines knowing that a relationship will not last forever, while at the same time never being certain of when it will end. Observe, however, that under this interpretation the distribution over possible lengths of the game has unbounded support. One does not have to believe that the game will last forever but must believe that it *could* last an arbitrarily long time. Moreover, the hazard rate is constant. No matter how old the relationship is, the expected number of additional periods before the relationship ends is unchanged. The potential for the relationship's lasting forever has thus been disguised but not eliminated.

If the stage game has more than one Nash equilibrium, then intertemporal incentives can be constructed even without the infinite-repetition possibility embedded in a constant continuation probability. Indeed, as we saw in the discussion of the game in figure 1.1.1 and return to in section 4.4, the prospect of future play can have an effect even if the continuation probability falls to 0 after the second period, that is, even if the game is played only twice. However, if the stage game has a unique Nash equilibrium, such as the prisoners' dilemma, then the possibility of an infinite horizon is crucial for constructing intertemporal incentives. To see this, suppose that the horizon may be uncertain, but it is *commonly known* that the game will certainly end no later than after $T + 1$ periods (where T is some possibly very large but finite number). Then we again face a backward-induction argument. The game may not last so long, but should period T ever be reached, the players will know it is the final period, and there is only one possible equilibrium action profile. Hence, if the game reaches period $T - 1$, the players will know that their current behavior has no effect on the future, either because the game ends and there is no future play or because the game continues for (only) one more round of play, in which case the unique stage-game Nash equilibrium again appears. The argument continues, allowing us to conclude that only the stage-game Nash equilibrium can appear in any period.

The assumption in the last paragraph that it is commonly known that the game will end by some T is critical. Suppose for example, that although both players know the game must end by period T, player 1 does *not* know that player 2 knows this. In that case, if period $T-1$ were to be reached, player 1 may now place significant probability on player 2 not knowing that the next period is the last possible period, and so 1 may exert effort rather than shirk. This argument can be extended, so that even if both players know the game will end by T, and both players know that both know, if it is not commonly known (i.e., at some level of iteration, statements of the form "i knows that j knows that i knows ... that the game will end by T" fails), then it may be possible to support effort in the repeated prisoners' dilemma (see Neyman 1999). To do so, however, requires that, for any arbitrarily large T, player i believes that j believes that i believes ... that the game could last longer than T periods.

In the presence of a unique stage-game Nash equilibrium, nontrivial outcomes in the repeated game thus hinge on at least the possibility of an infinite horizon. However, one might then seek to soften the role of the infinite horizon by allowing the continuation probability to decline as the game is played. We may never be certain that a particular period will be the last but may think it more likely to be the last if the game has already continued for a long time. In the next section, we construct nontrivial intertemporal incentives for stage games with a unique Nash equilibrium, when the continuation probability (discount factor) converges to 0, as long as the rate at which this occurs is not too fast.

4.3 Declining Discount Factors

Suppose players share a sequence of one-period discount factors $\{\delta_t\}_{t=0}^{\infty}$, where δ_t is the rate applied to period $t+1$ payoffs in period t. Hence, payoffs in period t are discounted to the beginning of the game at rate

$$\prod_{\tau=0}^{t-1} \delta_\tau.$$

We can interpret these discount factors as representing either time preferences or uncertainty as to the length of the game. In the latter case, δ_τ is the probability that play continues to period $\tau + 1$, conditional on play having reached period τ. We are interested in the case where $\lim_{\tau \to \infty} \delta_\tau = 0$. In terms of continuation probabilities, this implies that as τ gets large, the game ends after the current period with probability approaching 1.

One natural measure for evaluating payoffs is the average discounted value of the payoff stream $\{u_i^t\}_{t=0}^{\infty}$,

$$\sum_{t=0}^{\infty} \left(\prod_{\tau=0}^{t-1} \delta_\tau \right) (1 - \delta_t) u_i^t,$$

where $\prod_{\tau=0}^{-1} \delta_\tau \equiv 1$. When the discount factor is constant, this coincides with the average discounted value. However, unlike with a constant discount factor, the implied intertemporal preferences are different from the preferences without averaging, given by

$$\sum_{t=0}^{\infty} \left(\prod_{\tau=0}^{t-1} \delta_\tau \right) u_i^t.$$

We view the unnormalized payoffs u_i^t as the true payoffs and, because the factor $(1 - \delta_t)$ is here not simply a normalization, we work without it.

We assume the players do discount, in the sense that

$$\sum_{t=0}^{\infty} \prod_{\tau=0}^{t-1} \delta_\tau \equiv \Delta < \infty. \qquad (4.3.1)$$

When (4.3.1) is satisfied, a constant payoff stream $u_i^t = u_i$ for all t has a well-defined value in the repeated game, Δu_i. Condition (4.3.1) is trivially satisfied if $\delta_t \to 0$.

Our interest in the case $\lim_{\tau \to 0} \delta_\tau = 0$ imposes some limitations. Intuitively, intertemporal incentives work whenever the myopic incentive to deviate is less than the size of the discounted future cost incurred by the deviation. For small δ, the intertemporal incentives are necessarily weak, and so can only work if the myopic incentive to deviate is also small, such as would arise for action profiles close to a stage-game Nash equilibrium.

The most natural setting for considering weak myopic incentives to deviate is found in infinite games, such as the oligopoly example of section 2.6.2.

Assumption 4.3.1 *Each player's action space A_i is an interval $[\underline{a}_i, \bar{a}_i] \subset \mathbb{R}$. The stage game has a Nash equilibrium profile a^N contained in the interior of the set of action profiles A. Each payoff function u_i is twice continuously differentiable in a and strictly quasi-concave in a_i, and each player's best reply function $\phi_i : \prod_{-i} A_j \to A_i$ is continuously differentiable in a neighborhood of a^N.[2] The Jacobian matrix of partial derivatives $Du(a^N)$ has full rank.*

This assumption is satisfied, for example, in the oligopoly model of section 2.6.2, with an upper bound on quantities of output so as to make the strategy sets compact.

We now describe, following Bernheim and Dasgupta (1995), how to construct nontrivial intertemporal incentives. Though the discount factors can converge to 0, intertemporal incentives require that this not occur too quickly. We assume that there exist constants $c, \Lambda > 0$ such that,

$$\prod_{k=0}^{\tau-1} \delta_k^{2^{\tau-1-k}} \geq c \Lambda^{2^\tau}, \forall \tau \geq 1. \qquad (4.3.2)$$

This condition holds for a constant discount factor δ (take $c = 1/\delta$ and $\Lambda = \delta$). By taking logs and rearranging, (4.3.2) can be seen to be equivalent to

$$\lim_{\tau \to \infty} \sum_{k=0}^{\tau-1} \frac{1}{2^{k+1}} \ln \delta_k > -\infty, \qquad (4.3.3)$$

implying that 2^k must grow faster than $|\ln \delta_k|$ (if $\delta_k \to 0$, then $\ln \delta_k \to -\infty$).

---

2. Because u_i is assumed strictly quasi-concave in a_i, $\phi_i(a_{-i}) \equiv \arg\max_{a_i} u_i(a_i, a_{-i})$ is unique, for all a_{-i}.

4.3 ∎ Declining Discount Factors

Proposition 4.3.1 *Suppose $\{\delta_k\}_{k=0}^{\infty}$ is a sequence of discount factors satisfying (4.3.1) and (4.3.2), but possibly converging to 0. There exists a subgame-perfect equilibrium of the repeated game in which in every period, every player receives a higher payoff than that produced by playing the (possibly unique) stage-game Nash equilibrium, a^N.*

Proof Because $Du(a^N)$ has full rank, and a^N is interior, there exists a constant $\eta > 0$, a vector $z \in \mathbb{R}^N$ and an $\varepsilon' > 0$ such that, for all $\varepsilon < \varepsilon'$ and all i,

$$u_i(a^N + \varepsilon z) - u_i(a^N) \geq \eta \varepsilon. \tag{4.3.4}$$

Letting $u_i^d(a)$ denote player i's maximal deviation payoff,

$$u_i^d(a) = \max_{a_i \in A_i} u_i(a_i, a_{-i}),$$

we have $u_i^d(a^N) = u_i(\phi_i(a_{-i}^N), a_{-i}^N)$. Because $\partial u_i(a_i^N)/\partial a_i = 0$,

$$\frac{\partial u_i^d(a^N)}{\partial a_j} = \frac{\partial u_i(\phi_i(a_{-i}^N), a_{-i}^N)}{\partial a_i} \frac{\partial \phi_i(a_{-i}^N)}{\partial a_j} + \frac{\partial u_i(\phi_i(a_{-i}^N), a_{-i}^N)}{\partial a_j}$$

$$= \frac{\partial u_i(\phi_i(a_{-i}^N), a_{-i}^N)}{\partial a_j} = \frac{\partial u_i(a^N)}{\partial a_j}.$$

(This is an instance of the envelope theorem.) By Taylor's formula, there exists a $\beta > 0$ and ε'' such that, for all i and $\varepsilon < \varepsilon''$,

$$u_i^d(a^N + \varepsilon z) - u_i(a^N + \varepsilon z)$$

$$\leq u_i^d(a^N) - u_i(a^N) + \varepsilon \sum_{j=1}^n z_j \left(\frac{\partial u_i^d(a^N)}{\partial a_j} - \frac{\partial u_i(a^N)}{\partial a_j} \right) + \varepsilon^2 \beta$$

$$= \beta \varepsilon^2. \tag{4.3.5}$$

In the oligopoly example of section 2.6.2, this is the observation that the payoff increment from the optimal deviation from proposed output levels, as a function of those levels, has a zero slope at the stage-game Nash equilibrium (in figure 2.6.1, μ^d is tangential to μ at $q = a^N$). As we have noted, this is an envelope theorem result. The most one can gain by deviating from the stage-game equilibrium strategies is 0, and the most one can gain by deviating from nearby strategies is nearly 0.

Summarizing, moving actions away from the stage-game Nash equilibrium in the direction z yields first-order payoff gains (4.3.4) while prompting only second-order increases in the temptation to defect (4.3.5). This suggests that the threat of Nash reversion should support the non-Nash candidate actions as an equilibrium, if these actions are sufficiently close to the Nash profile. The discount factor will determine the effectiveness of the threat, and so how close the candidate actions must be to the Nash point. As the discount factor gets lower, the punishment becomes relatively less severe. As a result, the proposed actions may have to move closer to the stage-game equilibrium to reinforce the weight of the punishment compared to the temptation of a deviation.

Consider, then, the following strategies. We construct a sequence ε_t for $t = 0, \ldots,$ and let the period t equilibrium outputs be

$$a^N + \varepsilon_t z,$$

with any deviation prompting Nash reversion (i.e., subsequent, permanent play of a^N). These strategies will be an equilibrium if every ε_t is less than $\min\{\varepsilon', \varepsilon''\}$ and if, for all t,

$$\beta \varepsilon_t^2 \leq \delta_t \eta \varepsilon_{t+1}. \qquad (4.3.6)$$

The left side of this inequality is an upper bound on the immediate payoff gain produced by deviating from the proposed equilibrium sequence, and the right side is a lower bound on the discounted value of next period's payoff loss from reverting to Nash equilibrium play. To establish the existence of an equilibrium, we then need only verify that we can construct a sequence $\{\varepsilon_t\}_{t=0}^{\infty}$ satisfying (4.3.6) while remaining below the bounds ε' and ε'' for which (4.3.4) and (4.3.5) hold.

To do this, fix ε_0 and recursively define a sequence $\{\varepsilon_t\}_{t=0}^{\infty}$ by setting

$$\varepsilon_{t+1} = \frac{\beta \varepsilon_t^2}{\eta \delta_t}.$$

This leads to the sequence, for $t \geq 1$,

$$\varepsilon_t = \left(\frac{\beta}{\eta}\right)^{2^t - 1} (\varepsilon_0)^{2^t} \left(\prod_{k=0}^{t-1} \delta_k^{2^{t-1-k}}\right)^{-1}.$$

By construction, this sequence satisfies (4.3.6). We need only verify that the sequence can be constructed so that no term exceeds $\min\{\varepsilon', \varepsilon''\}$. To do this, notice that, from (4.3.2), we have, for all $t \geq 1$,

$$\varepsilon_t \leq \frac{\eta}{\beta c} \left(\frac{\beta}{\eta} \frac{\varepsilon_0}{\Lambda}\right)^{2^t}.$$

It now remains only to choose ε_0 sufficiently small that $\varepsilon_0 < \min\{\varepsilon', \varepsilon''\}$, that $\beta \varepsilon_0^2 / [\eta c \Lambda^2] < \min\{\varepsilon', \varepsilon''\}$ (ensuring $\varepsilon_1 < \min\{\varepsilon', \varepsilon''\}$), and that $\beta \varepsilon_0 / \eta \Lambda \leq 1$ (so that $\varepsilon_{t+1} \leq \varepsilon_t$ for $t \geq 1$). ∎

It is important to note that the above result covers the case where the stage game has a unique equilibrium. Multiple stage game equilibria only make it easier to construct nontrivial intertemporal incentives.

Bernheim and Dasgupta (1995) provide conditions under which (4.3.3) is necessary as well as sufficient for there to exist any subgame-perfect equilibrium other than the continuous play of a unique stage-game Nash equilibrium. This necessity result is not intuitively obvious. The optimality condition given by (4.3.6) assumes that deviations are punished by Nash reversion and makes use only of the first period of such a punishment when verifying the optimality of equilibrium behavior. Observe, however, that as the discount factors tend to 0, eventually all punishments are effectively single-period punishments. From the viewpoint of period t, period $t+1$ is discounted

4.3 ■ Declining Discount Factors

at rate δ_t, and period $t+2$ is discounted at rate $\delta_t \delta_{t+1}$. Because

$$\lim_{t\to\infty} \frac{\delta_t \delta_{t+1}}{\delta_t} = \lim_{t\to\infty} \delta_{t+1} = 0,$$

it is eventually the case that any payoffs occurring two periods in the future are discounted arbitrarily heavily compared to next period's payoffs, ensuring that effectively only the next period matters in any punishment.

There may well be more severe punishments than Nash reversion, and much of the proof of the necessity of (4.3.3) is concerned with such punishments. The intuition is straightforward. Equation (4.3.6) is the optimality condition for punishments by Nash reversion. The prospect of a more serious punishment allows us to replace (4.3.6) with a counterpart that has a larger penalty on the right side. But if (4.3.6) holds for some η and this large penalty, then it must also hold for some larger value η' and the penalty of Nash reversion (then proportionately reducing both η' and β, if needed, so that (4.3.4) and (4.3.5) hold). In a sense, the Nash reversion and optimal punishment cases differ only by a constant. Translating (4.3.3) into the equivalent (4.3.2), the existence of nontrivial subgame-perfect equilibria based on either Nash reversion or optimal punishments both imply the existence of a (possibly different) constant for which (4.3.2) holds.

This result shows that nontrivial subgame-perfect equilibria can be supported with discount factors that decline to 0. However, it only establishes that we can support *some* equilibrium path of outcomes that does not always coincide with the stage-game Nash equilibrium, leaving open the possibility that the equilibrium path is very close to continual play of the stage-game Nash equilibrium. We can apply the insights from finitely repeated games (discussed in the next section) to construct additional equilibria that are "far" from the Nash equilibrium, as long as there is an initial sequence of sufficiently large discount factors.

Proposition 4.3.2 *Suppose the set of feasible payoffs $\mathscr{F}$ has full dimension and $\{\tilde{\delta}_t\}_{t=0}^{\infty}$ is a sequence of discount factors satisfying (4.3.1) and (4.3.2). Suppose the action profile a is strictly individually rational with $u(a)$ in the interior of $\mathscr{F}$. Then there exist a $\bar{\delta} \in (0,1)$ and an integer T^* such that for all $T \geq T^*$, if $\delta_t \geq \bar{\delta}$ for all $t \leq T$ and $\delta_t \geq \tilde{\delta}_{t-T-1}$ for all $t > T$, then there exists a subgame-perfect equilibrium of the game with discount factors $\{\delta_t\}_{t=0}^{\infty}$ in which a is played in the first $T - T^*$ periods.*

Proof Consider a repeated game characterized by $(T, \bar{\delta})$, where $\delta_t \geq \bar{\delta} > \tilde{\delta}_0$ for all $t \leq T$ and $\delta_t \geq \tilde{\delta}_{t-T-1}$ for all $t \geq T+1$. The subgame beginning in period $T+1$ of the repeated game has (at least) two subgame-perfect equilibria, the infinite repetition of a^N and an equilibrium σ from proposition 4.3.1 with payoffs $\bar{v}$ satisfying $\bar{v}_i > \Delta u_i(a^N)$ for all i (recall (4.3.1)).

Let $\bar{a}$ be the action profile in the first period of σ, and note that $\min_i u_i(\bar{a}) - u_i(a^N) > 0$. Suppose $u(a)$ strictly dominates $u(a^N)$, and define $\varepsilon \equiv \min\{u_i(a) - u_i(a^N), u_i(\bar{a}) - u_i(a^N)\} > 0$. For $T^* < T$, let σ^{T^*} be the profile that specifies a in every period $t < T - T^*$, $\bar{a}$ in periods $T - T^* \leq t \leq T$, and then σ as the continuation profile in period $T+1$. A deviation in any period results in Nash reversion. Because $\delta_t \geq \tilde{\delta}_0$ for $t \leq T$, and σ is a subgame-perfect equilibrium of the subgame beginning in period $T+1$, no player wishes to deviate in any period $t \geq T - T^*$. If player i deviates in a period $t < T - T^*$, there is a one-period gain of $u_i^d(a) - u_i(a)$, which must be compared to the flow losses of $u_i(a) - u_i(a^N) \geq \varepsilon$ in periods $t+1, \ldots, T - T^* - 1$, and $u_i(\bar{a}) - u_i(a^N) \geq \varepsilon$ in periods $T - T^*, \ldots, T$, as well as the continuation value loss of $\bar{v}_i - \Delta u_i(a^N) > 0$ in period $T+1$. By choosing T^* large enough and $\bar{\delta}$ close enough to 1, the discounted value of a loss of at least ε for at least T^* periods dominates the one-period gain.

Constructing equilibria for a with payoffs that do not strictly dominate $u(a^N)$ is more complicated and is accomplished using the techniques of Benoit and Krishna (1985), discussed in section 4.4.

∎

These results depend on the assumption that the stage-game strategy sets are compact intervals of the reals, rather than finite, so that actions can eventually become arbitrarily close to stage-game Nash equilibrium actions. They do not hold for finite stage games. In particular, if the stage game is finite, then there is a lower bound on the amount that can be gained by at least some player by deviating from any pure-strategy outcome that is not a Nash equilibrium. As a result, if the stage-game has a unique strict Nash equilibrium, then when discount factors decline to 0, the only subgame-perfect equilibria of the repeated game will feature perpetual play of the stage-game Nash equilibrium.

4.4 Finitely Repeated Games

This section shows that the unique equilibrium of the finitely repeated prisoners' dilemma is not typical of finitely repeated games in general. The discussion of figure 1.1.1 provides a preview of the results presented here. If the stage game has multiple Nash equilibrium payoffs, then it is possible to construct effective intertemporal incentives. Consequently, sufficiently long but finitely repeated games feature sets of equilibrium payoffs very much like those of infinitely repeated games. In contrast, a straightforward backward-induction argument shows that the only subgame-perfect equilibrium of a finitely repeated stage game with a unique stage-game Nash equilibrium has this equilibrium action profile played in every period.

We assume that players in the finitely repeated game maximize average payoffs over the finite horizon, that is, if the stage game characterized by payoff function $u : \prod_i A_i \to \mathbb{R}^n$ is played T times, then the payoff to player i from the outcome $(a^0, a^1, \ldots, a^{T-1})$ is

$$\frac{1}{T} \sum_{t=0}^{T-1} u_i(a^t). \tag{4.4.1}$$

It is perhaps more natural to retain the discount factor δ, and use the discounted finite sum

$$\frac{1-\delta}{1-\delta^T} \sum_{t=0}^{T-1} \delta^t u_i(a^t) \tag{4.4.2}$$

4.4 ■ Finitely Repeated Games

	A	B	C
A	4,4	0,0	0,0
B	0,0	3,1	0,0
C	2,2	0,0	1,3

Figure 4.4.1 The game for example 4.4.1. The game has three strict Nash equilibria, *AA*, *BB*, and *CC*.

as the T period payoff. However, this payoff converges to the average given by (4.4.1) as $\delta \to 1$, and it is simpler to work with (4.4.1) and concentrate on the limit as T gets large than to work with (4.4.2) and the limits as both T gets large and δ gets sufficiently close to 1.

One difficulty in working with finitely repeated games is that they obviously do not have the same recursive structure as infinitely repeated games. The horizons of the continuation subgame become shorter after longer histories.

We begin with an example to illustrate that even in some short games, it can be easy to construct player-specific punishments.

Example 4.4.1 The stage game is given in figure 4.4.1. The two action profiles *BB* and *CC* are player specific punishments, in the sense that *BB* minimizes player 2's payoffs over the Nash equilibria of the stage game, and *CC* does for player 1. If the game is played twice, then it is possible to support the profile *CA* in the first period in a subgame-perfect equilibrium: If *CA* is played in the first period, play *AA* in the second period; if *AA* is played in the first period (i.e., 1 deviated to *A*), play *CC* in the second; if *CC* is played in the first period (i.e., 2 deviated to *C*), play *BB* in the second; and after all other first-period action profiles, play *AA* in the second period. The second-period Nash equilibrium specified by this profile depends on the identity of the first period deviator.

If the game is played T times, then the profile can be extended, maintaining subgame perfection, so that *CA* is played in the first $T - 1$ periods, with any unilateral deviation by player i being immediately followed by *CC* if $i = 1$ and *BB* if $i = 2$ till the end of the game. It is important for this profile that *CA* yields higher payoffs to both players than their respective punishing Nash equilibria of the stage game. If the payoff profile from *CA* were $(0, 0)$ rather than $(2, 2)$, then player 1, for example, prefers to trigger the punishment (which gives him 1 in each period) than to play *CA* (which gives him 0).[3]

Finally, note that the "rewarding" Nash profile of *AA* is not needed. Suppose the payoff profile from *AA* were $(3, -2)$ rather than $(4, 4)$. Player 1 still has a

---

3. When the payoff to *CA* is $(0, 0)$, player 1 prefers the outcome *CA, CA, AA, AA* to the outcome *AA, CC, CC, CC*. For $T = 3, 4$, it is easily seen that there is then a subgame-perfect equilibrium with *AA* played in the last two periods and *CA* played in the first one or two periods.

myopic incentive to play A rather than C, when 2 is playing A. Because AA is not a Nash equilibrium of the stage game, it cannot be used in the last periods to reward players for playing CA in earlier periods. However, observe that each player prefers an outcome that alternates between BB and CC to perpetual play of their punishment profile. Hence, for $T = 3$, the following profile is a subgame-perfect equilibrium: Play CA in the first period; if CA is played in the first period, play BB in the second period and CC in the third period; if AA is played in the first period, play CC in the second and third periods; if CC is played in the first period, play BB in the second and third periods; and after all other first-period action profiles, play BB in the second period and CC in the third period.

●

Let $\mathcal{N}$ be set of Nash equilibria of the stage game. For each player i, define

$$w_i \equiv \min_{\alpha \in \mathcal{N}} \{u_i(\alpha)\},$$

player i's worst Nash equilibrium payoff in the stage game. We first state and prove a Nash-reversion folk theorem, due to Friedman (1985).

Proposition 4.4.1 *Suppose there exists a payoff $v \in \operatorname{co}(\mathcal{N})$ with $w_i < v_i$ for all i. For all $a^* \in A$ satisfying $u_i(a^*) > w_i$ for all i, for any $\varepsilon > 0$, there exists T_ε such that for every $T > T_\varepsilon$, the T period repeated game has a subgame-perfect equilibrium with payoffs within ε of $u(a^*)$.*

Proof Because the inequality $w_i < v_i$ is strict, there exists a rational convex combination of Nash equilibria, $\tilde{v} \in \operatorname{co}(\mathcal{N})$, with $\tilde{v}_i > w_i$. That is, there are m equilibria $\{\alpha^k \in \mathcal{N} : k = 1, \ldots, m\}$ and m positive integers, $\{\lambda^k\}$, such that, for all i,

$$w_i < \tilde{v}_i = \frac{1}{L} \sum_{k=1}^{m} \lambda^k u_i(\alpha^k),$$

where $L = \sum \lambda^k$. Set $\eta \equiv \min_i \tilde{v}_i - w_i > 0$, and $\Delta \equiv \max_{i,a_i} |u_i(a_i, a^*_{-i}) - u_i(a^*)| \geq 0$. Let t^* be the smallest multiple of L larger than Δ/η, so that $t^* = \ell L$ for some positive integer ℓ, and consider games of length T at least t^*.

The strategy profile specifies a^* in the first $T - t^*$ periods, provided that there has been no deviation from such behavior. In periods $T - t^*, \ldots, T$, the equilibrium α^k is played $\ell \lambda^k$ times (the order is irrelevant because payoffs are averaged). After the first unilateral deviation by player i, the stage-game equilibrium yielding payoffs w_i is played in every subsequent period.

It is straightforward to verify that this is a subgame-perfect equilibrium. Finally, by choosing T sufficiently large, the payoffs from the first $T - t^*$ periods will dominate. There is then a length T_ε such that for any $T > T_\varepsilon$, the payoffs from the constructed equilibrium strategy profile are within ε of $u(a^*)$.

■

This construction uses the ability to switch between stage-game Nash equilibria near the end of the game to create incentives supporting a target action profile a^* in earlier periods that is not a Nash equilibrium of the stage game. As we discussed in

4.4 ■ Finitely Repeated Games

	A	B	C	D
A	4, 4	0, 0	18, 0	1, 1
B	0, 0	6, 6	0, 0	1, 1
C	0, 18	0, 0	13, 13	1, 1
D	1, 1	1, 1	1, 1	0, 0

Figure 4.4.2 A game with repeated-game payoffs below 4.

example 4.4.1, it is essential for this construction that for each player i, there exists a Nash equilibrium of the stage game giving i a lower payoff than does the target action a^*. If this were not the case, players would be eager to trigger the punishments supposedly providing the incentives to play a^*.

With somewhat more complicated strategies, we can construct equilibria of a repeated game with payoffs lower than the payoffs in *any* stage-game Nash equilibrium.

Example 4.4.2 In the stage game of figure 4.4.2, there are two stage-game Nash equilibria, AA and BB. The lowest payoff in any stage-game pure-strategy Nash equilibrium is 4. Suppose this game is played twice. There is an equilibrium with an average payoff of 3 for each player: Play DD in the first period, followed by BB in the second, and after any deviation play AA in the second period.

We can now use this two-period equilibrium to support outcomes in the three-period game. Observe first that in the three-period game, CC cannot be supported in the first period using the threat of two periods of AA rather than two periods of BB, because the myopic incentive to deviate is 5, and the total size of the punishment is 4 (to get the repeated-game payoffs, divide everything by 3). On the other hand, we can support CC in the three-period game as follows: The equilibrium outcome is (CC, BB, BB); after any first period profile $a^0 \neq CC$, play the two-period equilibrium outcome (DD, BB).

●

If $\hat{\mathbf{a}}$ is a subgame-perfect outcome path of the $\hat{T}$ period game, and $\tilde{\mathbf{a}}$ is a subgame-perfect outcome path of the $\tilde{T}$ period game, then by "restarting" the game after the first $\hat{T}$ periods, it is clear that $\hat{\mathbf{a}}\tilde{\mathbf{a}}$ is a subgame-perfect outcome path of the $\hat{T} + \tilde{T}$ period game. Hence, i's worst punishment (equilibrium) payoff in the $\hat{T} + \tilde{T}$ period game is at least as low as the average of his punishment payoffs in the $\hat{T}$ period game and the $\tilde{T}$ period game. Typically, it will be strictly lower. In example 4.4.2, the worst two-period punishment (with outcome path (DD, BB)) is strictly worse than the repetition of the worst one-period punishment (of AA). This subadditivity allows us to construct equilibria with increasingly severe punishments as the length of the game increases, in turn allowing us to achieve larger sets of equilibrium payoffs. Benoit and Krishna (1985) prove the following proposition.

Proposition 4.4.2 *Suppose for every player i, there is a stage-game Nash equilibrium α^i such that $u_i(\alpha^i) > w_i$, and that $\mathscr{F}^*$ has full dimension. Then for any payoff vector $v \in \mathscr{F}^{\dagger p}$ and any $\varepsilon > 0$, there exists T^* such that, for all $T \geq T^*$, there exists a perfect equilibrium path $(a^0, a^1, \ldots, a^{T-1})$ with*

$$\left| \frac{1}{T} \sum_{t=0}^{T-1} u(a^t) - v \right| < \varepsilon.$$

We omit the involved proof, confining ourselves to some comments and an example. The full-dimensionality assumption appears for the same reason as it does in the case of infinitely repeated games. It allows us the freedom to construct punishments that treat players differently. The key insight is Benoit and Krishna's (1985) lemma 3.5. This lemma essentially demonstrates that given punishments in a T period game, for sufficiently larger T', it is possible to construct for each player a punishment with a lower average payoff. The punishment shares some similarities to those used in the proof of proposition 3.4.1 and uses three phases. In the first phase of player i's punishment, player i is pure-action minmaxed.[4] In the second phase, players $j \neq i$ are rewarded (relative to the T period punishments that could be imposed) for minmaxing i during the first phase, whereas i is held to his T period punishment level. Finally, in the last phase, stage-game Nash equilibria are played, so that the average payoff in this phase exceeds w_i for all i. This last phase is chosen sufficiently long that no player has an incentive to deviate in the second phase. The lengths of the first two phases are simultaneously determined to maintain the remaining incentives, and to ensure that i's average payoff over the three phases is less than his T-period punishment.

Example 4.4.3 As in the infinitely repeated game (see section 3.3), the ability to mutually minmax in the two-player case can allow a more transparent approach. We illustrate taking advantage of the particular structure of the game in figure 4.4.2. The pure (and mixed) minmax payoff for this game is 1 for each player, with profile DD minmaxing each player.

We define a player 1 punishment lasting T periods, for any $T \geq 0$. This punishment gives player 1 a payoff very close to his minmax level (for large T) while using mutual minmaxing behavior to sustain the optimality of the behavior producing this punishment payoff. First, we define the outcome for the punishment. Fix Q and S, satisfying conditions to be determined. First, for any $T > Q + S$, define the outcome path

$$\mathbf{a}(T) \equiv (\underbrace{DD, DD, \ldots, DD}_{Q \text{ times}}, \underbrace{AD, AD, \ldots, AD}_{T-Q-S \text{ times}}, \underbrace{BB, BB, \ldots, BB}_{S \text{ times}}).$$

For T satisfying $S < T \leq Q + S$, define

$$\mathbf{a}(T) \equiv (\underbrace{DD, DD, \ldots, DD}_{T-S \text{ times}}, \underbrace{BB, BB, \ldots, BB}_{S \text{ times}}),$$

---

4. It seems difficult (if not impossible) to replicate the arguments from section 3.8, extending the analysis to mixed minmax payoffs, for the finite horizon case. However, there is a mixed minmax folk theorem (proved using different techniques), see Gossner (1995).

4.4 ■ Finitely Repeated Games

and for $T \leq S$,
$$\mathbf{a}(T) \equiv \underbrace{(BB, BB, \ldots, BB)}_{T \text{ times}}.$$

Observe that for large T, most of the periods involve action profile AD and hence player 1's minmax payoff of 1. This will play a key role in allowing us to push the average player 1 payoff over the course of the punishment close to the minmax payoff of 1. We must specify a punishment for all values of T, not simply large values of T, because the shorter punishments will play a role in ensuring that we can obtain the longer punishment outcomes as outcomes of equilibrium strategies.

We now describe a strategy profile σ^T with outcome path $\mathbf{a}(T)$, and then provide conditions on Q and S that guarantee that σ^T is subgame perfect. We proceed recursively, as a function of T.

Suppose first that $T \leq S$. Because the punishment outcome $\mathbf{a}(T)$ specifies a stage-game Nash equilibrium in each of the last T periods for any $T \leq S$, no player has a myopic incentive to deviate. The strategy σ^T then ignores all deviations, specifying BB after every history.

For T satisfying $S < T \leq 2Q + S$, σ^T specifies that after the first unilateral deviation from $\mathbf{a}(T)$ in a period $t \leq T - S - 1$, AA is played in every subsequent period (with further deviations ignored). Deviations from the stage-game action profile in the final S periods are ignored.

For $T > 2Q + S$, any unilateral deviation from $\mathbf{a}(T)$ in any period t results in the continuation profile σ^{T-t-1}.

The critical feature of this profile is that for short horizons, a switch from BB to AA provides incentives. For longer horizons, however, the incentive to carry on with the punishment is provided by the fact that a deviation triggers Q new periods of mutual minmax. In an infinite horizon game, this would be the only punishment needed. In a finite horizon, we cannot continue with the threat of Q new periods of minmaxing forever, and so the use of this threat early in the game must be coupled with a transition to other threats in later periods.

In establishing conditions under which these strategies constitute an equilibrium, it is important that both players' payoffs under the outcome $\mathbf{a}(T)$ are identical, and that under the candidate profile, player 2's incentives to deviate are always at least as large as those of player 1. These properties appear because the stage game features the action profile AD, under which both players earn their minmax payoffs (though this is not a mutual-minmax action profile).

For $S < T \leq Q + S$, the most profitable deviation is in the first period, and this is not profitable if $1 + (T - 1)4 \leq 6S$, that is, if $4T - 3 \leq 6S$. The constraint is most severe if $T = Q + S$, and so the deviation is not profitable if $4Q - 3 \leq 2S$.

Suppose $Q + S < T \leq 2Q + S$. There are two classes of deviations we need to worry about, deviations in the first Q periods, and those after the first Q periods but before the last S periods. The most profitable deviation in the first class occurs in the first period, and is not profitable if (where $R = T - Q - S$), $1 + (Q - 1 + R + S)4 \leq R + 6S$. This constraint is most severe when $R = Q$, and so is satisfied if $7Q - 3 \leq 2S$. Because the payoff from AD is less than that of AA, the most profitable deviation in the second class is in period $Q + 1$, and is not profitable if (because player 2 has the greater incentive to deviate)

$4 + (R - 1 + S)4 \leq R + 6S$, which is most severe when $R = Q$. For $Q \geq 1$, this constraint is implied by $7Q - 3 \leq 2S$.

Suppose now $T > 2Q + S$. Observe first that deviating in a period $t \leq Q$ results in the deviator being minmaxed, and so at best reorders payoffs, but brings no increase. The only remaining deviations are from AD by player 2, and the most profitable is in period $Q + 1$. Such a deviation yields a total payoff of $4 + Q \times 0 + (R - 1 - Q) \times 1 + 6S$, whereas the total payoff from not deviating is $R \times 1 + 6S$. The deviation is not profitable if

$$4 + (R - 1 - Q) + 6S \leq R + 6S,$$

that is,

$$3 \leq Q.$$

Hence, if $Q \geq 3$ and $S \geq (7Q - 3)/2$, the strategy profile σ^T is a subgame-perfect equilibrium for any T. We can then fix Q and S and, by choosing T sufficiently large, make player 1's punishment payoff arbitrarily close to 1. The same can be done for player 2. These punishments can then be used to construct equilibria sustaining any strictly individually rational outcome path, in a finitely repeated game of sufficient length.

●

4.5 Approximate Equilibria

We have noted that the sole Nash equilibrium of the finitely repeated prisoners' dilemma features mutual shirking in every period. For example, there is no Nash-equilibrium outcome in which both players exert effort in every period, because it is a superior response to exert effort until the last period and then shirk. However, this opportunity to shirk may have a tiny impact on payoffs compared to the total stakes of the game. What if we weakened our equilibrium concept by asking only for approximate best responses?

Let h be a history and let σ^h be the strategy profile σ modified at (only) histories preceding h to ensure that σ^h generates history h. Notice that $\sigma|_h$ is the continuation strategy induced by σ and history h, which may or may not appear under strategy σ, while σ^h is a potentially different strategy, in the original game, under which history h certainly occurs. The strategies $\sigma|_h$ and $\sigma^h|_h$ are identical.

Definition 4.5.1 *A strategy profile $\hat{\sigma}$ is an ε-Nash equilibrium if, for each player i and strategy σ_i, we have*

$$U_i(\hat{\sigma}) \geq U_i(\sigma_i, \hat{\sigma}_{-i}) - \varepsilon.$$

A strategy profile $\hat{\sigma}$ is an ex ante perfect ε-equilibrium if, for each player i, history h and strategy σ_i, we have

$$U_i(\hat{\sigma}^h) \geq U_i(\sigma_i^h, \hat{\sigma}_{-i}^h) - \varepsilon.$$

If we set $\varepsilon = 0$ in the first definition, we have the definition of a Nash equilibrium of the repeated game. Setting ε equal to 0 in the second gives the definition of subgame perfection. Any ex ante perfect ε-equilibrium must be an ε-Nash equilibrium.

4.5 ■ Approximate Equilibria

Example 4.5.1 Consider the finitely repeated prisoners' dilemma of figure 1.2.1, with horizon T and payoffs evaluated according to the average-payoff criterion $\frac{1}{T}\sum_{t=0}^{T-1} u(a^t)$. For every $\varepsilon > 0$, there is a finite T_ε such that for every $T \geq T_\varepsilon$, there is an ex ante perfect ε-equilibrium in which the players exert effort in every period (Radner 1980).[5]

The verification of this statement is a simple calculation. Let the equilibrium strategies be given by grim trigger, prescribing effort after every history featuring no shirking and prescribing shirking otherwise. The best response to such a strategy is to exert effort until the final period and then shirk. The equilibrium strategy is then an ex ante perfect ε-equilibrium if $2 \geq \frac{1}{T}[2(T-1) + 3] - \varepsilon$, or $\frac{1}{T} \leq \varepsilon$. For any ε, the proposed strategies are thus an ε-equilibrium in any game of length at least $T_\varepsilon = 1/\varepsilon$.

●

Consistent mutual effort is an ex ante perfect ε-equilibrium because the final-period deviation to shirking generates a payoff increment that averaged over T_ε periods is less than ε. If $\varepsilon = 0$, of course, then the finitely repeated prisoners' dilemma admits only shirking as a Nash-equilibrium outcome. These statements are reconciled by noting the order of the quantifiers in example 4.5.1. If we fix the length of the game T, then, for sufficiently small values of ε, the only ε-Nash equilibrium calls for universal shirking. However, for any fixed value of ε, we can find a value of T sufficiently large as to support effort as an ε-equilibrium.

Remark 4.5.1 The payoff increment to final-period shirking may be small in the context of a T period average, but may also loom large in the final period. A strategy profile σ^* is a *contemporaneous perfect ε-equilibrium* if, for each player i, history h and strategy σ_i, we have $U_i(\sigma^*|h) \geq U_i(\sigma_i|h, \sigma^*_{-i}|h) - \varepsilon$. A contemporaneous perfect ε-equilibrium thus evaluates the strategy σ^{*h} not in the original game but in the continuation game induced by h.[6]

Radner (1980) shows that for every $\varepsilon > 0$, there is a finite T_ε such that there is a contemporaneous perfect ε-equilibrium of the T-period prisoners' dilemma with the average-payoff criterion $\frac{1}{T}\sum_{t=0}^{T-1} u(a^t)$, for any $T \geq T_\varepsilon$, in which players exert effort in all but the final T_ε periods. As T increases, we can thus sustain effort for an arbitrarily large proportion of the periods. The strategies prescribe effort in the initial period and effort in every subsequent period that is preceded by no shirking and that occurs no later than period $T - T_\varepsilon - 1$ (for some T_ε to be determined), with shirking otherwise. The best response to such a strategy exerts effort through period $T - T_\varepsilon - 2$ and then shirks. The profile is a contemporaneous perfect ε-equilibrium if $2/(T_\varepsilon + 1) \geq 3/(T_\varepsilon + 1) - \varepsilon$, or $T_\varepsilon \geq 1/\varepsilon - 1$. The key to this result is that the average payoff criterion can cause payoff differences in distant periods to appear small in the current period, and can also cause payoff

---

5. While Radner (1980) examines finitely repeated oligopoly games, the issues can be more parsimoniously presented in the context of the prisoners' dilemma. Radner (1981) explores analogous equilibria in the context of a repeated principal-agent game, where the construction is complicated by the inability of the principal to perfectly monitor the agent's actions.

6. Lehrer and Sorin (1998) and Watson (1994) consider concepts that require contemporaneous ε-optimality conditional only on those histories that are reached along the equilibrium path.

differences in the current period to appear small, as long as there are enough periods left in the continuation game. Mailath, Postlewaite, and Samuelson (2005) note that if payoffs are discounted instead of simply averaged, then for sufficiently small ε, the only contemporaneous perfect equilibrium outcome is to always shirk, no matter how long the game. In this respect, the average and discounted payoff criteria of (4.4.1) and (4.4.2) have different implications.

◆

Proposition 4.4.2 shows that the set of equilibrium payoffs, for very long finitely repeated games with multiple stage-game Nash equilibria, is close to its counterpart in infinitely repeated games. Considering ε-equilibria allows us to establish a continuity result that also applies to stage games with a unique Nash equilibrium.

Fix a finite stage game G and let G^T be its T-fold repetition, with G^∞ being the corresponding infinitely repeated game. Let payoffs be discounted, and hence given by (4.4.2). Let σ^∞ and σ^T denote strategy profiles for G^∞ and G^T. We would like to speak of a sequence of strategy profiles $\{\sigma^T\}_{T=0}^\infty$ as converging to σ^∞. To do so, first convert each strategy σ^T to a strategy $\hat{\sigma}^T$ in game G^∞ by concatenating σ^T with an arbitrary strategy for G^∞. We then say that $\{\sigma^T\}_{T=0}^\infty$ converges to σ^∞ if the sequence $\{\hat{\sigma}^T\}_{T=0}^\infty$ converges to σ^∞ in the product topology (i.e., for any history h^t, $\hat{\sigma}^T(h^t)$ converges to $\sigma^\infty(h^t)$ as $T \to \infty$). Fudenberg and Levine's (1983) theorem 3.3 implies:

Proposition *The strategy profile σ^∞ is a subgame-perfect equilibrium of G^∞ if and only if*
4.5.1 *there exist sequences $\varepsilon(n)$, $T(n)$, and $\sigma^{T(n)}$, $n = 1, 2, \ldots$, with $\sigma^{T(n)}$ an ex ante perfect ε-equilibrium of $G^{T(n)}$ and with $\lim_{n\to\infty} \varepsilon(n) = 0$, $\lim_{n\to\infty} T(n) = \infty$, and $\lim_{n\to\infty} \sigma^{T(n)} = \sigma^\infty$.*

Fudenberg and Levine (1983) establish this result for games that are continuous at infinity, a class of games that includes repeated games with discounted payoffs as a common example. Intuitively, a game is continuous at infinity if two strategy profiles yield nearly the same payoff profile whenever they generate identical behavior on a sufficiently long initial string of periods. Hence, behavior differences that occur sufficiently far in the future must have a sufficiently small impact on payoffs. At the cost of a more complicated topology on strategies (simplified by Harris 1985), Fudenberg and Levine (1983) extend the result beyond finite games (see also Fudenberg and Levine 1986 and Börgers 1989). Mailath, Postlewaite, and Samuelson's (2005) observation that the finitely repeated prisoners' dilemma with discounting has a unique contemporaneous perfect ε-equilibrium for sufficiently small ε and for any length, shows that a counterpart of proposition 4.5.1 does not hold for contemporaneous perfect ε-equilibrium.

4.6 Renegotiation

For sufficiently high discount factors, it is an equilibrium outcome for the players in a repeated prisoners' dilemma to exert effort in every period with the temptation to shirk deterred by a subsequent switch to perpetual, mutual shirking. This equilibrium

4.6 ■ Renegotiation

	L	R
T	9,9	0,8
B	8,0	7,7

Figure 4.6.1 Coordination game in which efficiency and risk dominance conflict.

is subgame perfect, so that carrying out the punishment is itself an equilibrium of the continuation game facing the players after someone has shirked.

Suppose the punishment phase has been triggered. Why doesn't one player approach the other and say, "Let's just forget about carrying on with this punishment and start over with a new equilibrium, in which we exert effort. It's an equilibrium to do so, just as it's an equilibrium to continue with the punishment, but starting over gives us a *better* equilibrium." If the other player is convinced, the punishment path is not implemented.

Renegotiation-proof equilibria limit attention to equilibria that survive this type of renegotiation credibility test. However, if the punishment path indeed fails this renegotiation test, then it is no longer obvious that we can support the original equilibrium in which effort was exerted, in which case the renegotiation challenge to the punishment may not be convincing, in which case the punishment is again available, and so on. Any notion of renegotiation proofness must resolve such self-references.

Renegotiation is a compelling consideration for credibility to the extent that the restriction to efficiency is persuasive. There is a large body of research centered on the premise that economic systems do *not* always yield efficient outcomes, to the extent that events such as bank failures (Diamond and Dybvig 1983), discrimination (Coate and Loury 1993), or economic depressions (Cooper and John 1988) are explained as events in which players have coordinated on an inefficient equilibrium.

At the same time, it is frequently argued that we should restrict attention to efficient equilibria of games. The study of equilibrium is commonly justified by a belief that there is some process resulting in equilibrium behavior. The argument then continues that this same process should be expected to produce not only an equilibrium but an efficient one. Suppose, for example, the process is thought to involve (nonbinding) communication among the players in which they agree on a plan of play before the game is actually played. If this communcation leads the players to agree on an equilibrium, why not an efficient one?

This case for efficiency is not always obvious. Consider the game shown in figure 4.6.1, taken from Aumann (1990). The efficient outcome is TL, for payoffs of $(9, 9)$. However, equilibrium BR is less risky, in the sense that B and R are best responses unless one is virtually certain that L and T will be played.[7] Even staunch believers in efficiency might entertain sufficient doubt about their opponents as to give rise to BR. Suppose now that we put some preliminary communication into the mix in an attempt

---
7. Formally, (B, R) is the *risk-dominant* equilibrium (Harsanyi and Selten 1988).

to banish these doubts. One can then view player 2 as urging 1 to play T, as part of equilibrium TL, with player 1 all the while thinking, "Player 2 is better off if I choose T no matter what she plays, so there is nothing to be learned from her advocacy of T. In addition, if she entertains doubts as to whether she has convinced me to play T, she will optimally play R, suggesting that I should protect myself by playing B." The same holds in reverse, suggesting that the players may still choose BR, despite their protests to the contrary.

The concept of a renegotiation-proof equilibrium pushes the belief in efficiency further. We are asked to think of the players being able to coordinate on an efficient equilibrium not only at the beginning of the game but also at any time during the game. Hence, should they ever find themselves facing an inefficient continuation equilibrium, whether on or off the equilibrium path, they can renegotiate to achieve an efficient equilibrium.

4.6.1 Finitely Repeated Games

The logic of renegotiation proofness seems straightforward: Agents should not settle for a continuation equilibrium if it is strictly dominated by some other equilibrium. However, we should presumably exclude an equilibrium only if it gives rise to a continuation equilibrium that is strictly dominated by some other *renegotiation-proof* equilibrium. This self-reference in the definition makes an obvious formulation of renegotiation proofness elusive.

Many of the attendant difficulties are eliminated in finitely repeated games, where backward induction allows us to avoid the ambiguity. We accordingly start with a discussion of finitely repeated games (based on Benoit and Krishna 1993).

We consider a normal-form two-player game G that is played T times, to yield the game G^T.[8] Normalized payoffs are given by (4.4.1). As in section 4.4, it simplifies the analysis to work with the limiting case of no discounting. We work throughout without public correlation.

Fixing a stage game G, let $\mathscr{E}^t$ be the set of *unnormalized* subgame-perfect equilibrium payoff profiles for game G^t.[9] Given any set W^t of unnormalized payoff profiles, let $\mathscr{G}(W^t)$ be the subset consisting of those profiles in W^t that are not strictly dominated by any other payoff profile in W^t. We refer to such payoffs throughout as efficient in W^t, or often simply as efficient, with the context providing the reference set. Finally, given a set of (unnormalized) payoff profiles W^t for game G^t, let $Q(W^t)$ be the set of unnormalized payoff profiles for game G^{t+1} that can be decomposed on W^t. Hence, the payoff profile v lies in $Q(W^t)$ if there exists an action profile α and a function $\gamma: A \to W^t$ such that, for all i and $a_i \in A_i$,

$$v_i = u_i(\alpha) + \sum_{a \in A} \gamma_i(a)\alpha(a)$$
$$\geq u(a_i, \alpha_{-i}) + \sum_{a_{-i} \in A_{-i}} \gamma(a_i, a_{-i})\alpha_{-i}(a_{-i}),$$

---

8. Wen (1996) extends the analysis to more than two players.
9. Hence, $\mathscr{E}^t$ consists of payoffs of the form $\sum_{\tau=0}^{t-1} u^\tau$.

4.6 ∎ Renegotiation

	A	B	C
A	0,0	4,1	0,0
B	1,4	0,0	5,3
C	0,0	3,5	0,0

Figure 4.6.2 Stage game with two inefficient pure-strategy Nash equilibria.

with a corresponding definition restricted to pure strategies for infinite games. We will be especially interested in this function in the case in which W^t is a subset of $\mathscr{E}^t$. The set $Q(W^t)$ will then identify subgame-perfect equilibrium payoffs in G^{t+1}.

Definition 4.6.1 Let

$$Q^1 = \mathscr{E}^1 \quad \text{and} \quad \mathscr{W}^1 = \mathscr{G}(Q^1),$$

and define, for $t = 2, \ldots, T$,

$$Q^t = Q(\mathscr{W}^{t-1}) \quad \text{and} \quad \mathscr{W}^t = \mathscr{G}(Q^t).$$

Then $R^T \equiv \frac{1}{T}\mathscr{W}^T$ is the set of renegotiation-proof subgame-perfect equilibrium payoffs in the repeated game of length T.

This construction begins at the end of the game by choosing the set of efficient stage-game Nash equilibria as candidates for play in the final period. These are the renegotiation-proof equilibria in the final period. In the penultimate period, the renegotiation-proof equilibria are those subgame-perfect equilibria that are efficient in the set of equilibria whose final-period continuation paths are renegotiation-proof. Continuing, in any period t, the renegotiation-proof equilibria are those subgame-perfect equilibria that are efficient in the set of equilibria whose next-period continuation paths are renegotiation-proof. Working our way back to the beginning of the T period game, we obtain the set R^T of normalized renegotiation-proof equilibrium payoffs.

We are interested in the limit (under the Hausdorff metric) of R^T as $T \to \infty$, if it exists.[10] Denote the limit by R^∞; this set is closed and nonempty by definition.

Example 4.6.1 Consider the stage game shown in figure 4.6.2. We restrict attention in this example to pure strategies.

Pure-strategy minmax payoffs are 1, imposed by playing A. Proposition 4.4.1 then tells us that the set of subgame-perfect equilibrium payoffs approaches

$$\mathscr{F}^{\dagger p} = \text{co}\{(0, 0), (1, 4), (3, 5), (5, 3), (4, 1)\} \cap \{v \in \mathbb{R}^2 : v_i > 1, i = 1, 2\}$$

---

10. The Hausdorff distance between two nonempty compact sets, $\mathscr{A}$ and $\mathscr{A}'$ is given by

$$d(\mathscr{A}, \mathscr{A}') = \max\left\{\max_{a \in \mathscr{A}} \min_{a' \in \mathscr{A}'} |a - a'|, \max_{a' \in \mathscr{A}'} \min_{a \in \mathscr{A}} |a - a'|\right\}.$$

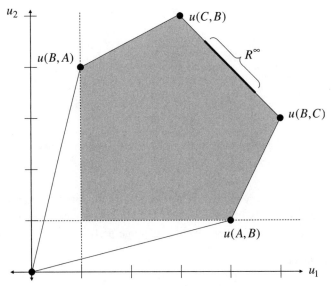

Figure 4.6.3 The set $\mathscr{F}^{\dagger p}$ (shaded) for the game in figure 4.6.2, along with the limiting set $\lim_{T \to \infty} \frac{1}{T} \mathscr{W}^T = R^\infty$ of renegotiation-proof equilibrium payoff profiles.

for sufficiently long finitely repeated games. Figure 4.6.3 shows the sets $\mathscr{F}^{\dagger p}$ and R^∞.

We now examine renegotiation-proof equilibria. We have

$$Q^1 = \{(1,4), (4,1)\} \quad \text{and} \quad \mathscr{W}^1 = \{(1,4), (4,1)\}.$$

The set Q^2 includes equilibria constructed by preceding either of the elements of $\mathscr{W}^1$ with first-period action profiles that yield either $(1, 4)$ or $(4, 1)$. In addition, we can play BC in the first period followed by BA in the second, with first-period deviations followed by AB, or can reverse the roles in this construction. We thus have

$$Q^2 = \{(2,8), (6,7), (5,5), (7,6), (8,2)\}$$
$$\text{and} \quad \mathscr{W}^2 = \{(2,8), (6,7), (7,6), (8,2)\}.$$

The sets Q^{t+1} quickly grow large, and it is helpful to note the following shortcuts in calculating $\mathscr{W}^{t+1}$. First, $\mathscr{W}^{t+1}$ will never include an element whose first-period payoff is $(0, 0)$ (even though Q^{t+1} may), because this is strictly dominated by an equilibrium whose first-period action profile gives payoff $(1, 4)$ or $(4, 1)$, with play then continuing as in the candidate equilibrium. We can thus restrict attention to elements of Q^{t+1} whose first-period payoffs are drawn from the set $\{(1, 4), (3, 5), (5, 3), (4, 1)\}$. Second, there is an equilibrium in Q^{t+1} whose first period payoff is $(5, 3)$ and whose continuation payoff is any element of $\mathscr{W}^t$ other than the one that minimizes player 2's payoff, because the latter leaves no opportunity to create the incentives for player 2 to choose C in period 1. Similarly, there is an equilibrium in Q^{t+1} whose first period payoff is $(3, 5)$ and whose

4.6 ■ Renegotiation

continuation payoff is any element of $\mathscr{W}^t$ other than the one that minimizes player 1's payoff. Third, this in turn indicates that any element of Q^{t+1} that begins with payoff (4, 1) and continues with a payoff in $\mathscr{W}^t$ that does *not* maximize player 1's payoff will not appear in W^{t+1}, being strictly dominated by an element of Q^{t+1} with initial payoff (5, 3) and identical continuation. A similar observation holds for payoffs beginning with (1, 4). Hence, we can simplify the presentation by restricting attention to elements of Q^{t+1} that begin with payoff (4, 1) only if they continue with the element $\mathscr{W}^t$ that maximizes player 1's payoff and similarly to equilibria beginning with (1, 4) only if the continuations maximize player 2's payoff. Letting $\hat{Q}^{t+1}$ be the subset these three rules identify, we have

$$\hat{Q}^3 = \{(3, 12), (9, 12), (10, 11), (11, 7), (7, 11), (11, 10), (12, 9), (12, 3)\},$$

and so

$$\mathscr{W}^3 = \{(3, 12), (9, 12), (10, 11), (11, 10), (12, 9), (12, 3)\}.$$

The next iteration gives

$$\mathscr{W}^4 = \{(12, 17), (13, 16), (14, 15), (15, 14), (16, 13), (17, 12)\}.$$

None of these equilibrium payoffs have been achieved by preceding an element of $\mathscr{W}^3$ with either payoff (1, 4) or (4, 1). Instead, the equilibrium payoffs constructed by beginning with these payoffs are inefficient, and are excluded from $\mathscr{W}^4$. We next calculate

$$\mathscr{W}^5 = \{(13, 21), (16, 21), (17, 20), (18, 19), (19, 18),$$
$$(20, 17), (21, 16), (21, 13)\}.$$

In this case, the equilibrium payoffs (13, 21) and (21, 13) are obtained by preceding continuation payoffs (12, 17) and (17, 12) with (1, 4) and (4, 1). At our next iteration, we have

$$\mathscr{W}^6 = \{(19, 26), (20, 25), (21, 24), (22, 23), (23, 22),$$
$$(24, 21), (25, 20), (26, 19)\}.$$

Here again, all of the equilibria in $\hat{Q}^6$ constructed by beginning with payoffs (1, 4) or (4, 1) are strictly dominated. Continuing in this fashion, we find that for even values of $t \geq 4$,

$$\mathscr{W}^t = \{(\underline{v}^t, \bar{v}^t), (\underline{v}^t + 1, \bar{v}^t - 1), (\underline{v}^t + 2, \bar{v}^t - 2), \ldots, (\bar{v}^t - 1, \underline{v}^t + 1), (\bar{v}^t, \underline{v}^t)\},$$

where

$$\underline{v}^t = 12 + 7\tfrac{t-4}{2} \quad \text{and} \quad \bar{v}^t = 17 + 9\tfrac{t-4}{2},$$

whereas for odd values of $t \geq 4$ we have

$$\mathscr{W}^t = \{(\underline{v}^{t-1} + 1, \bar{v}^t), (\underline{v}^t, \bar{v}^t), (\underline{v}^t + 1, \bar{v}^t - 1), (\underline{v}^t + 2, \bar{v}^t - 2),$$
$$\ldots, (\bar{v}^t - 1, \underline{v}^t + 1), (\bar{v}^t, \underline{v}^t), (\bar{v}^t, \underline{v}^{t-1} + 1)\},$$

where

$$\underline{v}^t = \underline{v}^{t-1} + 4 \quad \text{and} \quad \bar{v}^t = \bar{v}^{t-1} + 4.$$

In particular, it is straightforward to verify that for each $t \geq 4$, $\mathscr{W}^{t+1}$ is obtained from $\mathscr{W}^t$ by constructing equilibria that precede the continuation payoff profile from $\mathscr{W}^t$ that minimizes player 1's payoff with payoff $(1, 4)$, that precede every other continuation profile with $(3, 5)$, that precede the profile that minimizes player 2's payoff with $(4, 1)$, and that precede every other with $(5, 3)$, and then eliminating inefficient profiles.

It is then apparent that we have

$$\lim_{T \to \infty} R^T = \text{co}\{(\tfrac{7}{2}, \tfrac{9}{2}), (\tfrac{9}{2}, \tfrac{7}{2})\}.$$

Hence, the set of renegotiation-proof equilibrium payoffs converges to a subset of the set of efficient payoffs.

●

Example 4.6.2 Consider the game shown in figure 4.6.4. We again consider pure-strategy equilibria. The first step in the construction of renegotiation-proof equilibria is straightforward:

$$Q^1 = \{(2, 4), (3, 3), (4, 2)\}$$
$$\text{and} \quad \mathscr{W}^1 = \{(2, 4), (3, 3), (4, 2)\}.$$

We now note that no element of Q^2 can begin with a first-period payoff of $(0, 0)$. Every such outcome either presents at least one player with a deviation that increases his first-period payoff by at least 3 or presents both players with deviations that increase their first-period payoffs by at least 2. The continuation payoffs presented by $\mathscr{W}^1$ cannot be arranged to deter all such deviations (nor, as will be apparent from (4.6.1)–(4.6.2), will the continuation payoffs contained in any $\mathscr{W}^t$ be able to do so). However, Q^2 does contain an equilibrium in which the payoffs $(7, 7)$ are attained in the first period, followed by $(3, 3)$ in the second,

	A	B	C	D
A	0,0	2,4	0,0	8,0
B	4,2	0,0	0,0	0,0
C	0,0	0,0	3,3	0,0
D	0,8	0,0	0,0	7,7

Figure 4.6.4 Game whose renegotiation-proof payoffs exhibit an oscillating pattern, converging to a single inefficient payoff as the horizon gets large.

with any first-period deviation deterred by a switch to a second-period payoff of (2, 4) or (4, 2), allocating the lower payoff to the first-period miscreant. This just suffices to deter such deviations. We then have

$$Q^2 = \{(4, 8), (5, 7), (6, 6), (7, 5), (8, 4), (10, 10)\}$$
$$\text{and} \quad \mathscr{W}^2 = \{(10, 10)\}.$$

The finding that $\mathscr{W}^2$ is a singleton ensures that in the three-period game, first-period play must constitute a stage-game Nash equilibrium, because there is no opportunity to arrange continuation payoffs to punish deviations. Hence,

$$Q^3 = \{(12, 14), (13, 13), (14, 12)\}$$
$$\text{and} \quad \mathscr{W}^3 = \{(12, 14), (13, 13), (14, 12)\}.$$

This again opens the possibility for a first-period payoff of (7, 7) in the four-period game, giving

$$Q^4 = \{(14, 18), (15, 17), (16, 16), (17, 15), (18, 14), (20, 20)\}$$
$$\text{and} \quad \mathscr{W}^4 = \{(20, 20)\}.$$

It should now be apparent that we have an oscillating structure, with

$$\mathscr{W}^{2t} = \{(10t, 10t)\} \tag{4.6.1}$$

and

$$\mathscr{W}^{2t+1} = \{(10t + 2, 10t + 4), (10t + 3, 10t + 3), (10t + 4, 10t + 2)\}. \tag{4.6.2}$$

It is then immediate that

$$\lim_{T \to \infty} R^T = \{(5, 5)\}.$$

●

This last example shows that the set of renegotiation-proof payoffs can be quite sensitive to precisely how many periods remain until the end of the game. Hence, although the backward-induction argument renders the concept of renegotiation proofness unambiguous, it raises just the sort of end-game effects that often motivate a preference for infinite horizon games.

When will renegotiation-proof equilibria be efficient? Let b_i^1 be the largest stage-game equilibrium payoff for player i.

Proposition 4.6.1 *If $(b_1^1, b_2^1) \notin \mathscr{F}^\dagger$ and R^∞ exists, then each element of R^∞ is efficient.*

Remark 4.6.1 Proposition 4.6.1 provides sufficient but not necessary conditions for the set of renegotiation-proof equilibrium payoffs R^T to be nearly efficient for large T. We have not offered conditions ensuring that R^∞ exists, which remains in general an open question. The condition $(b_1^1, b_2^1) \notin \mathscr{F}^\dagger$ is not necessary: The coordination game shown in figure 4.6.5 satisfies $(b_1^1, b_2^1) = (2, 2) \in \mathscr{F}^\dagger$, thus failing the

	L	R
T	2,2	0,0
B	0,0	1,1

Figure 4.6.5 A coordination game.

conditions of proposition 4.6.1, but R^∞ nonetheless uniquely contains the efficient payoff $(2, 2)$.

◆

The proof requires several lemmas, maintaining the hypotheses of proposition 4.6.1 throughout. Let b_i^T be the best payoff for player i in the set R^T.

Lemma 4.6.1 *For player $i \in \{1, 2\}$ and length $T \geq 1$,*

$$b_i^T \geq b_i^1.$$

Hence, as the game gets longer, each player's best renegotiation-proof outcome can never dip below the payoff produced by repeating his best equilibrium payoff in the one-shot game.

Proof Let w_i^T be the worst payoff for player i in the set R^T. We proceed by induction. When $T = 1$, we have the tautological $b_i^1 \geq b_i^1$. Hence, suppose that $b_i^{T-1} \geq b_i^1$, and consider a game of length T. For convenience, consider player 1. We first note that

$$(b_1^1, w_2^1) + (T-1)(b_1^{T-1}, w_2^{T-1}) \in Q^T.$$

This statement follows from two observations. First, because no two elements of R^t can be strictly ranked (i.e., neither strictly dominates the other), there is an equilibrium payoff in R^t giving player 1 his best payoff and player 2 her worst payoff, or (b_1^t, w_2^t), for any t. Second, we can construct an equilibrium of the T period game by playing the stage-game equilibrium featuring payoffs (b_1^1, w_2^1) in the first period, followed by the continuation equilibrium of the $T - 1$ period remaining game that gives payoffs (b_1^{T-1}, w_2^{T-1}). Then, because R^T consists of the normalized undominated elements of Q^T, we have

$$\begin{aligned} Tb_1^T &\geq b_1^1 + (T-1)b_1^{T-1} \\ &\geq b_1^1 + (T-1)b_1^1 \\ &= Tb_1^1, \end{aligned}$$

where the second inequality follows from the induction hypothesis.

∎

Let b_i and w_i be the best and worst payoff for player i in R^∞. The fact that R^∞ is closed ensures existence. Then the primary intuition behind proposition 4.6.1 is contained in the following.

Lemma 4.6.2 If $b_1 > w_1$ and $b_2 > w_2$, then each payoff $v \in R^\infty$ with $v_i > w_i$, $i = 1, 2$, is efficient.

The idea behind this proof is that if there is a feasible payoff profile v'' that strictly dominates one of the payoffs v' in R^∞, then we can append this payoff to the beginning of an "equilibrium" in R^∞ with payoff v', using the worst equilibria in R^∞ for the two players to deter deviations. But this yields a new equilibrium that strictly dominates v', a contradiction. The conclusion is then that there must be no such v' and v'', that is, that the payoffs in R^∞ must be efficient. To turn this intuition into a proof, we must recognize that R^∞ is the *limit* of sets of equilibria in finitely repeated games, and make the translation from "an equilibrium in R^∞" to equilibria along this sequence.

Proof Observe first that the payoff profiles $(w_1, b_2) \equiv v^2$ and $(b_1, w_2) \equiv v^1$ are both contained in R^∞.

Suppose there is a payoff profile $v' \in R^\infty$, with $v'_i > w_i$, and with another payoff profile $v'' \in \mathscr{F}^\dagger$ with v'' strictly dominating v'. We now seek a contradiction.

Because $v'' \in \mathscr{F}^\dagger$ strictly dominates v', v' must also be strictly dominated by some payoff profile in $\mathscr{F}^\dagger$ close to v'' (for which we retain the notation v'') for which there exists a finite K and (not necessarily distinct) pure-action profiles $a^1, \ldots, a^K$ such that $\frac{1}{K}\sum_{k=1}^K u(a^k) = v''$. Fix $\varepsilon > 0$ so that

$$v_i^j + \varepsilon < v'_i, \qquad i, j = 1, 2, j \neq i, \tag{4.6.3}$$

and, for any such ε, fix a length T_ε for the finitely repeated game and an equilibrium $\sigma^{v'}(T_\varepsilon)$ such that for $i = 1, 2$,

$$\left|U_i(\sigma^{v'}(T_\varepsilon)) - v'_i\right| < \frac{\varepsilon}{3}, \quad \text{and} \tag{4.6.4}$$

$$\frac{\varepsilon}{3}T_\varepsilon > K(M - m) \tag{4.6.5}$$

(where, as usual, $U_i(\sigma)$ is the *average* payoff in G^t of the G^t-profile σ and M (m, respectively) is the maximum (minimum, respectively) stage game payoff), and there is a pair of sequences of equilibria $\{\sigma^j(T)\}_{T \geq T_\varepsilon}$, $j = 1, 2$, such that, for any $T \geq T_\varepsilon$, $i \neq j$,

$$U_i(\sigma^j(T)) < v_i^j + \frac{\varepsilon}{3}. \tag{4.6.6}$$

We now recursively define a sequence of equilibria. The idea is to build a sequence of equilibria for ever longer games, appending in turn each of the actions $a^1, \ldots, a^K$ to the beginning of the equilibrium, starting the cycle anew every K periods. Deviations by player i are punished by switching to the equilibrium $\sigma^j(T)$.

We begin with the equilibrium for $G^{T_\varepsilon+1}$, which plays a^1 in period 1, followed by $\sigma^{v'}(T_\varepsilon)$, with a first-period deviation by player i followed by play of $\sigma^j(T_\varepsilon)$. Checking that this is a subgame-perfect equilibrium requires showing that first-period deviations are not optimal, for which it suffices that

$$m + T_\varepsilon\left(v'_i - \frac{\varepsilon}{3}\right) \geq M + T_\varepsilon\left(v_i^j + \frac{\varepsilon}{3}\right),$$

or
$$T_\varepsilon(v'_i - v^j_i - \tfrac{2}{3}\varepsilon) \geq M - m,$$

which is implied by (4.6.3) and (4.6.5). We thus have an equilibrium payoff in $Q^{T_\varepsilon+1}$, which is either itself contained in $R^{T_\varepsilon+1}$ or is strictly dominated by some equilibrium payoff profile in $R^{T_\varepsilon+1}$. Denote by $\sigma(T_\varepsilon + 1)$ the equilibrium with payoff profile in $R^{T_\varepsilon+1}$. Note that $(T_\varepsilon + 1)U_i(\sigma(T_\varepsilon + 1)) \geq u_i(a^1) + T_\varepsilon U_i(\sigma^{v'}(T_\varepsilon))$.

Given an equilibrium $\sigma(T_\varepsilon + k)$ with payoffs in $R^{T_\varepsilon+k}$ for $k = 1, \ldots, K-1$, consider the profile that plays a^{k+1} in the first period, with deviations by player i punished by continuing with $\sigma^j(T_\varepsilon + k)$, and with $\sigma(T_\varepsilon + k)$ played in the absence of a deviation. To verify that the profile is a subgame-perfect equilibrium, it is enough to show that there are no first-period profitable deviations. In doing so, however, we are faced with the difficulty that we have little information about the payoff $u_i(a^{k+1})$ or the potential benefit from a deviation. We do know that each complete cycle $a^1, \ldots, a^K$ has average payoff v'' (which will prove useful for periods $T > T_\varepsilon + K$). Accordingly, we proceed by first noting that a lower bound on i's equilibrium payoff is

$$(k+1)m + T_\varepsilon\left(v'_i - \frac{\varepsilon}{3}\right), \qquad (4.6.7)$$

obtained by placing no control on payoffs during the first $k + 1$ periods and then using (4.6.4). Conditions (4.6.3) and (4.6.5) give the first inequality in the following, the second follows from the definition of M, and the third from (4.6.6),

$$(k+1)m + T_\varepsilon\left(v'_i - \frac{\varepsilon}{3}\right) \geq (k+1)M + T_\varepsilon\left(v^j_i + \frac{\varepsilon}{3}\right)$$
$$\geq M + (T_\varepsilon + k)\left(v^j_i + \frac{\varepsilon}{3}\right)$$
$$> M + (T_\varepsilon + k)U_i(\sigma^j(T_\varepsilon + k)).$$

Because the first term is a lower bound on i's equilibrium payoff, first-period deviations are suboptimal. We thus have an equilibrium payoff profile in $Q^{T(\varepsilon)+k+1}$, which is again either contained in or strictly dominated by some profile in $R^{T_\varepsilon+k+1}$. Denote by $\sigma(T_\varepsilon + k + 1)$ the equilibrium with payoff profile in $R^{T_\varepsilon+k+1}$. Note that $(T_\varepsilon + k + 1)U_i(\sigma(T_\varepsilon + k + 1)) \geq u_i(a^{k+1}) + (T_\varepsilon + k)U_i(\sigma^{v'}(T_\varepsilon + k))$.

This argument yields an equilibrium for $T = T_\varepsilon + K$, $\sigma(T_\varepsilon + K)$ with payoffs in $R^{T_\varepsilon+K}$, and satisfying

$$(T_\varepsilon + K)U_i(\sigma(T_\varepsilon + K)) \geq T_\varepsilon U_i(\sigma^{v'}(T_\varepsilon)) + \sum_k u(a^k)$$
$$\geq T_\varepsilon\left(v'_i - \tfrac{\varepsilon}{3}\right) + K v''_i.$$

Suppose $T = T_\varepsilon + \ell K + k$ for some $\ell \in \{1, 2, \ldots\}$ and $k \in \{1, \ldots, K-1\}$, and $\sigma(T-1)$ is an equilibrium with payoffs in R^{T-1}, and satisfying

$$(T-1)U_i(\sigma(T-1)) \geq T_\varepsilon\left(v'_i - \tfrac{\varepsilon}{3}\right) + \ell K v''_i + \sum_{h=1}^{k-1} u_i(a^h). \qquad (4.6.8)$$

4.6 ■ Renegotiation

Consider the profile that plays a^k in the first period, with deviations by player i punished by continuing with $\sigma^j(T-1)$, and with $\sigma(T-1)$ played in the absence of a deviation. As before, deviations in the first period are unprofitable:

$$TU_i(\sigma(T)) \geq km + \ell K v_i'' + T_\varepsilon\left(v_i' - \tfrac{\varepsilon}{3}\right)$$
$$\geq M + (T-1)\left(v_i^j + \tfrac{\varepsilon}{3}\right)$$
$$> M + (T-1)U_i(\sigma^j(T-1)).$$

We thus have an equilibrium payoff profile in Q^T, which is again either contained in or strictly dominated by some profile in R^T. Denote by $\sigma(T)$ the equilibrium with payoff profile in R^T. Note also that $\sigma(T)$ satisfies (4.6.8).

We thus have two sequences of equilibria, $\sigma^{v'}(T)$ and $\sigma(T)$, each yielding payoffs contained in R^T. The average payoff of each element of the former sequence lies within $\varepsilon/3$ of v', whereas the average payoff of the latter eventually strictly dominates $v'' - \varepsilon/3$. For sufficiently small ε, the latter eventually strictly dominates the former, a contradiction.

■

Proving proposition 4.6.1 requires extending lemma 4.6.2 to the boundary cases in which $v_i = w_i$. We begin with an intermediate result. Let M_i be the largest stage-game payoff for player i.

Lemma 4.6.3 Suppose v^1 and v^2 are two profiles in R^∞ with $v_i^2 < v_i^1$ and with no $v'' \in R^\infty$ satisfying $v_i^2 < v_i'' < v_i^1$. Then $\max\{v_j : (v_i^\ell, v_j) \in R^\infty\} = M_j$ for $\ell = 1, 2$.

Proof Without loss of generality, we take $i = 1$. Under the hypothesis of the lemma, the set R^∞ has a gap between two equilibria. Because $v_1^1 > v_1^2$, $\max\{v_2 : (v_1^1, v_2) \in R^\infty\} \leq \max\{v_2 : (v_1^2, v_2) \in R^\infty\}$. We let a be an action profile with $u_2(a) = M_2$ and derive a contradiction from the assumption that $\max\{v_2 : (v_1^1, v_2) \in R^\infty\} < M_2$. Without loss of generality, we assume $v_2^1 = \max\{v_2 : (v_1^1, v_2) \in R^\infty\}$.

Let $\varepsilon = \tfrac{1}{4}(v_1^1 - v_1^2)$ and choose T_ε so that for all $T \geq T_\varepsilon$,

$$M + T(v_1^2 + \varepsilon) < m + T(v_1^1 - 2\varepsilon). \tag{4.6.9}$$

We consider two cases. First, suppose $u_1(a) < v_1^1 - \varepsilon$, and let $\lambda \in (0, 1)$ and the payoff profile $\tilde{v}$ solve

$$\tilde{v}_1 = v_1^1 - \varepsilon,$$
and $$\tilde{v} = \lambda u(a) + (1 - \lambda)v^1.$$

Let $\eta < \min\{\varepsilon, (\tilde{v}_2 - v_2^1)/2\}$. Because $v_2^1 = \max\{v_2 : (v_1, v_2) \in R^\infty, v_1 \geq v_1^1\}$, $R^\infty \cap V = \emptyset$, where $V \equiv \{v : v_1 \geq v_1^1 - 2\varepsilon, v_2 \geq v_2^1 + \eta\}$. Note that $\tilde{v} \in V$. The proof proceeds by showing that, for large T, $R^T \cap V \neq \emptyset$, which suffices for the contradiction in the first case.

For every v with $\|v^1 - v\| < \eta/2$,[11] $\|\lambda u(a) + (1 - \lambda)v - \tilde{v}\| < \eta/2$. Choose a game length $T_\eta > T_\varepsilon$ such that for all $T > T_\eta$, the Hausdorff distance between

---

11. We use $\|\cdot\|$ to denote the max or sup norm, reserving $|\cdot|$ for Euclidean distance.

R^T and R^∞ is less than $\eta/2$. Hence, for each $T > T_\eta$, there are equilibria $\sigma^1(T)$ and $\sigma^2(T)$ with payoffs in R^T, such that $\|U(\sigma^i(T)) - v^i\| < \eta/2$. Moreover, there are sequences of game lengths $T(n)$ and integers $K(n) < T(n) - T_\eta$, such that

$$\left\| \frac{1}{T(n)} \left\{ K(n)u(a) + (T(n) - K(n))v^1 \right\} - \tilde{v} \right\| < \eta.$$

For each $T(n)$ in the sequence, we construct an equilibrium with payoffs in $R^{T(n)} \cap V$ (which is the desired contradiction), using a payoff obtained from playing $u(a)$ for $K(n)$ periods and (approximately) v^1 for the remaining $T - K(n)$ periods. Notice that $K(n)/T(n)$ will be approximately λ, so there is no difficulty in requiring $T(n) - K(n) > T_\eta$.

For each $T(n)$, we recursively construct a sequence of equilibria, beginning with the game of length $T(n) - K(n) + 1$. In the first period, profile a is played, followed by the continuation equilibrium $\sigma^1(T(n) - K(n))$. Player 1 deviations in the first period are punished by continuation equilibrium $\sigma^2(T(n) - K(n))$. Player 2 has no incentive to deviate in the first period, because a gives her the maximum stage-game payoff M_2, and so can be ignored. This strategy profile is subgame perfect if 1 has no incentive to deviate, which is ensured by (4.6.9). We denote by $\sigma(T(n) - K(n) + 1)$ an equilibrium whose payoff profile is in $R^{T(n)-K(n)+1}$ and which either equals or strictly dominates the equilibrium payoff we have just constructed.

Given such an equilibrium $\sigma(T(n) - K(n) + m)$ for some $m = 1, \ldots, K(n) - 1$, construct a profile for $G^{T(n)-K(n)+m+1}$ by prescribing play of a in the first period, followed by $\sigma(T(n) - K(n) + m)$, with player 1 deviations punished by $\sigma^2(T(n) - K(n) + m)$. Condition (4.6.9) again provides the necessary condition for this to be an equilibrium with payoffs in $Q^{T(n)-K(n)+m+1}$. Denote by $\sigma(T(n) - K(n) + m + 1)$ an equilibrium with payoffs in $R^{T(n)-K(n)+m+1}$ that equal or strictly dominate those of the equilibrium we have constructed. Proceeding in this fashion until reaching length $T(n)$ provides an equilibrium with payoffs in $R^{T(n)}$; it is straightforward to verify that the payoffs also lie in V.

The remaining case, $u_1(a) > v_1^1 - \varepsilon$, is easier to handle, because adding payoff $u(a)$ to an equilibrium increases rather than decreases 1's payoff and hence makes it less likely to raise incentive problems for player 1. In this case, we can take $\tilde{v}$ to be an arbitrary nontrivial convex combination of M and v^1 and then proceed as before.

∎

We now extend lemma 4.6.2 to the boundary cases in which $v_i = w_i$.

Lemma 4.6.4 *Let $b_1 > w_1$ and $b_2 > w_2$. Then each payoff in $v \in R^\infty$ is efficient.*

Proof Lemma 4.6.2 establishes the result for any profile that does not give one player the worst payoff in R^∞. By hypothesis, R^∞ contains at least two points.

Suppose $(w_1, w_2) \notin R^\infty$, and let $v_j > w_j$ be the smallest player j playoff for which $(w_i, v_j) \in R^\infty$. If $v_j = M_j$, then (w_i, v_j) is efficient. Suppose $v_j < M_j$. Then, by lemma 4.6.3, there is path in R^∞ between (w_i, v_j) and (b_i, v'_j), for some $v'_j \geq w_j$. Hence, there is a sequence of payoff profiles $\{v^n\}_{n=0}^\infty$ in R^∞ that converges to (w_i, v_j) and, by the definition of v_j, $v_i^n > w_i$ and $v_j^n > w_j$ for all sufficiently large n. By lemma 4.6.2, each v^n is efficient, and hence so is (w_i, v_j), as are all points $(w_i, v_j'') \in R^\infty$ with $v_j'' > v_j$.

It remains to rule out the possibility that $(w_1, w_2) \in R^\infty$. We suppose $(w_1, w_2) \in R^\infty$ and argue to a contradiction. Lemma 4.6.3, coupled with the observation that R^∞ can contain no profile that strictly dominates another, ensures that R^∞ consists of two line segments, one joining (w_1, b_2) with (w_1, w_2), and one joining (w_1, w_2) with (b_1, w_2). Let a be an action profile with $u_2(a) = M_2$, and fix $\lambda \in (0, 1)$ so that

$$w_1 < \tilde{v}_1 \equiv \lambda u_1(a) + (1 - \lambda) b_1,$$
$$\text{and} \quad w_2 < \tilde{v}_2 \equiv \lambda M_2 + (1 - \lambda) w_2.$$

Let $\varepsilon = \min\{\tilde{v}_1 - w_1, \tilde{v}_2 - w_2\}/4 > 0$ and choose T_ε so that for all $T \geq T_\varepsilon$,

$$M + T(w_1 + \varepsilon) < m + T(b_1 - 2\varepsilon).$$

The proof proceeds by showing that for large T, there are payoffs in R^T within ε of $\tilde{v}$, contradicting the assumption that $(w_1, w_2) \in R^\infty$. Because the details of the rest of the argument are almost identical to the proof of lemma 4.6.3, they are omitted.

∎

Proof of Proposition 4.6.1 Suppose that $(b_1^1, b_2^1) \notin \mathscr{F}^\dagger$. Lemma 4.6.1 ensures that R^∞ contains one profile whose payoff to player 1 is at least b_1^1 and one whose payoff to player 2 is at least b_2^2. Because it is impossible to do both simultaneously, it must be that $b_1 > w_1$ and $b_2 > w_2$. But then lemma 4.6.4 implies that R^∞ consists of efficient equilibria.

∎

The payoffs b_1^1 and b_2^1 are the best stage-game Nash equilibrium payoffs for players 1 and 2. The criterion provided by proposition 4.6.1 is then to ask whether it is feasible (though not necessarily an equilibrium) to simultaneously obtain these payoffs in the stage game. If not, then renegotiation-proof equilibria must be nearly efficient. The payoff profile (b_1^1, b_2^1) is infeasible in example 4.6.1, and hence the limiting renegotiation-proof equilibria are efficient. Example 4.6.2 accommodates such payoffs, thus failing the sufficient condition for efficiency, and featuring inefficient limiting payoffs.

Remark 4.6.2 Benoit and Krishna (1993) extend lemma 4.6.4 to allow one of the inequalities $b_1 > w_1$ and $b_2 > w_2$ to be weak. By doing so, they obtain the result that either the set R^∞ is a singleton or it is efficient. The only case that is not covered by our proof is that in which R^∞ is either a vertical or horizontal line segment, to which analogous arguments apply. Example 4.6.2 illustrates the case in which R^∞ is inefficient and is hence a singleton. Example 4.6.1 illustrates a case in which R^∞ contains only efficient profiles and contains an interval of such payoffs. There remains the possibility that R^∞ may contain only a single payoff which is efficient. This is the case for the coordination game of figure 4.6.5.

◆

	E	S
E	2, 2	−1, 3
S	3, −1	0, 0

Figure 4.6.6 The prisoners' dilemma.

4.6.2 Infinitely Repeated Games

We now consider infinite horizon games.[12] The point of departure is that a continuation equilibrium is not "credible" if it is strictly dominated by an alternative continuation equilibrium. In making this determination, however, one is not interested in all alternative equilibria, but only in those that themselves survive the credibility test.

Definition 4.6.2 *A subgame-perfect equilibrium* $(\mathcal{W}, w^0, f, \tau)$ *of the infinitely repeated game is weakly renegotiation proof if the payoffs at any pair of states are not strictly ranked (i.e., if for all $w', w'' \in \mathcal{W}$, if $V_i(w') > V_i(w'')$ for some i, then $V_j(w') \leq V_j(w'')$ for some j).*

Intuitively, if the payoffs beginning with the initial state w' strictly dominate the payoffs beginning with the initial state w'', then the latter are vulnerable to renegotiation, because the strategy profile $(\mathcal{W}, w', f, \tau)$ is both unanimously preferred and part of equilibrium behavior.

Weakly renegotiation proof equilibria always exist, because for any stage-game Nash equilibrium, the repeated-game profile that plays this equilibrium after every history is weakly renegotiation proof.

Example 4.6.3 Consider the prisoners' dilemma of figure 1.2.1, reproduced in figure 4.6.6. Grim trigger is not a weakly renegotiation-proof equilibrium. The strategies following the null history feature effort throughout the remainder of the equilibrium path. The resulting payoffs strictly dominate those produced by the strategies following a history featuring an instance of shirking, which call for perpetual shirking.

This may suggest that perpetual defection is the only renegotiation-proof equilibrium of the repeated prisoners' dilemma, because any other equilibrium must involve a punishment that will be strictly dominated by the original equilibrium. This is the case for strongly symmetric strategy profiles, that is, profiles in which the two players always choose the same actions, because any punishment must then

---

12. We illustrate the issues by first presenting some results taken from Evans and Maskin (1989) and Farrell and Maskin (1989), followed by an alternative perspective due to Abreu, Pearce, and Stacchetti (1993). Bernheim and Ray (1989) comtemporaneously presented closely related work. Similar issues appear in the idea of a coalition-proof equilibrium (Bernheim, Peleg, and Whinston 1987; Bernheim and Whinston 1987), where one seeks equilibria that are robust to alternative choices that are jointly coordinated by the members of a coalition. Here again, one seeks robustness to deviations that are themselves robust, introducing the self-reference that makes an obvious definition elusive.

4.6 ◾ Renegotiation

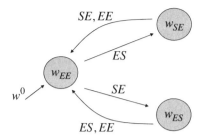

Figure 4.6.7 Weakly renegotiation-proof strategy profile supporting mutual effort in every period, in the prisoners' dilemma. Unspecified transitions leave the state unchanged.

push both players below their original equilibrium payoff. However, asymmetric punishments open other possibilities.

Consider the strategy profile in figure 4.6.7 (and taken from van Damme 1989). Intuitively, these strategies begin with both players choosing effort. If player i shirks, then he must pay a penance of exerting effort while player j shirks. Once i has done so, all is forgiven and the players return to mutual effort. This is a renegotiation-proof subgame-perfect equilibrium, for sufficiently high discount factors. Beginning with subgame perfection, we require that neither player prefer to shirk in the first period of the game, or the first period of a punishment phase. The accompanying incentive constraints are

$$2 \geq (1-\delta)[3 + \delta(-1)] + \delta^2 2,$$
$$\text{and} \quad (1-\delta)(-1) + \delta 2 \geq \delta[(1-\delta)(-1) + \delta 2].$$

Both are satisfied for $\delta \geq 1/3$. Our next task is to check renegotiation proofness. Continuation payoff profiles are $(2, 2)$, $(3\delta - 1, 3 - \delta)$, and $(3 - \delta, 3\delta - 1)$. For $\delta \in [1/3, 1)$, no pair of these profiles is strictly ranked.

◾

This result extends. The key to constructing nontrivial renegotiation-proof equilibria is to select punishments that reward the player doing the punishing, ensuring that the punisher is not tempted by the prospect of renegotiating. We consider repeated games with two players.[13]

Proposition 4.6.2 *Suppose the game has two players and $\tilde{a} \in A$ is a strictly individually rational pure action profile, with $v = u(\tilde{a})$. If there exist pure-action profiles a^1 and a^2 such that, for $i = 1, 2$,*

$$\max_{a_i \in A_i} u_i(a_i, a_j^i) < v_i, \qquad (4.6.10)$$

$$\text{and} \qquad u_j(a^i) \geq v_j, \qquad (4.6.11)$$

then, for sufficiently large δ, there exists a pure-strategy weakly renegotiation-proof equilibrium with payoffs v. If v is the payoff of a pure-strategy weakly

---

13. An extension to more than two players remains to be done.

renegotiation-proof equilibrium, then there exist pure actions a^1 and a^2 satisfying the weak inequality version of (4.6.10) and satisfying (4.6.11).

The action profiles a^1 and a^2 allow us to punish one player while rewarding the other. In the case of the prisoners' dilemma of figure 4.6.6, letting $a^1 = ES$ and $a^2 = SE$ ensures that for any payoff vector $v \in \mathscr{F}^*$, we have, for $i = 1$ in (4.6.10)–(4.6.11)

$$\max_{a_1 \in \{E,S\}} u_1(a_1, S) = 0 < v_1$$

and
$$u_2(E, S) = 3 \geq v_2.$$

Hence, any feasible, strictly individually rational payoff vector can be supported by a weakly renegotiation-proof equilibrium in the prisoners' dilemma. Notice that in the case of the prisoners' dilemma, a single pair of action profiles a^1 and a^2 provide the punishments needed to support any payoff $v \in \mathscr{F}^*$. In general, this need not be the case, with different equilibrium payoffs requiring different punishments.

Proof of Proposition 4.6.2 We prove the sufficiency of (4.6.10)–(4.6.11). For necessity, see Farrell and Maskin (1989), whose proof of necessity in their theorem 1 holds for our class of equilibria and games (i.e., pure-strategy equilibria in games with unobservable mixtures).

The equilibrium strategy profile is a simple strategy profile (definition 2.6.1), $\sigma(\mathbf{a}(0), \mathbf{a}(1), \mathbf{a}(2))$, where $\mathbf{a}(0)$ is the constant outcome path in which $\tilde{a}$ is played in every period, and $\mathbf{a}(i)$, $i = 1, 2$, is i's *punishment* outcome path of L periods of a^i followed by a return to $\tilde{a}$ in every period (where L is to be determined). An automaton for this profile has a set of states

$$\{w(0)\} \cup \{w(i, k), i = 1, 2, k = 1, \ldots, L\},$$

initial state $w(0)$, output function

$$f(w) = \begin{cases} \tilde{a}, & \text{if } w = w(0), \\ a^i, & \text{if } w = w(i, k), k = 1, \ldots, L, \end{cases}$$

and transitions

$$\tau(w(0), a) = \begin{cases} w(i, 1), & \text{if } a_i \neq \tilde{a}_i, a_{-i} = \tilde{a}_{-i}, \\ w(0), & \text{otherwise,} \end{cases}$$

and

$$\tau(w(i, k), a) = \begin{cases} w(j, 1), & \text{if } a_j \neq a^i_j, a_{-j} = a^i_{-j}, \\ w(i, k+1), & \text{otherwise,} \end{cases}$$

where we take $w(i, L+1)$ to be $w(0)$.

Choose L so that

$$L \min_i (v_i - u_i(a^i)) \geq M - \min_i v_i,$$

where M is, as usual, the largest stage game payoff.

4.6 ■ Renegotiation

	h	ℓ
H	2, 3	0, 2
L	3, 0	1, 1

Figure 4.6.8 The product-choice game.

Standard arguments show that for sufficiently high δ, no one-shot deviations are profitable from $w(0)$ nor by player i in any state $w(i, k)$ (see the proof of proposition 3.3.1). It is immediate that player i has no incentive to deviate from punishing player j, because i is rewarded for punishing j.

It is also straightforward to verify $V_i(w(i,k)) < V_i(w(i, k+1)) < v_i \leq V_i(w(j, k+1)) \leq V_i(w(j,k))$, and so the profile is weakly renegotiation proof.

∎

Example 4.6.4 Consider the product-choice game of figure 1.5.1, reproduced in figure 4.6.8. Suppose both players are long-lived. For sufficiently large discount factors, we have seen that equilibria exist in which Hh is played in every period. However, it is clear that no such equilibrium can be renegotiation proof. In particular, this equilibrium gives player 2 the largest feasible payoff in the stage game. There is then no way to punish player 1, who faces the equilibrium incentive problem, without also punishing player 2. If there exists an equilibrium in which Hh is always played, then this equilibrium path must yield payoffs strictly dominating some of its own punishments, precluding renegotiation proofness.

●

Remark 4.6.3 **Public correlation** In the presence of a public correlating device, there is an ex ante and an ex post notion of weak renegotiation proofness. In the latter, renegotiation is possible after the realization of the correlating device, whereas in the former, renegotiation is only possible before the realization. The analysis easily extends to ex ante (but not to ex post) weak renegotiation proofness with public correlation and allows a weakening of the sufficient conditions (4.6.10)–(4.6.11) by replacing $\tilde{a}$, a^1 and a^2 with correlated action profiles. In doing so, we require (4.6.10) to hold for every realization of the public correlating device, whereas (4.6.11) is required only in expectation.

Farrell and Maskin (1989) allow a^1 and a^2 to be (independent) mixtures and allow v to be the payoff of any publicly correlated action profile. They then show that payoff v can be obtained from a deterministic sequence of (possibly mixed but uncorrelated) profiles, using an argument similar to that of section 3.7 but assuming observable mixed strategies and complicated by the need to show weak renegotiation proofness.

◆

Every feasible, strictly individually rational payoff in the prisoners' dilemma can be supported as the outcome of a weakly renegotiation-proof equilibrium, including

	A	B	C	D
A	4, 2	z, z	z, 0	z, z
B	z, z	0, 3	z, 0	z, z
C	5, z	0, z	3, 0	0, z
D	z, z	z, z	z, 5	2, 4

Figure 4.6.9 Stage game for repeated game in which a strongly renegotiation proof equilibrium does not exist, where z is negative and large in absolute value.

the payoffs provided by persistent shirking and the payoff provided by persistent effort. Why isn't the former equilibrium disrupted by the opportunity to renegotiate to the latter?

The difficulty is that weak renegotiation proofness ensures that no two continuation paths, *that potentially arise as part of a single equilibrium*, are strictly ranked. However, our intuition is that renegotiation-proofness should also restrict attention to a *set of equilibria* with the property that no two equilibria in the set are strictly ranked. Toward this end, Farrell and Maskin (1989) offer:

Definition 4.6.3 *A strategy profile σ is* strongly renegotiation proof *if it is weakly renegotiation proof and no continuation payoff is strictly dominated by the payoff in another weakly renegotiation-proof equilibrium.*

The strategy profile shown in figure 4.6.7 is strongly renegotiation proof. The following example, adapted from Bernheim and Ray (1989), shows that strongly renegotiation-proof equilibria need not exist.

Example 4.6.5 We consider the game shown in figure 4.6.9. We let $\delta = 1/5$ and consider only pure-strategy equilibria. When doing so, it suffices for nonexistence that $z < -5/4$.[14] The minmax values for each agent are 0.

The first observation is that no pure-strategy equilibrium can ever play an action profile that does not lie on the diagonal of the stage game. Equivalently, any pure equilibrium must be strongly symmetric. Off the diagonal, at least one player receives a stage-game payoff of z and hence a repeated game payoff at most $(1-\delta)z + \delta 5$. Given $\delta = 1/5$ and $z < -5/4$, this is less than the minmax value of 0, a contradiction.

The strategy profile that specifies *BB* after every history is weakly renegotiation proof, as is the profile that specifies *CC* after each history, with payoffs (0, 3) and

---

14. Bernheim and Ray (1989) note that the argument extends to mixed equilibria if z is sufficiently large and negative. The key is that in that case, any equilibrium mixtures must place only very small probability on outcomes off the diagonal, allowing reasoning similar to that of the pure-strategy case to apply.

(3, 0) respectively. Similarly, paths that vary between *BB* and *CC* can generate convex combinations of (0, 3) and (3, 0) as outcomes of weakly renegotiation-proof equilibria.

We now consider an equilibrium in which *AA* is played. Such an equilibrium exists only if we can find continuation payoffs $\gamma(AA)$ and $\gamma(CA)$ from the interval $[0, 4]$ such that

$$(1 - \delta)4 + \delta\gamma_1(AA) \geq (1 - \delta)5 + \delta\gamma_1(CA).$$

Given $\delta = 1/5$, this inequality can be satisfied only if $\gamma_1(AA) = 4$ and $\gamma_1(CA) = 0$. Hence, there exists a unique equilibrium outcome in which *AA* is played, which features *AA* in every period and with deviations followed by a continuation value (0, 3). Similarly, there exists an equilibrium in which *DD* is played in every period, with deviations followed by continuation values (3, 0). Both equilibria are weakly renegotiation proof.

It is now immediate that there are no strongly renegotiation-proof equilibria. Any equilibrium that features only actions *BB* and *CC* is strictly dominated either by the equilibrium that always plays *AA* or the one that always plays *DD*. However, each of the latter includes a punishment phase that always plays either *BB* or *CC*, ensuring that the latter equilibria are also not strongly renegotiation proof.

●

Strongly renegotiation-proof equilibria fail to exist because of a lack of efficient punishments. Sufficient conditions for existence are straightforward when $\mathscr{F} = \mathscr{F}^\dagger$, as is often (but not always, see figure 3.1.3) the case when actions sets are continua.[15] First, note that taking $\tilde{a}$ in proposition 4.6.2 to be efficient gives sufficient conditions for the existence of an efficient weakly renegotiation-proof equilibrium.[16]

Proposition 4.6.3 *Suppose $\mathscr{F} = \mathscr{F}^\dagger$ and an efficient (in $\mathscr{F}^\dagger$) pure-strategy weakly renegotiation-proof equilibrium exists. Let v^i be the payoff of the efficient weakly renegotiation-proof equilibrium that minimizes player i's payoff over the set of such equilibria. If there are multiple such equilibria, choose the one maximizing j's payoff. For each i, assume there is an action profile a^i satisfying*

$$\max_{a_i} u_i(a_i, a^i_j) < v^i_i$$

and $\quad u_j(a^i) \geq v^i_j.$

Then for every efficient action profile a^* with $u_i(a^*) \geq v^i_i$ for each i, there is a $\underline{\delta}$ such that for all $\delta \in (\underline{\delta}, 1)$ there is a pure-strategy strongly renegotiation-proof equilibrium yielding payoffs $u(a^*)$.

---

15. In the presence of public correlation and ex ante weak renegotiation proofness (remark 4.6.3), we can drop the assumption that $\mathscr{F} = \mathscr{F}^\dagger$. Farrell and Maskin (1989) do not assume $\mathscr{F} = \mathscr{F}^\dagger$ or the existence of public correlation (but see remark 4.6.3).
16. Evans and Maskin (1989) show that efficient weakly renegotiation-proof equlibria generically exist, though again assuming observable mixtures.

Proof Let $\breve{a}^i$ be a pure-action profile with $u(\breve{a}^i) = v^i$ (recall $\mathscr{F} = \mathscr{F}^\dagger$). The equilibrium strategy profile is, again, a simple strategy profile (definition 2.6.1), $\sigma(\mathbf{a}(0), \mathbf{a}(1), \mathbf{a}(2))$, where $\mathbf{a}(0)$ is the constant outcome path in which a^* is played in every period, and $\mathbf{a}(i)$, $i = 1, 2$, is i's *punishment* outcome path of L periods of a^i followed by the perpetual play of $\breve{a}^i$, where L satisfies

$$L \min_i (u_i(a^*) - u_i(a^i)) \geq M - \min_i u_i(a^*),$$

and M is the largest stage game payoff.

As in the proof of proposition 4.6.2, standard arguments show that for sufficiently high δ, no one-shot deviations are profitable (see the proof of proposition 3.3.1).

It thus remains to show that none of the continuation equilibria induced by this strategy profile are strictly dominated by other weakly renegotiation-proof equilibria. There are five types of continuation path to consider. Three of these are immediate. The continuation paths consisting of repeated stage-game action profiles with payoffs $u(a^*)$ and v^i cannot be strictly dominated, because each of these payoff profiles is by construction efficient. The remaining two cases are symmetric, and we present the argument for only one of them. Consider a continuation path in which player 1 is being punished, consisting of between 1 and L initial periods in which a^1 is played, followed by the perpetual play of $\breve{a}^1$. Let $\tilde{v}$ be the payoff from this path. Given the properties of a^1, we know that

$$\tilde{v}_1 < v_1^1 \quad \text{and} \quad \tilde{v}_2 \geq v_2^1.$$

Suppose $\hat{v}$ is a weakly renegotiation-proof payoff profile dominating $\tilde{v}$. Then, of course, we must have

$$\hat{v}_2 > \tilde{v}_2 \geq v_2^1.$$

Now suppose $\hat{v}_1 > v_1^1$. Then $\hat{v}$ strictly dominates v^1, contradicting the assumption that v^1 is an efficient weakly renegotiation-proof equilibrium. Suppose next $\hat{v}_1 = v_1^1$. Then $\hat{v}$ must be efficient and, because $\hat{v}_2 > v_2^1$, this contradicts the definition of v^1. Suppose finally that $\hat{v}_1 < v_1^1$ (and hence is inefficient, because we otherwise have a contradiction to the choice of v^1 as the efficient weakly renegotiation-proof equilibrium that minimizes player 1's payoff). Let $(v_1', \hat{v}_2)$ be efficient. Then applying the necessity portion of proposition 4.6.2 to $\hat{v}$, using $\mathscr{F} = \mathscr{F}^\dagger$ to obtain $\hat{a}^1$, and recalling the properties of a^2, we have

$$\max_{a_1 \in A_1} u_1(a_1, \hat{a}_2^1) \leq \hat{v}_1 < v_1',$$

$$u_2(\hat{a}^1) \geq \hat{v}_2,$$

$$\max_{a_2 \in A_2} u_2(a_1^2, a_2) \leq v_2^2 < \hat{v}_2,$$

$$\text{and} \quad u_1(a^2) \geq v_1^2 \geq v_1'.$$

Then $\hat{a}^1, a^2$ and $(v_1', \hat{v}_2)$ satisfy the sufficient conditions from proposition 4.6.2 for there to exist a pure-strategy weakly renegotiation-proof equilibrium with payoff $(v_1', \hat{v}_2)$. By construction, this payoff is efficient. This contradicts the efficiency of v^1 (if $v_1' > v_1^1$) or the other elements of the definition of v^1 (if $v_1' \leq v_1^1$). ∎

4.6 ■ Renegotiation

Renegotiation-proof equilibria are so named because they purportedly survive any attempt by the players to switch to another equilibrium. In general, the players will have different preferences over the possible alternative equilibria, giving rise to a bargaining problem. The emphasis on efficiency in renegotiation proofness reflects a presumption that the players could agree on replacing an equilibrium with another that gives everyone a higher payoff. However, given the potential conflicts of interest, this belief in agreement may be optimistic. Might not a deal on a superior equilibrium be scuttled by player i's attempt to secure an equilibrium that is yet better for him? It is not clear we can answer without knowing more about how bargaining proceeds.

Abreu, Pearce, and Stacchetti (1993) take an alternative approach to renegotiation proofness that focuses on the implied bargaining. Their view is that if player i can point to an alternative equilibrium in which every player (including i) earns a higher payoff after every history than i currently earns, then i can make a compelling case for switching to the alternative equilibrium. For an equilibrium σ define $\underline{v}_i(\sigma) = \inf\{U_i(\sigma|_{h^t}) : h^t \in \mathcal{H}\}$.

Definition 4.6.4 *An equilibrium σ^* is a* consistent bargaining equilibrium *if there is no alternative subgame-perfect equilibrium σ with $\min_i\{\underline{v}_i(\sigma)\} > \min_i\{\underline{v}_i(\sigma^*)\}$.*

Hence, the consistent bargaining criterion rejects an equilibrium if it could reach a continuation game with one player whose continuation payoff is lower than every continuation payoff in some alternative equilibrium. Notice the comparisons embedded in this notion are no longer based on dominance. It suffices to reject an equilibrium that it fails the test for one player, though this player must make comparisons with every other player's payoffs in the proposed alternative equilibrium. A player for whom there is such an alternative equilibrium promising a higher payoff in every circumstance is viewed as having a winning case for abandoning the current one. Abreu, Pearce, and Stacchetti (1993) apply this concept only to symmetric games. Symmetry is not required for this concept to be well defined or to exist, but helps in interpreting the payoff comparisons across players as natural comparisons the players might make.

Example 4.6.6 Consider the battle of the sexes game of figure 4.6.10. With the help of a public correlating device, we can construct an equilibrium in which *TL* and *BR* are each played with equal probability in each period, for expected payoffs of (2, 2). It is also clear that there is no equilibrium with a higher total payoff, so this equilibrium is the unique consistent bargaining equilibrium.

We use this game to explain the need for a comparison across players' payoffs in this notion. Suppose that we suggested that an equilibrium σ^* should

	L	R
T	1, 3	0, 0
B	0, 0	3, 1

Figure 4.6.10 A battle-of-the-sexes game.

be a consistent bargaining equilibrium if there is no alternative subgame-perfect equilibrium σ and player i with $v_i(\sigma) > v_i(\sigma^*)$. Then notice that there exists a subgame-perfect equilibrium σ^1 in which BR is played after every history, and hence $v_1(\sigma^1) = 3$. There exists another equilibrium σ^2 that features TL after every history and hence $\underline{v}_2(\sigma^2) = 3$. Given the infeasibility of payoffs $(3, 3)$, this ensures that there would be no consistent bargaining equilibrium. This result illustrates the intuition surrounding a consistent bargaining equilibrium. A player i has a persuasive objection to an equilibrium, not simply when there is an alternative equilibrium under which that player invariably fares better, but when there is an alternative under which *every* player invariably fares better than i currently does. In the former case, the player is arguing that he would prefer an equilibrium in which things work out better for him, whereas in the latter he can argue that he is being asked to endure an outcome that no one need endure.

●

Example 4.6.7 We consider the three versions of the prisoners' dilemmas, shown in figure 4.6.11. The first two are familiar, and the third makes it most attractive to shirk when the opponent is exerting effort. Consider an equilibrium path with a value v_i to player i. We use the possibility of public correlation to assume that this value is received in every period, simplifying the calculations. Let Δ_i be the increase in player i's current payoff that can be achieved by deviating from equilibrium play. Restrict attention to simple strategies, and let $\tilde{v}_i$ be the value of the resulting punishment. We can further assume that this punishment takes the form of a finite number of periods of mutual shirking, followed by a return to the equilibrium path. This ensures that the continuation payoff provided by the punishment is lowest in its initial period, and that the only nontrivial incentive constraint is the condition that play along the equilibrium path be optimal, which can be rearranged to give

$$\tilde{v}_i \leq v_i - \frac{1-\delta}{\delta} \Delta_i.$$

Let us focus on the case of $\delta = 1/2$, so that the incentive constraint becomes

$$\tilde{v}_i \leq v_i - \Delta_i.$$

	E	S
E	2, 2	−1, 3
S	3, −1	0, 0

	E	S
E	3, 3	−1, 4
S	4, −1	1, 1

	E	S
E	2, 2	−1, 4
S	4, −1	0, 0

Figure 4.6.11 Three prisoners' dilemma games. In the left game, the incentives to shirk rather than exert effort are independent of the opponent's action. The incentive to shirk is strongest when the opponent shirks in the middle game, and strongest when the opponent exerts effort in the right game.

4.6 ■ Renegotiation

The search for a consistent bargaining equilibrium thus becomes the search for an equilibrium featuring a pair of values $(v_i, \tilde{v}_i)$ with the property that (because $\tilde{v}_i \leq v_i$) no other equilibrium features continuation values that all exceed $\tilde{v}_i$.

Now consider the prisoners' dilemma in the left panel of figure 4.6.11. Here, Δ_i is fixed at 1, independently of the opponents' actions. This fixes the relationship between v_i and $\tilde{v}_i$, for any equilibrium. Finding a consistent bargaining equilibrium now becomes a matter of finding the equilibrium with the largest value of $\tilde{v}_i$. Given the fixed relationship between v_i and $\tilde{v}_i$, this is equivalent to finding the equilibrium that maximizes v_i. This calls for permanent effort and equilibrium payoffs (2, 2). This is the unique consistent bargaining equilibrium outcome in this game.

In the middle game, the incentives to shirk now depend on the equilibrium actions, so that Δ_i is no longer fixed. Finding a consistent bargaining equilibrium again requires finding the equilibrium with the largest value of $\tilde{v}_i$. Moreover, Δ_i is larger the more likely one's opponent is to shirk. This reinforces the fact that maximizing $\tilde{v}_i$ calls for maximizing v_i. As a result, a consistent bargaining equilibrium again requires that v_i be maximized, and hence the unique consistent bargaining equilibrium again calls for payoffs (2, 2).

In the right game, the incentives to shirk are strongest when the opponent exerts effort. Hence, Δ_i becomes smaller the more likely is the opponent to shirk. It is then no longer the case that the equilibrium maximizing $\tilde{v}_i$, and hence providing our candidate for a consistent bargaining equilibrium, is also the equilibrium that maximizes v_i. Instead, the unique consistent bargaining equilibrium here calls for the two players to attach equal probability (again, with the help of a public correlating device) to *ES* and *SE*. Notice that equilibrium is inefficient, with expected payoffs of 3/2 rather than the expected payoffs (2, 2) provided by mutual effort. To see that this is the unique consistent bargaining equilibrium, notice that the incentive constraint for the proposed equilibrium must deter shirking when the public correlation has selected the agent in question to exert effort and the opponent to shirk. A deviation to shirking under these circumstances brings a payoff gain of 1, so that the punishment value $\tilde{v}_i$ must satisfy

$$\tilde{v}_i \leq v_i - \Delta_i = \tfrac{3}{2} - 1 = \tfrac{1}{2}.$$

The proposed equilibrium thus has continuation values of 3/2 and 1/2. Any equilibrium featuring higher continuation values must sometimes exhibit mutual effort. Here, the incentive constraint is

$$\tilde{v}_i \leq v_i - \Delta_i = v_i - 2.$$

Because this incentive constraint must hold for both players, at least one player must face a continuation payoff of 0. This ensures that no such equilibrium can yield a collection of continuation values which are all strictly larger than 1/2. As a result, the equilibrium that mixes over outcomes *SE* and *ES* is the unique consistent bargaining equilibrium.

●

5 Variations on the Game

An important motivation for work with repeated games, as with all economic models, is that the repeated game is an analytically convenient model of a more complicated reality. This chapter explores some variations of the canonical repeated-game model in which the strategic interactions are not literally identical from period to period.

5.1 Random Matching

We begin with a setting in which every player shares some of the characteristics of a short-lived player and some of a long-lived player.[1] Our analysis is based on the familiar prisoners' dilemma of figure 1.2.1, reproduced in figure 5.1.1. The phrase "repeated prisoners' dilemma" refers to a single pair of long-lived players who face each other in each period. Here, we study a model with an even number $M \geq 4$ of players. In each period, the M players are matched into $M/2$ pairs to play the game. Matchings are independent across periods, with each possible configuration of pairs equally likely in each period. Each matched pair then plays the prisoners' dilemma, at which point we move to the next period with current matches dissolved and the players entered into a new matching process. We refer to such a game as a *matching game*. A common interpretation is a market where people are matched to complete bilateral trades in which they may either act in good faith or cheat.

Players in the matching game discount their payoffs at rate δ and maximize the average discounted value of payoffs. In any given match, both players appear to be effectively short-lived, in the sense that they play one another once and then depart. Were this literally the end of the story, we would have a collection of ordinary prisoners' dilemma stage games with the obvious (and only) equilibrium outcome of shirking at every opportunity.

However, the individual interactions in the matching game are not perfectly isolated. With positive probability, each pair of matched players has a subsequent rematch. Given enough time, the matching process will almost certainly bring them together again. But if the population is sufficiently large, the probability of a rematch in the near future will be sufficiently small as to have a negligible effect on current incentives. In that event, mutual shirking appears to be the only equilibrium outcome.

---

1. The basic model is due to Kandori (1992a). Our elaboration is due to Ellison (1994), which allows for a public correlating device.

	E	S
E	2, 2	−1, 3
S	3, −1	0, 0

Figure 5.1.1 Prisoners' dilemma.

5.1.1 Public Histories

There may still be hope for outcomes other than mutual shirking. Much hinges on the information players have about others' past play. We first consider the *public-history matching* game, in which in every period each player observes the actions played in *every* match. A period t history is then a t-tuple of outcomes, each of which consists of an $(M/2)$-tuple of pairs of actions. Because histories are public, just as in the standard perfect-monitoring repeated game, every history leads to a subgame. Consequently, the appropriate equilibrium concept is, again, subgame perfection.

Proposition 5.1.1 *An outcome* $\mathbf{a} = (a^0, a^1, \ldots) \in A^\infty$ *is an equilibrium outcome in the repeated prisoners' dilemma if and only if there is an equilibrium in the public-history matching game in which action profile a^t is played by every pair of matched players in period t.*

The idea of the proof is that any deviation from $\mathbf{a}$ in the public-history matching game can prompt the most severe punishments possible from the repeated game. This can be done despite the fact that partners are scrambled each period, because histories are public. Because deviations from outcome $\mathbf{a}$ are deterred by the punishments of the equilibrium supporting outcome $\mathbf{a}$ in the repeated game, they are also deterred by the (possibly more severe) punishments of the matching game. This is essentially the argument lying behind optimal penal codes (section 2.6.1). Though interest often focuses on the prisoners' dilemma, the argument holds in general for two-player games.

Proof Let $\mathbf{a}$ be an equilibrium outcome in the repeated prisoners' dilemma and σ the corresponding equilibrium profile. Let σ' be a strategy profile that duplicates σ except that at any history $h^t \neq \mathbf{a}^t$, σ' prescribes that every player shirk. Because perpetual mutual shirking is an optimal punishment in the prisoners' dilemma, from Corollary 2.6.1, σ' is also a subgame-perfect equilibrium of the repeated game with outcome $\mathbf{a}$.

We now define a strategy profile σ'' for the public-history matching game. A pair of players matched in period t chooses action profile a^t if every pair of matched players in every period $\tau < t$ has chosen profile a^τ. Otherwise, the players shirk. Notice that the publicness of the histories in the public-history matching game makes this strategy profile feasible. Notice also that these strategies accomplish the goal of producing profile a^t in each match in each period t. It remains to show that the proposed strategies are a subgame-perfect equilibrium of the public-history matching game.

5.1 ■ Random Matching

Consider a player in the public-history matching game, in period t, with a^τ having been played in every match in every period $\tau < t$. Let the resulting history be called $\hat{h}^t$ and let h^t be the corresponding history (i.e., $\mathbf{a}^t$), in the repeated game. Then we need only note that the optimality of action profile a^t in the matching game is equivalent to

$$U_i(\sigma'|_{h^t}) \geq U_i(\sigma_i, \sigma'_{-i}|_{h^t}),$$

for all repeated-game strategies σ_i, where U_i is the repeated-game payoff function. This is simply the observation that the player faces precisely the same future play, with or without a deviation, in the matching game as in the repeated game. The optimality conditions for the repeated game thus imply the corresponding optimality conditions for the full-information matching game. Similar reasoning gives the converse.

■

There is thus nothing special about having the same players matched with each other in each period of a repeated game, as long as sufficient information about previous play is available. What matters in the matching game is that player i be punished in the future for current deviations, regardless of whether the punishment is done by the current partner or someone else. In the equilibrium we have constructed, a deviation from the equilibrium path is observed by all and trips everyone over to the punishment. Having one's partner enter the punishment suffices to deter the deviation in the repeated game, and having the remainder of the population do so suffices in the matching game. Why would someone against whom a player has never played punish the player? Because it is an equilibrium to do so, and failing to do so brings its own punishment.

5.1.2 Personal Histories

We now consider the *personal-history* game. We consider the extreme case in which at the beginning of period t, player i's history consists only of the actions played in each of the previous t matches in which i has been involved. Player i does not know the identities of the partners in the earlier matches, nor does he know what has transpired in any other match.[2] Hence, a period t personal history is an element of A^t, where $A = \{EE, SE, ES, SS\}$.

In analyzing this game, we cannot simply study subgame-perfect equilibria, because there are no nontrivial subgames, and so subgame perfection is no more restrictive than Nash. The appropriate notion of sequential rationality in this setting is *sequential equilibrium*. Though originally introduced for finite extensive form games by Kreps and Wilson (1982b), the notion has a straightforward intuitive description in this setting: A strategy profile is a sequential equilibrium if, after every personal history, player i is best responding to the behavior of the other players, given beliefs over the personal histories of the other players that are "consistent" with the personal history that player i has observed.

We are interested in the possible existence of a sequential equilibrium of the personal-history game in which players always exert effort. Figure 5.1.2 displays the candidate equilibrium strategies, where $q \in [0, 1]$ is to be determined; denote the

---

2. In particular, player i does not know if he has been previously matched with his current partner.

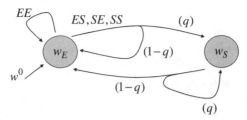

Figure 5.1.2 The individual automaton describing a single player i's strategy $\hat{\sigma}_i$ for the personal history game. Transitions not labeled by a strategy profile occur for all strategy profiles. Transitions labeled by $(\cdot)$ are random and are conditioned on ω, the realization of a public random variable, returning the automaton to state w_E if $\omega > q$; the probability of that transition is $1 - q$.

strategy profile by $\hat{\sigma}$. In contrast to the other automata we use in part I but in a preview of our work with private strategies and private-monitoring games, an automaton here describes the strategy of a single player rather than a strategy profile, with one such automaton associated with each player (see remark 2.3.1). The actions governing the transitions in the automaton are those of the player and the partner in the current stage game. The transitions are also affected by a public random variable, which allows the players in the population to make perfectly correlated transitions.

Each player's automaton begins in state w_E, where the player exerts effort. If the player or the player's partner shirks, then a transition is made to state w_S, and to shirking, unless the realization $\omega \in [0, 1]$ of the uniformly distributed public random variable is larger than q, in which case no transition is made. The automaton remains in state w_S until at some future date $\omega > q$. The public randomization thus ensures that in every period, there is probability $1 - q$ that every player returns to state w_E of their automaton. With probability q, every player proceeds with a transition based on his personal history.

Under these strategies, any deviation to shirking sets off a contagion of further shirking. In the initial stages of this contagion, the number of shirkers approximately doubles every period (as long as $\omega < q$), as the relatively small number of shirkers tend to meet primarily partners who have not yet been contaminated by shirking. We think of the public correlation device as occasionally (and randomly) pushing a "reset button" that ends the contagion.

Consider first a player who has observed only mutual effort. Then, he is in state w_E. Because he does not observe the personal history of his current match, he also does not *know* her current state. However, the only belief about her personal history that is consistent with his personal history and the profile $\hat{\sigma}$ is that she also has observed only mutual effort and so is in state w_E.

Suppose now that a player has observed his partner in the last period play S, and $\omega > q$. Then again, because ω is public, player i must believe that every other players' current state is w_E. On the other hand, if $\omega \leq q$, then the player's current state is w_S. If the previous period was the initial period, sequentiality requires player i to believe that his period 0 partner unilaterally deviated, and at the end of the period 0, all players except player i and his period 0 partner are in state w_E, whereas the remaining two players are in state w_S. Consequently, in period 1, he is rematched with his previous partner with probability $1/(M-1)$, and with complementary probability, is matched with a player in state w_E.

However, if this observation occurs much later in the game, sequentiality would allow player i to believe that the original deviation occurred in period 0 and had spread throughout most of the population. At the same time, if in every period after observing S, $\omega \leq q$, player i must believe that at least one other player has observed a personal history leaving that player in state w_S. As will be clear from the proof, the most difficult incentive constraint to satisfy is that player i, in state w_S, shirk when he believes that exactly one other player is in state w_S.

Proposition 5.1.2 *There exists $\underline{\delta} < 1$ such that, for any $\delta \in (\underline{\delta}, 1)$, there exists a value of $q = q(\delta)$ such that $\hat{\sigma}$ (the profile of strategies in figure 5.1.2) is a sequential equilibrium of the personal-history matching game.*

Proof Let $V(\mathscr{S}, \delta, q)$ be the expected value of a player $i \in \mathscr{S} \subset \{1, \ldots, M\}$ when the players in $\mathscr{S}$ are in state w_S and the remaining players are in state w_E, given discount factor δ and probability $q(\delta) = q$. Given the uniformity of the matching process, this value depends only on the number of players in state w_S and is independent of their identities and the identity of player $i \in \mathscr{S}$. We take i to be player 1 unless otherwise specified and often take $\mathscr{S}$ to be $\mathscr{S}_k \equiv \{1, \ldots, k\}$.

We begin by identifying two sufficient conditions for the proposed strategy profile to be an equilibrium. The first condition is that player 1 prefer not to shirk when in state w_E. Player 1 then believes every other player (including his partner) to also be in state w_E, as specified by the equilibrium strategies, either because no shirking has yet occurred or because no shirking has occurred since the last realization $\omega > q$. This gives an incentive constraint of

$$2 \geq (1-\delta)3 + \delta[(1-q)2 + qV(\mathscr{S}_2, \delta, q)],$$

or

$$1 \leq \frac{\delta q}{1-\delta}[2 - V(\mathscr{S}_2, \delta, q)]. \qquad (5.1.1)$$

The second condition is that player 1 must prefer not to exert effort while in state w_S. Because exerting effort increases the likelihood of effort from future partners, this constraint is not trivial. In this case, player 1 cannot be certain of the state of his partner and may be uncertain as to how many other players are in state w_S of their automata. Notice, however, that if player 1's partner shirks in the current interaction, then exerting effort only reduces player 1's current payoff without retarding the contagion to shirking and hence is suboptimal. If there are to be any gains from exerting effort when called on to shirk, they must come against a partner who exerts effort. If player 1 shirks against a partner who exerts effort, the number of players in state w_S next period (conditional on $\omega < q$) is larger by 1 (player 1's partner) than it would be if player 1 exerts effort. Because we are concerned with the situation in which player 1 is in state w_S and meets a partner in state w_E, there must be at least two players in the population in state w_S (player

1 and his previous partner). Hence, it suffices for shirking to be optimal that, for all $k \geq 2$,[3]

$$(1-\delta)3 + \delta[(1-q)2 + qV(\mathscr{S}_{k+1}, \delta, q)]$$
$$\geq (1-\delta)2 + \delta[(1-q)2 + qV(\mathscr{S}_k, \delta, q)],$$

where $\mathscr{S}_k$ is the set of players in state w_S next period if player 1 exerts effort, and $\mathscr{S}_{k+1}$ is the set in state w_S if player 1 shirks. Rearranging, this is

$$1 \geq \frac{\delta q}{1-\delta}[V(\mathscr{S}_k, \delta, q) - V(\mathscr{S}_{k+1}, \delta, q)]. \quad (5.1.2)$$

We now proceed in two steps. We first show that there exists $\underline{\delta}$ such that for every $\delta \in (\underline{\delta}, 1)$, we can find a $q(\delta) \in [0, 1]$ such that (5.1.1) holds with equality, that is, players are indifferent between shirking and exerting effort on the induced path. Note that in that case, (5.1.2) holds for $k = 1$, because $V(\mathscr{S}_1, \delta, q) = 2$ is the value to a player when all other players are in state w_E, and the player deviates to S. We then show that if (5.1.2) holds for any value k, it also holds strictly for any larger value k'. In particular, when a player is indifferent between shirking and exerting effort on the induced path, after deviating, that player now strictly prefers to shirk because the earlier deviation has triggered the contagion.

Step 1. Fix $q = 1$ and consider how the right side of (5.1.1) varies in the discount factor,

$$\lim_{\delta \to 0} \frac{\delta}{(1-\delta)}[2 - V(\mathscr{S}_2, \delta, 1)] = 0,$$

and $\quad \lim_{\delta \to 1} \frac{\delta}{(1-\delta)}[2 - V(\mathscr{S}_2, \delta, 1)] = \infty.$

The first equality follows from noting that $V(\mathscr{S}_2, \delta, 1) \in [0, 3]$ for all δ. The second follows from noting that $V(\mathscr{S}_2, \delta, 1)$ approaches 0 as δ approaches 1, because under $\hat{\sigma}$ with $q = 1$, eventually every player is in state w_S, and hence the expected stage-game payoffs converge to 0, at a rate independent of δ. Because $\frac{\delta}{(1-\delta)}[2 - V(\mathscr{S}_2, \delta, 1)]$ is continuous, there is then (by the intermediate value theorem) a value of δ at which $\frac{\delta}{1-\delta}[2 - V(\mathscr{S}_2, \delta, 1)] = 1$, and hence at which (5.1.1) holds with equality when $q = 1$. Denote this value by $\underline{\delta}$, and then take $q(\underline{\delta}) = 1$. Summarizing,

$$1 = \frac{\delta}{(1-\underline{\delta})}q(\underline{\delta})[2 - V(\mathscr{S}_2, \underline{\delta}, 1)]. \quad (5.1.3)$$

We now show that (5.1.1) holds for every $\delta \in (\underline{\delta}, 1)$, for suitable $q(\delta)$. We first acquire the tools for summarizing key aspects of a history of play. Fix a period and a history of play, and consider the continuation game induced by that history. Let ξ be a specification of which players are matched in the period 0 (of the continuation

---

3. In fact, it would suffice to have $k \geq 3$ because there are at least three players in state w_S next period (including the current partner of the other shirker). However, as will become clear, it is convenient to work with the stronger requirement, $k \geq 2$.

5.1 ■ Random Matching

game) and in each future period. Let $i(j, t, \xi)$ denote the partner of player j in period t under the matching ξ. We now recursively construct a sequence of sets, beginning with an arbitrary set $\mathscr{S} \subset \{1, \ldots, M\}$. For any set of players $\mathscr{S}$, let

$$T_0(\mathscr{S}, \xi) = \{1, \ldots, M\} \setminus \mathscr{S}$$

and $\quad T_{t+1}(\mathscr{S}, \xi) = \{j \in T_t(\mathscr{S}, \xi) \mid i(j, t, \xi) \in T_t(\mathscr{S}, \xi)\}.$

In our use of this construction, $\mathscr{S}$ is the set of players who enter the continuation game in state w_S, with the remaining players in state w_E. The state $T_t(\mathscr{S}, \xi)$ then identifies the players that are still in state w_E of their automata in period t, given that the public reset to state w_E has not occurred. Define $V(\mathscr{S}, \delta, q \mid \xi)$ to be the continuation value of player $i \in \mathscr{S}$ when the players in set $\mathscr{S}$ (only) are in state w_S, given matching realization ξ. Hence, $V(\mathscr{S}, \delta, q)$ is the expectation over ξ of $V(\mathscr{S}, \delta, q \mid \xi)$. Let $\mathscr{S}_k = \{1, \ldots, k\}$ denote the first k players, and suppose we are calculating player 1's ($\in \mathscr{S}$) value in $V(\mathscr{S}_k, \delta, q \mid \xi)$. Then,

$$V(\mathscr{S}_k, \delta, q \mid \xi) - V(\mathscr{S}_{k+1}, \delta, q \mid \xi)$$
$$= \sum_{t=0}^{\infty} (1-\delta) q^t \delta^t 3 \chi_{\{\tau : i(1,\tau,\xi) \in T_\tau(\mathscr{S}_k,\xi) \setminus T_\tau(\mathscr{S}_{k+1},\xi)\}}(t), \quad (5.1.4)$$

where χ_C is the indicator function, $\chi_C(t) = 1$ if $t \in C$ and 0 otherwise. Equation (5.1.4) gives the loss in continuation value to player 1, when entering a continuation game in state w_S, from having $k+1$ players rather than k players in state w_S.

It is immediate from (5.1.4) that $\frac{\delta}{(1-\delta)} q(\delta)[V(\mathscr{S}_1, \delta, q) - V(\mathscr{S}_2, \delta, q)]$ depends only on the product $q\delta$. For every $\delta \in [\underline{\delta}, 1)$, we can then define

$$q(\delta) = \underline{\delta}/\delta.$$

This ensures that (5.1.3) holds for every $\delta \in (\underline{\delta}, 1)$ and $q(\delta)$, and hence that (5.1.1) holds for every such pair. This completes our first step.

Step 2. We finish the argument by showing that (5.1.2) holds for all values $k \geq 2$, for all $\delta \in (\underline{\delta}, 1)$ and $q = q(\delta)$. For this, it suffices to show that for all $k \geq 2$ and $s = 1, \ldots, M - k - 1$,

$$V(\mathscr{S}_k, \delta, q \mid \xi) - V(\mathscr{S}_{k+1}, \delta, q \mid \xi) > V(\mathscr{S}_{k+s}, \delta, q \mid \xi) - V(\mathscr{S}_{k+s+1}, \delta, q \mid \xi),$$

for any realization ξ, that is, slowing the contagion to shirking is more valuable when fewer people are currently shirking. This statement follows immediately from (5.1.4) and the observation that, with strict containment for some t,

$$T_t(\mathscr{S}_{k+s}, \xi) \setminus T_t(\mathscr{S}_{k+s+1}, \xi) \subset T_t(\mathscr{S}_k, \xi) \setminus T_t(\mathscr{S}_{k+1}, \xi).$$

∎

Kandori (1992a) works without a public correlating device, thus examining the special case of this strategy profile in which $q = 1$ and punishments are therefore permanent. As Ellison (1994) notes, the purpose of the reset button is to make punishments less severe, and hence make it less tempting for a player who has observed an incidence of shirking to continue exerting effort in an attempt to retard the contagion to shirking. Without public correlation, a condition on payoffs is needed to ensure that such an attempt to slow the contagion is not too tempting (Kandori 1992a, theorem 1). This condition requires that the payoff lost by exerting effort (instead of shirking) against a shirking partner be sufficiently large and requires this loss to grow without bound as the population size grows. If this payoff is fixed, then effort will become impossible to sustain, no matter how patient the players, when the population is sufficiently large. In contrast, public correlation allows us to establish a lower bound $\underline{\delta}$ for the discount factor that increases as does the size of the population, but with the property that for any population size there exist discount factors sufficiently large (but short of unity) that support equilibrium effort.

Remark 5.1.1 Although we have assumed that players cannot remember the identities of earlier partners, the strategy profile $\hat{\sigma}$ is still a sequential equilibrium when players are not so forgetful, with essentially the same proof. Harrington (1995) considers this extension as well as more general matching technologies. Okuno-Fujiwara and Postlewaite (1995) examine a related model in which each player is characterized by a history-dependent status that allows direct but partial transfer of information across encounters. Ahn and Suominen (2001) examine a model with indirect information transmission.

◆

5.2 Relationships in Context

In many situations, players are neither tied permanently to one another (as in a repeated game) nor destined to part at first opportunity (as in the matching games of section 5.1). In addition, players often have some control over how long their relationship lasts and have views about the desirability of continuing that depend on the context in which the relationship is embedded. One may learn about one's partner over the course of a relationship, making it either more or less attractive to continue the relationship. One might be quite willing to leave one's job or partner if it is easy to find another, but not if alternatives are scarce. One may also pause to reflect that alternative partners may be available because they have come from relationships whose continuation was not profitable.

This section presents some simple models that allow us to explore some of these issues. These models share the feature that in each period, players have the opportunity to terminate their relationship. Although many of the models in this section are not games (because there is no initial node), the notions of strategy and equilibrium apply in the obvious manner.

The ability to end a relationship raises issues related to the study of renegotiation in section 4.6. Renegotiating to a new and better equilibrium may not be the only recourse available to players who find themselves facing an undesirable continuation equilibrium. What if, instead, the players have the option of quitting the game? This possibility becomes more interesting if there is a relationship between the value of quitting a game and play in the game itself. Suppose, for example, that quitting a game allows one to begin again with a new partner. We must then jointly examine the equilibrium of the game and the implications of this equilibrium for the environment in which the game is played.

5.2.1 A Frictionless Market

As our point of departure, we consider a market in which the stage game is the prisoners' dilemma of figure 5.1.1 and players who terminate one relationship encounter no obstacles in forming a new one.

We can view this model as a version of the personal-history matching model of section 5.1 with three modifications. First, we assume that there is an infinite number of players, most commonly taken to correspond to a continuum such as the unit interval, with individual behavior unobserved (see remark 2.7.1). This precludes any possibility of using the types of contagion arguments explored in section 5.1 to support nontrivial equilibrium outcomes.

Second, at the end of each period the players in each match simultaneously decide whether to continue or terminate their match. If both continue, they proceed to the next period. If either decides to terminate, then both enter a pool of unmatched players where they are paired with a new partner to begin play in the next period. The prospect for continuing play restores some hope of creating intertemporal incentives.

Third, similar to section 4.2, at the end of each period (either before or after making their continuation decisions), each pair of matched players either "dies" (with probability $1 - \delta$) or survives until the next period (with probability δ).[4] Agents who die are replaced by new players who join the pool of unmatched players.[5] We assume that players do not discount, with the specter of death filling the role of discounting (section 4.2), though it costs only extra notation to add discounting as well.

At the beginning of each period, each player is either matched or unmatched. Unmatched players are matched into pairs, and each (new or continuing) pair plays the prisoners' dilemma. Agents in a continuing match observe the history of play in their own match, but players receive no information about new partners they meet in the matching pool. In particular, they cannot observe whether a partner is a new entrant to the unmatched pool or has come from a previous relationship.

The flow of new players into the pool of unmatched players ensures that this pool is never empty, and hence players who terminate a match can always anonymously rematch. Otherwise, the possibility arises that no player terminates a match in equilibrium, because there would be no possibility of finding a new match (because the matching pool is empty...).

In this market, players who leave a match can wipe away any record of their past behavior and instantly find a new partner. What effect does this have on equilibrium play?

---

4. It simplifies the calculations to assume that if one player in a pair dies, so does the other. At the cost of somewhat more complicated expressions, we could allow the possibility that one player in a pair leaves the market while the other remains, entering the matching pool, without affecting the conclusions.
5. From the point of view of an individual player, the probability of death is random, but we exploit the continuum of players to assume that there is no aggregate uncertainty.

We begin with two immediate observations. First, there exists an equilibrium in which every player shirks at every opportunity. Second, there is no equilibrium in which the equilibrium path of play features only effort. Suppose that there were such an equilibrium. A player could then increase his payoff by shirking in the current period. This shirking results in a lower continuation if the player remained matched with his current partner, but the player can instead terminate the current match, immediately find a new partner, and begin play anew with a clean history and with a continuation equilibrium that prescribes continued effort. Consequently, shirking is a profitable deviation.

The difficulty here is that punishments can be carried out only within a match, whereas there is no penalty attached to abandoning a match to seek a new one. The public-history matching model of section 5.1 avoids this difficulty by allowing a player's past to follow the player into the matching pool. The personal-history matching game provides players with less information about past play but exploits the finiteness of the set of players to allow enough information transmission to support effort (with sufficient patience). In the current model, no information survives the termination of a match, and there are no punishments that can sustain effort.

5.2.2 Future Benefits

Players may be less inclined to shirk and run if new matches are not as valuable as current ones. Moreover, these differences in value may arise endogenously as a product of equilibrium play. In the simplest version of this, consider equilibrium strategies in which matched players shirk during periods $0, \ldots, T-1$, begin to exert effort in period T, and exert thereafter. Any deviation from these strategies prompts both players to abandon the match and enter the matching pool. If a deviation from such strategies is ever to be optimal, it must be optimal in period T, the first period of effort. It then suffices that it is optimal to exert effort in period T (for a continuation payoff 2) instead of shirking, which requires:

$$2 \geq (1-\delta)3 + \delta^{T+1}2,$$

which we can solve for

$$2 \geq \frac{1-\delta}{1-\delta^{T+1}}3.$$

For $\delta > 1/2$, the inequality is satisfied for $T \geq 1$. Taking the limit as $T \to \infty$, we find that this inequality can be satisfied for finite T if the familiar inequality $\delta > 1/3$ holds. Hence, as long as the effective continuation probability is sufficiently large to support the strict optimality of effort in the ordinary repeated prisoners' dilemma, we can make the introductory shirking phase sufficiently long to support *some* effort when an exit option is added to the repeated game. Moreover, as the continuation probability δ approaches 1, we can set $T=1$ and still preserve incentives. Hence, in the limit as $\delta \to 1$, the equilibrium value of $\delta 2$ converges to 2. Very patient players can come arbitrarily close to the payoff they could obtain by always exerting effort.

An alternative interpretation is that the introductory phase may take place outside of the current relationship. Suppose that whenever a player returns to the matching pool, he must wait some number of periods T before beginning another match. Then the previous calculations apply immediately, indicating that effort can be supported within a match as long as T is sufficiently long relative to the incentives to shirk. This trade-off is the heart of the efficiency wage model of Shapiro and Stiglitz (1984). Efficiency wages, in the form of wages that offer a matched worker a continuation value higher than the expected continuation value of entering the pool of unemployed workers, create the incentives to exert effort rather than accept the risk of terminating the relationship that accompanies shirking. In Shapiro and Stiglitz (1984), the counterparts of the value of T as well as the payoffs in the prisoners' dilemma are determined as part of a market equilibrium.

Carmichael and MacLeod (1997) note that any technology for paying sunk costs can take the place of the initial T periods of shirking. Suppose players can burn a sum of money θ. Consider strategies prescribing that matched players initially exert effort if both have burned the requisite sum θ, with any deviation prompting termination to enter the matching pool. Once a pair of players have matched and burned their money, continued effort is optimal if

$$2 \geq (1-\delta)3 + \delta(2 - (1-\delta)\theta),$$

or

$$\theta \geq \frac{1}{\delta}.$$

Hence, the sum to be burned must exceed the payoff premium to be earned by shirking rather than exerting effort against a cooperator. For these strategies to be an equilibrium, it must in turn be the case that the amount of money to be burned is less than the benefits of the resulting effort, or

$$(1-\delta)\theta \leq 2.$$

These two constraints on θ can be simultaneously satisfied if $\delta \geq 1/3$. Notice that the efficiency loss of burning the money can be as small as $(1-\delta)/\delta$, which becomes arbitrarily small as the continuation probability approaches 1.

We seldom think of people literally burning money. Instead, this is a metaphor for an activity that is costly but creates no value. In some contexts, advertising is offered as an example. In the case of people being matched to form partnerships, it suffices for the players to exchange gifts that are costly to give but have no value (other than sentimental) to the recipient. Wedding rings are often mentioned as an example.

5.2.3 Adverse Selection

Ghosh and Ray (1996) offer a model in which returning to the matching pool is costly because it exposes players to adverse selection. We consider a simplified version of their argument.

We now assume that players discount as well as face a probability that the game does not continue. There are two types of players, "impatient" players characterized by a discount factor of 0 and "patient" players characterized by a discount factor $\delta > 0$. The pool of unmatched players will contain a mixture of patient and impatient players. Matching is random, with players able to neither affect the type of partner with whom they are matched nor observe this type. Once matched, their partners' plays may allow them to draw inferences about their partner that will affect the relative payoffs of continuing or abandoning the match. Denote the continuation probability by ρ. The effective discount factor is then $\delta\rho$, and we normalize accordingly payoffs by the factor $(1 - \delta\rho)$.

We begin by assuming that the proportion of impatient types in the matching pool is exogenously fixed at λ. Players who enter the matching pool are matched with a new partner in the next period.

As usual, one configuration of equilibrium strategies is that every player shirks at every opportunity, coupled with any configuration of decisions about continuing or abandoning matches. We are interested in an alternative equilibrium in which impatient players again invariably shirk but patient players initially exert effort, continuing to do so as long as they are matched with a partner who exerts effort. Patient players respond to any shirking by terminating the match. Patient players thus screen their partners, abandoning impatient ones to seek new partners while remaining and exerting effort with patient ones. The temptation to shirk against a patient partner is deterred by the fact that such shirking requires a return to the matching pool, where one may have to sort through a number of impatient players before finding another patient partner.

The incentive facing the impatient players in this equilibrium are trivial, and they must always shirk in any equilibrium. Suppose a patient player is in the middle of a match with a partner who is exerting effort and hence who is also patient. Equilibrium requires that continued effort be optimal, rather than shirking and then returning to the matching pool. Let V be the value of entering the pool of unmatched players. Then the optimality of effort requires:

$$2 \geq (1 - \delta\rho)3 + \delta\rho V. \tag{5.2.1}$$

Now consider a patient player at the beginning of a match with a partner of unknown type, being impatient with probability λ. If it is to be optimal for the patient player to exert effort, we must have

$$(1 - \lambda)2 + \lambda[(1 - \delta\rho)(-1) + \delta\rho V] \geq (1 - \lambda)[(1 - \delta\rho)3 + \delta\rho V] + \lambda\delta\rho V,$$

or

$$2 - \frac{\lambda}{1 - \lambda}(1 - \delta\rho) \geq (1 - \delta\rho)3 + \delta\rho V.$$

The latter constraint is clearly more stringent than that given by (5.2.1), for exerting effort in a continuing relationship. Hence, if it is optimal to exert in the first period, continued effort against a patient partner is optimal. We thus need only investigate initial effort.

Let V_E and V_S be the expected payoff to a player who exerts effort and shirks, respectively, in the initial period of a match. Then we have

$$V_E = 2(1 - \lambda) + \lambda[(1 - \delta\rho)(-1) + \delta\rho V] \tag{5.2.2}$$

and

$$V_S = (1 - \lambda)[(1 - \delta\rho)3 + \delta\rho V] + \lambda\delta\rho V. \tag{5.2.3}$$

If we are to have an equilibrium in which patient players exert effort in the initial period, then $V_E = V$, and we can solve (5.2.2) to find

$$V_E = \frac{2 - 3\lambda + \delta\rho\lambda}{1 - \delta\rho\lambda}. \qquad (5.2.4)$$

A necessary and sufficient condition for the optimality of exerting effort is obtained by inserting $V = V_E$ in (5.2.2)–(5.2.3) to write $V_E \geq V_S$ as

$$V_E \geq (1 - \lambda)[(1 - \delta\rho)3 + \delta\rho V_E] + \lambda\delta\rho V_E,$$

and then using (5.2.4) to solve for

$$3\delta\rho\lambda^2 - 4\delta\rho\lambda + 1 \leq 0.$$

If $\delta\rho < 3/4$, there are no real values of λ that satisfy this inequality. Effort cannot be sustained without sufficient patience. If $\delta\rho \geq 3/4$, then this condition is satisfied, and effort is optimal, for values of $\lambda \in [\underline{\lambda}, \bar{\lambda}]$, for some

$$0 < \underline{\lambda} < \bar{\lambda} < 1.$$

Hence, if effort is to be optimal, the proportion of impatient players λ must be neither too high nor too low. If λ is low, and hence almost every player is patient, then there is little cost to entering the matching pool in search of a new partner, making it optimal to shirk in the first round of a match and then abandon the match. If λ is very high, so that almost all players are impatient, then the probability that one's new partner is patient is so low as to not make it worthwhile risking initial effort, no matter how bleak the matching pool.

What fixes the value of λ, the proportion of impatient types in the matching pool? We illustrate one possibility here. First, we assume that departing players are replaced by new players of whom ϕ are impatient and $1 - \phi$ patient.[6] Let ℓ and h be the mass of players in the unmatched pool in each period who are impatient and patient, respectively. We then seek values of ℓ and h, and hence $\lambda = \ell/(\ell + h)$, for which we have a steady state, in the sense that the values ℓ and h remain unchanged from period to period, given the candidate equilibrium strategies.[7]

Let z_{HH}, z_{HL}, and z_{LL} be the mass of matches in each period that are between two patient players, one patient and one impatient player, or between two impatient players. Notice that these must sum to $1/2$ because there are half as many matches as players. In a steady state,

$$\ell = \phi(1 - \rho) + \rho(z_{HL} + 2z_{LL}) \qquad (5.2.5)$$

and

$$h = (1 - \phi)(1 - \rho) + \rho z_{HL}, \qquad (5.2.6)$$

6. Recall that there is a continuum of players, of mass 1, and we assume there is no aggregate uncertainty.
7. A more ambitious analysis would fix ℓ and h at arbitrary initial values and allow them to evolve. This is significantly more difficult, because the value of λ is then no longer constant, and hence our preceding analysis inapplicable.

reflecting the flow of players into the unmatched pool from replacing departing players (the first term in each case) or from terminating matches (the second term).

In each period, surviving matches between patient players as well as newly formed matches from players in the unmatched pool form matches in proportions,

$$z_{HH} = \rho z_{HH} + \frac{1}{2}(h+\ell)\frac{h^2}{(h+\ell)^2},$$

$$z_{HL} = \frac{1}{2}(h+\ell)\frac{2h\ell}{(h+\ell)^2},$$

and

$$z_{LL} = \frac{1}{2}(h+\ell)\frac{\ell^2}{(h+\ell)^2}.$$

Substituting into (5.2.5), we obtain

$$\ell = \phi(1-\rho) + \rho\ell$$
$$\implies \ell = \phi,$$

whereas substituting into (5.2.6), we obtain

$$h = (1-\phi)(1-\rho) + \rho\frac{h\ell}{h+\ell}$$
$$= (1-\phi)(1-\rho) + \rho\frac{h\phi}{h+\phi} \equiv f(h).$$

Because $f(0) > 0$, $f(1-\phi) < 1-\phi$, and $f'(h) \in (0,1)$, f has a unique fixed point, and this point is strictly smaller than $1-\phi$. It remains to verify that $\lambda \in [\underline{\lambda}, \bar{\lambda}]$, that is, that the steady state is indeed an equilibrium. As for the exogenous λ case, extreme values of ϕ preclude equilibrium. Fix ρ and δ so that $\delta\rho > 3/4$. As $\phi \to 0$, $\ell \to 0$ while $h \to 1 - \rho$ and so $\lambda \to 0$. On the other hand, as $\phi \to 1$, $\ell \to 1$ while $h \to 0$ and so $\lambda \to 1$. From continuity, there will be intermediate values of ϕ for which the implied steady-state values of h and ℓ are consistent with equilibrium.

In equilibrium, the proportion of patient players in the unmatched pool falls short of the proportion of new entrants who are patient, ensuring that the unmatched pool is biased toward impatient players ($\lambda > \phi$). This reflects adverse selection in the process by which players reach the unmatched pool. Patient players tend to lock themselves into partnerships that keep them out of the unmatched pool, whereas impatient players continually flow back into the pool.

5.2.4 Starting Small

In section 5.2.2, we saw that patient players have an incentive to not behave opportunistically in the presence of attractive outside options when the value of the relationship increases over time. In addition, the possibility that potential partners may be impatient can reduce the value of the outside option sufficiently to again provide patient players with an incentive to not behave opportunistically (section 5.2.3). Here we explore an adverse selection motivation, studied by Watson (1999, 2002), for "starting small."[8]

---

8. Similar ideas appear in Diamond (1989).

5.2 ■ Relationships in Context

We continue with the prisoners' dilemma of figure 5.1.1. As in section 5.2.3, we assume that players can come in two types, impatient and patient. The impatient type has a discount factor $\underline{\delta} < 1/3$, and the patient type has a discount factor $\bar{\delta} > 1/3$. Because our focus is on the players' response to this adverse selection within a relationship (rather than how their behavior responds to the adverse selection within the population), we assume each player is impatient with exogenous probability λ. Each player knows his own type. Similarly, at the end of each period, each player has the option of abandoning the relationship and, on doing so, receives an exogenous outside option. The impatient type's option is worth 0, and a patient type's is worth $\bar{V} \in (0, 2)$. As in section 5.1.2, the relevant equilibrium notion is sequential equilibrium.

Because $\underline{\delta} < 1/3$, even if an impatient player knows he is facing a patient player playing grim trigger, the impatient player's unique best reply is to shirk immediately. In other words, the immediate reward from shirking is too tempting for the impatient player. Consequently, the only equilibrium outcome of the repeated prisoner's dilemma of figure 5.1.1, when one player is known to be impatient, is perpetual shirking. Thus, if the probability of the impatient type is sufficiently large, the only equilibrium outcome is again perpetual shirking.

The new aspect considered in this section is that the scale of the relationship may change over time. Early periods involve confidence building, whereas in later periods the players hope to reap the rewards of the relationship. To capture this, we study the game in figure 5.2.1. This partnership game extends the prisoners' dilemma by adding a moderate effort choice that lowers the cost from having a partner shirk while lowering the benefit from effort.

We consider separating strategies of the following form. Patient players choose M for T periods, and then exert effort E as long as their partner also does so, abandoning the match after any shirking. Impatient players always shirk and terminate play after the first period of any game.

What is required for these strategies to be optimal? There are two strategies that could be optimal for an impatient player. He could shirk in the first period of a match, thereafter abandoning the match, or he could choose M until period T (if the partner does, abandoning the match otherwise), and then shirk in period T. Under the latter strategy, the player shirks at high stakes against a patient opponent, at the cost of

	E	M	S
E	2, 2	0, 1	−1, 3
M	1, 0	2z, 2z	−z, 3z
S	3, −1	3z, −z	0, 0

Figure 5.2.1 A partnership game extending the prisoners' dilemma, where $z \in [0, 1]$.

delaying the shirking and the risk of being exploited by an impatient partner. The incentive constraint is given by

$$(1-\lambda)(1-\delta)3z \geq (1-\lambda)(1-\delta)\left(\sum_{t=0}^{T-1}\underline{\delta}^t 2z + \underline{\delta}^T 3\right) + \lambda(1-\delta)(-z).$$

The left side is the payoff produced by immediate shirking. If $z = 1$, then the impatient player will always shirk immediately. There is then no gain from waiting for the chance to shirk at higher stakes. Alternatively, if $z = 0$, then the impatient player will surely wait until period T to shirk, because this is the only opportunity available to secure a positive payoff.

For the patient player, the relevant choices are either to shirk immediately and then abandon the match or follow the equilibrium strategy, for an incentive constraint of

$$(1-\lambda)[(1-\bar{\delta})3z + \bar{\delta}\bar{V}] + \lambda\bar{\delta}\bar{V}$$
$$\leq (1-\lambda)\left[(1-\bar{\delta})\sum_{t=0}^{T-1}\bar{\delta}^t 2z + \bar{\delta}^T 2\right] + \lambda[(1-\bar{\delta})(-z) + \bar{\delta}\bar{V}].$$

The left side is again the payoff from immediate shirking. For any T, if $\bar{\delta}$ is sufficiently large, this incentive constraint will hold for any $z \in (0, 1)$. Suppose $\bar{\delta}$ is sufficiently large that it holds for $T = 1$.

We now evaluate the impact of varying z and T on the payoff of the patient player. In doing so, we view z as a characteristic of the partnership (perhaps one that could be designed in an effort to nurture the relationship), whereas T, of course, is a parameter describing the strategy profile. We are thus asking what relationship the patient type prefers, noting that the answer depends on the attendant equilibrium.

The derivative of the patient player's payoff under the candidate equilibrium profile with respect to z is given by

$$(1-\bar{\delta})\left[2(1-\lambda)\frac{(1-\bar{\delta}^T)}{(1-\bar{\delta})} - \lambda\right].$$

If $\lambda < 2/3$, then this derivative is positive, for any T. Patient players then prefer z to be as large as possible, leading to an optimum of $z = 1$ or $T = 0$ (and reinforcing the incentives of impatient players in the process). In this case, impatient players are sufficiently rare that it is optimal to simply bear the consequences of meeting such players, rather than taking steps to minimize the losses they inflict that also reduce the gains earned against impatient players. As a result, patient players would prefer that the interaction start at full size.

If $\lambda > 2/3$, the derivative is negative for $T = 1$. In this case, an impatient player is sufficiently likely that the patient player prefers a smaller value of z and $T = 1$ to $T = 0$ to minimize the losses imposed by meeting an impatient type. The difficulty is that a small value of z, with $T = 1$, may violate the incentive constraint for the impatient player. Because the constraint holds strictly with $z = 1$, there will exist an interval $[z_1, 1)$ for which the constraint holds. The patient player will prefer the equilibrium

	E	S
E	2, 2	−1, 3
S	3, −1	0, 0

	E	S
E	3, 3	−1, 4
S	4, −1	0, 0

Figure 5.3.1 Prisoners' dilemma stage games for two repeated games between players 1 and 2.

$T = 1$ for any $z \in [z_1, 1)$ to the equilibrium with $T = 0$. We thus have a basis for starting small.[9]

5.3 Multimarket Interactions

Suppose that players 1 and 2 are engaged in more than one repeated game. The players may be firms who produce a variety of products and hence repeatedly compete in several markets. They may be agents who repeatedly bargain with one another on behalf of a variety of clients. How does this multitude of interactions affect the set of equilibrium payoffs? For example, if the players are firms, does the multimarket interaction enhance the prospects for collusion?

One possibility is to view the games in isolation, conditioning behavior in each game only on the history of play in that game. In this case, any payoff that can be sustained as an equilibrium in one of the constituent repeated games can also be sustained as the equilibrium outcome of that game in the combined interaction. Putting the games together can thus never decrease the set of equilibrium payoffs.

An alternative is to treat the constituent games as a single metagame, now allowing behavior in one game to be conditioned on actions in another game. It initially appears as if this must enlarge the set of possible payoffs. A deviation in one game can be punished in each of the other games, allowing the construction of more severe punishments that should seemingly increase the set of equilibrium payoffs. At the same time, however, the prospect arises of simultaneously deviating from equilibrium play in each of the constituent games, making deviations harder to deter.

It is immediate that if the constituent games are identical, then the set of equilibrium payoffs in the metagame, averaged across constituent games, is precisely that of any single constituent game. If the constituent games differ, then the opportunity for cross-subsidization arises.

We illustrate these points with an example. Consider the pair of prisoner's dilemmas in figure 5.3.1. It is a familiar calculation that an expected payoff of (2, 2) can be achieved in the left game, featuring an equilibrium path of persistent effort, whenever $\delta \geq 1/3$, and that otherwise a payoff of (0, 0) is the only equilibrium possibility

---

9. There may exist yet better combinations of z and $T > 1$ for the patient player.

(see section 2.5.2). A payoff of (3, 3), the result of persistent effort, is available in the right game if and only if $\delta \geq 1/4$.

Suppose now that $\delta = 3/10$. If the games are played separately, the largest symmetric equilibrium payoff, summed across the two games, is (3, 3), reflecting the fact that the left game features only mutual shirking as an equilibrium outcome, whereas effort can be supported in the right game. Now suppose that the two games are combined, with an equilibrium strategy of exerting effort in both games as long as no prior shirking in either game has occurred, and shirking in both games otherwise. To verify these are equilibrium strategies, we need only show that the most lucrative deviation, to simultaneously shirking in both games, is unprofitable. The equilibrium payoff from the two games is 5, and the deviation brings an immediate payoff of 7 followed by a future of zero payoffs, for an incentive constraint of

$$5 \geq (1 - \delta)7,$$

which we solve for $\delta \geq 2/7$ (<3/10).

The mechanism at work here is that the right game satisfies the incentive constraint for exerting effort with some slack when $\delta = 3/10$. This slack can be used to subsidize effort in the left game, allowing effort to be an equilibrium outcome in both.

The potential gains to be realized from linking heterogeneous games are only available when the discount factor is sufficiently large that at least one of the constituent games has a nontrivial subgame-perfect equilibrium. Multimarket interactions cannot manufacture new possibilities out of thin air. But if there is such a game, and the incentive constraints in that game hold with strict inequality, then new possibilities arise in games that in isolation allow only trivial equilibria. Bernheim and Whinston (1990) pursue this issue in the context of oligopoly markets.

5.4 Repeated Extensive Forms

This section discusses repeated games in which the players play an extensive-form stage game in each period. For our purposes, we can restrict attention to extensive-form games without nature and continue to use subgame perfection as the equilibrium notion. A stage-game action then specifies, for the player in question, a choice at each of the player's information sets in the extensive-form stage game. At the end of the period, the players observe the terminal node y reached as a result of that period's play. In any particular period, some of a player's information sets may be reached in the course of play, but others may not be. As a result, observing a terminal node y will not in general reveal the players' actions. That is, and as our examples make clear, if the game has a nontrivial dynamic structure, several action profiles can lead to the same terminal node. It is then natural to think of a repeated extensive-form game as a repeated game of imperfect public monitoring with the terminal node being the signal. Section 9.6 characterizes the set of subgame-perfect equilibrium payoffs for repeated extensive-form games and proves a folk theorem using tools developed for the general analysis of public-monitoring repeated games in chapters 7–9.[10] This section presents

---

10. The approach in sections 3.4.2 and 3.7 can also be used to immediately prove a folk theorem for repeated extensive-form games. See the end of section 5.4.1.

some examples illustrating the new issues raised by repeated extensive-form (rather than normal-form) games.[11]

Given an action profile a of the extensive form, because there are no moves of nature, a unique terminal node $y(a)$ is reached under the path of play implied by a. A sequence of action profiles $(a^0, a^1, \ldots, a^{t-1})$ induces a public history of terminal nodes $h^t = (y(a^0), y(a^1), \ldots, y(a^{t-1}))$, and a strategy for player i specifies in period t, as a function of the history of terminal nodes, h^t, the action to be taken in period t (where this action is a contingent plan of behavior within the extensive form).

A repeated extensive-form game often has more subgames than does the repeated game constructed from the corresponding normal form. These additional subgames consist of, for some node ξ of the extensive form, the subgame of the extensive form with that initial node, followed by the infinite repetition of the entire extensive form. A profile is *subgame perfect* if, for every subgame (including those whose initial node is not at the beginning of the extensive form), the continuation strategy profile is a Nash equilibrium of the subgame (see section 9.6 for a formal treatment). We are interested in comparing the set of subgame-perfect equilibrium payoffs in a repeated extensive-form game with the set of equilibrium payoffs in the repeated game constructed from the corresponding normal form. We continue to use $\mathscr{E}(\delta)$ for the set of subgame-perfect equilibrium payoff profiles in the repeated normal-form game, given discount factor δ, and let $\mathscr{E}^E(\delta)$ be the corresponding set for the repeated extensive form.

5.4.1 Repeated Extensive-Form Games Have More Subgames

The additional subgames in repeated extensive forms can have a significant impact. Consider the chain store game shown in figure 5.4.1.[12] We focus on equilibria that support an outcome in which player 2 chooses *Out* in every period.

Consider first the repeated normal-form game. Here, the action profile (Out, F) is a Nash equilibrium of the normal form. As a result, strategies that specify (Out, F) after every history are a subgame-perfect equilibrium of the infinitely repeated game, for every discount factor $\delta \in [0, 1]$.

Now consider the repeated extensive-form version of this game. Regardless of the discount factor, strategies specifying (Out, F) after every history do not constitute

---

11. An alternative approach to extensive form games, which we will not develop, restricts attention to extensive-form games that are *playable*, in the sense that there is a mapping from information sets to a set of numbers $\{1, 2, \ldots, T\}$ with the property that no information set mapped into t precedes an information set mapped into some $\tau < t$ (Mailath, Samuelson, and Swinkels 1994). For a playable extensive-form game, we can think of the corresponding repeated game as a dynamic game (sections 5.5–5.7), with each play of the extensive-form game broken into T periods, each of which is now considered a period in the repeated game. In many of these periods, only some players will have choices, and in many the payoff function will be trivial. There may also be imperfect monitoring, as some choices may not be observed by other players. We can then apply results for dynamic games to the study of repeated extensive forms.

12. The chain store game is commonly interpreted as a game between a long-lived incumbent firm and a sequence of short-lived entrants. In the stage game, the entrant first decides between entering the market (*In*) and staying out (*Out*); if the entrant does enter, the incumbent then chooses between acquiescing to the entry (*A*) or fighting (*F*) by, for example, pricing below cost. We examine the finitely repeated version (where it became famous) in chapter 17. Here, we consider an infinitely repeated game between two long-lived players but refer to the first mover as player 2, because this player is the entrant and is accordingly commonly referred to player 2 when modeled as a short-lived player.

164 Chapter 5 ■ Variations on the Game

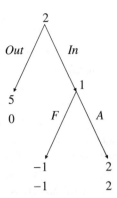

Figure 5.4.1 Chain-store game. Player 1's payoff is listed first at each terminal node of the extensive-form game.

a subgame-perfect equilibrium. Fix a strategy profile specifying such behavior, and suppose that a history h^t occurs after which player 2 has chosen *In*, with player 1 now to choose between F and A. Given that all subsequent histories prescribe the actions (*Out, F*), subgame perfection calls for player 1 to play A, no matter what the discount factor.

The force at work here is that playing the stage game in extensive form expands the set of histories after which choices must be made, and hence expands the list of conditions required for sequential rationality. The incentives for player 1 to choose F simultaneously with player 2's choice of *Out*, as in the normal-form game, differ from the incentives for player 1 to choose F after 2 has chosen *In*. The latter situation can arise only in the extensive-form game and suffices to ensure that invariably playing (*Out, F*) is not a subgame-perfect equilibrium.

This does not preclude the possibility of an equilibrium in the extensive-form game in which player 2 always chooses *Out*, though we must look for other strategies to support such an outcome. To minmax player 2, player 1 chooses F, with 2's minmax value being 0. Player 2 receives precisely this minmax value in any equilibrium in which she plays *Out* in every period. Hence, if player 2 is to choose *Out* in every period, it must be that *In* causes player 1 to choose F (with sufficiently high probability) in the current period, because this (rather than relying on continuation payoffs) is the only effective way to punish player 2. However, because F is not a best response in the stage game, we must use appropriately designed continuation payoffs to create the incentives for 1 to choose F.

Player 1 is minmaxed by 2's choice of *In*, giving player 1 a minmax value of 2. Our best hope for sustaining *Out* along the equilibrium path is then given by the strategy profile in which 2 chooses *Out* after every history in which every period has featured either *Out* or (*In, F*), and in which player 1 chooses F after any history in which all previous periods have featured either *Out* or (*In, F*) and player 2 has chosen *In* in the current period. Otherwise, (*In, A*) is played. Hence, player 2 stays out, deterred from entry by the threat of F. Player 1 chooses F whenever entry occurs, deterred by the threat of a switch to the subsequent play of the stage-game equilibrium (*In, A*), which minmaxes player 1.

5.4 ■ Repeated Extensive Forms

Because the equilibrium path involves a best response for player 2 and the punishment involves repeated play of a subgame-perfect equilibrium of the stage game, the only incentive problem involves the prescription that player 1 choose F, should 2 enter and the punishment not have been triggered. The optimality condition in this case is

$$(1-\delta)(-1) + \delta 5 \geq 2$$

or $\delta \geq 1/2$. We conclude that there exist equilibria of the repeated extensive-form game in which player 2 always chooses Out, but that these exist for a subset of the discount factors under which this outcome can be supported in the repeated normal-form game. For $\delta < 1/2$, $\mathscr{E}^E(\delta)$ is a strict subset of $\mathscr{E}(\delta)$.

We can construct a folk theorem result for the repeated extensive-form chain store game. The equilibrium outcome Out in every period minmaxes player 2, whereas the stage-game Nash equilibrium (A, In) minmaxes player 1. For any feasible and strictly individually rational payoff profile, we thus need only specify stage-game behavior that yields this payoff with deviations by player i punished by reversion to the equilibrium that minmaxes player i.

In general, we will not have the convenience of being able to minmax players by playing stage-game Nash equilibria. However, this construction illustrates why the extra subgames created by an extensive-form stage game are generally no obstacle to obtaining a folk theorem. The difficulty raised by the extensive form is that we must provide incentives for players to take appropriate actions at the subgames that arise within a period's play of the extensive-form game. If δ is small, all incentives must effectively be created within a period, which may be impossible. For large δ, however, we can freely rely on continuation play to create incentives, even for subgames that do not appear in the repeated normal form. The extra subgames created by the extensive form can thus fade into insignificance as δ approaches 1.[13]

5.4.2 Player-Specific Punishments in Repeated Extensive-Form Games

We concluded the previous subsection with a sketch of why a folk theorem holds for repeated extensive-form games. We now describe an example from Rubinstein and Wolinksy (1995), that illustrates the scope of that result. In this example, $\mathscr{E}^E$ is a strict subset of $\mathscr{E}$, even in the limit as δ approaches 1.

Consider the normal-form game shown in the left panel of figure 5.4.2. There are two pure Nash equilibria, (T, R) and (B, L). The former of these minmaxes each of the players. The set $\mathscr{F}^*$ of feasible, strictly individually rational payoffs is the convex combination of the stage-game Nash equilibrium payoffs $(1, 1)$ and $(2, 2)$ with positive weight on $(2, 2)$. By appropriately alternating between these two stage-game equilibria, we can construct repeated game equilibria for any sufficiently large discount factor that attain as a payoff any such convex combination of $(1, 1)$ and $(2, 2)$.

---

13. This discussion concerns strictly individually rational payoffs, and so does not contradict our observation that in supporting (Out, F) as an equilibrium outcome of the repeated extensive-form chain store game, player 2 must be deterred from choosing In by attaching a sufficiently high probability to F in that period (rather than by adjusting payoffs). In the latter case, at the beginning of every period, player 2 faces a continuation equilibrium payoff that is precisely her minmax value, ensuring that we cannot punish her by reducing continuation payoffs. The folk theorem, restricted to strict individual rationality, does not apply to such outcomes.

	L	R
T	1,1	1,1
B	2,2	0,0

	L	R
T	1,1	1,1
B	2,3	0,0

Figure 5.4.2 Normal-form game that does not allow player-specific punishments (left game) and one that does (right game).

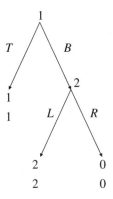

Figure 5.4.3 Extensive-form stage game accompanying the left normal form from figure 5.4.2.

Now consider the extensive form in figure 5.4.3, which has this normal form. We find a lower bound on equilibrium payoffs. Suppose player 1 chooses B at the first opportunity. In the repeated normal-form game, such an action could be quite unhelpful. For example, if the equilibrium calls for a first-period profile of (T, R) and player 2 has accordingly chosen R, playing B sacrifices current payoffs, possibly without bringing any future gains. In the extensive form, in contrast, player 2 has not made a choice and is faced with player 1's fait accompli of B. What are the continuation payoffs?

Let v^* be the smallest subgame-perfect equilibrium payoff. Given the agreement between the two players' stage-game payoffs, this minimum payoff must be the same for both players and hence needs no subscript. Following player 1's choice of B, a choice of L by player 2 ensures both players a payoff of at least $(1 - \delta)2 + \delta v^*$. The equilibrium continuation payoff must then be at least this high, and hence so must the equilibrium payoff. This gives

$$v^* \geq (1 - \delta)2 + \delta v^*,$$

which implies

$$v^* = 2.$$

No matter what the discount factor, $\mathscr{E}^E(\delta)$ is thus a strict subset of $\mathscr{E}(\delta)$.

The difficulty here is traced to the perfectly correlated payoffs or, equivalently, the lack of player-specific punishments. In a two-player repeated normal-form game, the

	L	R
T	4, 2	−4, 0
M	−2, 0	2, 2
B	−1, 10	−1, 10

Figure 5.4.4 Normal-form game.

feasibility of mutual minmaxing renders player-specific punishments unnecessary. For a repeated extensive-form game, we can no longer be assured of mutual minmaxing, even with two players. In this case, mutual minmaxing requires the play of *TR*. What if player 1 plays *B* instead? In the extensive form, this presents player 2 with a choice. If player 2 is to choose *R*, ensuring the effectiveness of the mutual minmaxing, then *L* must bring player 2 a lower continuation payoff. If payoffs are perfectly aligned and we are already at the minimum equilibrium payoff, no such punishment is available, giving rise to the unraveling argument leading to the conclusion that (2, 2) is the smallest subgame-perfect equilibrium payoff.

If payoffs are not perfectly aligned, then we can shift continuation payoffs along the lower frontier of subgame-perfect equilibrium payoffs, being able to always punish either player 1 for choosing *B* or player 2 for choosing *L*. This suffices to generate a folk theorem result as the players become patient. To illustrate, consider the right game in figure 5.4.2, which corresponds to changing player 2's payoff at the node reached by *BL* from 2 to 3. The set of feasible payoffs now has a nonempty interior. It is now straightforward that we can construct subgame-perfect equilibria with payoffs smaller than (2, 3) and that in the limit as $\delta \to 1$ the set of equilibrium payoffs approaches co$\{(1, 1), (1, 3/2), (2, 3)\}$, with a basic construction mimicking the proof of proposition 3.4.1.

5.4.3 Extensive-Form Games and Imperfect Monitoring

The extensive form of a game can conceal information from the players. We illustrate with an example adapted from Sorin (1995).[14] Consider the normal-form game shown in figure 5.4.4. The minmax utilities for the two players are given by (0, 1), which are produced by the stage-game Nash equilibrium in which player 1 mixes equally over *T* and *M* and player 2 mixes equally over *L* and *R*.

Suppose now that the game is played twice (without discounting). In the normal-form version of the game, the players can choose strategies in which player 1 plays *B* in the first period while 2 mixes equally between *L* and *R*. If *L* is chosen in the

---

14. Since the extensive-form game is nongeneric, the generality of the phenomenon remains to be explored.

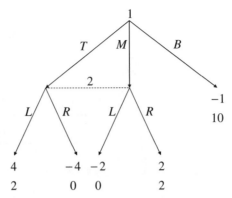

Figure 5.4.5 Extensive-form counterpart of game from figure 5.4.4.

first period, second-period play is (T, L), otherwise it is (M, R). Expected payoffs are $(2, 12)$. If player 1 deviates in the first period, the mixed stage-game equilibrium is played.

Consider the extensive-form version of this game, shown in figure 5.4.5 and again played twice. Suppose player 1 chooses B. Then player 2 has no opportunity to move, and hence player 1 collects no information about player 2's action. As a result, the payoff $(2, 12)$ cannot be obtained in the twice-played extensive form. The extensive form conceals information that the players can use to coordinate behavior in the normal form.

This difference would disappear, of course, if the players had access to a public correlating device. In its absence, the players are using 2's action in the first period of the normal-form game to coordinate their second-period actions.

5.4.4 Extensive-Form Games and Weak Individual Rationality

As we have noted, if player-specific punishments are possible, we obtain a folk theorem result for repeated extensive-form games. This nongeneric example, from Rubinstein and Wolinksy (1995), shows that there may still exist payoff profiles that can be obtained as subgame-perfect equilibria in the repeated normal-form game for sufficiently patient players, but cannot be obtained as equilibrium outcomes in the repeated extensive-form game, no matter how patient the players. If this is to be the case, the payoff profiles must lie in $\overline{\mathscr{F}^*} \setminus \mathscr{F}^*$, which is to say that they must be weakly but not strictly individually rational.

Consider the game in figure 5.4.6. The minmax utility level for each player is 0. The three players can be simultaneously minmaxed by player 1 choosing L and 2 choosing R. The set of feasible payoffs has a nonempty interior, allowing player-specific punishments. However, consider the payoff $(1, 0, 0)$. Players 2 and 3 receive their minmax utilities in this profile, removing it from the purview of the folk theorem. Nonetheless, this payoff profile can be obtained as the outcome of a subgame-perfect equilibrium in the repeated normal-form game, for any discount factor, because (R, L, L) is a Nash equilibrium of the stage game.

There is no subgame-perfect equilibrium of the repeated extensive-form game, for any discount factor, with payoff $(1, 0, 0)$. Once again, the key innovation is the

5.4 ■ Repeated Extensive Forms

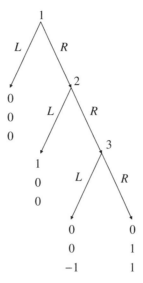

Figure 5.4.6 The weakly individually rational payoff $(1, 0, 0)$ can be supported in the repeated normal form but not in the repeated extensive form version of this game.

appearance of a new history, namely, one in which player 2 has chosen R and player 3 is called on to play. Suppose that we have a candidate equilibrium with payoffs $(1, 0, 0)$ and a history in which 2 chooses R. Then it must either be the case that (*i*) player 3 responds by choosing R or (*ii*) player 3's continuation payoff following L exceeds 0. In either case, the continuation payoff to player 3 is positive. However, player 3's payoffs along any continuation path can be positive only if player 2's payoffs are also positive because 2's stage-game payoffs are at least as large as player 3's. Player 2 can thus ensure a positive payoff by choosing R, in turn ensuring that there is no subgame-perfect equilibrium featuring actions $(R, L, \cdot)$ in each period. However, this is the only outcome yielding the payoff $(1, 0, 0)$, yielding the result.

5.4.5 Asynchronous Moves

Lagunoff and Matsui (1997) examine *asynchronous* repeated games. We follow their lead here in concentrating on two-player games with the simplest asynchronous timing structure, though they note that their results hold for a more general class of specifications. In period 0, players 1 and 2 simultaneously choose actions. Thereafter, player 1 chooses an action in every odd-numbered period and player 2 an action in every even-numbered period. If player i chooses action a_i in period t (other than player 1 in period 0), then i is constrained to also play action a_i in period $t + 1$, having the option of making a new choice in period $t + 2$.[15]

Consider the coordination game shown in figure 5.4.7. Suppose that an odd period t has been reached, so that player 1 must choose an action (to be played in periods t and $t + 1$), and that player 2's period t action (chosen in period $t - 1$) is L. What should

---

15. Maskin and Tirole (1988a,b) study oligopoly games with an asynchronous timing structure.

	L	R
T	2,2	0,0
B	0,0	1,1

Figure 5.4.7 Coordination game.

player 1 choose? From the standpoint of the current period, T is most attractive. In principle, the possibility remains that T may prompt sufficiently unfavorable continuation play to make B a superior choice. However, one can imagine player 1 reasoning, "For the same reasons that I now find T attractive, player 2 will find L attractive next period if I choose T now, and player 2 will do so secure in the knowledge that I will find T attractive in the next period...." Taking this logic one step further, suppose that player 1 has a chance to move, with R being player 2's previous-period move. Then B is player 1's myopic optimum. However, if player 1 is sufficiently patient, then player 1 might choose T instead, in anticipation that this will prompt player 2 to choose L, initiating a string of subsequent periods in which TL is played.

These considerations suggest that if one of the players ever chooses T or L, we should thereafter see TL, and that this process should allow patient players to secure an outcome in which TL is virtually always played. Notice that the situation is very much like the one arising in figure 5.4.3. The set of stage-game payoffs is $\mathscr{F} = \{(0,0), (1,1), (2,2)\}$ (and hence does not allow player-specific punishments). If the players move simultaneously, either because the game of figure 5.4.3 is played in normal form each period or the game of figure 5.4.7 is played as a synchronous repeated game, then there exists a subgame-perfect equilibrium of the repeated game with payoff $(1, 1)$. If player 1 can move first, either in the extensive form of figure 5.4.3 or the asynchronous version of figure 5.4.7, player 1 can present player 2 with 1's having chosen an action consistent with payoffs $(2, 2)$.

We consider asynchronous games with two-player normal-form pure coordination stage games (i.e., $u_1(a) = u_2(a)$ for all $a \in A$). Let a^* uniquely solve $\max_{a \in A} u_i(a) \equiv v^*$.

In the case of asynchronous games, we have:

Proposition 5.4.1 *Consider an asynchronous move game with two-player normal-form pure-coordination stage game. Fix a history h^t. Then for any $\varepsilon > 0$, there exists $\underline{\delta} < 1$ such that for all $\delta \in (\underline{\delta}, 1)$, every subgame-perfect equilibrium of the continuation game induced by history h^t gives an equilibrium payoff in excess of $v^* - \varepsilon$. If h^t is the null history, this statement holds for $\varepsilon = 0$.*

Proof We proceed in three steps.

First, let h^t be a history ending in a choice of a_i^* for player i (and hence $t \geq 1$), with player j now called on to make a choice in period t. We show that continuation payoffs are at least v^* for each player. To establish this result, let $\tilde{v}$ be the infimum over the set of subgame-perfect equilibrium payoffs, in a continuation game beginning in a period $t \geq 1$, in which player j must choose a

5.4 ■ Repeated Extensive Forms

current action and player i's period $t-1$ choice is a_i^*. Because the game is one of pure coordination, this value must be the same for both players. One option open to player j is to choose a_j^* in the current period, earning a payoff of v^* in the current period, followed by a continuation payoff no lower than $\tilde{v}$. We thus have

$$\tilde{v} \geq (1-\delta)v^* + \delta\tilde{v},$$

allowing us to solve for

$$\tilde{v} \geq v^*.$$

Hence, if the action a_i^* or a_j^* is ever played, then the continuation payoffs to both players must be v^*.

Second, we fix an arbitrary history h^t, including possibly the null history, with player j called on to move (or, in the case of the null history, with j being player 2, who does not move in the next period) and with no restriction on player i's previous move (or on player 1's concurrent move, in the case of the null history). Then one possibility is for player j to choose a_j^*. We have shown that this brings a continuation payoff, beginning in period $t+1$, of v^*. Hence, the payoff to this action must be at least

$$(1-\delta)m + \delta v^*,$$

where m is the smallest stage-game payoff. Choosing

$$\underline{\delta} = \frac{v^* - m - \varepsilon}{v^* - m}$$

then ensures that every subgame-perfect contniuation equilibrium gives a payoff at least $v^* - \varepsilon$.

Third, we consider the null history. We show that player 2 prefers to choose a_2^* in the first period, regardless of 1's first-period action, given subgame-perfect continuation play. The initial choice of a_2^* ensures that payoff v^* will be received in every period except the first, for a payoff of at least

$$(1-\delta)m + \delta v^*.$$

Choosing any other action can give a payoff at most

$$(1-\delta)[v' + \delta v'] + \delta^2 v^*,$$

where v' is the second-largest stage-game payoff. Choosing

$$\underline{\delta} = \frac{v' - m}{v^* - v'}$$

then suffices for the optimality of player 2's initial choice of a_2^*. Given that player 1 makes a new choice in period 1, it is immediate that 1 finds a_1^* optimal in the first period. This suffices for the result.

■

This finding contrasts with chapter 3's folk theorem for repeated normal-form games. The key to the result is again the appearance of histories in which player i can choose an action, knowing that player j will not be able to make another choice until after observing i's current choice. This allows player i to lead player j to coordination on the efficient outcome.

The asynchronous games considered here are a special case of the dynamic games examined in section 5.5. There, we developed a folk theorem for dynamic games. Why do we find a different result here? The folk theorem for dynamic games makes use of player-specific punishments that are not available in the asynchronous games considered here. Given that there are only two players, why not rely on mutual minmaxing, rather than player-specific punishments, to create incentives? In the asynchronous game, player i's current choice of a_i^* ensures future play of a^*. This in turn implies that player i will invariably choose a_i^* rather than minmax player j. Our only hope for inducing i to do otherwise would be to arrange future play to punish i for not minmaxing j, and the coordination game does not allow sufficient flexibility to do so. Conversely, the efficiency result obtained here is tied to the special structure of the coordination game, with the folk theorem for dynamic games appearing once one has the ability to impose player-specific punishments.

5.4.6 Simple Strategies

We are often interested in characterizing equilibrium behavior as well as payoffs. Mailath, Nocke, and White (2004) note that simple strategies may no longer suffice in extensive-form games. Instead, punishments may have to be tailored not only to the identity of a deviator but to the nature of the deviation as well.

We must first revisit what it means for a strategy to be simple. Definition 2.6.1, for normal-form games, defined simple strategies in terms of outcome paths, consisting of an equilibrium outcome path and a punishment path for each player. The essence of the definition is that any unilateral deviation by player i, whether from the equilibrium path or one of the punishments, triggered the *same* player i punishment. Punishments thus depend on the identity of the transgressor, but neither the circumstances nor the nature of the transgression.

As we saw in section 5.4.1, repeated extensive-form games have more information sets than repeated normal-form games, and so there are two candidates for the notion of simplicity. First observe that we cannot simply apply definition 2.6.1 because action profiles are not observed. We can still refer to an infinite sequence of actions **a** as a (potential) punishment path for i. Let $Y^d(a)$ be the set of terminal nodes reached by (potentially a sequence of) unilateral deviations from the action profile a, and let $i(y)$ be the last player who deviated on the path (within the extensive form) to $y \in Y^d(a)$ (there is a unique last player by the definition of $Y^d(a)$). Then, we say that a strategy profile is *uniformly simple* if i's punishment path after nodes $y \in Y^d(a)$ is constant on $Y_i^d(a) \equiv \{y' : i(y') = i, y' \in Y^d(a)\}$. This notion requires that i's intertemporal punishments not be tailored to either when the deviation occurred or the nature of the deviation.

A weaker notion requires that the intertemporal punishments not be tailored to the nature of the deviation but can depend on when the deviation occurred. The set $Y_i^d(a)$ can be partitioned by the information sets h at which i's deviation occurs; let $Y_h^d(a)$ denote a typical member of the partition. We say that a strategy profile is *agent simple* if i's punishment path after nodes $y \in Y^d(a)$ is constant on $Y_h^d(a)$ for all h.[16] In other

---

16. The term is chosen by analogy with the agent-normal form, where each information set of an extensive form is associated with a distinct agent.

5.4 ■ Repeated Extensive Forms

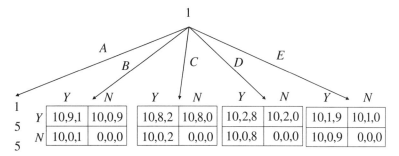

Figure 5.4.8 Representation of an extensive-form game. Each normal-form game represents a simultaneous move subgame played by players 2 and 3.

words, the punishment path is independent of the nature of the deviation at h but can depend on h.

We now describe an extensive form, illustrated in figure 5.4.8, and a subgame-perfect equilibrium outcome of its repetition that cannot be supported using agent-simple strategy profiles. Player 1 moves first. If player 1 chooses either $B, C, D,$ or E, then players 2 and 3 play a simultaneous-move subgame, each choosing Y or N. We are interested in equilibria in which player 1 chooses A in the first period. One such equilibrium calls for players 2 and 3 to both choose N after choices of $B, C, D,$ or E. These choices make player 1's first-period choice of A a best response in the stage game, but are not a stage-game subgame-perfect equilibrium. To support A in period 0 as a choice for player 1, we proceed as follows. In period 0, player 1 chooses A and players 2 and 3 respond to any other choice with NN. From period 1 on, a single stage-game subgame-perfect equilibrium is played in each period, whose identity depends on first-period play. Let "equilibrium x" for $x \in \{B, C, D, E\}$ call for player 1 to choose x and players 2 and 3 to choose Y in each period. The continuation play as a function of first-period actions is given in figure 5.4.9. Because A is a best response for player 1 in the first period, and because every first-period history leads to a continuation path consisting of stage-game subgame-perfect equilibria, verifying subgame perfection in the repeated game requires only ensuring that players 2 and 3 behave optimally out of equilibrium in the first period. This is ensured by punishing player 2 and 3 for deviations, by selecting the relatively unattractive continuation equilibria B and E, respectively. If player 2 and 3 behave as prescribed in period 1, after a choice of B, C, D or E, then the relatively low-payoff player is rewarded by the selection from continuation equilibria C or D that is relatively lucrative for that player.

If $\delta = 2/3$, these strategies are a subgame-perfect equilibrium. However, the strategy profile is not agent-simple because 1's punishment path depends on the nature of 1's deviation. Deviations by player 1 to either B or C are followed, given equilibrium play by players 2 and 3, by continuation equilibrium C. Deviations to D or E are followed by continuation equilibrium D. This absence of simplicity is necessary to obtain the payoffs provided by this equilibrium. Consider a deviation by player 1 to B, and consider player 2's incentives. Choosing Y produces a payoff of 9 in the current period, followed by an equilibrium giving player 2 her minmax value of 1, and hence imposing the most severe possible punishment on player 2. If 2's choice of N, for a current payoff of 0, is to be optimal, the resulting continuation payoff v_2 must satisfy

$$\delta v_2 \geq (1-\delta)9 + \delta,$$

First-period profile	Continuation	First-period profile	Continuation
A	B		
BNN, BYY	C	DNN, DYY	D
BYN	E	DYN	E
BNY	B	DNY	B
CNN, CYY	C	ENN, EYY	D
CYN	E	EYN	E
CNY	B	ENY	B

Figure 5.4.9 The continuation play of the equilibrium for the repeated game.

or
$$v_2 \geq \tfrac{11}{2}.$$

Player 1's deviation to B must then be followed by a continuation path giving player 2 a payoff of at least $11/2$. A symmetric argument shows that the same must be true for player 3 after 1's deviation to E. However, there is no stage-game action profile that gives players 2 and 3 both at least $11/2$. Hence, if player 1 is ever to play A, then deviations to B and E must lead to different continuation paths—the equilibrium strategy cannot be agent simple. Finally, player 1's choice of A gives payoffs of $(1, 5, 5)$, a feat that is impossible if 1 confines himself to choices in $\{B, C, D, E\}$. The payoffs provided by this equilibrium can thus be achieved only via strategies that are not agent simple.

5.5 Dynamic Games: Introduction

This section allows the possibility that the stage game changes from period to period for a fixed set of players, possibly randomly and possibly as a function of the history of play. Such games are referred to as *dynamic games* or, when stressing that the stage game may be a random function of the game's history, *stochastic games*.

The analysis of a dynamic game typically revolves around a set of *game states* that describe how the stage game varies from period to period. Unless we need to distinguish between game states and states of an automaton (*automaton states*), we refer to game states simply as states (see remark 5.5.2). Each state determines a stage game, captured by writing payoffs as a function of states and actions. The specification of the game is completed by a rule for how the state changes over the course of play.

In many applications, the context in which the game arises suggests what appears to be a natural candidate for the set of states. It is accordingly common to treat

5.5 ■ Dynamic Games: Introduction

the set of states as an exogenously specified feature of the environment. This section proceeds in this way. However, the appropriate formulation of the set of states is not always obvious. Moreover, the notion of a state is an intermediate convention that is not required for the analysis of dynamic games. Instead, we can define payoffs directly as functions of current and past actions, viewing states as tools for describing this function. This suggests that instead of inspecting the environment and asking which of its features appear to define states, we begin with the payoff function and identify the states that implicitly lie behind its structure. Section 5.6 pursues this approach. With these tools in hand, section 5.7 examines equilibria in dynamic games.

5.5.1 The Game

There are n players, numbered $1, \ldots, n$. There is a set of states S, with typical state s. Player i has the compact set of actions $A_i \subset \mathbb{R}^k$, for some k. Player i's payoffs are given by the continuous function $u_i : S \times A \to \mathbb{R}$. Because payoffs are state dependent, the assumption that A_i is state independent is without loss of generality: If state s has action set A_i^s, define $A_i \equiv \prod_s A_i^s$ and set $\tilde{u}_i(s, a) = u_i(s, a^s)$. Players discount at the common rate δ.

The evolution of the state is given by a continuous transition function $q : S \times A \cup \{\varnothing\} \to \Delta(S)$, associating with each current state and action profile a probability distribution from which the next state is drawn; $q(\varnothing)$ is the distribution over initial states. This formulation captures a number of possibilities. If S is a singleton, then we are back to the case of a repeated game. If $q(s, a)$ is nondegenerate but constant in a, then we have a game in which payoffs are random variables whose distribution is constant across periods. If $S = \{0, 1, 2, \ldots\}$ and $q(s_\ell, a)$ puts probability one on $s_{\ell+1}$, we have a game in which the payoff function varies deterministically across periods, independently of behavior.

We focus on two common cases. In one, the set of states S is finite. We then let $q(s' \mid s, a)$ denote the probability that the state s' is realized, given that the previous state was s and the players chose action profile a (the initial state can be random in this case). We allow equilibria to be either pure or, if the action sets are finite, mixed. In the other case, A_i and S are infinite, in which case $S \subset \mathbb{R}^m$ for some m. We then take the transition function to be deterministic, so that for every s and a, there exists s' with $q(s' \mid s, a) = 1$ and $q(s'' \mid s, a) = 0$ for all $s'' \neq s'$ (the initial state is deterministic and given by s^0 in this case). As is common, we then restrict attention to pure-strategy equilibria.

In each period of the game, the state is first drawn and revealed to the players, who then simultaneously choose their actions. The set of period t ex ante histories $\mathcal{H}^t$ is the set $(S \times A)^t$, identifying the state and the action profile in each period.[17] The set of period t ex post histories is the set $\tilde{\mathcal{H}}^t = (S \times A)^t \times S$, giving state and

---

17. Under the state transition rule q, many of the histories in this set may be impossible. If so, the specification of behavior at these histories will have no effect on payoffs.

action realizations for each previous period and identifying the current state. Let $\tilde{\mathscr{H}} = \cup_{t=0}^{\infty}\tilde{\mathscr{H}}^t$; we set $\tilde{\mathscr{H}}^0 = \{\varnothing\}$, so that $\tilde{\mathscr{H}}^0 = S$. Let $\mathscr{H}^\infty$ be the set of outcomes.

A pure strategy for player i is a mapping $\sigma_i : \tilde{\mathscr{H}} \to A_i$, associating an action with each ex post history. A pure strategy profile σ, together with the transition function q, induces a probability measure over $\mathscr{H}^\infty$. Player i's expected payoff is then

$$U_i(\sigma) = E^\sigma \left\{ (1-\delta) \sum_{t=0}^{\infty} \delta^t u_i(s^t, a^t) \right\},$$

where the expectation is taken with respect to the measure over $\mathscr{H}^\infty$ induced by $q(\varnothing)$ and σ. Note that the expectation may be nontrivial, even for a pure strategy profile, because the transition function q may be random. As usual, this formulation has a straightforward extension to behavior strategies. For histories other than the null history, we let $U_i(\sigma \mid \tilde{h}^t)$ denote i's expected payoffs induced by the strategy profile σ in the continuation game that follows the ex post history $\tilde{h}^t$.

An ex ante history ending in (s, a) gives rise to a continuation game matching the original game but with $q(\varnothing)$ replaced by $q(\cdot \mid s, a)$ and the transition function otherwise unchanged. An ex post history $\tilde{h}^t$, ending with state s, gives rise to the continuation game again matching the original game but with the initial distribution over states now attaching probability 1 to state s. It will be convenient to denote this game by $G(s)$.

Definition 5.5.1 *A strategy profile σ is a* Nash equilibrium *if $U_i(\sigma) \geq U_i(\sigma_i', \sigma_{-i})$ for all σ_i' and for all players i. A strategy profile σ is a* subgame-perfect equilibrium *if, for any ex post history $\tilde{h}^t \in \tilde{\mathscr{H}}$ ending in state s, the continuation strategy $\sigma|_{\tilde{h}^t}$ is a Nash equilibrium of the continuation game $G(s)$.*

Example 5.5.1 Suppose there are two equally likely states, independently drawn in each period, with payoffs given in figure 5.5.1. It is a quick calculation that strategies specifying effort after every ex post history in which there has been no previous shirking (and specifying shirking otherwise) are a subgame-perfect equilibrium if and only if $\delta \geq 2/7$. Strategies specifying effort after any ex post history ending in state 2 while calling for shirking in state 1 (as long as there have been no deviations from this prescription, and shirking otherwise) are an equilibrium if and only if $\delta \geq 2/5$. The deviation incentives are the same in both games, whereas exerting

	E	S			E	S
E	2, 2	−1, 3		E	3, 3	−1, 4
S	3, −1	0, 0		S	4, −1	0, 0

State 1   State 2

Figure 5.5.1 Payoff functions for states 1 and 2 of a dynamic game.

5.5 ■ Dynamic Games: Introduction

effort in only one state reduces the equilibrium continuation value, and hence requires more patience to sustain effort in the other state.[18]

Remark 5.5.1 **Repeated games with random states** The previous example illustrates the special case in which the probability of the current state is independent of the previous state and the players' actions, that is, $q(s \mid s', a') = q(s \mid s'', a'') \equiv q(s)$. We refer to such dynamic games as *repeated games with random states*. These games are a particularly simple repeated game with imperfect public monitoring (section 7.1.1). Player i's pure action set in the stage game is given by the set of functions from S into A_i, and players simultaneously choose such actions. The pure-action profile σ then gives rise to the signal $(s, \sigma_1(s), \ldots, \sigma_n(s))$ with probability $q(s)$, for each $s \in S$. Player i's payoff $u_i(s, a)$ can then be written as $u_i((s, \sigma_{-i}(s)), a_i(s))$, giving i's payoff as a function of i's action and the public signal. We describe a simpler approach in remark 5.7.1.

◆

5.5.2 Markov Equilibrium

In principle, strategies in a dynamic game could specify each period's action as a complicated function of the preceding history. It is common, though by no means universal, to restrict attention to Markov strategies:

Definition 5.5.2
1. *The strategy profile σ is a* Markov strategy *if for any two ex post histories $\tilde{h}^t$ and $\tilde{h}^\tau$ of the same length and terminating in the same state, $\sigma(\tilde{h}^t) = \sigma(\tilde{h}^\tau)$. The strategy profile σ is a* Markov equilibrium *if σ is a Markov strategy profile and a subgame-perfect equilibrium.*
2. *The strategy profile σ is a* stationary Markov strategy *if for any two ex post histories $\tilde{h}^t$ and $\tilde{h}^\tau$ (of equal or different lengths) terminating in the same state, $\sigma(\tilde{h}^t) = \sigma(\tilde{h}^\tau)$. The strategy profile σ is a* stationary Markov equilibrium *if σ is a stationary Markov strategy profile and a subgame-perfect equilibrium.*

It is sometimes useful to reinforce the requirement of subgame perfection by referring to a Markov equilibrium as a *Markov perfect equilibrium*. Some researchers also refer to game states as *Markov states* when using Markov equilibrium (but see remark 5.5.2).

Markov strategies ignore all of the details of a history except its length and the current state. Stationary Markov strategies ignore all details except the current state.

Three advantages for such equilibria are variously cited. First, Markov equilibria appear to be simple, in the sense that behavior depends on a relatively small set of variables, often being the simplest strategies consistent with rationality. To some, this simplicity is appealing for its own sake, whereas for others it is an analytical or computational advantage. Markov equilibria are especially common in applied work.[19]

---

18. In contrast to section 5.3, we face here one randomly drawn game in each period, instead of both games.
19. It is not true, however, that Markov equilibria are always simpler than non-Markov equilibria. The proof of proposition 18.4.4 goes to great lengths to construct a Markov equilibrium featuring high effort, in a version of the product choice game, that would be a straightforward calculation in non-Markov strategies.

Second, the set of Markov equilibrium outcomes is often considerably smaller than the set of all equilibrium outcomes. This is a virtue for some and a vice for others, but again contributes to the popularity of Markov equilibria in applied work.

Third, Markov equilibria are often viewed as having some intuitive appeal for their own sake. The source of this appeal is the idea that only things that are "payoff relevant" should matter in determining behavior. Because the only aspect of a history that affects current payoff functions is the current state, then a first step in imposing payoff relevance is to assume that current behavior should depend only on the current state.[20] Notice, however, that there is no reason to limit this logic to dynamic games. It could just as well be applied in a repeated game, where it is unreasonably restrictive, because *nothing* is payoff relevant in the sense typically used when discussing dynamic games. Insisting on Markov equilibria in the repeated prisoners' dilemma, for example, dooms the players to perpetual shirking. More generally, a Markov equilibrium in a repeated game must play a stage-game Nash equilibrium in every period, and a stationary Markov equilibrium must play the same one in every period.

Remark 5.5.2 **Three types of state** We now have three notions of a state to juggle. One is *Markov state*, an equivalence class of histories in which distinctions are payoff irrelevant. The second is *game state*, an element of the set S of states determining the stage game in a dynamic game. Though it is often taken for granted that the set of Markov states can be identified with the set of game states, as we will see in section 5.6, these are distinct concepts. Game states may not always be payoff relevant and, more important, we can identify Markov states without any a priori specification of a game state. Finally, we have *automaton states*, states in an automaton representing a strategy profile. Because continuation payoffs in a repeated game depend on the current automaton state, and only on this state, some researchers take the set of automata states as the set of Markov states. This practice unfortunately robs the Markov notion and payoff relevance of any independent meaning. The particular notion of payoff relevance inherent in labeling automaton states Markov is much less restrictive than that often intended to be captured by Markov perfection. For example, Markov perfection then imposes *no* restrictions beyond subgame perfection in repeated games, because any subgame-perfect equilibrium profile has an automaton representation, in contrast to the trivial equilibria that appear if we at least equate Markov states with game states.

◆

5.5.3 Examples

Example 5.5.2 Suppose that players 1 and 2 draw fish from a common pool. In each period t, the pool contains a stock of fish of size $s^t \in \mathbb{R}_+$. In period t, player i extracts $a_i^t \geq 0$ units of fish, and derives payoff $\ln(a_i^t)$ from extracting a_i^t.[21] The remaining

---

20. If the function $u(s, a)$ is constant over some values of s, then we could impose yet further restrictions.

21. We must have $a_1^t + a_2^t \leq s^t$. We can model this by allowing the players to choose extraction levels $\bar{a}_1^t$ and $\bar{a}_2^t$, with these levels realized if feasible and with a rationing rule otherwise determining realized extraction levels. This constraint will not play a role in the equilibrium, and so we leave the rationing rule unspecified and treat a_i^t and $\bar{a}_i^t$ as identical.

5.5 ■ Dynamic Games: Introduction

(depleted) stock of fish doubles before the next period. This gives a dynamic game with actions and states drawn from the infinite set $[0, \infty)$, identifying the current quantity of fish extracted (actions) and the current stock of fish (states), and with the deterministic transition function $s^{t+1} = 2(s^t - a_1^t - a_2^t)$. The initial stock is fixed at some value s^0.

We first calculate a stationary Markov equilibrium of this game, in which the players choose identical strategies. That is, although we assume that the players choose Markov strategies in equilibrium, the result is a strategy profile that is optimal in the full strategy set—there are no superior strategies, Markov or otherwise. We are thus calculating not an equilibrium in the game in which players are restricted to Markov strategies but Markov strategies that are an equilibrium of the full game.

The restriction to Markov strategies allows us to introduce a function $V(s)$ identifying the equilibrium value (conditional on the equilibrium strategy profile) in any continuation game induced by an ex post history ending in state s. Let $g^t(s^0)$ be the amount of fish extracted by each player at time t, given the period 0 stock s^0 and the (suppressed, in the notation) equilibrium strategies. The function $V(s^0)$ identifies equilibrium utilities, and hence must satisfy

$$V(s^0) = (1 - \delta) \sum_{t=0}^{\infty} \delta^t \ln(g^t(s^0)). \tag{5.5.1}$$

Imposing the Markov restriction that current actions depend only on the current state, let each player's strategy be given by a function $a(s)$ identifying the amount of fish to extract given that the current stock is s. We then solve jointly for the function $V(s)$ and the equilibrium strategy $a(s)$. First, the one-shot deviation principle (which we describe in section 5.7.1) allows us to characterize the function $a(s)$ as solving, for any $s \in S$ and for each player i, the Bellman equation,

$$a(s) \in \arg\max_{\tilde{a} \in A_i} (1 - \delta) \ln(\tilde{a}) + \delta V(2(s - \tilde{a} - a(s))),$$

where $\tilde{a}$ is player i's consumption and the $a(s)$ in the final term captures the assumption that player j adheres to the candidate equilibrium strategy. If the value function V is differentiable, the implied first-order condition is

$$\frac{(1 - \delta)}{a(s)} = 2\delta V'(2(s - 2a(s))).$$

To find an equilibrium, suppose that $a(s)$ is given by a linear function, so that $a(s) = ks$. Then we have $s^{t+1} = 2(s^t - 2ks^t) = 2(1 - 2k)s^t$. Using this and $a(s) = ks$ to recursively replace $g^t(s)$ in (5.5.1), we have

$$V(s) = (1 - \delta) \sum_{t=0}^{\infty} \delta^t \ln[k(2(1 - 2k))^t s],$$

and so V is differentiable with $V'(s) = 1/s$. Solving the first-order condition, $k = (1-\delta)/(2-\delta)$, and so

$$a(s) = \frac{1-\delta}{2-\delta}s$$

and $$V(s) = (1-\delta)\sum_{t=0}^{\infty} \delta^t \ln\left(\frac{1-\delta}{2-\delta}s\left(\frac{2\delta}{2-\delta}\right)^t\right).$$

We interpret this expression by noting that in each period, proportion

$$1 - 2a(s) = 1 - 2\frac{1-\delta}{2-\delta} = \frac{\delta}{2-\delta}$$

of the stock is preserved until the next period, where it is doubled, so that the stock grows at rate $2\delta/(2-\delta)$. In each period, each player consumes fraction $(1-\delta)/(2-\delta)$ of this stock.

Notice that in this solution, the stock of resource grows without bound if the players are sufficiently patient ($\delta > 2/3$), though payoffs remain bounded, and declines to extinction if $\delta < 2/3$. As is expected from these types of common pool resource problems, this equilibrium is inefficient. Failing to take into account the externality that their extraction imposes on their partner's future consumption, each player extracts too much (from an efficiency point of view) in each period.

This stationary Markov equilibrium is not the only equilibrium of this game. To construct another equilibrium, we first calculate the largest symmetric payoff profile that can be achieved when the firms choose identical Markov strategies. Again representing the solution as a linear function $a = ks$, we can write the appropriate Bellman equation as

$$a(s) = \underset{\tilde{a} \in A_i}{\operatorname{argmax}} \, 2(1-\delta)\ln(\tilde{a}) + \delta(1-\delta)\sum_{t=0}^{\infty} \delta^t 2\ln(k(2(1-2k))^t(s - 2\tilde{a})).$$

Taking a derivative with respect to $\tilde{a}$ and simplifying, we find that the efficient solution is given by

$$a(s) = \frac{1-\delta}{2}s.$$

As expected, the efficient solution extracts less than does the Markov equilibrium. The efficient solution internalizes the externality that player i's extraction imposes on player j, through its effect on future stocks of fish.

Under the efficient solution, we have

$$s^{t+1} = 2s^t\left(1 - 2\frac{1-\delta}{2}\right) = 2\delta s^t,$$

and hence

$$s^t = (2\delta)^t s^0.$$

The stock of the resource grows without bound if $\delta > 1/2$. Notice also that just as we earlier solved for Markov strategies that are an equilibrium in the complete strategy set, we have now found Markov strategies that maximize total expected payoffs over the set of all strategies.

5.5 ■ Dynamic Games: Introduction

We can support the efficient solution as a (non-Markov) equilibrium of the repeated game, if the players are sufficiently patient. Let strategies prescribe the efficient extraction after every history in which the quantity extracted has been efficient in each previous period, and prescribe the Markov equilibrium extraction $a(s) = [(1 - \delta)/(2 - \delta)]s$ otherwise. Then for sufficiently large δ, we have an equilibrium.

●

Example 5.5.3 Consider a market with a single good, produced by a monopoly firm facing a continuum of small, anonymous consumers.[22] We think of the firm as a long-lived player and interpret the consumers as short-lived players.

The good produced by the firm is durable. The good lasts forever, subject to continuous depreciation at rate η, so that 1 unit of the good purchased at time 0 depreciates to $e^{-\eta t}$ units of the good at time t. This durability makes this a dynamic rather than repeated game.

Time is divided into discrete periods of length Δ, with the firm making a new production choice at the beginning of each period. We will subsequently be interested in the limiting case as Δ becomes very short. The players in the model discount at the continuously compounded rate r. For a period of length Δ, the discount factor is thus $e^{-r\Delta}$.

The stock of the good in period t is denoted $x(t)$. The stock includes the quantity that the firm has newly produced in period t, as well as the depreciated remnants of past production. Though the firm's period t action is the period t quantity of production, it is more convenient to treat the firm as choosing the stock $x(t)$. The firm thus chooses a sequence of stocks $\{x(0), x(1), \ldots\}$, subject to $x(t) \geq e^{-\eta\Delta}x(t-1)$.[23] Producing a unit of the good incurs a constant marginal cost of c.

Consumers take the price path as given, believing that their own consumption decisions cannot influence future prices. Rather than modeling consumers' maximization behavior directly, we represent it with the inverse demand curve $f(x) = 1 - x$. We interpret $f(x)$ as the *instantaneous* valuation consumers attach to x units of the durable good. We must now translate this into our setting with periods of length Δ. The value *per unit* a consumer assigns to acquiring a quantity x at the beginning of a period and used *only* throughout that period, with no previous or further purchases, is

$$F(x) = \int_0^\Delta f(xe^{-\eta s})e^{-(r+\eta)s}ds$$
$$= \frac{1}{r+\eta}(1 - e^{-(r+\eta)\Delta}) - x\frac{1}{r+2\eta}(1 - e^{-(r+2\eta)\Delta})$$
$$\equiv \theta - \beta x.$$

---

22. For a discussion of durable goods monopoly problems, see Ausubel, Cramton, and Deneckere (2002). The example in this section is taken from Bond and Samuelson (1984, 1987). The introduction of depreciation simplifies the example, but is not essential to the results (Ausubel and Deneckere 1989).

23. Because this constraint does not bind in the equilibrium we construct, we can ignore it.

Because the good does not disappear at the end of the period, the period t price (reflecting current and future values) given the sequence $\mathbf{x}(t) \equiv \{x(t), x(t+1), \ldots\}$ of period t and future stocks of the good, is given by

$$p(t, \mathbf{x}(t)) = \sum_{s=0}^{\infty} e^{-(r+\eta)\Delta s} F(x(t+s)).$$

The firm's expected payoff in period t, given the sequence of actions $x(t-1)$ and $\mathbf{x}(t)$, is given by[24]

$$\sum_{\tau=t}^{\infty} (x(\tau) - x(\tau-1)e^{-\eta\Delta})(p(\tau, \mathbf{x}(\tau)) - c)e^{-r\Delta(\tau-t)}.$$

If the good were perishable, this would be a relatively straightforward intertemporal price discrimination problem. The durability of the good complicates the relationship between current prices and future actions. We begin by seeking a stationary Markov equilibrium. The state variable in period t is the stock $x(t-1)$ chosen in the previous period. The firm's strategy is described by a function $x(t) = g(x(t-1))$, giving the period t stock as a function of the previous period's stock. However, a more flexible description of the firm's strategy is more helpful. We consider a function

$$g(s, t, x),$$

identifying the period s stock, given that the stock in period $t \leq s$ is x. Hence, we build into our description of the Markov strategy the observation that if period t's stock is a function of period $t-1$'s, then so is period $t+2$'s stock a (different) function of $x(t-1)$, and so is $x(t+3)$, and so on. To ensure this representation of the firm's strategy is coherent, we impose the consistency condition that $g(s', t, x) = g(s', s, g(s, t, x))$ for $s' \geq s \geq t$. In addition, $g(s+\tau, s, x)$ must equal $g(t+\tau, t, x)$ for $s \neq t$, so that the same state variable produces identical continuation behavior, regardless of how it is reached and regardless of when it is reached.

The firm's profit maximization problem, in any period t, is to choose the sequence of stocks $\{x(t), x(t+1), \ldots, \}$ to maximize

$$V(\mathbf{x}(t), t \mid x(t-1))$$

$$= \sum_{\tau=t}^{\infty} [x(\tau) - x(\tau-1)e^{-\eta\Delta}](p(\tau, \mathbf{x}(\tau)) - c)e^{-r\Delta(\tau-t)} \quad (5.5.2)$$

$$= \sum_{\tau=t}^{\infty} [x(\tau) - x(\tau-1)e^{-\eta\Delta}]$$

$$\times \left(\sum_{s=0}^{\infty} e^{-(r+\eta)\Delta s}(\theta - \beta g(s+\tau, \tau, x(\tau))) - c \right) e^{-r\Delta(\tau-t)}. \quad (5.5.3)$$

In making the substitution for $p(\tau, x(\tau))$ that brings us from (5.5.2) to (5.5.3), $g(s+\tau, \tau, x(\tau))$ describes consumers' expectations of the firm's future stocks and

---

24. Notice that we must specify the stock in period $t-1$ because this combines with $x(t)$ to determine the quantity produced and sold in period t.

5.5 ■ Dynamic Games: Introduction

so their own future valuations. To find an optimal strategy for the firm, we differentiate $V(\mathbf{x}(t), t \mid x(t-1))$ with respect to $x(t')$ for $t' \geq t$ to obtain a first-order condition for the latter. In doing so, we hold the values $x(\tau)$ for $\tau \neq t'$ fixed, so that the firm chooses $x(\tau)$ and $x(t')$ independently. However, consumer expectations are given by the function $g(s + \tau, \tau, x(\tau))$, which builds in a relationship between the current stock and anticipated future stocks that determines current prices.

Fix $t' \geq t \geq 0$. The first-order condition $dV(\mathbf{x}(t), t \mid x(t-1))/dx(t') = 0$ is, from (5.5.3),

$$[\theta - \beta x(t') - c(1 - e^{-(r+\eta)\Delta})]$$
$$- \beta[x(t') - e^{-\eta\Delta}x(t'-1)] \sum_{s=0}^{\infty} e^{-(r+\eta)\Delta s} \left. \frac{dg(t'+s, t', x)}{dx} \right|_{x=x(t')} = 0.$$

Using $x(t') = g(t', t, x(t))$, we rewrite this as

$$[\theta - \beta g(t', t, x(t)) - c(1 - e^{-(r+\eta)\Delta})]$$
$$- \beta[g(t', t, x(t)) - e^{-\eta\Delta}g(t'-1, t, x(t))]$$
$$\times \left[\sum_{s=0}^{\infty} e^{-(r+\eta)\Delta s} \frac{dg(t'+s, t', g(t', t, x(t)))}{dx} \right] = 0.$$

As is typically the case, this difference equation is solved with the help of some informed guesswork. We posit $g(s, t, x)$ takes the form

$$g(s, t, x) = \bar{x} + \mu^{s-t}(x - \bar{x}),$$

where we interpret $\bar{x}$ as a limiting stock of the good and μ as identifying the rate at which the stock adjusts to this limit. With this form for g, it is immediate that the first-order conditions characterize the optimal value of $x(t')$.

Substituting this expression into our first-order condition gives,

$$\left\{ \theta - c(1 - e^{-(r+\eta)\Delta}) - \beta\bar{x}\left(1 + \frac{1 - e^{-\eta\Delta}}{1 - e^{-(r+\eta)\Delta}\mu}\right) \right\}$$
$$- \beta\mu^{t'-t-1}(x - \bar{x})\left\{ \mu + \frac{\mu - e^{-\eta\Delta}}{1 - e^{-(r+\eta)\Delta}\mu} \right\} = 0.$$

Because this equation must hold for all t', we conclude that each expression in braces must be 0. We can solve the second for μ and then insert in the first to solve for $\bar{x}$, yielding

$$\mu = \frac{1 - \sqrt{1 - e^{-(r+2\eta)\Delta}}}{e^{-(r+\eta)\Delta}}$$

and

$$\bar{x} = \frac{[\theta - c(1 - e^{-(r+\eta)\Delta})]}{\beta} \frac{\sqrt{1 - e^{-(r+2\eta)\Delta}}}{(\sqrt{1 - e^{-(r+2\eta)\Delta}} + 1 - e^{-\eta\Delta})}.$$

Notice first that $\mu < 1$. The stock of good produced by the monopoly thus converges monotonically to the limiting stock $\bar{x}$. In the expression for the limit $\bar{x}$,

$$\frac{[\theta - c(1 - e^{-(r+\eta)\Delta})]}{\beta}$$

is the competitive stock. Maintaining the stock at this level in every period gives $p(t, \mathbf{x}) = c$, and hence equality of price and marginal cost. The term

$$\frac{\sqrt{1 - e^{-(r+2\eta)\Delta}}}{(\sqrt{1 - e^{-(r+2\eta)\Delta}} + 1 - e^{-\eta\Delta})} \equiv \gamma(\Delta, \eta) \tag{5.5.4}$$

then gives the ratio between the monopoly's limiting stock of good and the competitive stock. The following properties follow immediately from (5.5.4):

$$\gamma(\Delta, \eta) < 1, \tag{5.5.5}$$

$$\lim_{\eta \to \infty} \gamma(\Delta, \eta) = \tfrac{1}{2}, \tag{5.5.6}$$

$$\gamma(\Delta, 0) = 1, \tag{5.5.7}$$

and

$$\lim_{\Delta \to 0} \gamma(\Delta, \eta) = 1. \tag{5.5.8}$$

Condition (5.5.5) indicates that the monopoly's limiting stock is less than that of a competitive market. Condition (5.5.6) shows that as the depreciation rate becomes arbitrarily large, the limiting monopoly quantity is half that of the competitive market, recovering the familiar result for perishable goods.

Condition (5.5.7) indicates that if the good is perfectly durable, then the limiting monopoly quantity approaches that of the competitive market. As the competitive stock is approached, the price-cost margin collapses to 0. With a positive depreciation rate, it pays to keep this margin permanently away from 0, so that positive profits can be earned on selling replacement goods to compensate for the continual deprecation. As the rate of depreciation goes to 0, however, this source of profits evaporates and profits are made only on new sales. It is then optimal to extract these profits, to the extent that the price-cost margin is pushed to 0.

Condition (5.5.8) shows that as the period length goes to 0, the stock again approaches the competitive stock. If the stock stops short of the competitive stock, every period brings the monopoly a choice between simply satisfying the replacement demand, for a profit that is proportional to the length of the period, or pushing the price lower to sell new units to additional consumers. The latter profit is proportional to the price and hence must overwhelm the replacement demand for short time periods, leading to the competitive quantity as the length of a period becomes arbitrarily short.

More important, because the adjustment factor μ is also a function of Δ, by applying l'Hôpital's rule to $\Delta^{-1} \ln \mu$, one can show that

$$\lim_{\Delta \to 0} \mu^{\frac{1}{\Delta}} = 0.$$

Hence, as the period length shrinks to 0, the monopoly's output path comes arbitrarily close to an instantaneous jump to the competitive quantity.[25] Consumers

---

25. At calender time T, T/Δ periods have elapsed, and so, as $\Delta \to 0$, $x(T/\Delta) \to \bar{x}$.

5.5 ■ Dynamic Games: Introduction

build this behavior into their pricing behavior, ensuring that prices collapse to marginal cost and the firm's profits collapse to 0. This is the Coase conjecture in action.[26]

How should we think about the Markov restriction that lies behind this equilibrium? The key question facing a consumer, when evaluating a price, concerns how rapidly the firm is likely to expand the stock and depress the price in the future. The firm will firmly insist that there is nary a price reduction in sight, a claim that the consumer would do well to treat with some skepticism. One obvious place for the consumer to look in assessing this claim is the firm's past behavior. Has the price been sitting at nearly its current level for a long time? Or has the firm been racing down the demand curve, having charged ten times as much only periods ago? Markov strategies insist that consumers ignore such information. If consumers find the information relevant, then we have moved beyond Markov equilibria.

To construct an alternative equilibrium with quite different properties, let

$$x_R = \frac{[\theta - c(1 - e^{-(r+\eta)\Delta})]}{2\beta}.$$

This is half the quantity produced in a competitive market. Choosing this stock in every period maximizes the firm's profits over the set of Nash equilibria of the repeated game.[27] Now let $\mathcal{H}^*$ be the set of histories in which the stock x_R has been produced in every previous period. Notice that this includes the null history. Then consider the firm's strategy $x(h^t)$, giving the current stock as a function of the history h^t, given by

$$x(h^t) = \begin{cases} x_R, & \text{if } h_t \in \mathcal{H}^*, \\ g(t, t-1, x(t-1)), & \text{otherwise,} \end{cases}$$

where $g(\cdot)$ is the Markov equilibrium strategy calculated earlier. In effect, the firm "commits" to produce the profit-maximizing quantity x_R, with any misstep prompting a switch to continuing with the Markov equilibrium. Let σ^R denote this strategy and the attendant best response for consumers.

It is now straightforward that σ^R is a subgame-perfect equilibrium, as long as the length of a period Δ is sufficiently short. To see this, let $U(\sigma^R \mid x)$ be the monopoly's continuation payoff from this strategy, given that the current stock is $x \leq x_R$. We are interested in the continuation payoffs given stock x_R, or

$$U(\sigma^R \mid x_R) = \sum_{\tau=0}^{\infty} e^{-r\Delta\tau}(1 - e^{-\eta\Delta})x_R\left(-c + \sum_{s=0}^{\infty} e^{-(r+\eta)\Delta s}F(x_R)\right)$$

$$= \frac{(1 - e^{-\eta\Delta})[F(x_R) - c(1 - e^{-(r+\eta)\Delta})]}{(1 - e^{-r\Delta})(1 - e^{-(r+\eta)\Delta})}x_R.$$

26. Notice that in examining the limiting case of short time periods, depreciation has added nothing other than extra terms to the model.
27. This quantity maximizes $[\theta - \beta x - c(1 - e^{-(r+\eta)\Delta})]x$, and hence is the quantity that would be produced in each period by a firm that retained ownership of the good and rented its services to the customers. A firm who sells the good can earn no higher profits.

The key observation now is that

$$\lim_{\Delta \to \infty} U(\sigma^R \mid x_R) > 0.$$

Even as time periods become arbitrarily short, there are positive payoffs to be made by continually replacing the depreciated portion of the profit-maximizing quantity x_R. In contrast, as we have seen, as time periods shorten, the continuation payoff of the Markov equilibrium, from any initial stock, approaches 0. This immediately yields:

Proposition 5.5.1 *There exists Δ^* such that, if $\Delta < \Delta^*$, then strategies σ^R are a subgame-perfect equilibrium.*

This example illustrates that a Markov restriction can make a great difference in equilibrium outcomes. One may or may not be convinced that a focus on Markov equilibria is appropriate, but one cannot rationalize the restriction simply as an analytical convenience.

●

5.6   Dynamic Games: Foundations

What determines the set S of states for a dynamic game? At first the answer seems obvious—states are things that affect payoffs, such as the stock of fish in example 5.5.2 or the stock of durable good in example 5.5.3. However, matters are not always so clear.

For example, our formulation of repeated games in chapter 2 allows players to condition their actions on a public random variable. Are these realizations states, in the sense of a dynamic game, and does Markov equilibrium allow behavior to be conditioned on such realizations? Alternatively, consider the infinitely repeated prisoners' dilemma. It initially appears as if there are no payoff-relevant states, so that Markov equilibria must feature identical behavior after every history and hence must feature perpetual shirking. Suppose, however, that we defined two states, an effort state and a shirk state. Let the game begin in the effort state, and remain there as along as there has been no shirking, being otherwise in the shirk state. Now let players' strategies prescribe effort in the effort state and shirking in the shirk state. We now have a Markov equilibrium (for sufficiently patient players) featuring effort. Are these states real, or are they a sleight of hand?

In general, we can define payoffs for a dynamic game as functions of current and past actions, without resorting to the idea of a state. As a result, it can be misleading to think of the set of states as being exogenously given. Instead, if we would like to work with the notions of payoff relevance and Markov equilibrium, we must endogenously infer the appropriate set of states from the structure of payoffs.

This section pursues this notion of a state, following Maskin and Tirole (2001). We examine a game with players $1, \ldots, n$. Each player i has the set A_i of stage-game actions available in each period. Hence, the set of feasible actions is again independent

5.6 ■ Dynamic Games: Foundations

of history.[28] Player i's payoff is a function of the outcome path $\mathbf{a} \in A^\infty$. This formulation is sufficiently general to cover the dynamic games of section 5.6. Deterministic transitions are immediately covered because the history of actions h^t determines the state s reached in period t. For stochastic transitions, such as example 5.5.1, introduce an artificial player 0, nature, with action space $A_0 = S$ and constant payoffs; random state transitions correspond to the appropriate fixed behavior strategy for player 0. In what follows, the term *players* refers to players $i \geq 1$, and *histories* include nature's moves.

Let σ be a pure strategy profile and h^t a history. Then we write $U_i^*(\sigma \mid h^t)$ for player i's payoffs given history h^t and the subsequent continuation strategy profile $\sigma|_{h^t}$. This is in general an expected value because future utilities may depend randomly on past play, for the same reason that the current state in a model with exogenously specified states may depend randomly on past actions.

Note that $U_i^*(\sigma \mid h^t)$ is *not*, in general, the continuation payoff from $\sigma|_{h^t}$. For example, for a repeated game in the class of chapter 2,

$$U_i^*(\sigma \mid (a^0, a^1, \ldots, a^{t-1})) = \sum_{\tau=0}^{t-1} \delta^\tau u_i(a^\tau) + \delta^t U_i(\sigma|_{(a^0, a^1, \ldots, a^{t-1})}),$$

where u_i is the stage game payoff and U_i is given by (2.1.2). In this case, $U_i^*(\sigma \mid h^t)$ and $U_i^*(\sigma' \mid \hat{h}^t)$ differ by only a constant if the continuation strategies $\sigma|_{h^t}$ and $\sigma'|_{\hat{h}^t}$ are identical, and history is important only for its role in coordinating future behavior.

This dual role of histories in a dynamic game gives rise to ambiguity in defining states. Suppose two histories induce different continuation payoffs. Do these differences arise because differences in future play are induced, in which case the histories would not satisfy the usual notion of being payoff relevant (though it can still be critical to take note of the difference), or because identical continuation play gives rise to different payoffs? Can we always tell the difference?

5.6.1 Consistent Partitions

Let $\mathcal{H}^t$ be the set of period t histories. Notice that we have no notion of a state in this context, and hence no distinction between ex ante and ex post histories. A period t history is an element of A^t. A partition of $\mathcal{H}^t$ is denoted $\mathbb{H}^t$, and $\mathbb{H}^t(h^t)$ is the partition element containing history h^t.

A sequence of partitions $\{\mathbb{H}^t\}_{t=0}^\infty$ is denoted $\mathbb{H}$; viewed as $\cup_t \mathbb{H}^t$, $\mathbb{H}$ is a partition of the set of all histories $\mathcal{H} = \cup_t \mathcal{H}^t$. We often find it convenient to work with several such sequences, one associated with each player, denoting them by $\mathbb{H}_1, \ldots, \mathbb{H}_n$. Given such a collection of partitions, we say that two histories h^t and $\hat{h}^t$ are i-equivalent if $h^t \in \mathbb{H}_i^t(\hat{h}^t)$.

A strategy σ_i is *measurable* with respect to $\mathbb{H}_i$ if, for every pair of histories h^t and $\hat{h}^t$ with $h^t \in \mathbb{H}_i^t(\hat{h}^t)$, the continuation strategy $\sigma_i|_{h^t}$ equals $\sigma_i|_{\hat{h}^t}$. Let $\Sigma_i(\mathbb{H})$ denote the set of pure strategies for player i that are measurable with respect to the partition $\mathbb{H}$.

---

28. If this were not the case, then we would first partition the set of period t histories $\mathcal{H}^t$ into subsets that feature the same feasible choices for each player i in period t and then work throughout with refinements of this partition, to ensure that our subsequent measurability requirements were feasible.

A collection of partitions $\{\mathbb{H}_1^t, \ldots, \mathbb{H}_n^t\}_{t=0}^{\infty}$ is *consistent* if for every player i, whenever other players' strategies σ_{-i} are measurable with respect to their partition, then for any pair of i-equivalent period t histories h^t and $\hat{h}^t$, player i has the same preferences over i's continuation strategies. Hence, consistency requires that for any player i, pure strategies $\sigma_j \in \Sigma_j(\mathbb{H}_j)$ for all $j \neq i$, and i-equivalent histories h^t and $\hat{h}^t$, there exist constants θ and $\beta > 0$ such that

$$U_i^*((\sigma_i, \sigma_{-i}) \mid h^t) = \theta + \beta U_i^*((\sigma_i, \sigma_{-i}) \mid \hat{h}^t).$$

If this relationship holds, conditional on the (measurable) strategies of the other players, i's utilities after histories h^t and $\hat{h}^t$ are affine transformations of one another. We represent this by writing

$$U_i^*((\cdot, \sigma_{-i}) \mid h^t) \sim U_i^*((\cdot, \sigma_{-i}) \mid \hat{h}^t). \tag{5.6.1}$$

We say that two histories h^t and $\hat{h}^t$, with the property that player i has the same preferences over continuation payoffs given these histories (as just defined) are *i-payoff equivalent*. Consistency of a partition is thus the condition that equivalence (under the partition) implies payoff equivalence.

The idea now is to define a Markov equilibrium as a subgame-perfect equilibrium that is measurable with respect to a consistent collection of partitions. To follow this program through, two additional steps are required. First, we establish conditions under which consistent partitions have some intuitively appealing properties. Second, there may be many consistent partitions, some of them more interesting than others. We show that a maximal consistent partition exists, and use this one to define Markov equilibria.

5.6.2 Coherent Consistency

One might expect a consistent partition to have two properties. First, we might expect players to share the same partition. Second, we might expect the elements of the period t partition to be subsets of partition in period $t - 1$, so that the partition is continually refined. Without some additional mild conditions, both of these properties can fail.

Lemma 5.6.1 *Suppose that for any players i and j, any period t, and any i-equivalent histories h^t and $\hat{h}^t$, there exists a repeated-game strategy profile σ and stage-game actions a_j and a_j' such that*

$$U_i^*((\cdot, \sigma_{-i}) \mid (h^t, a_j)) \not\sim U_i^*((\cdot, \sigma_{-i}) \mid (\hat{h}^t, a_j')). \tag{5.6.2}$$

Then if $(\mathbb{H}_1, \ldots, \mathbb{H}_n)$ is a consistent collection of partitions, then in every period t and for all players i and j, $\mathbb{H}_i^t = \mathbb{H}_j^t$.

The expression $U_i^*((\sigma_i, \sigma_{-i}) \mid (h^t, a_j))$ gives player i's payoffs, given that h^t has occurred and given that player j chooses a_j in period t, with behavior otherwise specified by (σ_i, σ_{-i}). Condition (5.6.2) then requires that player i's preferences, given (h^t, a_j) and $(\hat{h}^t, a_j')$, *not* be affine transformations of one another.

Proof Let h^t and $\hat{h}^t$ be i-equivalent. Choose a player j and suppose the strategy profile σ and actions a_j and a'_j satisfy (5.6.2). We suppose h^t and $\hat{h}^t$ are not j-equivalent (and derive a contradiction). Then player j's strategy of playing as in σ, except playing a_j after histories in $\mathbb{H}_j(h^t)$ and a'_j after histories in $\mathbb{H}_j(\hat{h}^t)$ is measurable with respect to $\mathbb{H}_j$. But then (5.6.2) contradicts (5.6.1): Player i's partition is not consistent (condition (5.6.2)), as assumed (condition (5.6.1)). ∎

To see the argument behind this proof, suppose that a period t arrives in which player i and j partition their histories differently. We exploit this difference to construct a measurable strategy for player j that differs across histories within a single element of player i's partition, in a way that affects player i's preferences over continuation play. This contradicts the consistency of player i's partition. There are two circumstances under which such a contradiction may not arise. One is that all players have the same partition, precluding the construction of such a strategy. This leads to the conclusion of the theorem. The other possibility is that we may not be able to find the required actions on the part of player j that affect i's preferences. In this case, we have reached a point at which, given a player i history $h^t \in \mathbb{H}_i(h^t)$, there is nothing player j can do in period t that can have any effect on how player i evaluates continuation play. Such degeneracies are possible (Maskin and Tirole, 2001, provide an example), but we hereafter exclude them, assuming that the sufficient conditions of lemma 5.6.1 hold throughout.

We can thus work with a single consistent partition $\mathbb{H}$ and can refer to histories as being "equivalent" and "payoff equivalent" rather than i-equivalent and i-payoff equivalent.

We are now interested in a similar link between periods.

Lemma 5.6.2 *Suppose that for any players i and j, and period t, any equivalent histories h^t and $\hat{h}^t$, and any stage-game action profile a^t, there exists a repeated-game strategy profile σ and player j actions a_j^{t+1} and $\tilde{a}_j^{t+1}$ such that*

$$U_i^*((\cdot, \sigma_{-i}) \mid (h^t, a_{-i}^t, a_j^{t+1})) \not\sim U_i^*((\cdot, \sigma_{-i}) \mid (\hat{h}^t, a_{-i}^t, \tilde{a}_j^{t+1})). \tag{5.6.3}$$

If $(\mathbb{H}_1, \ldots, \mathbb{H}_n)$ is a consistent collection of partitions under which h^t and $\hat{h}^t$ are equivalent histories, then for any action profile a^t, (h^t, a^t) and $(\hat{h}^t, a^t)$ are equivalent.

Proof Fix a consistent collection $\mathbb{H} = (\mathbb{H}_1, \ldots, \mathbb{H}_n)$ and an action profile a^t. Suppose h^t and $\hat{h}^t$ are equivalent, and the action profile a^t, strategy σ, and player j actions a_j^{t+1} and $\tilde{a}_j^{t+1}$ satisfy (5.6.3). The strategy for every player k other than i and j that plays according to σ_k, except for playing a_k^t in period t, is measurable with respect to $\mathbb{H}$. We now suppose that (h^t, a^t) and $(\hat{h}^t, a^t)$ are not equivalent and derive a contradiction. In particular, the player j strategy of playing a_j^{t+1} and then playing according to σ_j, after any history $h^{t+1} \in \mathbb{H}((h^t, a^t))$, and otherwise playing $\tilde{a}_j^{t+1}$ (followed by σ_j) is then measurable with respect to $\mathbb{H}$ and from (5.6.3), allows us to conclude that h^t and $\hat{h}^t$ are not equivalent (recall (5.6.1)), the contradiction. ∎

The conditions of this lemma preclude cases in which player j's behavior in period $t + 1$ has no effect on player i's period t continuation payoffs. If the absence of such an effect, the set of payoff-relevant states in period $t + 1$ can be coarser than the set in period t.[29]

We say that games satisfying the conditions of lemmas 5.6.1 and 5.6.2 are *nondegenerate* and hereafter restrict attention to such games.

5.6.3 Markov Equilibrium

There are typically many consistent partitions. The trivial partition, in which every history constitutes an element, is automatically consistent. There is clearly nothing to be gained in defining a Markov equilibrium to be measurable with respect to this collection of partitions, because every strategy would then be Markov. Even if we restrict attention to nontrivial partitions, how do we know which one to pick?

The obvious response is to examine the maximally coarse consistent partition, meaning a consistent partition that is coarser than any other consistent partition.[30] This will impose the strictest version of the condition that payoff-irrelevant events should not matter. Does such a partition exist?

Proposition 5.6.1 *Suppose the game is nondegenerate (i.e., satisfies the hypotheses of lemmas 5.6.1 and 5.6.2). A maximally coarse consistent partition exists. If the stage game is finite, then this maximally coarse consistent partition is unique.*

Proof Let Φ be the set of all consistent partitions of histories. Endow this set with the partial order $\prec$ defined by $\hat{\mathbb{H}} \prec \mathbb{H}$ if $\mathbb{H}$ is a coarsening of $\hat{\mathbb{H}}$. We show that there exists a maximal element under this partial order, unique for finite games.

This argument proceeds in two steps. The first is to show that there exist maximal elements. This in turn follows from Zorn's lemma (Hrbacek and Jech 1984, p. 171), if we can show that every chain (i.e., totally ordered subset) $\mathscr{C} = \{\mathbb{H}_{(m)}\}_{m=1}^{\infty}$ admits an upper bound. Let $\mathbb{H}_{(\infty)}$ denote the finest common coarsening (or meet) of $\mathscr{C}$, that is, for each element $h \in \mathscr{H}$, $\mathbb{H}_{(\infty)}(h) = \cup_{m=1}^{\infty} \mathbb{H}_{(m)}(h)$. Because $\mathscr{C}$ is a chain, $\mathbb{H}_{(m)}(h) \subset \mathbb{H}_{(m+1)}(h)$, and so $\mathbb{H}_{(\infty)}$ is a partition that is coarser than every partition in $\mathscr{C}$. It remains to show that $\mathbb{H}_{(\infty)}$ is consistent. To do this, suppose that two histories h^t and $\hat{h}^t$ are contained in a common element of $\mathbb{H}_{(\infty)}$. Then they must be contained in some common element of $\mathbb{H}_{(m)}$ for some m, and hence must satisfy (5.6.1). This ensures that $\mathbb{H}_{(\infty)}$ is consistent, and so is an upper bound for the chain. Hence by Zorn's lemma, there is a maximally coarse consistent partition.

The second step is to show that there is a unique maximal element for finite stage games. To do this, it suffices to show that for any two consistent partitions, their meet (i.e., finest common coarsening) is consistent. Let $\mathbb{H}$ and $\hat{\mathbb{H}}$ be consistent partitions, and let $\bar{\mathbb{H}}$ be their meet. Suppose h^t and $\hat{h}^t$ are contained in a single element of $\bar{\mathbb{H}}$. Because the stage game is finite, $\mathbb{H}^t$ and $\hat{\mathbb{H}}^t$ are both finite partitions. Then, by the definition of meet, there is a finite sequence of histories

---

29. Again, Maskin and Tirole (2001) provide an example.
30. A partition $\mathbb{H}'$ is a *coarsening* of another partition $\mathbb{H}$ if for all $H \in \mathbb{H}$ there exists $H' \in \mathbb{H}'$ such that $H \subset H'$.

$\{h^t, h^t(1), \ldots, h^t(n), \hat{h}^t\}$ such that each adjacent pair is contained in either the same element of $\mathbb{H}^t$ or the same element of $\hat{\mathbb{H}}^t$, and hence satisfy payoff equivalence. But then h^t and $\hat{h}^t$ must be payoff equivalent, which suffices to conclude that $\bar{\mathbb{H}}$ is consistent.

∎

Denote the maximally coarse consistent partition by $\mathbb{H}^*$.

Definition 5.6.1 *A strategy profile σ is a* Markov strategy *profile if it is measurable with respect to the maximally coarse consistent partition $\mathbb{H}^*$. A strategy profile is a* Markov equilibrium *if it is a subgame-perfect equilibrium and it is Markov. Elements of the partition $\mathbb{H}^*$ are called* Markov states *or* payoff-relevant histories.

No difficulty arises in finding a Markov equilibrium in a repeated game, because one can always simply repeat the Nash equilibrium of the stage game, making history completely irrelevant. This is a reflection of the fact that in a repeated game, the maximally coarse partition is the set of all histories. Indeed, all Markov equilibria feature a Nash equilibrium of the stage game in every period.

Repeated games have the additional property that every history gives rise to an identical continuation game. As we noted in section 5.5.2, the Markov condition on strategies is commonly supplemented with the additional requirement that identical continuation games feature identical continuation strategies. Such strategies are said to be *stationary*. A stationary Markov equilibria in a repeated game must feature the same stage-game Nash equilibrium in every period.

More generally, it is straightforward to establish the existence of Markov equilibria in dynamic games with finite stage games and without private information. A backward induction argument ensures that finite horizon versions of the game have Markov equilibria, and discounting ensures that the limit of such equilibria, as the horizon approaches infinity, is a Markov equilibrium of the infinitely repeated game (Fudenberg and Levine 1983).

Now consider dynamic games G in the class described in section 5.5. The set S induces a partition on the set of ex post histories in a natural manner, with two ex post histories being equivalent under this partition if they are histories of the same length and end with the same state $s \in S$. Refer to this partition as $\mathbb{H}^S$. Because the continuation $G(s)$ is identical, regardless of the history terminating in s, the following is immediate:

Proposition 5.6.2 *Suppose G is a dynamic game in the class described in section 5.5. Suppose $\mathbb{H}^S$ is the partition of $\mathscr{H}$ with $h \in \mathbb{H}^S(h')$ if h and h' are of the same length and both result in the same state $s \in S$. Then, $\mathbb{H}^S$ is finer than $\mathbb{H}^*$. If for every pair of states $s, s' \in S$, there is at least one player i for which $u_i(s, a)$ and $u_i(s', a)$ are not affine transformations of one another, then $\mathbb{H}^* = \mathbb{H}^S$.*

The outcomes ω and ω' of a public correlating device have no effect on players' preferences and hence fail the condition that there exist a player i for whom $u_i(\omega, a)$ and $u_i(\omega', a)$ are not affine transformations of one another. Therefore, the outcomes of a public correlating device in a repeated game do not constitute states.

Markov equilibrium precludes the use of public correlation in repeated games and restricts the players in the prisoners' dilemma to consistent shirking. Alternatively, much of the interest in repeated games focuses on non-Markov equilibria.[31] In our view, the choice of an equilibrium is part of the construction of the model. Different choices, including whether Markov or not, may be appropriate in different circumstances, with the choice of equilibrium to be defended not within the confines of the model but in terms of the strategic interaction being modeled.

Remark 5.6.1 **Games of incomplete information** The notion of Markov strategy also plays an important role in incomplete information games, where the beliefs of uninformed players are often treated as Markov states (see, for example, section 18.4.4). At an intuitive level, this is the appropriate extension of the ideas in this section. However, determining equivalence classes of histories that are payoff-equivalent is now a significantly more subtle question. For example, because the inferences that players draw from histories depend on the beliefs that players have about past play, the equivalence classes now must satisfy a complicated fixed point property. Moreover, Markov equilibria (as just defined) need not exist, and this has led to the notion of a weak Markov equilibrium in the literature on bargaining under incomplete information (see, for example, Fudenberg, Levine, and Tirole, 1985). We provide a simple example of a similar phenomenon in section 17.3.

◆

5.7 Dynamic Games: Equilibrium

5.7.1 The Structure of Equilibria

This section explores some of the common ground between ordinary repeated games and dynamic games. Recall that we assume either that the set of states S is finite, or that the transition function is deterministic. The proofs of the various propositions we offer are straightforward rewritings of their counterparts for repeated games in chapter 2 and hence are omitted.

We say that strategy $\hat{\sigma}_i$ is a one-shot deviation for player i from strategy σ_i if there is a unique ex post history $\tilde{h}^t$ such that

$$\hat{\sigma}_i(\tilde{h}^t) \neq \sigma_i(\tilde{h}^t).$$

It is then a straightforward modification of proposition 2.2.1, substituting ex post histories for histories and replacing payoffs with expected payoffs to account for the potential randomness of the state transition function, to establish a one-shot deviation principle for dynamic games:

Proposition 5.7.1 *A strategy profile σ is subgame perfect in a dynamic game if and only if there are no profitable one-shot deviations.*

Given a dynamic game, an automaton is $(\mathcal{W}, \mathbf{w}^0, \tau, f)$, where $\mathcal{W}$ is the set of automaton states, $\mathbf{w}^0 \colon S \to \mathcal{W}$ gives the initial automaton state as a function of the initial game state, $\tau \colon \mathcal{W} \times A \times S \to \mathcal{W}$ is the transition function giving the automaton

---

[31]. This contrast may not be so stark. Maskin and Tirole (2001) show that most non-Markov equilibria of repeated games are limits of Markov equilibria in nearby dynamic games.

5.7 ■ Dynamic Games: Equilibrium

state in the next period as a function of the current automaton state, the current action profile, and the next draw of the game state. Finally, $f : \mathscr{W} \to \prod_i \Delta(A_i)$ is the output function. (Note that this description agrees with remark 2.3.3 when S is the space of realizations of the public correlating device.)

The initial automaton state is determined by the initial game state, through the function $\mathbf{w}^0$. We will often be interested in the strategy induced by the automaton beginning with an automaton state w. As in the case of repeated games, we write this as $(\mathscr{W}, w, \tau, f)$.[32]

Let $\tau(\tilde{h}^t)$ denote the automaton state reached under the ex post history $\tilde{h}^t \in \tilde{\mathscr{H}}^t = (S \times A)^t \times S$. Hence, for a history $\{s\}$ that identifies the initial game-state s, we have

$$\tau(\{s\}) = \mathbf{w}^0(s)$$

and for any ex post history $\tilde{h}^t = (\tilde{h}^{t-1}, a, s)$,

$$\tau(\tilde{h}^t) = \tau(\tau(\tilde{h}^{t-1}), a, s).$$

Given an ex post history $\tilde{h}^t \in \tilde{\mathscr{H}}$, let $s(\tilde{h}^t)$ denote the current game state in $\tilde{h}^t$. Given a game state $s \in S$, the set of automaton states *accessible in game state s* is $\mathscr{W}(s) = \{w \in \mathscr{W} : \exists \tilde{h}^t \in \tilde{\mathscr{H}}, w = \tau(\tilde{h}^t), s = s(\tilde{h}^t)\}$.[33]

Proposition 5.7.2 *Suppose the strategy profile σ is described by the automaton $\{\mathscr{W}, \mathbf{w}^0, \tau, f\}$. Then σ is a subgame-perfect equilibrium of the dynamic game if and only if for any game state $s \in S$ and automaton state w accessible in game state s, the strategy profile induced by $\{\mathscr{W}, w, \tau, f\}$, is a Nash equilibrium of the dynamic game $G(s)$.*

Our next task is to develop the counterpart for dynamic games of the recursive methods for generating equilibria introduced in section 2.5 for repeated games. Restrict attention to finite sets of signals (with $|S| = m$) and pure strategies. For each game state $s \in S$, and each state $w \in \mathscr{W}(s)$, associate the profile of values $V_s(w)$, defined by

$$V_s(w) = (1 - \delta)u(s, f(w)) + \delta \sum_{s' \in S} V_{s'}(\tau(w, f(w), s'))q(s' \mid s, f(w)).$$

As is the case with repeated games, $V_s(w)$ is the profile of expected payoffs when beginning the game in game state s and automaton state w. Associate with each game state $s \in S$, and each state $w \in \mathscr{W}(s)$, the function $g^{(s,w)}(a) : A \to \mathbb{R}^n$, where

$$g^{(s,w)}(a) = (1 - \delta)u(s, a) + \delta \sum_{s' \in S} V_{s'}(\tau(w, a, s'))q(s' \mid s, a).$$

Proposition 5.7.3 *Suppose the strategy profile σ is described by the automaton $(\mathscr{W}, \mathbf{w}^0, \tau, f)$. Then σ is a subgame-perfect equilibrium if and only if for all game states $s \in S$ and all $w \in \mathscr{W}(s)$, $f(w)$ is a Nash equilibrium of the normal-form game with payoff function $g^{(s,w)}$.*

---

32. Hence, $(\mathscr{W}, \mathbf{w}^0, \tau, f)$ is an automaton whose initial state is specified as a function of the game state, $(\mathscr{W}, \mathbf{w}^0(s), \tau, f)$ is the automaton whose initial state is given by $\mathbf{w}^0(s)$, and $(\mathscr{W}, w, \tau, f)$ is the automaton whose initial state is fixed at an arbitrary $w \in \mathscr{W}$.

33. Note that $\mathbf{w}^0(s)$ is thus accessible in game state s, even if game state s does not have positive probability under $q(\cdot \mid \varnothing)$.

Let $\mathscr{W}^s$ be a subset of $\mathbb{R}^n$, for $s = 1, \ldots, m$. We interpret this set as a set of feasible payoffs in dynamic game $G(s)$. We say that the pure action profile a^* is *pure-action enforceable* on $(\mathscr{W}^1, \ldots, \mathscr{W}^m)$ given $s \in S$ if there exists a function $\gamma : A \times S \to \mathscr{W}^1 \cup \ldots \cup \mathscr{W}^m$ with $\gamma(a, s') \in \mathscr{W}^{s'}$ such that for all players i and all $a_i \in A_i$,

$$(1 - \delta)u_i(s, a^*) + \delta \sum_{s' \in S} \gamma_i(a^*, s')q(s' \mid s, a^*)$$
$$\geq (1 - \delta)u_i(s, a_i, a^*_{-i}) + \delta \sum_{s' \in S} \gamma_i(a_i, a^*_{-i}, s')q(s' \mid s, a_i, a^*_{-i}).$$

We say that the payoff profile $v \in \mathbb{R}^n$ is *pure-action decomposable* on $(\mathscr{W}^1, \ldots, \mathscr{W}^m)$ given $s \in S$ if there exists a pure action profile a^* that is pure-action enforceable on $(\mathscr{W}^1, \ldots, \mathscr{W}^m)$ given s, with the enforcing function γ satisfying, for all players i,

$$v_i = (1 - \delta)u_i(s, a^*) + \delta \sum_{s' \in S} \gamma_i(a^*, s')q(s' \mid s, a^*).$$

A vector of payoff profiles $\mathbf{v} \equiv (v(1), \ldots, v(m)) \in \mathbb{R}^{nm}$, with $v(s)$ interpreted as a payoff profile in game $G(s)$, is *pure-action decomposable* on $(\mathscr{W}^1, \ldots, \mathscr{W}^m)$ if for all $s \in S$, $v(s)$ is pure-action decomposable on $(\mathscr{W}^1, \ldots, \mathscr{W}^m)$ given s. Finally, $(\mathscr{W}^1, \ldots, \mathscr{W}^m)$ is *pure-action self-generating* if every vector of payoff profiles in $\prod_{s \in S} \mathscr{W}^s$ is pure-action decomposable on $(\mathscr{W}^1, \ldots, \mathscr{W}^m)$. We then have:[34]

Proposition 5.7.4 *Any self-generating set of payoffs $(\mathscr{W}^1, \ldots, \mathscr{W}^m)$ is a set of pure-strategy subgame-perfect equilibrium payoffs.*

As before, we have the corollary:

Corollary 5.7.1 *The set $(\mathscr{W}^{1*}, \ldots, \mathscr{W}^{m*})$ of pure-strategy subgame-perfect equilibrium payoff profiles is the largest pure-action self-generating collection $(\mathscr{W}^1, \ldots, \mathscr{W}^m)$.*

Remark 5.7.1 **Repeated games with random states** For these games (see remark 5.5.1), the set of ex ante feasible payoffs is independent of last period's state and action profile. Consequently, it is simpler to work with ex ante continuations in the notions of enforceability, decomposability, and pure-action self-generation. A pure action profile a^* is *pure-action enforceable* in state $s \in S$ on $\mathscr{W} \subset \mathbb{R}^n$ if there exists a function $\gamma : A \to \mathscr{W}$ with such that for all players i and all $a_i \in A_i$,

$$(1 - \delta)u_i(s, a^*) + \delta \gamma_i(a^*) \geq (1 - \delta)u_i(s, a_i, a^*_{-i}) + \delta \gamma_i(a_i, a^*_{-i}).$$

The ex post payoff profile $v^s \in \mathbb{R}^n$ is *pure-action decomposable* in state s on $\mathscr{W}$ if there exists a pure-action profile a^* that is pure-action enforceable in s on $\mathscr{W}$ with the enforcing function γ satisfying $v^s = (1 - \delta)u(s, a^*) + \delta \gamma(a^*)$. An ex ante payoff profile $v \in \mathbb{R}^n$ is *pure-action decomposable* on $\mathscr{W}$ if there exist ex post payoffs $\{v^s : s \in S\}$, v^s pure-action decomposable in s on $\mathscr{W}$, such that $v = \sum_s v^s q(s)$. Finally, $\mathscr{W}$ is *pure-action self-generating* if every payoff profile in $\mathscr{W}$ is pure-action decomposable on $\mathscr{W}$. As usual, any self-generating set of ex ante

---

34. See section 9.7 on games of symmetric incomplete information (in particular, proposition 9.7.1 and lemma 9.7.1) for an application.

payoffs is a set of subgame-perfect equilibrium ex ante payoffs, with the set of subgame-perfect equilibrium ex ante payoffs being the largest such set.

Section 6.3 analyzes a repeated game with random states in which players have an opportunity to insure one another against endowment shocks. Interestingly, this game is an example in which the efficient symmetric equilibrium outcome is sometimes necessarily nonstationary. It is no surprise that the equilibrium itself might not be stationary, that is, that efficiency might call for nontrivial intertemporal incentives and their attendant punishments. However, we have the stronger result that efficiency cannot be obtained with a stationary-outcome equilibrium.

◆

5.7.2 A Folk Theorem

This section presents a folk theorem for dynamic games. We assume that the action spaces and the set of states are finite.

Let Σ be the set of pure strategies in the dynamic game and let Σ^M be the set of pure Markov strategies. For any $\sigma \in \Sigma$, let

$$U^\delta(\sigma) = E^\sigma \left\{ (1-\delta) \sum_{t=0}^{\infty} \delta^t u(s^t, a^t) \right\}$$

be the expected payoff profile under strategy σ, given discount factor δ. The expectation accounts not only for the possibility of private randomization and public correlation but also for randomness in state transitions, including the determination of the initial state. We let

$$U^\delta(\sigma \mid h^t) = E^{\sigma, h^t} \left\{ (1-\delta) \sum_{\tau=t}^{\infty} \delta^{\tau-t} u(a^\tau, s^\tau) \right\}$$

be the analogous expectation conditioned on having observed the history h^t, where continuation play is given by $\sigma|_{h^t}$. As a special case of this, we have the expected payoff $U^\delta(\sigma \mid s)$, which conditions on the initial state realization s.

Let

$$\mathscr{F}(\delta) = \{\mathbf{v} \in \mathbb{R}^{nm} : \exists \sigma \in \Sigma^M \text{ s.t. } U^\delta(\sigma \mid s) = v(s) \, \forall s \}.$$

This is our counterpart of the set of payoff profiles produced by pure stage-game actions in a repeated game. There are two differences here. First, we identify functions that map from initial states to expected payoff profiles. Second, we now work directly with the collection of repeated-game payoffs rather than with stage-game payoffs because we have no single underlying stage game. In doing so, we have restricted attention only to payoffs produced by pure Markov strategies. We comment shortly on the reasons for doing so, and in the penultimate paragraph of this section on the sense in which this assumption is not restrictive.

We let

$$\mathscr{F} = \lim_{\delta \to 1} \mathscr{F}(\delta). \tag{5.7.1}$$

This is our candidate for the set of feasible pure-strategy payoffs available to patient players. We similarly require a notion of minmax payoffs, which we define contingent on the current state,

$$\underline{v}_i^\delta(s) = \inf_{\sigma_{-i} \in \Sigma_{-i}} \sup_{\sigma_i \in \Sigma_i} U_i^\delta(\sigma \mid s)$$

with

$$\underline{v}_i(s) = \lim_{\delta \to 1} \underline{v}_i^\delta(s). \tag{5.7.2}$$

Dutta (1995, lemma 2, lemma 4) shows that the limits in (5.7.1) and (5.7.2) exist. The restriction to Markov strategies in defining $\mathscr{F}(\delta)$ is useful here, as it is relatively easy to show that the payoffs to a pure Markov strategy, in the presence of finite sets of actions and states, are continuous as $\delta \to 1$.

Finally, we say that the collection of pure strategies $\{\sigma^1, \ldots, \sigma^n\}$ is a *player-specific punishment* for **v** if the limits $U(\sigma^i \mid s) = \lim_{\delta \to 1} U^\delta(\sigma^i \mid s)$ exist and the following hold for all s, s', and s'' in S:

$$U_i(\sigma^i \mid s') < v_i(s) \tag{5.7.3}$$

and

$$\underline{v}_i(s'') < U_i(\sigma^i \mid s') < U_i(\sigma^j \mid s). \tag{5.7.4}$$

We do not require the player-specific punishments to be Markov. The inequalities in conditions for player-specific punishments are required to hold uniformly across states. This imposes a tremendous amount of structure on the payoffs involved in these punishments. We comment in the final paragraph of this section on sufficient conditions for the existence of such punishments.

We then have the pure-strategy folk theorem.

Proposition 5.7.5 *Let* **v** $\in \mathscr{F}$ *be strictly individually rational, in the sense that for all players i and pairs of states s and s' we have*

$$v_i(s) > \underline{v}_i(s'),$$

let **v** *admit a player specific punishment, and suppose that the players have access to a public correlating device. Then, for any $\varepsilon > 0$, there exists $\underline{\delta}$ such that for all $\delta \in (\underline{\delta}, 1)$, there exists a subgame-perfect equilibrium σ whose payoffs $U(\sigma \mid s)$ are within ε of $v(s)$ for all $s \in S$.*

The proof of this proposition follows lines that are familiar from proposition 3.4.1 for repeated games. The additional complication introduced by the dynamic game is that there may now be two reasons to deviate from an equilibrium strategy. One is to obtain a higher current payoff. The other, not found in repeated games, is to affect the transitions of the state. In addition, this latter incentive potentially becomes more powerful as the players become more patient, and hence the benefits of affecting the future state become more important. We describe the basic structure of the proof. Dutta (1995) can be consulted for more details.

5.7 ■ Dynamic Games: Equilibrium

Proof We fix a sequence of values of δ approaching 1 and corresponding strategy profiles $\sigma(\delta)$ with the properties that

$$\lim_{\delta \to 1} U^\delta(\sigma(\delta) \mid s) = v(s).$$

This sequence allows us to approach the desired equilibrium payoffs. Moreover, Dutta (1995, proposition 3) shows that there exists a strategy profile $\hat{\sigma}$ such that for any $\eta > 0$, there exists an integer $L(\eta)$ such that for all $T \geq L(\eta)$ and $s \in S$,

$$E^{\hat{\sigma}, s} \frac{1}{T} \sum_{t=0}^{T-1} u_i(s^t, a^t) \leq \underline{v}_i^\delta(s) + \frac{\eta}{2},$$

and hence a value $\delta_1 < 1$ such that for $\delta \in (\delta_1, 1)$,

$$E^{\hat{\sigma}, s} \frac{1}{T} \sum_{t=0}^{T-1} u_i(s^t, a^t) \leq \underline{v}_i(s) + \eta.$$

Let $\{\sigma^1, \ldots, \sigma^n\}$ be the player-specific punishment for **v**. Conditions (5.7.3) and (5.7.4) ensure that we can fix $\delta_2 \in (\delta_1, 1)$, $\overline{\underline{v}}_i \equiv \max_{s \in S} \underline{v}_i(s)$ and η sufficiently small that, for all $\delta \in (\delta_2, 1)$, all i, and for any states s, s' and s'',

$$\underline{v}_i(s) + \eta \leq \overline{\underline{v}}_i + \eta < U_i^\delta(\sigma^i \mid s') < v_i(s'')$$

and

$$U_i^\delta(\sigma^i \mid s) < U_i^\delta(\sigma^j \mid s').$$

We now note that we can assume, for each player i, that the player-specific punishment σ^i has the property that there exists a length of time T_i such that for each $t, t' = 0, T_i, 2T_i, \ldots$, and for all ex ante histories h^t and $h^{t'}$ and states s, $\sigma^i|_{\{h^t, s\}} = \sigma^i|_{\{h^{t'}, s\}}$. Hence, σ^i has a cyclical structure, erasing its history and starting from the beginning every T_i periods.[35] We can also assume that σ^i provides a payoff to player i that is independent of its initial state.[36] We hereafter retain these properties for the player specific punishments.

The strategy profile now mimics that used to prove proposition 3.4.1, the corresponding result for repeated games. It begins with play following the strategy profile $\sigma(\delta)$; any deviation by player i from $\sigma(\delta)$, and indeed any deviation from any subsequent equilibrium prescription other than deviations from being

---

35. Suppose σ^i does not have this property. Because each inequality in (5.7.3) and (5.7.4) holds by at least ε, for some ε, we need only choose T_i sufficiently large that the average payoff from any strategy profile over its first T_i periods is within at least $\varepsilon/3$ of its payoff. Now construct a new strategy by repeatedly playing the first T_i periods of σ^i, beginning each time with the null history. This new strategy has the desired cyclic structure and satisfies (5.7.3) and (5.7.4). Dutta (1995, section 6.2) provides details.

36. Suppose this is not the case. Let s maximize $U_i(\sigma^i \mid s)$. Define a new strategy as follows: In each period $0, T_i, 2T_i, \ldots$, conduct a state-contingent public correlation that mixes between σ^i and a strategy that maximizes player i's repeated-game payoff, with the correlation set so as to equate the continuation payoff for each state s' with $U_i(\sigma^i \mid s)$. The uniform inequalities of the player-specific punishments ensure this is possible.

minmaxed, which are ignored, prompts the first L periods of the corresponding minmax strategy $\hat{\sigma}^i(\delta)$, followed by the play of the player-specific punishment σ^i.

Our task now is to show that these strategies constitute a subgame-perfect equilibrium, given the freedom to restrict attention to large discount factors and choose $L \geq L(\eta)$. As usual, let M and m be the maximum and minimum stage-game payoffs.

The condition for deviations from the equilibrium path to be unprofitable is that, for any state $s \in S$ (suppressing the dependence of strategies on δ)

$$(1-\delta)M + \delta(1-\delta^L)(\underline{v}_i + \eta) + \delta^{L+1}U_i^\delta(\sigma^i) \leq U_i^\delta(\sigma \mid s),$$

which, as δ converges to one for fixed L, becomes $U_i(\sigma^i) \leq v_i(s)$, which holds with strict inequality by virtue of our assumption that $\mathbf{v}$ admits player-specific punishments. There is then a value $\delta_3 \in [\delta_2, 1)$ such that this constraint holds for any $\delta \in (\delta_3, 1)$ and $L \geq L(\eta)$.

For player j to be unwilling to deviate while minmaxing i, the condition is

$$(1-\delta)M + \delta(1-\delta^L)(\underline{v}_j + \eta) + \delta^{L+1}U_j^\delta(\sigma^j) \leq (1-\delta^L)m + \delta^L U_j^\delta(\sigma^i).$$

Rewrite this condition as

$$(1-\delta)M + (1-\delta^L)(\delta(\underline{v}_j + \eta) - m) + \delta^L[\delta U_j^\delta(\sigma^j) - U_j^\delta(\sigma^i)] \leq 0.$$

The term $[\delta U_j^\delta(\sigma^j) - U_j^\delta(\sigma^i)]$ converges to $U_j(\sigma^j) - U_j(\sigma^i) < 0$ as $\delta \to 1$. We can then find a value $\delta_4 \in [\delta_3, 1)$ and an increasing function $L(\delta)$ $(\geq L(\eta))$ such that this constraint holds for any $\delta \in (\delta_4, 1)$ and the associated $L(\delta)$, and such that $\delta^{L(\delta)} < 1 - \gamma$, for some $\gamma > 0$. We hereafter take L to be given by $L(\delta)$.

Now consider the postminmaxing rewards. For player i to be willing to play σ^i, a sufficient condition is that for any ex post history $\tilde{h}^t$ under which current play is governed by σ^i,

$$(1-\delta)M + \delta(1-\delta^{L(\delta)})(\underline{v}_i + \eta) + \delta^{L(\delta)+1}U_i^\delta(\sigma^i) \leq U_i^\delta(\sigma^i \mid \tilde{h}^t).$$

This inequality is not obvious. The difficulty here is that we cannot exclude the possibility that $U_i^\delta(\sigma^i) > U_i^\delta(\sigma^i \mid \tilde{h}^t)$. There is no reason to believe that player i's payoff from strategy σ^i is constant across time or states. Should player i find himself at an ex post history (h^t, s) in which this strategy profile gives a particularly low payoff, i may find it optimal to deviate, enduring the resulting minmaxing to return to the relatively high payoff of beginning σ^i from the beginning. This is the incentive to deviate to affect the state that does not appear in an ordinary repeated game. In addition, this incentive seemingly only becomes stronger as the player gets more patient, and hence the intervening minmaxing becomes less costly.

A similar issue arises in the proof of proposition 3.8.1, the folk theorem for repeated games without public correlation, where we faced the fact that the deterministic sequences of payoffs designed to converge to a target payoff may feature continuation values that differ from the target. In the case of proposition 3.8.1,

5.7 ■ Dynamic Games: Equilibrium

the response involved a careful balancing of the relative sizes of δ and L. Here, we can use the cyclical nature of the strategy σ^i to rewrite this constraint as

$$(1 - \delta)M + \delta(1 - \delta^{L(\delta)})(v_i + \eta) + \delta^{L(\delta)+1}U_i^\delta(\sigma^i) \leq (1 - \delta^{T_i})m + \delta^{T_i}U_i^\delta(\sigma^i),$$

where $(1 - \delta^{T_i})m$ is a lower bound on player i's payoff from σ^i over T_i periods, and then the strategy reverts to payoff $U_i^\delta(\sigma^i)$. The key to ensuring this inequality is satisfied is to note that T_i is fixed as part of the specification of σ^i. As a result, $\lim_{\delta \to 1} \delta^{T_i} = 1$ while $\delta^{L(\delta)}$ remains bounded away from 1.

A similar argument establishes that a sufficiently patient player i has no incentive to deviate when in the middle of strategy σ^j. This argument benefits from the fact that a deviation trades a single-period gain for L periods of being minmaxed followed by a return to the less attractive payoff $U_i^\delta(\sigma^i)$. Letting $\delta_5 \in (\delta_4, 1)$ be the bound on the discount factor to emerge from these two arguments, these strategies constitute a subgame-perfect equilibrium for all $\delta \in (\delta_5, 1)$.

■

We have worked throughout with pure strategies and with pure Markov strategies when defining feasible payoffs. Notice first that these are pure strategies in the dynamic game. We are thus not restricting ourselves to the set of payoffs that can be achieved in pure stage-game actions. This makes the pure strategy restriction less severe than it may first appear. In addition, any feasible payoff can be achieved by publicly mixing over pure Markov strategies (Dutta 1995, lemma 1), so that the Markov restriction is also not restrictive in the presence of public correlation (which we used in modifying the player-specific punishments in the proof).

Another aspect of this proposition can be more directly traced to the dynamic structure of the game. We have worked with a function $\mathbf{v}$ that specifies a payoff profile for each state. Suppose instead we defined, for each $s \in S$,

$$\mathscr{F}(\delta, s) = \{v \in \mathbb{R}^n : \exists \sigma \in \Sigma^M \text{ s.t. } U^\delta(\sigma \mid s) = v\},$$

the set of feasible payoffs (in pure Markov strategies) given initial state s, with $\mathscr{F}(s) = \lim_{\delta \to 1} \mathscr{F}(\delta, s)$. Let us say that a stochastic game is *communicating* if for any pair of states s and s', there is a strategy σ and a time t such that if the game begins in state s, there is positive probability under strategy σ that the game is in state s' in period t. Dutta (1995, lemma 12) shows that in communicating games, $\mathscr{F}(s)$ is independent of s. If the game communicates independently of the actions of player i, for each i, then minmax values will also be independent of the state. In this case, we can formulate the folk theorem for stochastic games in terms of payoff profiles $v \in \mathbb{R}^n$ and minmax profiles v_i that do not depend on the initial state. In addition, full dimensionality of the convex hull of $\mathscr{F}$ then suffices for the existence of player-specific punishments for interior v. We can establish a similar result in games in which $\mathscr{F}(\delta, s)$ depends on s (and hence which are not communicating), in terms of payoffs that do not depend on the initial state by concentrating on those payoffs in the set $\cap_{s \in S}\mathscr{F}(s)$.

6 Applications

This chapter offers three examples of how repeated games of perfect monitoring have been used to address questions of economic behavior.

6.1 Price Wars

More stage-game outcomes can be supported as repeated-game equilibrium actions in some circumstances than in others. This section exploits this insight in a simple model of collusion between imperfectly competitive firms. We are interested in whether successful collusion is likely to be procyclical or countercyclical.[1]

6.1.1 Independent Price Shocks

We consider a market with n firms who costlessly produce a homogeneous output. In each period of the infinitely repeated game, a state s is first drawn from the finite set S and revealed to the firms. The state s is independently drawn in each period, with $q(s)$ being the probability of state s. We thus have a repeated game with random states (remarks 5.5.1 and 5.7.1).

After observing the states, the firms simultaneously choose prices for that period. If the state is s and the firms set prices $p_1, \ldots, p_n$, then the quantity demanded is given by $s - \min\{p_1, \ldots, p_n\}$. This quantity is split evenly among those firms who set the minimum price.[2]

The stage game has a unique Nash equilibrium outcome in which the minimum price is set at 0 in every period and each firm earns a 0 payoff. This is the mutual minmax payoff for this game. In contrast, the myopic monopoly price in state s is $s/2$, for a total payoff of $(s/2)^2$. Higher states feature higher monopoly profits—it is more valuable to (perfectly) collude when demand is high. However, it is also more lucrative to undercut the market price slightly, capturing the entire market at only a slightly smaller price, when demand is high.

---

1. This example is motivated by Rotemberg and Saloner (1986), who establish conditions under which oligopolistic firms will behave more competitively when demand is high rather than low.
2. This game violates assumption 2.1.1, because action sets are continua while the payoff function has discontinuities. The existence of a stage-game Nash equilibrium is immediate, so the discontinuity poses no difficulties.

In the repeated game, the firms have a common discount factor δ. We consider the strongly symmetric equilibrium that maximizes the firms' expected payoffs, which we refer to as the most collusive equilibrium. Along the equilibrium path, the firms set a price denoted by $p(s)$ whenever the state is s. Given our interest in an equilibrium that maximizes the firms' expected payoffs, we can immediately assume that $p(s) \leq s/2$, and can assume that deviations from equilibrium play are punished by perpetual minmaxing.

Let v^* be the expected payoff from such a strategy profile. Necessary and sufficient conditions for the strategy profile to be an equilibrium are, for each state s,

$$\frac{1}{n}(1-\delta)p(s)(s - p(s)) + \delta v^* \geq (1-\delta)p(s)(s - p(s)), \qquad (6.1.1)$$

where

$$v^* = \frac{1}{n} \sum_{\hat{s} \in S} p(\hat{s})(\hat{s} - p(\hat{s}))q(\hat{s}). \qquad (6.1.2)$$

Condition (6.1.1) ensures that the firm would prefer to set the prescribed price $p(s)$ and receive the continuation value v^* rather than deviate to a slightly smaller price (where the right side is the supremum over such prices) followed by subsequent minmaxing. Condition (6.1.2) gives the continuation value v^*, which is independent of the current state.

We are interested in the function $p(s)$ that maximizes v^*, among those that satisfy (6.1.1)–(6.1.2). For sufficiently large δ, the expected payoff v^* is maximized by setting $p(s) = s/2$, the myopic profit maximizing price, for every state s. This reflects the fact that the current state of demand fades into insignificance for high discount factors.

Suppose that the discount factor is too low for this to be an equilibrium. In the most collusive equilibrium, the constraint (6.1.1) can be rewritten as

$$\frac{n\delta v^*}{(n-1)(1-\delta)} \geq p(s)(s - p(s)). \qquad (6.1.3)$$

Hence, there exists $\bar{s} < \max S$ such that for all $s > \bar{s}$, $p(s)$ is the smaller of the two roots solving, from (6.1.3),[3]

$$p(s)(s - p(s)) = \frac{n\delta v^*}{(n-1)(1-\delta)}, \qquad (6.1.4)$$

and if $\bar{s} \in S$, then for all $s \leq \bar{s}$, $p(s)$ is the myopic monopoly price

$$p(s) = \frac{s}{2}.$$

It is immediate from (6.1.4) that for $s > \bar{s}$, $p(s)$ is decreasing in s, giving countercyclical collusion in high states. The most profitable deviation open to a firm is to

---

3. One of the solutions to (6.1.4) is larger than the myopic monopoly price and one smaller, with the latter being the root of interest. Deviations from any price higher than the monopoly price are as profitable as deviations from the monopoly price, so that the inability to enforce the monopoly price ensures that no higher price can be enforced.

undercut the equilibrium price by a minuscule margin, jumping from a $1/n$ share of the market to the entire market, at essentially the current price. The higher the state, the more tempting this deviation. For states $s > \bar{s}$, this deviation is profitable if firms set the monopoly price. They must accordingly set a price lower than the monopoly price, reducing the value of the market and hence the temptation to capture the entire market. The higher the state, and hence the higher the quantity demanded at any given price, the lower must the equilibrium price be to render this deviation unprofitable. The function $p(s)$ is thus decreasing for $s > \bar{s}$.[4] In periods of relatively high demand, colluding firms set lower prices.

6.1.2 Correlated Price Shocks

The demand shocks in section 6.1.1 are independently and identically distributed over time. Future values of the equilibrium path are then independent of the current realization of the demand shock, simplifying the incentive constraint given by (6.1.1). However, this assumption creates some tension with the interpretation of the model as describing how price wars might arise during economic booms or how pricing policies might change over the course of a business cycle.

Instead of independent distributions, one might expect the shock describing a business cycle to exhibit some persistence. If this persistence is sufficiently strong, then collusion may no longer be countercyclical. It is still especially profitable to cheat on a collusive agreement when demand is high, but continuation payoffs are now also especially lucrative in such states, making cheating more costly.

We illustrate these issues with a simple example. Let there be two firms. Demand can be either high ($s = 2$) or low ($s = 1$). The firms have discount factor $\delta = 11/20$.

When demand is low, joint profit maximization requires that each firm set a price of $1/2$, earning a payoff of $1/8$. When demand is high, joint profit maximization calls for a price of 1, with accompanying payoff of $1/2$.

Consider the most collusive equilibrium with independent equally likely shocks. When demand is low, each firm sets price $1/2$. When demand is high, each firm sets price $\tilde{p}$, the highest price for which the high-demand incentive constraint,

$$\sup_{p<\tilde{p}}[(1-\delta)p(2-p)] \leq (1-\delta)\tfrac{1}{2}\tilde{p}(2-\tilde{p}) + \delta\left(\tfrac{1}{2}\tfrac{1}{8} + \tfrac{1}{2}\tfrac{1}{2}\tilde{p}(2-\tilde{p})\right),$$

is satisfied. The left side of this constraint is the largest payoff that can be secured by undercutting $\tilde{p}$ to capture the entire high-demand market, whereas the right side is the value of the equilibrium strategy, conditional on high demand and noting that price $\tilde{p}$ is set in high-demand states. We solve this expression for

$$\tilde{p} = 0.22.$$

Hence, collusion is countercyclical in the sense that lower prices appear in the high-demand state. High demand coupled with collusion gives rise to especially strong incentives to cheat. These can be deterred only by making the collusion less valuable, leading to lower prices than those set in the low-demand state.

---

4. If $\bar{s} \in S$, for $s \leq \bar{s}$, the (monopoly) prices $p(s)$ may exceed the prices attached to some states $s > \bar{s}$. Section 6.1.2 presents an example.

Now suppose that the state is given by a Markov process. The prior distribution makes high and low demand equally likely in the first period, and thereafter the period t state is identical to the period $t-1$ state with probability $1-\phi$ and switches to the other state with probability ϕ. Suppose ϕ is quite small, so that rather than being almost independent draws, we have persistent states.

Consider a strategy profile in which the myopic monopoly price is set in each period, with deviations again prompting perpetual minmaxing. As ϕ approaches 0, the continuation value under this profile approaches $1/8$ when the current state is low and $1/2$ when the current state is high, each reflecting the value of receiving half of the monopoly profits in every future period, with the state held fixed at its current value. The incentive constraints for equilibrium approach

$$\sup_{p<\frac{1}{2}} (1-\delta)p(1-p) \leq \tfrac{1}{8}$$

and

$$\sup_{p<1} (1-\delta)p(2-p) \leq \tfrac{1}{2}.$$

These constraints hold if and only if $\delta \geq 1/2$. Given our discount factor of $\delta = 11/20$, there is a value $\phi^* > 0$ such that if $\phi \leq \phi^*$, there exists an equilibrium in which firms set the myopic monopoly price in each period. Collusion is now procyclical in the sense that higher prices and profits appear when demand is high.[5] Repeated games can thus serve as a framework for assessing patterns of collusion over the business cycle, but the conclusions depend importantly on the nature of the cycle.

6.2 Time Consistency

6.2.1 The Stage Game

The stage game is a simple model of an economy. Player 1 is a government. The role of player 2 is played by a unit continuum of small and anonymous investor/consumers. As we have noted in section 2.7 (in particular in remark 2.7.1), this has the effect of ensuring that player 2 is a short-lived player when we consider the repeated game.

Each consumer is endowed with one unit of a consumption good. The consumer divides this unit between consumption c and capital $1-c$. Capital earns a gross return of R, so that the consumer amasses $R(1-c)$ units of capital. The government sets a tax rate t on capital, collecting revenue $tR(1-c)$, with which it produces a public good. One unit of revenue produces $\gamma > 1$ of the public good, where $R - 1 < \gamma < R$. Untaxed capital is consumed.[6]

---

5. Bagwell and Staiger (1997), Haltiwanger and Harrington (1991), and Kandori (1991b) examine variations on this model with correlated shocks, establishing conditions under which collusive behavior is either more or less likely to occur during booms or recessions.

6. In a richer economic environment, period t untaxed capital is the period $t+1$ endowment, and so investors face a nontrivial intertemporal optimization problem. The essentials of the analysis are unchanged, because small and anonymous investors (while intertemporally optimizing) assume their individual behavior will not affect the government's behavior. Small and anonymous players who intertemporally optimize appear in example 5.5.3.

6.2 ■ Time Consistency

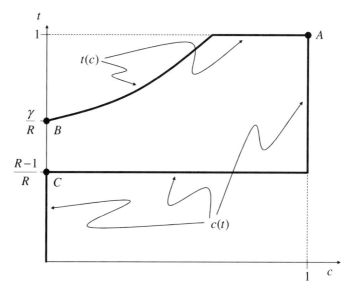

Figure 6.2.1 Consumer best response $c(t)$ as a function of the tax rate t, and government best response $t(c)$ as a function of the consumption c. The unique stage-game equilibrium outcome features $c = 1$, as at A, whereas the efficient allocation is B and the constrained (by consumer best responses) efficient allocation is C.

The consumer's utility is given by

$$c + (1 - t)R(1 - c) + 2\sqrt{G}, \tag{6.2.1}$$

where G is the quantity of public good. The government chooses its tax rates to maximize the representative consumer's utility (i.e., the government is benevolent). We examine equilibria in which the consumers choose identical strategies.

Each individual consumer makes a negligible contribution to the government's tax revenues, and accordingly treats G as fixed. The consumer thus chooses c to maximize $c + (1 - t)R(1 - c)$. The consumer's optimal behavior, as a function of the government's tax rate t, is then given by:

$$c = \begin{cases} 0, & \text{if } t < \frac{R-1}{R}, \\ 1, & \text{if } t > \frac{R-1}{R}. \end{cases}$$

When every consumer chooses consumption level c, the government's best response is to choose the tax rate maximizing

$$c + (1 - t)R(1 - c) + 2\sqrt{\gamma t R(1 - c)},$$

where the government recognizes that the quantity of the public good depends on its tax rate. The government's optimal tax rate as a function of c is

$$t = \min\left\{\frac{\gamma}{R(1 - c)}, 1\right\}. \tag{6.2.2}$$

Figure 6.2.1 illustrates the best responses of consumers and the government.[7]

---

7. We omit in figure 6.2.1 the fact that if consumers set $c = 1$, investing none of their endowment, then the government is indifferent over all tax rates, because all raise a revenue of 0.

Because the government's objective is to maximize the consumers' utilities, it appears as if there should be no conflict of interest in this economy. The efficient outcome calls for consumers to set $c = 0$. Because $R > 1$, investing in capital is productive, and because the option remains of using the accumulated capital for either consumption or the public good, this ensures that it is efficient to invest all of the endowment. The optimal tax rate (from (6.2.2)) is $t = \gamma/R$. This gives the allocation B in figure 6.2.1.

6.2.2 Equilibrium, Commitment, and Time Consistency

It is apparent from figure 6.2.1 that the stage game has a unique Nash equilibrium outcome in which consumers invest none of their endowment ($c = 1$) and the government tax rate is set sufficiently high as to make investments suboptimal. Outcome A in figure 6.2.1 is an example.[8] No matter what consumers do, the government's best response is a tax rate above $(R-1)/R$, the maximum rate at which consumers find it optimal to invest. No equilibrium can then feature positive investment by consumers. Notice that the stage-game equilibrium minmaxes both the consumer and the government.

It is impossible to obtain the efficient allocation (B in figure 6.2.1) as an equilibrium outcome, even in a repeated game. The consumers are not choosing best responses at B, whereas small and anonymous players must play best responses either in a stage-game equilibrium or in each period of a repeated-game equilibrium (section 2.7). Constraining consumers to choose best responses, the allocation that maximizes the government's (and hence also the consumers') payoffs is C in figure 6.2.1. The government sets the highest tax rate consistent with consumers' investing, given by $(R-1)/R$, and the latter invest all of their endowment. Let $\bar{v}_1$ denote the resulting payoff profile for the government (this is the pure-action Stackelberg payoff of section 2.7.2).

If the government could choose its tax rate first with this choice observed by consumers before they make their investment decisions, then it can guarantee a payoff (arbitrarily close to) $\bar{v}_1$. In the absence of the ability to do so, we can say that the government has a *commitment problem*—its payoff could be increased by the ability to commit to a tax rate before consumers make their choices.

Alternatively, this is often described as a *time consistency problem*, or the government is described as having a tax rate $((R-1)/R)$ that is optimal but *time inconsistent* (Kydland and Prescott, 1977). In keeping with the temporal connotation of the phrase, it is common to present the game with a sequential structure. For example, we might let consumers choose their allocations first, to be observed by the government, who then chooses a tax rate. The unique subgame-perfect equilibrium of this sequential stage game again features no investment, whereas a better outcome could be obtained if the government could commit to the time-inconsistent tax rate of $(R-1)/R$. Alternatively, we might precede either the simultaneous or sequential stage game with a stage at which the government makes a cheap-talk announcement of what its tax rate

---

8. There are other Nash equilibria in which the government sets a tax rate less than 1 (because the government is indifferent over all tax rates when $c = 1$), but they all involve $c = 1$.

6.2 ■ Time Consistency

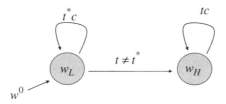

Figure 6.2.2 Candidate equilibrium for repeated game. In the low-tax/investment state w_L, the government chooses $t = t^*$ and consumers invest everything, and in the high-tax/no-investment state w_H, the government chooses $t = 1$ and consumers invest nothing.

is going to be. Here, time inconsistency appears in the fact that the government would then prefer an announcement that would induce investment, if there were any, but any such announcement would be followed by an optimal tax rate that rendered the investment suboptimal. Again, this is ultimately the observation that the government's payoff could be increased by the ability to commit to an action.

6.2.3 The Infinitely Repeated Game

Now suppose that the stage game is infinitely repeated, with the government discounting at rate δ.[9] As expected, repeated play potentially allows the government to effectively commit to more moderate tax rates. Let $t^* \equiv (R-1)/R$ be the optimal tax rate and consider the strategy profile described by the automaton $(\mathscr{W}, w_L, f, \tau)$, with states $\mathscr{W} = \{w_L, w_H\}$, output function

$$f(w) = \begin{cases} (t^*, 0), & \text{if } w = w_L, \\ (1, 1), & \text{if } w = w_H, \end{cases}$$

and transition function (where c is the average consumption),

$$\tau(w, (t, c)) = \begin{cases} w_L, & \text{if } w = w_L \text{ and } t = t^*, \\ w_H, & \text{otherwise.} \end{cases}$$

This is a grim-trigger profile with a low-tax/investment state w_L and a high-tax/no-investment state w_H. The profile is illustrated in figure 6.2.2. The government begins with tax rate $t^* = (R-1)/R$ and consumers begin by investing all of their endowment. These actions are repeated, in the absence of deviations, and any deviation prompts a reversion to the permanent play of the (minmaxing) stage-game equilibrium. Because it involves the play of stage-game Nash equilibrium, reversion to perpetual minmaxing is a continuation equilibrium of the repeated game and is also the most severe punishment that can be inflicted in the repeated game. The government is not playing a best response along the proposed equilibrium path but refrains from setting higher tax rates to avoid triggering the punishment.

---

9. The stage game is unchanged, so that in each period capital is either taxed away to support the public good or consumed but cannot be transferred across periods.

Proposition 6.2.1 *There exists $\underline{\delta}$ such that, for all $\delta \in [\underline{\delta}, 1)$, the strategy profile $(\mathscr{W}, w_L, f, \tau)$ of figure 6.2.2 is a subgame-perfect equilibrium of the repeated game in which the constrained efficient allocation (C in figure 6.2.1) is obtained in every period.*

Proof Given $c = 0$, the most profitable deviation by the government is to the optimal tax rate of γ/R, for a payoff gain of

$$\left(R\left(1 - \frac{\gamma}{R}\right) + 2\gamma\right) - \left(R\left(1 - \frac{R-1}{R}\right) + 2\sqrt{\gamma(R-1)}\right) \equiv \Delta > 0.$$

We then need only note that the strategy presented in figure 6.2.2 is an equilibrium if

$$(1 - \delta)\Delta \leq \delta(\bar{v}_1 - \underline{v}_1),$$

where $\underline{v}_1$ is the government's minmax payoffs. This inequality holds for sufficiently large δ. ∎

The tax policy of the government described by the automaton of figure 6.2.2 is sometimes called a *sustainable plan* (Chari and Kehoe, 1990). We have presented the simplest possible economic environment consistent with time consistency being an issue. Similar issues arise in more complicated economic environments, where agents (both private and the government) may have contracting opportunities, investment decisions may have intertemporal consequences, and markets may be incomplete. In such settings, the model will not be a repeated game. However, the model typically shares some critical features with the example here: The government is a large long-lived player, whereas the private agents are small and anonymous. The literature on sustainable plans exploits the strategic similarity between such economic models and repeated games.[10]

6.3 Risk Sharing

We now examine consumption dynamics in a simplification of a model introduced by Kocherlakota (1996) (and discussed by Ljungqvist and Sargent, 2004, chapters 19–20).[11]

---

10. For example, Chari and Kehoe (1993a,b) examine the question of why governments repay their debts and why governments are able to issue debt in the first place, given the risk that they will not repay. Domestic debt can be effectively repudiated by inflation or taxation, but there is seemingly no obstacle to an outright repudiation of foreign debt. Why is debt repaid? The common explanation is that its repudiation would preclude access to future borrowing. Although intuitive, a number of subtleties arise in making this explanation precise. Chari and Kehoe (1993b) show that a government can commit to issuing and repaying debt, on the strength of default triggering a reversion to a balanced budget continuation equilibrium, only if the government cannot enforce contracts in which it lends to its citizens. Such contracts can be enforced in Chari and Kehoe (1993a), and hence simple trigger strategies will not sustain equilibria with government debt. More complicated strategies do allow equilibria with debt. These include a case in which multiple balanced budget equilibria can be used to sustain equilibria with debt in finite horizon games.

11. Thomas and Worrall (1988) and Ligon, Thomas, and Worrall (2002) present related models. Koeppl (2003) qualifies Kocherlakota's (1996) analysis, see note 14 on page 219.

Interest in consumption dynamics stems from the following stylized facts: Conditional on the level of aggregate consumption, individual consumption is positively correlated with current and lagged values of individual income. People consume more when they earn more, and people consume more when they have earned more in the past.

At first glance, nothing could seem more natural—the rich consume more than the poor. On closer inspection, however, the observation is more challenging. The pattern observed in the data is not simply that the rich consume more than the poor but that the consumption levels of the poor and rich alike are sensitive to their current and past income levels (controlling for a variety of factors, such as aggregate consumption, so that we are not simply making the observation that everyone consumes more when more is available). If a risk-averse agent's income varies, there are gains to be had from smoothing the resulting consumption stream by insuring against the income fluctuations. Why aren't consumption fluctuations perfectly insured?

One common answer, to which we return in chapter 11, is based on the adverse selection that arises naturally in games of imperfect monitoring. It can be difficult to insure an agent against income shocks if only that agent observes the shocks. Here, we examine an alternative possibility based on moral hazard constraints that arise even in the presence of perfectly monitored shocks to income.

We work with a model in which agents are subject to perfectly observed income shocks. In the absence of any impediments to contracting on these shocks, the agents should enter into insurance contracts with one another, with each agent i making transfers to others when i's income is relatively high and receiving transfers when i's income is relatively low. In the simple examples we consider here, featuring no fluctuations in aggregate income, each agent's equilibrium consumption would be constant across states (though perhaps with some agents consuming more than others). We refer to this as a *full insurance* allocation. The conventional wisdom is that consumption fluctuates more than is warranted under a full insurance allocation.

This excess consumption sensitivity must represent some difficulties in conditioning consumption on income. We focus here on one such difficulty, an inability to write contracts committing to future payments. In particular, in each period, and after observing the current state, each agent is free to abandon the current insurance contract.[12] As a result, any dependence of current behavior on current income must satisfy incentive constraints. The dynamics of consumption and income arise out of this restriction.

6.3.1 The Economy

The stage game features two consumers, 1 and 2. There is a single consumption good. A random draw first determines the players' endowments of the consumption good to

---

12. Ljungqvist and Sargent (2004, chapter 19) examine a variation on this model in which one side of the insurance contract can be bound to the contract, whereas the other is free to abandon the contract at any time. Such a case would arise if an insurance company can commit to insurance policies with its customers, who can terminate their policies at will. We follow Kocherlakota (1996), Ljungqvist and Sargent (2004, chapter 20), and Thomas and Worrall (1988) in examining a model in which either party to an insurance contract has the ability to abandon the contract at will.

be either $e(1) \equiv (\bar{y}, \underline{y})$ or $e(2) \equiv (\underline{y}, \bar{y})$, with player 1's endowment listed first in each case and with $\bar{y} = 1 - \underline{y} \in (1/2, 1)$. Player i thus fares relatively well in state i. Each state is equally likely.

After observing the endowment $e(i)$, players 1 and 2 simultaneously transfer nonnegative quantities of the consumption good to one another and then consume the resulting net quantities $c_1(i)$ and $c_2(i)$, evaluated according to the utility function $u(\cdot)$. The function u is strictly increasing and strictly concave, so that players are risk averse.

This stage game obviously has a unique Nash equilibrium outcome in which no transfers are made. Because the consumers are risk averse, this outcome is inefficient.

Suppose now that the consumers are infinitely lived, playing the game in each period $t = 0, 1, \ldots$ The endowment draws are independent across periods, with the two endowments equally likely in each period. We thus again have a repeated game with random states (remarks 5.5.1 and 5.7.1). Players discount at the common rate δ.

An ex ante period t history is a sequence identifying the endowment and the transfers in each previous period. However, smaller transfers pose less stringent incentive constraints than do larger ones, and hence we need never consider cases in which the agents make simultaneous transfers, replacing them with an equivalent outcome in which one agent makes the net transfer and the other agent makes none. We accordingly take an ex ante period t history to be a sequence $((e^0, c_1^0, c_2^0), \ldots, (e^{t-1}, c_1^{t-1}, c_2^{t-1}))$ identifying the endowment and the consumption levels of the two agents in each previous period. An ex post period t history is the combination of an ex ante history and a selection of the period t state. We let $\tilde{\mathcal{H}}^t$ denote the set of period t ex post histories, with typical element $\tilde{h}^t$. Let $\tilde{\mathcal{H}}$ denote the set of ex post histories.

A pure strategy for player i is a function $\sigma_i : \tilde{\mathcal{H}} \to [0, 1]$, identifying the amount player i transfers after each ex post history (i.e., after each ex ante history and realized endowment). Transfers must satisfy the feasibility requirement of not exceeding the agent's endowment, that is, $\sigma_i(h^t, e) \leq y_i$ where y_i is i's endowment in e. When convenient, we describe strategy profiles in terms of the consumption levels rather than transfers that are associated with a history. We consider pure-strategy subgame-perfect equilibria of the repeated game.

6.3.2 Full Insurance Allocations

The consumers in this economy can increase their payoffs by insuring each other against the endowment risk. An ex post history is a *consistent history under* a strategy profile σ if, given the implied endowment history, in each period the transfers in the history are those specified by σ.

Definition 6.3.1 *A strategy profile σ features* full insurance *if there is a quantity c such that after every consistent ex post history under σ, player 1 consumes c and 2 consumes $1 - c$.*

As the name suggests, player 1's (and hence player 2's) consumption does not vary across histories in a full insurance outcome. Note that under this outcome, we have full *intertemporal consumption smoothing* as well as full insurance with respect to the uncertain realization of endowments. Risk-averse players strictly prefer smooth consumption profiles over time as well as states. We refer to the resulting payoffs as

6.3 ■ Risk Sharing

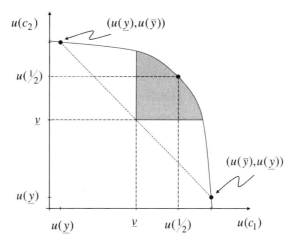

Figure 6.3.1 The set $\mathscr{F}^*$ of feasible, strictly individually rational payoffs consists of those payoff pairs (weakly) below the frontier connecting $(u(\underline{y}), u(\bar{y}))$ and $(u(\bar{y}), u(\underline{y}))$, and giving each player a payoff in excess of $\underline{v}$.

full insurance payoffs. There are many full insurance profiles that differ in how they distribute payoffs across the two agents.

When can full insurance strategy profiles be subgame-perfect equilibria? As usual, we start by examining the punishments that can be used to construct an equilibrium. The minmax payoff for each agent i is given by

$$\underline{v} = \tfrac{1}{2}u(\bar{y}) + \tfrac{1}{2}u(\underline{y}).$$

The players have identical minmax payoffs, and hence no subscript is required. Each player i can ensure at least this payoff simply by never making any transfers to player j. Figure 6.3.1 illustrates the set $\mathscr{F}^*$ of feasible, strictly individually rational payoffs for this game. Any of the payoff profiles along the frontier of $\mathscr{F}^*$ is a potential full insurance payoff, characterized by different allocations of the surplus between the two players.

The unique Nash equilibrium of the stage game features no transfers, and hence strategies that never make transfers are a subgame-perfect equilibrium of the repeated game. Because this equilibrium produces the minmax payoffs, it is the most severe punishment that can be imposed. We refer to this as the autarkic equilibrium.

Now fix a full insurance strategy profile and let (c_1, c_2) be the corresponding consumption of agents 1 and 2 after any (and every) history. Let the strategy profile specify play of the autarkic equilibrium after any nonequilibrium history. For this strategy profile to be a subgame-perfect equilibrium, it must satisfy the incentive constraints given by

$$u(c_1) \geq (1-\delta)u(\bar{y}) + \delta\underline{v}$$

and $\quad u(c_2) \geq (1-\delta)u(\bar{y}) + \delta\underline{v}.$

These ensure that deviating from the equilibrium to keep the entire relatively high endowment $\bar{y}$, when faced with the opportunity to do so, does not dominate the equilibrium payoff for either player. These constraints are most easily satisfied (i.e., are satisfied for the largest set of discount factors, if any) when $c_1 = c_2 = 1/2$. Hence, there exists at least one full-insurance equilibrium as long as

$$\frac{1-\delta}{\delta} \leq \frac{u(\frac{1}{2}) - v}{u(\bar{y}) - u(\frac{1}{2})}.$$

Let δ^* denote the value of the discount factor that solves this relationship with equality, and hence the minimum discount factor for supporting a full insurance equilibrium.

When $\delta = \delta^*$, this equal-payoff equilibrium is the only full insurance equilibrium outcome. Full insurance outcomes with asymmetric payoffs create stronger incentives for deviation on the part of the player receiving the relatively small payoff and hence require larger discount factors. The set of full insurance outcomes expands as the discount factor increases above δ^*, encompassing in the limit as $\delta \to 1$ any full insurance payoff that gives each player a payoff strictly larger than $\underline{v}$.

6.3.3 Partial Insurance

Suppose that δ falls short of δ^*, so that full insurance is impossible. If δ is not too small, we can nonetheless support equilibrium outcomes that do not simply repeat the stage-game Nash equilibrium in each period. To show this, we consider a class of equilibria with stationary outcomes in which the agents consume $(\bar{y} - \varepsilon, \underline{y} + \varepsilon)$ after any ex post history ending in endowment $e(1)$ and $(\underline{y} + \varepsilon, \bar{y} - \varepsilon)$ in endowment $e(2)$. Hence, the high-income player transfers ε of his endowment to the other player. In the full insurance equilibrium $\varepsilon = \bar{y} - \frac{1}{2} = \frac{1}{2} - \underline{y} \equiv \varepsilon^*$.

The incentive constraint for the high-endowment agent to find this transfer optimal is

$$(1-\delta)u(\bar{y}) + \delta \underline{v} \leq (1-\delta)u(\bar{y} - \varepsilon) + \delta \frac{1}{2}[u(\bar{y} - \varepsilon) + u(\underline{y} + \varepsilon)], \qquad (6.3.1)$$

or

$$\frac{1-\delta}{\delta} \leq \frac{\frac{1}{2}[u(\bar{y} - \varepsilon) + u(\underline{y} + \varepsilon)] - \underline{v}}{u(\bar{y}) - u(\bar{y} - \varepsilon)}.$$

Now consider the derivative of the right side of this expression in ε, evaluated at $\varepsilon = \varepsilon^*$. The derivative has the same sign as

$$-\left(u\left(\tfrac{1}{2}\right) - \underline{v}\right)u'\left(\tfrac{1}{2}\right) < 0.$$

Hence, reducing ε below ε^* increases the upper bound on $(1-\delta)/\delta$ for satisfying the incentive constraint, thereby decreasing the bound on values of δ for which the incentive constraint can be satisfied. This implies that there are values of the discount factor that will not support full insurance but will support stationary-outcome equilibria featuring partial insurance.

For a fixed value of $\delta < \delta^*$, with δ sufficiently large, there is a largest value of ε for which the incentive constraint given by (6.3.1) holds with equality. This value of ε describes the efficient, (ex ante) strongly symmetric, stationary-outcome equilibrium given δ, in which consumption is $(\bar{y} - \varepsilon, \underline{y} + \varepsilon)$ in endowment 1 and $(\underline{y} + \varepsilon, \bar{y} - \varepsilon)$ in endowment 2. It will be helpful to refer to this largest value of ε as $\hat{\varepsilon}$ and to refer to the associated equilibrium as $\hat{\sigma}$.

There is a collection of additional stationary-outcome equilibria in which the high-endowment agent makes transfer ε to the low-endowment agent for any $\varepsilon \in [0, \hat{\varepsilon})$. These equilibria give symmetric payoffs that are strictly dominated by the payoffs produced by $\hat{\sigma}$. There also exist equilibria with stationary outcomes that give asymmetric payoffs, with agent i making a transfer ε_i to agent j whenever i has a high endowment, and with $\varepsilon_1 \neq \varepsilon_2$.

6.3.4 Consumption Dynamics

We continue to suppose that $\delta < \delta^*$, so that players are too impatient to support a full insurance outcome, but that the discount factor nonetheless allows them to support a nontrivial subgame-perfect equilibrium. The equilibrium $\hat{\sigma}$ is the efficient stationary-outcome, strongly symmetric equilibrium. We now characterize the frontier of efficient equilibria.

Consider an ex ante history that induces equilibrium consumption profiles $c(1)$ and $c(2)$ and equilibrium continuation payoff profiles $\gamma(1)$ and $\gamma(2)$ in states 1 and 2. Then these profiles satisfy

$$(1-\delta)u(c_1(1)) + \delta\gamma_1(1) \geq (1-\delta)u(\bar{y}) + \delta\underline{v}, \quad (6.3.2)$$

$$(1-\delta)u(c_1(2)) + \delta\gamma_1(2) \geq (1-\delta)u(\underline{y}) + \delta\underline{v}, \quad (6.3.3)$$

$$(1-\delta)u(1-c_1(2)) + \delta\gamma_2(2) \geq (1-\delta)u(\bar{y}) + \delta\underline{v}, \quad (6.3.4)$$

and

$$(1-\delta)u(1-c_1(1)) + \delta\gamma_2(1) \geq (1-\delta)u(\underline{y}) + \delta\underline{v}. \quad (6.3.5)$$

These incentive constraints require that each agent in each endowment prefer the equilibrium payoff to the punishment of entering the autarkic equilibrium. As is typically the case with problems of this type, we expect only two of these constraints to bind, namely, those indicating that the high-income agent be willing to make an appropriate transfer to the low-income agent.

Fix a pure strategy profile σ. For each endowment history $\mathbf{e}^t \equiv (e^0, e^1, \ldots, e^t)$, only one period t ex post history is consistent under σ. Accordingly, we can write $\sigma[\mathbf{e}^t]$ for the period t transfers after the ex post history consistent under σ with endowment history $\mathbf{e}^t$. This allows us to construct a useful tool for examining equilibria. Consider two pure strategy profiles σ and $\check{\sigma}$. We recursively construct a strategy profile $\bar{\sigma}$ as follows. To begin, let $\bar{\sigma}(e^0) = \frac{1}{2}\sigma(e^0) + \frac{1}{2}\check{\sigma}(e^0)$. Then fix an ex post history $\bar{h}^t$ and let $\mathbf{e}^t$ be the sequence of endowments realized under this history. For each of σ and $\check{\sigma}$, there is a unique consistent ex post history associated with $\mathbf{e}^t$, denoted by h^t and $\check{h}^t$. If $\bar{h}^t$ is a consistent history under $\bar{\sigma}$, then let $\bar{\sigma}(\bar{h}^t) = \frac{1}{2}\sigma(h^t) + \frac{1}{2}\check{\sigma}(\check{h}^t)$, i.e., $\bar{\sigma}[\mathbf{e}^t] = \frac{1}{2}\sigma[\mathbf{e}^t] + \frac{1}{2}\check{\sigma}[\mathbf{e}^t]$ for all $\mathbf{e}^t$. Otherwise, the autarkic equilibrium actions are played. We refer to $\bar{\sigma}$ as the average of σ and $\check{\sigma}$.

Lemma *If σ and $\check{\sigma}$ are equilibrium strategy profiles, then so is their average, $\bar{\sigma}$. Moreover,*
6.3.1 $U_i(\bar{\sigma}) \geq (U_i(\sigma) + U_i(\check{\sigma}))/2$ *for* $i = 1, 2$.

Proof The average $\bar{\sigma}$ specifies the stage-game equilibrium at any history that is not consistent under $\bar{\sigma}$, so we need only consider histories consistent under $\bar{\sigma}$. We know that σ and $\check{\sigma}$ satisfy (6.3.2)–(6.3.5) for every consistent history and must show that $\bar{\sigma}$ does so. Notice that the right side of each constraint is independent of the strategy profile, and therefore we need show only that $\bar{\sigma}$ gives a left side of each constraint at least as large as the average of the values obtained from σ and $\check{\sigma}$.

The critical observation is that for any endowment history, the period t consumption profile under $\bar{\sigma}$ is the average of the corresponding consumption profiles under σ and $\check{\sigma}$, and so

$$u_i(\bar{\sigma}[\mathbf{e}^t]) \geq \frac{u_i(\sigma[\mathbf{e}^t]) + u_i(\check{\sigma}[\mathbf{e}^t])}{2}.$$

Because this observation holds for all endowment histories, and i's expected payoff from a profile σ is

$$U_i(\sigma) = (1 - \delta) \sum_{t=0}^{\infty} \delta^t E[u_i(\sigma[\mathbf{e}^t])],$$

where the expectation is taken over endowment histories, we then have

$$U_i(\bar{\sigma}) \geq \frac{U_i(\sigma) + U_i(\check{\sigma})}{2}.$$

Moreover, similar inequalities hold conditional on any endowment history, and so if σ and $\check{\sigma}$ satisfy (6.3.2)–(6.3.5), so does $\bar{\sigma}$.
∎

This result allows us to identify one point on the efficient frontier. The stationarity and strong symmetry constraints that we imposed when constructing the equilibrium profile $\hat{\sigma}$ are not binding:

Lemma *The efficient, strongly symmetric, stationary-outcome equilibrium strategy profile*
6.3.2 *$\hat{\sigma}$ is the strongly efficient, symmetric-payoff equilibrium.*

Proof Suppose that σ' is an equilibrium strategy profile with symmetric payoffs that strictly dominate those of $\hat{\sigma}$. Let σ'' be the strategy profile obtained from σ' that reverses the roles of players 1 and 2. Then, σ'' also has symmetric payoffs that strictly dominate those of $\hat{\sigma}$. Lemma 6.3.1 then implies that $\bar{\sigma}$, the average of σ' and σ'', is a strongly symmetric strategy profile, whose payoffs strictly dominate those of $\hat{\sigma}$.

We now construct a stationary-outcome strongly symmetric equilibrium, with payoffs at least as high as $\bar{\sigma}$, by replacing any low-payoff continuation with the highest payoff continuation. Because $c_i(s)$ is then the same after each ex ante history, the profile $\bar{\sigma}$ is a stationary-outcome, strongly symmetric equilibrium with payoffs higher than those of $\hat{\sigma}$, a contradiction.
∎

6.3 ■ Risk Sharing

We now characterize efficient equilibria as solutions to a maximization problem. Recall that $\mathscr{E}^p$ is the set of pure-strategy subgame-perfect equilibrium payoffs. The task is to choose current consumption levels $c_1(1)$ and $c_1(2)$, giving agent 1's consumption in states 1 and 2, and continuation payoffs $\gamma(1)$ and $\gamma(2)$ in states 1 and 2, to maximize

$$\tfrac{1}{2}\{(1-\delta)[u(c_1(1))+u(c_1(2))]+\delta[\gamma_1(1)+\gamma_1(2)]\}$$

subject to (6.3.2)–(6.3.5),

$$\tfrac{1}{2}((1-\delta)[u(1-c_1(1))+u(1-c_1(2))]+\delta[\gamma_2(1)+\gamma_2(2)]) \geq v_2$$

and

$$\gamma(1), \gamma(2) \in \mathscr{E}^p.$$

We describe this constrained optimization problem as MAX1. The problem MAX1 maximizes player 1's payoff given a fixed payoff for player 2 over the current consumption plan, identified by $c_1(1)$ and $c_1(2)$, and the continuation payoffs $\gamma(1)$ and $\gamma(2)$. Each of the latter is a subgame-perfect equilibrium payoff profile, as ensured by the final constraint. The first displayed constraint ensures that player 2's target utility is realized.

Lemma 6.3.3 *Fix v_2 and suppose that problem* MAX1 *has a solution in which (6.3.2)–(6.3.5) do not bind. Then*

$$c_1(1) = c_1(2)$$
$$\text{and}\quad \gamma_2(1) = \gamma_2(2) = v_2.$$

This in turn implies that there exists an equilibrium featuring $c_1(1) = c_1(2)$ after every history, and hence full insurance.

Proof Suppose that we have a solution to the optimization problem MAX1 in which none of the incentive constraints bind. Then $c_1(1) = c_1(2)$, that is, consumption does not vary with the endowment. (Suppose this were not the case, so that $c_1(1) > c_1(2)$. Then the current consumption levels can be replaced by a smaller value of $c_1(1)$ and larger $c_1(2)$ while preserving their expected value and the incentive constraints. But given the concavity of the utility function, this increases both players' utilities, ensuring that the candidate solution was not optimal.)

Next, suppose that $\gamma_2(1) \neq \gamma_2(2)$, so that continuation utilities vary with the state. Then states 1 and 2 must be followed by different continuation equilibria. The average of these equilibria is an equilibrium with higher average utility (lemma 6.3.1) and hence yields a contradiction.

Finally, suppose that $\gamma_2(1) = \gamma_2(2) \neq v_2$. Then the equilibrium features a consumption allocation for player 2 that is not constant over equilibrium histories, and the current expected utility v_2 is a convex combination of the consumption levels attached to these histories. Let c_2^* solve $u(c_2) = (1-\delta)u(1-c_1) + \delta\gamma_2$, where c_1 is 1's state-independent consumption and γ_2 is 2's state-independent continuation. Denote by σ^* the strategy profile that gives player 2 consumption c_2^* in the current period and after every consistent history and otherwise prescribes the stage-game Nash equilibrium. By construction, (6.3.4) and (6.3.5) are satisfied, with $\gamma_2^* = u(c_2)$. Because the utility function is concave, player 1's payoff in this newly constructed strategy profile is larger than in the equilibrium under consideration, and so (6.3.2) and (6.3.3) are satisfied, with $\gamma_1^* = u(1 - c_2^*)$. Finally, because (6.3.2)–(6.3.5) are simply the enforceability constraints for this game, we have just shown that $\{(\underline{v}, \underline{v}), (u(1 - c_2^*), u(c_2^*))\}$ is self-generating, and so $(u(1 - c_2^*), u(c_2^*)) \in \mathscr{E}^p$ (recall remark 5.7.1). We thus have an equilibrium that offers player 1 a higher payoff, a contradiction.

Hence, the solution to MAX1 has state-independent consumption, c_1 for player 1, and state-independent continuations for player 2, $\gamma_2 = v_2$ (so that $u(1 - c_1) = v_2$). Hence, as before, the set of payoffs $\{(\underline{v}, \underline{v}), (u(c_1), u(1 - c_1))\}$ is self-generating, and so $(u(c_1), u(1 - c_1)) \in \mathscr{E}^p$. The profile yielding payoffs $(u(c_1), u(1 - c_1))$ is clearly a full insurance equilibrium. ∎

The conclusion is that in any efficient equilibrium σ, if the incentive constraints do not bind at some ex ante consistent history under σ, then the continuation equilibrium is a full insurance equilibrium. Because this requires full insurance to be consistent with equilibrium, for the discount factors $\delta < \delta^*$, after every consistent history at least one incentive constraint must bind.

Now let us examine the subgame-perfect equilibrium that maximizes player 1's payoff. A first observation is immediate.

Lemma 6.3.4 *Let σ^* be a subgame-perfect equilibrium. Then player 1 receives at least as high a payoff from an equilibrium that specifies consumption $(\bar{y}, \underline{y})$ after any ex post history in which only state 1 has been realized, and otherwise specifies equilibrium σ^*.*

As a result, the equilibrium maximizing player 1's payoff must feature nonstationary outcomes, and must begin with $(\bar{y}, \underline{y})$ after any ex post history which only state 1 has been realized.

Proof We first note that consumption bundle $(\bar{y}, \underline{y})$ is incentive compatible in state 1, because no transfers are made. As a result, the prescription of playing $(\bar{y}, \underline{y})$ after any ex post history in which only state 1 has been realized, and otherwise playing equilibrium σ^*, is itself a subgame-perfect equilibrium. Because there is no subgame-perfect equilibrium in which player 1 earns $u(\bar{y})$, the constructed equilibrium must give at least as high a payoff as σ^*, for any equilibrium σ^*, and must give a strictly higher payoff if σ^* does not itself prescribe $(\bar{y}, \underline{y})$ after any ex post history in which only state 1 has been realized. ∎

The task of maximizing player 1's payoff now becomes one of finding that equilibrium that maximizes player 1's payoff, conditional on endowment 2 having been drawn in the first period.

Lemma 6.3.5 *The equilibrium maximizing player 1's payoff, conditional on state 2 having been drawn in the first period, is the efficient symmetric-payoff stationary-outcome equilibrium $\hat{\sigma}$.*

This result gives us a complete characterization of the equilibrium maximizing player 1's (and, reversing the roles, player 2's) equilibrium payoff. Consumption is

6.3 ■ Risk Sharing

given by $(\bar{y}, \underline{y})$ as long as state 1 is realized, with the first draw of state 2 switching play to the efficient, symmetric-payoff stationary-outcome equilibrium.

Proof We begin by noting that the payoffs provided by equilibrium $\hat{\sigma}$ are the equally weighted convex combination of two continuation payoffs, one following the draw of state 1 and one following the draw of state 2. In the first of these continuation equilibria, the incentive constraint for player 1 binds,

$$(1 - \delta)u(c_1(1)) + \delta\gamma_1(1) = (1 - \delta)u(\bar{y}) + \delta\underline{v},$$

whereas in the second the incentive constraint for player 2 binds,

$$(1 - \delta)u(1 - c_1(2)) + \delta\gamma_2(2) = (1 - \delta)u(\bar{y}) + \delta\underline{v}.$$

Refer to these continuation payoffs as $U(\hat{\sigma} \mid e(1))$ and $U(\hat{\sigma} \mid e(2))$, respectively, where player i draws the relatively high share of the endowment in endowment $e(i)$. Figure 6.3.2 illustrates the payoffs $U(\hat{\sigma} \mid e(1))$ and $U(\hat{\sigma} \mid e(2))$. These constraints ensure that in each case, the player drawing the high endowment earns an expected payoff of $(1 - \delta)u(\bar{y}) + \delta\underline{v} \equiv \bar{v}$. Notice also that

$$(1 - \delta)u(\bar{y}) + \delta\underline{v} = \bar{v} > u\left(\tfrac{1}{2}\right),$$

because otherwise the sufficient conditions hold for a full-insurance equilibrium in which consumption is $(1/2, 1/2)$ after every history, contrary to our hypothesis.

Now consider the equilibrium that maximizes player 1's payoff, conditional on state 2. Because we are maximizing player 1's payoff, the incentive constraint for player 2 to make a transfer to player 1 in state 2 must bind. Player 2 must then earn a continuation payoff of $(1 - \delta)u(\bar{y}) + \delta\underline{v}$. One equilibrium delivering such a payoff to player 2 is $\hat{\sigma}$. Is there another equilibrium that respects player 2's incentive constraint and provides player 1 a higher payoff than that which he

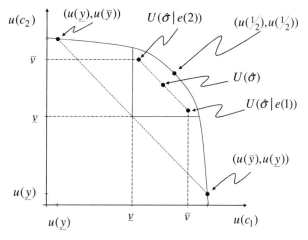

Figure 6.3.2 Representation of efficient, symmetric-payoff equilibrium payoff $U(\hat{\sigma})$ as the equally likely combination of $U(\hat{\sigma} \mid e(1))$ and $U(\hat{\sigma} \mid e(2))$. The utility $\bar{v}$ equals $(1 - \delta)u(\bar{y}) + \delta\underline{v}$.

earns under $\hat\sigma$? If so, then we would have an equilibrium that weakly dominates $\hat\sigma$, conditional on drawing state 2. Reversing the roles of the players, we could also construct an equilibrium weakly dominating $\hat\sigma$, conditional on endowment 1 having been drawn. Combining these two, we would find a symmetric-payoff equilibrium strictly dominating $\hat\sigma$, yielding a contradiction.

∎

We have thus constructed the equilibrium that maximizes player 1's payoff and can do similarly for player 2, as a convex combination of an initial segment, in which a consumption binge is supported by fortunate endowment draws, followed by continuation paths of the efficient symmetric-payoff equilibrium. Let σ^1 and σ^2 denote these equilibria. Figure 6.3.3 illustrates. Because

$$U(\sigma^1) = \tfrac{1}{2}U(\hat\sigma \mid e(2)) + \tfrac{1}{2}((1-\delta)(u(\bar y), u(\underline y)) + \delta U(\sigma^1)),$$

the payoff vector $U(\sigma^1)$ is a convex combination of $U(\hat\sigma \mid e(2))$ and $(u(\bar y), u(\underline y))$. Moreover, player 2 earns a payoff of $\underline v$. This follows from the observation that 2 earns $u(\underline y)$ after any initial history featuring only state 1, and on the first draw of state 2 earns a continuation payoff equivalent to receiving $u(\bar y)$ in the current period and $\underline v$ thereafter. Hence, 2's payoff equals that of receiving $\bar y$ whenever endowment 2 is drawn and $\underline y$ when endowment 1 is drawn, which is $\underline v$. Similarly, player 1 earns $\underline v$ in equilibrium σ^2.

This argument can be extended to show that every payoff profile on the efficient frontier consists of an initial segment in which the player drawing the high-income endowment does relatively well, with the first switch to the other endowment prompting a switch to $\hat\sigma$, with a continuation payoff of either $U(\hat\sigma \mid e(1))$ or $U(\hat\sigma \mid e(2))$, as appropriate. The most extreme version of this initial segment allows the high-income player i to consume his entire endowment (generating equilibrium σ^i), and the least extreme simply begins with the play of $\hat\sigma$. By varying the transfer made from the

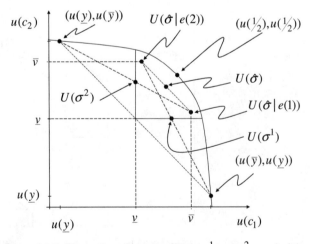

Figure 6.3.3 Illustration of how equilibria σ^1 and σ^2, maximizing player 1's and player 2's payoffs, respectively, are constructed as convex combinations of $(u(\bar y), u(\underline y))$ and $U(\hat\sigma \mid e(2))$ in the case of σ^1 and of $(u(\underline y), u(\bar y))$ and $U(\hat\sigma \mid e(1))$ in the case of σ^2.

high-income to the low-income player in this initial segment, we can generate the intervening frontier.

To see how the argument works, consider a payoff profile on the efficient frontier in which player 2 earns a payoff v_2 larger than $\underline{v}$ but smaller than $U_2(\hat{\sigma})$. One equilibrium producing a payoff of v_2 for player 2 calls for continuation equilibrium $\hat{\sigma}$ with value $U(\hat{\sigma} \mid e(2))$ on the first draw of state 2, with any history in which only state 1 has been drawn prompting a transfer of ε from player 1 to player 2, where ε is calculated so as to give player 2 an expected payoff of v_2.[13] Call this equilibrium profile σ'. Now let σ'' be an equilibrium that maximizes player 1's payoff, subject to player 2 receiving at least v_2. We argue that $U_1(\sigma'') = U_1(\sigma')$. Suppose instead that $U_1(\sigma'') > U_1(\sigma')$. Then equilibrium σ'' must yield a higher expected continuation payoff for player 1 from histories in which only state 1 has been realized (because σ' has the largest payoff for player 1 after state 2 is first realized), and hence a lower payoff for player 2. To preserve player 2's expected payoff of v_2, there must then be some consistent history under σ'' in which state 2 has been drawn and in which player 2 receives a higher payoff (and hence player 1 a lower payoff) than under $\hat{\sigma}$. This, however, is a contradiction, as the continuation payoff that σ'' prescribes after the history in question can be replaced by $\hat{\sigma}$ and payoff $U(\hat{\sigma} \mid e(2))$ without disrupting incentives, thereby increasing both players' payoffs.

This model captures part of the stylized consumption behavior of interest, namely, that current consumption fluctuates in current income more than would be the case if risk could be shared efficiently. However, beyond a potential initial segment preceding the first change of state, it fails to capture a link between past income and current consumption.

The stationarity typified by $\hat{\sigma}$ and built into every efficient strategy profile after some initial segment is not completely intuitive. Because the discount factor falls below δ^*, complete risk sharing within a period is impossible. However, players face the same expected continuation payoffs at the end of every period. Why not share some of the risk over time by eliciting a somewhat larger transfer from the high-income agent in return for a somewhat lower continuation value? One would expect such a trade-off to appear as a natural feature of the first-order conditions used to solve the maximization problem characterizing the efficient solution. The difficulty is that the frontier of efficient payoffs is not differentiable at the point $U(\hat{\sigma})$.[14] The failure of differentiability suffices to ensure that in no direction is there such a favorable trade-off between current and future risk.

6.3.5 Intertemporal Consumption Sensitivity

A richer model (in particular, more possible states) allows links between past income and current consumption to appear. We present an example with three states. We leave many of the details to Ljungqvist and Sargent (2004, chapter 20), who provide an

---

13. Such an ε exists, because setting $\varepsilon = 0$ produces expected payoff $\underline{v}$ for player 2, whereas $\varepsilon = \hat{\varepsilon}$ produces $U_2(\hat{\sigma})$. Notice also that the transfer of ε is incentive compatible for player 1, with deviations punished by reversion to autarky, because the transfer of $\hat{\varepsilon}$ is incentive compatible in equilibrium $\hat{\sigma}$.

14. The nondifferentiability of the frontier of repeated-game payoffs is demonstrated in Ljungqvist and Sargent (2004, section 20.8.2).

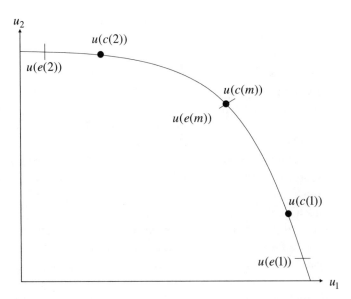

Figure 6.3.4 The payoff profiles for our three-state example and the first equilibrium we construct.

analysis of the general case of finitely many states, finding properties analogous to those of our three-state example.

The three equally likely endowments are $e(1)$, $e(m)$, and $e(2)$, with, as before, $e(1) \equiv (\bar{y}, \underline{y})$, $e(2) \equiv (\underline{y}, \bar{y})$, where $\bar{y} \in (1/2, 1]$ and $\underline{y} = 1 - \bar{y}$. The new endowment $e(m)$ splits the endowment evenly, $e(m) = (1/2, 1/2)$. Hence, we again have a symmetric model in which player i receives a relatively large endowment in state i. Figure 6.3.4 illustrates these endowments.

The most severe punishment available is again an autarkic equilibrium in which no transfers are made, giving each player their minmax value of

$$\underline{v} = \tfrac{1}{3}\bigl(u(\bar{y}) + u(\tfrac{1}{2}) + u(\underline{y})\bigr).$$

Incentive constraints are given by the fact that, in state i, player i must receive a payoff at least $\bar{v} = (1-\delta)u(\bar{y}) + \delta\underline{v}$. In state m, each player must receive at least $(1-\delta)u(\tfrac{1}{2}) + \delta\underline{v}$.

The symmetric-payoff full insurance outcome is straightforward. In state m, no transfers are made (and hence incentive constraints are automatically satisfied). In state i, the high-income player i transfers $\varepsilon^* = \bar{y} - \tfrac{1}{2}$ to the low-income player. Hence, consumption after every ex post history is $(1/2, 1/2)$. As in the two-state case, if there is any full insurance equilibrium outcome, then the symmetric-payoff full insurance outcome is an equilibrium outcome. If the discount factor is sufficiently large, then there also exist other full insurance equilibrium outcomes in which the surplus is split asymmetrically between the two players.

We now suppose the discount factor is sufficiently large that there exist subgame-perfect equilibria that do not simply repeat the autarkic equilibrium but not so large that there are full insurance equilibrium outcomes. Hence, it must be the case that $\bar{v} \equiv (1-\delta)u(\bar{y}) + \delta\underline{v} > u(\tfrac{1}{2})$.

6.3 ■ Risk Sharing

We examine the efficient, symmetric-payoff equilibrium for this case. We can construct a likely candidate for such an equilibrium by choosing ε to satisfy

$$(1-\delta)u(\bar{y}-\varepsilon) + \delta\tfrac{1}{3}\bigl(u(\bar{y}-\varepsilon) + u\bigl(\tfrac{1}{2}\bigr) + u(\underline{y}+\varepsilon)\bigr) = \bar{v}.$$

Given our presumption that the discount factor is large enough to support more than the autarky equilibrium but too small to support full insurance, this equation is solved by some $\varepsilon \in (0, \varepsilon^*)$. This equilibrium leaves consumption untouched in state m, while treating states 1 and 2 just as in the two-state case of the previous subsection. We refer to this as equilibrium σ. Figure 6.3.4 illustrates.

This equilibrium provides some insurance, but we can provide more. Notice first that in state m, there is slack in the incentive constraint of each agent, given by

$$(1-\delta)u\bigl(\tfrac{1}{2}\bigr) + \delta U(\sigma) \geq (1-\delta)u\bigl(\tfrac{1}{2}\bigr) + \delta\underline{v}.$$

Now choose some small ζ and let consumption in state m be given by $\bigl(\tfrac{1}{2}+\zeta, \tfrac{1}{2}-\zeta\bigr)$ with probability $1/2$ (conditional on state m occurring), and by $\bigl(\tfrac{1}{2}-\zeta, \tfrac{1}{2}+\zeta\bigr)$ with the complementary probability. The slack in the incentive constraints ensures that this is feasible. The derivative of an agent's utility, as ζ increases and evaluated at $\zeta=0$, is signed by $u'\bigl(\tfrac{1}{2}\bigr) - u'\bigl(\tfrac{1}{2}\bigr) = 0$. Hence, increasing ζ above 0 has only a second-order effect (a decrease) in expected payoffs.

Let us now separate ex ante histories into two categories, category 1 and category 2. A history is in category i if agent i is the most recent one to have drawn a high endowment. Hence, if the last state other than m was state 1, then we have a category 1 history. Now fix $\zeta > 0$, and let the prescription for any ex post history in which the agents find themselves in state m prescribe consumption $\bigl(\tfrac{1}{2}+\zeta, \tfrac{1}{2}-\zeta\bigr)$ if this is a category 1 history and consumption $\bigl(\tfrac{1}{2}-\zeta, \tfrac{1}{2}+\zeta\bigr)$ if this is a category 2 history. In essence, we are using consumption in state m to reward the last agent who has had a large endowment and transferred part of it to the other agent.

This modification of profile σ has two effects. We are introducing risk in state m but with a second-order effect on total expected payoffs. However, because we now allocate state m consumption as a function of histories, rather than randomly, this adjustment gives a first-order increase in the expected continuation payoff to agent i after a history of category i. This relaxes the incentive constraints facing agents in states 1 and 2. We can thus couple the increase in ζ with an increase in ε, where the latter is calculated to preserve equality in the incentive constraints in states 1 and 2, thereby allowing more insurance in states 1 and 2. The increased volatility of consumption in state m thus buys reduced volatility in states 1 and 2, allowing a first-order gain on the latter at a second-order cost on the former.

Figure 6.3.5 illustrates the resulting consumption pattern. The remaining task is to calculate optimal values of ζ and ε. For any value of ζ, we choose $\varepsilon(\zeta)$ to preserve equality in the incentive constraints in states 1 and 2. The principle here is that insurance is always valuable in states 1 and 2, where income fluctuations are relatively large, and hence we should insure up to the limits imposed by incentive constraints in those states.

As ζ increases from 0, this adjustment increases expected payoffs. We can continue increasing ζ, and hence the equilibrium payoffs, until one of two events occurs. First,

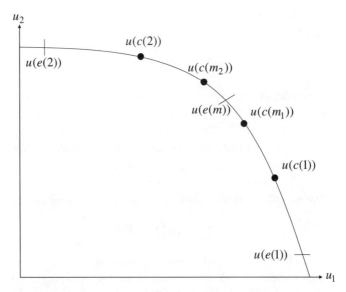

Figure 6.3.5 Payoff profiles for the efficient, symmetric equilibrium. Consumption bundle $c(m_i)$ follows a history in which state m has been drawn, and the last state other than m to be drawn was state i.

the state m incentive constraints may bind. At this point, we have reached the limits of our ability to insure across states m and states 1 and 2, and the optimal value of ζ is that which causes the state m incentive constraints to hold with equality. This gives the consumption pattern illustrated in figure 6.3.5, with four distinct consumption profiles. Second, it may be that the cost of the increased consumption risk in state m overwhelms the gains from increased insurance in states 1 and 2. This will certainly be the case if ζ becomes so large that

$$\bar{y} - \varepsilon(\zeta) = \tfrac{1}{2} + \zeta \quad \text{and} \quad \underline{y} + \varepsilon(\zeta) = \tfrac{1}{2} - \zeta.$$

In this last case, the optimal value of ζ is that which satisfies these two equalities. We now have a consumption pattern in which volatility in state m consumption matches that of states 1 and 2, and hence there are no further opportunities to smooth risk across states. Then, $c(1) = c(m_1)$ and $c(2) = c(m_2)$, and only two consumption bundles arise in equilibrium.

Together, these two possibilities fix the value of ζ that gives the efficient symmetric-payoff equilibrium. Notice, however, that this symmetric-payoff equilibrium does not feature stationary outcomes (nor is it strongly symmetric). Current consumption depends on whether the history is in category 1 or category 2, in addition to the realization of the current state. Intuitively, we are now spreading risk across time as well as states within a period, exchanging a relatively large transfer from a high-endowment agent for a relatively lucrative continuation payoff. In terms of consumption dynamics, agents with high endowments in their history are now more likely to have high current consumption.

Part II

Games with (Imperfect) Public Monitoring

7 The Basic Structure of Repeated Games with Imperfect Public Monitoring

The first part of the book has focused on games with perfect monitoring. In these games, deviations from the equilibrium path of play can be detected and punished. As we saw, it is then relatively straightforward to provide incentive for players to not myopically optimize (play stage-game best replies).

In this chapter, we begin our study of games with *imperfect monitoring*: games in which players have only noisy information about past play. The link between current actions and future play is now indirect, and in general, deviations cannot be unambiguously detected. However, equilibrium play will affect the distribution of the signals, allowing intertemporal incentives to be created by attaching "punishments" to signals that are especially likely to arise in the event of a deviation. This in turn will allow us to again support behavior in which players do not myopically optimize. Because the direct link between deviations and signals is broken, these punishments may sometimes occur on the equilibrium path. The equilibria we construct will thus involve strategies that are similar in spirit to those we have examined in perfect monitoring games, but significantly different in some of their details and implications.

Throughout this second part, we maintain the important assumption that any signals of past play, however imprecise and noisy, are invariably observed by all players. This is often stressed by using the phrase imperfect *public* monitoring to refer to such games. These commonly observed signals allow players to coordinate their actions in a way that is not possible if the signals observed by some players are not observed by others. The latter case, referred to as one of *private monitoring*, is deferred until chapter 12.

7.1 The Canonical Repeated Game

7.1.1 The Stage Game

The specification of the stage game follows closely that of perfect monitoring games, allowing for the presence of short-lived players (section 2.7). Players $1, \ldots, n$ are long-lived and players $n+1, \ldots, N$ are short-lived, with player i having a set of pure actions A_i. We explicitly allow $n = N$, so that there may be no short-lived players. As for perfect monitoring games, we assume each A_i is a compact subset of the Euclidean space $\mathbb{R}^k$ for some k. Players choose actions simultaneously. The correspondence mapping any mixed-action profile for the long-lived players to the corresponding set of static Nash equilibria for the short-lived players is denoted

$B: \prod_{i=1}^{n} \Delta(A_i) \Rightarrow \prod_{i=n+1}^{N} \Delta(A_i)$, with its graph denoted by $\mathbf{B} \subset \prod_{i=1}^{N} \Delta A_i$. If there are no short-lived players, $\mathbf{B} = \prod_{i=1}^{n} \Delta(A_i)$.

At the end of the stage game, players observe a *public signal* y, drawn from a signal space Y. The signal space Y is finite (except in section 7.5). The probability that the signal y is realized, given the action profile $a \in A \equiv \prod_i A_i$, is denoted by $\rho(y \mid a)$. The function $\rho: Y \times A \to [0, 1]$ is continuous (so that ex ante payoffs are continuous functions of actions). We have the obvious extension $\rho(y \mid \alpha)$ to mixed-action profiles. We say ρ has *full support* if $\rho(y \mid a) > 0$ for all y and a. We invoke full support only when needed and are explicit when doing so.

The players receive no information about opponents' play beyond the signal y. If players receive payoffs at the end of each period, player i's payoff after the realization (y, a) is given by $u_i^*(y, a_i)$.[1] Ex ante stage game payoffs are then given by

$$u_i(a) = \sum_{y \in Y} u_i^*(y, a_i) \rho(y \mid a). \tag{7.1.1}$$

For ease of reference, we now list the maintained assumptions on the stage game.

Assumption 7.1.1
1. A_i *is either finite or a compact and convex subset of the Euclidean space* $\mathbb{R}^k$ *for some* k. *As in part I, we refer to compact and convex action spaces as continuum action spaces.*
2. Y *is finite, and, if* A_i *is a continuum action space, then* $\rho: Y \times A \to [0, 1]$ *is continuous.*
3. *If* A_i *is a continuum action space, then* $u_i^*: Y \times A_i \to \mathbb{R}$ *is continuous, and* u_i *is quasiconcave in* a_i.

Remark 7.1.1 **Pure strategies** As for perfect monitoring games (see remark 2.1.1), when the action spaces are a continuum, we avoid some tedious measurability details by considering only pure strategies. We use α_i to both denote pure or mixed strategies in finite games and pure strategies only in continuum action games.

Abreu, Pearce, and Stacchetti (1990) do allow for a continuum of signals but assume A is finite and restrict attention to pure strategies. We discuss their bang-bang result, which requires a continuum of signals, in section 7.5. ◆

7.1.2 The Repeated Game

In the repeated game, the only public information available in period t is the t-period history of public signals, $h^t \equiv (y^0, y^1, \ldots, y^{t-1})$. The set of *public histories* is

$$\mathscr{H} \equiv \cup_{t=0}^{\infty} Y^t,$$

where we set $Y^0 \equiv \varnothing$.

---

1. The representation of ex ante stage-game payoffs as the expected value of ex post payoffs is typically made for interpretation and is not needed for any of the results in this chapter and the next ((7.1.1) is only used in lemma 9.4.1 and those results, propositions 9.4.1 and 9.5.1, that depend on it). An alternative (but less common) assumption is to view discounting as reflecting the probability of the end of the game (as discussed in section 4.2), with payoffs, a simple sum of stage-game payoffs, awarded at the end of play (so players cannot infer anything from intermediate stage-game payoffs).

7.1 ■ The Canonical Repeated Game

A history for a long-lived player i includes both the public history and the history of actions that he has taken, $h_i^t \equiv (y^0, a_i^0; y^1, a_i^1; \ldots; y^{t-1}, a_i^{t-1})$. The set of histories for player i is

$$\mathcal{H}_i \equiv \cup_{t=0}^{\infty} (A_i \times Y)^t.$$

A pure strategy for player i is a mapping from all possible histories into the set of pure actions,

$$\sigma_i : \mathcal{H}_i \to A_i.$$

A mixed strategy is, as usual, a mixture over pure strategies, and a behavior strategy is a mapping

$$\sigma_i : \mathcal{H}_i \to \Delta(A_i).$$

As for perfect-monitoring games, the role of a short-lived player $i = n+1, \ldots, N$ is filled by a countable sequence of players. We refer simply to a short-lived player i, rather than explicitly referring to the sequence of players. We assume that each short-lived player in period t only observes the public history h^t.[2]

In general, a pure strategy profile does not induce a deterministic outcome path (i.e., an infinite sequence in $(A \times Y)^\infty$), because the public signals may be random. As usual, long-lived players have a common discount factor δ. Following the notation for games with perfect monitoring, the vector of long-lived players' payoffs from a strategy profile σ is denoted $U(\sigma)$. To give a flavor of $U(\sigma)$, we write a partial sum for $U(\sigma)$ when σ is pure, letting $a^0 = \sigma(\varnothing)$:

$$U_i(\sigma) = (1-\delta)u_i(a^0)$$
$$+ (1-\delta)\delta \sum_{y^0 \in Y} u_i(\sigma_1(a_1^0, y^0), \ldots, \sigma_n(a_n^0, y^0))\rho(y^0 \mid a^0) + \cdots$$

Public monitoring games include the following as special cases.

1. **Perfect monitoring games.** Because we restrict attention to finite signal spaces (with the exception of section 7.5, which requires A finite), perfect monitoring games are only immediately covered when A is finite. In that case, simply set $Y = A$ and $\rho(y \mid a) = 1$ if $y = a$, and 0 otherwise. However, the analysis in this chapter also covers perfect monitoring games with continuum action spaces, given our restriction to pure strategies (see remark 2.1.1).[3]
2. **Games with a nontrivial extensive form.** In this case, the signal is the terminal node reached. There is imperfect observability, because only decisions on the path of play are observed. We have already discussed this case in section 5.4 and return to it in section 9.6.
3. **Games with noisy monitoring.** When the prisoners' dilemma is interpreted as a partnership game, it is natural to consider an environment where output

---

2. Although this assumption is natural, the analysis in chapters 7–9 is unchanged when short-lived players observe predecessors' actions (because the analysis restricts attention to public strategies).

3. While the signal space is a continuum in this case, its cardinality does not present measurability problems because, with pure strategies, all expectations over the signals are trivial, since for any pure action profile only one signal can arise. Compare, for example, the proof of proposition 2.5.1, which restricts attention to pure strategies, and proposition 7.3.1, which does not.

(the product of the partnership) is a random function of the choices of the partners. Influential early examples are Radner's (1985), Rubinstein's (1979b), and Rubinstein and Yaari's (1983) repeated principal-agent model with noisy monitoring (discussed in section 11.4.5) and Green and Porter's (1984) oligopoly with noisy prices (section 11.1).

4. Games of repeated adverse selection. In these games, the moves of players are public information, but moves are taken after players learn some *private* information. For example, in Athey and Bagwell's (2001) model of a repeated oligopoly, firm prices are public, but firm costs are subject to privately observed i.i.d. shocks. Firm i's price is a public signal of firm i's action, which is the mapping from possible costs into prices. We discuss this in section 11.2.

5. Games with incomplete observability. Games with semi-standard information (Lehrer 1990), where each player's action space is partitioned and other players only observe the element of the partition containing that action.

7.1.3 Recovering a Recursive Structure: Public Strategies and Perfect Public Equilibria

In perfect monitoring games, there is a natural isomorphism between histories and information sets. Consequently, in perfect monitoring games, every history, h^t, induces a continuation game that is strategically identical to the original repeated game, and for every strategy σ_i in the original game, h^t induces a well-defined continuation strategy $\sigma_i|_{h^t}$. Moreover, any Nash equilibrium induces Nash equilibria on the induced equilibrium outcome path.

Unfortunately, none of these observations hold for public monitoring games. Because long-lived players potentially have private information (their own past action choices), a player's information sets are naturally isomorphic to the set of their own private histories, $\mathcal{H}_i$, not to the set of public histories, $\mathcal{H}$. Thus there is no continuation game induced by *any* history—a public history is clearly insufficient, and i's private history will not be known by the other players. There are examples of Nash equilibria in public monitoring games whose continuation play resembles that of a correlated and not Nash equilibrium (see section 10.3). This lack of a recursive structure is a significant complication, not just in calculating Nash equilibria but in formulating a tractable notion of sequential rationality.[4]

A recursive structure does hold, however, on a restricted strategy space.

Definition 7.1.1 *A behavior strategy σ_i is* public *if, in every period t, it depends only on the public history $h^t \in Y^t$ and not on i's private history: for all $h_i^t, \hat{h}_i^t \in \mathcal{H}_i$ satisfying $y^\tau = \hat{y}^\tau$ for all $\tau \le t-1$,*

$$\sigma_i(h_i^t) = \sigma_i(\hat{h}_i^t).$$

A behavior strategy σ_i is private *if it is not public.*

---

4. The problem lies not in defining sequential rationality, because the notion of Kreps and Wilson (1982b) is the natural definition, adjusting for the infinite horizon. Rather, the difficulty is in applying the definition.

We can thus take $\mathcal{H}$ to be the domain of public strategies. Note that short-lived players are necessarily playing public strategies.[5] When all players but i are playing public strategies, player i essentially faces a Markov decision problem with states given by the public histories, and so has a Markov best reply. We consequently have the following result.

Lemma 7.1.1 *If all players other than i are playing a public strategy, then player i has a public strategy as a best reply.*

Because the result is obvious, we provide some intuition rather than a proof. Let $h_i^t \in \mathcal{H}_i$ be a private history for player i. Player i's actions before t may well be relevant in determining i's beliefs over the actions chosen by the other players before t. However, because the other players' continuation behavior in period t is only a function of the public history h^t and not their own past behavior, player i's expected payoffs are independent of i's beliefs over the past actions of the other players, and so i has a best reply in public strategies. Note that i need not have a public best reply to a nonpublic strategy profile.

We provide some examples in chapter 10 illustrating behavior that is ruled out by restricting attention to public strategies. However, to a large extent the restriction to public strategies is not troubling. First, every pure strategy is realization equivalent to a public pure strategy (lemma 7.1.2). Second, as chapter 9 shows, the folk theorem holds under quite general conditions with public strategies. Finally, for games with a product structure (discussed in section 9.5), the set of equilibrium payoffs is unaffected by the restriction to public strategies (proposition 10.1.1).

Restricting attention to public strategy profiles, every public history h^t induces a continuation game that is strategically identical (in terms of public strategies) to the original repeated public monitoring game, and for any public strategy σ_i in the original game, h^t induces a well-defined continuation public strategy $\sigma_i|_{h^t}$. Moreover, any Nash equilibrium in public strategies induces Nash equilibria (in public strategies) on the induced equilibrium outcome path.

Two strategies, σ_i and $\hat{\sigma}_i$, are *realization equivalent* if, for all strategies for the other players, σ_{-i}, the distributions over outcomes induced by (σ_i, σ_{-i}) and $(\hat{\sigma}_i, \sigma_{-i})$ are the same.

Lemma 7.1.2 *Every pure strategy in a public monitoring game is realization equivalent to a public pure strategy.*

Proof Let σ_i be a pure strategy. Let $a_i^0 = \sigma_i(\varnothing)$ be the first-period action. In the second period, after the signal y^0, the action $a_i^1(y^0) \equiv \sigma_i(y^0, a_i^0) = \sigma_i(y^0, \sigma_i(\varnothing))$ is played. Proceeding recursively, in period t after the public history, $h^t = (y^0, y^1, \ldots, y^{t-1}) = (h^{t-1}, y^{t-1})$, the action $a_i^t(h^t) \equiv \sigma_i(h^t; a_i^0, a_i^1(y^0), \ldots, a_i^{t-1}(h^{t-1}))$ is played. Hence, for any public outcome $h \in Y^\infty$, the pure strategy σ_i

---

5. This is a consequence of our assumption that a period t short-lived player i only observes h^t, the public history. We refer to such public monitoring games as *canonical public monitoring games*. There is no formal difficulty in assuming that a period t short-lived player i knows the choices of previous short-lived player i's, in which case short-lived players could play private strategies. However, in most situations, one would not expect short-lived players to have such private information.

puts probability one on the action path $(a^0, a^1(y^0), \ldots, a^t(h^t), \ldots)$. Consequently, the pure strategy σ_i is realization equivalent to the public strategy that plays $a_i^t(h^t)$, after the public history h^t.

∎

For example, let there be two actions available to player 1, T and B, and two signals, y' and y''. Let strategy σ_1 specify T in the first period, and then in each period t, specify T if $(a^{t-1}, y^{t-1}) \in \{(T, y''), (B, y')\}$, and B if $(a^{t-1}, y^{t-1}) \in \{(B, y''), (T, y')\}$. This is a private strategy. However, it is realization equivalent to a public strategy that plays action T in the first period and thereafter plays T after any history with an even number of y' signals and B after any history with an odd number of y' signals.

Remark 7.1.2 **Automata** The recursive structure of public strategies is most clearly seen by observing that a public strategy profile has an automaton representation very similar to that of strategy profiles in perfect monitoring games (discussed in section 2.3): Every public behavior strategy profile can be represented by a set of states $\mathscr{W}$, an initial state w^0, a decision rule $f : \mathscr{W} \to \prod_i \Delta(A_i)$ associating mixed-action profiles with states, and a transition function $\tau : \mathscr{W} \times Y \to \mathscr{W}$.

As for perfect monitoring games, we extend τ to the domain $\mathscr{W} \times \mathscr{H} \setminus \{\varnothing\}$ by recursively defining

$$\tau(w, h^t) = \tau(\tau(w, h^{t-1}), y^{t-1}).$$

A state $w' \in \mathscr{W}$ is *accessible* from another state $w \in \mathscr{W}$ if there exists a sequence of public signals such that beginning at w, the automaton transits eventually to w', that is, there exists h^t such that $w' = \tau(w, h^t)$.

In contrast, the automaton representation is more complicated for private strategies, requiring a separate automaton for each player (recall remark 2.3.1): Each behavior strategy σ_i can be represented by a set of states $\mathscr{W}_i$, an initial state w_i^0, a decision rule $f_i : \mathscr{W}_i \to \Delta(A_i)$ specifying a distribution over action choices for each state, and a transition function $\tau_i : \mathscr{W}_i \times A_i \times Y \to \mathscr{W}_i$. Note that the transitions are potentially private because they depend on the realized action choice, which is not public. It is also sometimes convenient to use mixed rather than behavior strategies when calculating examples with private strategies (see section 10.4.2 for an example). A private mixed strategy can also be represented by an automaton $(\mathscr{W}_i, w_i^0, f_i, \tau_i)$, where $f_i : \mathscr{W}_i \to \Delta(A_i)$ is the decision rule (as usual), but where transitions are potentially random, that is, $\tau_i : \mathscr{W}_i \times A_i \times Y \to \Delta(\mathscr{W}_i)$.

◆

Remark 7.1.3 **Minimal automata** An automaton $(\mathscr{W}, w^0, \tau, f)$ is *minimal* if every state $w \in \mathscr{W}$ is accessible from $w^0 \in \mathscr{W}$ and, for every pair of states $w, \hat{w} \in \mathscr{W}$, there exists a sequence of signals h^t such that for some i, $f_i(\tau(w, h^t)) \neq f_i(\tau(\hat{w}, h^t))$.

Every public profile has a minimal representing automaton. Moreover, this automaton is essentially unique: Suppose $(\mathscr{W}, w^0, \tau, f)$ and $(\tilde{\mathscr{W}}, \tilde{w}^0, \tilde{\tau}, \tilde{f})$ are two minimal automata representing the same public strategy profile. Define a mapping $\varphi : \mathscr{W} \to \tilde{\mathscr{W}}$ as follows: Set $\varphi(w^0) = \tilde{w}^0$. For $\hat{w} \in \mathscr{W} \setminus \{w^0\}$, let h^t be a public history reaching $\hat{w}$ (i.e., $\hat{w} = \tau(w^0, h^t)$), and set $\varphi(\hat{w}) = \tilde{\tau}(\tilde{w}^0, h^t)$. Because both automata are minimal and represent the same profile, φ does not depend on

7.1 ■ The Canonical Repeated Game

the choice of public history reaching $\hat{w}$. It is straightforward to verify that φ is one-to-one and onto. Moreover, $\tilde{\tau}(\tilde{w}, y) = \varphi(\tau(\varphi^{-1}(\tilde{w}), y))$, and $f(w) = \tilde{f}(\varphi(w))$.

◆

Remark 7.1.4 **Public correlation** If the game has public correlation, the only change in the automaton representation of public strategy profiles is that the initial state is determined by a probability distribution μ^0 and the transition function maps into probability distributions over states, that is, $\tau : \mathcal{W} \times Y \to \Delta(\mathcal{W})$ (see remark 2.3.3).

◆

Attention in applications is often restricted to strongly symmetric public strategy profiles, in which after every history, the same action is chosen by all long-lived players:[6]

Definition 7.1.2 *Suppose $A_i = A_j$ for all long-lived i and j. A public profile σ is strongly symmetric if, for all public histories h^t, $\sigma_i(h^t) = \sigma_j(h^t)$ for all long-lived i and j.*

Once we restrict attention to public profiles, there is an attractive formulation of sequential rationality, because every public history h^t does induce a well-defined continuation game (in public strategies).

Definition 7.1.3 *A perfect public equilibrium (PPE) is a profile of public strategies σ that for any public history h^t, specifies a Nash equilibrium for the repeated game, that is, for all t and all $h^t \in Y^t$, $\sigma|_{h^t}$ is a Nash equilibrium. A PPE is strict if each player strictly prefers his equilibrium strategy to every other public strategy.*

When the public monitoring has *full support*, that is, $\rho(y \mid a) > 0$ for all y and a, every public history arises with positive probability, and so every Nash equilibrium in public strategies is a PPE.

We denote the set of PPE payoff vectors of the long-lived players by $\mathscr{E}(\delta) \subset \mathbb{R}^n$.

The one-shot deviation principle plays as useful a role here as it does for subgame-perfect equilibria in perfect monitoring games. A *one-shot deviation* for player i from the public strategy σ_i is a strategy $\hat{\sigma}_i \neq \sigma_i$ with the property that there exists a unique public history $\tilde{h}^t \in Y^t$ such that for all $h^\tau \neq \tilde{h}^t$,

$$\sigma_i(h^\tau) = \hat{\sigma}_i(h^\tau).$$

The proofs of the next two propositions are straightforward modifications of their perfect-monitoring analogs (propositions 2.2.1 and 2.4.1), and so are omitted.

Proposition 7.1.1 **The one-shot deviation principle** *A public strategy profile σ is a PPE if and only if there are no profitable one-shot deviations, that is, if and only if for all public histories $h^t \in Y^t$, $\sigma(h^t)$ is a Nash equilibrium of the normal-form game with payoffs*

---

6. Because it imposes no restriction on short-lived players' behavior, the concept of strong symmetry is most useful in games without short-lived players. Section 11.2.6 presents a strongly symmetric equilibrium of a game with short-lived players that is most naturally described as asymmetric.

$$g_i(a) = \begin{cases} (1-\delta)u_i(a) + \delta \sum_{y \in Y} U_i(\sigma|_{h^t,y})\rho(y \mid a), & \text{for } i = 1, \ldots, n, \\ u_i(a), & \text{for } i = n+1, \ldots, N. \end{cases}$$

Proposition 7.1.2 *Suppose the public strategy profile σ is described by the automaton $(\mathcal{W}, w^0, f, \tau)$, and let $V_i(w)$ be the long-lived player i's average discounted value from play that begins in state w. The strategy profile σ is a PPE if and only if for all $w \in \mathcal{W}$ accessible from w^0, $f(w)$ is a Nash equilibrium of the normal-form game with payoffs*

$$g_i^w(a) = \begin{cases} (1-\delta)u_i(a) + \delta \sum_{y \in Y} V_i(\tau(w,y))\rho(y \mid a), & \text{for } i = 1, \ldots, n, \\ u_i(a), & \text{for } i = n+1, \ldots, N. \end{cases} \quad (7.1.2)$$

Equivalently, σ is a PPE if and only if for all $w \in \mathcal{W}$ accessible from w^0, $f(w) \in \mathbf{B}$ and $(f_1(w), \ldots, f_n(w))$ is a Nash equilibrium of the normal-form game with payoffs $(g_1^w, \ldots, g_n^w)$, where $g_i^w(\cdot, f_{n+1}(w), \ldots, f_N(w)) : \prod_{i=1}^n A_i \to \mathbb{R}$, for $i = 1, \ldots, n$, is given by (7.1.2).

We also have a simple characterization of strict PPE under full-support public monitoring: Because every public history is realized with positive probability, strictness of a PPE is equivalent to the strictness of the induced Nash equilibria of the normal form games of propositions 7.1.1 and 7.1.2.

Corollary 7.1.1 *Suppose $\rho(y \mid a) > 0$ for all $y \in Y$ and $a \in A$. The profile σ is a strict PPE if and only if for all $w \in \mathcal{W}$ accessible from w^0, $f(w)$ is a strict Nash equilibrium of the normal-form game with payoffs g^w.*

Clearly, strict PPE must be in pure strategies, and so we can define:

Definition 7.1.4 *Suppose $\rho(y \mid a) > 0$ for all $y \in Y$ and $a \in A$. A pure public strategy profile described by the automaton $(\mathcal{W}, w^0, f, \tau)$ is a uniformly strict PPE if and only if there exists $\upsilon > 0$ such that for all $w \in \mathcal{W}$ accessible from w^0, for all i,*

$$g_i^w(f(w)) \geq g_i^w(a_i, f_{-i}(w)) + \upsilon, \quad a_i \neq f_i(w),$$

where g^w is defined by (7.1.2).

7.2 A Repeated Prisoners' Dilemma Example

This section illustrates some key issues that arise in games of imperfect monitoring. We again study the repeated prisoners' dilemma. The imperfect monitoring is captured by two signals $\bar{y}$ and $\underline{y}$, whose distribution is given by

$$\rho(\bar{y} \mid a) = \begin{cases} p, & \text{if } a = EE, \\ q, & \text{if } a = SE \text{ or } ES, \\ r, & \text{if } a = SS, \end{cases} \quad (7.2.1)$$

7.2 ■ A Repeated Prisoners' Dilemma Example

	$\bar{y}$	$\underline{y}$
E	$\frac{(3-p-2q)}{(p-q)}$	$-\frac{(p+2q)}{(p-q)}$
S	$\frac{3(1-r)}{(q-r)}$	$-\frac{3r}{(q-r)}$

	E	S
E	$2,2$	$-1,3$
S	$3,-1$	$0,0$

Figure 7.2.1 The left matrix describes the ex post payoffs for the prisoners' dilemma with the public monitoring of (7.2.1). The implied ex ante payoff matrix is on the right and agrees with figure 1.2.1.

where $0 < q < p < 1$ and $0 < r < p$. If we interpret the prisoners' dilemma as a partnership game, then the effort choices of the players stochastically determine whether output is high ($\bar{y}$) or low ($\underline{y}$). High output, the "good" signal, has a higher probability when both players exert effort. Output thus provides some noisy information about whether both players exerted effort and may or may not provide information distinguishing the three stage-game outcomes in which at least one player shirks (depending on q and r). The ex post payoffs and implied ex ante payoffs (that agree with the payoffs from figure 1.2.1) are in figure 7.2.1.

In conducting comparative statics with respect to the monitoring distribution, we fix the ex ante payoffs and so are implicitly adjusting the ex post payoffs as well as the monitoring distribution. Although it is more natural for ex post payoffs to be fixed, with changes in monitoring reflected in changes in ex ante payoffs, fixing ex ante payoffs significantly simplifies calculations (without altering the substance of the results).

7.2.1 Punishments Happen

As with perfect monitoring games, any strategy profile prescribing a stage-game Nash equilibrium in each period, independent of history, constitutes a PPE of the repeated game of imperfect monitoring. In this case, players can simply ignore any signals they see, secure in the knowledge that every player is choosing a best response in every period. In the prisoners' dilemma, this implies that both shirk in every period.

Equilibria in which players exert effort require intertemporal incentives, and a central feature of such incentives is that some realizations of the signal must be followed by low continuation values. As such, they have the flavor of punishments, but unlike the case with perfect monitoring, these low continuation values need not arise from a deviation. As will become clear, they are needed to provide appropriate incentives for players to exert effort.

One of the simplest profiles in which signals matter calls for the players to exert effort in the first period and continue to exert effort until the first realization of low output $\underline{y}$, after which players shirk forever. We refer to this profile as *grim trigger* because of its similarity to its perfect monitoring namesake. This strategy has a simple representation as a two-state automaton. The state-space is $\mathcal{W} = \{w_{EE}, w_{SS}\}$, initial state w_{EE}, output function

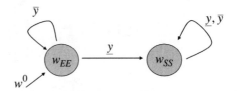

Figure 7.2.2 The grim trigger automaton for the prisoners' dilemma with public monitoring.

$$f(w_{EE}) = EE,$$
and $$f(w_{SS}) = SS,$$

and transition function

$$\tau(w, y) = \begin{cases} w_{EE}, & \text{if } w = w_{EE} \text{ and } y = \bar{y}, \\ w_{SS}, & \text{otherwise}, \end{cases}$$

where y is the previous-period signal. The automaton is illustrated in figure 7.2.2.

We associate with each state a value describing the expected payoff when play begins in the state in question. The value function is given by (omitting subscripts, because the setting is symmetric)

$$V(w_{EE}) = (1-\delta)2 + \delta\{pV(w_{EE}) + (1-p)V(w_{SS})\}$$

and

$$V(w_{SS}) = (1-\delta) \times 0 + \delta V(w_{SS}).$$

We immediately have $V(w_{SS}) = 0$, so that

$$V(w_{EE}) = \frac{2(1-\delta)}{1-\delta p}. \tag{7.2.2}$$

The strategies will be an equilibrium if and only if in each state, the prescribed actions constitute a Nash equilibrium of the normal-form game induced by the current payoffs and continuation values (proposition 7.1.2). Hence, the conditions for equilibrium are

$$V(w_{EE}) \geq (1-\delta)3 + \delta\{qV(w_{EE}) + (1-q)V(w_{SS})\} \tag{7.2.3}$$

and

$$V(w_{SS}) \geq (1-\delta)(-1) + \delta V(w_{SS}).$$

It is clear that the incentive constraint for defecting in state w_{SS} is trivially satisfied, because the state w_{SS} is absorbing. The incentive constraint in the state w_{EE} can be rewritten as

$$V(w_{EE}) \geq \frac{3(1-\delta)}{1-\delta q},$$

7.2 ■ A Repeated Prisoners' Dilemma Example

or using (7.2.2) to substitute for $V(w_{EE})$,

$$\frac{2(1-\delta)}{1-\delta p} \geq \frac{3(1-\delta)}{1-\delta q},$$

that is,

$$\delta(3p - 2q) \geq 1. \tag{7.2.4}$$

If $\bar{y}$ is a sufficiently good signal that both players had exerted effort, in the sense that

$$p > \tfrac{1}{3} + \tfrac{2}{3}q, \tag{7.2.5}$$

so that $3p - 2q > 1$, then grim trigger is an equilibrium, provided the players are sufficiently patient. For δ sufficiently large (and $p > \tfrac{1}{3} + \tfrac{2}{3}q$), (7.2.3) holds strictly, in which case the equilibrium is strict (in the sense of definition 7.1.3).

If the incentive constraint (7.2.3) holds, then the value of the equilibrium is given by (7.2.2). Notice that as p approaches 1, so that the action profile EE virtually guarantees the signal $\bar{y}$, the equilibrium value approaches 2, the perfect monitoring value. Moreover, if $p = 1$ and $q = 0$ (so that $\bar{y}$ is a perfect signal of effort when the opponent exerts effort), grim trigger here looks very much like grim trigger in the game with perfect monitoring, and (7.2.4) is equivalent to the bound $\delta \geq 1/3$ obtained in example 2.4.1.

As in the case of perfect monitoring, we conclude that grim trigger is an equilibrium strategy profile if the players are sufficiently patient. However, there is an important difference. Players are initially willing to exert effort under the grim trigger profile in the presence of imperfect monitoring because shirking triggers the transition to the absorbing state w_{SS} with too high a probability. Unlike the perfect monitoring case, playing EE does not guarantee that w_{SS} will not be reached. Players receive positive payoffs in this equilibrium only from the initial segment of periods in which both exert effort, before the inevitable and irreversible switch to mutual shirking. The timing of this switch is independent of the discount factor. As a result, as the players become more patient and the importance of the initial periods declines, so does their expected payoff (because $V(w_{EE}) \to 0 = V(w_{SS})$ as $\delta \to 1$ from (7.2.2)).

This profile illustrates the interaction between continuation values in their role of providing incentives (where a low continuation value after certain signals, such as y, strengthens incentives) and the contribution such values make to current values. As players become patient, the myopic incentive to deviate to the stage-game best reply is reduced, but at the same time, the impact of the low continuation values may be increased (we return to this issue in remark 7.2.1).

7.2.2 Forgiving Strategies

We now consider a profile that provides incentives to exert effort without the use of an absorbing state. Players exert effort after the signal $\bar{y}$ and shirk after the signal y, and exert effort in the first period. The two-state automaton representing this profile

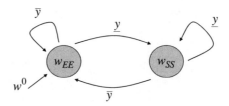

Figure 7.2.3 A simple profile with one-period memory.

has the same state space, initial state, and output function as grim trigger. Only the transition function differs from the previous example. It is given by

$$\tau(w, y) = \begin{cases} w_{EE}, & \text{if } y = \bar{y}, \\ w_{SS}, & \text{if } y = \underline{y}, \end{cases}$$

where y is the previous period signal. The automaton is illustrated in figure 7.2.3. We can think of the players as rewarding good signals and punishing bad ones. In this profile, punishments are attached to bad signals even though players may know these signals represent no shirking (such as $\underline{y}$ following $\bar{y}$).

The values in each state are given by

$$V(w_{EE}) = (1 - \delta)2 + \delta\{pV(w_{EE}) + (1 - p)V(w_{SS})\} \tag{7.2.6}$$

and

$$V(w_{SS}) = (1 - \delta) \times 0 + \delta\{rV(w_{EE}) + (1 - r)V(w_{SS})\}. \tag{7.2.7}$$

Solving (7.2.6)–(7.2.7),

$$\begin{bmatrix} V(w_{EE}) \\ V(w_{SS}) \end{bmatrix} = (1 - \delta) \begin{bmatrix} 1 - \delta p & -\delta(1 - p) \\ -\delta r & 1 - \delta(1 - r) \end{bmatrix}^{-1} \begin{bmatrix} 2 \\ 0 \end{bmatrix}$$

$$= \frac{1}{1 - \delta(p - r)} \begin{bmatrix} 2(1 - \delta(1 - r)) \\ 2\delta r \end{bmatrix}. \tag{7.2.8}$$

Notice that $V(w_{EE})$ and $V(w_{SS})$ are both independent of q, because q identifies the signal distribution induced by profiles SE and ES, which do not arise in equilibrium. We again have $\lim_{p \to 1} V(w_{EE}) = 2$, so that we approach the perfect monitoring value as the signals in state EE become arbitrarily precise. For any specification of the parameters, we have

$$\frac{2(1 - \delta)}{1 - \delta p} < \frac{2(1 - \delta(1 - r))}{1 - \delta(p - r)},$$

and hence the forgiving strategy of this section yields a higher payoff than the grim trigger strategy of the previous example.

As with grim trigger, the strategies will be an equilibrium if and only if in each state, the prescribed actions constitute a Nash equilibrium of the normal-form game induced

7.2 ■ A Repeated Prisoners' Dilemma Example

by the current payoffs and continuation values. Hence, the profile is an equilibrium if and only if

$$V(w_{EE}) \geq (1-\delta)3 + \delta\{qV(w_{EE}) + (1-q)V(w_{SS})\} \tag{7.2.9}$$

and

$$V(w_{SS}) \geq (1-\delta)(-1) + \delta\{qV(w_{EE}) + (1-q)V(w_{SS})\}.$$

Using (7.2.6) to substitute for the equilibrium value $V(w_{EE})$ of state w_{EE}, the incentive constraint (7.2.9) can be rewritten as

$$(1-\delta)2 + \delta\{pV(w_{EE}) + (1-p)V(w_{SS})\}$$
$$\geq (1-\delta)3 + \delta\{qV(w_{EE}) + (1-q)V(w_{SS})\},$$

which simplifies to

$$\delta(p-q)\{V(w_{EE}) - V(w_{SS})\} \geq (1-\delta). \tag{7.2.10}$$

From (7.2.8),

$$V(w_{EE}) - V(w_{SS}) = \frac{2(1-\delta)}{1-\delta(p-r)}. \tag{7.2.11}$$

From (7.2.10), the incentive constraint for state w_{EE} requires

$$2\delta(p-q) \geq 1 - \delta(p-r)$$

or

$$\delta \geq \frac{1}{3p - 2q - r}. \tag{7.2.12}$$

A similar calculation for the incentive constraint in state w_{SS} yields

$$(1-\delta) \geq \delta(q-r)\{V(w_{EE}) - V(w_{SS})\},$$

and substituting from (7.2.11) gives

$$\delta \leq \frac{1}{p + 2q - 3r}. \tag{7.2.13}$$

Conditions (7.2.12) and (7.2.13) are in tension: For the w_{EE} incentive constraint to be satisfied, players must be sufficiently patient (δ is sufficiently large) and the signals sufficiently informative ($p - q$ is sufficiently large). This ensures that the myopic incentive to play S is less than the continuation reward from the more favorable distribution induced by EE rather than that induced by SE. At the same time, for the w_{SS} incentive constraint to be satisfied, players must not be too patient (δ is not too large) relative to the signals. This ensures that the myopic cost from playing E is more than the continuation reward from the more favorable distribution induced by *ES* rather than that induced by *SS*. Conditions (7.2.12) and (7.2.13) are consistent as long as

$$p \geq 2q - r.$$

Moreover, (7.2.13) is trivially satisfied if q is sufficiently close to r.[7]

When (7.2.12) and (7.2.13) are satisfied, the one-period memory profile supports effort as an equilibrium choice in a period by "promising" players a low continuation after $\underline{y}$, which occurs with higher probability after shirking. It is worth emphasizing that the specification of S after $\underline{y}$ is *not* a punishment of a player for having deviated or shirked. For example, in period 0, under the profile, both players play E. Nonetheless, the players shirk in the next period after observing $\underline{y}$. However, if the profile did not specify such a negative repercussion from generating bad signals, players have no incentive to exert effort. The role of *SS* after $\underline{y}$ is analogous to that of the deductible in an insurance policy, which encourages due care.

It is also worth observing, that when (7.2.12) and (7.2.13) hold strictly, then the PPE is strict: Each player finds E a *strict* best reply after $\bar{y}$ and S a *strict* best reply after $\underline{y}$. In this sense, history is coordinating continuation play (as it commonly does in equilibria of perfect-monitoring games).

Remark 7.2.1 **Patient incentives** Suppose q is sufficiently close to r that the upper bound on δ, (7.2.13), is satisfied for all δ. Then, the profile of figure 7.2.3 is a strict PPE for all δ satisfying (7.2.12) strictly. The outcome path produced by these strategies can be described by a Markov chain on the state space $\{w_{EE}, w_{SS}\}$, with transition matrix

	w_{EE}	w_{SS}
w_{EE}	p	$1 - p$
w_{SS}	r	$1 - r$.

The process is ergodic and the stationary distribution puts probability $r/(1 - p + r)$ on EE and $(1 - p)/(1 - p + r)$ on SS.[8] Because $p < 1$, some shirking must occur in equilibrium. When starting in state w_{EE}, the current payoff is thus higher than the equilibrium payoff, and increasing the discount factor only makes the relatively low-payoff future more important, decreasing the expected payoff from the game. When in state w_{SS}, the current payoff is relatively low, and putting more weight on the future increases expected payoffs.

The myopic incentive to play a stage-game best reply becomes small as players become patient. At the same time, as in grim trigger and many other strategy profiles,[9] the size of the penalty from a disadvantageous signal ($V(w_{EE}) - V(w_{SS})$) also becomes smaller. In the limit, as the players get arbitrarily patient, expected

---

7. Note that $q = r$ is inconsistent with the assumption that the stage game payoff $u_1(a)$ is the expectation of ex post payoffs (7.1.1), because $q = r$ would imply $u_1(SE) = u_1(SS)$ (see also figure 7.2.1). A similar comment applies if $p = q$. We assume $q = r$ or $p = q$ in some examples to ease calculations; the conclusions hold for $q - r$ or $p - q$ close to 0.
8. For grim trigger, the transition matrix effectively has $r = 0$, so the stationary distribution in that case puts probability one on w_{SS}.
9. More specifically, this claim holds for profiles that are *connected*, that is, profiles with the property that there is a common finite sequence of signals taking any state into a common state (lemma 13.5.1). We discuss this issue in some detail in section 13.5.

7.2 ■ A Repeated Prisoners' Dilemma Example

payoffs are independent of the initial states. For a fixed signal distribution given by p and r, we have

$$\lim_{\delta \to 1} V(w_{EE}) = \lim_{\delta \to 1} V(w_{SS}) = \frac{2r}{1 - p + r}.$$

◆

7.2.3 Strongly Symmetric Behavior Implies Inefficiency

We now investigate the most efficient symmetric pure strategy equilibria that the players can support under imperfect monitoring. More precisely, we focus on *strongly symmetric pure-strategy equilibria*, equilibria in which after every history, the same action is chosen by both players (definition 7.1.2). It turns out that efficiency cannot typically be attained with a strongly symmetric equilibrium under imperfect monitoring (see proposition 8.2.1). Moreover, for this example, if $2p - q < 1$ and $q > 2r$, the best strongly symmetric PPE payoff is strictly smaller than the best symmetric PPE payoff, which is achieved using SE and ES (see remark 7.7.2 and section 8.4.3). As we have seen, imperfect monitoring ensures that punishments will occur along the equilibrium path, whereas symmetry ensures that these punishments reduce the payoffs of both players, precluding efficiency. We will subsequently see that asymmetric strategies, in which one player is punished while the other is rewarded, play a crucial role in achieving nearly efficient payoffs. At the same time, it will be important that the monitoring is sufficiently "rich" to allow this differential treatment (we return to this issue for this example in section 8.4.3 and in general in chapter 9).

We assume here that players can publicly correlate (leaving to section 7.7.1 the analysis without public correlation). As we will see, it suffices to consider strategies implemented by automata with two states, so that $\mathscr{W} = \{w_{EE}, w_{SS}\}$ with $f(w_{EE}) = EE$ and $f(w_{SS}) = SS$. Letting $\tau(w, y)$ be the probability of a transition to state w_{EE}, given that the current state $w \in \{w_{EE}, w_{SS}\}$ and signal $y \in \{\bar{y}, \underline{y}\}$, calculating the maximum payoff from a symmetric equilibrium with public correlation is equivalent to finding the largest ϕ for which the following transition function supports an equilibrium:

$$\tau(w, y) = \begin{cases} 1, & \text{if } w = w_{EE} \text{ and } y = \bar{y}, \\ \phi, & \text{if } w = w_{EE} \text{ and } y = \underline{y}, \\ 0, & \text{if } w = w_{SS}. \end{cases}$$

These strategies make use of a public correlating device, because the players perfectly correlate the random movement to state w_{SS} that follows the bad signal. The automaton is illustrated in figure 7.2.4.

The most efficient symmetric outcome is permanent EE, yielding both players a payoff of 2. The imperfection in monitoring precludes this outcome from being an equilibrium outcome. The "punishment" state of w_{SS} is the worst possible state and plays a similar role here as in grim trigger of section 7.2.1, which is captured by $\phi = 0$. For large δ, the incentive constraint in state w_{EE} when $\phi = 0$ (described by (7.2.3)) holds strictly, indicating that the continuation value after $\underline{y}$ can be slightly increased without disrupting the incentives players have to play E. Moreover, by increasing

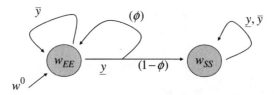

Figure 7.2.4 The grim trigger profile with public correlation. With probability ϕ play remains in w_{EE} after a bad signal, and with probability $1 - \phi$, play transits to the absorbing state w_{SS}.

that continuation valuation, the value in the previous period has been increased. By appropriately choosing ϕ, we can maximize the symmetric payoff while still just satisfying the incentive constraint.[10]

The value $V(w_{SS})$ in state w_{SS} equals 0, reflecting the fact that state w_{SS} corresponds to permanent shirking. The value for state w_{EE} is given by

$$V(w_{EE}) = (1-\delta)2 + \delta\{pV(w_{EE}) + (1-p)(\phi V(w_{EE}) + (1-\phi)V(w_{SS}))\}$$
$$= (1-\delta)2 + \delta(p + (1-p)\phi)V(w_{EE}),$$

and solving,

$$V(w_{EE}) = \frac{2(1-\delta)}{1 - \delta(p + (1-p)\phi)}. \quad (7.2.14)$$

Incentives for equilibrium behavior in state w_{SS} are trivial because play then consists of a stage-game Nash equilibrium in every subsequent period. Turning to state w_{EE}, the incentive constraint is (using $V(w_{SS}) = 0$):

$$V(w_{EE}) \geq (1-\delta)3 + \delta(q + (1-q)\phi)V(w_{EE}).$$

As explained, we need the largest value of ϕ for which this constraint holds. Clearly, such a value must cause the constraint to hold with equality, leading us to solve for

$$V(w_{EE}) = \frac{3(1-\delta)}{1 - \delta(q + (1-q)\phi)}.$$

Making the substitution for $V(w_{EE})$ from (7.2.14), and solving,

$$\phi = \frac{\delta(3p - 2q) - 1}{\delta(3p - 2q - 1)}. \quad (7.2.15)$$

This expression is nonnegative as long as[11]

$$\delta(3p - 2q) \geq 1. \quad (7.2.16)$$

Once again, equilibrium requires that players be sufficiently patient and that p and q not be too close. It is also worth noting that this condition is the same as (7.2.4), the

---

10. It is an implication of proposition 7.5.1 that this bang-bang behavior yields the most efficient symmetric pure strategy profile.
11. The expression could also be positive if both numerator and denominator are negative, but then it necessarily exceeds 1, a contradiction.

lower bound ensuring grim trigger is an equilibrium. If (7.2.16) fails, then even making the punishment as severe as possible, by setting $\phi = 0$, does not create sufficiently strong incentives to deter shirking. Conditional on being positive, the expression for ϕ will necessarily be less than 1. This is simply the observation that one can never create incentives for effort by setting $\phi = 1$, and hence dispensing with all threat of punishment.

A value of $\delta < 1$ will exist satisfying inequality (7.2.16) as long as

$$p > \tfrac{1}{3} + \tfrac{2}{3}q.$$

This is (7.2.5) (the necessary condition for grim trigger to be a PPE) and is implied by the necessary condition (7.2.12) for the one-period memory strategy profile of section 7.2.2 to be an equilibrium. If $\delta = 1/(3p - 2q)$, then we have $\phi = 0$, and hence grim trigger. In this case, the discount factor is so low as to be just on the boundary of supporting such strategies as an equilibrium, and sufficient incentives can be obtained only by having a bad signal trigger a punishment with certainty. On the other hand, $\phi \to 1$ as $\delta \to 1$. As the future swamps the present, bad signals need only trigger punishments with an arbitrarily small probability. However, this does not suffice to achieve the efficient outcome. Substituting the value of ϕ from (7.2.15) in (7.2.14), we find, for all $\delta \in (0, 1)$ satisfying the incentive constraint given by (7.2.16),[12]

$$V(w_{EE}) = \frac{3p - 2q - 1}{p - q} = 2 - \frac{(1 - p)}{(p - q)} < 2. \qquad (7.2.17)$$

Intuitively, punishments can become less likely as the discount factor climbs, but incentives will be preserved only if punishments remain sufficiently likely that the expected value of the game following a bad signal falls sufficiently short of the expected value following a good signal. Because bad signals occur with positive probability along the equilibrium path, this suffices to bound payoffs away from their efficient levels.

It is intuitive that the upper bound for the symmetric equilibrium payoff with public correlation is also an upper bound without public correlation. We will argue in section 7.7.1 that the bound of (7.2.17) is in fact tight (even in the absence of public correlation) for large δ.

7.3 Decomposability and Self-Generation

In this section, we describe a general method (introduced by Abreu, Pearce, and Stacchetti 1990) for characterizing $\mathscr{E}(\delta)$, the set of perfect public equilibrium payoffs. We have already seen a preview of this work in section 2.5.1. The essence of this approach is to view a PPE as describing after each history the specified action to be taken after the history and continuation promises. The continuation promises are themselves of course required to be equilibrium values. The recursive properties of PPE described in section 7.1.3 provide the necessary structure for this approach.

---

12. Notice that the maximum payoff is independent of δ, once δ is large enough to provide the required incentives.

The first two definitions are the public monitoring versions of the notions in section 2.5.1. Recall that $\mathbf{B} \subset \prod_{i=1}^{N} \Delta(A_i)$ is the set of feasible action profiles when players $i = n+1, \ldots, N$ are short-lived. If $n = N$ (there are no short-lived players), then $\mathbf{B} = \prod_{i=1}^{n} \Delta(A_i)$.

Definition 7.3.1 *For any $\mathscr{W} \subset \mathbb{R}^n$, a mixed action profile $\alpha \in \mathbf{B}$ is enforceable on $\mathscr{W}$ if there exists a mapping $\gamma : Y \to \mathscr{W}$ such that, for all $i = 1, \ldots, n$, and $a_i' \in A_i$,*

$$V_i(\alpha, \gamma) \equiv (1-\delta)u_i(\alpha) + \delta \sum_{y \in Y} \gamma_i(y) \rho(y \mid \alpha) \qquad (7.3.1)$$
$$\geq (1-\delta)u_i(a_i', \alpha_{-i}) + \delta \sum_{y \in Y} \gamma_i(y) \rho(y \mid a_i', \alpha_{-i}).$$

The function γ enforces α (on $\mathscr{W}$).

Remark 7.3.1 **Hidden short-lived players** The incentive constraints on the short-lived players are completely captured by the requirement that the action profile be an element of $\mathbf{B}$. We often treat $u(\alpha)$ as the vector of stage-game payoffs for the long-lived players, that is, $u(\alpha) = (u_1(\alpha), \ldots, u_n(\alpha))$, rather than the vector of payoffs for all players—the appropriate interpretation should be clear from context. ◆

Notice that the function V, introduced in definition 7.3.1, is continuous. Figure 7.3.1 illustrates the relationship between $u(\alpha)$, γ, and $V(\alpha, \gamma)$.

We interpret the function γ as describing expected payoffs from future play ("continuation promises") as a function of the public signal y. Enforceability is then essentially an incentive compatibility requirement. The profile α is enforceable if it is optimal for each player to choose α, given some γ describing the implications of current signals for future payoffs. Phrased differently, the profile α is enforceable if it is a Nash equilibrium of the one-shot game $g^\gamma(a) \equiv (1-\delta)u(a) + \delta E[\gamma(y) \mid a]$ (compare with proposition 7.1.2 and the discussion just before proposition 2.4.1).

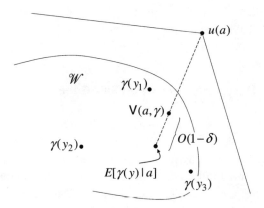

Figure 7.3.1 The relationship between $u(\alpha)$, γ, and $V(\alpha, \gamma)$. There are three public signals, $Y = \{y_1, y_2, y_3\}$, and $E[\gamma(y) \mid a] = \sum_\ell \gamma(y_\ell) \rho(y_\ell \mid a)$. Because $V(a, \gamma) = (1-\delta)u(a) + \delta E[\gamma(y) \mid a]$, the distance from $V(a, \gamma)$ to $E[\gamma \mid a]$ is of order $1 - \delta$.

Definition 7.3.2 *A payoff vector $v \in \mathbb{R}^n$ is* decomposable *on $\mathscr{W}$ if there exists a mixed action profile $\alpha \in \mathbf{B}$, enforced by γ on $\mathscr{W}$, such that*

$$v_i = V_i(\alpha, \gamma).$$

The payoff v is decomposed *by the pair (α, γ) (on $\mathscr{W}$).*

It will be convenient to identify for any set $\mathscr{W}$, the set of payoffs that can be decomposed on $\mathscr{W}$, as well as to have a function that identifies the action profile and associated enforcing γ decomposing any value in that set. When there are several enforceable profiles and enforcing promises for any decomposable payoff, the selection can be arbitrary.

Definition 7.3.3 *For all $\mathscr{W} \subset \mathbb{R}^n$, let $\mathscr{B}(\mathscr{W}) \equiv \{v \in \mathbb{R}^n : v = V(\alpha, \gamma)$ for some α enforced by γ on $\mathscr{W}\}$, and define the pair*

$$\mathsf{Q} : \mathscr{B}(\mathscr{W}) \to \mathbf{B} \quad \text{and} \quad \mathsf{U} : \mathscr{B}(\mathscr{W}) \to \mathscr{W}^Y$$

so that $\mathsf{Q}(v)$ is enforced by $\mathsf{U}(v)$ on $\mathscr{W}$ and $V(\mathsf{Q}(v), \mathsf{U}(v)) = v$. A payoff $v \in \mathscr{B}(\mathscr{W})$ is decomposed *by $(\mathsf{Q}(v), \mathsf{U}(v))$.*

We can think of $\mathscr{B}(\mathscr{W})$ as the set of equilibrium payoffs (for the long-lived players), given that $\mathscr{W}$ is the set of continuation equilibrium payoffs. In particular, $\mathscr{B}(\mathscr{W})$ contains any payoff vector that can be obtained from (decomposed) using an enforceable choice α in the current period and with the link γ between current signals and future equilibrium payoffs. Although $\mathscr{B}(\mathscr{W})$ depends on δ, we only make that dependence explicit when necessary, writing $\mathscr{B}(\mathscr{W}; \delta)$.

If $\mathscr{B}(\mathscr{W})$ is the set of payoffs that one can support given the set $\mathscr{W}$ of continuation payoffs, then a set $\mathscr{W}$ for which $\mathscr{W} \subset \mathscr{B}(\mathscr{W})$ should be of special interest. We have already seen in proposition 2.5.1 and remark 2.5.1 (where $\cup_{a \in A} \mathscr{B}_a(\mathscr{W})$ is the pure-action version of $\mathscr{B}(\mathscr{W})$) that a similar property is sufficient for the payoffs to be pure-strategy subgame-perfect equilibrium payoffs in games with perfect monitoring. Because the structure of the game is stationary (for public strategies), the set of equilibrium payoffs and equilibrium continuation payoffs coincide. The function $\mathscr{B}$ plays an important role in the investigation of equilibria of the repeated game and is commonly referred to as the *generating function* for the game. We first need the critical notion of self-generation:

Definition 7.3.4 *A set of payoffs $\mathscr{W} \subset \mathbb{R}^n$ is* self-generating *if $\mathscr{W} \subset \mathscr{B}(\mathscr{W})$.*

Our interest in self-generation is due to the following result (compare with proposition 2.5.1).

Proposition 7.3.1 **Self-generation** *For any bounded set $\mathscr{W} \subset \mathbb{R}^n$, if $\mathscr{W}$ is self-generating, $\mathscr{B}(\mathscr{W}) \subset \mathscr{E}(\delta)$ (and hence $\mathscr{W} \subset \mathscr{E}(\delta)$).*

Note that no explicit feasibility restrictions are imposed on $\mathscr{W}$ (in fact, no restrictions are placed on the set $\mathscr{W}$ beyond boundedness).[13] As a result, one would think that by

---

13. The space $\mathbb{R}^n$ is trivially self-generating. To generate an arbitrary payoff $v \in \mathbb{R}^n$, one need only couple the current play of a Nash equilibrium of the stage game (thus ensuring enforceability),

choosing a sufficiently large set $\mathscr{W}$, payoffs infeasible in the repeated game could be generated. The discipline is imposed by the fact that $\mathscr{W}$ must be capable of generating a superset of itself. Coupled with discounting, the assumption that $\mathscr{W}$ is bounded and the requirement that the first period's payoffs be given by $u(\alpha)$ for some $\alpha \in \mathbf{B}$ excludes infeasible payoffs.

As in the proof of proposition 2.5.1, the idea is to view $\mathscr{B}(\mathscr{W})$ as the set of states for an automaton, to which we apply proposition 7.1.2. Decomposability allows us to associate an action profile and a transition function describing continuation values to each payoff profile in $\mathscr{B}(\mathscr{W})$. Note that although vectors of continuations for the long-lived players are states for the automaton, the output function specifies actions for both long- and short-lived players.

Remark 7.3.2 **Equilibrium behavior** Though self-generation naturally directs attention to the set of equilibrium payoffs, the proof of proposition 7.3.1 constructs an equilibrium profile. In the course of this construction, however, one often has multiple choices for the decomposing actions and continuations (i.e., Q and U are selections). The resulting equilibrium can depend importantly on the choices one makes. Remark 7.7.1 discusses an example, including profiles with Nash reversion and with bounded recall. Chapter 13 gives one reason why this difference is important, showing that profiles with permanent punishments are often not robust to the introduction of private monitoring, whereas profiles with bounded recall always are.
◆

Proof For $v \in \mathscr{B}(\mathscr{W})$, we construct an automaton yielding the payoff vector v and satisfying the condition in proposition 7.1.2, so that the implied strategy profile σ is a PPE. Consider the collection of automata $\{(\mathscr{B}(\mathscr{W}), v, f, \tau) : v \in \mathscr{B}(\mathscr{W})\}$, where the common set of states is given by $\mathscr{B}(\mathscr{W})$, the common decision function by

$$f(v) = \mathsf{Q}(v)$$

for all $v \in \mathscr{B}(\mathscr{W})$, and the common transition function by

$$\tau(v, y) = \mathsf{U}(v)(y)$$

for all $y \in Y$ (recall that for $\mathsf{U}(v) \in \mathscr{W}^Y$, so that $\mathsf{U}(v)(y) \in \mathscr{W}$). Because $\mathscr{W}$ is self-generating, the decision and transition functions are well defined for all $v \in \mathscr{B}(\mathscr{W})$. These automata differ only in their initial state $v \in \mathscr{B}(\mathscr{W})$.

We need to show that for each $v \in \mathscr{B}(\mathscr{W})$, the automaton $(\mathscr{B}(\mathscr{W}), v, f, \tau)$ describes a PPE with payoff v. This will be an implication of proposition 7.1.2 and the decomposability of v on $\mathscr{W}$, once we have shown that

$$v_i = V_i(v), \qquad i = 1, \ldots, n,$$

where $V_i(v)$ is the value to long-lived player i of being in state v.

For any $v \in \mathscr{B}(\mathscr{W})$, we recursively define the implied sequence of continuations $\{\gamma^t\}_{t=0}^{\infty}$, where $\gamma^t : Y^t \to \mathscr{W}$, by setting $\gamma^0 = v$ and $\gamma^t(h^{t-1}, y^{t-1}) =$

with payoff profile v^N, with the function $\gamma(y) = v'$ for all y, where $(1-\delta)v^N + \delta v' = v$. Of course, the unboundedness of the reals plays a key role in this construction.

$U(\gamma^{t-1}(h^{t-1}))(y^{t-1})$. Let σ be the public strategy profile described by the automaton $(\mathscr{B}(\mathscr{W}), \gamma^0, f, \tau)$, so that $\sigma(\varnothing) = Q(\gamma^0)$, and for any history, h^t, $\sigma(h^t) = Q(\gamma^t(h^t))$.

Then, by construction,

$$v = V(Q(v), U(v)) = V(\sigma(\varnothing), U(v))$$
$$= (1-\delta)u(\sigma(\varnothing)) + \delta \sum_{y^0 \in Y} \gamma^1(y^0)\rho(y^0 \mid \sigma(\varnothing))$$
$$= (1-\delta)u(\sigma(\varnothing)) + \delta \sum_{y^0 \in Y} \Bigg\{ (1-\delta)u(\sigma(y^0))$$
$$+ \delta \sum_{y^1 \in Y} \gamma^2(y^0, y^1)\rho(y^1 \mid \sigma(y^0)) \Bigg\} \rho(y^0 \mid \sigma(\varnothing))$$
$$= (1-\delta)\sum_{s=0}^{t-1} \delta^s \sum_{h^s \in Y^s} u(\sigma(h^s)) \Pr\nolimits_\sigma(h^s) + \delta^t \sum_{h^t \in Y^t} \gamma^t(h^t) \Pr\nolimits_\sigma(h^t),$$

where $\Pr_\sigma(h^s)$ is the probability that the sequence of public signals h^s arises under σ. Because $\gamma^t(h^t) \in \mathscr{W}$ and $\mathscr{W}$ is bounded, $\sum_{h^t \in Y^t} \gamma^t(h^t) \Pr_\sigma(h^t)$ is bounded. Taking $t \to \infty$ yields

$$v = (1-\delta)\sum_{s=0}^{\infty} \delta^s \sum_{h^s \in Y^s} u(\sigma(h^s)) \Pr\nolimits_\sigma(h^s),$$

and so the value of σ is v. Hence, for all $v \in \mathscr{W}$ and the automaton, $(\mathscr{W}, v, f, \tau)$, $v = V(v)$.

Let $g^v : \prod_{i=1}^n A_i \to \mathbb{R}^n$ be given by

$$g^v(a) = (1-\delta)u(a) + \delta \sum_{y \in Y} U(v)(y)\rho(y \mid a)$$
$$= (1-\delta)u(a) + \delta \sum_{y \in Y} V(\tau(v, y))\rho(y \mid a),$$

where $a \in A$ is an action profile with long-lived players' actions unrestricted and short-lived players' actions given by $(Q_{n+1}(v), \ldots, Q_N(v))$. Because v is enforced by $(Q(v), U(v))$, $Q(v) \in \mathbf{B}$ is a Nash equilibrium of the normal-form game described by the payoff function g^v, and so proposition 7.1.2 applies. ∎

Proposition 7.3.1 gives us a criterion for identifying subsets of the set of PPE payoffs, because any self-generating set is such a subset. We say that a set of payoffs can be *factorized* if it is a fixed point of the generating function $\mathscr{B}$. The next proposition indicates that the set of PPE payoffs can be factorized. From proposition 7.3.1, the set of PPE payoffs is the largest such fixed point. Abreu, Pearce, and Stacchetti (1990) refer to the next proposition as *factorization*.

Proposition 7.3.2 $\mathscr{E}(\delta) = \mathscr{B}(\mathscr{E}(\delta))$.

Proof If $\mathscr{E}(\delta)$ is self-generating (i.e., $\mathscr{E}(\delta) \subset \mathscr{B}(\mathscr{E}(\delta))$), then (because $\mathscr{E}(\delta)$ is clearly bounded, being a subset of $\mathscr{F}^*$) by the previous proposition, $\mathscr{B}(\mathscr{E}(\delta)) \subset \mathscr{E}(\delta)$, and so $\mathscr{B}(\mathscr{E}(\delta)) = \mathscr{E}(\delta)$. It thus suffices to prove $\mathscr{E}(\delta) \subset \mathscr{B}(\mathscr{E}(\delta))$.

Suppose $v \in \mathscr{E}(\delta)$ and σ is a PPE with value $v = U(\sigma)$. Let $\alpha \equiv \sigma(\varnothing)$ and $\gamma(y) = U(\sigma|_y)$.

It is enough to show that α is enforced by γ on $\mathscr{E}(\delta)$ and $\mathsf{V}(\alpha, \gamma) = v$. But,

$$\mathsf{V}(\alpha, \gamma) = (1 - \delta)u(\alpha) + \delta \sum_y \gamma(y)\rho(y \mid \alpha)$$
$$= (1 - \delta)u(\alpha) + \delta \sum_y U(\sigma|_y)\rho(y \mid \alpha)$$
$$= U(\sigma) = v.$$

Because σ is a PPE, $\sigma|_y$ is also a PPE, and so $\gamma : Y \to \mathscr{E}(\delta)$. Finally, because σ is a PPE, there are no profitable one-shot deviations, and so α is enforced by γ on $\mathscr{E}(\delta)$, and so $v \in \mathscr{B}(\mathscr{E}(\delta))$. ∎

Lemma 7.3.1 $\mathscr{B}$ *is a monotone operator, that is,* $\mathscr{W} \subset \mathscr{W}' \Longrightarrow \mathscr{B}(\mathscr{W}) \subset \mathscr{B}(\mathscr{W}')$.

Proof Suppose $v \in \mathscr{B}(\mathscr{W})$. Then, $v = \mathsf{V}(\alpha, \gamma)$ for some α enforced by $\gamma : Y \to \mathscr{W}$. But then γ also decomposes v using α on $\mathscr{W}'$, and hence $v \in \mathscr{B}(\mathscr{W}')$. ∎

Lemma 7.3.2 *If* $\mathscr{W}$ *is compact,* $\mathscr{B}(\mathscr{W})$ *is closed.*

Proof Suppose $\{v^k\}_k$ is a sequence in $\mathscr{B}(\mathscr{W})$ converging to v, and (α^k, γ^k) is the associated sequence of enforceable action profiles and enforcing continuations, with $v^k = \mathsf{V}(\alpha^k, \gamma^k)$. Because $(\alpha^k, \gamma^k) \in \prod_i \Delta(A_i) \times \mathscr{W}^Y$ and $\prod_i \Delta(A_i) \times \mathscr{W}^Y$ is compact,[14] without loss of generality (taking a subsequence if necessary), we can assume $\{(\alpha^k, \gamma^k)\}_k$ is convergent, with limit (α, γ). The action profile α is clearly enforced by γ with respect to $\mathscr{W}$. Moreover, $\mathsf{V}(\alpha, \gamma) = v$ and so $v \in \mathscr{B}(\mathscr{W})$. ∎

The set of feasible payoffs $\mathscr{F}^\dagger$ is clearly compact. Moreover, every payoff that can be decomposed on the set of feasible payoffs must itself be feasible, that is, $\mathscr{B}(\mathscr{F}^\dagger) \subset \mathscr{F}^\dagger$. Because $\mathscr{E}(\delta)$ is a fixed point of $\mathscr{B}$ and $\mathscr{B}$ is monotonic,

$$\mathscr{E}(\delta) \subset \mathscr{B}^m(\mathscr{F}^\dagger) \subset \mathscr{F}^\dagger, \forall m.$$

In fact, $\{\mathscr{B}^m(\mathscr{F}^\dagger)\}_m$ is a decreasing sequence. Let

$$\mathscr{F}^\dagger_\infty \equiv \bigcap_m \mathscr{B}^m(\mathscr{F}^\dagger).$$

---

14. Continuing our discussion from note 3 on page 227 when A_i is a continuum action space for some i, $\mathscr{W}^Y$ is not sequentially compact under perfect monitoring. However, we can proceed as in the second part of the proof of proposition 2.5.2 to nonetheless obtain a convergent subsequence.

7.3 ■ Decomposability and Self-Generation

Each $\mathscr{B}^m(\mathscr{F}^\dagger)$ is compact and so $\mathscr{F}_\infty^\dagger$ is compact and nonempty (because $\mathscr{E}(\delta) \subset \mathscr{F}_\infty^\dagger$). Therefore, we have

$$\mathscr{E}(\delta) \subset \mathscr{F}_\infty^\dagger \subset \cdots \subset \mathscr{B}^2(\mathscr{F}^\dagger) \subset \mathscr{B}(\mathscr{F}^\dagger) \subset \mathscr{F}^\dagger. \tag{7.3.2}$$

The following proposition implies that the algorithm of iteratively calculating $\mathscr{B}^m(\mathscr{F}^\dagger)$ computes the set of PPE payoffs. See Judd, Yeltekin, and Conklin (2003) for an implementation.

Proposition 7.3.3 $\mathscr{F}_\infty^\dagger$ *is self-generating and so* $\mathscr{F}_\infty^\dagger = \mathscr{E}(\delta)$.

Proof We need to show $\mathscr{F}_\infty^\dagger \subset \mathscr{B}(\mathscr{F}_\infty^\dagger)$. For all $v \in \mathscr{F}_\infty^\dagger$, $v \in \mathscr{B}^m(\mathscr{F}^\dagger)$ for all m, and so there exists (α^m, γ^m) such that $v = \mathsf{V}(\alpha^m, \gamma^m)$ and $\gamma^m(y) \in \mathscr{B}^{m-1}(\mathscr{F}^\dagger)$ for all $y \in Y$.

By extracting convergent subsequences if necessary, we can assume the sequence $\{(\alpha^m, \gamma^m)\}_m$ converges to a limit (α^*, γ^*). It remains to show that α^* is enforced by γ^* on $\mathscr{F}_\infty^\dagger$ and $v = \mathsf{V}(\alpha^*, \gamma^*)$. We only verify that $\gamma^*(y) \in \mathscr{F}_\infty^\dagger$ for all $y \in Y$ (since the other parts are trivial). Suppose then that there is some $y \in Y$ such that $\gamma^*(y) \notin \mathscr{F}_\infty^\dagger$. As $\mathscr{F}_\infty^\dagger$ is closed, there is an $\varepsilon > 0$ such that

$$\bar{B}_\varepsilon(\gamma^*(y)) \cap \mathscr{F}_\infty^\dagger = \varnothing,$$

where $\bar{B}_\varepsilon(v)$ is the closed ball of radius ε centered at v. But, there exists m' such that for any $m > m'$, $\gamma^m(y) \in \bar{B}_\varepsilon(\gamma^*(y))$, which implies

$$\bar{B}_\varepsilon(\gamma^*(y)) \cap (\cap_{m \leq M} \mathscr{B}^m(\mathscr{F}^\dagger)) \neq \varnothing, \quad \forall M > m'.$$

Thus, the collection $\{\bar{B}_\varepsilon(\gamma^*(y))\} \cup \{\mathscr{B}^m(\mathscr{F}^\dagger)\}_{m=1}^\infty$ has the finite intersection property, and so (by the compactness of $\bar{B}_\varepsilon(\gamma^*(y)) \cup \mathscr{F}^\dagger$),

$$\bar{B}_\varepsilon(\gamma^*(y)) \cap \mathscr{F}_\infty^\dagger \neq \varnothing,$$

a contradiction. Thus, $\mathscr{F}_\infty^\dagger$ is self-generating, and because it is bounded, from proposition 7.3.1 $\mathscr{F}_\infty^\dagger \subset \mathscr{E}(\delta)$. But from (7.3.2), $\mathscr{E}(\delta) \subset \mathscr{F}_\infty^\dagger$, completing the argument.

■

The proposition immediately implies the compactness of $\mathscr{E}(\delta)$ (which can also be directly proved along the lines of proposition 2.5.2).

Corollary 7.3.1 *The set of perfect public equilibrium payoffs, $\mathscr{E}(\delta)$, is compact.*

We now turn to the monotonicity of PPE payoffs with respect to the discount factor. Intuitively, as players become more patient, it should be easier to enforce an action profile because myopic incentives to deviate are now less important. Consequently, we should be able to adjust continuation promises so that incentive constraints are still satisfied, and yet players' total payoffs have not been affected by the change in weighting between flow and continuation values. However, as we discussed near the end of section 2.5.4, there is a discreteness issue: If the set of available continuations is disconnected, it may not be possible to adjust the continuation value by a sufficiently small amount that the incentive constraint is not violated. On the other hand, if continuations can be chosen from a convex set, then the above intuition is valid. In particular, it is often valid for large δ, where the set of available continuations can often be taken to be convex.[15] The available continuations are also convex if players use a public correlating device.

It will be convenient to denote the set of payoffs that can be decomposed by α on $\mathscr{W}$, when the discount factor is δ, by $\mathscr{B}(\mathscr{W}; \delta, \alpha)$.

Proposition 7.3.4 *Suppose $0 < \delta_1 < \delta_2 < 1$, $\mathscr{W} \subset \mathscr{W}'$, and $\mathscr{W} \subset \mathscr{B}(\mathscr{W}'; \delta_1, \alpha)$ for some α. Then $\mathscr{W} \subset \mathscr{B}(\mathrm{co}(\mathscr{W}'); \delta_2, \alpha)$.*

In particular, if $\mathscr{W} \subset \mathscr{W}'$ and $\mathscr{W} \subset \mathscr{B}(\mathscr{W}'; \delta_1)$, then $\mathscr{W} \subset \mathscr{B}(\mathrm{co}(\mathscr{W}'); \delta_2)$.

Proof Fix $v \in \mathscr{W}$, and suppose v is decomposed by (α, γ) on $\mathscr{W}'$, given δ_1. Then define:

$$\bar{\gamma}(y) = \frac{(\delta_2 - \delta_1)}{\delta_2(1 - \delta_1)} v + \frac{\delta_1(1 - \delta_2)}{\delta_2(1 - \delta_1)} \gamma(y),$$

Because $v, \gamma(y) \in \mathscr{W}'$ for all y, we have $\bar{\gamma}(y) \in \mathrm{co}(\mathscr{W}')$ for all y.
Moreover,

$$(1 - \delta_2) u_i(a_i, \alpha_{-i}) + \delta_2 \sum_y \bar{\gamma}_i(y) \rho(y \mid a_i, \alpha_{-i})$$

$$= (1 - \delta_2) u_i(a_i, \alpha_{-i}) + \frac{(\delta_2 - \delta_1)}{(1 - \delta_1)} v + \frac{\delta_1(1 - \delta_2)}{(1 - \delta_1)} \sum_y \gamma_i(y) \rho(y \mid a_i, \alpha_{-i})$$

$$= \frac{(\delta_2 - \delta_1)}{(1 - \delta_1)} v + \frac{(1 - \delta_2)}{(1 - \delta_1)} \left\{ (1 - \delta_1) u_i(a_i, \alpha_{-i}) + \delta_1 \sum_y \gamma_i(y) \rho(y \mid a_i, \alpha_{-i}) \right\}.$$

Because α is enforced by γ and δ_1 with respect to $\mathscr{W}'$, it is also enforced by $\bar{\gamma}$ with respect to $\mathrm{co}(\mathscr{W}')$ and δ_2. Moreover, evaluating the term in $\{\cdot\}$ at α yields v, and so $V(\alpha, \bar{\gamma}; \delta_2) = v$. Hence, $v \in \mathscr{B}(\mathrm{co}(\mathscr{W}'); \delta_2, \alpha)$. ∎

Corollary 7.3.2 *Suppose $0 < \delta_1 < \delta_2 < 1$, and $\mathscr{W} \subset \mathscr{B}(\mathscr{W}; \delta_1)$. If, in addition, $\mathscr{W}$ is bounded and convex, then $\mathscr{W} \subset \mathscr{E}(\delta_2)$. In particular, if $\mathscr{E}(\delta_1)$ is convex, then for any $\delta_2 > \delta_1$, $\mathscr{E}(\delta_1) \subset \mathscr{E}(\delta_2)$.*

Proof $\mathscr{W}$ is self-generating, so the result follows from proposition 7.3.1. ∎

Remark 7.3.3 **Pure strategy restriction** Abreu, Pearce, and Stacchetti (1990) restrict attention to pure strategies but allow for a continuum of signals (see section 7.5). We use a superscript p to denote relevant expressions when we explicitly restrict to pure strategies of the long-lived players (we are already implicitly doing so for continuum action spaces—see remark 7.1.1). In particular, $\mathscr{B}^p(\mathscr{W})$ is the set of payoffs that can be decomposed on $\mathscr{W}$ using profiles in which the long-lived

---

15. More precisely, for any equilibrium payoff in the interior of $\mathscr{E}(\delta)$, under a mild condition, the continuations can be chosen from a convex set (proposition 9.1.2).

players play pure actions, and $\mathscr{E}^p(\delta)$ is the set of PPE payoffs when long-lived players are required to play pure strategies. Clearly, all the results of this section apply under this restriction. In particular, $\mathscr{E}^p(\delta)$ is the largest fixed point of $\mathscr{B}^p$ (see also remark 2.5.1).

With only a slight abuse of language, we call $\mathscr{E}^p(\delta)$ the set of pure-strategy PPE payoffs. We do not require short-lived players to play pure actions, because the static game played by the short-lived players implied by some long-lived player action profiles, $(a'_1, \ldots, a'_n)$, may not have a pure strategy Nash equilibrium.[16] In that event, there is no pure action profile $a \in \mathbf{B}$ with $a_i = a'_i$ for $i = 1, \ldots, n$. Allowing the short-lived players to randomize guarantees that for all long-lived player action profiles, $(a'_1, \ldots, a'_n)$, there exists $(\alpha_{n+1}, \ldots, \alpha_N)$ such that $(a'_1, \ldots, a'_n, \alpha_{n+1}, \ldots, \alpha_N) \in \mathbf{B}$.

◆

Proposition 7.3.4 also implies the following important corollary that will play a central role in the next chapter.

Definition 7.3.5 *A set $\mathscr{W} \subset \mathbb{R}^n$ is* locally self-generating *if for all $v \in \mathscr{W}$, there exists $\delta_v < 1$ and an open set $\mathscr{U}_v$ satisfying*

$$v \in \mathscr{U}_v \cap \mathscr{W} \subset \mathscr{B}(\mathscr{W}; \delta_v).$$

Corollary 7.3.3 *Suppose $\mathscr{W} \subset \mathbb{R}^n$ is compact, convex, and locally self-generating. There exists $\delta' < 1$ such that for $\delta \in (\delta', 1)$,*

$$\mathscr{W} \subset \mathscr{B}(\mathscr{W}; \delta) \subset \mathscr{E}(\delta).$$

Proof Because $\mathscr{W}$ is compact, the open cover $\{\mathscr{U}_v\}_{v \in \mathscr{W}}$ has a finite subcover. Let δ' be the maximum of the δ_v's on this subcover. Proposition 7.3.4 then implies that for any δ larger than δ', we have $\mathscr{W} \subset \mathscr{E}(\delta)$ for $\delta > \delta'$.

■

7.4 The Impact of Increased Precision

In this section, we present a result, due to Kandori (1992b), showing that improving the precision of the public signals cannot reduce the set of equilibrium payoffs.

A natural ranking of the informativeness of the public signals is provided by Blackwell's (1951) partial ordering of experiments. We can view the realized signal y as the result of an experiment about the underlying space of uncertainty, the space A of pure-action profiles. Two different public monitoring distributions (with different signal spaces) can then be viewed as two different experiments. Let R denote the $|A| \times |Y|$-matrix whose ath row corresponds to the probability distribution over Y conditional on the action profile a. We can construct a noisier "experiment" from ρ by assuming that when y is realized under ρ, y is observed with probability $1 - \varepsilon$ and a

---

16. This requires at least two short-lived players, because a single short-lived player always has a pure best reply.

uniform draw from Y is observed with probability ε. Denoting this distribution by ρ' and the corresponding probability matrix R', we have $R' = RQ$, where

$$Q = \begin{bmatrix} (1-\varepsilon) + \varepsilon/|Y| & \varepsilon/|Y| & \cdots & \varepsilon/|Y| \\ \varepsilon/|Y| & (1-\varepsilon) + \varepsilon/|Y| & & \vdots \\ \vdots & & \ddots & \varepsilon/|Y| \\ \varepsilon/|Y| & \cdots & \varepsilon/|Y| & (1-\varepsilon) + \varepsilon/|Y| \end{bmatrix}.$$

Note that Q is a *stochastic matrix*, a nonnegative matrix whose rows sum to 1.

More generally, we define

Definition 7.4.1 *The public monitoring distribution (Y', ρ') is a garbling of (Y, ρ) if there exists a stochastic matrix Q such that*

$$R' = RQ.$$

Note that there is no requirement that the signal spaces Y and Y' bear any particular relationship (in particular, Y' may have more or less elements than Y). The garbling partial order is not strict. For example, if Q is a permutation matrix, then Y' is simply a relabeling of Y and ρ and ρ' are garblings of each other (the inverse of a stochastic matrix is a stochastic matrix if and only if it is a permutation matrix).

It will be convenient to denote the set of payoffs that can be decomposed by α on $\mathscr{W}$, when the discount factor is δ and the public monitoring distribution is ρ, by $\mathscr{B}(\mathscr{W}; \delta, \rho, \alpha)$. Not surprisingly, the set of payoffs that can be decomposed on $\mathscr{W}$, when the discount factor is δ and the public monitoring distribution is ρ, will be written $\mathscr{B}(\mathscr{W}; \delta, \rho)$.

Proposition 7.4.1 *Suppose the public monitoring distribution (Y', ρ') is a garbling of (Y, ρ), and $\mathscr{W} \subset \mathscr{B}(\mathscr{W}'; \delta, \rho', \alpha)$ for some α. Then $\mathscr{W} \subset \mathscr{B}(\mathrm{co}\,(\mathscr{W}'); \delta, \rho, \alpha)$. In particular, if $\mathscr{W} \subset \mathscr{B}(\mathscr{W}'; \delta, \rho')$, then $\mathscr{W} \subset \mathscr{B}(\mathrm{co}(\mathscr{W}'); \delta, \rho)$.*

For large δ, the set of available continuations can often be taken to be convex (see note 15 on page 248). The available continuations are also convex if players use a public correlating device. Hence, the more precise signal y must give at least as large a set of self-generating payoffs. An immediate corollary is that the set of PPE payoffs is at least weakly increasing as the monitoring becomes more precise. Much of Kandori (1992b) is concerned with establishing conditions under which the monotonicity is strict and characterizing the nature of the increase.

Proof For any $\gamma : Y \to \mathbb{R}^n$, write $\gamma_i : Y \to \mathbb{R}$ as the vector in $\mathbb{R}^{|Y|}$ describing player i's continuation value under γ after different signals. Fix $v \in \mathscr{W}$ and suppose v is decomposed by (α, γ') under the public monitoring distribution ρ'. For any action profile α', denote the implied vector of probabilities on A also by α'. Player i's expected continuation value under ρ' and γ' from any action profile α', is then

$$E_{\rho'}[\gamma_i' \mid \alpha'] = \alpha' R' \gamma_i'.$$

Because (Y', ρ') is a garbling of (Y, ρ), there exists a stochastic matrix Q so that $R' = RQ$. Defining $\gamma_i = Q\gamma_i'$, we get

$$E_{\rho'}[\gamma_i' \mid \alpha'] = \alpha' RQ\gamma_i' = \alpha' R\gamma_i = E_\rho[\gamma_i \mid \alpha'].$$

In other words, for all action profiles α', player i's expected continuation value under ρ and γ is the same as that under ρ' and γ'. Hence, α is enforced by γ under ρ.

Given the vectors $\gamma_i \in \mathbb{R}^{|Y|}$ for all i, let $\gamma(y) \in \mathbb{R}^n$ describe the vector of continuation values for all the players for $y \in Y$. Since Q is independent of i,

$$\gamma(y) = \sum_{y' \in Y'} q_{yy'} \gamma'(y'),$$

where $Q = [q_{yy'}]$. Finally, because Q is a stochastic matrix, implying $\gamma(y)$ is a convex combination of the $\gamma'(y')$, $\gamma(y) \in \mathrm{co}(\mathscr{W}')$. ∎

7.5 The Bang-Bang Result

The analysis to this point has assumed the set of signals is finite. The results of the previous sections also hold when there is a continuum of signals, if the signals are continuously distributed, the spaces of stage-game actions are finite, and we restrict attention to pure strategies. The statements only change by requiring $\mathscr{W}$ to be Borel and the mapping γ to be measurable (the proofs are significantly complicated by measurability issues; see Abreu, Pearce, and Stacchetti 1990).

We now present the bang-bang result of Abreu, Pearce, and Stacchetti (1990, theorem 3). We take Y to be a subset of $\mathbb{R}^n$ with positive Lebesgue measure, and work with Lebesgue measurable functions $\gamma : Y \to \mathscr{W}$. A function has the *bang-bang property* if it takes on only extreme points of the set $\mathscr{W}$. The set of *extreme points* of $\mathscr{W} \subset \mathbb{R}^n$ is denoted $\mathrm{ext}\mathscr{W} \equiv \{v \in \mathscr{W} : \nexists v', v'' \in \mathrm{co}\mathscr{W}, \lambda \in (0,1), v = \lambda v' + (1-\lambda)v''\}$. Recall the notation from remark 7.3.3.

Definition 7.5.1 *The measurable function $\gamma : Y \to \mathscr{W}$ has the* bang-bang property *if $\gamma(y) \in \mathrm{ext}\mathscr{W}$ for almost all $y \in Y$. A pure-strategy PPE σ has the* bang-bang property *if after almost every public history h^t, the value of the continuation profile $V(\sigma|_{h^t})$ is in $\mathrm{ext}\mathscr{E}^p(\delta)$.*

The assumption of a continuum of signals is necessary for the result (see remark 7.6.3). Under that assumption, we can use Lyapunov's convexity theorem to guarantee that the range of the extreme points of enforcing continuations of an action profile lies in $\mathrm{ext}\mathscr{W}$. Lyapunov's theorem plays a similar role in the formulation of the bang-bang principle of optimal control theory.

Proposition 7.5.1 *Suppose A is finite. Suppose the signals are distributed absolutely continuously with respect to Lebesgue measure on a subset of $\mathbb{R}^k$, for some k. Suppose $\mathscr{W} \subset \mathbb{R}^n$ is compact and $a \in A$ is enforced by $\hat{\gamma}$ on the convex hull of $\mathscr{W}$. Then, there exists a measurable function $\bar{\gamma} : Y \to \mathrm{ext}\mathscr{W}$ such that a is enforced by $\bar{\gamma}$ on $\mathscr{W}$ and $V(a, \hat{\gamma}) = V(a, \bar{\gamma})$.*

Proof Let $L^\infty(Y, \mathbb{R}^n)$ be the space of bounded Lebesgue measurable functions from Y into $\mathbb{R}^n$ (as usual, we identify functions that agree almost everywhere). Define

$$\hat{\Gamma} = \{\gamma \in L^\infty(Y, \mathbb{R}^n) : a \text{ is enforced by } \gamma \text{ on } \mathrm{co}\mathscr{W}, \text{ and } V(a, \gamma) = V(a, \hat{\gamma})\}.$$

Because $\hat{\gamma} \in \hat{\Gamma}$, $\hat{\Gamma}$ is nonempty. It is also immediate that $\hat{\Gamma}$ is convex. If Y were finite, it would also be immediate that $\hat{\Gamma}$ is compact and so contains its extreme points (because it is a subset of a finite dimensional Euclidean space). We defer the proof that $\hat{\Gamma}$ has an extreme point under the current assumptions till the end of the section (lemma 7.5.1).

Let $\bar{\gamma}$ be an extreme point of $\hat{\Gamma}$. For finite Y, there is no expectation that the range of $\bar{\gamma}$ lies in $\text{ext}\,\mathscr{W}$ (see remark 7.6.3). It is here that the continuum of signals plays a role, because the finiteness of A together with an argument due to Aumann (1965, proposition 6.2) will imply that the range of $\bar{\gamma}$ lies in $\text{ext}\,\mathscr{W}$.

Suppose, en route to a contradiction, that for a positive measure set of signals, $\bar{\gamma}(y) \notin \text{ext}\,\mathscr{W}$. Then there exists $\gamma', \gamma'' \in L^{\infty}(Y, \mathbb{R}^n)$ taking values in $\text{co}\,\mathscr{W}$ such that $\bar{\gamma} = \frac{1}{2}(\gamma' + \gamma'')$ and for a positive measure set of signals, $\gamma'(y) \neq \gamma''(y)$.[17] Let $\gamma^* \equiv \frac{1}{2}(\gamma' - \gamma'')$, so that $\bar{\gamma} + \gamma^* = \gamma'$ and $\bar{\gamma} - \gamma^* = \gamma''$.

We now define a vector-valued measure μ by setting, for any measurable set $Y' \subset Y$,

$$\mu(Y') = \left(\int_{Y'} \gamma_i^*(y) \rho(dy \mid a') \right)_{i=1,\ldots,n; a' \in A} \in \mathbb{R}^{n|A|},$$

where $\rho(\cdot \mid a)$ is the probability measure on Y implied by $a \in A$. Because γ_i^* is bounded for all i (as $\mathscr{W}$ is compact), μ is a vector-valued finite nonatomic measure. By Lyapunov's convexity theorem (Aliprantis and Border 1999, theorem 12.33), $\{\mu(Y') : Y' \text{ a measurable subset of } Y\}$ is convex.[18] Hence, there exists Y' such that $\mu(Y') = \frac{1}{2}\mu(Y)$, with neither Y' nor $Y \setminus Y'$ having zero Lebesgue measure.

Now define $\bar{\gamma}', \bar{\gamma}'' \in L^{\infty}(Y, \mathbb{R}^n)$ by

$$\bar{\gamma}'(y) = \begin{cases} \gamma'(y), & y \in Y', \\ \gamma''(y), & y \notin Y', \end{cases} \quad \text{and} \quad \bar{\gamma}''(y) = \begin{cases} \gamma''(y), & y \in Y', \\ \gamma'(y), & y \notin Y'. \end{cases}$$

Note that both $\bar{\gamma}_i'$ and $\bar{\gamma}_i''$ take values in $\text{co}\,\mathscr{W}$. Because

$$\int_Y \bar{\gamma}_i'(y) \rho(dy \mid a')$$
$$= \int_{Y'} \gamma_i'(y) \rho(dy \mid a') + \int_{Y \setminus Y'} \gamma_i''(y) \rho(dy \mid a')$$
$$= \int_{Y'} [\bar{\gamma}_i(y) + \gamma_i^*(y)] \rho(dy \mid a') + \int_{Y \setminus Y'} [\bar{\gamma}_i(y) - \gamma_i^*(y)] \rho(dy \mid a')$$
$$= \int_Y \bar{\gamma}_i(y) \rho(dy \mid a') + \int_{Y'} \gamma_i^*(y) \rho(dy \mid a') - \int_{Y \setminus Y'} \gamma_i^*(y) \rho(dy \mid a')$$
$$= \int_Y \bar{\gamma}_i(y) \rho(dy \mid a') + \frac{1}{2} \int_Y \gamma_i^*(y) \rho(dy \mid a') - \frac{1}{2} \int_Y \gamma_i^*(y) \rho(dy \mid a')$$
$$= \int_Y \bar{\gamma}_i(y) \rho(dy \mid a'),$$

---

17. This step requires an appeal to a measurable selection theorem to ensure that γ' and γ'' can be chosen measurable, that is, in $L^{\infty}(Y, \mathbb{R}^n)$.
18. For an elementary proof of Lyapunov's theorem based on the intermediate value theorem, see Ross (2005).

7.5 ■ The Bang-Bang Result

$\bar{\gamma}' \in \hat{\Gamma}$. A similar calculation shows that we also have $\bar{\gamma}'' \in \hat{\Gamma}$. Finally, note that $\bar{\gamma} = \frac{1}{2}(\bar{\gamma}' + \bar{\gamma}'')$ and for a positive measure set of signals, $y \in Y$, $\bar{\gamma}'(y) \neq \bar{\gamma}''(y)$. But this contradicts $\bar{\gamma}$ being an extreme point of $\hat{\Gamma}$.

■

Remark 7.5.1 Adding a public correlating device to a finite public monitoring game with a finite signal space yields a public monitoring game whose signals satisfy the hypotheses of proposition 7.5.1. Suppose $Y = \{\underline{y}, \bar{y}\}$ in the finite public monitoring game. Denoting the realization of the correlating device by $\omega \in [0, 1]$, a signal is now

$$\tilde{y} = \begin{cases} \omega, & \text{if } y = \underline{y}, \\ \omega + 1, & \text{if } y = \bar{y}. \end{cases}$$

If the distribution of ω is absolutely continuous on $[0, 1]$, such as for the public correlating device described in definition 2.1.1, then the distribution of $\tilde{y}$ is absolutely continuous on $[0, 2]$.

The analysis in section 7.2.3 provides a convenient illustration of the bang-bang property in a symmetric setting. For $\delta \geq 1/(3p - 2q)$, the two extreme values are 0 and $2 - (1 - p)/(p - q)$, and the optimal PPE determines a set of values for ω (though the set is not unique, its probability is and is given by ϕ), which leads to a bang-bang equilibrium with value $2 - (1 - p)/(p - q)$.

More generally, any particular equilibrium payoff can be decomposed using a correlating distribution with finite support (because, from Carathéodory's theorem [Rockafellar 1970, theorem 17.1], any point in co $\mathcal{W}$ can be written as a finite convex combination of points in ext $\mathcal{W}$). To ensure that *any* equilibrium payoff can be decomposed, however, the correlating device must have a continuum of values, so that any finite distribution can be constructed.

Even in the absence of a public correlating device, a continuously distributed public signal (via Lyapunov's theorem) effectively allows us to construct any finite support public randomizing device "for free." For example, if the support of y is the unit interval, by dividing the interval into small subintervals, and identifying alternating subintervals with Heads and Tails, we obtain a coin flip.

◆

Corollary 7.5.1 *Under the hypotheses of proposition 7.5.1, if $\mathcal{W} \subset \mathbb{R}^n$ is compact, $\mathcal{B}^p(\mathcal{W}) = \mathcal{B}^p(\text{co}\mathcal{W})$.*

Proof By monotonicity of $\mathcal{B}^p$, we immediately have $\mathcal{B}^p(\mathcal{W}) \subset \mathcal{B}^p(\text{co}\mathcal{W})$. Proposition 7.5.1 implies $\mathcal{B}^p(\text{co}\mathcal{W}) \subset \mathcal{B}^p(\mathcal{W})$.

■

Corollary 7.5.2 *Under the hypotheses of proposition 7.5.1, if $0 < \delta_1 < \delta_2 < 1$, $\mathcal{E}^p(\delta_1) \subset \mathcal{E}^p(\delta_2)$.*

Proof Let $\mathcal{W} = \mathcal{E}^p(\delta_1)$. From proposition 7.3.4, $\mathcal{W} \subset \mathcal{B}^p(\text{co}\mathcal{W}; \delta_2)$. Because $\mathcal{W}$ is compact, corollary 7.5.1 implies $\mathcal{W} \subset \mathcal{B}^p(\mathcal{W}; \delta_2)$. Because $\mathcal{W}$ is self-generating (with respect to δ_2), proposition 7.3.1 implies $\mathcal{W} \subset \mathcal{E}^p(\delta_2)$.

■

Remark 7.5.2 **Symmetric games** Recall that an equilibrium of a symmetric game is *strongly symmetric* if all players choose the same action after every public history. Proposition 7.5.1 applies to strongly symmetric PPE.[19] Consequently, strongly symmetric equilibria have a particulary simple structure in this case, because there are only *two* extreme points in the convex hull of the set of strongly symmetric PPE payoffs (we discuss an example in detail in section 11.1.1). However, proposition 7.5.1 does not immediately imply a similar simple structure for general PPE. As we will see in chapter 9, we are often interested in self-generating sets that are balls, whose set of extreme points is a circle with two players and the surface of a sphere with three players. Restricting continuations to this set is not a major simplification. ◆

We now complete the proof of proposition 7.5.1.

Lemma 7.5.1 *Under the hypotheses of proposition 7.5.1,*

$$\hat{\Gamma} = \{\gamma \in L^\infty(Y, \mathbb{R}^n) : a \text{ is enforced by } \gamma \text{ on } \operatorname{co}\mathscr{W}, \text{ and } \mathsf{V}(a, \gamma) = \mathsf{V}(a, \hat{\gamma})\},$$

has an extreme point.

Proof We denote by $L^1(Y, \mathbb{R}^m)$ the collection of functions $f = (f_1, \ldots, f_m)$, with each f_i Lebesgue integrable. Writing e_i for the ith standard basis vector (i.e., the vector whose ith coordinate equals 1 and all other coordinates are 0), any function $f \in L^1(Y, \mathbb{R}^m)$ can be written as $\sum_i f_i e_i$, with $f_i \in L^1(Y, \mathbb{R})$. For any continuous linear functional F on $L^1(Y, \mathbb{R}^m)$, let F_i be the implied linear functional on $L^1(Y, \mathbb{R})$ defined by, for any function $h \in L^1(Y, \mathbb{R})$, $F_i(h) \equiv F(he_i)$. From the Riesz representation theorem (Royden 1988, theorem 6.13), there exists $g_i \in L^\infty(Y, \mathbb{R})$, such that for all $h \in L^1(Y, \mathbb{R})$, $F_i(h) = \int h g_i$. Then,

$$F(f) = \sum_i F(f_i e_i) = \sum_i F_i(f_i) = \sum_i \int f_i g_i = \int \langle f, g \rangle,$$

where $g = (g_1, \ldots, g_m) \in L^\infty(Y, \mathbb{R}^m)$, and $\langle \cdot, \cdot \rangle$ is the standard inner product on $\mathbb{R}^m$.

Hence, $L^\infty(Y, \mathbb{R}^m)$ is the dual of $L^1(Y, \mathbb{R}^m)$, and so the set of functions $\|g\|_\infty \leq 1$ (the "unit ball") is weak-* compact (by Alaoglu's theorem; Aliprantis and Border 1999, theorem 6.25). But this immediately implies the weak-* compactness of

$$\hat{\Gamma}^\dagger \equiv \{\gamma \in L^\infty(Y, \mathbb{R}^n) : \gamma(y) \in \operatorname{co}\mathscr{W} \ \forall y \in Y\},$$

because for some $m \leq n$, it is the image of the unit ball in $L^\infty(Y, \mathbb{R}^m)$ under a continuous function. (Because $\operatorname{co}\mathscr{W}$ is compact and convex, it is homeomorphic to $\{w \in \mathbb{R}^m : |w| \leq 1\}$ for some m; let $\varphi : \{w \in \mathbb{R}^m : |w| \leq 1\} \to \operatorname{co}\mathscr{W}$ denote the homeomorphism. Then, $\hat{\Gamma}^\dagger$ is the image of the unit ball in $L^\infty(Y, \mathbb{R}^m)$ under the continuous map J, where $J(g) = \varphi \circ g$.)

---

19. Abreu, Pearce, and Stacchetti (1986) develops the argument for strongly symmetric equilibria.

We now turn to $\hat{\Gamma}$, and argue that it is a weak-* closed subset of $\hat{\Gamma}^\dagger$. Suppose $\gamma \in L^\infty(Y, \mathbb{R}^n)$ is the weak-* limit of a net $\{\gamma^\beta\}$. Hence,

$$\int \langle \gamma^\beta(y), f(y) \rangle dy \to \int \langle \gamma(y), f(y) \rangle dy \quad \forall f \in L^1(Y, \mathbb{R}^n),$$

and, in particular, for all i,

$$\int \gamma_i^\beta(y) f(y) dy \to \int \gamma_i(y) f(y) dy \quad \forall f \in L^1(Y, \mathbb{R}).$$

Because the signals are distributed absolutely continuously with respect to Lebesgue measure, for each action profile, $a' \in A$, there is a Radon-Nikodym derivative $d\rho(\cdot \mid a')/dy \in L^1(Y, \mathbb{R})$ with

$$\int \gamma_i'(y) \rho(dy \mid a') = \int \gamma_i'(y) \frac{d\rho(y \mid a')}{dy} dy, \quad \forall i, \quad \forall \gamma' \in L^\infty(Y, \mathbb{R}^n).$$

This implies that $\hat{\Gamma}$ is a weak-* closed (and so compact) subset of $\hat{\Gamma}^\dagger$, because the additional constraints on γ that define $\hat{\Gamma}$ only involve expressions of the form $\int \gamma_i(y) \rho(dy \mid a')$. Finally, the Krein-Milman theorem (Aliprantis and Border 1999, theorem 5.117) implies that $\hat{\Gamma}$ has an extreme point. ∎

7.6 An Example with Short-Lived Players

We revisit the product-choice example of example 2.7.1. The stage game payoffs are reproduced in figure 7.6.1. We begin with the game on the left, where player 1 is a long-lived and player 2 a short-lived player. We also begin with perfect monitoring. For a mixed profile α in the stage game, let $\alpha^H = \alpha_1(H)$ and $\alpha^h = \alpha_2(h)$. We denote a mixed action for player 1 by α^H and for player 2 by α^h. Then the relevant set of profiles incorporating the myopic behavior of player 2 is

$$\mathbf{B} = \{(\alpha^H, \ell) : \alpha^H \leq \tfrac{1}{2}\} \cup \{(\alpha^H, h) : \alpha^H \geq \tfrac{1}{2}\} \cup \{(\tfrac{1}{2}, \alpha^h) : \alpha^h \in [0,1]\}.$$

	h	ℓ
H	2,3	0,2
L	3,0	1,1

	h	ℓ
H	2,3	1,2
L	3,0	0,1

Figure 7.6.1 The games from figure 2.7.1. The left game is the product-choice game of figure 1.5.1. Player 1's action L is a best reply to 2's choice of ℓ in the left game, but not in the right.

Moreover, from section 2.7.2, player 1's minmax payoff, $\underline{v}_1$, equals 1, and the upper bound on his equilibrium payoffs, $\bar{v}_1$, equals 2. In other words, the set of possible PPE player 1 payoffs is the interval $[1, 2]$.

Because there is only one long-lived player, enforceability and decomposition occur on subsets of $\mathbb{R}$, and we drop the subscript on player 1 payoffs.

7.6.1 Perfect Monitoring

With perfect monitoring, the set of signals is simply $A_1 \times A_2$. The simplest candidate equilibrium in which Hh is played is the grim trigger profile in which play begins with Hh, and remains there until the first deviation, after which play is perpetual $L\ell$. This profile is described by the doubleton set of payoffs $\{1, 2\}$. It is readily verified that this set is self-generating for $\delta \geq 1/2$, and so the trigger strategy profile is a PPE, and so a subgame-perfect equilibrium, for $\delta \geq 1/2$.

We now ask when the set $[1, 2]$ is self-generating, and begin with the pure profiles in **B**, namely, Hh and $L\ell$. Because $L\ell$ is a stage-game Nash equilibrium, $L\ell$ is trivially enforceable using a constant continuation. Player 1's payoffs must lie in the interval $[1, 2]$, so the set of payoffs decomposed by $L\ell$ on $[1, 2]$, $\mathscr{W}^{L\ell}$, is given by

$$v \in \mathscr{W}^{L\ell} \iff \exists \gamma \in [1, 2] \text{ such that } v = \mathsf{V}(L\ell, \gamma) = (1 - \delta) + \delta \gamma.$$

Hence, $\mathscr{W}^{L\ell} = [1, 1 + \delta]$.

The enforceability of Hh on $[1, 2]$ is straightforward. The pair of continuations $\gamma = (\gamma(Lh), \gamma(Hh))$ enforces Hh if

$$(1 - \delta)2 + \delta \gamma(Hh) \geq (1 - \delta)3 + \delta \gamma(Lh),$$

that is,

$$\gamma(Hh) \geq \gamma(Lh) + \frac{(1 - \delta)}{\delta}. \tag{7.6.1}$$

Hence, the set of payoffs decomposed by Hh on $[1, 2]$, $\mathscr{W}^{Hh}$, is given by

$$v \in \mathscr{W}^{Hh} \iff \exists \gamma(Lh), \gamma(Hh) \in [1, 2] \text{ satisfying (7.6.1)}$$
$$\text{such that } v = \mathsf{V}(Hh, \gamma) = (1 - \delta)2 + \delta \gamma(Hh).$$

Hence, $\mathscr{W}^{Hh} = [3 - 2\delta, 2]$ if $\delta \geq 1/2$ (it is empty if $\delta < 1/2$).

The set of possible PPE payoffs $[1, 2]$ is self-generating if $\mathscr{W}^{L\ell} \cup \mathscr{W}^{Hh} \supset [1, 2]$. Thus, $[1, 2]$ is self-generating, and so every payoff in $[1, 2]$ is a subgame-perfect equilibrium payoff, if $\delta \geq 2/3$. This requirement on δ is tighter than the requirement ($\delta \geq 1/2$) for grim trigger to be an equilibrium. Just as in section 2.5.4, we could describe the equilibrium-outcome path for any $v \in [1, 2]$. Of course, here player 2's action is always a myopic best reply to that of player 1.

Although $\mathscr{W}^{L\ell} \cup \mathscr{W}^{Hh} \supset [1, 2]$ is sufficient for the self-generation of $[1, 2]$, it is not necessary, as we now illustrate. In fact, the set $[1, 2]$ *is* self-generating for $\delta \in [1/2, 2/3)$ as well as for higher δ once we use mixed strategies. Consider the

7.6 ■ An Example with Short-Lived Players

mixed profile $\alpha = (1/2, \alpha^h)$ for fixed $\alpha^h \in [0, 1]$. Because both actions for player 1 are played with strictly positive probability, the continuations γ must satisfy

$$v = \alpha^h[(1 - \delta)2 + \delta\gamma(Hh)] + (1 - \alpha^h)[(1 - \delta) \times 0 + \delta\gamma(H\ell)]$$
$$= \alpha^h[(1 - \delta)3 + \delta\gamma(Lh)] + (1 - \alpha^h)[(1 - \delta) + \delta\gamma(L\ell)],$$

where the first expression is the expected payoff from H and the second is that from L. Rearranging, we have the requirement

$$\delta\alpha^h[\gamma(Hh) - \gamma(H\ell) - (\gamma(Lh) - \gamma(L\ell))] = (1 - \delta) + \delta(\gamma(L\ell) - \gamma(H\ell)).$$

We now try enforcing α with continuations that satisfy $\gamma(Hh) = \gamma(H\ell) \equiv \gamma^H$ and $\gamma(Lh) = \gamma(L\ell) \equiv \gamma^L$. Imposing this constraint yields

$$\gamma^H = \gamma^L + \frac{(1 - \delta)}{\delta}. \tag{7.6.2}$$

Because we can choose $\gamma^H, \gamma^L \in [1, 2]$ to satisfy this constraint as long as $\delta \geq 1/2$, we conclude that α is enforceable on $[1, 2]$. Moreover, the set of payoffs decomposed by $\alpha = (1/2, \alpha^h)$ on $[1, 2]$, $\mathscr{W}^\alpha$, is given by

$$v \in \mathscr{W}^\alpha \iff \exists \gamma^L, \gamma^H \in [1, 2] \text{ satisfying (7.6.2)}$$
$$\text{such that } v = V(\alpha, \gamma) = 2\alpha^h(1 - \delta) + \delta\gamma^H.$$

Hence,

$$\mathscr{W}^\alpha = [2\alpha^h(1 - \delta) + 1, \ 2\alpha^h(1 - \delta) + 2\delta].$$

Finally,

$$\cup_{\{\alpha:\alpha^h \in [0,1]\}} \mathscr{W}^\alpha = [1, 2].$$

Hence, the set of payoffs $[1, 2]$ is self-generating, and therefore is the maximal set of subgame-perfect equilibrium payoffs, for all $\delta \geq 1/2$. The lower bound on δ agrees with that for the trigger strategy profile to be an equilibrium, and is lower than for self-generation of $[1, 2]$ under pure strategies. Moreover, we have shown that for any $v \in [1, 2]$, a subgame-perfect equilibrium with that expected payoff has the long-run player randomizing in every period, putting equal probability on H and L, *independent of history*. The randomizations of the short-lived player, on the other hand, do depend on history.

We now ask whether there is a simple strategy of this form. In particular, is the doubleton set $\{\gamma^L, \gamma^H\} \equiv \{\gamma^H - (1 - \delta)/\delta, \ \gamma^H\}$ self-generating, where the value for γ^L was chosen to satisfy (7.6.2)? The payoff γ^H is decomposed by $\alpha = (1/2, \alpha^h(H))$ on $\{\gamma^L, \gamma^H\}$, with $\alpha^h(H) = \gamma^H/2$. The payoff γ^L is decomposed by $\alpha = (1/2, \alpha^h(L))$ on $\{\gamma^L, \gamma^H\}$ if

$$\gamma^L = \alpha^h(L)2(1 - \delta) + \delta\gamma^H.$$

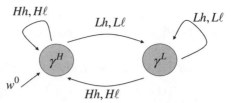

Figure 7.6.2 The automaton representing the profile with payoffs $\{\gamma^L, \gamma^H\}$. The associated behavior is $f(\gamma^H) = (1/2, \alpha^h(H))$ and $f(\gamma^L) = (1/2, \alpha^h(L))$. Building on the intuition of section 7.3, states are labeled with their values. State transitions only depend on player 1's realized action and player 1 plays identically in the two states, whereas player 2 plays differently.

Solving for the implied $\alpha^h(L)$ gives

$$\alpha^h(L) = \frac{\gamma^L - \delta \gamma^H}{2(1-\delta)} = \frac{\gamma^H(1-\delta)\delta - (1-\delta)}{2\delta(1-\delta)}$$
$$= \frac{\delta \gamma^H - 1}{2\delta} = \alpha^h(H) - \frac{1}{2\delta}.$$

The quantities $\alpha^h(H)$ and $\alpha^h(L)$ are well-defined probabilities for $\gamma^H \in [1/\delta, 2]$ and $\delta \geq 1/2$. Note that the implied range for γ^L is $[1, 2 - (1-\delta)/\delta]$. Hence, we have a one-dimensional manifold of mixed equilibria. The associated profile is displayed in figure 7.6.2. Player 1 is indifferent between L, the myopically dominant action, and H, because a play of H is rewarded in the next period by h with probability α^H, rather than with the lower probability α^L.

Remark 7.6.1 Because player 1's behavior in these mixed strategy equilibria is independent of history, based on such profiles, it is possible to construct nontrivial equilibria when signals are private, precluding the use of histories to coordinate continuation play. We construct such an equilibrium in section 12.5. ♦

We turn now to the stage game on the right in figure 7.6.1, maintaining perfect monitoring and our convention on α^H and α^h. The relevant set of profiles is still given by the set **B** calculated earlier (since player 2's payoffs are identical in the two games). Moreover, the bounds on player 1's equilibrium payoffs are unchanged: Player 1's payoffs must lie in the interval $[1, 2]$. The crucial change is that player 1's choice L (facilitating 2's minmaxing of 1) is no longer a myopic best reply to that action of ℓ.

As before, we begin with the pure profiles in **B**, namely, Hh and $L\ell$. Because $L\ell$ is not a stage-game Nash equilibrium, the enforceability of $L\ell$ must be dealt with similarly to that of Hh. The enforceability of both $L\ell$ and Hh on $[1, 2]$ is, as before, immediate; it is enough that $\gamma(L\ell) - \gamma(H\ell), \hat{\gamma}(Hh) - \hat{\gamma}(Lh) \geq (1-\delta)/\delta$, where γ and $\hat{\gamma}$ are the continuation functions for $L\ell$ and Hh, respectively. Hence, the set of payoffs decomposed by $L\ell$ is the interval $[1, 2\delta]$, whereas the set of payoffs decomposed by Hh is the interval $[3 - 2\delta, 2]$. Thus, $[1, 2]$ is self-generating for $\delta \geq 3/4$.

Because $L\ell$ is not a Nash equilibrium of the stage game, there are no pure strategy equilibria in trigger strategies. However, as we saw in example 2.7.1, there are simple

7.6 ■ An Example with Short-Lived Players

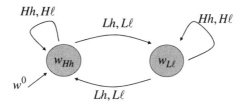

Figure 7.6.3 The automaton representing the profile with payoffs $\{2\delta, 2\}$. The associated behavior is $f(w_a) = a$.

profiles in which Hh is played. For example, the doubleton set of payoffs $\{2\delta, 2\}$ is easily seen to be self-generating if $\delta \geq 1/2$, because in that case $2 - 2\delta \geq (1 - \delta)/\delta$. This self-generating set corresponds to the equilibrium described in example 2.7.1, and illustrated in figure 7.6.3.

We now turn to enforceability of mixed-action profiles, $\alpha = (1/2, \alpha^h)$ for fixed $\alpha^h \in [0, 1]$. Similarly to before, the continuations γ must satisfy

$$v = \alpha^h[(1-\delta)2 + \delta\gamma(Hh)] + (1-\alpha^h)[(1-\delta) + \delta\gamma(H\ell)]$$
$$= \alpha^h[(1-\delta)3 + \delta\gamma(Lh)] + (1-\alpha^h)[(1-\delta) \times 0 + \delta\gamma(L\ell)].$$

Rearranging, we have

$$\alpha^h[(1-\delta)2 + \delta(\gamma(L\ell) - \gamma(L\ell) - (\gamma(Hh) - \gamma(H\ell)))]$$
$$= (1-\delta) + \delta(\gamma(H\ell) - \gamma(L\ell)). \quad (7.6.3)$$

We now try enforcing α with continuations that satisfy $\gamma(Hh) = \gamma(H\ell) \equiv \gamma^H$ and $\gamma(Lh) = \gamma(L\ell) \equiv \gamma^L$. Imposing this constraint, and rearranging, yields

$$\gamma^H = \gamma^L + (2\alpha^h - 1)\frac{(1-\delta)}{\delta}. \quad (7.6.4)$$

Because for $\delta \geq 1/2$ and any $\alpha^h \in [0, 1]$ we can choose $\gamma^H, \gamma^L \in [1, 2]$ to satisfy this constraint, α is enforceable on $[1, 2]$. Note that, unlike for the product choice game, here the continuations depend on the current randomizing behavior of player 2. This is a result of player 1's myopic incentive to play L depending on the current play of player 2. An alternative would be to choose continuations so that the coefficient of α^h equaled 0 in (7.6.3), in which case continuations depend on the realized action of player 2.

The set of payoffs decomposed by $\alpha = (1/2, \alpha^h)$ on $[1, 2]$, $\mathscr{W}^\alpha$, is given by

$$v \in \mathscr{W}^\alpha \iff \exists \gamma^L, \gamma^H \in [1, 2] \text{ satisfying (7.6.4)}$$
$$\text{such that } v = 3\alpha^h(1-\delta) + \delta\gamma^L.$$

Hence, for $\alpha^h \geq 1/2$,

$$\mathscr{W}^\alpha = [3\alpha^h(1-\delta) + \delta, \ \alpha^h(1-\delta) + 1 + \delta],$$

and for $\alpha^h \leq 1/2$,

$$\mathscr{W}^\alpha = [\alpha^h(1-\delta) + 1, \ 3\alpha^h(1-\delta) + 2\delta].$$

Finally,
$$\cup_{\{\alpha:\alpha^h \in [0,1]\}} \mathscr{W}^\alpha = [1, 2].$$
Hence, the set of payoffs $[1, 2]$ is self-generating for $\delta \geq 1/2$. Consequently, every payoff in $[1, 2]$ is a subgame-perfect equilibrium payoff for $\delta \geq 1/2$.

7.6.2 Imperfect Public Monitoring of the Long-Lived Player

We now discuss the impact of public monitoring on the product-choice game (the left game in figure 7.6.1). The actions of the long-lived player are now not public, but there is a public signal $y_1 \in \{\underline{y}, \bar{y}\}$ of his action. We interpret $\bar{y}$ as high quality and $\underline{y}$ as low quality. The actions of the short-lived player remain public, and so the space of public signals is $Y \equiv \{\underline{y}, \bar{y}\} \times A_2$.[20] The distribution of y_1 is given by

$$\rho_1(\bar{y} \mid a) = \begin{cases} p, & \text{if } a_1 = H, \\ q, & \text{if } a_1 = L, \end{cases} \quad (7.6.5)$$

with $0 < q < p < 1$. The joint distribution ρ over Y is given by $\rho(y_1 y_2 \mid a) = \rho_1(y_1 \mid a)$ if $y_2 = a_2$, and 0 otherwise.

The relevant set of profiles continues to be given by the set **B** calculated earlier. Moreover, the bounds on player 1's equilibrium payoffs are still valid: Player 1's payoffs must lie in the interval $[1, 2]$. Denote player 1's maximum PPE payoff by $v^* \geq 1$ (recall that because there is only one long-lived player, we drop the subscript on player 1's payoff).

As before, we begin with the pure profiles in **B**, namely, Hh and $L\ell$. Because $L\ell$ is a stage-game Nash equilibrium, $L\ell$ is trivially enforceable using a constant continuation. Moreover, we consider first only continuations implemented by pure strategy PPE. Let v^{*p} denote the maximum pure strategy PPE player 1 payoff. Player 1's payoffs must lie in the interval $[1, v^{*p}]$, so the set of payoffs decomposed by $L\ell$ on $[1, v^{*p}]$, $\mathscr{W}^{L\ell}$, is given by

$$v \in \mathscr{W}^{L\ell} \iff \exists \gamma \in [1, v^{*p}] \text{ such that } v = (1 - \delta) + \delta\gamma.$$

Hence,
$$\mathscr{W}^{L\ell} = [1, 1 + \delta(v^{*p} - 1)].$$

It is worth noting that if $v^{*p} > 1$, then $v^{*p} \notin \mathscr{W}^{L\ell}$.

The continuations $\gamma : \{\underline{y}, \bar{y}\} \times A_2 \to [1, v^{*p}]$ enforce Hh if
$$2(1 - \delta) + \delta\{p\gamma(\bar{y}h) + (1 - p)\gamma(\underline{y}h)\} \geq 3(1 - \delta) + \delta\{q\gamma(\bar{y}h) + (1 - q)\gamma(\underline{y}h)\},$$

that is,
$$\gamma(\bar{y}h) \geq \gamma(\underline{y}h) + \frac{(1 - \delta)}{\delta(p - q)}. \quad (7.6.6)$$

Hence, the payoff following the good signal $\bar{y}$ must exceed that of the bad signal $\underline{y}$ by the difference $(1 - \delta)/[\delta(p - q)]$. This payoff difference shrinks as the discount

---

20. It is worth noting that the analysis is unchanged when the short-lived players' actions are not public. This corresponds to taking $\{\underline{y}, \bar{y}\}$ as the space of public signals.

7.6 ■ An Example with Short-Lived Players

factor increases, and hence the temptation of a current deviation diminishes, and as $p - q$ increases, and hence signals become more responsive to actions.

Let $\mathscr{W}^{Hh}$ denote the set of payoffs that can be decomposed by Hh on $[1, v^{*p}]$. Then,

$$v \in \mathscr{W}^{Hh} \iff \exists \gamma(\underline{y}h), \gamma(\bar{y}h) \in [1, v^{*p}] \text{ satisfying (7.6.6) such that}$$
$$v = \mathsf{V}(Hh, \gamma) = (1-\delta)2 + \delta\{p\gamma(\bar{y}h) + (1-p)\gamma(\underline{y}h)\}.$$

The maximum value of $\mathscr{W}^{Hh}$ is obtained by setting $\gamma(\bar{y}h) = v^{*p}$ and having (7.6.6) hold with equality. We thus have (if $v^{*p} > 1$)

$$v^{*p} = \max \mathscr{W}^{Hh} = (1-\delta)2 - \frac{(1-\delta)(1-p)}{(p-q)} + \delta v^{*p},$$

and solving for v^{*p} gives

$$v^{*p} = 2 - \frac{(1-p)}{(p-q)} < 2 = \bar{v}.$$

We need to verify that $v^{*p} > 1$, which is equivalent to $(1-p) < (p-q)$, that is, $2p - q > 1$. If $2p - q \leq 1$, then the only pure strategy PPE is $L\ell$ in every period.

Remark 7.6.2 **Inefficiency due to binding moral hazard** Player 1 is subject to *binding moral hazard* (definition 8.3.1). As a result, all PPE are inefficient (proposition 8.3.1). We have just shown that the maximum pure strategy PPE player 1 payoff is strictly less than 2. We next demonstrate that for large δ, *all* PPE in this example are bounded away from efficiency (we return to the general case in section 8.3.2). This is in contrast to both the perfect monitoring version of this example (where we have already seen that for a sufficiently patient long-lived player, there are efficient PPE), and the general case of only long-lived players (where, under appropriate regularity conditions and for patient players, there are approximately efficient PPE, proposition 9.2.1).

Intuitively, the imperfection in the monitoring implies that on the equilibrium path player 1 must face low continuations with positive probability. Because player 2 is short-lived, it is impossible to use intertemporal transfers of payoffs to maintain efficiency while still providing incentives. ◆

The minimum value of $\mathscr{W}^{Hh}$ is obtained by setting $\gamma(\underline{y}h) = 1$ and having (7.6.6) hold with equality. We thus have

$$\min \mathscr{W}^{Hh} = 1 + \frac{(1-\delta)(2p-q)}{(p-q)}.$$

The set of payoffs $[1, v^{*p}]$ is self-generating using pure strategies if $\mathscr{W}^{L\ell} \cup \mathscr{W}^{Hh} \supset [1, v^{*p}]$, which is implied by

$$\min \mathscr{W}^{Hh} \leq \max \mathscr{W}^{L\ell}.$$

Substituting and solving for the bound on δ gives

$$\frac{(2p-q)}{(4p-2q-1)} \leq \delta. \tag{7.6.7}$$

The bound $2p - q > 1$ implies the left side is less than 1.

When $\mathscr{W}^{L\ell} \cup \mathscr{W}^{Hh} \supset [1, v^{*p}]$, v^{*p} can be achieved in a pure strategy equilibrium, because the continuation promises $\gamma(yh)$ and $\gamma(\bar{y}h)$ decomposing v^{*p} using Hh are elements of $\mathscr{W}^{L\ell} \cup \mathscr{W}^{Hh}$, and the continuation promises supporting $\gamma(yh)$ and $\gamma(\bar{y}h)$ are themselves in $\mathscr{W}^{L\ell} \cup \mathscr{W}^{Hh}$, and so on.

A strategy profile that achieves the payoff v^{*p} can be constructed as follows. First, set $\gamma(v) = (v - (1-\delta))/\delta$ (the constant continuation decomposing v using $L\ell$), and let $\gamma^{\bar{y}}, \gamma^{\underline{y}} : [0, v^{*p}] \to \mathbb{R}$ be the functions solving $v = \mathbf{V}(Hh, (\gamma^{\bar{y}}(v), \gamma^{\underline{y}}(v)))$ when (7.6.6) holds with equality:

$$\gamma^{\bar{y}} = \frac{v}{\delta} - \frac{2(1-\delta)}{\delta} + \frac{(1-p)(1-\delta)}{\delta(p-q)},$$

and
$$\gamma^{\underline{y}} = \frac{v}{\delta} - \frac{2(1-\delta)}{\delta} - \frac{p(1-\delta)}{\delta(p-q)}.$$

Now, define $\zeta : \mathscr{H} \to [1, v^{*p}]$ as follows: $\zeta(\varnothing) = v^{*p}$ and for $h^t \in \mathscr{H}^t$,

$$\zeta(h^t, y_1 a_2) = \begin{cases} \gamma^{\bar{y}}(\zeta(h^t)), & \text{if } y_1 = \bar{y} \text{ and } \zeta(h^t) \in \mathscr{W}^{Hh}, \\ \gamma^{\underline{y}}(\zeta(h^t)), & \text{if } y_1 = \underline{y} \text{ and } \zeta(h^t) \in \mathscr{W}^{Hh}, \\ \gamma(\zeta(h^t)), & \text{if } \zeta(h^t) \in \mathscr{W}^{L\ell} \setminus \mathscr{W}^{Hh}. \end{cases}$$

The strategies are then given by

$$\sigma_1(\varnothing) = H, \quad \sigma_2(\varnothing) = h,$$

$$\sigma_1(h^t) = \begin{cases} H, & \text{if } \zeta(h^t) \in \mathscr{W}^{Hh}, \\ L, & \text{if } \zeta(h^t) \in \mathscr{W}^{L\ell} \setminus \mathscr{W}^{Hh}, \end{cases}$$

and

$$\sigma_2(h^t) = \begin{cases} h, & \text{if } \zeta(h^t) \in \mathscr{W}^{Hh}, \\ \ell, & \text{if } \zeta(h^t) \in \mathscr{W}^{L\ell} \setminus \mathscr{W}^{Hh}. \end{cases}$$

We thus associate with every history of signals h^t a continuation payoff $\zeta(h^t)$. This continuation payoff allows us to associate an action with the history, either Hh or $L\ell$, depending on whether the continuation payoff lies in the set $\mathscr{W}^{Hh}$ or $\mathscr{W}^{L\ell} \setminus \mathscr{W}^{Hh}$. We then associate with the signals continuation payoffs that decompose the current continuation payoff into the payoff of the currently prescribed action and a new continuation payoff. If the currently prescribed action is $L\ell$, this decomposition is trivial, as the current payoff is 1 and there are no incentives to be created. We can then take the continuation payoff to be simply $(\zeta(h^t) - (1-\delta))/\delta$. When the currently prescribed action is Hh, we assign continuation payoffs according to the functions $\gamma^{\bar{y}}$ and $\gamma^{\underline{y}}$ that allow us to enforce payoffs in the set $\mathscr{W}^{Hh}$.

There is still the possibility that mixing by the players will allow player 1 to achieve a higher PPE payoff or achieve additional PPE payoffs for lower δ. In what follows,

7.6 ■ An Example with Short-Lived Players

recall that v^* is the maximum player 1 PPE payoff, allowing for mixed strategies, and so $v^* \geq v^{*p}$. Consider the mixed profile $\alpha = (1/2, \alpha^h)$ for fixed $\alpha^h \in [0, 1]$. Because both actions for player 1 are played with strictly positive probability, the continuations γ must satisfy

$$\begin{aligned} v &= \alpha^h[(1-\delta)2 + \delta\{p\gamma(\bar{y}h) + (1-p)\gamma(\underline{y}h)\}] \\ &\quad + (1-\alpha^h)[(1-\delta) \times 0 + \delta\{p\gamma(\bar{y}\ell) + (1-p)\gamma(\underline{y}\ell)\}] \\ &= \alpha^h[(1-\delta)3 + \delta\{q\gamma(\bar{y}h) + (1-q)\gamma(\underline{y}h)\}] \\ &\quad + (1-\alpha^h)[(1-\delta) + \delta\{q\gamma(\bar{y}\ell) + (1-q)\gamma(\underline{y}\ell)\}], \end{aligned}$$

where the first expression is the expected payoff from H and the second is that from L. Rearranging, the continuations must satisfy the requirement

$$\delta(p-q)[\alpha^h\gamma(\bar{y}h) + (1-\alpha^h)\gamma(\bar{y}\ell) - \{\alpha^h\gamma(\underline{y}h) + (1-\alpha^h)\gamma(\underline{y}\ell)\}] = (1-\delta). \tag{7.6.8}$$

Letting

$$\gamma^{\bar{y}}(\alpha^h) = \alpha^h\gamma(\bar{y}h) + (1-\alpha^h)\gamma(\bar{y}\ell)$$

and

$$\gamma^{\underline{y}}(\alpha^h) = \alpha^h\gamma(\underline{y}h) + (1-\alpha^h)\gamma(\underline{y}\ell),$$

requirement (7.6.8) can be rewritten as

$$\gamma^{\bar{y}}(\alpha^h) = \gamma^{\underline{y}}(\alpha^h) + \frac{(1-\delta)}{\delta(p-q)}. \tag{7.6.9}$$

If we can choose $\gamma(y_1 a_2) \in [1, v^*]$ to satisfy this constraint, then α is enforceable on $[1, v^*]$. A sufficient condition to do so is

$$\frac{(1-\delta)}{\delta(p-q)} \leq v^{*p} - 1$$

because $v^{*p} \leq v^*$. The above inequality is implied by

$$\frac{1}{(2p-q)} \leq \delta. \tag{7.6.10}$$

If α is enforceable on $[1, v^*]$, the set of payoffs decomposed by $\alpha = (1/2, \alpha^h)$ on $[1, v^*]$, $\mathscr{W}^\alpha$, is given by

$$\begin{aligned} v \in \mathscr{W}^\alpha &\iff \exists \gamma(y_1 a_2) \in [1, v^*] \text{ satisfying (7.6.9)} \\ &\quad \text{such that } v = V(\alpha, \gamma) \\ &\quad = 2\alpha^h(1-\delta) + \delta\{p\gamma^{\bar{y}}(\alpha^h) + (1-p)\gamma^{\underline{y}}(\alpha^h)\}. \end{aligned}$$

Hence,

$$\mathscr{W}^\alpha = \left[2\alpha^h(1-\delta) + \delta + \frac{p(1-\delta)}{(p-q)},\ 2\alpha^h(1-\delta) + \delta v^* - \frac{(1-p)(1-\delta)}{(p-q)}\right],$$

which is nonempty if (7.6.10) holds.

We can now determine v^*: Because $v^* = \sup_\alpha \max \mathscr{W}^\alpha$, and the supremum is achieved by $\alpha^h = 1$, it is straightforward to verify $v^* = v^{*p}$. Hence, under (7.6.10),

$$\cup_{\{\alpha:\alpha^h \in [0,1]\}} \mathscr{W}^\alpha = [1, v^*] = [1, v^{*p}],$$

and the set of payoffs $[1, v^*]$ is self-generating, and so is the maximal set of PPE payoffs, for all δ satisfying (7.6.10). Moreover, this lower bound on δ is lower than that in (7.6.7), the bound for self-generation of $[1, v^*]$ under pure strategies.

Remark 7.6.3 **Failure of bang-bang** The payoff v^* can only be decomposed by $(Hh, \hat{\gamma})$ on $[1, v^*]$, where $\hat{\gamma}(\bar{y}) = v^*$ and $\hat{\gamma}(\underline{y})$ solves (7.6.6) with equality. Consequently, the set $\hat{\Gamma}$ in the proof of proposition 7.5.1 is the singleton $\{\hat{\gamma}\}$, and so (trivially) $\hat{\gamma}$ is an extreme point of $\hat{\Gamma}$. Because $\text{ext}[1, v^*] = \{1, v^*\}$, $\hat{\gamma}(\underline{y}) \notin \text{ext}[1, v^*]$, and the bang-bang property fails. The problem is that after the signal $\underline{y}$, decomposability of v^* requires a lower payoff than v^*, but not as low as 1. If there was a public correlating device, then the bang-bang property could trivially be achieved by specifying a public correlation over continuations in $\{1, v^*\}$ whose expected value equaled $\hat{\gamma}(\underline{y})$. ◆

7.7 The Repeated Prisoners' Dilemma Redux

7.7.1 Symmetric Inefficiency Revisited

Let $\bar{\gamma}$ be the largest payoff in any strongly symmetric pure-strategy PPE of the repeated prisoners' dilemma from section 7.2. From (7.2.17), we know

$$\bar{\gamma} \leq 2 - \frac{(1-p)}{(p-q)}. \tag{7.7.1}$$

We now show that this upper bound can be achieved without the correlating device used in section 7.2.3. The strategy profile in section 7.2.3 used grim trigger, with a bad signal triggering permanent SS with a probability smaller than 1 (using public correlation). Here we show that by appropriately keeping track of the history of signals, we can effectively do the same thing without public correlation. We are interested in strongly symmetric pure-strategy equilibria, so we need only be concerned with the enforceability of EE and SS, and we will be enforcing on diagonal subsets of $\mathbb{R}^2$, that is, subsets where $\gamma_1 = \gamma_2$. We treat such subsets as subsets of $\mathbb{R}$.

We are interested in constructing a profile similar to grim trigger, but where the first observation of $\underline{y}$ need not trigger the switch to SS. (Two examples of such profiles are displayed in figures 13.4.1 and 13.4.2.) We show that a set $\mathscr{W} \equiv \{0\} \cup [\gamma, \bar{\gamma}]$, with $\gamma > 0$, is self-generating. Note that $\mathscr{W}$ is *not* an interval. Observe first that 0 is trivially decomposed on this set by SS, because SS is a Nash equilibrium of the stage game and no intertemporal considerations are needed to enforce SS. Hence a constant continuation payoff of 0 suffices.

We decompose the other payoffs using EE, with two different specifications of continuations. Let $\mathscr{W}'$ be the set of payoffs that can be decomposed using EE on $\mathscr{W}$ and imposing the continuation value $\gamma^{\underline{y}} = 0$, whereas $\mathscr{W}''$ will be the set of payoffs that can be decomposed using EE on $[\gamma, \bar{\gamma}]$ (so that $\gamma^{\underline{y}} > 0$).

7.7 ■ The Repeated Prisoners' Dilemma Redux

The pair of values $\gamma = (\gamma^{\underline{y}}, \gamma^{\bar{y}})$ enforces EE if

$$(1-\delta)2 + \delta\{p\gamma^{\bar{y}} + (1-p)\gamma^{\underline{y}}\} \geq (1-\delta)3 + \delta\{q\gamma^{\bar{y}} + (1-q)\gamma^{\underline{y}}\},$$

that is,

$$\gamma^{\bar{y}} \geq \gamma^{\underline{y}} + \frac{(1-\delta)}{\delta(p-q)} \equiv \gamma^{\underline{y}} + \Delta. \tag{7.7.2}$$

The set of payoffs that can be decomposed using EE on $\mathscr{W}$ and imposing the continuation value $\gamma^{\underline{y}} = 0$ is given by

$$v \in \mathscr{W}' \iff \exists \gamma^{\bar{y}} \in [\underline{\gamma}, \bar{\gamma}] \text{ satisfying (7.7.2) such that}$$
$$v = \mathsf{V}(EE, \gamma) = (1-\delta)2 + \delta p \gamma^{\bar{y}}.$$

Observe that if $\mathscr{W}$ is self-generating, then the minimum of $\mathscr{W}'$ must equal $\underline{\gamma}$. To find that value, we set $\gamma^{\bar{y}} = \underline{\gamma}$, and solve

$$\underline{\gamma} = (1-\delta)2 + \delta p \underline{\gamma}$$

to obtain

$$\underline{\gamma} = \frac{(1-\delta)2}{1-\delta p}.$$

To decompose $\underline{\gamma}$ in this way, (7.7.2) must be feasible, that is,

$$\underline{\gamma} \geq \frac{(1-\delta)}{\delta(p-q)},$$

or

$$\delta \geq \frac{1}{(3p-2q)} \equiv \delta_0. \tag{7.7.3}$$

When this condition holds, the two-point set $\{0, \underline{\gamma}\}$ is self-generating (and corresponds to grim trigger), whereas the set $\mathscr{W}'$ is empty if (7.7.3) fails. The bound (7.7.3) is the same as that for grim trigger to be an equilibrium without (see (7.2.4)) or with (see (7.2.16)) public correlation.

The maximum value of $\mathscr{W}'$, $\bar{\gamma}'$ is obtained by setting $\gamma^{\bar{y}} = \bar{\gamma}$, and so we have

$$\mathscr{W}' = [\underline{\gamma}, \bar{\gamma}'] = \left[\frac{(1-\delta)2}{1-\delta p}, 2(1-\delta) + \delta p \bar{\gamma}\right].$$

We now consider $\mathscr{W}''$. To find the maximum $v \in \mathscr{W}''$, we set $v = \bar{\gamma} = \gamma^{\bar{y}}$ and suppose (7.7.2) holds with equality:

$$\bar{\gamma} = (1-\delta)2 + \delta\left\{p\bar{\gamma} + (1-p)\left(\bar{\gamma} - \frac{(1-\delta)}{\delta(p-q)}\right)\right\}$$
$$= 2(1-\delta) + \delta\bar{\gamma} - \frac{(1-p)(1-\delta)}{(p-q)},$$

so that

$$\bar{\gamma} = 2 - \frac{(1-p)}{(p-q)},$$

matching the value in (7.7.1). If condition (7.7.3) holds with equality, then $\gamma = \bar{\gamma}$. We must also verify that (7.7.2) is consistent with decomposing $\bar{\gamma}$ on $[\gamma, \bar{\gamma}]$, which requires

$$\gamma \leq \bar{\gamma} - \Delta = \bar{\gamma} - \frac{(1-\delta)}{\delta(p-q)}. \tag{7.7.4}$$

Because $\bar{\gamma}$ is independent of δ, and both Δ and γ converge to 0 as $\delta \to 1$, there exists a $\delta_1 < 1$ such that for $\delta > \delta_1$, (7.7.4) is satisfied. Moreover, because $\gamma = \bar{\gamma}$ for $\delta = \delta_0$ (the lower bound in (7.7.3)), $\delta_1 > \delta_0$. This implies that for $\delta \in [\delta_0, \delta_1]$, $\mathscr{W}'''$ is empty.

To find the minimum value $\gamma'' \equiv \min \mathscr{W}'''$, we set $\gamma^y = \gamma$ and again suppose (7.7.2) holds with equality. We thus have

$$\mathscr{W}''' = [\gamma'', \bar{\gamma}] = \left[2(1-\delta) + \delta\gamma + \delta p\Delta, 2 - \frac{(1-p)}{(p-q)} \right].$$

For self-generation, we need

$$\mathscr{W} = \{0\} \cup \mathscr{W}' \cup \mathscr{W}''',$$

which is implied by $\gamma'' \leq \bar{\gamma}'$. This is equivalent to

$$\gamma \leq p(\bar{\gamma} - \Delta).$$

Similarly to (7.7.4), there exists $\delta_2 > \delta_1$ such that the above inequality holds for all larger δ. If $\delta \geq \delta_2$, the upper bound $\bar{\gamma}$ can be achieved in a pure strategy symmetric equilibrium. The implied strategy profile has a structure similar to grim trigger with public correlation. For all strictly positive continuation values, the players continue to exert effort, but the first time a continuation γ^y equals 0, SS is played permanently thereafter.

Remark 7.7.1 **Multiple self-generating sets** We could instead have determined the conditions under which the entire interval $[0, \bar{\gamma}]$ is self-generating. Because the analysis for that case is very similar to that in section 7.6.2, we simply observe that $[0, \bar{\gamma}]$ is indeed self-generating if

$$\delta \geq \frac{3p - 2q}{2(3p - 2q) - 1}. \tag{7.7.5}$$

Notice that the bound δ_2 and the bound given by (7.7.5) exceed the bound (7.2.16) for obtaining $\bar{\gamma}$ with public correlation.

The self-generating sets $[0, \bar{\gamma}]$ and $\{0\} \cup [\gamma, \bar{\gamma}]$ allow Nash reversion (because both include the payoff 0). Because PPE need not rely on Nash reversion (for example, the profile in section 7.2.2 has one-period recall), there are self-generating sets that exclude 0 and indeed that exclude a neighborhood of 0. For example, it is easily verified that if $q < 2r$, the set $[1, \bar{\gamma}]$ is self-generating for sufficiently large δ. Finally, the set of continuation values generated by a strongly symmetric PPE with bounded recall (see definition 13.3.1) is a finite self-generating set and, if EE can be played, excludes 0.

◆

As $p - q \to 0$, the bounds on δ implied by (7.7.3)–(7.7.5) become increasingly severe and the maximum strongly symmetric PPE payoff converges to 0. This is not surprising, because if $p - q$ is small, it is very difficult to detect a deviation when the opponent is choosing E. In particular, if $p - q$ is sufficiently close to 0 that (7.7.2) is violated, even for $\gamma^{\bar{y}} = 2$ and $\gamma^{\underline{y}} = 0$, the pure-action profile EE cannot be enforced. On the other hand, if $q - r$ is large, then the distribution over signals under ES is very different than under SS. Thus, if player i is playing S with positive probability, then E may now be enforceable. (This possibility also underlies the superiority of the private strategy examples of sections 10.2 and 10.4.) Consequently, there may be strongly symmetric mixed equilibria with a payoff strictly larger than 0, a possibility we explore in section 7.7.2. The observation that even though a pure-action profile a is unenforceable, a mixed-action profile that assigns positive probability to a may be enforceable, is more general.

Remark 7.7.2 **Symmetric payoffs from asymmetric profiles** If $q > 2r$, it is possible to achieve the symmetric payoff of $(1, 1)$ using the asymmetric action profiles ES and SE. For sufficiently large δ, familiar calculations verify that the largest set of pure-strategy PPE payoffs in which only ES and SE are played in each period is $\mathscr{W} = \{(v \in \mathbb{R}^2 : v_1 + v_2 = 2, v_i \geq r/(q - r), i = 1, 2\}$ (i.e., $\mathscr{W}$ is the largest self-generating set using only ES and SE). Because $q > 2r$, we have $r/(q - r) < 1 < 2 - r/(q - r)$, and there is an asymmetric pure-strategy PPE in which each player has payoff 1. Hence, for p sufficiently close to q, although $(1, 1)$ cannot be achieved in a strongly symmetric PPE, it can be achieved in an asymmetric PPE. Finally, there is an asymmetric equilibrium in which EE is never played, and player 2's payoff is strictly larger than $2 - r/(q - r)$ (this PPE uses SS to relax incentive constraints, see note 6 on page 289).

♦

7.7.2 Enforcing a Mixed-Action Profile

Let α denote a *completely mixed* symmetric action profile in the prisoners' dilemma. We first argue that α can in some situations be enforced even when EE cannot. When q is close to p, player 1 still has a (weak) incentive to play E when player 2 is randomizing because with positive probability (the probability that 2 plays S) a change from E to S will substantially change the distribution over signals. To conserve on notation, we denote the probability that each player assigns to E in the symmetric profile by α. Let $\gamma^{\underline{y}}, \gamma^{\bar{y}} \in \mathscr{F}^\dagger$ be the continuations after $\underline{y}$ and $\bar{y}$. The payoff to player 1 from E when player 2 is randomizing with probability α on E is

$$g_1^\gamma(E, \alpha) = \alpha\{2(1-\delta) + \delta[p\gamma_1^{\bar{y}} + (1-p)\gamma_1^{\underline{y}}]\}$$
$$+ (1-\alpha)\{(-1)(1-\delta) + \delta[q\gamma_1^{\bar{y}} + (1-q)\gamma_1^{\underline{y}}]\}$$
$$= (1-\delta)(3\alpha - 1) + \delta[\alpha(p\gamma_1^{\bar{y}} + (1-p)\gamma_1^{\underline{y}})$$
$$+ (1-\alpha)(q\gamma_1^{\bar{y}} + (1-q)\gamma_1^{\underline{y}})].$$

The payoff from S is

$$g_1^\gamma(S,\alpha) = \alpha\{3(1-\delta) + \delta[q\gamma_1^{\bar{y}} + (1-q)\gamma_1^{\underline{y}}]\}$$
$$+ (1-\alpha)\{\delta[r\gamma_1^{\bar{y}} + (1-r)\gamma_1^{\underline{y}}]\}.$$

For player 1 to be willing to randomize, we need $g_1^\gamma(E,\alpha) = g_1^\gamma(S,\alpha)$, that is,

$$\left(\gamma_1^{\bar{y}} - \gamma_1^{\underline{y}}\right) = \frac{(1-\delta)}{\delta[\alpha(p-q) + (1-\alpha)(q-r)]}. \tag{7.7.6}$$

We now investigate when there exists a value of α and feasible continuations solving (7.7.6). Let $f(\alpha, \gamma_1^{\bar{y}})$ be the payoff to player 1 when player 2 puts α weight on E and (7.7.6) is satisfied:

$$f(\alpha, \gamma_1^{\bar{y}}) = (1-\delta)(3\alpha - 1) - \frac{(\alpha(1-p) + (1-\alpha)(1-q))(1-\delta)}{(\alpha(p-q) + (1-\alpha)(q-r))} + \delta\gamma_1^{\bar{y}}.$$

Self-generation suggests solving $\bar{\gamma}(\alpha) = f(\alpha, \bar{\gamma}(\alpha))$, yielding

$$\bar{\gamma}(\alpha) = (3\alpha - 1) - \frac{(\alpha(1-p) + (1-\alpha)(1-q))}{(\alpha(p-q) + (1-\alpha)(q-r))}. \tag{7.7.7}$$

We need to know if there is a value of $\alpha \in [0,1]$ for which $\bar{\gamma}(\alpha)$ is sufficiently large that (7.7.6) can be satisfied with $\gamma_1^{\bar{y}} = \bar{\gamma}(\alpha)$ for some α and $\gamma_1^{\underline{y}} \geq 0$.

To simplify calculations, we first consider $p = q$.[21] In this case, the function $\bar{\gamma}(\alpha)$ is maximized at

$$\bar{\alpha} = 1 - \sqrt{\frac{(1-q)}{3(q-r)}}$$

with a value

$$\bar{\gamma}(\bar{\alpha}) = 2 - 2\sqrt{\frac{3(1-q)}{(q-r)}}.$$

From (7.7.6), α is enforceable if

$$\bar{\gamma}(\bar{\alpha})(1 - \bar{\alpha}) \geq \frac{(1-\delta)}{\delta(q-r)}. \tag{7.7.8}$$

Inequality (7.7.8) is satisfied for δ sufficiently close to 1, if $\bar{\gamma}(\bar{\alpha})(1-\bar{\alpha}) > 0$. The payoff $\bar{\gamma}(\bar{\alpha})$ is strictly positive if

$$3 + r < 4q, \tag{7.7.9}$$

and this inequality also implies $\bar{\alpha} \in (2/3, 1)$. Given q and r satisfying (7.7.9), let $\underline{\delta}$ be a lower bound on δ so that (7.7.8) holds.

---

21. As we discuss in note 7 on page 238, assuming $p = q$ is inconsistent with the representation of stage-game payoffs as expected ex post payoffs. We relax the assumption $p = q$ at the end of the example.

Suppose (7.7.9) holds and $\delta \in (\underline{\delta}, 1)$. Because $\bar{\gamma}(\alpha)(1-\alpha)$ is a continuous function of α, with $\bar{\gamma}(0) < 0$, there exists $\alpha(\delta)$ such that

$$\bar{\gamma}(\alpha(\delta))(1-\alpha(\delta)) = \frac{(1-\delta)}{\delta(q-r)}.$$

Hence, the set of payoffs $\{(0,0), (\bar{\gamma}(\alpha(\delta)), \bar{\gamma}(\alpha(\delta)))\}$ is self-generating for $\delta \in (\underline{\delta}, 1)$. The associated PPE is strongly symmetric and begins with each player randomizing with probability $\alpha(\delta)$ on E, and continually playing this mixed action until the first observation of $\underline{y}$. After the first observation of $\underline{y}$, play switches to permanent S.

Indeed, for sufficiently high δ, the symmetric set of payoffs described by the interval $[0, \bar{\gamma}(\tilde{\alpha})]$ is self-generating: Every payoff in the interval $[0, \delta\bar{\gamma}(\tilde{\alpha})]$ can be decomposed by SS and some constant continuation in $[0, \bar{\gamma}(\tilde{\alpha})]$. The function f is increasing in $\gamma_1^{\bar{y}}$ for fixed α, as well as in α for fixed $\gamma_1^{\bar{y}}$. Moreover, consistent with $\gamma_1^{\underline{y}} \geq 0$ and recalling the role of (7.7.6) in the definition of f, f is minimized at $(\alpha, \gamma_1^{\bar{y}}) = (0, (1-\delta)/[\delta(q-r)])$ with a value $(1-\delta)r/(q-r)$. Hence, every payoff in $[(1-\delta)r/(q-r), \bar{\gamma}(\tilde{\alpha})]$ can be decomposed by some α and continuations in $[0, \bar{\gamma}(\tilde{\alpha})]$.[22]

We now return to the case $p > q$. By continuity, for p sufficiently close to q, $\bar{\gamma}(\alpha)$ will be maximized at some $\tilde{\alpha} \in (0, 1)$, with $\bar{\gamma}(\tilde{\alpha}) > 0$. Hence for δ sufficiently close to 1, the set $[0, \bar{\gamma}(\tilde{\alpha})]$ is self-generating.

7.8 Anonymous Players

In section 2.7, we argued that short-lived players can be interpreted as a continuum of small, anonymous, long-lived players when histories include only aggregate outcomes. Because a small and anonymous player, though long-lived, has no effect on the aggregate outcome and hence on future play, he should choose a myopic best response. As we discussed in remark 2.7.1, this interpretation opens a potentially troubling discontinuity: Equilibrium behavior with a large but finite number of small players may be very different under perfect monitoring than with small anonymous players.

Under imperfect monitoring, this discontinuity disappears. Consider, for example, a version of the prisoners' dilemma with a single long-lived player 1 and a finite number $N - 1$ of long-lived players in the role of player 2. Player 1's actions are perfectly monitored, whereas the actions of the other players are only imperfectly monitored. For each player i, $i = 2, \ldots, N$, there are two signals e_i and s_i, independently distributed according to

$$\rho_i(e_i \mid a_i) = \begin{cases} 1 - \varepsilon, & a_i = E, \\ \varepsilon, & a_i = S, \end{cases}$$

where $\varepsilon < 1/2$. Player 1's stage game payoff is the average of the payoff from the play with each player $i \in \{2, \ldots, N\}$. Figure 7.8.1 gives the ex post payoffs from a single interaction as a function of player 1's action E or S and the signal e and s generated

---

22. For some parameter values ($p \approx q$ and $13q - r > 12$), the best strongly symmetric mixed-strategy profile achieves a higher symmetric payoff than the asymmetric profile of remark 7.7.2, which in turn achieves a higher symmetric payoff than any pure-strategy strongly symmetric PPE.

	e_i	s_i
E	$\frac{2-\varepsilon}{1-2\varepsilon}, \frac{2-5\varepsilon}{1-2\varepsilon}$	$-\frac{(1+\varepsilon)}{1-2\varepsilon}, \frac{3-5\varepsilon}{1-2\varepsilon}$
S	$\frac{3-3\varepsilon}{1-2\varepsilon}, -\frac{(1-\varepsilon)}{1-2\varepsilon}$	$-\frac{3\varepsilon}{1-2\varepsilon}, \frac{\varepsilon}{1-2\varepsilon}$

	E	S
E	$2,2$	$-1,3$
S	$3,-1$	$0,0$

Figure 7.8.1 Ex post payoffs from a single interaction between player 1 and a player $i \in \{2, \ldots, N\}$ (left panel) as a function of player 1's action E or S and the signal e_i or s_i, and the ex ante payoffs (right panel), the payoffs in figure 7.2.1.

by player 2's action, as well as the ex ante payoff (which agrees with our standard prisoners' dilemma payoff matrix, figure 7.2.1).

We consider a class of automata with two states and public correlation:[23] Given a public signal $y \in \{E, S\} \times \prod_{i \geq 2} \{e_i, s_i\}$, we define $\#(y) = |\{i \geq 2 : y_i = s_i\}|$, that is, $\#(y)$ is the number of "shirk" signals in y (apart from player 1, who is perfectly monitored). The set of states is $\mathcal{W} = \{w_E, w_S\}$, the initial state is $w^0 = w_E$, the output function is $f_i(w_E) = E$ and $f_i(w_S) = S$ for $i = 1, \ldots, N$, and the transition function τ is given by, where $\tau(w, y)$ is the probability of a transition to state w_E,

$$\tau(w, y) = \begin{cases} 1, & \text{if } w = w_E, y_1 = E \text{ and } \#(y) \leq \kappa, \\ \phi, & \text{if } w = w_E, y_1 = E \text{ and } \#(y) > \kappa, \\ 0, & \text{otherwise.} \end{cases}$$

Any deviation by player 1 (which is detected for sure) triggers immediate permanent shirking, and more than κ signals of s_i for the other players only randomly triggers permanent shirking. Consequently, whenever the incentive constraints for a player $i, i = 2, \ldots, N$, are satisfied, so are player 1's, and so we focus on player i.

The analysis of this profile is identical to that of the profile in section 7.2.3, once we set p equal to the probability of the event that $\#(y) \leq \kappa$ when no player deviates and q equal to the probability of that event when exactly one player deviates. Thus,

$$p = \Pr\{\#(y) \leq \kappa \mid a_i = E, i = 1, \ldots, N\}$$
$$= \sum_{\tau=0}^{\kappa} \frac{(N-1)!}{\tau!(N-1-\tau)!} \varepsilon^\tau (1-\varepsilon)^{N-1-\tau}$$

and

$$q = \Pr\{\#(y) \leq \kappa \mid a_N = S, a_i = E, i = 1, \ldots, N-1\}$$
$$= (1-\varepsilon) \sum_{\tau=0}^{\kappa-1} \frac{(N-2)!}{\tau!(N-2-\tau)!} \varepsilon^\tau (1-\varepsilon)^{N-2-\tau}$$
$$+ \varepsilon \sum_{\tau=0}^{\kappa} \frac{(N-2)!}{\tau!(N-2-\tau)!} \varepsilon^\tau (1-\varepsilon)^{N-2-\tau},$$

---

23. In sections 7.6.2 and 7.7.1, we saw that public correlation serves only to simplify the calculation of upper bounds on payoffs.

7.8 ■ Anonymous Players

where the first sum in the expression for q equals 0 if $\kappa = 0$. Because all players play identically under the profile, the payoff to player 1 *under the profile* equals that of any player i, $i = 2, \ldots, N$.

If (7.2.16) is satisfied, the profile describes a PPE, and the payoff in state w_E is, from (7.2.17),

$$V(w_E) = 2 - \frac{1-p}{(p-q)}. \tag{7.8.1}$$

Suppose we first fix N. Then, for sufficiently small ε, (7.2.16) is only satisfied for $\kappa = 0$ (if $\kappa \geq 1$, $\lim_{\varepsilon \to 0} q = 1$) and $\lim_{\varepsilon \to 0} V(w_E) = 2$. In other words, irrespective of the number of players, for arbitrarily accurate monitoring all players essentially receive a payoff of 2 under this profile.

Suppose we now fix $\varepsilon > 0$, and consider increasing N. Observe first that if $\kappa = 0$, then for large N it is too easy to "fail the test," that is, $\lim_{N \to \infty} p = 0$. From (7.8.1), we immediately get an upper bound less than 1. By appropriately increasing κ as N increases, one could hope to keep p not too small. However, although that is certainly possible, it turns out that irrespective of how κ is determined, $p - q$ will go to 0. For fixed κ and player i, we have

$$p - q = \sum_{\tau=1}^{\kappa-1} \left\{ \frac{(N-1)}{(N-1-\tau)} - 1 \right\} \frac{(N-2)!}{\tau!(N-2-\tau)!} \varepsilon^\tau (1-\varepsilon)^{N-1-\tau}$$

$$- \sum_{\tau=1}^{\kappa+1} \frac{(N-2)!}{(\tau-1)!(N-1-\tau)!} \varepsilon^\tau (1-\varepsilon)^{N-1-\tau}$$

$$= \frac{(N-2)!}{\kappa!(N-2-\kappa)!} \varepsilon^\kappa (1-\varepsilon)^{N-2-\kappa} (1-2\varepsilon)$$

$$= \Pr(\text{player } i \text{ is pivotal})(1-2\varepsilon).$$

Observe that $\Pr(\text{player } i \text{ is pivotal})$ is the probability $b(\kappa, N-2, \varepsilon)$ that there are exactly κ successes under a binomial distribution with $N-2$ draws. It is bounded above by $\max_k b(k, N-2, \varepsilon)$, and therefore for fixed ε, $p - q \to 0$ as $N \to \infty$.

Consequently, for sufficiently large N, (7.2.16) fails and the only symmetric equilibrium (of the type considered here) for large N has players choosing S in every period (as predicted by the model with small anonymous players, see remark 2.7.1). Given the imperfection in monitoring, with a large population, even though there is a separate signal for each player, the chance that any single player will be pivotal in determining the transition to permanent shirking is small, and so each imperfectly monitored player can effectively ignore the future and myopically optimizes.

We restricted attention to strongly symmetric PPE, but a similar result holds in general. We can thus view short-lived players as a continuum of small, anonymous, long-lived players, recognizing that this is an approximation of a world in which such players are relatively plentiful relative to the imperfection in the monitoring. A more complete discussion can be found in Al-Najjar and Smorodinsky (2001); Fudenberg, Levine, and Pesendorfer (1998); Levine and Pesendorfer (1995); Green (1980); and Sabourian (1990).

8 Bounding Perfect Public Equilibrium Payoffs

The observation that the set of perfect public equilibrium payoffs $\mathscr{E}(\delta)$ is the largest self-generating set does not provide a simple technique for characterizing the set. We now describe a technique that allows us to bound, and for large δ completely characterize, the set of PPE payoffs.[1]

8.1 Decomposing on Half-Spaces

It is helpful to begin with figure 8.1.1. We have seen in section 7.3 that every payoff $v \in \mathscr{E}(\delta)$ is decomposed by some action profile α (such as the pure-action profile a in the figure) and continuation payoffs $\gamma(y) \in \mathscr{E}(\delta)$. Recall from remark 7.3.1 that $u(\alpha)$ will often denote the vector of stage-game payoffs for long-lived players, a practice we follow in this chapter.

To bound $\mathscr{E}(\delta)$, we are clearly interested in the boundary points of $\mathscr{E}(\delta)$ (unlike the v in the figure). Consider a boundary point $v' \in \mathscr{E}(\delta)$ decomposed by a non–stage-game Nash equilibrium action profile (such as a in the figure). Moreover, for simplicity, suppose that v' is in fact a boundary point of the convex hull of $\mathscr{E}(\delta)$, $\text{co}\mathscr{E}(\delta)$. Then, the continuations (some of which must differ from v') lie in a convex set ($\text{co}\mathscr{E}(\delta)$) that can be separated from both v' and $u(a)$ by the supporting hyperplane to v' on $\text{co}\mathscr{E}(\delta)$. That is, the continuations lie in a half-space whose boundary is given by the supporting hyperplane.

Because $\mathscr{F}^\dagger$ is the convex hull of $\mathscr{F}$, a compact set, it is the intersection of the closed half-spaces containing $\mathscr{F}$ (Rockafellar 1970, corollary 11.5.1 or theorem 18.8). Now, for any direction $\lambda \in \mathbb{R}^n \setminus \{\mathbf{0}\}$, the smallest half-space containing $\mathscr{F}^\dagger$ is given by $\{v' : \lambda \cdot v' \leq \max_{\alpha \in \mathbf{B}} \lambda \cdot u(\alpha)\}$. In other words, because $\mathscr{E}(\delta) \subset \mathscr{F}^\dagger$, $\mathscr{E}(\delta)$ is contained in the intersection over λ of the half-spaces $\{v' : \lambda \cdot v' \leq \max_{\alpha \in \mathbf{B}} \lambda \cdot u(\alpha)\}$.

This bound on $\mathscr{E}(\delta)$ ignores decomposability constraints. The set of payoffs decomposable on a general set $\mathscr{W}$ (such as $\mathscr{E}(\delta)$) is complicated to describe (as we discuss just before proposition 7.3.4, even the weak property of monotonicity of $\mathscr{E}(\delta)$ is not guaranteed). On the other hand, the set of payoffs decomposable on a half-space has a particularly simple dependence on δ, allowing us to describe the smallest half-space respecting the decomposability constraints independently of δ.

---

1. The essentials of the characterization are due to Fudenberg, Levine, and Maskin (1994) and Fudenberg and Levine (1994), with a refinement by Kandori and Matsushima (1998). Some similar ideas appear in Matsushima (1989).

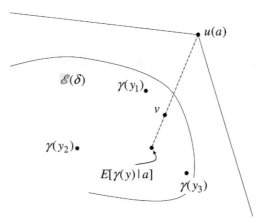

Figure 8.1.1 The payoff $v \in \mathscr{E}(\delta)$ is decomposable by a on $\mathscr{E}(\delta)$. There are three public signals, $Y = \{y_1, y_2, y_3\}$, and $E[\gamma(y) \mid a] = \sum_\ell \gamma(y_\ell) \rho(y_\ell \mid a)$. Note that $u(a) \notin \mathscr{E}(\delta)$, and so a is not a Nash equilibrium of the stage game.

Given a direction $\lambda \in \mathbb{R}^n$ and constant $k \in \mathbb{R}$, $H(\lambda, k)$ denotes the half-space $\{v \in \mathbb{R}^n : \lambda \cdot v \leq k\}$. In figure 8.1.2, we have chosen a direction λ so that v is *decomposable on the half-space* $H(\lambda, \lambda \cdot v)$ using the equilibrium continuations from figure 8.1.1. In general, of course, when a payoff is decomposed on a half-space there is no guarantee that the continuations are equilibrium, or even feasible, continuations.

Recall that $\mathscr{B}(\mathscr{W}; \delta, \alpha)$ is the set of payoffs that can be decomposed by α on $\mathscr{W}$, when the discount factor is δ. For fixed λ and α, let[2]

$$k^*(\alpha, \lambda, \delta) = \max_v \lambda \cdot v \qquad (8.1.1)$$
$$\text{subject to} \quad v \in \mathscr{B}(H(\lambda, \lambda \cdot v); \delta, \alpha).$$

The pair (v, γ) illustrated in figure 8.1.2 does not solve this linear program for the indicated λ and $\alpha = a$, because moving all quantities toward $u(a)$ would increase $\lambda \cdot v$ while not losing enforceability. Intuitively, v's in the interior of $\mathscr{E}(\delta)$ cannot solve this problem. We will see in section 9.1, in the limit (as $\delta \to 1$), boundary points of $\mathscr{E}(\delta)$ do solve this problem in many cases.

The linear programming nature of (8.1.1) can be seen most clearly by noting that the constraint $v \in \mathscr{B}(H(\lambda, \lambda \cdot v); \delta, \alpha)$ is equivalent to the existence of $\gamma : Y \to \mathbb{R}^n$ satisfying

$$v_i = (1 - \delta) u_i(\alpha) + \delta E[\gamma_i(y) \mid \alpha], \quad \forall i,$$
$$v_i \geq (1 - \delta) u_i(a_i, \alpha_{-i}) + \delta E[\gamma_i(y) \mid (a_i, \alpha_{-i})], \quad \forall a_i \in A_i, \forall i,$$

and

$$\lambda \cdot v \geq \lambda \cdot \gamma(y), \quad \forall y \in Y.$$

---

2. As usual, if the constraint set is empty, the value is $-\infty$. It is clear from the characterization of $\mathscr{B}(H(\lambda, \lambda \cdot v); \delta, \alpha)$ given by (8.1.2)–(8.1.4) that both $\mathscr{B}(H(\lambda, \lambda \cdot v); \delta, \alpha)$ is closed and $\{\lambda \cdot v : v \in \mathscr{B}(H(\lambda, \lambda \cdot v); \delta, \alpha)\}$ is bounded above, ensuring the maximum exists.

8.1 ■ Decomposing on Half-Spaces

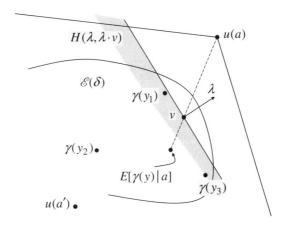

Figure 8.1.2 Illustration of decomposability on half-spaces. The payoff $v \in \mathscr{E}(\delta)$ is decomposable with respect to a and the half space $H(\lambda, \lambda \cdot v)$.

By setting $x_i(y) = \delta(\gamma_i(y) - v_i)/(1-\delta)$, we can characterize $\mathscr{B}(H(\lambda, \lambda \cdot v); \delta, \alpha)$ independently of the discount factor δ, that is, $v \in \mathscr{B}(H(\lambda, \lambda \cdot v); \delta, \alpha)$ if and only if there exists $x : Y \to \mathbb{R}^n$ satisfying

$$v_i = u_i(\alpha) + E[x_i(y) \mid \alpha], \quad \forall i \tag{8.1.2}$$

$$v_i \geq u_i(a_i, \alpha_{-i}) + E[x_i(y) \mid (a_i, \alpha_{-i})], \quad \forall a_i \in A_i, \forall i, \tag{8.1.3}$$

and

$$0 \geq \lambda \cdot x(y), \quad \forall y \in Y. \tag{8.1.4}$$

We call $x : Y \to \mathbb{R}^n$ the *normalized continuations*. If x satisfies (8.1.2) and (8.1.3) for some v, we say x *enforces* α, and if, in addition, (8.1.4) is satisfied for some λ, we say x *enforces* α *in the direction* λ. If x satisfies (8.1.2)–(8.1.4) with $\lambda \cdot x(y) = 0$ for all y, we say x *orthogonally enforces* α (in the direction λ). Suppose $\lambda_i > 0$ for all i and α is orthogonally enforced by x. Then for each y, $x_i(y)$ cannot be the same sign for all i; if $x_i(y) > 0$, so that i is "rewarded" after signal y, then there must be some other player j who is "punished," $x_j(y) < 0$. In other words, orthogonal enforceability typically involves a transfer of continuations from some players to other players, with λ determining the "transfer price."

The unnormalized continuations are given by

$$\gamma_i(y) = v_i + x_i(y)(1-\delta)/\delta, \quad \forall y \in Y, \tag{8.1.5}$$

where $v = u(\alpha) + E[x \mid \alpha]$. If x enforces α in the direction λ, then the associated γ from (8.1.5) enforces α with respect to the hyperplane $H(\lambda, \lambda \cdot v)$.

Recall that e_j denotes the jth standard basis vector. A direction λ is a *coordinate direction* if $\lambda = \lambda_i e_i$ for some constant $\lambda_i \neq 0$ and some i. A direction λ is *ij-pairwise* if there are two players, $i \neq j$, such that $\lambda = \lambda_i e_i + \lambda_j e_j$, $\lambda_i \neq 0$, $\lambda_j \neq 0$; denote such a direction by λ^{ij}.

Lemma 8.1.1
1. $k^*(\alpha, \lambda, \delta)$ is independent of δ, and so can be written as $k^*(\alpha, \lambda)$.
2. $k^*(\alpha, \lambda) \leq \lambda \cdot u(\alpha)$, so that $u(\alpha) \notin \operatorname{int} H(\lambda, k^*(\alpha, \lambda))$; and
3. $k^*(\alpha, \lambda) = \lambda \cdot u(\alpha)$ if α is orthogonally enforced in the direction λ. Moreover, when $\rho(y \mid \alpha) > 0$ for all $y \in Y$, this sufficient condition is also necessary.
4. If, for all pairwise directions λ^{ij}, α is orthogonally enforceable in the direction λ^{ij}, then α is orthogonally enforceable in all noncoordinate directions.

Proof
1. Immediate from the equivalence of the constraint in (8.1.1) and (8.1.2)–(8.1.4).
2. Let v^* solve (8.1.1). Then, $k^*(\alpha, \lambda) = \lambda \cdot v^* = \lambda \cdot (u(\alpha) + E[x^*(y) \mid \alpha])$, where x^* satisfies (8.1.2)–(8.1.4), so that in particular, $\lambda \cdot x^*(y) \leq 0$.
3. Sufficiency is immediate. For necessity, observe that if $k^*(\alpha, \lambda) = \lambda \cdot u(\alpha)$, then $E[\lambda \cdot x^*(y) \mid \alpha] = 0$, where x^* is an enforcing normalized continuation solving (8.1.2)–(8.1.4). Necessity then follows from (8.1.4) and $\rho(y \mid \alpha) > 0$ for all $y \in Y$.
4. Consider first a direction λ with four nonzero coordinates, $\{i, j, k, \ell\}$. Let $\lambda^{ij} \equiv \lambda_i e_i + \lambda_j e_j$ be the pairwise direction reflecting the $i - j$ coordinates, and $\lambda^{k\ell}$ the pairwise direction reflecting the $k - \ell$ coordinates. Let x^{ij} and $x^{k\ell}$ be the enforcing normalized continuations for the pairwise directions λ^{ij} and $\lambda^{k\ell}$. By assumption, $\lambda^{ij} \cdot x^{ij} = 0$ and $\lambda^{k\ell} \cdot x^{k\ell} = 0$. Let x denote the normalized continuation,

$$x_m(y) = \begin{cases} x_m^{ij}, & \text{if } m = i, j, \\ x_m^{k\ell}, & \text{if } m = k, \ell, \\ x_m^{ij}, & \text{otherwise.} \end{cases}$$

Then, (8.1.2) and (8.1.3) are satisfied, and $\lambda \cdot x = (\lambda^{ij} + \lambda^{k\ell}) \cdot x = 0$. This argument can clearly be extended to cover an even number of nonzero coordinates.

Suppose now the direction λ has three nonzero coordinates, $\{i, j, k\}$. Now let $\lambda^{ij} = \lambda_i e_i + \frac{1}{2}\lambda_j e_j$ and $\lambda^{jk} = \frac{1}{2}\lambda_j e_j + \lambda_k e_k$. Let x^{ij} and x^{jk} be the enforcing normalized continuations for the pairwise directions λ^{ij} and λ^{jk}. In this case, we let x denote the normalized continuation,

$$x_m(y) = \begin{cases} x_i^{ij}, & \text{if } m = i, \\ \frac{1}{2}(x_j^{ij} + x_j^{jk}), & \text{if } m = j, \\ x_k^{jk}, & \text{if } m = k, \\ x_m^{ij}, & \text{otherwise.} \end{cases}$$

Then, we again have (8.1.2) and (8.1.3) satisfied and $\lambda \cdot x = (\lambda^{ij} + \lambda^{jk}) \cdot x = 0$. For a general odd number of nonzero coordinates, we apply this argument for three nonzero coordinates, and the previous argument to the remaining even number of nonzero coordinates. ∎

Intuitively, we would like to approximate $\mathscr{E}(\delta)$ by the half-spaces $H(\lambda, k^*(\alpha, \lambda))$. However, as is clear from figure 8.1.2, we need to choose α appropriately. In particular, we can only move v toward $u(a)$ (and so include more of $\mathscr{E}(\delta)$) because $u(a)$ is separated from the half-space containing the enforcing continuations γ. The action profile a' does not yield an appropriate approximating half-space for the displayed choice of λ, because by lemma 8.1.1(2), the enforcing continuations need to lie in a *lower* half-space, excluding much of $\mathscr{E}(\delta)$. Accordingly, for each direction, we use the action profile that maximizes $k^*(\alpha, \lambda)$, that is, set $k^*(\lambda) \equiv \sup_{\alpha \in \mathbf{B}} k^*(\alpha, \lambda)$ and $H^*(\lambda) \equiv H(\lambda, k^*(\lambda))$.[3] We refer to $H^*(\lambda)$ as the *maximal half-space* in the direction λ.

The coordinate direction $\lambda = -e_j$ corresponds to minimizing player j's payoff, because $-e_j \cdot v = -v_j$. The next lemma shows that any payoff profile in the half-space $H^*(-e_j)$ is weakly individually rational for player j. Note that $\underline{v}_j$ is the minmax payoff of (2.7.1).

Lemma 8.1.2 *For all j,*

$$k^*(-e_j) \leq -\underline{v}_j = -\min_{\alpha \in \mathbf{B}} \max_{a_j} u_j(a_j, \alpha_{-j}).$$

Proof For $\lambda = -e_j$, constraint (8.1.4) becomes $x_j(y) \geq 0$ for all $y \in Y$. Constraint (8.1.3) then implies $v_j \geq u_j(a_j, \alpha_{-j})$ for all a_j, and so $v_j \geq \max_{a_j} u_j(a_j, \alpha_{-j})$. In other words, $v \in \mathscr{B}(H(-e_j, -\underline{v}_j); \delta, \alpha)$ implies $v_j \geq \max_{a_j} u_j(a_j, \alpha_{-j})$. Then,

$$\begin{aligned}
k^*(-e_j) &= \sup_{\alpha \in \mathbf{B}} k^*(\alpha, -e_j) \\
&= \sup_{\alpha \in \mathbf{B}} \max_v \{-v_j : v \in \mathscr{B}(H(-e_j, -\underline{v}_j); \delta, \alpha)\} \\
&\leq -\min_{\alpha \in \mathbf{B}} \max_{a_j} u_j(a_i, \alpha_{-i}) \\
&= -\underline{v}_j.
\end{aligned}$$
∎

Proposition 8.1.1 *For all δ,*

$$\mathscr{E}(\delta) \subset \cap_\lambda H^*(\lambda) \equiv \mathscr{M} \subset \overline{\mathscr{F}^*}.$$

Proof Suppose the first inclusion fails. Then there exists a half-space $H(\lambda, k^*(\lambda))$ and a point $v' \in \mathscr{E}(\delta)$ with $\lambda \cdot v' > k^*(\lambda)$. Let v^* maximize $\lambda \cdot v$ over $v \in \mathscr{E}(\delta)$. Because $\lambda \cdot v \leq \lambda \cdot v^*$ for all $v \in \mathscr{E}(\delta)$, v^* is decomposable on $H(\lambda, \lambda \cdot v^*)$, implying $k^*(\lambda) \geq \lambda \cdot v^*$, a contradiction.

It is a standard result from convex analysis that the convex hull of a compact set is the intersection of the closed half-spaces containing the set (Rockafellar 1970, corollary 11.5.1 or theorem 18.8). The inclusion of $\mathscr{M}$ in $\overline{\mathscr{F}^*}$, the set of feasible and weakly individually rational payoffs, is then an immediate implication of lemmas 8.1.1 and 8.1.2, because $k^*(\lambda) \equiv \sup_{\alpha \in \mathbf{B}} k^*(\alpha, \lambda) \leq \sup_{\alpha \in \mathbf{B}} \lambda \cdot u(\alpha)$.
∎

Each half-space is convex and the arbitrary intersection of convex sets is convex, so $\mathscr{M}$ also bounds the convex hull of $\mathscr{E}(\delta)$. The bound $\mathscr{M}$ is crude for small δ, because it is independent of δ. However, as we will see in section 9.1, provided $\mathscr{M}$ has non-empty interior, $\lim_{\delta \to 1} \mathscr{E}(\delta) = \mathscr{M}$.

---

3. As half-spaces are not compact, it is not obvious that $k^*(\alpha, \lambda)$ is upper semicontinuous in α. Note that $k^*(\lambda) > -\infty$, and so $H^*(\lambda) \neq \varnothing$ for all λ: let v^* maximize $\lambda \cdot v$ over $v \in \mathscr{E}(\delta)$. Because $\lambda \cdot v \leq \lambda \cdot v^*$ for all $v \in \mathscr{E}(\delta)$, v^* is decomposable on $H(\lambda, \lambda \cdot v^*)$.

Remark **Interpretation of $k^*(\lambda)$** For a direction λ, we can interpret $\lambda \cdot u$ as the "average"
8.1.1 utility of all players in the direction λ. Given lemma 8.1.1, we can interpret $k^*(\lambda)$ as a bound on the "average" utility consistent with providing each player appropriate incentives (and section 9.1 shows that this bound is typically achieved for patient players). If $k^*(\lambda) = k^*(\alpha, \lambda)$ for some α orthogonally enforceable in the direction λ, then it is costless to provide the appropriate incentives in aggregate, in the sense that $\lambda \cdot x(y) = 0$ for all y (lemma 8.1.1). That is, it is possible to simultaneously provide incentives to all players by appropriately transferring continuations without lowering average utility below $\lambda \cdot u(\alpha)$. On the other hand, if there is no α orthogonally enforceable in the direction λ for which $k^*(\lambda) = k^*(\alpha, \lambda)$, then it is costly to provide the appropriate incentives in aggregate, in the sense that $\lambda \cdot x(y) < 0$ for some y. Moreover, if y has positive probability under α, this cost has an impact in that it is impossible to simultaneously provide incentives to all players without lowering average utility below $\lambda \cdot u(\alpha)$. ◆

Remark **Pure strategy PPE** Recall that $\mathscr{E}^p(\delta)$ is the set of pure strategy PPE payoffs
8.1.2 (remark 7.3.3). The analysis in this section applies to pure strategy PPE. In particular, we immediately have the following pure strategy version of lemma 8.1.2 and proposition 8.1.1.

Proposition *Let $k^{*p}(\lambda) \equiv \sup\{k^*(\alpha, \lambda) : \alpha_i \text{ is pure for } i = 1, \ldots, n\}$ and $H^{*p} \equiv H(\lambda, k^{*p}(\lambda))$.*
8.1.2 *Then,*

$$k^{*p}(-e_j) \leq -\underline{v}_i^p$$

and, for all δ,

$$\mathscr{E}^p(\delta) \subset \cap_\lambda H^{*p}(\lambda) \equiv \mathscr{M}^p.$$

◆

8.2 The Inefficiency of Strongly Symmetric Equilibria

The bound in proposition 8.1.1 immediately implies the pervasive inefficiency of strongly symmetric equilibria in symmetric games with full-support public monitoring.

Definition *Suppose there are no short-lived players. The stage game is* symmetric *if all*
8.2.1 *players have the same action space and same ex post payoff function ($A_i = A_j$ and $u_i^* = u_j^*$ for all i, j), and the distribution ρ over the public signal is unaffected by permutation of player indices.*

From (7.1.1), ex ante payoffs in a symmetric game are also appropriately symmetric. Recall from definition 7.1.2 that an equilibrium is *strongly symmetric* if after every public history, each player chooses the same action.

In a strongly symmetric PPE, $x_i(y) = x_j(y)$ for all $y \in Y$. For the remainder of this subsection, we restrict attention to such symmetric normalized continuations. Lemma 8.1.1 and proposition 8.1.1 hold under this restriction ($\mathscr{E}(\delta)$ is now the set of strongly symmetric PPE payoffs). Because the continuations are symmetric, without loss of generality, we can restrict attention to directions $\mathbf{1}$ and $-\mathbf{1}$, where $\mathbf{1}$ is an n-vector with 1 in each coordinate.

8.2 ■ Strongly Symmetric Equilibria

Let $\bar{u} \equiv \max_{\alpha_1} u_1(\alpha_1 \mathbf{1})$ denote the maximum symmetric payoff.

Proposition 8.2.1 *Suppose there are no short-lived players, the stage game is finite and symmetric, ρ has full support, and $\bar{u}$ is not a Nash equilibrium payoff of the ex ante stage game. The maximum strongly symmetric PPE payoff, $\bar{v}(\delta)$, is bounded away from $\bar{u}$, independently of δ (i.e., $\exists \varepsilon > 0$ such that $\forall \delta \in [0, 1)$, $\bar{v}(\delta) + \varepsilon < \bar{u}$).*

The inefficiency of strongly symmetric equilibria should not be surprising: If $\bar{u}$ is not a stage-game equilibrium payoff, for any $\bar{\alpha}$ satisfying $\bar{u} = u(\bar{\alpha})$, each player i needs an incentive to play $\bar{\alpha}_i$, which requires a lower continuation after some signal. But strong symmetry implies that every player must be punished at the same time, and full support that the punishment occurs on the equilibrium path, so inefficiency arises.

A little more formally, suppose $k^*(\bar{\alpha}_1 \mathbf{1}, \mathbf{1}) = n\bar{u}$ for some $\bar{\alpha}_1$. Inequality (8.1.4) and strong symmetry imply $x_i(y) \leq 0$ for all y. The definition of $\bar{u}$ then implies $E[x_i(y) \mid \bar{\alpha}_1 \mathbf{1}] = 0$. Full support of ρ finally implies $x_i(y) = 0$ for all y, implying $\bar{\alpha}_1$ is a best reply to $\bar{\alpha}_1 \mathbf{1}$, contradicting the hypothesis that $\bar{u}$ is not a Nash equilibrium payoff of the ex ante stage game.

The argument just given, though intuitive, relies on $\sup_{\alpha_1} k^*(\alpha_1 \mathbf{1}, \mathbf{1})$ being attained by some α_1. The proof must deal with the general possibility that the supremum is not attained (see note 3 on page 277).

Proof We now suppose the supremum in the definition of k^* is not achieved, and derive a contradiction to $k^*(\mathbf{1}) = n\bar{u}$. Because $k^*(\mathbf{1}) = \sup_{\alpha_1} k^*(\alpha_1 \mathbf{1}, \mathbf{1})$, for all $\ell \in \mathbb{N}$, there exists α_1^ℓ such that $k^*(\alpha_1^\ell \mathbf{1}, \mathbf{1}) \geq n\bar{u} - n/\ell$. We write $\alpha^\ell = \alpha_1^\ell \mathbf{1}$. Because A_1 is finite, by extracting a subsequence if necessary, without loss of generality we can assume $\{\alpha_1^\ell\}_\ell$ is a convergent sequence with limit α_1^∞. Let a_1 be an action minimizing $u_1(a_1', \alpha_{-1}^\infty)$ over $a_1' \in \text{supp}(\alpha_1^\infty)$. By the definition of $\bar{u}$, $u_1(a_1, \alpha_{-1}^\infty) \leq u_1(\alpha_1^\infty \mathbf{1}) \leq \bar{u}$. For sufficiently large ℓ, $\text{supp}(\alpha_1^\infty) \subset \text{supp}(\alpha_1^\ell)$, so we can also assume a_1 is in the support α_1^ℓ. Finally, by extracting a subsequence and renumbering if necessary, $u_1(a_1, \alpha_{-1}^\ell) \leq u_1(a_1, \alpha_{-1}^\infty) + 1/\ell$.

Because $\alpha_1^\ell(a_1) > 0$,

$$\bar{u} - \frac{1}{\ell} \leq \frac{1}{n} k^*(\alpha_1^\ell \mathbf{1}, \mathbf{1}) = u_1(a_1, \alpha_{-1}^\ell) + E[x_1^\ell(y) \mid (a_1, \alpha_{-1}^\ell)], \quad (8.2.1)$$

where x^ℓ is the normalized enforcing continuation in the direction $\mathbf{1}$ (so that $x_1^\ell(y) \leq 0$ for all y). Then, for all ℓ,

$$-\frac{2}{\ell} \leq E[x_1^\ell(y) \mid (a_1, \alpha_{-1}^\ell)] \leq 0. \quad (8.2.2)$$

For fixed ℓ, (8.1.3) is

$$u_1(a_1, \alpha_{-1}^\ell) + E[x_1^\ell(y) \mid (a_1, \alpha_{-1}^\ell)]$$
$$\geq u_1(a_1', \alpha_{-1}^\ell) + E[x_1^\ell(y) \mid (a_1', \alpha_{-1}^\ell)], \quad \forall a_1' \in A_1. \quad (8.2.3)$$

Inequality (8.2.2) implies $\lim_\ell E[x_1^\ell(y) \mid (a_1, \alpha_{-1}^\ell)] = 0$ and (using $x_1^\ell(y) \leq 0$), for all ℓ,

$$-\frac{2}{\ell} \leq x_1^\ell(y) \sum_{a_{-1}} \rho(y \mid (a_1, a_{-1})) \alpha_{-1}^\ell(a_{-1}) \leq 0.$$

As ρ has full support (by hypothesis), and $\underline{\rho} \equiv \min_{y,a}\{\rho(y \mid a)\} > 0$ (by the finiteness of Y and A), we thus have for all ℓ and all $a_1' \in A_1$,

$$-\frac{2}{\ell\underline{\rho}} \leq x_1^\ell(y) \sum_{a_{-1}} \rho(y \mid (a_1', a_{-1})) \alpha_{-1}^\ell(a_{-1}) \leq 0.$$

Hence, $\lim_\ell E[x_1^\ell(y) \mid (a_1', \alpha_{-1}^\ell)] = 0$ for all $a_1' \in A_1$, and so taking $\ell \to \infty$ in (8.2.1) and (8.2.3) yields

$$\bar{u} = u_1(a_1, \alpha_{-1}^\infty) \geq u_1(a_1', \alpha_{-1}^\infty), \quad \forall a_1' \in A_1.$$

Consequently, $\alpha_1^\infty \mathbf{1}$ is a Nash equilibrium of the stage game, and so $\bar{u}$ is a Nash equilibrium payoff, contradiction. ∎

8.3 Short-Lived Players

8.3.1 The Upper Bound on Payoffs

The techniques of this chapter provide an alternative proof of proposition 2.7.2.

Lemma 8.3.1 *For $j = 1, \ldots, n$,*

$$k^*(e_j) \leq \bar{v}_j \equiv \max_{\alpha \in \mathbf{B}} \min_{a_j \in \mathrm{supp}(\alpha_j)} u_j(a_j, \alpha_{-j}).$$

Proof For $\lambda = e_j$, constraint (8.1.4) becomes $x_j(y) \leq 0$ for all $y \in Y$. Constraints (8.1.2) and (8.1.3) imply, for all $a_j \in \mathrm{supp}(\alpha_j)$,

$$v_j = u_j(a_j, \alpha_{-j}) + E[x_j(y) \mid (a_j, \alpha_{-j})],$$

and so

$$v_j \leq u_j(a_j, \alpha_{-j}).$$

Hence,

$$k^*(e_j) = \sup_{\alpha \in \mathbf{B}} k^*(\alpha, e_j) = \sup_{\alpha \in \mathbf{B}} \max\{v_j : v \in \mathscr{B}(H(e_j, v_j); \delta, \alpha)\}$$
$$\leq \sup_{\alpha \in \mathbf{B}} \min_{a_j \in \mathrm{supp}(\alpha_j)} u_j(a_j, \alpha_{-j}) = \max_{\alpha \in \mathbf{B}} \min_{a_j \in \mathrm{supp}(\alpha_j)} u_j(a_j, \alpha_{-j}).$$
∎

Note that in the absence of short-lived players, $\mathbf{B} = \prod_i \Delta(A_i)$ and $k^*(e_j) = \max_{\alpha \in \prod_i \Delta(A_i)} \min_{a_j \in \mathrm{supp}(\alpha_j)} u_j(a_j, \alpha_{-j}) = \max_{\alpha \in \prod_i \Delta(A_i)} u_j(\alpha)$.

Lemma 8.3.1, with proposition 8.1.1 (implying that every PPE equilibrium payoff for a long-lived player j is bounded above by $k^*(e_j)$), implies proposition 2.7.2: Because $\mathscr{E}(\delta) \subset H^*(e_j)$, in every PPE no long-run player can earn a payoff greater than $\bar{v}_j$. Moreover, in the presence of imperfect public monitoring, the bound $\bar{v}_i$ may not be tight. It is straightforward to show for the example of section 7.6.2, that $k^*(e_1) = v^* < \bar{v}_1$, an instance of the next section's result.

It is worth emphasizing that with short-lived players, the case of one long-lived player is interesting and the analysis of this chapter (and the next) applies. In that case, of course, the only directions to be considered are e_1 and $-e_1$.

8.3.2 Binding Moral Hazard

From lemma 8.3.1, the presence of short-lived players constrains the set of long-lived players' payoffs that can be achieved in equilibrium. Moreover, as we saw in section 7.6.2, for the product-choice game with a short-lived player 2 and imperfect public monitoring, the imperfection in the monitoring necessarily further reduced the maximum equilibrium payoff that player 1 could receive. This phenomenon is not special to that example but arises more broadly and for reasons similar to those leading to the inefficiency of strongly symmetric equilibria (proposition 8.2.1).

We provide a simple sufficient condition implying that player i's maximum equilibrium payoff for all δ is bounded away from $\bar{v}_i$, the upper bound from lemma 8.3.1.

Definition 8.3.1 *Player i is subject to* binding moral hazard *if for all a_{-i}, for all a_i, a_i', $\mathrm{supp}\rho(y \mid (a_i, a_{-i})) = \mathrm{supp}\rho(y \mid (a_i', a_{-i}))$, and for any $\alpha \in \mathbf{B}$ satisfying $u_i(\alpha) = \bar{v}_i$, α_i is not a best reply to α_{-i}.*

The first condition implies there is no signal that unambiguously indicates that player i has deviated. The second condition then guarantees that player i's incentive constraint due to moral hazard is binding. Fudenberg and Levine (1994) say a game with short-lived players is a *moral-hazard mixing game* if it has a product structure (see section 9.5) and all long-lived players are subject to binding moral hazard.

Proposition 8.3.1 *Suppose a long-lived player i is subject to binding moral hazard, and A is finite. Then, $k^*(e_i) < \bar{v}_i$. Consequently, there exists $\kappa > 0$ such that for all δ and all $v \in \mathscr{E}(\delta)$, $v_i \leq \bar{v}_i - \kappa$.*

Proof Though the result is intuitive, the proof must deal with the general possibility that $\sup_\alpha k^*(\alpha, e_i)$ is not attained by any α (see note 3 on page 277). We assume here the supremum is attained; the proof for the more general case is a straightforward modification of the proof of proposition 8.2.1.

So, we suppose $k^*(\bar{\alpha}, e_i) = \bar{v}_i$ for some $\bar{\alpha} \in \mathbf{B}$. Let a_i minimize $u_i(a_i', \bar{\alpha}_{-i})$ over $a_i' \in \mathrm{supp}(\bar{\alpha}_i)$. Then, by the definition of $\bar{v}_i$, $\bar{v}_i \geq u_i(a_i, \bar{\alpha}_{-i})$. Because $\bar{\alpha}$ is enforced in the direction e_i by some x, $x_i(y) \leq 0$ for all y and

$$\bar{v}_i = k^*(\bar{\alpha}, e_i) = u_i(a_i, \bar{\alpha}_{-i}) + E[x_i(y) \mid (a_i, \bar{\alpha}_{-i})]$$
$$\leq u_i(a_i, \bar{\alpha}_{-i}) \leq \bar{v}_i.$$

Hence, $x_i(y) = 0$ for all $y \in \mathrm{supp}\rho(\cdot \mid (a_i, \bar{\alpha}_{-i}))$. Because $\mathrm{supp}\rho(\cdot \mid (a_i', \bar{\alpha}_{-i}))$ is independent of $a_i' \in A_i$, $\bar{\alpha}_i$ must be a best reply to $\bar{\alpha}_{-i}$, contradicting the hypothesis that player i is subject to binding moral hazard.
■

The inefficiency due to binding moral hazard arises for similar reasons to the inefficiency of strongly symmetric PPE (indeed, the proofs are very similar). In both cases, in the "target" action profile $\bar{\alpha}$, player i's action is not a best reply to $\bar{\alpha}_{-i}$, and achieving the target payoff requires orthogonal enforceability of the target profile in the relevant direction, a contradiction.

Remark 8.3.1 *Role of short-lived players* In the absence of short-lived players, $\mathbf{B} = \prod_i \Delta(A_i)$, and $\bar{v}_i = \max_{a \in A} u_i(a)$, so that player i can never be subject to binding moral

	E	S
E	2,2	$-c,b$
S	$b,-c$	0,0

Figure 8.4.1 The prisoners' dilemma from figure 2.5.1, where $b > 2$, $c > 0$, and $0 < b - c < 4$.

hazard. If player i *is* subject to binding moral hazard, the payoff $\bar{v}_i$ can only be achieved if player i is *not* playing a best reply to the behavior of the other players, and so incentives must be provided. But this involves a lower normalized continuation with positive probability, and so player i's payoffs in any PPE are bounded away from $\bar{v}_i$.

◆

8.4 The Prisoners' Dilemma

8.4.1 Bounds on Efficiency: Pure Actions

Our first application of these tools is to investigate the potential inefficiency of PPE in the repeated prisoners' dilemma. We return to the prisoners' dilemma from figure 2.5.1 (reproduced in figure 8.4.1) and the monitoring distribution (7.2.1),

$$\rho(\bar{y} \mid a) = \begin{cases} p, & \text{if } a = EE, \\ q, & \text{if } a = SE \text{ or } ES, \\ r, & \text{if } a = SS, \end{cases}$$

where $0 < q < p < 1$ and $0 < r < p$. An indication of the potential inefficiency of PPE can be obtained by considering $H^*(\lambda)$ for $\lambda_i \geq 0$. From proposition 8.1.1, $H^*(\lambda)$ is a bound on $\mathscr{E}(\delta)$, and in particular, for all $v \in \mathscr{E}(\delta)$, $\lambda \cdot v \leq k^*(\lambda)$. In particular, for $\hat{\lambda} = (1, 1)$, if $k^*(\hat{\lambda}) < 4$, then all PPE are bounded away from the maximum aggregate payoff. Moreover, if $k^*(\hat{\lambda}) < 2(b + c)/(2 + c)$, then all PPE are necessarily inefficient, and are bounded away from efficiency for all δ (see figure 8.4.2).

We first consider the pure action profile ES. Equations (8.1.2) and (8.1.3) reduce to

$$x_1(\bar{y}) \geq x_1(\underline{y}) + \frac{c}{q - r}$$

and

$$x_2(\bar{y}) \leq x_2(\underline{y}) + \frac{b - 2}{p - q}.$$

Setting $x_1(\bar{y}) = c/[2(q - r)]$ and $x_1(\underline{y}) = -c/[2(q - r)]$, with $x_2(y) = -\lambda_1 x_1(y)/\lambda_2$ for $y = \underline{y}, \bar{y}$, yields normalized continuations that orthogonally enforce ES. Hence, $k^*(ES, \lambda) = \lambda_2 b - \lambda_1 c$. Symmetrically, we also have $k^*(SE, \lambda) = \lambda_1 b - \lambda_2 c$.

8.4 ■ The Prisoners' Dilemma

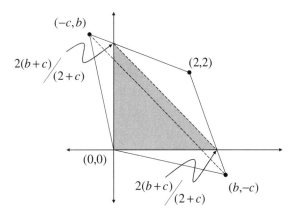

Figure 8.4.2 The set of feasible payoffs for the prisoners' dilemma of figure 8.4.1. If $v_1 + v_2 < 2(b+c)/(2+c)$, then v is in the shaded region and is necessarily inefficient.

For EE, (8.1.2) and (8.1.3) imply $x_i(\bar{y}) - x_i(\underline{y}) \geq (b-2)/(p-q)$, and consequently, the profile cannot be orthogonally enforced. Then,

$$\begin{aligned}
\lambda \cdot (u(EE) + E[x \mid EE]) &= (\lambda_1 + \lambda_2)2 + \lambda \cdot (px(\bar{y}) + (1-p)x(\underline{y})) \\
&= (\lambda_1 + \lambda_2)2 + \lambda \cdot (x(\bar{y}) - (1-p)(x(\bar{y}) - x(\underline{y}))) \\
&\leq (\lambda_1 + \lambda_2)2 - (1-p)\lambda \cdot (x(\bar{y}) - x(\underline{y})) \\
&\leq (\lambda_1 + \lambda_2)\left\{2 - \frac{(1-p)(b-2)}{(p-q)}\right\} \\
&\equiv (\lambda_1 + \lambda_2)\bar{v}.
\end{aligned} \qquad (8.4.1)$$

Setting $x_i(\bar{y}) = 0$ and $x_i(\underline{y}) = -(b-2)/(p-q)$ for $i = 1, 2$ enforces EE and achieves the upper bound just calculated. Hence, $k^*(EE, \lambda) = (\lambda_1 + \lambda_2)\bar{v}$.

We are now in a position to study the inefficiency of pure-strategy PPE. From proposition 8.1.2, for all pure strategy PPE payoffs v,

$$\lambda \cdot v \leq k^{*p}(\lambda) = \max\{k^*(EE, \lambda), k^*(ES, \lambda), k^*(SE, \lambda)\}$$

(we omit the action profile SS, because $k^*(SS, \lambda) = 0$ and this is clearly less than one of the other three expressions for $\lambda_i > 0$).

Because $k^*(ES, \lambda) \geq k^*(SE, \lambda)$ if and only if $\lambda_2 \geq \lambda_1$, we restrict our analysis to $\lambda_2 \geq \lambda_1$ and the comparison of $k^*(ES, \lambda)$ with $k^*(EE, \lambda)$ (with symmetry covering the other case). If

$$\frac{b-c}{2} \geq \bar{v}, \qquad (8.4.2)$$

then for all $\lambda_2 \geq \lambda_1$, $k^*(ES, \lambda) \geq k^*(EE, \lambda)$, and all pure strategy PPE payoffs must fall into that part of the shaded region below the line connecting $(-c, b)$ and $(b, -c)$, in figure 8.4.2.

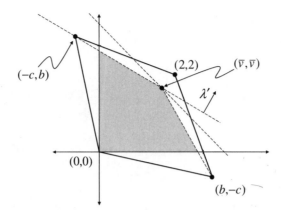

Figure 8.4.3 A bounding set for $\mathscr{E}^p(\delta)$ when (8.4.2) does not hold. Every pure strategy PPE payoff vector must fall into the shaded region.

On the other hand, if (8.4.2) does not hold, then $k^*(ES, \hat{\lambda}) < k^*(EE, \hat{\lambda})$ for some $\hat{\lambda}$. Because $k^*(ES, \lambda) > k^*(EE, \lambda)$ for λ_1 close to 0, there is a direction λ' such that $k^*(ES, \lambda') = k^*(EE, \lambda')$, and so all pure strategy PPE must fall into the shaded region in figure 8.4.3.

Clearly, independently of (8.4.2), pure strategy PPE payoffs are bounded away from the Pareto frontier and so are necessarily inefficient. The reason for the failure is intuitive: Note first that efficiency requires EE be played. With only two signals and a symmetric monitoring structure, it is impossible to distinguish between a deviation to S by player 1 from that of player 2, leading to a failure of orthogonal enforceability. Hence when providing incentives when the target action profile is EE, the continuations are necessarily inefficient.

The pure strategy folk theorems for repeated games with public monitoring exclude this example (as they must) by assuming there are sufficient signals to distinguish between deviations of different players. We illustrate this in section 8.4.4.

8.4.2 Bounds on Efficiency: Mixed Actions

We now consider mixed-action profiles and, for tractability, focus on the direction $\hat{\lambda} = (1, 1)$. Just like EE, many (but not necessarily all) mixed profiles fail to be orthogonally enforceable. The critical feature is whether action choices by 1 can be distinguished from those by 2. Given a mixed action α_j for player j, player i's choice of S implies

$$\rho(\bar{y} \mid S\alpha_j) = \alpha_j q + (1 - \alpha_j)r,$$

and i's choice of E implies

$$\rho(\bar{y} \mid E\alpha_j) = \alpha_j p + (1 - \alpha_j)q.$$

Thus, if $\rho(\bar{y} \mid S\alpha_j) < \rho(\bar{y} \mid E\alpha_j)$, then $\underline{y}$ is a signal that player i had played S. Suppose that for some α_i, we also have $\rho(\bar{y} \mid S\alpha_i) > \rho(\bar{y} \mid E\alpha_i)$ (this is only possible if $q < r$). Then, $\underline{y}$ is *not* also a signal that j had played S, and so it should be possible at $\alpha = (\alpha_i, \alpha_j)$ to separately provide incentives for i and j (similar to the discussion for ES).

8.4 ■ The Prisoners' Dilemma

Let $\mathscr{A}$ be the set of mixed-action profiles α that have full support and under which the signals distinguish between the players, that is,

$$\rho(\bar{y} \mid S\alpha_i) < \rho(\bar{y} \mid E\alpha_i) \quad \text{and} \quad \rho(\bar{y} \mid S\alpha_j) > \rho(\bar{y} \mid E\alpha_j), \quad i \neq j. \quad (8.4.3)$$

The action profiles in $\mathscr{A}$ are asymmetric: The asymmetry in the interpretation of signals, captured by (8.4.3), requires asymmetry in the underlying behavior. The set $\mathscr{A}$ is nonempty if and only if $q < r$.

We now argue that some (but not all) $\alpha \in \mathscr{A}$ can be orthogonally enforced in the direction $\hat{\lambda}$. Because α has full support, (8.1.2) and (8.1.3) imply the equations $i, j = 1, 2, j \neq i$,

$$\begin{aligned}\alpha_j(b-2) + (1-\alpha_j)c &= [\rho(\bar{y} \mid E\alpha_j) - \rho(\bar{y} \mid S\alpha_j)](x_i(\bar{y}) - x_i(\underline{y})) \\ &= [\alpha_j(p-q) + (1-\alpha_j)(q-r)](x_i(\bar{y}) - x_i(\underline{y})). \quad (8.4.4)\end{aligned}$$

The left side is necessarily positive, so the right side must also be positive. Orthogonality implies that $x_i(y)$ and $x_j(y)$ are either both 0 or of opposite sign. When (8.4.3) holds, it is also possible to choose the normalized continuations so that the sign of $x_i(\bar{y}) - x_i(\underline{y})$ is the reverse of $[\rho(\bar{y} \mid E\alpha_j) - \rho(\bar{y} \mid S\alpha_j)]$ for both players.

Orthogonal enforceability in the direction $\hat{\lambda}$ requires more than a sign reversal, because it requires $x_2(y) = -x_1(y)$, or

$$\frac{\alpha_1(b-2) + (1-\alpha_1)c}{\alpha_1(p-q) + (1-\alpha_1)(q-r)} = -\frac{\alpha_2(b-2) + (1-\alpha_2)c}{\alpha_2(p-q) + (1-\alpha_2)(q-r)}. \quad (8.4.5)$$

Let $g(\alpha_i)$ be the reciprocal of the term on the left side, that is,

$$g(\alpha_i) = \frac{\alpha_i(p-q) + (1-\alpha_i)(q-r)}{\alpha_i(b-2) + (1-\alpha_i)c}. \quad (8.4.6)$$

From (8.4.3), $q < r$, and so $g(0) < 0$ and $g(1) > 0$. Therefore there is some value of α_i, say $\hat{\alpha}_i$, for which $g(\hat{\alpha}_i) = 0$. Because the denominator is always positive, and the numerator is affine in α_i, there is a function $h(\alpha_i)$, $h(\hat{\alpha}_i) = \hat{\alpha}_i$, such that for a range of values of α_i containing $\hat{\alpha}_i$, $g(h(\alpha_i)) = -g(\alpha_i)$. Hence, (8.4.5) holds for all $\alpha \in \mathscr{A}^* \equiv \{\alpha \in \mathscr{A} : h(\alpha_i) = \alpha_j\}$, and so for such α,

$$k^*(\alpha, \hat{\lambda}) = u_1(\alpha) + u_2(\alpha).$$

Therefore if $\mathscr{A} \neq \varnothing$, $\mathscr{A}^* \neq \varnothing$ and

$$k^*(\hat{\lambda}) \geq \sup_{\alpha \in \mathscr{A}^*} u_1(\alpha) + u_2(\alpha).$$

Section 8.4.5 illustrates the use of this lower bound. On the other hand, we trivially have the upper bound,

$$k^*(\alpha, \hat{\lambda}) \leq u_1(\alpha) + u_2(\alpha).$$

For later reference, we define

$$\kappa^* = \sup_{\alpha \in \mathscr{A}} u_1(\alpha) + u_2(\alpha), \quad (8.4.7)$$

where (as usual) $\kappa^* = -\infty$ if $\mathscr{A}$ is empty.

Example Suppose $b=3$ and $c=2$, and $p-q=r-q>0$. It is easy to calculate that
8.4.1 $\mathscr{A} = \{(\alpha_1, \alpha_2) : (\frac{1}{2}-\alpha_1)(\frac{1}{2}-\alpha_2) < 0\}, \hat{\alpha}_i = \frac{1}{2}$,

$$h(\alpha_i) = \frac{4-5\alpha_i}{5-4\alpha_i},$$

and $\mathscr{A}^* = \{\alpha \in \mathscr{A} : \alpha_j = h(\alpha_i)\}$. While $(1/2, 1/2) \notin \mathscr{A}$, $\lim_{\alpha_i \to 1/2} h(\alpha_i) = 1/2$, and we have

$$\sup_{\alpha \in \mathscr{A}^*} u_1(\alpha) + u_2(\alpha) = u_1(\tfrac{1}{2}, \tfrac{1}{2}) + u_2(\tfrac{1}{2}, \tfrac{1}{2}) = \tfrac{3}{2}.$$

Note that if (8.4.2) holds, then the lower bound on $k^*(\hat{\lambda})$ from asymmetric mixing in $\mathscr{A}^*$ is larger than the bound from any pure action profile.

•

Finally, we consider mixed-action profiles not in $\mathscr{A}$. In this case, signals do not distinguish between players, and so intuitively it is impossible to separately provide incentives. More formally, it is not possible to orthogonally enforce such profiles (this is immediate from the discussion following (8.4.4)). Letting $\Delta x_i \equiv x_i(\bar{y}) - x_i(\underline{y})$, where $j \neq i$, we have

$$\sum_i \{u_i(\alpha) + E[x_i \mid \alpha]\} = \sum_i \{u_i(\alpha) + \alpha_i[\alpha_j(p-q) + (1-\alpha_j)(q-r)]\Delta x_i$$
$$+ (\alpha_j q + (1-\alpha_j)r)\Delta x_i + x_i(\underline{y})\}.$$

From (8.4.4), we thus have for full support α,

$$k^*(\alpha, \hat{\lambda}) = \sum_i \{u_i(\alpha) + \alpha_i[\alpha_j(b-2) + (1-\alpha_j)c]$$
$$+ (\alpha_j q + (1-\alpha_j)r)\Delta x_i + x_i(\underline{y})\}$$
$$= \sum_i \{\alpha_j b + (\alpha_j q + (1-\alpha_j)r)\Delta x_i + x_i(\underline{y})\}$$
$$\leq \sum_i \{\alpha_j b - (1-\alpha_j)q - (1-\alpha_j)r)\Delta x_i\},$$

where we used the inequality $\sum_i x_i(\underline{y}) = \sum_i \{x_i(\bar{y}) - \Delta x_i\} \leq -\sum_i \Delta x_i$. Because $x_i(\bar{y}) = 0$ is a feasible choice for the normalized continuations and (8.4.4) then determines $x_i(\underline{y})$, the inequality is an equality, and we have

$$k^*(\alpha, \hat{\lambda}) = \sum_j \left\{ \alpha_j b - \frac{(1-\alpha_j q - (1-\alpha_j)r)(\alpha_j(b-2) + (1-\alpha_j)c)}{\alpha_j(p-q) + (1-\alpha_j)(q-r)} \right\}.$$

Observe that the expression is separable in α_1 and α_2, so that maximizing $k^*(\alpha, \hat{\lambda})$ over $\alpha \notin \mathscr{A}$ is equivalent to maximizing the expression,

$$\alpha_j b - \frac{(1-\alpha_j q - (1-\alpha_j)r)(\alpha_j(b-2) + (1-\alpha_j)c)}{\alpha_j(p-q) + (1-\alpha_j)(q-r)},$$

which yields the maximum strongly symmetric payoff. For the case $b=3$ and $c=1$, this expression reduces to the function $\bar{\gamma}$ given in (7.7.7). Let v^{SSE} denote the maximum strongly symmetric payoff,

$$v^{\text{SSE}} = \sup_{\alpha_j} \alpha_j b - \frac{(1 - \alpha_j q - (1 - \alpha_j) r)(\alpha_j (b - 2) + (1 - \alpha_j) c)}{\alpha_j (p - q) + (1 - \alpha_j)(q - r)}.$$

Evaluating the expression at $\alpha_j = 1$ yields $\bar{v}$ of (8.4.1), and so $k^*(EE, \hat{\lambda}) = 2\bar{v} \leq 2v^{\text{SSE}}$. Section 7.7.2 discusses an example where $v^{\text{SSE}} > \bar{v}$.

Summarizing the foregoing discussion, if v is a PPE payoff, then[4]

$$v_1 + v_2 \leq \max \{b - c, \kappa^*, 2v^{\text{SSE}}\}, \tag{8.4.8}$$

where κ^* is defined in (8.4.7).

8.4.3 A Characterization with Two Signals

We return to the prisoners' dilemma of figure 7.2.1 (i.e., $b = 3$ and $c = 1$) with imperfect public monitoring. We now characterize $\mathscr{M}$ for the special case $p - q = q - r$ of the monitoring distribution (7.2.1). As we saw in section 8.4.2, mixed actions can play a role in the determination of $\mathscr{M}$. For the special case under study, however, the myopic incentives to shirk are independent of the actions of the other player. Moreover, because $p - q = q - r$, the impact on the distribution of signals is also independent of the actions of the other player. In terms of the analysis of the previous section, the expression determining v^{SSE} is affine in α_i, so the best strongly symmetric equilibrium payoff is achieved in pure actions. In addition, $\mathscr{A}$ is empty because $q > r$. Consequently, we can restrict attention to pure-action profiles.

Each player's minmax value is 0, so lemma 8.1.2 implies that $k^*(-e_i) \leq 0$. Moreover, because (S, S) is a Nash equilibrium of the stage game, $(0, 0)$ satisfies the constraint in (8.1.1) for $\alpha = SS$ and λ. Consequently, $k^*(-e_i) = 0$, and so $\mathscr{M} \subset \mathbb{R}_+^2$. By a similar argument, the maximal half-spaces in directions with $\lambda_1, \lambda_2 \leq 0$ impose no further restriction on $\mathscr{M}$.

We now consider directions λ with $\lambda_2 > 0 > \lambda_1$. If $-\lambda_1 \leq \lambda_2$, the action profile ES can be orthogonally enforced and so, by the third claim in lemma 8.1.1, $k^*(\lambda) = \lambda \cdot (-1, 3) = 3\lambda_2 - \lambda_1$. To obtain the normalized continuations satisfying (8.1.2)–(8.1.4), we first solve (8.1.2) and (8.1.3) for player 1 as an equality with $E[x_1 \mid ES] = 0$, giving

$$x_1(\bar{y}) = \frac{(1 - q)}{(q - r)}, \quad \text{and} \quad x_1(\underline{y}) = \frac{-q}{(q - r)}.$$

We then set

$$x_2(\bar{y}) = \frac{-\lambda_1}{\lambda_2} x_1(\bar{y}) \quad \text{and} \quad x_2(\underline{y}) = \frac{-\lambda_1}{\lambda_2} x_1(\underline{y}).$$

---

4. We have not discussed asymmetric profiles in which only one player is randomizing. Observe, however, κ^* also bounds any such profiles in the closure of $\mathscr{A}$. For any such profiles not in the closure of $\mathscr{A}$, $2v^{\text{SSE}}$ is still the relevant bound.

By construction, $\lambda \cdot x(y) = 0$. Finally,

$$(p-q)[x_2(\bar{y}) - x_2(\underline{y})] = (p-q)\left(\frac{-\lambda_1}{\lambda_2}\right)[x_1(\bar{y}) - x_1(\underline{y})]$$
$$\leq (q-r)[x_1(\bar{y}) - x_1(\underline{y})] = 1,$$

(using $p - q = q - r$ and $-\lambda_1 \leq \lambda_2$), and so (8.1.3) holds for player 2.

On the other hand, if $-\lambda_1 > \lambda_2$, the action profile ES cannot be orthogonally enforced. In this case, $k^*(ES, \lambda)$ is the maximum value of $\lambda_1 v_1 + \lambda_2 v_2$ for v with the property that there exists x satisfying

$$v_1 = -1 + qx_1(\bar{y}) + (1-q)x_1(\underline{y}) \geq 0 + rx_1(\bar{y}) + (1-r)x_1(\underline{y}),$$
$$v_2 = 3 + qx_2(\bar{y}) + (1-q)x_2(\underline{y}) \geq 2 + px_2(\bar{y}) + (1-p)x_2(\underline{y}),$$

and

$$0 \geq \lambda_1 x_1(y) + \lambda_2 x_2(y) \quad \text{for } y = \bar{y}, \underline{y}.$$

The first two constraints imply, because $\lambda_1 < 0$ and using $p - q = q - r$,

$$\lambda_1 x_1(\bar{y}) + \lambda_2 x_2(\bar{y}) \leq \lambda_1 x_1(\underline{y}) + \lambda_2 x_2(\underline{y}) + (\lambda_1 + \lambda_2)/(q-r),$$

so the last constraint implies

$$k^*(ES, \lambda) \leq -\lambda_1 + 3\lambda_2 + q(\lambda_1 + \lambda_2)/(q-r). \tag{8.4.9}$$

The continuations $x_1(\underline{y}) = x_2(\underline{y}) = 0$ and $x_1(\bar{y}) = x_2(\bar{y}) = 1/(q-r)$ satisfy the constraints, so the inequality in (8.4.9) is an equality. Because $q/(q-r) > 1$, $k^*(ES, \lambda) < 0$ for large $|\lambda_1|$, so $k^*(ES, \lambda) < k^*(SS, \lambda) = 0$. Consequently,[5]

$$k^*(\lambda) = \max\{k^*(ES, \lambda), k^*(SS, \lambda), k^*(EE, \lambda), k^*(SE, \lambda)\}$$
$$= \begin{cases} 3\lambda_2 - \lambda_1, & \text{if } 0 < -\lambda_1 \leq \lambda_2, \\ 3\lambda_2 - \lambda_1 + q(\lambda_1 + \lambda_2)/(q-r), & \text{if } 0 < \lambda_2 < -\lambda_1 \text{ and} \\ & \quad 3\lambda_2 - \lambda_1 + q(\lambda_1 + \lambda_2)/(q-r) > 0, \\ 0, & \text{otherwise.} \end{cases}$$

Hence, the upper left boundary of $\mathscr{M}$ is given by the line through $(0, 0)$ perpendicular to the normal λ satisfying $3\lambda_2 - \lambda_1 + q(\lambda_1 + \lambda_2)/(q-r) = 0$, that is, given by $x_2 = [(4q - 3r)/r]x_1 = [(2p-r)/r]x_1$ (the last equality uses $p - q = q - r$).

Symmetrically, for $0 < -\lambda_2 \leq \lambda_1$ we also have

$$k^*(\lambda) = \begin{cases} 3\lambda_1 - \lambda_2, & \text{if } 0 < -\lambda_2 \leq \lambda_1, \\ 3\lambda_1 - \lambda_2 + q(\lambda_2 + \lambda_1)/(q-r), & \text{if } 0 < \lambda_1 < -\lambda_2 \text{ and} \\ & \quad 3\lambda_1 - \lambda_2 + q(\lambda_1 + \lambda_2)/(q-r) > 0, \\ 0, & \text{otherwise.} \end{cases}$$

---

5. It is immediate that $k^*(EE, \lambda) \leq 2(\lambda_1 + \lambda_2) < 0$ and $k^*(SE, \lambda) \leq 3\lambda_1 - \lambda_2 < 0$.

8.4 ■ The Prisoners' Dilemma

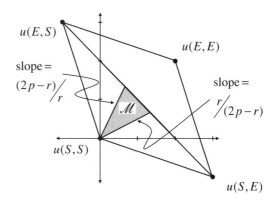

Figure 8.4.4 The bounding set $\mathscr{M}$ when $2p - q \leq 1$.

It remains to solve for $k^*(\lambda)$ when $\lambda_1, \lambda_2 \geq 0$. We apply the analysis from section 8.4.1. Condition (8.4.2) in the current setting reduces to $2p - q \leq 1$, and the maximal half-spaces can then always be obtained by using either ES or SE, that is, $k^*(\lambda) = \max\{3\lambda_2 - \lambda_1, 3\lambda_1 - \lambda_2\}$ for all λ. The bounding set $\mathscr{M}$ is illustrated in figure 8.4.4.[6]

On the other hand, if $2p - q > 1$, then some maximal half-spaces (in particular, those with λ_1 and λ_2 close) require EE. For example, $k^*(\lambda) = 2[2 - (1 - p)/(p - q)]$, when $\lambda_1 = \lambda_2 = 1$. This possibility is illustrated in figure 8.4.5. Observe that the symmetric upper bound we obtained here, $2 - (1 - p)/(p - q)$ on a player's payoff, agrees with our earlier calculation yielding (7.7.1) in section 7.7.1, as it should. Moreover, for p and q satisfying $2p - q < 1$, the best *strongly symmetric* PPE payoff is strictly smaller than the best symmetric PPE payoff (which is 1) for δ large (applying theorem 9.1.1).

8.4.4 Efficiency with Three Signals

We now consider a modification of the monitoring structure that does not preclude efficiency and (as an implication of theorem 9.1.1), yields all feasible and strictly individually rational payoffs as PPE payoffs for sufficiently large δ.

The new monitoring structure has three signals $\underline{y}$, y', and $\bar{y}$, whose distribution is given by

$$\rho(\bar{y} \mid a) = \begin{cases} p, & \text{if } a = EE, \\ q, & \text{if } a = SE \text{ or } ES, \\ r, & \text{if } a = SS, \end{cases} \quad \rho(y' \mid a) = \begin{cases} p', & \text{if } a = EE, \\ q'_1, & \text{if } a = SE, \\ q'_2, & \text{if } a = ES, \\ r', & \text{if } a = SS, \end{cases}$$

with $\underline{y}$ receiving the complementary probability.

We proceed as for the previous monitoring distribution to calculate $\mathscr{M}$. As before, for λ with $\lambda_1, \lambda_2 \leq 0$, we obtain the restriction $\mathscr{M} \subset \mathbb{R}_+^2$. Consider now directions λ

---

[6]. Proposition 9.1.1 implies that any $v \in \mathscr{M}$ is a PPE payoff for sufficiently high δ. It is worth observing that $\max_{v \in \mathscr{M}} v_2$ exceeds $2 - r/(q - r)$, the maximum payoff player 2 could receive in any PPE in which only ES and SE are played (a similar comment applies to player 1, of course). This is most easily seen by considering the normal λ determining the upper left boundary of $\mathscr{M}$. For such a direction, it is immediate that $\max_{v \in \mathscr{W}} \lambda \cdot v = 3\lambda_2 - \lambda_1 + q(\lambda_1 - \lambda_2)/(q - r) < 0$, where $\mathscr{W}$ is the set defined in remark 7.7.2. This is an illustration of the general phenomenon that the option of using inefficient actions, such as SS, relaxes incentive constraints.

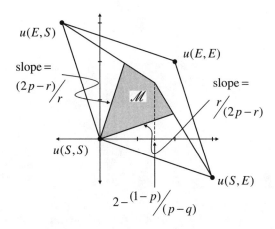

Figure 8.4.5 The bounding set $\mathscr{M}$ when $2p - q > 1$.

with $\lambda_2 > 0 > \lambda_1$. If $-\lambda_1 \leq \lambda_2$, ES can be orthogonally enforced by setting $x_i(y') = x_i(y)$ for $i = 1, 2$, and arguing as in the previous case. Hence, for such a direction λ, $k^*(\lambda) = 3\lambda_2 - \lambda_1$.

Consider now directions λ with $-\lambda_1 \geq \lambda_2$ (recall that it was not possible to orthogonally enforce ES in these directions under the public monitoring described by (7.2.1)). As above, $k^*(\lambda) = 3\lambda_2 - \lambda_1$ if ES can be orthogonally enforced. The orthogonal enforceability of ES requires, from equations (8.1.2) and (8.1.3),

$$-1 = -1 + qx_1(\bar{y}) + q_2'x_1(y') + (1 - q - q_2')x_1(\underline{y}) \qquad (8.4.10)$$
$$\geq rx_1(\bar{y}) + r'x_1(y') + (1 - r - r')x_1(\underline{y})$$

and

$$3 = 3 + qx_2(\bar{y}) + q_2'x_2(y') + (1 - q - q_2')x_2(\underline{y}) \qquad (8.4.11)$$
$$\geq 2 + px_2(\bar{y}) + p'x_2(y') + (1 - p - p')x_2(\underline{y}). \qquad (8.4.12)$$

Using $\lambda \cdot x = 0$ to eliminate x_2 from (8.4.11) and (8.4.12), and writing the inequalities as equalities ((8.4.11) is an implication of (8.4.10) and $\lambda \cdot x = 0$), we obtain the matrix equation

$$\begin{bmatrix} q & q_2' & 1-q-q_2' \\ r & r' & 1-r-r' \\ p & p' & 1-p-p' \end{bmatrix} \begin{bmatrix} x_1(\bar{y}) \\ x_1(y') \\ x_1(\underline{y}) \end{bmatrix} = \begin{bmatrix} 0 \\ -1 \\ -\lambda_2/\lambda_1 \end{bmatrix}.$$

Hence if the matrix of stacked distributions is invertible, then ES is orthogonally enforceable in the direction λ, so $k^*(EE, \lambda) = 3\lambda_2 - \lambda_1$. A similar argument shows that $k^*(SE, \lambda) = 3\lambda_1 - \lambda_2$.

For $\lambda_1, \lambda_2 \geq 0$, both ES and SE can be orthogonally enforced by setting $x_i(y') = x_i(y)$ for $i = 1, 2$ and arguing as in the previous case. Hence, for such λ, $k^*(ES, \lambda) = 3\lambda_2 - \lambda_1$ and $k^*(SE, \lambda) = 3\lambda_1 - \lambda_2$.

Finally, we show that invertibility of another matrix of stacked distributions is sufficient for EE to be orthogonally enforced in any direction λ with $\lambda_1, \lambda_2 \geq 0$.

8.4 ■ The Prisoners' Dilemma

The orthogonal enforceability of EE requires, from equations (8.1.2) and (8.1.3), for $i = 1, 2$,

$$2 = 2 + px_i(\bar{y}) + p'x_i(y') + (1 - p - p')x_i(\underline{y})$$
$$\geq 3 + qx_i(\bar{y}) + q'_i x_i(y') + (1 - q - q'_i)x_i(\underline{y}). \quad (8.4.13)$$

Using $\lambda \cdot x = 0$ as above to eliminate x_2 from (8.4.13) and writing the inequalities for $i = 1, 2$ as equalities, we obtain the matrix equation

$$\begin{bmatrix} p & p' & 1-p-p' \\ q & q'_1 & 1-q-q'_1 \\ q & q'_2 & 1-q-q'_2 \end{bmatrix} \begin{bmatrix} x_1(\bar{y}) \\ x_1(y') \\ x_1(\underline{y}) \end{bmatrix} = \begin{bmatrix} 0 \\ -1 \\ \lambda_2/\lambda_1 \end{bmatrix}.$$

Hence if the matrix of stacked distributions is invertible, then EE is orthogonally enforceable in the direction λ, and so $k^*(EE, \lambda) = (\lambda_1 + \lambda_2)2$.

Thus,

$$k^*(\lambda) = \begin{cases} 3\lambda_2 - \lambda_1, & \text{if } 3\lambda_1 \leq \lambda_2, \\ (\lambda_1 + \lambda_2)2, & \text{if } \lambda_2/3 < \lambda_1 < 3\lambda_2, \\ 3\lambda_1 - \lambda_2, & \text{if } \lambda_1 \geq 3\lambda_2, \end{cases}$$

and so $\mathscr{M} = \overline{\mathscr{F}^*}$ (note that the slope of the line connecting $u(E, S)$ and $u(E, E)$ is $-1/3$).

The requirement that the matrix of stacked distributions be invertible is an example of the property of *pairwise full rank*, which we explore in section 9.2.

8.4.5 Efficient Asymmetry

This section presents an example, based on Kandori and Obara (2003), in which players exploit the asymmetric mixed actions in $\mathscr{A}$ (described in section 8.4.2) to attach asymmetric continuations to the public signals. Paradoxically, if the monitoring technology is sufficiently noisy, the set of PPE payoffs for patient players is very close to $\mathscr{F}^*$.[7]

We return to the prisoners' dilemma of figure 8.4.1 with public monitoring distribution (7.2.1). Fix a closed set $\mathscr{C} \subset \text{int}\mathscr{F}^*$, and $1 > r > q \in (0, 1)$. We will show that for sufficiently small $\varepsilon > 0$, if $q < p < q + \varepsilon$, then $\mathscr{C} \subset \mathscr{M}$. From proposition 9.1.2, the set of PPE payoffs then contains $\mathscr{C}$ for sufficiently large δ. This is *not* a folk theorem result, because we first fix the set $\mathscr{C}$, then choose a monitoring distribution, and then allow δ to approach 1.

We first dispense with some preliminaries. Suppose p, r satisfy $p, r > q$ and

$$\frac{r-q}{p-q} > \frac{c}{b-2} \quad (8.4.14)$$

(for fixed q and r, (8.4.14) is always satisfied for p sufficiently close to q). It is straightforward to check that SS is orthogonally enforced in all directions, SE is orthogonally

---

7. This result is not inconsistent with section 7.4, because the increased noisiness does not come in the form of a less informative signal in the sense of Blackwell.

enforced in all directions λ such that $\lambda_1 > 0$ and $\lambda_1 \geq \lambda_2$, and ES is orthogonally enforced in all directions λ such that $\lambda_2 > 0$ and $\lambda_2 \geq \lambda_1$. Consequently, $\mathcal{M}$ contains $\mathcal{D} \equiv \text{co}\{(0,0), (b,-c), (-c,b)\}$.

Let λ^1 be the direction orthogonal to the line connecting $(2,2)$ and $(b,-c)$, and λ^2 the direction orthogonal to the line connecting $(2,2)$ and $(-c,b)$. Let $\Lambda^* \equiv \{\lambda : \lambda_i > 0, \lambda_2^1/\lambda_1^1 \leq \lambda_2/\lambda_1 \leq \lambda_2^2/\lambda_1^2\}$ be the set (cone) of directions "between" λ^1 and λ^2. It is immediate that for any direction not in Λ^*, $k^*(\lambda) = k^*(a, \lambda)$ for some $a \in \{SS, SE, ES\}$. Consequently, if $\mathcal{M}$ is larger than $\mathcal{D}$, it is because $k^*(\lambda) > \max_{a \in \{SS, SE, ES\}} k^*(a, \lambda)$ for $\lambda \in \Lambda^*$.

It remains to argue that for p sufficiently close to q, $\mathcal{M}$ contains any closed set in the interior of the convex hull of $(2,2)$ and $\mathcal{D}$. This is an implication of the following lemma.

Lemma 8.4.1 *There exists $\varepsilon > 0$ such that if $q < p < q + \varepsilon$, then for all directions $\lambda \in \Lambda^*$ and all $\eta > 0$, there is an asymmetric mixed profile α η-close to EE that can be orthogonally enforced in the direction λ.*

Proof Fix a direction $\lambda \in \Lambda^*$. We begin by observing that, as for the direction $\hat{\lambda}$ in section 8.4.2, orthogonal enforceability in the direction λ requires, from (8.4.4) and (8.4.6),

$$\lambda_2 g(\alpha_2) = -\lambda_1 g(\alpha_1). \tag{8.4.15}$$

Just as in section 8.4.2, this equation implies a relationship between α_1 and α_2 for α orthogonally enforceable in the direction λ. In particular, if for $\alpha_1 = 1$, the implied value of α_2 from (8.4.15) is a probability, then $(1, \alpha_2)$ is orthogonally enforced in the direction λ. Setting $\alpha_1 = 1$ in (8.4.15) and solving for α_2 gives

$$\alpha_2 = \frac{\lambda_2(r-q)(b-2) - \lambda_1(p-q)c}{\lambda_2(p-q-(q-r))(b-2) + \lambda_1(p-q)(b-2-c)}.$$

Because λ_2/λ_1 is bounded for $\lambda \in \Lambda^*$, there exists $\varepsilon > 0$ such that $1 - \eta < \alpha_2 < 1$ if $p < q + \varepsilon$, which is what we were required to prove. ∎

9 The Folk Theorem with Imperfect Public Monitoring

The previous chapters showed that intertemporal incentives can be constructed even when there is imperfect monitoring. In particular, nonmyopically optimal behavior can be supported as equilibrium behavior by appropriately specifying continuation play, so that a deviation by a player adversely changes the probability distribution over continuation values.

This chapter shows that the set $\mathscr{M}$, from proposition 8.1.1, completely describes the set of limit PPE payoffs, and explores conditions under which $\mathscr{M}$ is appropriately "large." As we saw in section 8.4, the nature of the dependence of the signal distribution on actions determines the size of $\mathscr{M}$ and so plays a crucial role in obtaining nearly efficient payoffs as PPE payoffs.

This analysis leads to several folk theorems. Folk theorems can (and have) given rise to a variety of interpretations. We refer to section 3.2.1 for a discussion.

9.1 Characterizing the Limit Set of PPE Payoffs

In section 8.1, we introduced the notion of decomposability with respect to half-spaces and defined the maximal half-space in the direction λ, $H^*(\lambda)$. Recall that the maximal half-space $H^*(\lambda)$ is the largest $H(\lambda, k)$ half-space with the property that a boundary point of the half-space can be decomposed with respect to that half-space. In section 8.1, we showed that the set of PPE payoffs is contained in $\mathscr{M}$, the intersection of all the maximal half-spaces, $\cap_\lambda H^*(\lambda)$. In this section, we show that for large δ, the set of PPE payoffs is typically well approximated by $\mathscr{M}$.

Proposition 9.1.1 *Suppose the stage game satisfies assumption 7.1.1. If $\mathscr{M} \subset \mathbb{R}^n$ has nonempty interior (and so dimension n), then for all $v \in \text{int} \mathscr{M}$, there exists $\underline{\delta} < 1$ such that for all $\delta \in (\underline{\delta}, 1)$, $v \in \mathscr{E}(\delta)$, that is, there is a PPE with value v. Hence,*[1]

$$\lim_{\delta \to 1} \mathscr{E}(\delta) = \mathscr{M}.$$

This proposition is an implication of proposition 9.1.2, stated independently for ease of reference.

Definition 9.1.1 *A subset $\mathscr{W} \subset \mathbb{R}^n$ is* smooth *if it is closed and has nonempty interior with respect to $\mathbb{R}^n$ and the boundary of $\mathscr{W}$ is a $\mathscr{C}^2$-submanifold of $\mathbb{R}^n$.*

---

1. More precisely, the Hausdorff distance between $\mathscr{E}(\delta)$ and $\mathscr{M}$ converges to 0 as $\delta \to 1$.

294   Chapter 9 ■ The Folk Theorem

The set $\mathcal{M}$ is a compact convex set. If it has nonempty interior, it can be approximated arbitrarily closely by a smooth convex set $\mathcal{W}$ contained in its interior.[2] Moreover, for any fixed $v \in \operatorname{int}\mathcal{M}$, $\mathcal{W}$ can be chosen to contain v. Proposition 9.1.1 is proven once we have shown $\mathcal{W} \subset \mathcal{E}(\delta)$ for sufficiently large δ.

Proposition 9.1.2 *Suppose the stage game satisfies assumption 7.1.1. Suppose $\mathcal{W}$ is a smooth convex subset of the interior of $\mathcal{M}$. There exists $\underline{\delta} < 1$ such that for all $\delta \in (\underline{\delta}, 1)$,*

$$\mathcal{W} \subset \mathcal{B}(\mathcal{W}; \delta) \subset \mathcal{E}(\delta).$$

There are two key insights behind this result. The first we have already seen: The quantity $k^*(\lambda)$ measures the maximum "average" utility of all the players in the direction λ consistent with providing each player with appropriate incentives (remark 8.1.1). The second is that because $\mathcal{W}$ is smooth, locally its boundary is sufficiently flat that the unnormalized continuations γ required to enforce any action profile can be kept inside $\mathcal{W}$ by making players patient (thereby making the variation in continuations arbitrarily small, see (8.1.5)).

Proof   From corollary 7.3.3, it is enough to show that $\mathcal{W}$ is locally self-generating, that is, for all $v \in \mathcal{W}$, there exists $\delta_v < 1$ and an open set $\mathcal{U}_v$ such that

$$v \in \mathcal{U}_v \cap \mathcal{W} \subset \mathcal{B}(\mathcal{W}; \delta_v).$$

We begin with a payoff vector $v \in \operatorname{int}\mathcal{W}$. Because v is interior, there exists an open set $\mathcal{U}_v$ s.t. $v \in \mathcal{U}_v \subset \overline{\mathcal{U}_v} \subset \operatorname{int}\mathcal{W}$. Let $\tilde{\alpha}$ be a Nash equilibrium of the stage game. Consequently, there exists a discount factor δ_v such that for any $v' \in \mathcal{U}_v$, there is a payoff $v'' \in \mathcal{W}$ such that

$$v' = (1 - \delta_v)u(\tilde{\alpha}) + \delta_v v''. \tag{9.1.1}$$

Because $\tilde{\alpha}$ is an equilibrium of the stage game, the pair $(\tilde{\alpha}, \gamma')$ decomposes v' with respect to $\mathcal{W}$, where $\gamma'(y) = v''$ for all $y \in Y$.

The interior of $\mathcal{W}$ presents a potentially misleading picture of the situation. The argument just presented decomposes payoffs using the myopic Nash equilibrium of the stage game. Consequently, no intertemporal incentives were needed. However, it would be incorrect to conclude from this that nontrivial intertemporal incentives are only required for a *negligible* set of payoffs. The value of δ_v satisfying (9.1.1) approaches 1 as v approaches $\operatorname{bd}\mathcal{W}$, the boundary of $\mathcal{W}$. Local decomposability on the boundary of $\mathcal{W}$, which requires intertemporal incentives, guarantees that δ can be chosen strictly less than 1. The strategy implicitly constructed through an appeal to corollary 7.3.3 uses intertemporal incentives for payoffs in a neighborhood of the boundary.

We turn to the heart of the argument, decomposability of points on the boundary of $\mathcal{W}$, $\operatorname{bd}\mathcal{W}$. We now suppress the dependence of the set $\mathcal{U}$ and the discount factor δ on v. Fix a point v on the boundary of $\mathcal{W}$. Because $\mathcal{W}$ is smooth and convex, there is a unique supporting hyperplane to $\mathcal{W}$ at v, with normal λ. Let $k = \lambda \cdot v$. Because $\mathcal{W} \subset \operatorname{int}H^*(\lambda)$, $k < k^*(\lambda)$. Moreover, for all $\tilde{v} \in \mathcal{W}$, $\lambda \cdot \tilde{v} \leq k < k^*(\lambda)$.

---

2. More precisely, for any ε, there exists a smooth convex $\mathcal{W} \subset \operatorname{int}\mathcal{M}$ whose Hausdorff distance from $\mathcal{M}$ is less than ε.

9.1 ■ The Limit Set of PPE Payoffs

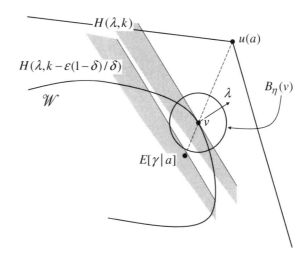

Figure 9.1.1 The hyperspaces $H(\lambda, k)$ and $H(\lambda, k - \varepsilon(1-\delta)/\delta)$.

Let $\alpha \in \mathbf{B}$ be an action profile that decomposes a point $v^* \in \operatorname{bd} H^*(\lambda)$ on $H^*(\lambda)$, so that $k < \lambda \cdot v^* = k^*(\lambda)$.[3] Let x^* denote the associated normalized continuations satisfying (8.1.2)–(8.1.4) for α and v^*. In general, all we know is that $\lambda \cdot x^*(y) \leq 0$, and it will be useful to decompose points using normalized continuations with a strictly negative value under λ. Accordingly, let v' satisfy $k = \lambda \cdot v < \lambda \cdot v' < \lambda \cdot v^*$, and let $x'(y) = x^*(y) - v^* + v'$. Because x' is x^* translated by a constant independent of y, v' is decomposed by (α, γ'), where $\gamma' = v' + x'(1-\delta)/\delta$. Moreover, $\lambda \cdot x' = \lambda \cdot x^*(y) - k^*(\lambda) + \lambda \cdot v' = \lambda \cdot x^*(y) - \varepsilon \leq -\varepsilon$ for all y, where $\varepsilon = k^*(\lambda) - \lambda \cdot v' > 0$.

Any $\tilde{v} \in \mathscr{W}$ can be decomposed by $(\alpha, \tilde{\gamma})$, where

$$\tilde{\gamma}(y) = \tilde{v} + [x'(y) - v' + \tilde{v}]\frac{(1-\delta)}{\delta}, \quad \forall y \in Y. \tag{9.1.2}$$

Because $\lambda \cdot \tilde{v} \leq k < \lambda \cdot v'$, $\lambda \cdot \tilde{\gamma}(y) \leq k - \varepsilon(1-\delta)/\delta$ (see figure 9.1.1). For $\tilde{v} = v$, the decomposing continuation is denoted by γ.

It remains to show that there exists $\eta > 0$ and $\delta < 1$ such that for all $\tilde{v} \in B_\eta(v) \cap \mathscr{W}$, we have $\tilde{\gamma}(y) \in \mathscr{W}$, for all y.

Now,

$$|\tilde{\gamma}(y) - E\gamma| \leq |\tilde{\gamma}(y) - E\tilde{\gamma}| + |E\tilde{\gamma} - E\gamma|$$
$$= |x'(y) - Ex'|\frac{(1-\delta)}{\delta} + |\tilde{v} - v|\frac{1}{\delta}.$$

In other words, by making η small, and δ sufficiently close to 1, we can make $\tilde{\gamma}(y)$ arbitrarily close to $E[\gamma \mid \alpha]$. From lemma 8.1.1(2), for δ sufficiently close to 1, $E[\gamma \mid \alpha] \in \mathscr{W}$. However, at the same time, $E[\gamma \mid \alpha]$ approaches the boundary

---

3. This assumes there exists an α for which $k^*(\lambda) = k^*(\alpha, \lambda)$. If there is no such α (see note 3 on page 277), the remainder of the proof in the text continues to hold with α and v^* chosen such that $k^*(\alpha, \lambda)$ is sufficiently close to $k^*(\lambda)$, which we can do because $k^*(\lambda) = \sup_{\alpha \in \mathbf{B}} k^*(\alpha, \lambda)$.

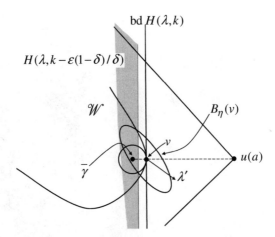

Figure 9.1.2 Figure 9.1.1 with the new coordinate system. The vector λ' is λ in the new system. Note that $B_\eta(v)$ is now an ellipse. The disc centered at $\bar{\gamma}$ is the ball of radius β' mentioned in the proof.

of $\mathscr{W}$. There is therefore no immediate guarantee that points close to $E[\gamma \mid \alpha]$ will also be in $\mathscr{W}$. However, as we now argue, the guarantee is an implication of the smoothness of the boundary of $\mathscr{W}$.

By construction, all continuations are no closer than $\varepsilon(1-\delta)/\delta$ to the supporting hyperplane of $\mathscr{W}$ at v, that is, the boundary of $H(\lambda, k)$. We now change bases to $\{f_i : 1 \leq i \leq n\}$, where $\{f_i : 1 \leq i \leq n-1\}$ is a set of spanning vectors for $\operatorname{bd} H(\lambda, k)$ and $f_n = u(\alpha) - v$. This change of basis allows us to apply a Pythagorean identity. Let $F : \mathbb{R}^n \to \mathbb{R}^n$ denote the linear change-of-basis map, and $|\cdot|_1$ denote the norm on $\mathbb{R}^n$ that makes F an isometry, that is, for all $x, x' \in \mathbb{R}^n$, $|x - x'|_1 = |Fx - Fx'|$, where $|x|$ is the Euclidean norm. The two norms $|\cdot|_1$ and $|\cdot|$ are *equivalent*, that is, there exist two constants $\kappa_1, \kappa_2 > 0$, so that for all $x \in \mathbb{R}^n$, $\kappa_1|x| \leq |x|_1 \leq \kappa_2|x|$. The image of the boundary of $\mathscr{W}$ under F is still a $\mathscr{C}^2$-submanifold. Hence, for some β', the ball of $|\cdot|_1$-radius β' centered at a point $\bar{\gamma}$ on the f_n-axis is contained in $\mathscr{W}$ and contains v (this is the reason for the new basis—see figure 9.1.2).

Let γ^δ denote the convex combination of the vectors v and $\bar{\gamma}$ satisfying $\gamma^\delta \in \operatorname{bd} H(\lambda, k - \varepsilon(1-\delta)/\delta)$, so that $|\gamma^\delta - v| = \varepsilon(1-\delta)/\delta$ (this is well defined for δ close to 1). Then, $|\bar{\gamma} - \gamma^\delta|_1 \leq \beta' - \kappa_1 \varepsilon(1-\delta)/(\delta|\lambda|)$.[4] For any point $\hat{\gamma} \in \operatorname{bd} H(\lambda, k - \varepsilon(1-\delta)/\delta)$ satisfying $|\hat{\gamma} - \gamma^\delta|_1 = \kappa\sqrt{1-\delta}$ (where $\kappa > 0$ is to be determined), we have,[5]

$$|\bar{\gamma} - \hat{\gamma}|_1^2 = |\bar{\gamma} - \gamma^\delta|_1^2 + |\hat{\gamma} - \gamma^\delta|_1^2$$
$$\leq (\beta' - \kappa_1\varepsilon(1-\delta)/(\delta|\lambda|))^2 + \kappa^2(1-\delta)$$
$$= (\beta')^2 - (1-\delta)\left[\frac{\kappa_1\varepsilon}{\delta|\lambda|}\left(2\beta' - \frac{\kappa_1\varepsilon(1-\delta)}{\delta|\lambda|}\right) - \kappa^2\right].$$

---

4. Letting $w = \gamma^\delta - v$, we have by the definition of γ^δ, $\varepsilon(1-\delta)/\delta = \lambda \cdot w$. Setting $\eta = \lambda \cdot w/\lambda \cdot \lambda$, we have $|w|_1 \geq \kappa_1|w| \geq \kappa_1\eta|\lambda| = \kappa_1\varepsilon(1-\delta)/(\delta|\lambda|)$. Finally, $|\bar{\gamma} - \gamma^\delta|_1 = |\bar{\gamma} - v|_1 - |w|_1$.

5. The first equality, a Pythagorean identity, follows from $F(\bar{\gamma} - \gamma^\delta) \cdot F(\gamma - \gamma^\delta) = 0$.

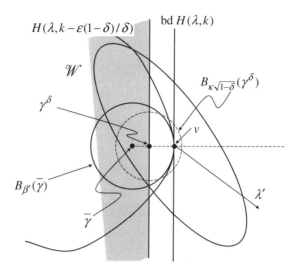

Figure 9.1.3 A "zoom-in" of figure 9.1.2 including the critical set $\{\tilde{\gamma} : |\tilde{\gamma} - \gamma^\delta|_1 < \kappa\sqrt{1-\delta}\} \cap H(\lambda, k - \varepsilon(1-\delta)/\delta) \subset \mathcal{W}$.

There exists a value for κ and a bound δ', such that for all $\delta > \delta'$, the term in square brackets is strictly positive,[6] and so

$$\{\tilde{\gamma} : |\tilde{\gamma} - \gamma^\delta|_1 < \kappa\sqrt{1-\delta}\} \cap H(\lambda, k - \varepsilon(1-\delta)/\delta) \subset \mathcal{W}. \quad (9.1.3)$$

This is illustrated in figure 9.1.3. Note that if $\mathcal{W}$ had empty interior, (9.1.3) must fail.

From (9.1.2), we already know that $\tilde{\gamma}(y) \in H(\lambda, k - \varepsilon(1-\delta)/\delta)$. Now,

$$|\tilde{\gamma}(y) - \gamma^\delta|_1 \leq |\tilde{\gamma}(y) - E\gamma|_1 + |E\gamma - v|_1 + |\gamma^\delta - v|_1$$
$$\leq |x'(y) - Ex'|\kappa_2\frac{(1-\delta)}{\delta} + |\tilde{v} - v|\kappa_2\frac{1}{\delta}$$
$$+ |Ex' - v' + v|_1\frac{(1-\delta)}{\delta} + \varepsilon\kappa_2\frac{(1-\delta)}{\delta}.$$

There exists κ' and δ'' such that, for $\delta > \delta''$, the above is bounded above by

$$\kappa'(1-\delta) + 2\kappa_2|\tilde{v} - v|.$$

Choose $\delta > \max\{\delta', \delta''\}$ so that $2\kappa'(1-\delta) < \kappa\sqrt{1-\delta}$ (this inequality is clearly satisfied for δ sufficiently close to 1). Finally, let $\eta < (\kappa/4\kappa_2)\sqrt{1-\delta}$. Then, for all $\tilde{v} \in B_\eta(v)$, the implied $\tilde{\gamma}$ satisfy

$$|\tilde{\gamma}(y) - \gamma^\delta|_1 < \kappa\sqrt{1-\delta},$$

and so, from (9.1.3), v is locally decomposed, completing the proof. ∎

6. As we will see soon, we need the radius of the ball centered at γ^δ to be of order strictly greater than $O(1-\delta)$ so that $\tilde{\gamma}$ will, for large δ, be in the ball. At the same time, if the ball had radius strictly greater than $O(\sqrt{1-\delta})$, then we could not ensure $|\tilde{\gamma} - \gamma|_1 < \beta'$ for large δ.

Remark 9.1.1 **Pure strategy PPE** The above analysis also applies to the case where long-lived players are restricted to pure strategies. Recalling the notation of remarks 7.3.3 and 8.1.2, we have the following result.

Proposition 9.1.3 *Suppose the game satisfies assumption 7.1.1. Suppose $\mathscr{M}^p$ has nonempty interior, and $\mathscr{W}$ is a smooth convex subset of the interior of $\mathscr{M}^p$. Then, there exists $\underline{\delta} < 1$ such that for all $\delta \in (\underline{\delta}, 1)$,*

$$\mathscr{W} \subset \mathscr{B}^p(\mathscr{W}; \delta) \subset \mathscr{E}^p(\delta).$$

Hence, for all $v \in \text{int}\mathscr{M}^p$, there exists $\underline{\delta} < 1$ such that for all $\delta \in (\underline{\delta}, 1)$, $v \in \mathscr{E}^p(\delta)$.

Proof The argument is almost identical to that of the proof of proposition 9.1.2, because every payoff in $\mathscr{W}$ can be decomposed using a pure action profile. This is obvious for points on the boundary of $\mathscr{W}$, and for interior points if the stage game has a pure strategy Nash equilibrium. If the stage game does not have a pure strategy Nash equilibrium, decompose points in the interior with respect to some pure action profile and then argue as for the boundary points. ■

◆

9.2 The Rank Conditions and a Public Monitoring Folk Theorem

When all players are long-lived and $\mathscr{M}$ has full dimension, from proposition 9.1.1, a folk theorem for games with imperfect public monitoring holds if $\mathscr{M} = \overline{\mathscr{F}^*}$. From lemma 8.1.1(3), this would occur if various pure action profiles (such as efficient profiles) are orthogonally enforced. Fudenberg, Levine, and Maskin's (1994) rank conditions deliver the requisite orthogonal enforceability. We assume action spaces are finite for the rest of this chapter.

If an action profile α is to be enforceable, and it is not an equilibrium of the stage game, then deviations must lead to different expected continuations. That is, the distribution over signals should be different if a player deviates. A sufficient condition is that the distribution over signals induced by α be different from the distribution induced by any profile (α'_i, α_{-i}) with $\alpha'_i \neq \alpha_i$. This is clearly implied by the following.

Definition 9.2.1 *Suppose A is finite. The profile α has* individual full rank for player i *if the $|A_i| \times |Y|$ matrix $R_i(\alpha_{-i})$ with elements $[R_i(\alpha_{-i})]_{a_i y} \equiv \rho(y \mid a_i, \alpha_{-i})$ has full row rank (i.e., the collection of probability distributions $\{\rho(\cdot \mid a_i, \alpha_{-i}) : a_i \in A_i\}$ is linearly independent). If this holds for all players i, then α has* individual full rank.

Individual full rank requires $|Y| \geq \max_i |A_i|$, and ensures that the signals generated by any (possibly mixed) action α_i are statistically distinguishable from those of any other mixture α'_i. Because both α_i and α'_i are convex combinations of pure actions, it would suffice that each $\rho(y \mid \alpha)$ is a unique convex combination of the distributions in $\{\rho(y \mid a_i, \alpha_{-i})\}_{a_i \in A_i}$ (such a condition is discussed in Kandori and Matsushima 1998).

9.2 ■ Public Monitoring Folk Theorem

If an action profile α has individual full rank, then for arbitrary v, the equations obtained from (8.1.2) and imposing equality in (8.1.3),

$$v_i = u_i(a_i, \alpha_{-i}) + E[x_i(y) \mid (a_i, \alpha_{-i})], \quad \forall i, \forall a_i \in A_i \qquad (9.2.1)$$

can be solved for enforcing normalized continuations x. Writing equation (9.2.1) in matrix form as, for each i,

$$R_i(\alpha_{-i})x_i = [(v_i - u_i(a_i, \alpha_{-i}))_{a_i \in A_i}],$$

the coefficient matrix $R_i(\alpha_{-i})$ has full row rank, so the equation can be solved for x_i (Gale 1960, theorem 2.5). In other words, we choose x so that each player has no incentive to deviate from α. Although each player also has no strict incentive to play α (a necessary property when α is mixed), there are many situations in which strict incentives can be given (see proposition 9.2.2).

Without further assumptions, there is no guarantee the enforcing x enforces α in any direction. For example, in the prisoners' dilemma example of section 8.4.3, $\alpha = EE$ has individual full rank for both players and is enforced by any x satisfying $x_i(\bar{y}) - x_i(\underline{y}) \geq 1/(p-q)$. However, any x satisfying $\lambda \cdot x(\bar{y}) > 0 > \lambda \cdot x(\underline{y})$ fails to enforce α in the direction λ. Moreover, if $0 = \lambda \cdot E[x(y) \mid \alpha]$, then necessarily $\lambda \cdot x(\bar{y}) > 0 > \lambda \cdot x(\underline{y})$. Therefore, $\alpha = EE$ cannot be orthogonally enforced in any direction (as we saw in section 8.4.3).

The coordinate directions, $\lambda = e_i$ and $-e_i$ for some i, impose the least restriction on the continuations, in that x_j is unconstrained for $j \neq i$. At the same time, $\lambda \cdot x(y) = 0$ implies $x_i(y) = 0$, for all y. In other words, only profiles in which player i is playing a static (or myopic) best reply to α_{-i} can be orthogonally enforced in the directions $-e_i$ or e_i. On the other hand, if all players are playing a static best reply (i.e., α is a static Nash equilibrium), $x_i(y) = 0$ for all i and y trivially satisfies (8.1.2)–(8.1.4), with $v_i = u_i(\alpha)$ and (8.1.4) holding with equality for any λ. This discussion implies the following lemma.

Lemma 1. *If α has individual full rank, then it is enforceable.*
9.2.1 2. *If α is a static Nash equilibrium, then it can be orthogonally enforced in any direction.*
3. *If and only if α is enforceable and α_i maximizes $u_i(a_i, \alpha_{-i})$ over A_i, then α is orthogonally enforceable in the directions e_i and $-e_i$.*

We turn now to the general conditions that allow orthogonal enforceability in noncoordinate directions. From lemma 8.1.1(4), it will be enough to obtain orthogonal enforceability in all pairwise directions, λ^{ij}. We begin with a strengthening of the individual full rank condition.

Definition *Suppose A is finite. The profile α has pairwise full rank for players i and j if the*
9.2.2 *$(|A_i| + |A_j|) \times |Y|$ matrix*

$$R_{ij}(\alpha) = \begin{bmatrix} R_i(\alpha_{-i}) \\ R_j(\alpha_{-j}) \end{bmatrix}$$

has rank $|A_i| + |A_j| - 1$.

Note that $|A_i| + |A_j| - 1$ is the maximum feasible rank for $R_{ij}(\alpha)$ because $\alpha_i R_i(\alpha_{-i}) = [\rho(y \mid \alpha)_{y \in Y}] = \alpha_j R_j(\alpha_{-j})$. Moreover, pairwise full rank for i and j implies individual full rank for i and j.

Lemma 9.2.2 *If α has pairwise full rank for players i and j and individual full rank, then it is orthogonally enforceable in any ij-pairwise direction.*

Proof Fix an ij-pairwise direction λ^{ij} and choose v satisfying $\lambda^{ij} \cdot v = \lambda^{ij} \cdot u(\alpha)$ (a necessary condition for orthogonal enforceability from lemma 8.1.1). Because $\lambda_j^{ij} \neq 0$, the constraint $\lambda^{ij} \cdot x(y) = 0$ for all y can be written as $x_j(y) = -\lambda_i^{ij} x_i(y)/\lambda_j^{ij}$ for all $y \in Y$. Imposing this equality, the equations obtained from (8.1.2) and imposing equality in (8.1.3) for i and j can be written as

$$R_{ij}(\alpha)x_i = \begin{bmatrix} (v_i - u_i(a_i, \alpha_{-i}))_{a_i \in A_i} \\ \left(-\frac{\lambda_j^{ij}}{\lambda_i^{ij}}\{v_j - u_j(a_j, \alpha_{-j})\}\right)_{a_j \in A_j} \end{bmatrix}. \quad (9.2.2)$$

Because $\alpha_i R_i(\alpha_{-i}) = \alpha_j R_j(\alpha_{-j})$ and $R_{ij}(\alpha)$ has rank $|A_i| + |A_j| - 1$, this system can be solved for x if (Gale 1960, theorem 2.4)

$$\alpha_i[(v_i - u_i(a_i, \alpha_{-i}))_{a_i \in A_i}] = -\frac{\lambda_j^{ij}}{\lambda_i^{ij}} \alpha_j[(v_j - u_j(a_j, \alpha_{-j}))_{a_j \in A_j}]. \quad (9.2.3)$$

The left side equals $v_i - u_i(\alpha)$, and the right side equals $-\frac{\lambda_j^{ij}}{\lambda_i^{ij}}(v_j - u_j(\alpha))$, so (9.2.3) is implied by $\lambda^{ij} \cdot v = \lambda^{ij} \cdot u(\alpha)$.

Finally, individual full rank for $k \neq i, j$ implies

$$R_k(\alpha_{-k})x_k = [(v_k - u_k(a_k, \alpha_{-k}))_{a_k \in A_k}],$$

can be solved for x_k. ∎

Recall from proposition 8.1.1 that $\cap_\lambda H^*(\lambda) \subset \overline{\mathscr{F}^*} \subset \mathscr{F}^\dagger$.

Corollary 9.2.1 *Suppose there are no short-lived players and all the pure action profiles yielding the extreme points of $\mathscr{F}^\dagger$, the set of feasible payoffs, have pairwise full rank for all pairs of players. Then, if Λ^n denotes the set of noncoordinate directions,*

$$\mathscr{F}^\dagger \subset \cap_{\lambda \in \Lambda^n} H^*(\lambda).$$

Proof Because $\mathscr{F}^\dagger$ is convex, for any noncoordinate direction λ, there is an extreme point $u(a^\lambda)$ of $\mathscr{F}^\dagger$ such that $\lambda \cdot v \leq \lambda \cdot u(a^\lambda)$ for all $v \in \mathscr{F}^\dagger$. The extreme points are orthogonally enforceable in all noncoordinate directions (lemmas 8.1.1 and 9.2.2), so $k^*(\lambda) = \lambda \cdot u(a^\lambda) \geq \lambda \cdot v$ for all $v \in \mathscr{F}^\dagger$, that is, $\mathscr{F}^\dagger \subset H^*(\lambda)$. ∎

Because $\mathscr{F}^\dagger$ is the intersection of the closed half-spaces with normals in Λ^n, the inclusion in corollary 9.2.1 can be strengthened to equality.

9.2 ■ Public Monitoring Folk Theorem

The characterization of $\mathscr{M}$ is completed by considering coordinate directions. Pairwise full rank implies individual full rank, for all i, so we have $k^*(e_i) = \max_a u_i(a)$, and e_i does not further restrict the intersection.[7] The coordinate directions $-e_i$ can impose new restrictions, despite being in the closure of the set of noncoordinate directions, Λ^n, because different constraints are involved: If player i is not best responding to α_{-i}, then as $\lambda_j^{ij} \to 0$ (so that λ^{ij} approaches an i-coordinate direction), the normalized continuations for player j can become arbitrarily large in magnitude (because $x_j(y) = -\lambda_i^{ij} x_i(y)/\lambda_j^{ij}$). Enforceability in the coordinate directions implies the normalized continuations used in all directions can be taken from a bounded set.

Proposition 9.2.1 **The public monitoring folk theorem** *Suppose there are no short-lived players, A is finite, $\mathscr{F}^\dagger \subset \mathbb{R}^n$ has nonempty interior, and all the pure action profiles yielding the extreme points of $\mathscr{F}^\dagger$, the set of feasible payoffs, have pairwise full rank for all pairs of players.*

1. *If $\tilde{\alpha}$ is an inefficient Nash equilibrium, then for all $v \in \text{int}\{v' \in \mathscr{F}^\dagger : v_i' \geq u_i(\tilde{\alpha}) \text{ for all } i\}$, there exists $\underline{\delta} < 1$ such that for all $\delta \in (\underline{\delta}, 1)$, $v \in \mathscr{E}(\delta)$.*
2. *If $\underline{v} = (\underline{v}_1, \ldots, \underline{v}_n)$ is inefficient and each player's minmax profile $\hat{\alpha}^i$ has individual full rank, then, for all $v \in \text{int}\mathscr{F}^*$, there exists $\underline{\delta} < 1$ such that for all $\delta \in (\underline{\delta}, 1)$, $v \in \mathscr{E}(\delta)$.*

Proof 1. Because $\tilde{\alpha}$ is a Nash equilibrium, it is orthogonally enforced in all directions, and so $k^*(-e_i) \geq k^*(\tilde{\alpha}, -e_i) = -u_i(\tilde{\alpha})$. Consequently, for any v satisfying $v_i \geq u_i(\tilde{\alpha})$ for all i, we have $-v_i \leq -u_i(\tilde{\alpha}) \leq k^*(-e_i)$, and so $v \in H^*(-e_i)$. Because $\mathscr{F}^\dagger \subset \mathbb{R}^n$ has nonempty interior, it has full dimension, and the inefficiency of $u(\tilde{\alpha})$ implies the same for $\mathscr{M}$, and the result follows from proposition 9.1.1.

2. Each player's minmax profile $\hat{\alpha}^i$ has individual full rank, so from lemma 9.2.1, for all i, $k^*(-e_i) = -\underline{v}_i$, and so $\mathscr{M} = \overline{\mathscr{F}^*}$. The inefficiency of $\underline{v}$ implies the full dimensionality of $\mathscr{M}$, and the result again follows from proposition 9.1.1. ∎

The conditions given in proposition 9.2.1 are stronger than necessary. Consider first the assumption that all pure-action profiles yielding the extreme points of $\mathscr{F}^\dagger$ have pairwise full rank for all players. Many games of interest fail such a condition. In particular, in symmetric games (definition 8.2.1) such as the oligopoly game of Green and Porter (1984), the distribution over public signals depends only on the number of players choosing different actions, not which players. Consequently, any profile in which all players choose the same action cannot have pairwise full rank: All profiles where a single player deviates to the same action induce the same distribution over signals. However, because $k^*(\lambda) = \sup_\alpha k^*(\alpha, \lambda)$, it would clearly be enough to have a dense set of actions within $\mathscr{F}^\dagger$, each of which has pairwise full rank. (See section 8.4.5 for an illustration.) Perhaps surprisingly, this seemingly very strong condition is an implication of the existence for each pair of players of just one profile with pairwise full rank for that pair (Fudenberg, Levine, and Maskin 1994, lemma 6.2). Similarly, the assumption that each players' minmax profile has individual full rank can be replaced

---

7. As we saw in lemma 8.3.1, and return to section 9.3, this is not true in the presence of short-lived players.

by the assumption that all pure strategy profiles have individual full rank (Fudenberg, Levine, and Maskin 1994, lemma 6.3).

Remark 9.2.1 **Identifiability and informativeness** It is worth emphasizing that the conditions in proposition 9.2.1 concern the statistical identifiability of deviations and make no mention of the informativeness of the signals. Pairwise full rank of a for players i and j asserts simply that the matrix $R_{ij}(a)$ has an invertible square submatrix with $|A_i| \times |A_j| - 1$ rows; there is no bound on how close to singular the matrix can be. For example, consider a sequence of monitoring distributions $\{\rho^\ell\}_\ell$, with all a having pairwise full rank under each ρ^ℓ and with $\rho^\ell(y \mid a) \to 1/|Y|$, the uniform distribution. In the limit, the public monitoring distribution is uniform on Y for all a, and under that monitoring distribution, all PPE are trivial. On the other hand, for each ℓ, proposition 9.2.1 holds. However, the relevant bound on δ becomes increasingly severe (i.e., the bound converges to 1 as $\ell \to \infty$): Because the determinant of any full row rank submatrix of R_{ij} in (9.2.2) converges to 0, the normalized continuations become arbitrarily large (as $\ell \to \infty$) and so in the proof of proposition 9.1.2, the necessary bound on δ to ensure the unnormalized continuations γ are in the appropriate set becomes increasingly severe.

♦

The equilibria implicitly constructed in proposition 9.2.1 are weak: All players are indifferent over all actions. However, for a pure profile a, observe that (8.1.2) and (8.1.3) are implied by, for all i,

$$v_i = u_i(a) + E[x_i(y) \mid a] \\ = u_i(a'_i, a_{-i}) + E[x_i(y) \mid (a'_i, a_{-i})] - 1, \quad \forall a'_i \in A_i, a'_i \neq a_i. \tag{9.2.4}$$

As with (9.2.1), if a has individual full rank, (9.2.4) can be solved for x. A similar modification to (9.2.2) yields, when a has pairwise full rank, continuations with strict incentives. This then gives us the following strengthening of proposition 9.2.1. Let $\mathcal{E}^s(\delta)$ be the set of *strict* PPE payoffs.

Proposition 9.2.2 **The public monitoring folk theorem in strict PPE** *Suppose there are no short-lived players, A is finite, and $\rho(y \mid a) > 0$ for all $y \in Y$ and $a \in A$. Suppose $\mathcal{F}^\dagger \subset \mathbb{R}^n$ has nonempty interior and all the pure-action profiles yielding the extreme points of $\mathcal{F}^\dagger$ have pairwise full rank for all pairs of players. If each player's pure-action minmax profile $\hat{a}^i$ has individual full rank and $\underline{v}^p = (\underline{v}_1^p, \ldots, \underline{v}_n^p)$ is inefficient, then*

$$\lim_{\delta \to 1} \mathcal{E}^s(\delta) = \overline{\mathcal{F}^{*p}}.$$

Proof Because strict PPE equilibria are necessarily in pure strategies, we begin, as in proposition 9.1.3, with a smooth convex set $\mathcal{W} \subset \text{int} \mathcal{M}^p$. We argue that $\mathcal{W} \subset \mathcal{E}^s(\delta)$ for sufficiently large δ (the argument is then completed as before). Moreover, observe that if we have local decomposability using continuations that imply strict incentives, then the resulting equilibrium will be strict (corollary 7.1.1). Note first that for the case of a point on the boundary of $\mathcal{W}$, the relevant part of the proof of proposition 9.1.2 carries through (with the new definitions), with the normalized continuations x^* chosen so that (9.2.4) holds (this is possible because each pure action profile has individual full rank). Turning to points in the interior of $\mathcal{W}$, if the stage game has a strict Nash equilibrium, then the obvious decomposition of (9.1.1) will give strict incentives, and the argument is completed. If, on the other hand, the stage game does not have a strict Nash equilibrium, we then decompose the point in the interior with respect to some pure-action profile and then argue as for boundary points. ∎

9.3 Perfect Monitoring Characterizations

9.3.1 The Folk Theorem with Long-Lived Players

Proposition 9.2.1 also implies the mixed-action minmax folk theorem for *perfect monitoring* repeated games with unobservable randomization. The condition that $\mathscr{F}^\dagger$ have nonempty interior is called the *full-dimensionality condition* and is stronger than the assumption of pure-action player-specific punishments of proposition 3.4.1.

Proposition 9.3.1 **The perfect monitoring folk theorem** *Suppose there are no short-lived players, A is finite, $\mathscr{F}^\dagger$ has nonempty interior, and $\underline{v} = (\underline{v}_1, \ldots, \underline{v}_n)$ is inefficient. For every $v \in \text{int}\mathscr{F}^*$, there exists $\underline{\delta} < 1$ such that for all $\delta \in (\underline{\delta}, 1)$, there exists a subgame-perfect equilibrium of the repeated game with perfect monitoring with value v. Moreover, if $v_i > \underline{v}_i^p$ for all i, v is a pure strategy subgame-perfect equilibrium payoff.*

Proof For perfect-monitoring games, $Y = A$ and $\rho(a' \mid a) = 1$ if $a' = a$ and 0 otherwise. Hence, every pure action profile has pairwise full rank for all pairs of players, and every action profile has individual full rank.[8]

The first claim then follows from proposition 9.2.1. The second claim follows from lemma 8.1.1(4) and proposition 9.1.3.
∎

9.3.2 Long-Lived and Short-Lived Players

We can also now easily characterize the set of subgame-perfect equilibrium payoffs of the perfect monitoring game with many patient long-lived players and one or more short-lived players. Recall from section 3.6 that for such games, $\mathscr{F}^\dagger$ denotes the payoffs in the convex hull of $\{(u_1(\alpha), \ldots, u_n(\alpha)) \in \mathbb{R}^n : \alpha \in \mathbf{B}\}$. Even with this change, corollary 9.2.1 does not hold in the presence of short-lived players, because some of the extreme points of $\mathscr{F}^\dagger$ may involve randomization by a long-lived player (see example 3.6.1). However, under perfect monitoring, all action profiles have pairwise full rank for all pairs of long-lived players:

Lemma 9.3.1 *Suppose A is finite and there is perfect monitoring (i.e., $Y = A$ and $\rho(a' \mid a) = 1$ if $a' = a$ and 0 otherwise). Then, all action profiles have pairwise full rank for all pairs of long-lived players.*

---

8. Indeed, under perfect monitoring, all action profiles have pairwise full rank (lemma 9.3.1), but we need only the weaker statement here.

Proof Fix a mixed profile α, and consider the matrix $R_{ij}(\alpha)$. Every column of this matrix corresponds to an action profile a and has no, one, or two, nonzero entries:

1. if the action profile a satisfies $\alpha(a) > 0$, then the entries corresponding to the a_i and a_j rows are $\alpha_{-i}(a_{-i})$ and $\alpha_{-j}(a_{-j})$ respectively, with every other entry 0;
2. if a satisfies $\alpha_{-k}(a_{-k}) > 0$ and $\alpha_k(a_k) = 0$ ($k = i$ or j), then the entry corresponding to the a_k-row is $\alpha_{-k}(a_{-k})$, with every other entry 0; and
3. if a satisfies $\alpha_i(a_i) = 0$ and $\alpha_j(a_j) = 0$, or $\alpha_{-ij}(a_{-ij}) = 0$, then every entry is 0.

Deleting one row corresponding to an action taken with positive probability by player i, say, a_i', yields a $(|A_i| + |A_j| - 1) \times |A|$ matrix, R'. We claim R' has full row rank. Observe that for any $a_{-i} \in \text{supp}(\alpha_{-i})$, the column in R' corresponding to $a_i' a_{-i}$ has exactly one nonzero term (corresponding to a_j). Moreover, every column of R' corresponding to a satisfying $\alpha_{-j}(a_{-j}) > 0$ and $\alpha_j(a_j) = 0$ also has exactly one nonzero term. Hence for any $\beta \in \mathbb{R}^{|A_i|+|A_j|-1}$ with $\beta R' = 0$, every entry corresponding to an action for player j equals 0. But this implies $\beta = 0$. ∎

We now proceed as in the proof of corollary 9.2.1. For any pairwise direction λ^{ij}, let α' maximize $\lambda^{ij} \cdot u(\alpha)$ over $\alpha \in \text{ext} \mathscr{F}^\dagger$. Then, $k^*(\lambda^{ij}) = \lambda^{ij} \cdot u(\alpha')$, and lemmas 9.2.2 and 8.1.1 imply

$$\cap_{\lambda \in \Lambda^n} H^*(\lambda) = \mathscr{F}^\dagger.$$

We now consider the coordinate directions, e_j and $-e_j$. Unlike the case of only long-lived players, the direction e_j does further restrict the intersection. From lemma 8.3.1 we have for long-lived player j,

$$k^*(e_j) \leq \bar{v}_j \equiv \max_{\alpha \in \mathbf{B}} \min_{a_j \in \text{supp}(\alpha_j)} u_j(a_j, \alpha_{-j}).$$

We now argue that we have equality. Fix $\alpha \in \mathbf{B}$, and set, for $i = 1, \ldots, n$, $v_i = \min_{a_i \in \text{supp}(\alpha_i)} u_i(a_i, \alpha_{-i})$. Let x be the normalized continuations given by

$$x_i(a_i, a_{-i}) = \begin{cases} v_i - u_i(a_i, \alpha_{-i}), & \text{if } \alpha_i(a_i) > 0, \\ -2 \max_\alpha |u_i(\alpha)|, & \text{if } \alpha_i(a_i) = 0. \end{cases}$$

Then, for all $j = 1, \ldots, n$, the normalized continuations x, with v, satisfy (8.1.2)–(8.1.4). So,

$$k^*(\alpha, e_j) \geq v_j = \min_{a_j \in \text{supp}(\alpha_j)} u_j(a_j, \alpha_{-j}),$$

and therefore

$$k^*(e_j) = \sup_{\alpha \in \mathbf{B}} k^*(\alpha, e_j) \geq \sup_{\alpha \in \mathbf{B}} \min_{a_j \in \text{supp}(\alpha_j)} u_j(a_j, \alpha_{-j}).$$

Finally, we need to argue that $k^*(-e_j) = -\underline{v}_j$. From lemma 8.1.2, $k^*(-e_j) \leq -\underline{v}_j$. We cannot apply lemma 9.2.1, because $\hat{\alpha}_j^j$ need not be a best reply to $\hat{\alpha}_{-j}^j$, the profile minmaxing j. Fix a long-lived player j and for $i = 1, \ldots, n$, set $v_i = \max_{a_i \in A_i} u_i(a_i, \hat{\alpha}_{-i}^j)$. Let x be the normalized continuations given by, for all $a \in A$,

$$x_i(a) = v_i - u_i(a_i, \hat{\alpha}^j_{-i}).$$

Then, the normalized continuations x, with v, satisfy (8.1.2)–(8.1.4). So,

$$k^*(\hat{\alpha}^j, -e_j) = -e_j \cdot v = -\max_{a_j \in A_j} u_j(a_j, \hat{\alpha}^j_{-j}),$$

and therefore

$$k^*(-e_j) = \sup_{\alpha \in \mathbf{B}} k^*(\alpha, -e_j) \geq k^*(\hat{\alpha}^j, -e_j) = -\underline{v}_j.$$

Summarizing this discussion, we have the following characterization.

Proposition 9.3.2 *Suppose A is finite and $\mathscr{F}^\dagger$ has nonempty interior. For every $v \in \operatorname{int} \mathscr{F}^*$ satisfying $v_i < \bar{v}_i$ for $i = 1, \ldots, n$, there exists $\underline{\delta}$ such that for all $\delta \in (\underline{\delta}, 1)$, there exists a subgame-perfect equilibrium of the repeated game with perfect monitoring with value v.*

Because intertemporal incentives need be provided for long-lived players only, it is sufficient that their actions be perfectly monitored. In particular, the foregoing analysis applies if the long-lived players' action spaces are finite, and the signal structure is given by $Y = \prod_{i=1}^n A_i \times Y'$, for some finite Y', and $\rho((a'_1, \ldots, a'_n, y') \mid a) = 0$ if for some $i = 1, \ldots, n, a'_i \neq a_i$.

As suggested by the development of the folk theorem for long-lived players in section 3.5, there is scope for weakening the requirement of a nonempty interior.

9.4 Enforceability and Identifiability

Individual full rank, which requires $|Y| > |A_i|$, is violated in some important games. For example, in repeated adverse selection, discussed in section 11.2, the inequality is necessarily reversed. Nonetheless, strongly efficient action profiles can still be enforced (lemma 9.4.1)—though not necessarily *orthogonally* enforced (for example, EE in the prisoners' dilemma of section 8.4.3).

The presence of short-lived players constrains strong efficiency in a natural manner: $\alpha \in \mathbf{B}$ is *constrained strongly efficient* if, for all $\alpha' \in \mathbf{B}$ satisfying $u_i(\alpha) < u_i(\alpha')$ for some $i = 1, \ldots, n$, there is some $\ell = 1, \ldots, n$, for which $u_\ell(\alpha) > u_\ell(\alpha')$. In the absence of short-lived players, the notions of constrained strongly efficient and strongly efficient coincide.

The enforceability of constrained strongly efficient actions is an implication of the assumption that player i's ex ante payoff is the expected value of his ex post payoffs (see (7.1.1)), which only depend on the realized signal and his own action. Intuitively, enforceability of a profile α fails if for some player i, there is another action α'_i such that α and (α'_i, α_{-i}) induce the same distribution over signals with (α'_i, α_{-i}) yielding higher stage-game payoffs. But the form of ex ante payoffs in (7.1.1) implies that the other players are then indifferent between α and (α'_i, α_{-i}), and so α is not constrained strongly efficient.

Lemma 9.4.1 *Suppose u_i is given by (7.1.1) for some u_i^*, $i = 1, \ldots, n$. Any constrained strongly efficient profile is enforceable in some direction.*

Proof If α' is constrained strongly efficient, then there is a vector $\lambda \in \mathbb{R}^n$, with $\lambda_i > 0$ for all i, such that $\alpha' \in \operatorname{argmax}_{\alpha \in \mathbf{B}} \lambda \cdot u(\alpha)$. For all $i \leq n$, let $M_i = \max_{y, a_i} u_i^*(y, a_i)$ and set

$$x_i(y) \equiv \sum_{\substack{j=1,\ldots,n,\\ j \neq i}} \lambda_j u_j^*(y, \alpha_j')/\lambda_i - (n-1) M_i, \quad i = 1, \ldots, n.$$

Then, $\lambda \cdot x(y) = \sum_{i=1,\ldots,n} \sum_{j \neq i} \lambda_j u_j^*(y, \alpha_j') - (n-1) \sum_i \lambda_i M_i \leq 0$. Moreover, for any α_i,

$$u_i(\alpha_i, \alpha_{-i}') + E[x_i(y) \mid (\alpha_i, \alpha_{-i}')] \tag{9.4.1}$$

$$= \frac{1}{\lambda_i} \left\{ \lambda_i u_i(\alpha_i, \alpha_{-i}') + \sum_y \sum_{\substack{j=1,\ldots,n,\\ j \neq i}} \lambda_j u_j^*(y, \alpha_j') \rho(y \mid \alpha_i, \alpha_{-i}') \right\} - (n-1) M_i$$

$$= \frac{1}{\lambda_i} \left\{ \lambda_i u_i(\alpha_i, \alpha_{-i}') + \sum_{j \neq i} \lambda_j u_j(\alpha_i, \alpha_{-i}') \right\} - (n-1) M_i$$

$$= \frac{1}{\lambda_i} \lambda \cdot u(\alpha_i, \alpha_{-i}') - (n-1) M_i.$$

Because $\alpha' \in \operatorname{argmax}_{\alpha \in \mathbf{B}} \lambda \cdot u(\alpha)$, α_i' maximizes (9.4.1), and so α' is enforceable in the direction λ. ∎

We now turn to Fudenberg, Levine, and Maskin's (1994) weakening of pairwise full rank to allow for enforceable profiles that fail individual full rank.

Definition 9.4.1 *A profile α is* pairwise identifiable *for players i and j if*

$$\operatorname{rank} R_{ij}(\alpha) = \operatorname{rank} R_i(\alpha_{-i}) + \operatorname{rank} R_j(\alpha_{-j}) - 1,$$

where $\operatorname{rank} R$ denotes the rank of the matrix R.

Note that pairwise full rank is simply individual full rank plus pairwise identifiability.

Lemma 9.4.2 *If an action profile with pure long-lived player actions is enforceable and pairwise identifiable for long-lived players i and j, then it is orthogonally enforceable in all ij-pairwise directions, λ^{ij}.*

Proof Suppose $\alpha^\dagger$ is an action profile with pure long-lived player actions, that is, $\alpha^\dagger = (a_1^\dagger, \ldots, a_n^\dagger, \alpha_{SL}^\dagger)$ for some $a_i^\dagger \in A_i$, $i = 1, \ldots, n$. For all $a_i \in A_i$, let $r_i(a_i)$ denote the vector of probabilities $[\rho(y \mid (a_i, \alpha_{-i}^\dagger)]_{y \in Y}$ and set $u_i^\dagger(a_i) \equiv u_i(a_i, \alpha_{-i}^\dagger)$. Say that the action a_i is *subordinate* to a set $A_i' \subset A_i$ if there exists subordinating scalars $\{\beta_i(a_i')\}_{a_i' \in A_i'}$ with $\beta_i(a_i) = 0$ such that

9.4 ■ Enforceability and Identifiability

$$\sum_{a_i' \in A_i'} \beta_i(a_i') r_i(a_i') = r_i(a_i) \tag{9.4.2}$$

and

$$\sum_{a_i' \in A_i'} \beta_i(a_i') u_i^\dagger(a_i') \geq u_i^\dagger(a_i). \tag{9.4.3}$$

From (9.4.2), $1 = \sum_y \rho(y \mid (a_i, a_{-i}^\dagger)) = \sum_y \sum_{a_i' \in A_i'} \beta_i(a_i') \rho(y \mid (a_i', a_{-i}^\dagger)) = \sum_{a_i' \in A_i'} \beta_i(a_i')$.

Claim *Suppose the set $\{r_i(a_i')\}_{a_i' \in A_i'}$ is linearly dependent for a set of actions A_i'. Then,*
9.4.1 *there exists an action $a_i \in A_i'$ subordinate to A_i'.*

Proof The linear dependence of $\{r_i(a_i')\}_{a_i' \in A_i'}$ implies the existence of scalars $\{\zeta_i(a_i')\}$, not all 0, such that $\sum_{a_i' \in A_i'} \zeta_i(a_i') r_i(a_i') = 0$. Let a_i be an action for which $\zeta_i(a_i) \neq 0$. Then,

$$\sum_{a_i' \in A_i' \setminus \{a_i\}} [-\zeta_i(a_i')/\zeta_i(a_i)] r_i(a_i') = r_i(a_i). \tag{9.4.4}$$

Suppose

$$\sum_{a_i' \in A_i' \setminus \{a_i\}} [-\zeta_i(a_i')/\zeta_i(a_i)] u_i^\dagger(a_i') < u_i^\dagger(a_i). \tag{9.4.5}$$

Because all elements of the vectors $r_i(a_i')$ are nonnegative, $-\zeta_i(a_i')/\zeta_i(a_i) > 0$ for at least one $a_i' \in A_i' \setminus \{a_i\}$; let a_i'' denote such an action. Multiplying (9.4.4) and (9.4.5) by the term $\zeta_i(a_i)/\zeta_i(a_i'')$, and rearranging, shows that a_i'' is subordinate to A_i'.

On the other hand, if the inequality in (9.4.5) fails, then a_i is subordinate to A_i'. □

Claim *If $a_i \in A_i'$ is subordinate to A_i' and $(a_i, a_{-i}^\dagger)$ is enforceable, then there exists*
9.4.2 *$a_i'' \neq a_i$, $a_i'' \in A_i'$, subordinate to A_i'.*

Proof Because $a_i \in A_i'$ is subordinate to A_i', there are subordinating scalars $\{\beta_i(a_i')\}$ with $\beta_i(a_i) = 0$ satisfying (9.4.2) and (9.4.3). Suppose (9.4.3) holds strictly. If $\beta_i(a_i') \geq 0$ for all $a_i' \in A_i'$, then $\{\beta_i(a_i')\}$ corresponds to a mixture over A_i' (recall that $\sum_{a_i' \in A_i'} \beta_i(a_i') = 1$). Given $a_{-i}^\dagger$, this mixture is both statistically indistinguishable from a_i and yields a higher payoff than a_i, so the profile $(a_i, a_{-i}^\dagger)$ cannot be enforceable. Hence, there is some $a_i'' \in A_i'$ for which $\beta_i(a_i'') < 0$. Dividing both (9.4.2) and (9.4.3) by $\beta_i(a_i'') < 0$ and rearranging shows that a_i'' is subordinate to A_i'.

Suppose now that (9.4.3) holds as an equality. Let $a_i'' \in A_i'$ be an action with $\beta_i(a_i'') \neq 0$. A familiar rearrangement, after dividing (9.4.2) and (9.4.3) by $\beta_i(a_i'')$, shows that a_i'' is subordinate to A_i'. □

Claim 9.4.3 *If a_i is subordinate to A'_i and a''_i is subordinate to $A'_i\setminus\{a_i\}$, then a_i is subordinate to $A'_i\setminus\{a''_i\}$.*

Proof Let $\{\beta_i(a'_i)\}$ be the scalars subordinating a_i to A'_i, and $\{\beta''_i(a'_i)\}$ be the scalars subordinating a''_i to $A'_i\setminus\{a_i\}$. It is straightforward to verify that $\{\tilde{\beta}_i(a'_i)\}$ given by

$$\tilde{\beta}_i(a'_i) = \begin{cases} \beta_i(a'_i) + \beta_i(a''_i)\beta''_i(a'_i), & \text{if } a'_i \neq a_i \text{ and } a'_i \neq a''_i, \\ 0, & \text{otherwise,} \end{cases}$$

subordinates a_i to $A'_i\setminus\{a''_i\}$. □

We claim that there is a set $A_i^\dagger \subset A_i$ containing $a_i^\dagger$ such that $\{r_i(a'_i)\}_{a'_i\in A_i^\dagger}$ is linearly independent and each $a''_i \notin A_i^\dagger$ is subordinate to $A_i^\dagger$: If $\{r_i(a'_i)\}_{a'_i\in A_i}$ is linearly dependent (i.e., $R_i(a_{-i}^\dagger)$ does not have full row rank), then claim 9.4.1 implies there is an action $a_i^1 \in A_i$ subordinate to A_i. By claim 9.4.2, we can assume $a_i^1 \neq a_i^\dagger$. Trivially, a_i^1 is subordinate to $A_i^1 \equiv A_i\setminus\{a_i^1\}$. If $\{r_i(a'_i)\}_{a'_i\in A_i^1}$ is linearly independent, then A_i^1 is the desired set of actions. If not, then claim 9.4.1 again implies there is an action $a_i^2 \in A_i^1$ subordinate to A_i^1. By claim 9.4.2, we can assume $a_i^2 \neq a_i^\dagger$. By claim 9.4.3, a_i^1 is subordinate to $A_i^2 \equiv A_i^1\setminus\{a_i^2\}$, and trivially, a_i^2 is subordinate to A_i^2. If $\{r_i(a'_i)\}_{a'_i\in A_i^2}$ is linearly independent, then A_i^2 is the desired set of actions. If not, then we apply the same argument. Because A_i is finite, we must eventually obtain the desired set $A_i^\dagger$.

Let $R_i^\dagger(\alpha_{-i}^\dagger)$ denote the corresponding matrix of probability distributions $\{r_i(a'_i)\}_{a'_i\in A_i^\dagger}$. Note that $R_i^\dagger(\alpha_{-i}^\dagger)$ has (full row) rank of $\text{rank } R_i(\alpha_{-i}^\dagger)$, and because every $a''_i \notin A_i^\dagger$ is subordinate to $A_i^\dagger$, $r_i(a''_i)$ is a linear combination of the vectors $\{r_i(a'_i)\}_{a'_i\in A_i^\dagger}$.

By the same argument, there is a corresponding set $A_j^\dagger$ and matrix of probability distributions $R_j^\dagger(\alpha_{-j}^\dagger)$ for player j. Pairwise identifiability thus implies that the matrix

$$\begin{bmatrix} R_i^\dagger(\alpha_{-i}^\dagger) \\ R_j^\dagger(\alpha_{-j}^\dagger) \end{bmatrix}$$

has rank equal to $\text{rank } R_{ij}(\alpha^\dagger)$. Hence, for all ij-pairwise directions λ^{ij}, there exists normalized continuations x satisfying $\lambda^{ij} \cdot x = 0$ and, for $k = i, j$,

$$\begin{aligned} v_k &\equiv u_k(\alpha^\dagger) + E[x_k(y) \mid \alpha^\dagger] \\ &= u_k(a'_k, \alpha_{-k}^\dagger) + E[x_k(y) \mid (a'_k, \alpha_{-k}^\dagger)], \quad \forall a'_k \in A_k^\dagger, \end{aligned} \quad (9.4.6)$$

(by the same argument as in the proof of lemma 9.2.2), as well as for $k \neq i, j$, $k \leq n$,

$$u_k(\alpha^\dagger) + E[x_k(y) \mid \alpha^\dagger] \geq u_k(a'_k, \alpha_{-k}^\dagger) + E[x_k(y) \mid (a'_k, \alpha_{-k}^\dagger)], \quad \forall a'_k \in A_k$$

(because $\alpha^\dagger$ is enforceable, and $\lambda_k^{ij} = 0$ for $k \neq i, j$ implies that our choice of x_k is unrestricted).

It remains to show that for $k = i, j$, and $a_k'' \notin A_k^\dagger$,

$$v_k \geq u_k(a_k'', \alpha_{-k}^\dagger) + E[x_k(y) \mid (a_k'', \alpha_{-k}^\dagger)].$$

Let $\{\beta_k(a_k')\}$ denote the scalars subordinating $a_k'' \notin A_k^\dagger$ to $A_k^\dagger$. Then we have, using $\sum_{a_k' \in A_k^\dagger} \beta_k(a_k') = 1$ and (9.4.6),

$$\begin{aligned}
v_k &= \sum_{a_k' \in A_k^\dagger} \beta_k(a_k') v_k \\
&= \sum_{a_k' \in A_k^\dagger} \beta_k(a_k') u_k(a_k', \alpha_{-k}^\dagger) + \sum_{a_k' \in A_k^\dagger} \beta_k(a_k') E[x_k(y) \mid (a_k', \alpha_{-k}^\dagger)] \\
&= \sum_{a_k' \in A_k^\dagger} \beta_k(a_k') u_k^\dagger(a_k') + \sum_{a_k' \in A_k^\dagger} \beta_k(a_k') x_k \cdot r_k(a_k') \\
&\geq u_k^\dagger(a_k'') + x_k \cdot r_k(a_k'') = u_k(a_k'', \alpha_{-k}^\dagger) + E[x_k(y) \mid (a_k'', \alpha_{-k}^\dagger)].
\end{aligned}$$

∎

Pairwise identifiability allows us to easily weaken pairwise full rank in the "Nash-threat" folk theorem as follows.

Proposition 9.4.1 **A weaker Nash-threat folk theorem** *Suppose there are no short-lived players, A is finite, every pure-action strongly efficient profile is pairwise identifiable for all players, and $\mathscr{F}^\dagger$ has nonempty interior. Let $\tilde{\mathscr{F}}$ denote the convex hull of $u(\tilde{\alpha})$, where $\tilde{\alpha}$ is an inefficient Nash equilibrium, and the set of pure-action strongly efficient profiles. Then, for all $v \in \text{int}\{v' \in \tilde{\mathscr{F}} : v_i' \geq u_i(\tilde{\alpha})\}$, there exists $\underline{\delta} < 1$ such that for all $\delta \in (\underline{\delta}, 1)$, $v \in \mathscr{E}(\delta)$.*

Proof By lemma 9.2.1(2), $\tilde{\alpha}$ is orthogonally enforceable in any direction, and from lemmas 9.4.1, 9.4.2, and 8.1.1(4), the pure-action strongly efficient profiles are orthogonally enforceable in all noncoordinate directions. Suppose $v \in \text{int}\{v' \in \tilde{\mathscr{F}} : v_i' \geq u_i(\tilde{\alpha})\}$. Because v is in the convex hull $\tilde{\mathscr{F}}$, for any noncoordinate direction λ, there is an extreme point $u(a)$ of $\tilde{\mathscr{F}}$ such that $\lambda \cdot v \leq \lambda \cdot u(a)$. Because the extreme points are orthogonally enforceable in noncoordinate directions, $\lambda \cdot v \leq k^*(\lambda)$, that is, $v \in H^*(\lambda)$. The coordinate direction e_i does not impose any further restriction, because at least one action profile maximizing $u_i(a)$ is strongly efficient, and so enforceable, implying $k^*(e_i) = \max_a u_i(a)$. Finally, for the coordinate direction $-e_i$, $-v_i \leq -u_i(\tilde{\alpha}) \leq k^*(-e_i)$, so $v \in H^*(-e_i)$. Therefore, $\text{int}\tilde{\mathscr{F}} \subset \mathscr{M}$ and the proof is then completed by an appeal to proposition 9.1.1.
∎

This is a Nash-threat folk theorem, because the equilibrium profiles that are implicitly constructed through the appeal to self-generation have the property that every continuation equilibrium has payoffs that dominate the payoffs in the Nash equilibrium $\tilde{\alpha}$ (see proposition 9.1.2). Fudenberg, Levine, and Maskin (1994) discuss minmax versions of the folk theorem under pairwise identifiability.

9.5 Games with a Product Structure

In many games of interest (such as games with repeated adverse selection), there is a separate independent public signal for each long-lived player.

Definition 9.5.1 *A game has a* product structure *if its space of public signals Y can be written as $\prod_{i=1}^{n} Y_i \times Y_{SL}$, and*

$$\rho(y \mid a) = \rho((y_1, \ldots, y_n, y_{SL}) \mid (a_1, \ldots, a_N))$$
$$= \rho_{SL}(y_{SL} \mid a_{n+1}, \ldots, a_N) \prod_{i=1}^{n} \rho_i(y_i \mid a_i, a_{n+1}, \ldots, a_N),$$

where ρ_{SL} is the marginal distribution of the short-lived player' signal y_{SL} and ρ_i is the marginal distribution of long-lived player i's signal y_i.

As we discuss in section 10.1, every sequential equilibrium payoff of a game with a product structure is also a PPE payoff (proposition 10.1.1).

The distribution determining the long-lived player i's public signal depends only on player i's action (and the actions of the short-lived players), so it is easy to distinguish between deviations by player i and player j.

Lemma 9.5.1 *If the game has a product structure, every action profile with pure long-lived player actions is pairwise identifiable for all pairs of long-lived players.*

Proof Suppose $\alpha^\dagger$ is an action profile with pure long-lived player actions, that is, $\alpha^\dagger = (a_1^\dagger, \ldots, a_n^\dagger, \alpha_{SL}^\dagger)$ for some $a_i^\dagger \in A_i, i = 1, \ldots, n$. Fix a pair of long-lived players i and j. For each $k = i, j$, there is a set of actions, $A_k^\dagger = \{a_k(1), \ldots, a_k(\ell_k)\} \subset A_k$ such that $a_k(\ell_k) = a_k^\dagger$, $|A_k^\dagger| = \ell_k = \operatorname{rank} R_k(a_{-k}^\dagger)$ and the collection of distributions $\{\rho(\cdot \mid a_k(h), \alpha_{-k}^\dagger) : h = 1, \ldots, \ell_k\}$ is linearly independent. We need to argue that the collection $\{\rho(\cdot \mid a_i(h), \alpha_{-i}^\dagger) : h = 1, \ldots, \ell_i\} \cup \{\rho(\cdot \mid a_j(h), \alpha_{-j}^\dagger) : h = 1, \ldots, \ell_j - 1\}$ is linearly independent, because this implies $\operatorname{rank} R_{ij}(\alpha^\dagger) = \ell_i + \ell_j - 1 = \operatorname{rank} R_i(\alpha_{-i}^\dagger) + \operatorname{rank} R_j(\alpha_{-j}^\dagger) - 1$. The rank of $R_{ij}(\alpha^\dagger)$ cannot be larger, because the collections of distributions corresponding to $A_i^\dagger$ and $A_j^\dagger$ contain the common distribution, $\rho(\cdot \mid \alpha^\dagger)$.

We argue to a contradiction. Suppose there are scalars $\{\beta_k(h)\}$ such that for all y, $\sum_{h=1}^{\ell_i} \beta_i(h) \rho(y \mid a_i(h), \alpha_{-i}^\dagger) + \sum_{h=1}^{\ell_j - 1} \beta_j(h) \rho(y \mid a_j(h), \alpha_{-j}^\dagger) = 0$. Because the collections of distributions for each player are linearly independent, at least one $\beta_i(h)$ and one $\beta_j(h)$ are nonzero. The game has a product structure, so we can sum over y_{-ij} to get, for all y_i and y_j (where we have suppressed the common α_{SL} term in ρ_i and ρ_j),

$$\sum_{h=1}^{\ell_i} \beta_i(h) \rho_i(y_i \mid a_i(h)) \rho_j(y_j \mid a_j^\dagger) + \sum_{h=1}^{\ell_j - 1} \beta_j(h) \rho_i(y_i \mid a_i^\dagger) \rho_j(y_j \mid a_j(h)) = 0. \tag{9.5.1}$$

For any pair of signals y_i and y_j with $\rho_i(y_i \mid a_i^\dagger) \rho_j(y_j \mid a_j^\dagger) > 0$, dividing by that product gives

$$\sum_{h=1}^{\ell_i} \beta_i(h) \frac{\rho_i(y_i \mid a_i(h))}{\rho_i(y_i \mid a_i^\dagger)} + \sum_{h=1}^{\ell_j - 1} \beta_j(h) \frac{\rho_j(y_j \mid a_j(h))}{\rho_j(y_j \mid a_j^\dagger)} = 0.$$

If the second summation depends on y_j, then we have a contradiction because the first does not. So, suppose the second summation is independent of y_j, that is,

$$\sum_{h=1}^{\ell_j-1} \beta_j(h)\rho_j(y_j \mid a_j(h)) = c\rho_j(y_j \mid a_j^\dagger)$$

for some c, for all y_j receiving positive probability under $a^\dagger$. Moreover, for y_j receiving zero probability under $a^\dagger$, (9.5.1) implies $\sum_{h=1}^{\ell_j-1} \beta_j(h)\rho_j(y_j \mid a_j(h)) = 0$, and so $\{\rho(\cdot \mid a_j(h), \alpha_{-j}^\dagger) : h = 1, \ldots, \ell_j\}$ is linearly dependent, a contradiction. ∎

As an immediate implication of proposition 9.4.1, we thus have the following folk theorem.

Proposition 9.5.1 *Suppose there are no short-lived players, A is finite, the game has a product structure, and $\mathscr{F}^\dagger$ has nonempty interior. Let $\tilde{\mathscr{F}}$ denote the convex hull of $u(\tilde{\alpha})$, where $\tilde{\alpha}$ is an inefficient Nash equilibrium, and the set of pure-action strongly efficient profiles. Then, for all $v \in \text{int}\{v' \in \tilde{\mathscr{F}} : v_i' \geq u_i(\tilde{\alpha})\}$, there exists $\underline{\delta} < 1$ such that for all $\delta \in (\underline{\delta}, 1)$, $v \in \mathscr{E}(\delta)$.*

9.6 Repeated Extensive-Form Games

We now consider repeated extensive-form games. In this section, there are no short-lived players. The stage game Γ is a finite extensive-form game with no moves of nature. We denote the infinite repetition of Γ by Γ^∞. Because the public signal of play within the stage game is the terminal node reached, the set of terminal nodes is denoted Y with typical element y. An action for player i specifies a *move* for player i at every information set owned by that player. Given an action profile a, because there are no moves of nature, a unique terminal node $y(a)$ is reached under the moves implied by a. The ex post payoff $u_i^*(y, a_i)$ depends only on y (payoffs are assigned to terminal nodes of the extensive form), and $u_i(a) = u_i^*(y(a)) \equiv u_i^*(y(a), a_i)$. A sequence of action profiles $(a^0, a^1, \ldots, a^{t-1})$ induces a public history of terminal nodes $h^t = (y(a^0), y(a^1), \ldots, y(a^{t-1}))$.

As we discuss in section 5.4, there are additional incentive constraints that must be satisfied when the stage game is an extensive-form game, arising from behavior at unreached subgames. In particular, a PPE profile need *not* be a subgame-perfect equilibrium of the repeated extensive-form game.

It is immediate that the approach in sections 3.4.2 and 3.7 can be used to prove a folk theorem for repeated extensive-form games.[9] Here we modify the tools from chapters 7

---

9. For example, one modifies the construction in proposition 3.4.1 by prescribing an equilibrium outcome path featuring a stage-game action profile a^0 that attains the desired payoff from $\mathscr{F}^*$, with play entering a player i punishment path in any period in which player i was the last player to deviate from equilibrium actions (given the appropriate history) in the previous period's (extensive-form) stage game. Wen (2002) presents a folk theorem for multistage games with observable actions, replacing the full dimensionality condition with a generalization of NEU.

and 8 to accommodate the additional incentive constraints raised by extensive-form stage games. There are only minor modifications to these tools, so we proceed quickly.

Let Ξ be the collection of initial nodes of the subgames of the extensive form, with ξ^0 being the initial node of the extensive form. We denote the subgame of Γ with initial node $\xi \in \Xi$ by Γ_ξ; note that $\Gamma_{\xi^0} = \Gamma$. If the extensive form has no nontrivial subgames, then $\Xi = \{\xi^0\}$. Given a node $\xi \in \Xi$, $u_i(a \mid \xi)$ is i's payoff in Γ_ξ given the moves in Γ_ξ implied by a. The terminal node reached by a conditional on ξ is denoted $y(a \mid \xi)$ (this is a terminal node of Γ_ξ), so that $u_i(a \mid \xi) = u_i^*(y(a \mid \xi))$ and $u_i(a \mid \xi^0) = u_i(a)$.

Every subgame of Γ^∞ is reached by some public history of past terminal nodes, $h \in \mathcal{H}$ (i.e., $h \in Y^t$ for some t), and a sequence of moves within Γ that reach some node $\xi \in \Xi$. For such a history (h, ξ), the subgame reached is strategically equivalent to the infinite horizon game that begins with Γ_ξ, followed by Γ^∞; we denote this game by $\Gamma^\infty(\xi)$. Note that $\Gamma^\infty(\xi^0) = \Gamma^\infty$, the game strategically equivalent to the subgames that are the focus of chapters 2 and 3. Denote the strategy profile induced by σ on $\Gamma^\infty(\xi)$ by $\sigma|_{(h,\xi)}$.

Definition 9.6.1 *A strategy profile σ is a subgame-perfect equilibrium if for every public history $h \in \mathcal{H}$ and every node $\xi \in \Xi$, the profile $\sigma|_{(h,\xi)}$ is a Nash equilibrium of $\Gamma^\infty(\xi)$.*

We then have the following analog of propositions 2.2.1 and 7.1.1 (the proof is essentially identical, mutatis mutandis).

Proposition 9.6.1 *Suppose there are no short-lived players. A strategy profile is a subgame-perfect equilibrium if and only if for all histories (h, ξ), $\sigma(h)$ is a Nash equilibrium of the normal-form game with payoffs*

$$g(a) = (1-\delta)u(a \mid \xi) + \delta U(\sigma|_{(h, y(a|\xi))}).$$

We now modify definitions 7.3.1 and 7.3.2.

Definition 9.6.2 *For any $\mathcal{W} \subset \mathbb{R}^n$, a pure action profile $a \in A$ is subgame enforceable on $\mathcal{W}$ if there exists a mapping $\gamma : Y \to \mathcal{W}$ such that, for all initial nodes $\xi \in \Xi$, for all i, and $a_i' \in A_i$,*

$$(1-\delta)u_i(a \mid \xi) + \delta\gamma_i(y(a \mid \xi)) \geq (1-\delta)u_i(a_i', a_{-i} \mid \xi) + \delta\gamma_i(y(a_i', a_{-i} \mid \xi)).$$

The function γ subgame enforces a (on $\mathcal{W}$).

Because we only use pure action decomposability in this section, we omit the adjective "pure-action" in the following notions.

Definition 9.6.3 *A payoff vector $v \in \mathbb{R}^n$ is subgame decomposable on $\mathcal{W}$ if there exists an action profile $a \in A$, subgame enforced by γ on $\mathcal{W}$, such that*

$$v_i = (1-\delta)u_i(a) + \delta\gamma_i(y(a)).$$

The payoff v is subgame decomposed by the pair (a, γ) (on $\mathcal{W}$).

A set of payoffs $\mathcal{W} \subset \mathbb{R}^n$ is subgame self-generating if every payoff profile in $\mathcal{W}$ is subgame decomposed by some pair (a, γ) on $\mathcal{W}$.

9.6 ■ Repeated Extensive-Form Games

We now have an analog to propositions 2.5.1 and 7.3.1 (the proof is identical, apart from the appeal to proposition 9.6.1, rather than proposition 2.2.1 or 7.1.1):

Proposition 9.6.2 *For any bounded set $\mathscr{W} \subset \mathbb{R}^n$, if $\mathscr{W}$ is subgame self-generating, then $\mathscr{W} \subset \mathscr{E}^E(\delta)$, that is, $\mathscr{W}$ is a set of subgame-perfect equilibrium payoffs.*

It is a straightforward verification that proposition 7.3.4 and corollaries 7.3.2 and 7.3.3 hold for the current notions (where *locally subgame self-generating* is the obvious notion).

We now consider the analog of (8.1.1) for the current scenario. Let $\kappa^*(a, \lambda, \delta)$ be the maximum value of $\lambda \cdot v$ over v for which there exists $\gamma : Y \to \mathbb{R}^n$ satisfying

$$v = (1 - \delta)u(a) + \delta\gamma(y(a)),$$

$$(1-\delta)u_i(a \mid \xi) + \delta\gamma_i(y(a \mid \xi)) \geq (1-\delta)u_i(a'_i, a_{-i} \mid \xi) + \delta\gamma_i(y(a'_i, a_{-i} \mid \xi)),$$

$$\forall \xi \in \Xi, \forall a'_i \in A_i, \forall i,$$

and

$$\lambda \cdot v \geq \lambda \cdot \gamma(y), \quad \forall y \in Y.$$

As in chapter 8, we can replace γ_i with $x_i(y) = \delta(\gamma_i(y) - v_i)/(1 - \delta)$ to obtain constraints independent of the discount factor. It is easy to verify that the subgame version of lemma 8.1.1 holds. Let $\kappa^*(\lambda) = \max_{a \in A} \kappa^*(a, \lambda, \delta)$ and

$$\mathscr{M}^E \equiv \cap_\lambda \{v : \lambda \cdot v \leq \kappa^*(\lambda)\}.$$

We then have the analog of proposition 9.1.2 (which is proved similarly):

Proposition 9.6.3 *Suppose the stage game is a finite extensive form. Suppose $\mathscr{W}$ is a smooth convex subset of the interior of $\mathscr{M}^E$. Then, $\mathscr{W}$ is locally subgame self-generating, so there exists $\underline{\delta} < 1$ such that for all $\delta \in (\underline{\delta}, 1)$, $\mathscr{W} \subset \mathscr{E}^E(\delta)$.*

We are now in a position to state and prove the folk theorem for repeated extensive-form games.

Proposition 9.6.4 *Suppose there are no short-lived players, and the stage game is a finite extensive-form game with no moves of nature. Suppose, moreover, that $\mathscr{F}^\dagger$ has nonempty interior and $\underline{v}^p = (\underline{v}_1^p, \ldots, \underline{v}_n^p)$ is inefficient. For every $v \in \text{int} \mathscr{F}^{\dagger p}$, there exists $\underline{\delta} < 1$ such that for all $\delta \in (\underline{\delta}, 1)$ there exists a subgame-perfect equilibrium of the repeated extensive-form game with value v.*

Proof Given proposition 9.6.3, we need only prove that $\mathscr{F}^{\dagger p} \subset \mathscr{M}^E$.

Step 1. We first show that any pure action profile a is subgame enforced, that is, there exists $x : Y \to \mathbb{R}^n$ such that, for all $\xi \in \Xi$,

$$u_i(a \mid \xi) + x_i(y(a \mid \xi)) \geq u_i(a'_i, a_{-i} \mid \xi) + x_i(y(a'_i, a_{-i} \mid \xi)),$$

$$\forall a'_i \in A_i, \forall i. \quad (9.6.1)$$

As Ξ is the set of initial nodes of subgames of Γ, it is partially ordered by precedence, where $\xi \prec \xi'$ if ξ' is a node in Γ_ξ. Let $\{\Xi_\ell\}_{\ell=0}^L$ be a partition of Ξ, where $\Xi_0 = \{\xi^0\}$, $\Xi_\ell = \{\xi \in \Xi : \xi' \prec \xi, \nexists \xi'', \xi' \prec \xi'' \prec \xi \text{ for some } \xi' \in \Xi_{\ell-1}\}$.

Because Ξ is finite, L is finite. Let Y_ξ be the set of terminal nodes in Γ_ξ. If ξ and ξ' are not ordered by $\prec$, then $Y_\xi \cap Y_{\xi'} = \emptyset$, whereas if $\xi \prec \xi'$ then $Y_{\xi'} \subsetneq Y_\xi$ (i.e., we have strict inclusion).[10] Of course, $Y_{\xi^0} = Y$.

Fix an initial node $\xi \in \Xi_L$. Any two unilateral deviations from a in Γ_ξ that result in a terminal node different from $y(a \mid \xi)$ being reached must reach distinct terminal nodes, that is, if $y(a'_i, a_{-i} \mid \xi) \neq y(a \mid \xi)$ and $y(a'_j, a_{-j} \mid \xi) \neq y(a \mid \xi)$, then $y(a'_i, a_{-i} \mid \xi) \neq y(a'_j, a_{-j} \mid \xi)$. Consequently, for $y \in Y_\xi$, we can specify $x^L(y)$ so that (9.6.1) is satisfied using $x = x^L$.

We now proceed by induction. Suppose we have specified, for $\ell \geq 1$, $x^\ell(y)$ for $y \in \cup_{\xi \in \Xi_\ell} Y_\xi$ so that (9.6.1) holds for all $\xi \in \cup_{k \geq \ell} \Xi_\ell$. Fix an initial node $\xi' \in \Xi_{\ell-1}$. If there is no node in Ξ_ℓ following ξ', then $x^\ell(y)$ for $y \in Y_{\xi'}$ is undefined. As for the case $\xi \in \Xi_L$, we can specify $x^{\ell-1}(y)$ for $y \in Y_{\xi'}$ so that (9.6.1) is satisfied using $x = x^{\ell-1}$ for ξ', without affecting (9.6.1) for $\xi \in \cup_{k \geq \ell} \Xi_\ell$.

Suppose now there is one (or more nodes) in Ξ_ℓ following ξ'. If $y(a \mid \xi') \notin \cup_{\xi \in \Xi_\ell} Y_\xi$ (i.e., none of these nodes are reached under a from ξ'), then set

$$x_i^{\ell-1}(y) = \begin{cases} \bar{M}^\ell, & y = y(a \mid \xi'), \\ x_i^\ell(y), & y \in \cup_{\xi \in \Xi_\ell} Y_\xi, \\ 0, & y \notin \cup_{\xi \in \Xi_\ell} Y_\xi, y \neq y(a \mid \xi'), \end{cases}$$

where $\bar{M}^\ell = \max u_i(a) - \min u_i(a) + 1 + \max x_i^\ell(y)$. This choice of $x^{\ell-1}(y)$ for $y \in Y_{\xi'}$ ensures that (9.6.1) is satisfied using $x = x^{\ell-1}$ for ξ', without affecting (9.6.1) for $\xi \in \cup_{k \geq \ell} \Xi_\ell$ (because the normalized continuations on the subgames are unaffected).

If $y(a \mid \xi') \in Y_{\xi''}$ for some (necessarily unique) $\xi'' \in \Xi_\ell$, set

$$x_i^{\ell-1}(y) = \begin{cases} x_i^\ell(y) + \bar{M}^\ell, & y \in Y_{\xi''}, \\ x_i^\ell(y), & y \in \cup_{\xi \in \Xi_\ell, \xi \neq \xi''} Y_\xi, \\ 0, & y \notin \cup_{\xi \in \Xi_\ell} Y_\xi. \end{cases}$$

This choice of $x^{\ell-1}(y)$ for $y \in Y_{\xi'}$ ensures that (9.6.1) is satisfied using $x = x^{\ell-1}$ for ξ', without affecting (9.6.1) for $\xi \in \cup_{k \geq \ell} \Xi_\ell$ (because the normalized continuations on the subgames are either unaffected, or modified by a constant). Hence, (9.6.1) holds for all $\xi \in \Xi$.

Consequently (recall lemma 9.2.1), the pure-action minmax profile $\hat{a}^i$ is orthogonally enforceable in the direction $-e_i$.

Step 2. It remains to argue that every pure action profile is orthogonally subgame enforced in all noncoordinate directions. Because a "subgame" version of lemma 8.1.1 holds with identical proof, it is enough to show that a is orthogonally subgame enforced in all pairwise directions. Denote by $\hat{x}$ the normalized continuations just constructed to satisfy (9.6.1). For a fixed ij-pairwise direction λ,

---

10. Without loss of generality, we assume each node has at least two moves.

9.6 ■ Repeated Extensive-Form Games

we now construct orthogonally enforcing continuations from $\hat{x}$. We first define new continuations that satisfy the orthogonality constraint for $y = y(a)$, that is,

$$\tilde{x}_k(y) = \begin{cases} \hat{x}_k(y), & \text{if } k \neq i, j, \\ \hat{x}_k(y) - \hat{x}_k(y(a)), & \text{if } k = i \text{ or } j. \end{cases}$$

Then, $\lambda \cdot \tilde{x}(y(a)) = 0$. Moreover, because a constant was subtracted from x_i independent of y (and similarly for j), (9.6.1) holds for all $\xi \in \Xi$ under $\tilde{x}$.

We first iteratively adjust $\tilde{x}$ so that on the outcome path of every subgame of Γ, orthogonality holds. Suppose $\xi \in \Xi_1$ and that ξ is reached from ξ^0 by a unilateral deviation by player i (hence, $\exists y \in Y_\xi$, $\exists a_i' \neq a_i$, $y = y(a_i', a_{-i})$).

For $y \in Y_\xi$, we subtract from each of player j's continuations the quantity $\tilde{x}_j(y(a \mid \xi)) + \lambda_i \tilde{x}_i(y(a \mid \xi))/\lambda_j$. Observe that these new continuations satisfy (9.6.1) for all $\xi'' \in \Xi$. Moreover, denoting the new continuations also by $\tilde{x}$, $\lambda \cdot \tilde{x}(y(a \mid \xi)) = 0$.

We proceed analogously for a node $\xi \in \Xi_1$ reached from ξ^0 by a unilateral deviation by player j. The only other possibility is that ξ cannot be reached by a unilateral deviation by i or j, in which case we can set $\tilde{x}_i(y) = \tilde{x}_j(y) = 0$ for all $y \in Y_\xi$ without any impact on incentives.

Proceeding inductively, we now assume $\lambda \cdot \tilde{x}(y(a \mid \xi')) = 0$ for all $\xi \in \cup_{m=0}^{\ell-1} \Xi_m$. Suppose $\xi \in \Xi_\ell$, and denote its immediate predecessor by $\xi^{\ell-1} \in \Xi_{\ell-1}$. If $y(a \mid \xi^{\ell-1}) = y(a \mid \xi)$, then $\lambda \cdot \tilde{x}(y(a \mid \xi)) = 0$, and no adjustment is necessary. Suppose $y(a \mid \xi^{\ell-1}) \neq y(a \mid \xi)$ and ξ is reached from $\xi^{\ell-1}$ by a unilateral deviation by either i or j (hence, either $\exists y \in Y_\xi$, $\exists a_i'$, $y = y(a_i', a_{-i} \mid \xi^{\ell-1})$, or $\exists y \in Y_\xi$, $\exists a_j'$, $y = y(a_j', a_{-j} \mid \xi^{\ell-1})$). Suppose it is player i. For $y \in Y_\xi$, we subtract $\tilde{x}_j(y(a \mid \xi)) + \lambda_i \tilde{x}_i(y(a \mid \xi))/\lambda_j$ from each of player j's continuations. These new continuations also satisfy (9.6.1) for all $\xi'' \in \Xi$. Moreover, denoting the new continuations also by $\tilde{x}$, $\lambda \cdot \tilde{x}(y(a \mid \xi)) = 0$.

As before, the only other possibility is that ξ cannot be reached from $\xi^{\ell-1}$ by a unilateral deviation by i or j, in which case we can set $\tilde{x}_i(y) = \tilde{x}_j(y) = 0$ for all $y \in Y_\xi$ without any impact on incentives.

Proceeding in this way yields continuations $\tilde{x}$ that satisfy (9.6.1) and for all $\xi \in \Xi$, $\lambda \cdot \tilde{x}(y(a \mid \xi')) = 0$. It remains to adjust $\tilde{x}$ for $y \neq y(a \mid \xi)$ for all ξ.

Fix such a y. There is a "last" node $\xi \in \Xi$ such that $y \in Y_\xi$ (that is, $\nexists \xi'$, $\xi \prec \xi'$, with $y \in Y_{\xi'}$). We partition Y_ξ into three sets, $\{y(a \mid \xi)\}$, Y_ξ^1, and Y_ξ^2, where $Y_\xi^1 = \{y \in Y_\xi : y = y(a_i', a_{-i} \mid \xi)$ for some $a_i' \neq a_i$, or $y = y(a_j', a_{-j} \mid \xi)$ for some $a_j' \neq a_j\}$ is the set of nodes reached via a unilateral deviation by either i or j from ξ, and $Y_\xi^2 = Y_\xi \setminus (Y_\xi^1 \cup \{y(a \mid \xi)\})$ are the remaining terminal nodes. If $y \in Y_\xi^2$, we can redefine $\tilde{x}_i(y) = \tilde{x}_j(y) = 0$ without affecting any incentive constraints, and obtaining $\lambda \cdot \tilde{x}(y) = 0$. Finally, suppose $y \in Y_\xi^1$ and $y = y(a_i', a_{-i})$ for some $a_i' \in A_i$. Then, player j cannot unilaterally deviate from a and reach y, so that the value of $x_j(y)$ is irrelevant for j's incentives, and we can set $x_j(y(a_i', a_{-i})) = -\lambda_i x_i(y(a_i', a_{-i}))/\lambda_j$.

■

9.7 Games of Symmetric Incomplete Information

This section examines an important class of dynamic games, games of symmetric incomplete information. Although there is incomplete information about the state of the world, because players have identical beliefs about the state after all histories, decomposability and self-generation are still central concepts.

The game begins with the draw of a state ξ from the finite set of possible states Ξ, according to the prior distribution μ^0, with $\mu^0(\xi) > 0$ for all $\xi \in \Xi$. The players know the prior distribution μ^0 but receive no further information about the state. Each player i, $i = 1, \ldots, n$, has a finite set of actions A_i, independent of ξ, with payoffs given by $u_i : A \times \Xi \to \mathbb{R}$.[11]

Every pair of states $\xi, \xi' \in \Xi$ are *distinct*, in the sense that there is an action profile a for which $u(a, \xi) \neq u(a, \xi')$. As usual, $u(\alpha, \mu) = \sum_\xi \sum_a u_i(a, \xi)\alpha(a)\mu(\xi)$.

The players share a common discount factor δ. At the end of each period t, each player observes the action profile a^t and the realized payoff profile $u(a^t, \xi)$.

Observing the action and payoff profiles may provide the players with considerable information about the state, but it does not ensure that the state is instantly learned or even eventually learned. For any pair of states ξ and ξ', there may be some action profiles that give identical payoffs, so that some observations may give only partial information about the state or may give no information at all. Because the set of states is finite and every pair of states is distinct, there are repeated-game strategy profiles ensuring that the players will learn the state in a finite number of periods. However, it is not obvious that an equilibrium will feature such behavior. The players in the game face a problem similar to a multiarmed bandit. Learning the payoffs of all the arms requires that they all be pulled. It may be optimal to always pull arm 1, never learning the payoff of arm 2.

As players become increasingly patient, experimenting in their action choices to collect information about the state becomes increasingly inexpensive. However, even arbitrarily patient players may optimally never learn the state. The players' bandit problem is interactive. The informativeness of player i's action depends on j's choice, so that i alone cannot ensure that the state is learned. In addition, the outcome that appears once a state is learned is itself an equilibrium phenomenon, raising the possibility that a player may prefer to *not* learn the state.

Example 9.7.1 **Players need not learn** Suppose there are two players and two equally likely states, with the stage games for the two states given in figure 9.7.1. Suppose first that $x = y = 0$, and consider an equilibrium in which players 1 and 2 each choose C in each period, regardless of history. These are stage-game best responses for any posterior belief about the state. In addition, no unilateral deviation results in a profile that reveals any information about the state (even though there exist profiles that would identify the state). We thus have an equilibrium in which the players never learn the state.

There are also equilibria in which players learn the state. One such profile has an outcome path with AA in the initial period, followed by BB in odd periods

---

11. As in section 5.5.1, this is without loss of generality.

9.7 ■ Symmetric Incomplete Information

	A	B	C
A	5,1	0,0	y,0
B	0,0	1,5	x,0
C	0,y	0,x	2,2

ξ'

	A	B	C
A	0,0	1,5	x,0
B	5,1	0,0	y,0
C	0,x	0,y	2,2

ξ''

Figure 9.7.1 Stage games for state ξ' and state ξ''.

and AA in even periods if payoffs reveal the state is ξ', and by AB in odd periods and BA in even periods if the state is revealed to be ξ''. Given the specification of AA in the initial period, the state is revealed at the end of period 0 even if a player unilaterally deviates to B. The profile is completed by specifying that any deviation to C in the initial period restarts the profile, that other deviations by player 2 after period 0 and all other deviations by player 1 are ignored, and finally that a deviation by player 2 in period 0 to B results in permanent AA in ξ' and permanent BA in ξ''. Player 1 is always playing a stage-game best response. Though A is not a best response for player 2 in period 0 (she is best responding in every subsequent period, given the revelation of the state in period 0), the specification ensures that she is optimizing for large δ.

Suppose now that $x = 6$ and $y = 7$. The profile in which CC is played in every period after every history is clearly no longer an equilibrium. Nonetheless, there is still an equilibrium in which players do not learn the state. In the equilibrium, CC is played after any history featuring only the outcomes CC. After any history in which the state becomes known as a result of a unilateral deviation by player 1 (respectively, 2), continuation play consists of the perpetual play of BB (respectively, AA) in state ξ' and AB (respectively, BA) in state ξ''. Continuation play thus penalizes the deviating player for learning the state, and sufficiently patient players will find it optimal to play CC, deliberately not learning the state.

●

Example 9.7.2 Learning facilitates punishment Continuing from example 9.7.1, suppose $x = y = 10$. We are still interested in the existence of equilibria in which learning does not occur, that is, in equilibria with outcome path $CC, CC, CC, \ldots$. In contrast to example 9.7.1, a deviation (though myopically profitable) does not reveal the state. However, we can use subsequent learning to create appropriate incentives. Consider the following *revealing* profile, σ^1. In the initial period, play BB, followed by BB in state ξ' and AB in state ξ'', in every subsequent period. Deviations to C in period 0 restart the profile, and other deviations are ignored. The actions are a stage-game Nash equilibrium in each period and only an initial deviation to C can affect information about the state, so this profile is an equilibrium, giving player 1 a payoff close to 1. Denote by σ^2 the analogous revealing profile that gives player 2 a payoff close to 1. Then, for large δ, there exists a pure-strategy nonrevealing equilibrium with outcome path $CC, CC, CC, \ldots$, simultaneous deviations ignored, and continuation play given by σ^i after unilateral deviation by player i.

◆

Remark 9.7.1 **Imperfect monitoring** Both actions and payoffs are perfectly monitored in the model just described. The canonical single-agent bandit problem considers the more complicated inference problem in which arms have random payoffs. Wiseman (2005) considers a game of symmetric incomplete information with a similar inference problem. For each action profile a and state ξ, there is a distribution $\rho(y \mid a, \xi)$ determining the draw of a public signal $y \in Y$ for some finite Y. Ex ante payoffs, as a function of the action profile and the state, are given by

$$u_i(a, \xi) = \sum_{y \in Y} u_i^*(y, a) \rho(y \mid a, \xi),$$

where $u_i^* : Y \times A \to \mathbb{R}$ is i's ex post payoff.

After each period, players observe both the actions chosen in that period and the realized payoff profile (or equivalently, the signal y). Our formulation is the special case in which the distribution $\rho(y \mid a, \xi)$ is degenerate for each a and ξ.

◆

9.7.1 Equilibrium

For each player i and state ξ, we let $\underline{v}_i^p(\xi)$ be the pure minmax payoff:

$$\underline{v}_i^p(\xi) = \min_{a_{-i}} \max_{a_i} u_i(a_i, a_{-i}, \xi).$$

We then let $\mathscr{F}(\xi)$, $\mathscr{F}^\dagger(\xi)$, and $\mathscr{F}^{\dagger p}(\xi)$ be the set of pure stage-game payoffs, the convex hull of this set, and the subset of the latter that is strictly individually rational, for state ξ. We assume that $\mathscr{F}^{\dagger p}(\xi)$ has dimension n, for each ξ.

A period t history $(\xi, \{a^\tau, u^\tau\}_{\tau=0}^{t-1})$ contains the state and the action profiles and payoffs realized in periods 0 through $t-1$. A period t public history h^t contains the action profiles and payoffs from periods 0 to $t-1$. The set of feasible period t public histories $\mathscr{H}^t$ is thus a subset of $(A \times \mathbb{R}^n)^t$, with $\{a^\tau, u^\tau\}_{\tau=0}^{t-1} \in \mathscr{H}^t$ if there exists $\xi \in \Xi$ with the property that, for each period $\tau = 0, \ldots, t-1, u^\tau = u(a^\tau, \xi)$.

Given a prior $\mu \in \Delta(\Xi)$, an action profile a and a payoff u may rule out some states but otherwise does not alter the odds ratio. That is, the posterior is given by

$$\varphi(\mu \mid a, u)(\xi) \equiv \begin{cases} \mu(\xi) / \sum_{\{\xi' : u(a, \xi') = u\}} \mu(\xi'), & \text{if } u(a, \xi) = u, \\ 0, & \text{if } u(a, \xi) \neq u. \end{cases}$$

We write $\varphi^*(\mu \mid a, \xi)$ for $\varphi(\mu \mid a, u(a, \xi))$.

The period t posterior distribution over Ξ, given the period t history h^t, is denoted by $\mu(\cdot \mid h^t)$ or $\mu(h^t)$. These posteriors are defined recursively, with

$$\mu(\varnothing) = \mu^0 \tag{9.7.1}$$

9.7 ■ Symmetric Incomplete Information

and
$$\mu(h^t) = \varphi(\mu(h^{t-1}) \mid a, u), \tag{9.7.2}$$
where $h^t = (h^{t-1}, a, u)$.

Fix a prior distribution μ^0 and let $\mathscr{K}$ be the set of possible posterior distributions, given prior μ^0. The set $\mathscr{K}$ is finite with $2^K - 1$ elements, where K is the number of states in Ξ. Each posterior is identified by the states that have not been ruled out, that is, that still command positive probability. It is clear from (9.7.1) and (9.7.2) that for any subset of states, every history under which beliefs attach positive probability to precisely that subset must give the same posterior distribution.

A behavior strategy σ for player i is a function mapping from the set of public histories $\mathscr{H}$ into $\Delta(A)$. We let $U_i(\sigma, \mu)$ be the expected payoff to player i given strategy profile σ and prior distribution over states μ.

We can treat this game of symmetric incomplete information as a dynamic game of perfect information by setting the set of *game states* equal to $\mathscr{K}$, setting the stage-game payoffs equal to the expected payoffs under the current state, and determining game-state transitions by φ.

Because Bayes' rule specifies beliefs after every public history, beliefs trivially satisfy the consistency requirement of sequential equilibrium.

Definition 9.7.1 *The strategy profile σ is a* sequential equilibrium *if, given the beliefs implied by (9.7.1) and (9.7.2), for all players i, histories h, and alternative strategies σ_i',*
$$U_i(\sigma \mid_h, \mu(h)) \geq U_i(\sigma_i', \sigma_{-i} \mid_h, \mu(h)).$$

The set of sequential equilibrium profiles of the game of symmetric incomplete information clearly coincides with the set of subgame-perfect equilibrium profiles of the implied dynamic game.

We now provide a self-generating characterization of sets of sequential equilibrium payoffs. Let $\mathscr{W} \subset \mathbb{R}^n$ be a set of payoff profiles. A pair $(v, \mu) \in \mathscr{W} \times \mathscr{K}$ identifies a payoff profile v for the repeated game whose prior probability over Ξ is μ. Let $\mathscr{W}_\mathscr{K}$ be a subset of $\mathscr{W} \times \mathscr{K}$ (where $\mathscr{W}_\mathscr{K}$ need not be a product set). A pair (v, μ) is *decomposed* by the profile $\alpha \in \prod_i \Delta(A_i)$ on the set $\mathscr{W}_\mathscr{K}$ if there exists a function $\gamma : A \times \mathscr{K} \to \mathbb{R}^n$ with $(\gamma(a, \hat{\mu}), \hat{\mu}) \in \mathscr{W}_\mathscr{K}$ for all a and $\hat{\mu} \in \mathscr{K}$ such that, for each player i and action a_i',

$$v_i = (1-\delta) u_i(\alpha, \mu) + \delta \sum_{a \in A} \gamma_i(a, \varphi^*(\mu \mid a, \xi)) \alpha(a) \mu(\xi) \tag{9.7.3}$$

$$\geq (1-\delta) u_i(a_i', \alpha_{-i}, \mu) + \delta \sum_{a_{-i} \in A_{-i}} \gamma(a_i', a_{-i}, \varphi^*(\mu \mid a_i, a_{-i}, \xi)) \alpha_{-i}(a_{-i}) \mu(\xi). \tag{9.7.4}$$

The first condition ensures that expected payoffs are given by v, and the second supplies the required incentive compatibility constraints. We cannot in general take the set $\mathscr{W}_\mathscr{K}$ to be the product $\mathscr{W} \times \mathscr{K}$. There may be payoff profiles that can be achieved in games with some prior distributions over states but not in games with other prior distributions.

The set $\mathscr{W}_\mathscr{K}$ is *self-generating* if every pair $(v, \mu) \in \mathscr{W}_\mathscr{K}$ can be decomposed on $\mathscr{W}_\mathscr{K}$. The following is a special case of proposition 5.7.4 for dynamic games:

Proposition 9.7.1 *Suppose $\mathscr{W}_{\mathscr{H}}$ is self-generating. Then for every $(v, \mu) \in \mathscr{W}_{\mathscr{H}}$, there is a sequential equilibrium of the repeated game with prior belief μ whose expected payoff is v.*

9.7.2 A Folk Theorem

Let $U(\sigma \mid \xi)$ be the repeated-game payoff from strategy profile σ, given that the state is in fact ξ. Let the states in Ξ be denoted by $\{\xi_1, \ldots, \xi_K\}$. The following is a special case of Wiseman's (2005) main result (see remark 9.7.1).

Proposition 9.7.2 *Fix a prior μ^0 on Ξ with $\mu^0(\xi_k) > 0$ for all $\xi_k \in \Xi$. For any vector of payoff profiles $(v^*(\xi_1), \ldots, v^*(\xi_K))$, with $v^*(\xi_k) \in \operatorname{int} \mathscr{F}^{\dagger p}(\xi_k)$ for all k, and any $\varepsilon > 0$, there is a discount factor $\underline{\delta}$ such that, for all $\delta \in (\underline{\delta}, 1)$, there exists a sequential equilibrium σ of the game of incomplete information with prior μ^0 such that*

$$|U(\sigma \mid \xi_k) - v^*(\xi_k)| < \varepsilon \quad \forall k.$$

We thus have a state-by-state approximate folk theorem. The profile constructed in the proof produces the appropriate payoffs given the state, by first inducing the players to learn the state. As we have seen in example 9.7.1, learning need not occur in equilibrium, and so specific incentives must be provided.

Proof From (9.7.1) and (9.7.2), each posterior in $\mathscr{H}$ is identified by its support. Let $\mathscr{H}(\ell)$ be the collection of posteriors in which ℓ states receive positive probability. Fix $(v^*(\xi_k))_k$, a vector of payoff profiles, with $v^*(\xi_k) \in \operatorname{int} \mathscr{F}^{\dagger p}(\xi_k)$ for all k. Choose $\varepsilon' < \varepsilon/2$ sufficiently small that $\overline{B}_{\varepsilon'}(v^*(\xi_k))$ is in the interior of $\mathscr{F}^{\dagger p}(\xi_k)$ for each ξ_k. For each posterior $\mu \in \mathscr{H}(\ell)$, let

$$C_\mu = \overline{B}_{\varepsilon' \frac{K-\ell+1}{K}} \left(\sum_{\xi_k \in \Xi} v^*(\xi_k) \mu(\xi_k) \right). \tag{9.7.5}$$

For each possible posterior distribution $\mu \in \mathscr{H}$, C_μ is a closed ball of payoffs, centered at the expected value of the target payoffs under μ. The radius decreases in the number of states contained in the support of μ, ranging from a radius of ε'/K when no states have been excluded to a radius of ε' when probability one is attached to a single state. This behavior of the radius of C_μ is critical in providing sufficient freedom in the choice of continuation values after revealing action profiles to provide incentives.

Our candidate for a self-generating set is the union of these sets,

$$\mathscr{C}_{\mathscr{H}} = \cup_{\mu \in \mathscr{H}} (C_\mu \times \{\mu\}).$$

The pair (a, μ), consisting of a pure action profile and a belief, is *revealing* if the belief μ attaches positive probability to two or more states that give different payoffs under a. Hence, a revealing pair (a, μ) ensures that if a is played given posterior μ, then the set of possible states will be refined. In equilibrium, there can only be a finite number of revealing profiles played, at which point beliefs either attach unitary probability to a single state or no further revealing profiles are played.

9.7 ■ Symmetric Incomplete Information

The proof relies on two lemmas. First, lemma 9.7.1 shows there exists a discount factor $\underline{\delta}'$ such that for all $\delta \in (\underline{\delta}', 1)$, the set $\mathscr{C}_{\mathscr{H}}$ is self-generating. This suffices to ensure that $\mathscr{C}_{\mathscr{H}}$ is a set of sequential equilibrium payoffs (proposition 9.7.1) for such δ. Second, lemma 9.7.2 shows that for all $\delta \in (\underline{\delta}', 1)$, there exists an equilibrium with every continuation-payoff/posterior pair (v, μ) in $\mathscr{C}_{\mathscr{H}}$ with the property that in every period, either a revealing profile is played, or $\mu \in \mathscr{H}(1)$.

As a result, for all $\delta \in (\underline{\delta}', 1)$, we are assured of the existence of an equilibrium such that within at most K periods, the continuation payoff lies within $\varepsilon' < \varepsilon/2$ of $v^*(\xi_k)$ for the true state ξ_k. To ensure that payoffs are within ε of v^*, it then suffices to restrict the discount factor to be high enough that

$$(1 - \delta^K)(M - m) < \frac{\varepsilon}{2},$$

where $M \equiv \max_{i,a,\xi} u_i(a, \xi)$ and $m \equiv \min_{i,a,\xi} u_i(a, \xi)$. This gives us the discount factor $\underline{\delta}$ such that for $\delta \in (\underline{\delta}, 1)$, there exists a sequential equilibrium with the desired payoffs.

■

Lemma *There exists $\underline{\delta}'$ such that for all $\delta \in (\underline{\delta}', 1)$, the set $\mathscr{C}_{\mathscr{H}}$ is self-generating.*
9.7.1

Proof The proof proceeds in four steps.

Step 1. Consider the sets C_μ for $\mu \in \mathscr{H}(1)$. Here, the identity of the state is known, and the continuation game is a repeated game of compete information and perfect monitoring. Letting ξ be the state to which μ attaches probability one, each set C_μ is a compact set contained in the interior of the set $\mathscr{F}^{\dagger p}(\xi)$, the set of feasible, pure-action strictly individually rational payoffs for this game. It is then an implication of proposition 9.1.3 and the observation that under perfect monitoring, every pure action profile has pairwise full rank, that there exists δ_μ such that any $(v, \mu) \in C_\mu \times \{\mu\}$ can be decomposed on $C_\mu \times \{\mu\}$, for any $\delta \in (\delta_\mu, 1)$.

Step 2. Suppose μ assigns positive probability to more than one state, and fix $\eta > 0$. Denote by C_μ^η the set of values $v \in C_\mu$ satisfying, for each player i,

$$v_i \geq \min_{\hat{v} \in C_\mu} \hat{v}_i + \eta. \tag{9.7.6}$$

In this step, we show that there exists δ_η such that every point in C_μ^η can be decomposed on $\mathscr{C}_{\mathscr{H}}$ using a revealing action profile, for all $\delta \in (\delta_\eta, 1)$.

Choose $\delta_\eta \in (0, 1)$ so that for all $\delta \in (\delta_\eta, 1)$,

$$\frac{1-\delta}{\delta}(M - m) < \min\left\{\eta, \frac{\varepsilon'}{2\sqrt{n}K}\right\} \equiv \zeta. \tag{9.7.7}$$

Because states are distinct, there is a revealing action profile a that discriminates between at least two of the states receiving positive probability under μ. Fix $v \in C_\mu^\eta$, and let v' be the payoff satisfying

$$v = (1-\delta)u(a,\mu) + \delta v'.$$

Rearranging, (9.7.7) implies $|v'_i - v_i| < \frac{\varepsilon'}{2\sqrt{n}K}, i = 1,\ldots,n$. Hence, letting $\Delta \equiv v' - v$, we have $|\Delta| < \frac{\varepsilon'}{2K}$. Let

$$\Delta_v \equiv v - \sum_{\xi_k \in \Xi} v^*(\xi_k)\mu(\xi_k),$$

so that $|\Delta_v| < \varepsilon'(K - \ell + 1)/K$ (from (9.7.5)). For all $\xi_j \in \text{supp}\mu$ (with the specification for other ξ being arbitrary), set

$$\gamma(a, \varphi^*(\mu \mid a, \xi_j)) = \sum_{\xi_k \in \Xi} v^*(\xi_k)\varphi^*(\mu \mid a, \xi_j)(\xi_k) + \Delta_v + \Delta. \tag{9.7.8}$$

The expected continuation under a and γ is

$$\sum_{\xi_j \in \Xi} \gamma(a, \varphi^*(\mu \mid a, \xi_j))\mu(\xi_j)$$

$$= \sum_{\xi_j \in \Xi} \left(\sum_{\xi_k \in \Xi} v^*(\xi_k)\varphi^*(\mu \mid a, \xi_j)(\xi_k) + \Delta_v + \Delta \right) \mu(\xi_j)$$

$$= \sum_{\xi_j \in \Xi} \left(\sum_{\xi_k \in \Xi} v^*(\xi_k)\varphi^*(\mu \mid a, \xi_j)(\xi_k) \right) \mu(\xi_j) + \Delta_v + \Delta$$

$$= \sum_{\xi_j \in \Xi} \left(\sum_{\xi_k \in \text{supp}\,\varphi^*(\mu \mid a, \xi_j)} v^*(\xi_k)\mu(\xi_k) \right) \left(\sum_{\xi_k \in \text{supp}\,\varphi^*(\mu \mid a, \xi_j)} \mu(\xi_k) \right)^{-1} \mu(\xi_j) + \Delta_v + \Delta$$

$$= \sum_{\xi_k \in \Xi} v^*(\xi_k)\mu(\xi_k) + \Delta_v + \Delta$$

$$= v + \Delta$$

$$= v',$$

so that we have the desired expected value. Because

$$|\Delta_v + \Delta| < |\Delta_v| + \frac{\varepsilon'}{2K} < \varepsilon'\left(K - \ell + \frac{3}{2}\right)/K,$$

the continuation-belief pair satisfy $(\gamma(a, \varphi^*(\mu \mid a, \xi_j)), \varphi^*(\mu \mid a, \xi_j)) \in \mathscr{C}_{\mathscr{H}}$.

We now specify the continuations for unilateral deviations from a (ignoring multiple deviations, as usual). For any unilateral deviation $a' = (a'_i, a_{-i})$ that is a revealing action profile, mimic (9.7.8) in setting, for all $\xi_j \in \text{supp}\mu$,

$$\gamma(a', \varphi^*(\mu \mid a', \xi_j)) = \sum_{\xi_k \in \Xi} v^*(\xi_k)\varphi^*(\mu \mid a', \xi)(\xi_j) + \Delta_v + \Delta - e_i\zeta. \tag{9.7.9}$$

9.7 ■ Symmetric Incomplete Information

Repeating the argument following (9.7.8), player i's payoff under a' is given by $v' - \zeta$. Invoking (9.7.7), a' is not a profitable deviation for player i. Because $\zeta < \varepsilon'/(2K)$, we have

$$|\Delta_v + \Delta - e_i \zeta| < \varepsilon'(K - \ell + 2)/K, \tag{9.7.10}$$

and so again $(\gamma(a', \varphi^*(\mu \mid a', \xi_j)), \varphi^*(\mu \mid a', \xi_j)) \in \mathscr{C}_{\mathscr{H}}$.

For any unilateral deviation $a' = (a'_i, a_{-i})$ that is not a revealing action profile, $\varphi^*(\mu \mid a', \xi_k)) = \mu$ for all $\xi_k \in \operatorname{supp}\mu$. In this case, set

$$\gamma(a', \mu) = \underset{\hat{v} \in C_\mu}{\arg \min} \, \hat{v}_i,$$

so that γ_i is no larger than $v_i - \eta$. From (9.7.7), the deviation is again unprofitable, and since $(\gamma(a', \mu), \mu) \in \mathscr{C}_{\mathscr{H}}$, we have decomposed every point in C_μ^η on $\mathscr{C}_{\mathscr{H}}$.

Step 3. As in Step 2, suppose the posterior μ attaches positive probability to more than one state. Set

$$\eta' \equiv \frac{\varepsilon'}{K}\left(1 - \frac{1}{\sqrt{2}}\right).$$

For any $\eta < \eta'$, if $v \notin C_\mu^\eta$, then (9.7.6) fails for one, and only one, player.[12] Fix an $\eta \in (0, \eta')$, and a payoff profile v in C_μ for which, for one player i,

$$v_i \leq \min_{\hat{v} \in C_\mu} \hat{v}_i + \eta. \tag{9.7.11}$$

The potential difficulty in decomposing v is that we now do not have sufficient punishments available in C_μ to deter player i's deviations.[13]

Let

$$\xi(i, \mu) \in \underset{\{\xi \in \Xi : \mu(\xi) > 0\}}{\arg \min} \, \underline{v}_i^p(\xi). \tag{9.7.12}$$

Of the states assigned positive probability by μ, $\xi(i, \mu)$ is that state giving player i the lowest minmax payoff, if the state were known. Let $\hat{a}_{-i}(\mu)$ be the associated minmax profile in state $\xi(i, \mu)$. This action profile need not minmax player i given posterior μ because it only necessarily minmaxes i in one of the states receiving positive probability under μ. However, $\underline{v}_i^p(\xi(i, \mu)) < v_i$ because v_i is the average (over states) of payoffs which exceed the corresponding minmax payoffs $\underline{v}_i^p(\xi)$, each of which is at least $\underline{v}_i^p(\xi(i, \mu))$.

The idea is to decompose v using the profile $a \equiv (\hat{a}_i, \hat{a}_{-i}(\mu))$, where $\hat{a}_i$ maximizes $u_i(a''_i, \hat{a}_{-i}(\mu), \mu)$. Suppose the action profile $(\hat{a}_i, \hat{a}_{-i}(\mu))$ is revealing. The analysis in step 2 shows that with the exception of nonrevealing deviations $a' = (a'_i, \hat{a}_{-i}(\mu))$, the continuations after any profile $\tilde{a}$ and payoff $u(\tilde{a}, \xi)$ can be

---

12. Inequality (9.7.6) can fail for two players only if $\sqrt{2(r-\eta)^2} < r$ or $\eta > r(1 - 1/\sqrt{2})$ (where r is the radius of C_μ). Finally, $r \geq \varepsilon'/K$ for all $\mu \in \mathscr{H}$.
13. If we use a revealing action to decompose v, then the analysis from step 2 can be used to construct continuations that deter any deviations to revealing actions. The difficulty arises with a deviation to a nonrevealing action profile a'.

chosen so deviations by i are not profitable and $(\gamma(\tilde{a}, \varphi^*(\mu \mid \tilde{a}, \xi)), \varphi^*(\mu \mid \tilde{a}, \xi))$ $\in \mathscr{C}_{\mathscr{H}}$. The difficulty with nonrevealing action profiles $a' = (a'_i, \hat{a}_{-i}(\mu))$ is that the continuation payoff must now come from the set C_μ and player i's payoff is already near its minimum in the set C_μ. Consequently, incentives cannot be provided via continuations. However, for the deviation to a' not to be profitable, it suffices that

$$u_i(a', \mu) \leq u_i((\hat{a}_i, \hat{a}_{-i}(\mu)), \mu).$$

We show that this inequality holds. Player i's expected payoff from $(a'_i, \hat{a}_{-i}(\mu))$ must be the same in every positive probability state (otherwise the profile is revealing). But player i's payoff from $(a'_i, \hat{a}_{-i}(\mu))$ in state $\xi(i, \mu)$ can be no larger than $\underline{v}_i^p(\xi(i, \mu))$, and hence his payoff in every state can be no larger. The deviation to a'_i thus brings a smaller current payoff than v_i. Choosing v as the continuation payoff then ensures that a'_i is suboptimal.

Hence, combining these last two steps, if μ has nonsingleton support, for all $\eta \in (0, \eta')$ and all $\delta \in (\delta_\eta, 1)$, every payoff in C_μ is decomposed on $\mathscr{C}_{\mathscr{H}}$, apart from $v \in C_\mu$ with $v_i \leq \min_{\hat{v} \in C_\mu} \hat{v}_i + \eta$ and $((\hat{a}_i, \hat{a}_{-i}(\mu)), \mu)$ nonrevealing.

Step 4. It remains to decompose payoffs $v \in C_\mu$ with $v_i \leq \min_{\hat{v} \in C_\mu} \hat{v}_i + \eta$ and $((\hat{a}_i, \hat{a}_{-i}(\mu)), \mu)$ nonrevealing, for small η. As we argued in the penultimate paragraph of step 3, $u_i(a, \mu) \equiv u_i((\hat{a}_i, \hat{a}_{-i}(\mu)), \mu) = \underline{v}_i^p(\xi(i, \mu))$. This value is strictly less than v_i. Let $v^i \in C_\mu$ be the unique point for which i's payoff in C_μ is minimized (the point is unique because C_μ is a strictly convex set). Because $v_i^i > u_i(a, \mu)$, there is a $\theta' > 1$ such that $(1 - \theta)u(a, \mu) + \theta v^i \in \text{int} C_\mu$ for all $\theta \in (1, \theta')$. Consequently, there exist $\eta'' \in (0, \eta')$ and $\theta'' > 1$ such that for all $v \in C_\mu$ satisfying (9.7.11), $(1 - \theta)u(a, \mu) + \theta v \in \text{int} C_\mu$ for all $\theta \in (1, \theta'')$. This implies that for all $\delta \in (1/\theta'', 1)$, there exists v' in the interior of C_μ such that

$$v = (1 - \delta)u(a) + \delta v'.$$

We decompose v with a profile $(\hat{a}_i, \hat{a}_{-i}(\mu))$ and continuation payoff v' after $(\hat{a}_i, \hat{a}_{-i}(\mu))$. Continuation payoffs for deviations by players other than i or by deviations on the part of player i to revealing actions are constructed as in step 2. Nonrevealing deviations by player i are ignored. Set $\delta_\mu = \max\{\delta_{\eta''}, 1/\theta''\}$, where $\delta_{\eta''}$ is defined in (9.7.7) for $\eta = \eta''$. It is now straightforward to verify that the required incentive constraints hold for all $\delta \in (\delta_\mu, 1)$.

Summarizing steps 2 through 4, for all μ with nonsingleton support, there exists δ_μ such that any $(v, \mu) \in \mathscr{C}_{\mathscr{H}}$ can be decomposed on $\mathscr{C}_{\mathscr{H}}$, for any $\delta \in (\delta_\mu, 1)$. The proof of the lemma is completed by taking $\underline{\delta}'$ as the maximum of δ_μ over all $\mu \in \mathscr{H}$ (a finite set).

∎

Lemma 9.7.2 *For every $\delta \in (\underline{\delta}', 1)$, there exists a sequential equilibrium with the properties that every history gives rise to a pair $(v, \mu) \in \mathscr{C}_{\mathscr{H}}$ and that in every period, either a revealing profile is played or $\mu \in \mathscr{H}(1)$.*

9.7 ■ Symmetric Incomplete Information

Proof Because δ satisfies (9.7.7) for $\eta = \eta''$ from step 4 of the proof of lemma 9.7.1, by step 2 the payoff $\sum_{\xi_k \in \Xi} v^*(\xi_k) \mu^0(\xi_k)$ can be decomposed by a revealing current action profile a and continuation payoffs $\gamma(a, \varphi^*(\mu \mid a, \xi_j))$ that lie within $\varepsilon'/2K$ of $\sum_{\xi_k \in \Xi} v^*(\xi_k) \varphi^*(\mu \mid a, \xi_j)(\xi_k)$.

For any $\mu = \varphi^*(\mu^0 \mid a, \xi_j)$, because a is revealing, $\mu \in \cup_{\ell < K} \mathcal{H}(\ell)$ and the ball C_μ has radius at least $2\varepsilon'/K$. Hence, given the belief $\mu = \varphi^*(\mu^0 \mid a, \xi_j)$ we can again apply step 2 with a new revealing action profile a' to decompose the continuations $\gamma(a, \varphi^*(\mu \mid a, \xi_j))$. Moreover, the application of step 2 yields continuation payoffs $\gamma(a', \varphi^*(\varphi^*(\mu^0 \mid a, \xi_j) \mid a', \xi_j))$ that lie within ε'/K of $\sum_{\xi_k \in \Xi} v^*(\xi_k) \varphi^*(\varphi^*(\mu^0 \mid a, \xi_j) \mid a', \xi_j)(\xi_k)$ and so satisfy (9.7.6) (because $\eta'' < \varepsilon'/K$).

Now, note that for $\mu \in \mathcal{H}(\ell)$, if $v \in C_\mu$ satisfies $|v - \sum_\xi v^*(\xi)\mu(\xi)| < [\varepsilon'(K - \ell)/(2K)]$, then

$$v_i - \min_{v' \in C_\mu} v'_i \geq \frac{\varepsilon'}{K}\left(K - \ell + 1 - \frac{K - \ell}{2}\right)$$
$$= \frac{\varepsilon'}{K}\left(\frac{K - \ell}{2} + 1\right)$$
$$\geq \frac{\varepsilon'}{K}$$
$$> \eta'',$$

and so satisfies (9.7.6).

Hence, continuing in this fashion, we construct a profile with a revealing action in each period and with the property that any continuation value on the path of play satisfies (9.7.6) (because at each stage, the continuation v is within $\varepsilon'(K - \ell)/(2K)$ of the center of C_μ). Consequently, within at most K periods, the posterior probability is in $\mathcal{H}(1)$ and continuation payoffs are within $\frac{K-1}{K}\varepsilon' < \varepsilon' < \frac{\varepsilon}{2}$ of $v^*(\xi_j)$ for that (single) state ξ_j to which the posterior attaches positive probability. Because the set $\mathcal{C}_\mathcal{H}$ is self-generating for $\delta > \underline{\delta}'$, and $\mathcal{H}$ is finite, for sufficiently large δ, the profile is a sequential equilibrium. ■

There are two key insights in this argument. The first is that by choosing the set $\mathcal{C}_\mathcal{H}$ to be the union of sets $C_\mu \times \{\mu\}$, where the size of C_μ is increasing in the confidence that players have about the state, revealing action profiles can be enforced (step 2 of the proof of lemma 9.7.1). The second is that when a nonrevealing action profile is played, because payoffs are uninformative, continuations are not needed to provide incentives (steps 3 and 4 in the proof of lemma 9.7.1).

These insights play a similarly important role in Wiseman's (2005) argument, which covers mixed minmax payoffs and games of imperfect monitoring (see remark 9.7.1). This noisy monitoring of payoffs poses two difficulties. First, the set of posteriors, and hence the set $\mathcal{H}$, is no longer finite. Second, we no longer have the sharp distinction between a revealing action and a nonrevealing action profile.

The argument replaces our finite union $\cup_{\mu \in \mathcal{H}}(C_\mu \times \{\mu\})$ with a function from the simplex of posteriors into sets of continuation payoffs, with the property that the sets of payoffs become larger as beliefs approach certainty. The next step is to replace the idea of a revealing profile with a profile that maximizes expected learning, that is, maximizes the expected distance between the current posterior and the future posterior. One then argues that every payoff/belief pair (v, μ) can be decomposed with a current action that maximizes expected learning, except those that minimize a player's payoff, which can be decomposed by forcing the player in question to choose an action that either induces some learning or yields a low payoff. The argument concludes by showing that there is a finite number of periods with the property that with very high probability, within this collection of periods the posterior converges to very near the truth and stays there forever.

9.8 Short Period Length

In section 3.2.3, we introduced the interpretation of patience as short period length. For perfect monitoring games, it is purely a question of taste whether one views high δ as patient players or short-period length. This is not the case, however, for public monitoring. Although interpreting payoffs on a flow basis poses no difficulty as period length goes to 0, the same is not true for imperfect public signals.

Consider, for example, the imperfect monitoring prisoners' dilemma of section 7.2. We embed this game in continuous time, interpreting the discount factor δ as $e^{-r\Delta}$, where r is the players' common discount rate and Δ is the length of the period. In each period of length Δ, a signal is generated according to the distribution 7.2.1. In a unit length of time, there are Δ^{-1} realizations of the public signal. For small Δ, the signals over the unit length of time thus become arbitrarily informative in distinguishing between the events that the players had always exerted effort in that time interval, or one player had always shirked.[14]

Once the repeated game is embedded in continuous time, the monitoring distribution should reflect this embedding. In this section, we illustrate using an example motivated by Abreu, Milgrom, and Pearce (1991). We consider two possibilities. First, assume the public monitoring distribution is parameterized by Δ as

$$\rho(\bar{y} \mid a) = \begin{cases} e^{-\beta \Delta}, & \text{if } a = EE, \\ e^{-\mu \Delta}, & \text{otherwise,} \end{cases}$$

with $0 < \beta < \mu$. When the period length Δ is small, we can view the probability distribution as approximating a Poisson process, where the signal $\underline{y}$ constitutes an arrival (and signal $\bar{y}$ denotes the absence of an arrival) of a "bad" signal, and where the instantaneous arrival rate is β if both players exert effort and μ otherwise. Observe that although taking $\Delta \to 0$ does make players more patient in the sense that $\delta \to 1$, this limit also makes the signals uninformative (both p and q converge to 1), and so

---

14. This can be interpreted as an intuition for the observation in remark 9.2.1 that the sufficient conditions for the public monitoring folk theorem concern only the statistical identifiability of deviations. There are no conditions on the informativeness of the signals.

9.8 ■ Short Period Length

the structural aspects of the model have changed. On the other hand, players unambiguously become patient as the *preference* parameter r goes to 0, because in this case $\delta \to 1$ whereas no other aspect of the model is affected.

We saw in section 7.7.1 that a necessary and sufficient condition for mutual effort to be played in a strongly symmetric equilibrium is (7.7.3), that is, $\delta[3\rho(\bar{y} \mid EE) - 2\rho(\bar{y} \mid SE)] \geq 1$, which becomes

$$e^{-r\Delta}(3e^{-\beta\Delta} - 2e^{-\mu\Delta}) > 1. \tag{9.8.1}$$

Taking $r \to 0$ yields

$$3e^{-\beta\Delta} - 2e^{-\mu\Delta} > 1,$$

which is simply $3p - 2q > 1$ (see (7.2.5)).

Suppose we now make the period length arbitrarily small, fixing the players' discount rate r. The left side of (9.8.1) equals 1 and its derivative equals $2\mu - 3\beta - r$ for $\Delta = 0$, so (9.8.1) holds for small $\Delta > 0$ if

$$r < 2\mu - 3\beta. \tag{9.8.2}$$

If players are sufficiently patient that (9.8.2) is satisfied, then from (7.7.1), the largest strongly symmetric PPE payoff converges to

$$\lim_{\Delta \to 0} 2 - \frac{1 - e^{-\beta\Delta}}{e^{-\beta\Delta} - e^{-\mu\Delta}} = 2 - \frac{\beta}{\mu - \beta} > 0.$$

Hence, in the "bad news" case, even for arbitrarily short period lengths, it is possible to support some effort in equilibrium.

Consider now the "good news" case, where the public monitoring distribution is parameterized by Δ as

$$\rho(\bar{y} \mid a) = \begin{cases} 1 - e^{-\beta\Delta}, & \text{if } a = EE, \\ 1 - e^{-\mu\Delta}, & \text{otherwise}, \end{cases}$$

with $0 < \mu < \beta$. The probability distribution approximates a Poisson process where the signal $\bar{y}$ constitutes an arrival (and signal $\underline{y}$ denotes the absence of an arrival) of a "good" signal, and where the instantaneous arrival rate is β if both players exert effort and μ otherwise. In the current context, (7.7.3) becomes

$$e^{-r\Delta}(1 - 3e^{-\beta\Delta} + 2e^{-\mu\Delta}) > 1.$$

Suppose we now make the period length arbitrarily small, fixing the players' discount rate r. Then,

$$\lim_{\Delta \to 0} e^{-r\Delta}(1 - 3e^{-\beta\Delta} + 2e^{-\mu\Delta}) = 0,$$

implying that there is *no* strongly symmetric perfect public equilibrium with strictly positive payoffs.

We see then a striking contrast between the "good news" and "bad news" cases. This contrast is exaggerated by the failure of pairwise identifiability. If, for example, the game has a product structure, with player i's signal $y_i \in \{\underline{y}_i, \bar{y}_i\}$, distributed as

$$\rho(\bar{y}_i \mid a_i) = \begin{cases} 1 - e^{-\beta \Delta}, & \text{if } a_i = E, \\ 1 - e^{-\mu \Delta}, & \text{if } a_i = S, \end{cases}$$

then it is possible to support some effort for small Δ even in the "good news" case.

An alternative to considering short period length is to directly analyze continuous time games with public monitoring. We content ourselves with the comment that there are appropriate analogs to many of the ideas in chapter 7 (see Sannikov 2004).

10 Private Strategies in Games with Imperfect Public Monitoring

The techniques presented in chapters 7–9 provide the tools for examining perfect public equilibria. It is common to restrict attention to such equilibria, and doing so is without loss of generality in many cases. For example, the public-monitoring folk theorem (proposition 9.2.1) only relies on public strategies. Moreover, if games have a product structure, all sequential equilibrium outcomes are PPE outcomes (proposition 10.1.1). In this chapter, we explore the impact of allowing for *private* strategies, that is, strategies that are nontrivial functions of private (rather than public) histories.

10.1 Sequential Equilibrium

The appropriate equilibrium notion when players use private strategies is *sequential equilibrium*, which we define at the end of this section. This equilibrium notion combines sequential rationality with consistency conditions on each player's beliefs over the private histories of other players.

Any PPE is a sequential equilibrium. Because any pure strategy is realization equivalent to a public strategy (lemma 7.1.2), any pure strategy sequential equilibrium outcome is also a PPE outcome. Moreover, every mixed strategy is clearly realization equivalent to a mixture over public pure strategies. Thus every Nash equilibrium is realization equivalent to a Nash equilibrium in which each player's mixture has only public pure strategies in its support.

If the analysis could be restricted to either pure strategies or Nash equilibria, there is no need to consider private strategies. However, sequential rationality with mixing requires us to work with behavior strategies. A mixture over pure strategies need not be realization equivalent to *any* public behavior strategy. For example, consider the twice-played prisoners' dilemma with imperfect monitoring and with the set of signals $\{\underline{y}, \bar{y}\}$, as in section 7.2. Let $\hat{\sigma}_1$ and $\tilde{\sigma}_1$ be the following pure strategies:

$$\hat{\sigma}_1(\varnothing) = \hat{\sigma}_1(\bar{y}) = E, \quad \hat{\sigma}_1(\underline{y}) = S,$$

and

$$\tilde{\sigma}_1(\varnothing) = \tilde{\sigma}_1(\bar{y}) = \tilde{\sigma}_1(\underline{y}) = S.$$

Each of these strategies is obviously public, because behavior is specified only as a function of the public signal. Now consider a mixed strategy that assigns probability

1/2 to each of these public pure strategies. This is by construction a mixture over public strategies. However, the corresponding behavior strategy, denoted by σ_1, is given by

$$\sigma_1(\varnothing) = \tfrac{1}{2} \circ E + \tfrac{1}{2} \circ S,$$

and

$$\sigma_1(a_1, y) = \begin{cases} E, & \text{if } (a_1, y) = (E, \bar{y}), \\ S, & \text{otherwise.} \end{cases}$$

This is a private strategy, as the second-period action following signal $\bar{y}$ depends on the player's (private) period 1 action. Notice in particular that it does not suffice to specify the second period action as a half/half mixture following signal $\bar{y}$. Doing so assigns positive probability to player 1 histories of the form $(E, \bar{y}, S, \cdot)$ or $(S, \bar{y}, E, \cdot)$, which the half-half mixture of $\hat{\sigma}_1$ and $\tilde{\sigma}_1$ does not allow. As a consequence, there are mixed strategy Nash equilibria that are not realization equivalent to any Nash equilibrium in public behavior strategies. Restricting attention to public behavior strategies may then exclude some equilibrium outcomes that can be achieved with private strategies.

Sequential equilibrium, originally defined for finite extensive form games (Kreps and Wilson 1982b), has a straightforward definition in games of public monitoring if the distribution of public signals generated by every action profile has full support. After any private history h_i^t, player i's belief over the private histories of the other players is then necessarily given by Bayes' rule (even if player i had deviated in the history h_i^t). Let $\sigma_{-i}|_{h_{-i}^t} = (\sigma_1|_{h_1^t}, \ldots, \sigma_{i-1}|_{h_{i-1}^t}, \sigma_{i+1}|_{h_{i+1}^t}, \ldots, \sigma_n|_{h_n^t})$.

Definition 10.1.1 *Suppose $\rho(y \mid a) > 0$ for all $y \in Y$ and all $a \in A$. A strategy profile σ is a sequential equilibrium of the repeated game with public monitoring if, for all private histories h_i^t, $\sigma_i|_{h_i^t}$ is a best reply to $E[\sigma_{-i}|_{h_{-i}^t} \mid h_i^t]$.*

A sequential equilibrium outcome can fail to be a PPE outcome only if players are randomizing and at least one player's choice is a nontrivial function of his own realized actions as well as his signals. Conditioning on his period $t-1$ action is of value to player i (because that action itself is private) only if player i's prediction of others' period t continuation play depends on i's action. Even if player i's best responses are independent of his beliefs about others' continuation play, appropriately specifying his period t best response as a function of his past actions may create new incentives beyond those achievable in public strategies. Both possibilities require the informativeness of the period $t-1$ public signal about others' period $t-1$ actions to depend on player i's period $t-1$ action. This is not the case in games with a product structure (section 9.5), suggesting the following result (Fudenberg and Levine 1994, theorem 5.2):[1]

Proposition 10.1.1 *Suppose A is finite, the game has a product structure, and the short-lived players' actions are observable. The set of sequential equilibrium payoffs is given by $\mathscr{E}(\delta)$, the set of PPE payoffs.*

[1] Intuitively, a sequential equilibrium in a public monitoring game without full-support monitoring is a profile such that after every private history, player i is best responding to the behavior of the other players, given beliefs over the private histories of the other players that are minimally consistent with his own private history.

10.2 A Reduced-Form Example

We first illustrate the potential advantages of private strategies in a two-period game. The first-period stage game is (yet again) the prisoners' dilemma stage game from figure 1.2.1, reproduced on the left in figure 10.2.1, and the second period game is given on the right. We think of the second-period game as representing the equilibrium continuation payoffs in an infinitely repeated game, here collapsed into a stage game to simplify the analysis and focus attention on the effects of private strategies. As the labeling and payoffs suggest, we will think of R as a rewarding action and P as a punishing action. The payoff ties in the second-period subgame are necessary for the type of equilibrium we construct. Section 10.4 shows that we can obtain such a pattern of continuation payoffs in the infinitely repeated prisoners' dilemma.

In keeping with this interpretation, we let payoffs in the two-period game, given period 1 and period 2 payoff profiles u^1 and u^2, be given by

$$(1-\delta)u^1 + \delta u^2.$$

The public monitoring distribution is given by (7.2.1), where the signals $\{\underline{y}, \bar{y}\}$ are distributed according to

$$\rho(\bar{y} \mid a) = \begin{cases} p, & \text{if } a = EE, \\ q, & \text{if } a = ES \text{ or } SE, \\ r, & \text{if } a = SS. \end{cases} \quad (10.2.1)$$

We assume

$$p = \tfrac{9}{10}, \quad q = \tfrac{4}{5}, \quad r = \tfrac{1}{5}, \quad \text{and} \quad \delta = \tfrac{25}{27}. \quad (10.2.2)$$

Observe that $p - q$ is small, so that any incentives based on shifting the distribution of signals from $p \circ \bar{y} + (1-p) \circ \underline{y}$ to $q \circ \bar{y} + (1-q) \circ \underline{y}$ are weak.

10.2.1 Pure Strategies

Suppose first that the players do not have access to a public correlating device. Then the best pure-strategy symmetric perfect public equilibrium payoff is obtained by playing

	E	S
E	2,2	−1,3
S	3,−1	0,0

	R	P
R	$\tfrac{8}{5},\tfrac{8}{5}$	$0,\tfrac{8}{5}$
P	$\tfrac{8}{5},0$	$0,0$

Figure 10.2.1 The two-period game for section 10.2, with the first-period game on the left, and the second period on the right.

EE in the first period, followed by RR in the second if the first-period signal is $\bar{y}$ and PP otherwise. This provides an expected payoff of $\frac{2}{27}(2) + \frac{25}{27}\frac{9}{10}\frac{8}{5} = \frac{40}{27}$.

The first-period incentive constraint in the equilibrium we have just presented, that players prefer effort to shirking, is

$$(1-\delta)2 + \delta p \tfrac{8}{5} \geq (1-\delta)3 + \delta q \tfrac{8}{5}.$$

Given the values from (10.2.2), this incentive constraint holds with strict inequality ($40/27 > 38/27$). There is then a sense in which too much of the potential expected payoff in the game is expended in creating incentives, suggesting that a higher payoff could be achieved if incentives could be more finely tuned.

10.2.2 Public Correlation

Suppose now players have access to a public correlating device, allowing for a finer tuning of incentives. The best pure-strategy strongly symmetric PPE payoff is now given by strategies that prescribe E in the first period, R in the second period following signal $\bar{y}$, and R with probability ϕ (with P otherwise) in the second period following signal y. The largest possible payoff is given by setting ϕ equal to the value that just satisfies the incentive constraint for effort in the first period, or

$$(1-\delta)(2) + \delta\left(p\tfrac{8}{5} + (1-p)\phi\tfrac{8}{5}\right) = (1-\delta)(3) + \delta\left(q\tfrac{8}{5} + (1-q)\phi\tfrac{8}{5}\right).$$

Under (10.2.2), this equality is solved by $\phi = 1/2$ and the corresponding expected payoff is $42/27$. Note that this exceeds the payoff $40/27$ achieved without the correlating device.

10.2.3 Mixed Public Strategies

The signals in this game are more informative about player 1's action when player 2 shirks than when she exerts effort. As a result, shirking with positive probability in the first period can improve incentives. We have seen examples of this possibility in sections 7.7.2, 8.4.1, and 8.4.5.

Let α be the probability that each player i plays E in the first period. Assume that RR is played in the second period after the first-period signal $\bar{y}$, and that ϕ is again the probability of playing RR after signal y. Then the incentive constraint for the first-period mixture between effort and shirking is

$$\begin{aligned}
\alpha\{(1-\delta)2 + \delta(p\tfrac{8}{5} + (1-p)\phi\tfrac{8}{5})\} & \\
+ (1-\alpha)\{(1-\delta)(-1) + \delta(q\tfrac{8}{5} + (1-q)\phi\tfrac{8}{5})\} & \\
= \alpha\{(1-\delta)(3) + \delta(q\tfrac{8}{5} + (1-q)\phi\tfrac{8}{5})\} & \\
+ (1-\alpha)\{\delta(r\tfrac{8}{5} + (1-r)\phi\tfrac{8}{5})\}. & \quad (10.2.3)
\end{aligned}$$

The left side of this equality is the payoff to E and the right side the payoff to S, where α and $(1-\alpha)$ are the probabilities that the opponent exerts effort and shirks.

10.2 ■ A Reduced-Form Example

We can solve for

$$\phi = \frac{11 - 10\alpha}{2(6 - 5\alpha)}. \tag{10.2.4}$$

This fraction exceeds $1/2$ whenever $\alpha < 1$. Hence, first-period mixtures allow incentives to be created with a smaller threat of second-period punishment, reflecting the more informative monitoring. This increased informativeness is purchased at the cost of some shirking in the first period.

Substituting (10.2.4) into (10.2.3) to obtain expected payoffs gives

$$\frac{224 - 152\alpha - 30\alpha^2}{27(6 - 5\alpha)}.$$

For $\alpha \in [0, 1]$, this is maximized at $\alpha = \frac{6}{5} - \frac{2}{15}\sqrt{3} = 0.969$, with a value $1.5566 > 42/27$.

The benefits of better monitoring outweigh the costs of shirking, and mixed strategies thus allow a (slightly) higher payoff than is possible under pure strategies.

10.2.4 Private Strategies

Because the signals about player 1 are particularly informative when 2 shirks, it is natural to ask whether we could do even better by punishing 1 only after a bad signal was observed when 2 shirked. This is a private strategy profile. Each player not only mixes but conditions future behavior on the outcome of the mixture.

Suppose players mix between E and S in the first period. After observing the signal $\bar{y}$, R is played in the second period. After signal $\underline{y}$, a player potentially mixes between R and P, but attaches positive probability to the punishment action P only if that player played S in the first period.

Let α again be the probability placed on E in the first period, and now let ξ denote the probability that a player chooses R in the second period after having chosen S in the first period and observed signal $\underline{y}$. Players choose R for sure after E in the first period, regardless of signal, or after S but observing signal $\bar{y}$. Figure 10.2.2 illustrates these strategies with an automaton. The automaton does *not* describe the strategy profile; rather, each players' behavior is described by an automaton (see remark 2.3.1). The profile we describe is symmetric, so the automaton description is common. However, because the strategy described is private, players may end up in different states.

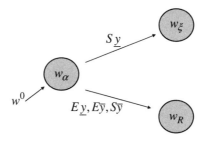

Figure 10.2.2 The automaton of the private strategy described in section 10.2.4. Subscripts on states indicate the specified behavior.

The incentive constraint for the first-period mixture is

$$\alpha\{(1-\delta)2 + \delta\tfrac{8}{5}\} + (1-\alpha)\{(1-\delta)(-1) + \delta(q\tfrac{8}{5} + (1-q)\xi\tfrac{8}{5})\}$$
$$= \alpha\{(1-\delta)(3) + \delta\tfrac{8}{5}\} + (1-\alpha)\{\delta(r\tfrac{8}{5} + (1-r)\xi\tfrac{8}{5})\}.$$

Solving for ξ gives

$$\xi = \frac{11 - 12\alpha}{12(1-\alpha)},$$

which we use to calculate the expected payoff as $\tfrac{2}{9}(\alpha + \tfrac{56}{9})$. Maximizing α subject to the constraint $\xi \in [0, 1]$ gives $\alpha = 11/12$ (and $\xi = 0$), with expected payoff $\tfrac{2}{9}(\tfrac{11}{12} + \tfrac{56}{9}) = 1.5864$. We thus have the following ranking of payoffs:

$$\underbrace{1.4815}_{\text{pure strategies}} < \underbrace{1.5556}_{\text{public correlation}} < \underbrace{1.5566}_{\text{mixed public strategies}} < \underbrace{1.5864}_{\text{private mixed strategies}}.$$

Private mixed strategies allow the players to randomize in the second period but not in a way that can be implemented by a public correlating device. From player 1's perspective, player 2's strategy depends on a history that he has not observed. However, the second-period incentive constraints are trivially satisfied by construction—each player is indifferent between R and P, no matter how he expects the opponent to play. With a finite horizon, this is clearly very special. We show in section 10.4 that it is possible to construct equilibria in the infinitely repeated game with this property.

10.3 Two-Period Examples

This section presents three examples from Mailath, Matthews, and Sekiguchi (2002), in which private strategies open up new equilibrium possibilities in a two-period game with *identical* stage games in each period.

10.3.1 Equilibrium Punishments Need Not Be Equilibria

We first examine a stage-game with a unique Nash and correlated equilibrium.[2] The set of perfect public equilibria for this game is then trivial—the stage-game Nash equilibrium must be played in each period. However, there exists an equilibrium in private strategies that does not play the stage-game Nash equilibrium in the first period and that offers payoffs superior to those of the perfect public equilibrium. The key to this result is that out-of-equilibrium first-period actions need not lead to equilibrium behavior in the second period.

---

2. A *correlated equilibrium* (Aumann 1974) of a stage game $u : \prod_i A_i \to \mathbb{R}^n$ is a probability distribution over all action profiles, $\mu \in \Delta(\prod_i A_i)$, such that when an action profile is drawn according to μ, and each player is *privately* recommended to play his part of the realized profile, that player has no incentive to deviate from the recommendation, assuming other players follow their own recommendations.

10.3 ■ Two-Period Examples

	c^1	c^2	c^3	c^4
r^1	6, 0	0, 1	0, 0	0, 0
r^2	5, 6	1, 5	11, 0	11, 1
r^3	0, 0	0, 0	10, 10	10, 10

Figure 10.3.1 The stage game for section 10.3.1.

	c^1	c^2	c^3	c^4
r^1	.5	.5	.5	.5
r^2	.5	.5	.9	.1
r^3	.5	.5	.5	.5

Figure 10.3.2 The probability of $\bar{y}$ under different action profiles.

The private equilibrium we construct yields payoffs dominating those of the perfect public equilibrium. Given the uniqueness of the stage-game Nash equilibrium, this perfect public equilibrium is also the unique subgame-perfect equilibrium outcome in the perfect-monitoring version of the game. Thus we have an example of a case in which the players gain higher payoffs as a result of a noisier monitoring scheme. Section 12.1.3 discusses an example, due to Kandori (1991a), illustrating a similar phenomenon in private-monitoring games.

The stage game of figure 10.3.1 is played twice. The stage game has a unique Nash and correlated equilibrium in which player 1 mixes equally over r^1 and r^2 and player 2 mixes equally over c^1 and c^2. Suppose that the set of public signals is $Y = \{y, \bar{y}\}$, with the monitoring technology $\rho(\bar{y} \mid r^i c^j)$ displayed in figure 10.3.2. The key feature is that player 1 learns something about player 2's action only if 1 plays r^2, in which case 1 receives information about the relative likelihoods of c^3 and c^4. The signals provide player 1 with no other information.

We construct an equilibrium that does not duplicate the unique PPE. The following strategies (σ_1, σ_2) are a sequential equilibrium that in the first period gives either $r^3 c^3$ or $r^3 c^4$. In the first period, player 1 chooses r^3 and player 2 mixes over c^3 and c^4:

$$\sigma_1(\varnothing) = r^3, \quad \text{and}$$
$$\sigma_2(\varnothing) = \frac{1}{2} \circ c^3 + \frac{1}{2} \circ c^4.$$

In the second period, regardless of the signal received in the first period, player 1 mixes equally between r^1 and r^2 if he took the first-period equilibrium action of r^3 or if he deviated to r^1, but plays r^2 if he deviated to r^2 in the first period:

$$\sigma_1(a_1 y) = \begin{cases} \frac{1}{2} \circ r^1 + \frac{1}{2} \circ r^2, & \text{if } a_1 = r^1 \text{ or } r^3, \\ r^2, & \text{if } a_1 = r^2. \end{cases}$$

Player 2 chooses c^2 after the histories $c^3 \bar{y}$ or $c^4 \underline{y}$, and otherwise plays c^1:

$$\sigma_2(a_2 y) = \begin{cases} c^2, & \text{if } a_2 y = c^3 \bar{y} \text{ or } c^4 \underline{y}, \\ c^1, & \text{otherwise.} \end{cases}$$

Along the equilibrium path, the second-period outcome is the unique Nash and correlated equilibrium of the stage game—player 1 mixes equally over r^1 and r^2, and player 2 mixes equally over c^1 and c^2. Player 2 is playing a stage-game best response in the first period, but player 1 is not. Why does player 1 choose r^3 instead of the myopically superior r^2? A deviation by player 1 to r^2 in the first period induces player 2 to play $0.1 \circ c^1 + 0.9 \circ c^2$ in the second period, whereas player 1 plays his best response of r^2. This is less lucrative than second-period equilibrium play, sufficiently so as to deter the first-period deviation.

If a first-period signal y is realized with positive probability in a Nash equilibrium, then the *equilibrium* distribution of second-period actions, conditional on y, must be a correlated equilibrium of the stage game.[3] If the equilibrium strategies are public, the uniqueness of the stage-game correlated equilibrium then ensures that every signal is followed by identical second-period behavior. First-period actions thus cannot affect second-period behavior, making it impossible to create incentives to play anything other than a myopic best response in the first period and leading to the unique perfect public equilibrium.

In contrast, player 2's private strategy allows the stage-game correlated equilibrium to be played in the second period in equilibrium, while still allowing second-period behavior to respond to player 1's first-period out-of-equilibrium actions. The key to achieving these two properties is that when strategies are private, second-period play for a given public signal need not be a stage-game Nash equilibrium. A deviation by player 1 in the first period can elicit player 2 behavior that is not a best response to player 1's second-period behavior, though it is a best response to player 2's equilibrium beliefs.

The example we constructed in section 10.2.4 similarly exploits a dependence between first-period and second-period behavior. In that game, however, there are

---

3. We cannot expect a Nash equilibrium in the second period, because players' actions may be conditioned on their own private histories. These histories take the role of the privately observed signals that lie behind the standard formulation of a correlated equilibrium. However, any first-period private history must lead to a second-period action that is a best response to the beliefs induced by that private history (and any second-period action played after two different private histories must be a best response to any average of the beliefs induced by those histories). This ensures that the realized distribution in the second period is a correlated equilibrium.

10.3 ■ Two-Period Examples

	c^1	c^2	c^3
r^1	0,0	1,2	2,1
r^2	2,1	0,0	1,2
r^3	1,2	2,1	0,0

Figure 10.3.3 The stage game for section 10.3.2.

multiple Nash equilibria of the second-period stage game, allowing us to construct links between the periods even with public strategies. Private strategies are then valuable in making these links stronger, and hence relaxing incentive constraints, by associating punishments with histories that are especially likely to appear if a player deviates from equilibrium play.

10.3.2 Payoffs by Correlation

Our next example again concerns a stage game in which the private strategies yield equilibrium payoffs dominating those of any PPE (or any subgame-perfect equilibrium of the perfect monitoring game). The stage game has a unique Nash equilibrium but also has a correlated equilibrium offering higher payoffs than the Nash equilibrium. The repeated game has a unique PPE that repeats the stage-game Nash equilibrium in each period. Private strategies allow the players to play the correlated rather than Nash equilibrium of the stage game in the second period, leading to higher payoffs.

The stage game is given in figure 10.3.3. This game has a unique Nash equilibrium, in which each player mixes equally over her three actions, for expected payoffs $(1, 1)$. It has a correlated equilibrium in which probability $1/6$ is placed on each off-diagonal outcome, for payoffs $(3/2, 3/2)$.

There are three signals y^1, y^2, and y^3, with conditional distribution given in figure 10.3.4. Note that a player's first-period action and the public signal allow that player to rule out one of the opponent's actions as a first-period possibility but provides no evidence as to which of the other two actions was taken. It should be apparent that this already brings us most of the way toward constructing a correlated equilibrium because the essence of the optimality conditions for the correlated equilibrium is that each player believe the opponent is mixing equally over two actions.

We now specify strategies in which each player mixes equally over all three actions in the first period, with second-period play given by (where we adopt the conventions that $r^4 = r^1$, $c^4 = c^1$, $r^0 = r^3$, and $c^0 = c^3$):

$$\sigma_1(r^\ell, y) = \begin{cases} r^\ell, & \text{if } y = y^1, \\ r^{\ell-1}, & \text{if } y = y^2, \\ r^{\ell+1}, & \text{if } y = y^3, \end{cases}$$

	c^1	c^2	c^3
r^1	$\frac{1}{2} \circ y^2 + \frac{1}{2} \circ y^3$	$\frac{1}{2} \circ y^1 + \frac{1}{2} \circ y^3$	$\frac{1}{2} \circ y^1 + \frac{1}{2} \circ y^2$
r^2	$\frac{1}{2} \circ y^1 + \frac{1}{2} \circ y^2$	$\frac{1}{2} \circ y^2 + \frac{1}{2} \circ y^3$	$\frac{1}{2} \circ y^1 + \frac{1}{2} \circ y^3$
r^3	$\frac{1}{2} \circ y^1 + \frac{1}{2} \circ y^3$	$\frac{1}{2} \circ y^1 + \frac{1}{2} \circ y^2$	$\frac{1}{2} \circ y^2 + \frac{1}{2} \circ y^3$

Figure 10.3.4 The monitoring distribution.

and

$$\sigma_2(c^\ell, y) = \begin{cases} c^\ell, & \text{if } y = y^1, \\ c^{\ell+1}, & \text{if } y = y^2, \\ c^{\ell-1}, & \text{if } y = y^3. \end{cases}$$

It is then straightforward to verify that second-period play constitutes a correlated equilibrium. For example, if player 1 has played r^3 and observed signal y^3, then his beliefs are that player 2 played either c^1 or c^3 in the first period, and therefore will mix equally over c^2 and c^3 in the second, to which the prescribed player 1 second-period action r^1 is a best response. In addition, the expected payoff from the second period is 3/2, no matter which first-period action a player takes, and hence the first-period mixture is optimal.

10.3.3 Inconsistent Beliefs

This section considers a stage game with two pure-strategy Nash equilibria. Once again an equilibrium in private strategies exists dominating every perfect public equilibrium. As in the first example, in section 10.3.1, the key to this result is that out-of-equilibrium first-period actions need not be followed by second-period equilibria. However, the second-period equilibrium construction now depends not on players being unable to detect deviations by their opponents, as in the first example, but on players being unable to detect the identity of the defector. For this to play a role, we need a game with at least three players.

The stage game is given in figure 10.3.5. Player 1 chooses rows, 2 chooses columns, and 3 chooses matrices. Notice first that R strictly dominates L for player 3, giving a payoff that is larger by 9, regardless of the other players' actions. The stage-game Nash equilibria consist of *LRR* and *RLR*, for payoffs of (1, 1, 12), and a mixture in which players 1 and 2 put probability 1/3 on L, for payoffs (1/3, 1/3, 74/9).

We are interested in equilibria that place significant probability on the jointly lucrative outcome *LLL*. The difficulty with a PPE is that the second period must feature a Nash equilibrium of the stage game. Switching between Nash equilibria in the second period of the game allows us to construct a difference in continuation payoffs for player 3 of only 34/9, which is not sufficient to overwhelm the payoff increment of 9 that

10.3 ■ Two-Period Examples

	L	R
L	15, 15, 17	−85, 5, 3
R	5, −85, 3	5, 5, −9

L

	L	R
L	−1, −1, 26	1, 1, 12
R	1, 1, 12	0, 0, 0

R

Figure 10.3.5 The stage game for section 10.3.3.

player 3 receives from choosing R instead of L. As a result, no arrangement of second-period play in a public equilibrium can induce player 3 to choose L in the first period. Given this, the best that can be accomplished in a public equilibrium, no matter what the monitoring technology (including perfect), is some sequence of LRR and RLR, for payoffs of $(2, 2, 24)$.

Suppose $Y = \{y^0, y^1, y^2, y^3\}$. If the number of players choosing L in the first period is ℓ ($\in \{0, 1, 2, 3\}$), then the probability of y^ℓ is $1 - 3\varepsilon$, whereas that of y^m, $m \neq \ell$, is ε. Now consider first-period strategies given by

$$\sigma_1(\varnothing) = \theta \circ L + (1 - \theta) \circ R,$$
$$\sigma_2(\varnothing) = \theta \circ L + (1 - \theta) \circ R,$$
$$\sigma_3(\varnothing) = L,$$

where
$$\theta = \frac{90 - 2\varepsilon}{100 - 3\varepsilon}.$$

These actions make LLL relatively likely in the first period. Second-period play depends on private histories only if signal y^2 is received, and is given by

$$(\sigma_1(a_1 y), \sigma_2(a_2 y), \sigma_3(a_3 y)) = (L, R, R) \quad \forall (a_1, a_2, a_3), \ y \neq y^2,$$
$$\sigma_1(Ly^2) = \sigma_2(Ly^2) = R,$$
$$\sigma_1(Ry^2) = \sigma_2(Ry^2) = L,$$
and $\quad \sigma_3(Ly^2) = \sigma_3(Ry^2) = R.$

If ε is sufficiently small, then play in the second period calls only for best responses. After any signal other than y^2, the stage-game equilibrium LRR is played. For sufficiently small ε, the signal y^2 almost certainly follows first-period behavior of LRL or RLL, in which cases the second-period Nash equilibria RLR and LRR are played, respectively. The strictness of these Nash equilibria ensures that second-period actions following y^2 are best responses, for small ε.

Now consider the first period. Player 1's and player 2's expected second-period payoffs following their first-period play, given player 3's choice of L in the first period, are given in figure 10.3.6. Adding these payoffs to the first-period payoffs, player 3's first-period choice of L induces a first-period game between players 1 and 2 whose equilibrium is given by $\theta \circ L + (1 - \theta) \circ R$.

	L	R
L	$1-\varepsilon, 1-\varepsilon$	$1, 1$
R	$1, 1$	$1-2\varepsilon, 1-2\varepsilon$

Figure 10.3.6 Expected continuation payoffs for players 1 and 2 as a function of their first-period actions.

If player 3 chooses L in period 1, the first-period signal is y^3 with very high probability (for small ε), for a second-period payoff of 12. Switching to R implies that with high probability, the signal will be y^2. Because both players 1 and 2 choose L in the first period with high probability, this realization of y^2 generates with high probability the second-period outcome RRR, for a player 3 payoff of 0. For small ε, this deviation is unprofitable.

The key to interpreting this equilibrium is to think in terms of the beliefs player 1 and 2 have about the origin of signal y^2, given that they played L. In equilibrium, the most likely cause of such a signal is that the other player's mixture generated a choice of R, and player 3 chose the equilibrium action L. Both players 1 and 2 have this belief and both switch to R in the second period, dooming player 3.

As $\varepsilon \to 0$, the expected payoff from this equilibrium approaches $(6, 6, 26.22)$, giving a private equilibrium that is superior to any public equilibrium. In this case, the advantage of the private equilibrium is in allowing players to condition second-period actions on different beliefs about the origins of first-period signals.

10.4 An Infinitely Repeated Prisoners' Dilemma

10.4.1 Public Transitions

The examination of private strategies in section 10.2 was simplified by the second period's game having been constructed to pose no incentive issues. We now present an analysis, from Kandori and Obara (2006), showing that equilibria in public and private strategies can differ in the infinitely repeated prisoners' dilemma.

The stage game is the prisoners' dilemma from the left panel of figure 10.2.1. There are two public signals, y and $\bar{y}$, distributed according to (10.2.1). We assume $q = 0$ here and analyze $q > 0$ in section 10.4.2. We will explain why this violation of the full support assumption does not pose difficulties for the application of sequential equilibrium (see definition 10.1.1). We assume $p > 0$ and make no assumptions about r at this point. The highest symmetric payoff from perfect public equilibrium is then (see (7.7.1) or section 8.4.1)

$$2 - \frac{(1-p)}{p} < 2.$$

We now argue that one can do better with the following private strategy, described as an automaton. The profile is symmetric, so the automaton description is common.

10.4 ■ An Infinitely Repeated Game

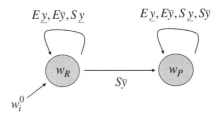

Figure 10.4.1 An automaton representation of player i's private strategy, with $f_i(w_R) = \alpha$ and $f_i(w_P) = S$.

However, because the strategy described is private, even with a common initial state, players may in general end up in different states. The set of states for player i is $\mathscr{W}_i = \{w_R, w_P\}$, with initial state w_R. The output function is $f_i(w_R) = \alpha \in (0, 1]$, where α is the probability of E, and $f_i(w_P) = S$. The transition function is given by

$$\tau_i(w_i, a_i y) = \begin{cases} w_R, & \text{if } w_i = w_R \text{ and either } a_i = E \text{ or } a_i y = S\underline{y}, \\ w_P, & \text{otherwise,} \end{cases}$$

where w_i is the current state of player i. The strategy is private because the transition function depends on a player's own actions. Figure 10.4.1 illustrates the automaton.

Although the strategy is private, the particular monitoring structure (i.e., $q = 0$) implies a very special property: If, for both $i = 1$ and 2, player i begins in the state w_R, then when both players follow the *private* strategy just described, both players are always in the same state. That is, even though the transition from state w_R to w_P is potentially private, both players necessarily transit at the same time. The transition only occurs when a player observes $\bar{y}$ (which is public) and has chosen S, and $q = 0$ implies that when $\bar{y}$ is observed after one player chose S, the other player must also have chosen S, and so will also transit. Consequently, when a player has transited to w_P, he believes his opponent has also transited to the same state. Moreover, because the transition to w_P is effectively public, when both players start in w_R, and a player remains in w_R as a result of his private history, so does his opponent. Finally, state w_P is absorbing (and hence specifies the myopic dominant action, as it must if the relevant incentive constraints are satisfied in w_P). In other words, under the profile, players are always in a common state (whether it is w_R or w_P). As a result, the application of sequential equilibrium is straightforward.

Suppose then that player i is in the state w_R, for $i = 1, 2$. Given that the two players are always in a common state, we can let $V(w_R)$ be the expected payoff to a player when both players are in state w_R. The value of exerting effort in state w_R is given by

$$V^E(w_R) \equiv (1-\delta)(2\alpha + (-1)(1-\alpha)) + \delta V(w_R). \tag{10.4.1}$$

Similarly, the value of shirking is given by

$$\begin{aligned} V^S(w_R) &= \alpha\{(1-\delta)3 + \delta V(w_R)\} \\ &\quad + (1-\alpha)\{(1-\delta) \times 0 + \delta[(1-r)V(w_R) + r \times 0]\} \\ &= (1-\delta)3\alpha + \delta\{1 - r(1-\alpha)\}V(w_R), \end{aligned} \tag{10.4.2}$$

where 0 is the value of state w_P. Indifference between actions E and S in state w_R requires
$$V^E(w_R) = V^S(w_R) = V(w_R),$$
and so, from (10.4.1)
$$V(w_R) = 3\alpha - 1.$$
and, from (10.4.2)
$$3\alpha - 1 = (1-\delta)3\alpha + \delta\{1 - r(1-\alpha)\}(3\alpha - 1).$$

Solving this for the probability of E in state w_R yields
$$\alpha = \frac{1}{3\delta r}\left(2\delta r + \sqrt{\delta^2 r^2 - 3\delta r + 3\delta^2 r}\right),$$
which is a well-defined probability for $r > 0$ and δ close to 1. As players become patient, we have
$$\lim_{\delta \to 1} \alpha = 1,$$
and hence
$$\lim_{\delta \to 1} V(w_R) = 2.$$

That is, the private strategy profile is efficient for patient players and in particular implies higher payoffs than the maximum attainable under public strategies.

This profile is *not* an equilibrium if $q > 0$, even if q is arbitrarily small. It is now not true that the two players are always in the same state when the profile is followed. Suppose $w_1 = w_2 = w_R$ and player 1, for example, stays in w_R after observing $\bar{y}$ having chosen E. When $q > 0$, there is some (even if small) probability that player 2 had chosen S, and so will transit to w_P. In other words, the transition from w_R to w_P is now truly private. This has implications for incentives. Consider the player 1 private history $E\bar{y}, (E\underline{y})^k$ for large k. Immediately after observing $E\bar{y}$, the probability assigned by player 1 to player 2 being in the state w_R is
$$\begin{aligned} \beta^0(q) &= \Pr\{w_2 = w_R \mid E\bar{y}\} \\ &= \Pr\{a_2 = E \mid E\bar{y}\} \\ &= \frac{\Pr\{\bar{y} \mid EE\}\Pr\{a_2 = E\}}{\Pr\{\bar{y} \mid a_1 = E\}} \\ &= \frac{p\alpha}{p\alpha + q(1-\alpha)} < 1. \end{aligned}$$

Note that $\beta^0(0) = 1$ and β^0 is continuous in q. Let
$$\beta^k(q) = \Pr\{w_2 = w_R \mid E\bar{y}, (E\underline{y})^k\}.$$

Then $\beta^k(0) = 1$ for all k, but $\beta^k(q) \searrow 0$ as $k \to \infty$ for all $q > 0$. Eventually, player 1 must conclude that player 2 is almost certainly in state w_P, disrupting the optimality of cooperation.

It is then trivial that the candidate profile is not an equilibrium. The value of α was calculated assuming the opponent was in state w_R with probability 1. This is no longer the case once the second period has been reached (when $q > 0$), violating the indifference required for optimality. In addition, adjusting the probability of E in period t for an agent in state w_R, to maintain indifference between E and S, will not rescue the equilibrium. Such a construction fails for large t because S is a strict best response for an agent in w_P, and hence the convergence of $\beta^k(q)$ to 0 ensures that eventually E must fail to be a best response. When transitions are private, some other approach to sustaining effort is required.

10.4.2 An Infinitely Repeated Prisoners' Dilemma: Indifference

In general, studying equilibria in private strategies is complicated by the potential need to keep track of players' beliefs of other players' private histories. Optimal actions will then depend on beliefs about beliefs about ..., quickly becoming prohibitively complicated. This difficulty surfaced at the end of the last section, where setting $q > 0$ and hence making transitions private disrupted the equilibrium we had constructed.

The two-period example in section 10.2 suggests an alternative to keeping track of players' beliefs. If a player has the same best replies for all beliefs, then beliefs are irrelevant. In section 10.2 we ensured that players had the same best responses for all period 2 beliefs by building payoff ties into the second-period stage game. Such indifferences arise endogenously in infinitely repeated games in *belief-free* equilibria (definition 14.1.2). Belief-free equilibria play an important role in repeated games with private monitoring, and are discussed in chapter 14.

This section presents an example of a belief-free equilibrium. In the course of doing so, we present a symmetric-payoff equilibrium in private strategies where the sum of the players' payoffs is higher than can be achieved in any equilibrium with public strategies.

We examine the repeated prisoners' dilemma of figure 10.4.2, where we will be interested in the comparative statics of b. There are two public signals, y and $\bar{y}$. The monitoring structure is given by (10.2.1). The equilibria we examine are described by an automaton, again describing an individual strategy rather than a strategy profile, with two states, w_R and w_P. Intuitively, we think of these as a reward state and a punishment state. The initial state is w_R. The output function is given by

$$f_i(w) = \begin{cases} \alpha^R \circ E + (1 - \alpha^R) \circ S, & w = w_R, \\ S, & w = w_P. \end{cases}$$

	E	S
E	2, 2	−b, b
S	b, −b	0, 0

Figure 10.4.2 The prisoners' dilemma from figure 8.4.1, with $c = b$.

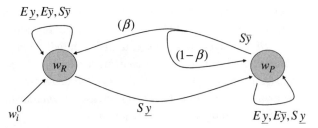

Figure 10.4.3 Private strategies under which each agent is indifferent in each state, regardless of the opponent's state. In state w_R, the player plays $\alpha^R \circ E + (1 - \alpha^R) \circ S$, and in state w_P, the player chooses S for sure.

The transition function is given by

$$\tau_i(w_i, a_i y) = \begin{cases} w_R, & \text{if } w_i = w_R \text{ and } a_i = E \text{ or } a_i y = S\bar{y}, \\ w_P, & \text{if } w_i = w_R \text{ and } a_i y = S\underline{y}, \text{ or} \\ & \text{if } w_i = w_P \text{ and } a_i = E \text{ or } a_i y = S\underline{y}, \\ \beta \circ w_R + (1 - \beta) \circ w_P, & \text{if } w_i = w_P \text{ and } a_i y = S\bar{y}. \end{cases}$$

Figure 10.4.3 illustrates the strategy.

The idea is to determine the two probabilities, β and α^R, so that each player is indifferent between E and S, irrespective of the current state of the opponent (in which case, beliefs are irrelevant). Let $V_i^x(a_i)$ denote player i's value under the profile, when player j's current state is w_x and i plays a_i this period. We are thus searching for probabilities so that $V_i^R(E) = V_i^R(S) \equiv V^R$ and $V_i^P(E) = V_i^P(S) \equiv V^P$ (the profile is symmetric, so we drop the player subscript). The equation $V^R = V_i^R(E)$ is

$$V^R = (1 - \delta)[2\alpha^R - b(1 - \alpha^R)] \\ + \delta[\alpha^R V^R + (1 - \alpha^R)\{qV^R + (1 - q)V^P\}]. \quad (10.4.3)$$

This value must equal $V_i^R(S)$,

$$V^R = (1 - \delta)[b\alpha^R] + \delta[\alpha^R V^R + (1 - \alpha^R)\{rV^R + (1 - r)V^P\}]. \quad (10.4.4)$$

Similarly, for V^P, we have

$$V^P = (1 - \delta)[-b] + \delta[q(\beta V^R + (1 - \beta)V^P) + (1 - q)V^P] \quad (10.4.5)$$
$$= \delta[r(\beta V^R + (1 - \beta)V^P) + (1 - r)V^P]. \quad (10.4.6)$$

If all four equations hold, then player i is always indifferent between exerting effort and shirking. This in turn implies that any strategy for player i is a best response, including the proposed strategy. It also implies that i's payoff depends only on j's state, as desired.

10.4 ■ An Infinitely Repeated Game

Simplifying (10.4.3) yields

$$(1-\delta)(V^R - (2+b)\alpha^R + b) = -\delta(1-\alpha^R)(1-q)(V^R - V^P), \qquad (10.4.7)$$

(10.4.4) yields

$$(1-\delta)(V^R - b\alpha^R) = -\delta(1-\alpha^R)(1-r)(V^R - V^P), \qquad (10.4.8)$$

(10.4.5) yields

$$(1-\delta)(V^P + b) = \delta\beta q(V^R - V^P), \qquad (10.4.9)$$

and finally, (10.4.6) simplifies to

$$(1-\delta)V^P = \delta\beta r(V^R - V^P). \qquad (10.4.10)$$

Eliminating $V^R - V^P$ from (10.4.7) and (10.4.8), we have

$$V^R = \left(\frac{2(1-r)}{(q-r)} + b\right)\alpha^R - \frac{b(1-r)}{(q-r)}. \qquad (10.4.11)$$

Eliminating $V^R - V^P$ from (10.4.9) and (10.4.10),

$$V^P = \frac{br}{(q-r)}. \qquad (10.4.12)$$

Eliminating V^P from (10.4.9) gives

$$V^R - V^P = \frac{b(1-\delta)}{\delta\beta(q-r)}. \qquad (10.4.13)$$

Inserting (10.4.11) and (10.4.13) in (10.4.8), and solving yields

$$\beta = \frac{b(1-\alpha^R)}{b - 2\alpha^R}. \qquad (10.4.14)$$

Substituting for β in (10.4.13), and using (10.4.11) and (10.4.12), we have

$$(1-\alpha^R)\delta[(2(1-r) + b(q-r))\alpha^R - b] - (1-\delta)(b - 2\alpha^R) = 0. \qquad (10.4.15)$$

For $\delta = 1$, $\alpha^R = 1$ is clearly a solution to (10.4.15). We now set $p = 1/2$, $q = 1/2 - \varepsilon$, and $r = \varepsilon$, and consider ε small and b close to 2. Then, we can apply the implicit function theorem to the function described by (10.4.15) to conclude that for $\delta < 1$ but close to 1, there is a value of $\alpha^R < 1$ but close to 1, satisfying (10.4.15). Moreover, the value of β from (10.4.14) is also a well-defined probability, because $b > 2$. Consequently, for δ close to 1, equations (10.4.3)–(10.4.6) can be simultaneously satisfied through an appropriate specification of β and α^R.

Remark 10.4.1 **Determinacy of equilibria** In this automaton, players play S for sure in state w_P, and state w_R has deterministic transitions, allowing us to determine both α^R and β. Kandori and Obara (2006) describe a two-dimensional manifold of belief-free equilibria using private strategies by allowing for randomization in both behavior at w_P and in the state transitions at w_R. This is typical of belief-free equilibria and arises even in games with perfect monitoring; see chapter 14.

◆

Applying (8.4.8) from section 8.4.1 for our parametrization, no PPE can provide a sum of payoffs exceeding 0 (the set $\mathscr{A}$ is empty, $b - c = 0$, and for small ε and b close to 2, $v^{\text{SSE}} < 0$). Nonetheless, for the *same* parameters, not only are there nontrivial equilibria in private strategies but the equilibrium just constructed is nearly efficient, because from (10.4.11), V^R is close to 2. The advantage of private strategies appears here in a setting in which there are too few signals for the public-monitoring folk theorem to hold. Kandori and Obara (2006) also show that for a fixed discount factor, private equilibria can be effective in increasing the set of perfect equilibrium payoffs even when the perfect public equilibrium folk theorem holds.

11 Applications

The study of imperfect public monitoring games began with work centered around economic applications, including noisy principal-agent models (e.g., Radner 1985; Rubinstein 1979b; Rubinstein and Yaari 1983; discussed in section 11.4) and imperfect competition with noisy prices (Green and Porter 1984; Porter 1983a; discussed in section 11.1). This chapter presents several examples.

11.1 Oligopoly with Imperfect Monitoring

11.1.1 The Game

In the stage game, each firm i, $i = 1, \ldots, n$, simultaneously chooses a quantity of output $a_i \in A_i$. Though it is common to take A_i to be all of $\mathbb{R}_+$, we assume that firms must choose from a common finite subset of $\mathbb{R}_+$. We say that the grid A_i becomes increasingly fine as the maximal distance between adjacent quantities approaches 0. We could interpret the restriction to A_i as reflecting an indivisibility that restricts outputs to be multiples of a smallest unit. Working with finite action sets allows us to restrict attention to bang-bang equilibria (see proposition 7.5.1).

The market price p is determined by the firms' quantities $(a_1, \ldots, a_n) \equiv a$ and a price shock $\theta \in \mathbb{R}_+$. The firms produce homogeneous products, so that $p : A \times \mathbb{R}_+ \to \mathbb{R}_+$ depends only on the aggregate quantity produced ($p(a, \theta) = p(a', \theta)$ if $\sum_{i=1}^n a_i = \sum_{i=1}^n a'_i$). The function $p(a, \theta)$ is strictly decreasing in a where it is positive, strictly increasing in θ, and continuously differentiable in a. There is an upper bound on expected revenues.

The price shock θ is randomly drawn after the firms have chosen their actions, according to the cumulative distribution F with continuously differentiable density f. The probability measure on prices implied by an action profile $a \in A$ is $\rho(\cdot \mid a)$. Because F has a density, $\rho(\cdot \mid a)$ is absolutely continuous with respect to Lebesgue measure. We assume the price shocks are sufficiently noisy that $\rho(\cdot \mid a)$ has full support on $\mathbb{R}_+$.

The price p is the signal in a game of imperfect public monitoring. We write firm i's ex post payoff, as a function of the action profile a and realized price p, as

$$u_i^*(a, p) = pa_i - c(a_i),$$

where $c(a_i)$ is an increasing, convex cost function. Firm i's ex ante payoff is

$$u_i(a) = \int_0^\infty u_i^*(a, p(a, \theta))dF(\theta).$$

We assume that the stage game has a Nash equilibrium in pure strategies, in which each firm i produces an identical quantity that we denote by $a^N > 0$, and that the symmetric action profile that maximizes the total payoffs for the n firms calls for each firm i to produce the identical quantity $a^m < a^N$, where $a^N \in A_i$. Both are satisfied if, for example, (11.1.8) holds and the demand curve $p(a)$ is a linear function of $\sum_{i=1}^n a_i$. We omit the subscripts on the quantities a^N and a^m.

Turning to the repeated game, with $\delta \in (0, 1)$, the price shock θ^t is independently and identically distributed across periods. In each period t, the firms choose their quantities simultaneously, yielding profile a^t, and then observe the market price. They are unable to observe either their rivals' quantities or (unlike the analysis of Rotemberg and Saloner 1986 in section 6.1) the shock θ^t, either before or after choosing their actions.

11.1.2 Optimal Collusion

We are interested in the equilibrium that earns the largest repeated-game payoff for the firm among pure-strategy strongly symmetric equilibria. The impact of restricting attention to pure strategies vanishes as the grid of feasible quantities becomes arbitrarily fine. We refer to the resulting equilibrium as the *most collusive* equilibrium.

In an equilibrium with $a_i < a^N$, a firm seeing a low market price cannot be certain whether this reflects a relatively large market output or simply an adverse demand shock. A "collusive" equilibrium in which the firms restrict their output levels to increase the market price then runs the risk that a firm may be tempted to cheat on the agreement, increasing its output while hoping that the deviation will be masked by the randomness in market prices.[1]

We can restrict attention to equilibria with the bang-bang property. Let $\bar{V}_i$ and $\tilde{V}_i$ be the maximum and minimum (strongly symmetric) pure perfect public equilibrium payoffs for player i. By proposition 7.5.1, there exists an equilibrium with value $\bar{V}_i$ and with continuation values drawn exclusively from the set $\{\bar{V}_i, \tilde{V}_i\}$. Let $\bar{a}$ denote the action profile played in the first period of this equilibrium. Similarly, there is an equilibrium with value $\tilde{V}_i$ and the same set of continuations. Let $\tilde{a}$ denote its first-period action profile. The equilibrium with value $\bar{V}_i$ consists of a reward state during which $\bar{a}$ is produced, and a punishment state during which $\tilde{a}$ is produced. Play begins in the former. Transitions are governed by a pair of sets of prices $\bar{P}$ and $\tilde{P}$. In the reward state, any price in $\bar{P}$ prompts a switch to the punishment state, with play otherwise continuing in the reward state. In the punishment state, any price in $\tilde{P}$ causes play to continue in the punishment state, with play otherwise switching to the reward state. Figure 11.1.1 illustrates.

Player i's continuation values satisfy

$$\bar{V}_i = (1-\delta)u_i(\bar{a}) + \delta[(1-\rho(\bar{P} \mid \bar{a}))\bar{V}_i + \rho(\bar{P} \mid \bar{a})\tilde{V}_i] \qquad (11.1.1)$$

---

1. Porter (1983b), with elaboration by Coslett and Lee (1985) and Ellison (1994), uses a repeated game of imperfect monitoring with this type of equilibrium to examine the pricing behavior of railroads in the 1880s.

11.1 ■ Oligopoly

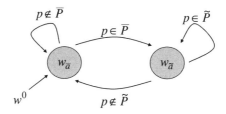

Figure 11.1.1 Equilibrium strategy profile for repeated oligopoly with imperfect monitoring.

and

$$\tilde{V}_i = (1-\delta)u_i(\tilde{a}) + \delta[(1-\rho(\tilde{P}\mid\tilde{a}))\bar{V}_i + \rho(\tilde{P}\mid\tilde{a})\tilde{V}_i]. \quad (11.1.2)$$

Solving (11.1.1) and (11.1.2), firm i's equilibrium payoff is

$$\bar{V}_i = \frac{u_i(\bar{a})(1-\delta\rho(\tilde{P}\mid\tilde{a})) + \delta u_i(\tilde{a})\rho(\bar{P}\mid\bar{a})}{1-\delta[\rho(\tilde{P}\mid\tilde{a}) - \rho(\bar{P}\mid\bar{a})]}.$$

There is always an equilibrium in which $\bar{a}_i = \tilde{a}_i = a^N$ for all i, duplicating the Nash equilibrium of the stage game in every period. In this case, the incentive constraints for equilibrium play are trivially satisfied. If the action sets A_i are sufficiently fine (see proposition 11.1.2), this equilibrium payoff falls short of $\bar{V}_i$.

To achieve payoff $\bar{V}_i$, it must be that $\bar{a}_i < a^N$ (again, for sufficiently fine A_i). Player i prefers to choose $\bar{a}_i$ rather than any alternative action a'_i, in either case thereafter continuing with the equilibrium strategies, if, for all feasible a'_i,

$$(1-\delta)[u_i(a'_i, \bar{a}_{-i}) - u_i(\bar{a})] \le \delta[\rho(\bar{P}\mid (a'_i, \bar{a}_{-i})) - \rho(\bar{P}\mid\bar{a})](\bar{V}_i - \tilde{V}_i). \quad (11.1.3)$$

We see the basic equilibrium trade-off in this inequality. The left side captures the immediate payoff gain of deviating from the proposed equilibrium strategy. The right side captures the loss in *future* payoffs caused by the attendant increase in the likelihood of triggering the punishment.

Green and Porter (1984) and Porter (1983a) consider strategies in which punishments consist of a finite number of periods of playing the stage-game Nash equilibrium, followed by a return to the reward phase. Optimal punishments are in general more severe, and hence allow equilibria that are "more collusive," in the sense that they allow a higher equilibrium payoff. These optimal punishments involve large quantities, that is, $\tilde{a}_i > a^N$, and give rise to the incentive constraint that, for all feasible a'_i,

$$(1-\delta)[u_i(a'_i, \tilde{a}_{-i}) - u_i(\tilde{a})] \le \delta[\rho(\tilde{P}\mid (a'_i, \tilde{a}_{-i})) - \rho(\tilde{P}\mid\tilde{a})](\bar{V}_i - \tilde{V}_i). \quad (11.1.4)$$

Remark 11.1.1 **Likelihood ratios** To develop intuition about the sets $\bar{P}$ and $\tilde{P}$, suppose there were only two actions available, the actions $\bar{a}_i$ and $\tilde{a}_i$. The incentive constraint (11.1.3) imposes a lower bound on the difference $\rho(\bar{P}\mid (\tilde{a}_i, \bar{a}_{-i})) - \rho(\bar{P}\mid\bar{a})$. Let h_a be the density (Radon-Nikodym derivative) of $\rho(\cdot\mid a)$. The bang-bang equilibrium strategy achieving payoff $\bar{V}_i$ must minimize $\rho(\bar{P}\mid\bar{a})$ subject to the incentive constraint (11.1.3) and holding fixed $\bar{a}$ and the continuation values $\bar{V}_i$ and $\tilde{V}_i$ in (11.1.3). This can be written as

$$\min_{\bar{P}} \int_{\bar{P}} h_{\bar{a}}(p)\, dp \tag{11.1.5}$$

such that

$$\int_{\bar{P}} h_{(\tilde{a}_i, \bar{a}_{-i})}(p)\, dp - \int_{\bar{P}} h_{\bar{a}}(p)\, dp \geq \frac{(1-\delta)[u_i(\tilde{a}_i, \bar{a}_{-i}) - u_i(\bar{a})]}{\delta(\bar{V}_i - \tilde{V}_i)}. \tag{11.1.6}$$

The solution to (11.1.5)–(11.1.6) calls for $\bar{P}$ to be a set of prices with the property that every price (signal) in $\bar{P}$ has a higher likelihood ratio than any price not in $\bar{P}$, where the *likelihood ratio* is

$$\frac{h_{(\tilde{a}_i, \bar{a}_{-i})}(p)}{h_{\bar{a}}(p)} \equiv \ell(p). \tag{11.1.7}$$

To see why, suppose there is a set of prices $\hat{P}$ contained in the complement of $\bar{P}$ that has positive measure under $\rho(\cdot \mid \bar{a})$, that is, $\int_{\hat{P}} h_{\bar{a}}(p)\, dp = \eta > 0$, and that the likelihood ratio is larger on $\hat{P}$ than on some set $P^\dagger \subset \bar{P}$ with $\int_{P^\dagger} h_{\bar{a}}(p)\, dp = \eta$. There exists $\kappa > 0$ such that for all $p \in \hat{P}$ and all $p' \in P^\dagger$, $\ell(p') < \kappa < \ell(p)$, and so

$$\int_{\hat{P}} h_{(\tilde{a}_i, \bar{a}_{-i})}(p)\, dp - \int_{P^\dagger} h_{(\tilde{a}_i, \bar{a}_{-i})}(p)\, dp$$
$$= \int_{\hat{P}} h_{\bar{a}}(p)\ell(p)\, dp - \int_{P^\dagger} h_{\bar{a}}(p)\ell(p)\, dp$$
$$> \int_{\hat{P}} h_{\bar{a}}(p)\kappa\, dp - \int_{P^\dagger} h_{\bar{a}}(p)\kappa\, dp = 0.$$

Hence, modifying $\bar{P}$ by replacing $P^\dagger$ with $\hat{P}$ leaves $\rho(\bar{P} \mid \bar{a})$ unchanged but increases $\rho(\bar{P} \mid (a'_i, \bar{a}_{-i}))$. This in turn introduces slack into the constraint (11.1.6) in the minimization problem (11.1.5), allowing $\bar{P}$ to be made smaller, reducing the equilibrium probability of a punishment while still preserving incentives.

The next section builds on this intuition, with additional assumptions on the game, to characterize $\bar{P}$ when players have many alternative actions.

◆

11.1.3 Which News Is Bad News?

Green and Porter (1984) and Porter (1983a) assume that the set of prices $\bar{P}$ resulting in a punishment while in the reward phase is of the form $[0, \bar{p}]$. It seems natural to associate punishments with low prices, because the latter are in turn associated with the increases in quantity that allow profitable deviations from collusive outputs. However, nothing from the bang-bang argument of section 7.5 suggests the set of signals triggering a punishment should reflect such intuition. This section establishes (restrictive) conditions under which the most collusive equilibrium entails $\bar{P} = [0, \bar{p}]$.

11.1 ■ Oligopoly

In general, the set of prices that maximizes the likelihood ratio (11.1.7) need not be of the form $[0, \bar{p}]$. A sufficient condition is that the density $h_a(p)$ satisfy the *monotone likelihood ratio property*: For any profiles a and a' with $\sum_i a_i < \sum_i a'_i$,

$$\frac{h_{a'}(p)}{h_a(p)}$$

is strictly decreasing in p. However, the monotone likelihood ratio alone does not ensure $\bar{P} = [0, \bar{p}]$, because we have more than the two actions of remark 11.1.1 to contend with. An incentive constraint must then be satisfied for each alternative action. By the argument in remark 11.1.1, if the monotone likelihood ratio property holds, then for each action $a'_i > \bar{a}_i$, $p(a'_i, \bar{a}_{-i}) < p(\bar{a})$, and so there is a price $\bar{p}_{a'_i}$ such that $[0, \bar{p}_{a'_i}]$ is the most effective way to discourage a deviation to a'_i. Hence, if no other constraints bind, then $\bar{P} = [0, \bar{p}]$ for $\bar{p} \equiv \sup\{p_{a'_i} | a'_i > \bar{a}_i\}$. Notice that under this $\bar{P}$, player i is just indifferent between $\bar{a}_i$ and some larger quantity.

What about deviations to smaller quantities? On the face of it, such deviations appear to be obviously counterproductive because they reduce current payoffs. However, such a deviation may be useful because it reduces the probability of a punishment.

A repetition of our previous argument, again under the monotone likelihood ratio property, suggests that deviations to smaller quantities are most effectively deterred by attaching punishments to large rather than small prices. Hence, if the optimal punishment set is to be of the form $[0, \bar{p}]$, it must be the case that the binding incentive constraints are those to larger quantities, and the incentive constraints to smaller quantities do not bind.

We establish sufficient conditions under which this is the case, and hence $\bar{P} = [0, \bar{p}]$. We now assume the price shock θ enters the function $p(a, \theta)$ multiplicatively, or

$$p(a, \theta) = \theta p(a). \tag{11.1.8}$$

We continue to use the notation p in this case, trusting to the context to indicate whether it is $p(a, \theta)$ or simply $p(a)$. Then,

$$\rho([0, \hat{p}] \mid a) = F\left(\frac{\hat{p}}{p(a)}\right) \quad \text{and} \quad h_a(\hat{p}) = \frac{1}{p(a)} f\left(\frac{\hat{p}}{p(a)}\right).$$

We let the distribution of shocks be given by

$$F(\theta) = \frac{\theta}{\theta + 1}, \quad \text{with density} \quad f(\theta) = \frac{1}{(\theta + 1)^2}. \tag{11.1.9}$$

Under (11.1.8) and (11.1.9), the monotone likelihood ratio property is satisfied. We further assume that demand is given by $p(a) = r - \sum_i a_i$ and for some $r > 0$, that costs are such that $a^N < r/n$, and (to ensure full-support signals and hence the bang-bang principle) that $A_i \subset [0, r/n)$.

If $\bar{P} = [0, \bar{p}]$ for some $\bar{p}$, then the payoff for firm i from profile $a' = (a_i, \bar{a}_{-i})$ is

$$(1 - \delta)u_i(a') + \delta \left(1 - F\left(\frac{\bar{p}}{p(a')}\right)\right) \bar{V}_i + \delta F\left(\frac{\bar{p}}{p(a')}\right) \tilde{V}_i.$$

Differentiating (recalling that $dp(a)/da_i = -1$) gives

$$(1 - \delta)\frac{du_i(a')}{da_i} - \delta f\left(\frac{\bar{p}}{p(a')}\right) \frac{\bar{p}}{(p(a'))^2}(\bar{V}_i - \tilde{V}_i). \tag{11.1.10}$$

Differentiating again, denoting $df(\cdot)/d\theta$ by $f'(\cdot)$ and using the specification of f given by (11.1.9) to infer the final inequality:

$$(1-\delta)\frac{d^2 u_i(a')}{da_i^2} - \delta f'\left(\frac{\bar{p}}{p(a')}\right) \frac{\bar{p}^2}{(p(a'))^4}(\bar{V}_i - \tilde{V}_i) - \delta f\left(\frac{\bar{p}}{p(a')}\right) \frac{2\bar{p}}{(p(a'))^3}(\bar{V}_i - \tilde{V}_i)$$

$$= (1-\delta)\frac{d^2 u_i(a')}{da_i^2} - \delta(\bar{V}_i - \tilde{V}_i)\frac{\bar{p}}{(p(a'))^3}\left(f'\left(\frac{\bar{p}}{p(a')}\right)\frac{\bar{p}}{p(a')} + 2f\left(\frac{\bar{p}}{p(a')}\right)\right)$$

$$< 0.$$

Hence, firm i's payoff function is concave. Coupled with player i's indifference between $\bar{a}_i$ and some $a_i > \bar{a}_i$, this ensures that deviations to quantities smaller than $\bar{a}_i$ are suboptimal, and hence $\bar{P} = [0, \bar{p}]$ for $\bar{p} = \sup\{p_{a_i'} | a_i' > \bar{a}_i\}$.

We summarize the discussion in the following proposition.

Proposition 11.1.1 *Suppose the demand shocks satisfy (11.1.9), the price function satisfies (11.1.8) with $p(a) = r - \sum_{i=1}^{n} a_i$ for some $r > 0$, and $A_i \subset [0, r/n)$. Suppose there is a nontrivial strongly symmetric PPE (so that $\bar{V}_i > \tilde{V}_i$). Then $\bar{V}_i$ is the value of a bang-bang equilibrium and $\bar{P} = [0, \bar{p}]$ for some $\bar{p} \in \mathbb{R}_+$.*

In the punishment phase of a nontrivial bang-bang equilibrium, $\tilde{a}_i > a^N$ and hence the firm has a myopic incentive to lower its quantity. Applying similar reasoning, one might then expect $\tilde{P}$ to be given by a set of the form $[\tilde{p}, \infty)$. This is indeed the most effective way to deter deviations to smaller quantities. However, the equilibrium payoff is now given by

$$\tilde{V}_i = (1-\delta)u_i(\tilde{a}) + \delta F\left(\frac{\tilde{p}}{p(\tilde{a})}\right) \bar{V}_i + \delta\left(1 - F\left(\frac{\tilde{p}}{p(\tilde{a})}\right)\right) \tilde{V}_i,$$

which is not obviously concave. If $\tilde{a}_i = \max A_i$, then deviations to larger quantities are impossible and hence $\tilde{P} = [\tilde{p}, \infty)$. Otherwise, our argument leaves open the possibility that $\tilde{P}$ has a more complicated structure.

11.1.4 Imperfect Collusion

If the choice sets A_i contain a sufficiently fine grid of points, then the most collusive equilibrium will be nontrivial. However, it will not entail perfect collusion:

Proposition 11.1.2 *Let $a^m \in A_i$. Under the assumptions of proposition 11.1.1, if the finite grid of feasible quantities A_i is sufficiently fine, then $a^m < \bar{a}_i < a^N$.*

11.1 ■ Oligopoly

Firms produce quantities larger than the monopoly output to reduce the incentives to deviate from equilibrium play, which in turn allows equilibrium punishments to be less severe. Because punishments actually occur in equilibrium, this reduced severity is valuable. The proof of this result makes precise the following intuition: If we take the joint profit maximizing outputs a^m as our point of departure, then an increase in quantities causes a second-order reduction in reward-phase payoffs, while attaining a first-order reduction in punishment severity, ensuring that $\bar{a}_i > a^m$.

Proof It is immediate that $\bar{a}_i \in [a^m, a^N]$, because quantities below a^m only make deviations more tempting and hence require more likely equilibrium punishments, without any flow payoff benefit. We suppose $\bar{a}_i = a^m \in A_i$ and derive a contradiction by constructing an alternative equilibrium, when A_i is sufficiently fine, with higher expected profits.

We begin by assuming every action in $[0, r/n)$ is available as a candidate for $\bar{a}_i$, while retaining the remaining structure of the equilibrium. From the proof of proposition 11.1.1, we know that firm i's payoff $\bar{V}_i(a')$ is concave in a_i and maximized at $\bar{a}_i$. In the limit as A_i approaches $[0, r/n)$, the first-order condition implied by (11.1.10) must approach zero. Denote the equilibrium value $\bar{p}/p(\bar{a})$ by $\bar{\theta}$ and rewrite this condition as

$$(1-\delta)\frac{du_i(\bar{a})}{da_i} - \delta f(\bar{\theta})\frac{\bar{\theta}}{p(\bar{a})}(\bar{V}_i - \tilde{V}_i) = 0. \tag{11.1.11}$$

From (11.1.9), $f(\bar{\theta})\bar{\theta}$ is single-peaked with a maximum at $\bar{\theta} = 1$. Because larger values of $\bar{\theta}$ make punishments more likely, and because (11.1.11) holds in the most collusive equilibrium (and hence must correspond to choosing $\bar{\theta}$ and $\bar{a}$ to maximize the firms' payoffs, holding fixed the continuation values $\bar{V}_i$ and $\tilde{V}_i$), its solution must correspond to a value $\bar{\theta} \leq 1$.

Consider marginally increasing $\bar{a}_i$ for all i from its current value of a^m, while adjusting $\bar{\theta}$ (or equivalently, $\bar{p}$) to preserve the equality in (11.1.11), holding fixed $\bar{V}_i - \tilde{V}_i$. This adjusts the first-period quantity of output and punishment criteria, holding fixed the remainder of the equilibrium. Because (11.1.11) captures the first-period incentive constraints, this adjustment gives us an alternative equilibrium. If $d\bar{\theta}/d\bar{a}_i < 0$, the probability of punishment falls and this alternative equilibrium gives a higher payoff, establishing the desired result. In particular, as we consider actions larger than but arbitrarily close to a^m, we couple a reduction in the probability of a punishment with a current payoff sacrifice that is (in the limit) an order of magnitude smaller, giving the result.

To determine $d\bar{\theta}/d\bar{a}_i$, implicitly differentiate (11.1.11) to obtain

$$0 = (1-\delta)\left[\frac{d^2 u_i(\bar{a})}{d\bar{a}_i d\bar{a}_i} + (n-1)\frac{d^2 u_i(\bar{a})}{d\bar{a}_i d\bar{a}_j}\right] - \delta f(\bar{\theta})\frac{\bar{\theta} n}{(p(\bar{a}))^2}(\bar{V}_i - \tilde{V}_i)$$
$$- \delta \left(f'(\bar{\theta})\frac{\bar{\theta}}{p(\bar{a})} + f(\bar{\theta})\frac{1}{p(\bar{a})}\right)(\bar{V}_i - \tilde{V}_i)\frac{d\bar{\theta}}{d\bar{a}_i}.$$

The linearity of the demand curve and convexity of cost functions ensures that the bracketed part of the first term in the first line is the negative, and the second term in the first line is also obviously negative. Condition (11.1.9) and $\bar{\theta} \leq 1$ ensure that the terms in the second line are positive except for $d\bar{\theta}/d\bar{a}_i$. The latter must then be negative. For a sufficiently fine grid for the finite set A_i, this implies that there exists an equilibrium with higher payoffs, giving a contradiction.

Similar ideas (leading to an argument reminiscent of those in section 4.3) allow us to show that $\bar{a}_i < a^N$.

∎

11.2 Repeated Adverse Selection

11.2.1 General Structure

In each period, each player i first learns his type, θ_i, which is drawn from a finite set Θ_i. The distribution of types is independent across players and time with π_i the commonly known distribution of player i's type.[2] After learning their types, players simultaneously choose their *moves*, with player i's move $y_i \in Y_i$ being publicly observable. The public outcome is the profile $y = (y_1, \ldots, y_n)$.

A pure action for player i is a mapping from his type space into his space of moves, $a_i : \Theta_i \to Y_i$. Assume the payoff structure is one of *private values*, so that player i's *realized* payoff is given by $r_i(y, \theta_i)$ when his type is θ_i and the public outcome profile is y.

Player i's expected payoff, given action profile a, is then given by

$$u_i(a) = \sum_{\theta \in \Theta} r_i(a(\theta), \theta_i) \prod_{j=1}^{n} \pi_j(\theta_j)$$
$$= \sum_{\theta_i \in \Theta_i} \pi_i(\theta_i) \sum_{y_{-i} \in Y_{-i}} r_i(a_i(\theta_i), y_{-i}, \theta_i) \prod_{j \neq i} \rho_j(y_j \mid a_j),$$

where

$$\rho_j(y_j \mid a_j) = \begin{cases} \sum_{\theta_j \in a_j^{-1}(y_j)} \pi_j(\theta_j), & \text{if } y_j \in a_j(\Theta_j), \\ 0, & \text{otherwise.} \end{cases}$$

Define

$$u_i^*(y, a_i) = \begin{cases} E[r_i(y_i, y_{-i}, \theta_i) \mid \theta_i \in a_i^{-1}(y_i)], & \text{if } y_i \in a_i(\Theta_i), \\ 0, & \text{otherwise.} \end{cases}$$

Because y_i arises with zero probability under a_i if $y_i \notin a_i(\Theta_i)$, $u_i^*(y, a_i)$ can then be defined arbitrarily. With this definition of u_i^*, player i's payoff can be written as an expectation of the ex post payoff u_i^* (see (7.1.1)). This description of i's payoffs is not natural here because u_i^* now involves taking expectations over player i's private information about his type.

---

2. This setting of a repeated stage game with private information contrasts with the reputation games that we study in part IV. Here, players' types are drawn anew in each period, in contrast to being fixed throughout the repeated game in reputation games.

Remark 11.2.1 It is immediate from definition 9.5.1 that games of repeated adverse selection have a product structure. Proposition 9.5.1 then applies, giving us a folk theorem for such games. As a result, adverse selection in repeated interactions is typically not an impediment to efficiency if players are patient. Section 11.2.6 provides an illustration.

◆

The notion of enforceability is unchanged from definition 7.3.1. This gives us *ex ante enforceability*, applied to actions $a_i : \Theta_i \to Y_i$ that map from types into moves and that can be viewed as describing players' behavior before they learn their realized type and choose their move.

Say that an action profile a is *interim enforceable* with respect to $\mathscr{W}$ if there exists a mapping $\gamma : Y \to \mathscr{W}$ such that for all i, for all $\theta_i \in \Theta_i$, and all $y'_i \in Y_i$,

$$\sum_{\theta_{-i}} \{(1-\delta) r_i(a_i(\theta_i), a_{-i}(\theta_{-i}), \theta_i) + \delta \gamma_i(a_i(\theta_i), a_{-i}(\theta_{-i}))\} \prod_{j \neq i} \pi_j(\theta_j)$$

$$\geq \sum_{\theta_{-i}} \{(1-\delta) r_i(y'_i, a_{-i}(\theta_{-i}), \theta_i) + \delta \gamma_i(y'_i, a_{-i}(\theta_{-i}))\} \prod_{j \neq i} \pi_j(\theta_j).$$

Players' types are private, so the function describing the continuation promises is independent of type (that is, the same function γ is used for all types).

It is immediate that if an action profile is interim enforceable, then it is ex ante enforceable. Taking expectations over the above inequality with respect to player i's type yields the relevant inequality in definition 7.3.1. Moreover, it is almost as immediate that ex ante enforceability implies interim enforceability: Suppose a is ex ante enforced by γ but that for player i there is some type θ'_i with a profitable deviation y'_i against γ. Then a cannot have been ex ante enforced by γ, because $\hat{a}_i$ given by $\hat{a}_i(\theta'_i) = y'_i$ and $\hat{a}_i(\theta_i) = a_i(\theta_i)$ for $\theta_i \neq \theta'_i$ is a profitable deviation.

11.2.2 An Oligopoly with Private Costs: The Game

This section introduces an example of repeated adverse selection, based on Athey and Bagwell (2001) and Athey, Bagwell, and Sanchirico (2004).

There are two firms, denoted 1 and 2. Nature first independently draws, for each firm, a constant marginal cost equal to either $\underline{\theta}$ or $\bar{\theta} > \underline{\theta}$, with the two values being equally likely. The firms then simultaneously choose prices in $\mathbb{R}_+$. There is a unit mass of consumers, with a reservation price of r. A consumer purchases from the firm setting the lower price if it does not exceed r. Consumers are indifferent between the two firms if they set identical prices, in which case we specify consumer decisions as part of the equilibrium. Consumers are thus small anonymous players. A firm from whom a mass μ of consumers purchases at price p, with cost θ, earns payoff $\mu(p-\theta)$.

The stage game has a unique symmetric Nash equilibrium. A firm whose cost level is $\bar{\theta}$ sets price $\bar{\theta}$ and earns a zero expected profit. A low-cost firm chooses a price according to a distribution $\alpha(p)$ with support on $[(\underline{\theta} + \bar{\theta})/2, \bar{\theta}]$.[3] If the low-cost firm sets price p, its expected payoff is

---
3. It is straightforward that prices above $\bar{\theta}$ are vulnerable to being undercut by one's rival. Prices above $\bar{\theta}$ thus will not appear in equilibrium, and high-cost firms will set price $\bar{\theta}$. The lower bound

$$\left[\tfrac{1}{2} + \tfrac{1}{2}(1 - \alpha(p))\right][p - \underline{\theta}],$$

because the probability that the customer buys from the firm is $1/2$ (from the event the other firm is high-cost) plus $(1 - \alpha(p)/2)$ (from the event that the other firm is low cost but sets a higher price). Differentiating, the equilibrium is completed by solving the differential equation

$$1 + (1 - \alpha(p)) - \frac{d\alpha(p)}{dp}(p - \underline{\theta}) = 0.$$

The expected payoff to each firm from this equilibrium is given by $(\bar{\theta} - \underline{\theta})/4$. The noteworthy aspect of this equilibrium is that it does not vary with r. If r is much larger than $\bar{\theta}$, the firms are falling far short of the monopoly profit.

Minmax payoffs are $(0, 0)$. Each firm can ensure the other earns a zero payoff by setting $p = \underline{\theta}$, whereas a firm can ensure at least this payoff by setting $p \geq \bar{\theta}$. To maximize firm i's payoff, we have firm i set $p = r$ and the other firm set a price higher than r. This gives expected payoff $r - (\underline{\theta} + \bar{\theta})/2$. An upper bound on the payoffs in a symmetric-payoff equilibrium would call for both firms to set price r, but with only low-cost firms (if there is such a firm) selling output, for an expected payoff to each firm of

$$\tfrac{1}{8}(r - \bar{\theta}) + \tfrac{3}{8}(r - \underline{\theta}) = \tfrac{1}{2}\left(r - \tfrac{3}{4}\underline{\theta} - \tfrac{1}{4}\bar{\theta}\right).$$

The firms can come arbitrarily close to this payoff by having a firm with high costs set price r and a low-cost firm set a slightly smaller price.

11.2.3 A Uniform-Price Equilibrium

We first study an equilibrium of the repeated game in which each firm sets price r in every period, as long as such behavior has been observed in the past, with any deviation triggering a switch to permanent play of the stage-game Nash equilibrium. We assume that half of the consumers buy from each firm in this case. The payoff from this strategy profile is

$$\tfrac{1}{2}\left(r - \tfrac{1}{2}(\underline{\theta} + \bar{\theta})\right). \tag{11.2.1}$$

A firm can earn a higher payoff in the stage game by setting a price slightly lower than r, allowing the firm to capture the entire market (rather than simply half of it). This undercutting is most attractive when the firm has drawn cost $\underline{\theta}$. In addition, the larger is the price that the firm sets, given that it falls below r, the more profitable the deviation, with a supremum on the set of deviation payoffs of $r - \underline{\theta}$. The condition

---

on the support of the low-cost firm's price distribution makes the firm indifferent between selling with probability 1 at that price and selling with probability $1/2$ at price $\bar{\theta}$, or $p - \underline{\theta} = (\bar{\theta} - \underline{\theta})/2$, giving $p = (\underline{\theta} + \bar{\theta})/2$. Though this example violates our practice of considering only pure strategies when examining continuum action spaces (remark 7.1.1), in the profiles we study, mixing only appears as behavior in an absorbing state of the automaton, so the measurability issues do not arise.

for the candidate strategies to be an equilibrium is that a deviation commanding this payoff, having drawn cost $\underline{\theta}$ be unprofitable, or

$$(1-\delta)(r-\underline{\theta}) + \delta\tfrac{1}{4}[\bar{\theta}-\underline{\theta}] \leq (1-\delta)\tfrac{1}{2}(r-\underline{\theta}) + \delta\tfrac{1}{2}\bigl(r - \tfrac{1}{2}(\underline{\theta}+\bar{\theta})\bigr).$$

We assume that δ is sufficiently large that this constraint holds, and therefore that we have at least one equilibrium providing payoffs larger than continual play of the stage-game Nash equilibrium. This is the best symmetric-payoff equilibrium when the firms are constrained to set prices that do not depend on their realized costs.

11.2.4 A Stationary-Outcome Separating Equilibrium

We now maximize the firms' equilibrium payoff, in a strongly symmetric equilibrium, where consumers treat firms identically when they choose the same price and firms choose a different price when drawing cost $\underline{\theta}$ than when drawing cost $\bar{\theta}$.[4] The resulting equilibrium has a stationary outcome, and so no intertemporal incentives on the equilibrium path. Let p_i denote the price set by firm i.

The first step in simplifying this problem is recognizing that in enforcing an action profile a, we can think of prices as falling in two categories, equilibrium and out of equilibrium. Equilibrium prices are those that are set with positive probability by a, and out-of-equilibrium prices are the remainder. Attaching a punishment to an out-of-equilibrium price has no effect on the equilibrium payoff, unlike the case of an equilibrium price.

Moreover, if we are examining the limit as the discount factor gets large, we can ignore out-of-equilibrium prices. For sufficiently large δ, the maximum symmetric equilibrium payoff exceeds that of the stage-game Nash equilibrium. It then follows that for any candidate equilibrium, attaching the punishment of perpetual play of the stage-game Nash equilibrium to out-of-equilibrium prices suffices to ensure (again, for large discount factors) that such prices will be suboptimal.

We thus restrict attention to equilibrium prices when evaluating the enforceability of the action profile a. Because firms price similarly in a symmetric profile, any symmetric action profile a is described by two prices, $\underline{p}$ and $\bar{p}$, set by a firm drawing cost $\underline{\theta}$ and $\bar{\theta}$, respectively. Separation requires $\underline{p} \neq \bar{p}$. Without loss of generality, we assume $\underline{p} < \bar{p}$. A separating symmetric action profile a is enforced if, for each player i,

$$(1-\delta)\tfrac{3}{4}(\underline{p}-\underline{\theta}) + \delta E_{(\underline{p},a_j)}\gamma(p_i,p_j) \geq (1-\delta)\tfrac{1}{4}(\bar{p}-\underline{\theta}) + \delta E_{(\bar{p},a_j)}\gamma(p_i,p_j) \tag{11.2.2}$$

and

$$(1-\delta)\tfrac{1}{4}(\bar{p}-\bar{\theta}) + \delta E_{(\bar{p},a_j)}\gamma(p_i,p_j) \geq (1-\delta)\tfrac{3}{4}(\underline{p}-\bar{\theta}) + \delta E_{(\underline{p},a_j)}\gamma(p_i,p_j), \tag{11.2.3}$$

where $E_{(\bar{p},a_j)}\gamma(p_i,p_j)$ is the expected value of $\gamma(p_i,p_j)$ with respect to $p_j = a_j(\theta_j)$, fixing $p_i = \bar{p}$ (given our focus on symmetry, $\gamma_i = \gamma_j$). The enforceability constraints have been written in interim form.

---

4. Strong symmetry is not a particularly restrictive notion in the current context. See note 6 on page 231 and remark 11.2.3.

We now reformulate the search for the best strongly symmetric separating equilibrium. Consider choosing a pair of prices $(\underline{p}, \bar{p})$ and a pair of values $\underline{x}$ and $\bar{x}$ to solve
$$\max_{\underline{p},\bar{p},\underline{x},\bar{x}} \tfrac{1}{2}\big(\tfrac{3}{4}(\underline{p}-\underline{\theta})+\underline{x}\big) + \tfrac{1}{2}\big(\tfrac{1}{4}(\bar{p}-\bar{\theta})+\bar{x}\big)$$
subject to
$$\tfrac{3}{4}(\underline{p}-\underline{\theta})+\underline{x} \geq \tfrac{1}{4}(\bar{p}-\underline{\theta})+\bar{x},$$
$$\tfrac{3}{4}(\underline{p}-\bar{\theta})+\underline{x} \leq \tfrac{1}{4}(\bar{p}-\bar{\theta})+\bar{x},$$
and
$$\underline{x},\bar{x} \leq 0.$$

To interpret this in terms of the repeated game, let $\bar{v}$ be the maximum strongly symmetric equilibrium payoff and let
$$\bar{x} = \frac{\delta}{1-\delta}[E_{(\bar{p},\alpha_j)}\gamma(p)-\bar{v}]$$
and
$$\underline{x} = \frac{\delta}{1-\delta}[E_{(\underline{p},\alpha_j)}\gamma(p)-\bar{v}].$$

Therefore, $\bar{x}$ and $\underline{x}$ are the normalized expected penalties a player faces, measured in terms of shortfall from the maximum possible payoff, for choosing the price consistent with cost level $\bar{\theta}$ and $\underline{\theta}$, respectively.

Simplifying the first two constraints, we find
$$2\underline{\theta} + 4(\bar{x}-\underline{x}) \leq 3\underline{p} - \bar{p} \leq 2\bar{\theta} + 4(\bar{x}-\underline{x}). \quad (11.2.4)$$

We now note that the prices $\bar{p}$ and $\underline{p}$ enter this constraint only in the difference $3\underline{p}-\bar{p}$. Because high prices bring high payoffs and we seek a payoff-maximizing equilibrium, we must set both prices as large as possible. We therefore set $\bar{p}$ at its maximum value
$$\bar{p} = r,$$
and then increase $\underline{p}$ until the upper constraint in (11.2.4) binds, giving constraints of
$$2\underline{\theta} + 4(\bar{x}-\underline{x}) \leq 3\underline{p} - \bar{p} = 3\underline{p} - r = 2\bar{\theta} + 4(\bar{x}-\underline{x}).$$
We can solve for
$$\underline{p} = \tfrac{1}{3}(2\bar{\theta}+r) + \tfrac{4}{3}(\bar{x}-\underline{x}).$$
Given the prices we have calculated, the payoff from the mechanism is given by
$$\tfrac{1}{2}\big(\tfrac{3}{4}(\underline{p}-\underline{\theta})+\underline{x}\big) + \tfrac{1}{2}\big(\tfrac{1}{4}(\bar{p}-\bar{\theta})+\bar{x}\big) = \tfrac{1}{4}r + \tfrac{1}{8}(\bar{\theta}-3\underline{\theta}) + \bar{x}.$$

We now note that this expected payoff does not depend on $\underline{x}$ and is increasing in $\bar{x}$. Hence, the problem has a solution in which we set $\bar{x}$ at its largest value of 0. Once we have made this choice, the value of $\underline{x}$ is arbitrary and can without loss of generality be taken to also equal 0. We then obtain prices of
$$\bar{p} = r \quad (11.2.5)$$
and
$$\underline{p} = \tfrac{1}{3}(2\bar{\theta}+r). \quad (11.2.6)$$

We imposed no constraints beyond nonpositivity on $\bar{x}$ and $\underline{x}$, so there is in general no guarantee that its solution leads to an equilibrium of the repeated game. In this case, however, we have found a solution with the convenient property that $\underline{x} = \bar{x} = 0$. This implies that we can maximize the expected payoff by choosing the prices given by (11.2.5)–(11.2.6) while respecting the incentive constraints and holding future payoffs constant. It then follows that setting the prices (11.2.5)–(11.2.6) in every period must be a stationary-outcome, strongly symmetric equilibrium of the repeated game.

In the course of solving this problem, we have seen ideas that are familiar from mechanism design. For example, the finding that only the upper constraint in (11.2.4) binds is the statement that high-cost firms may prefer to mimic low-cost firms, but not the other way around. Athey and Bagwell (2001) and Athey, Bagwell, and Sanchirico (2004) refer to this as the *mechanism design approach* to repeated games of adverse selection. The idea is to attach a punishment to out-of-equilibrium actions sufficient to deter them when players are patient. One then notes that the remaining optimality conditions for equilibrium actions look much like a mechanism design problem, with continuation payoffs in the repeated game taking the place of transfers in the mechanism. This allows us to bring the insights of mechanism design theory to work in solving for the equilibrium of the repeated game.

11.2.5 Efficiency

We assumed throughout section 11.2.4 that the two types of firm, low-cost and high-cost, will choose different prices. Together with section 11.2.3, this gives us two candidates for the stationary-outcome, strongly symmetric equilibrium (in which consumers treat firms identically when prices are equal) that maximizes the players' payoffs. In one of these equilibria, from section 11.2.4, players with different costs set different prices in each period, with prices given by (11.2.5)–(11.2.6). Expected payoffs are given by

$$\tfrac{1}{4}r + \tfrac{1}{8}(\bar{\theta} - 3\underline{\theta}).$$

We refer to this as the discriminatory-price equilibrium. The other equilibrium, from section 11.2.3, is the uniform-price equilibrium in which both first set price r in every period, for expected payoffs of (see (11.2.1))

$$\tfrac{1}{2}\left(r - \tfrac{1}{2}\underline{\theta} - \tfrac{1}{2}\bar{\theta}\right).$$

For sufficiently large r, the second payoff dominates.

The stage-game Nash equilibrium in this market has the virtue that production is efficient, in the sense that the low-cost firm always sells the output. However, competition between the firms ensures that they cannot set prices above $\bar{\theta}$. If r is much larger than θ, the incentive to improve on this outcome by colluding is tremendous.

The question then arises as to how the firms can use their repeated play to collude. It is immediate that they can achieve prices larger than $\bar{\theta}$ by attaching punishments to undercutting prices. If they set the same price regardless of their cost level, they sacrifice productive efficiency but all deviations are to out-of-equilibrium prices that can be deterred by Nash reversion punishments that never occur in equilibrium. If the firms are to produce efficiently, they must set different prices when drawing low and high cost. There are now two equilibrium prices, and the temptation for a high-cost firm to mimic a low-cost firm must now be countered by incentives that occur on the equilibrium path and thus are costly.

When r is sufficiently large, the payoff-maximizing strongly symmetric equilibrium of the repeated game abandons all attempts at productive efficiency and calls for both firms to set the monopoly price r. The inefficiency cost of not directing production to the low-cost firm is $(\bar{\theta} - \underline{\theta})/4$, which is independent of r. Arranging production efficiently brings an incentive for the high-cost firm to undercut the low-cost firm's price to obtain a payoff arbitrarily close to $r - \bar{\theta}$, which grows as does r. Counteracting this temptation requires a similarly sized incentive cost. When r is large, this incentive cost is larger than the productive inefficiency of uniform pricing, and the latter maximizes payoffs. The result is an equilibrium payoff bounded away from efficiency, regardless of the discount factor.

11.2.6 Nonstationary-Outcome Equilibria

We once again examine equilibria with symmetric payoffs, but we no longer limit attention to stationary-outcome equilibria and no longer assume that consumers treat the firms identically when they charge identical prices. Recall that the efficient stage-game payoff profile is

$$\left(\tfrac{1}{2}\bigl(r - \tfrac{3}{4}\underline{\theta} - \tfrac{1}{4}\bar{\theta}\bigr),\ \tfrac{1}{2}\bigl(r - \tfrac{3}{4}\underline{\theta} - \tfrac{1}{4}\bar{\theta}\bigr)\right) \equiv v^*.$$

We are interested in the existence of an equilibrium with payoff near profile v^*. Recall from section 11.2.5 that if r is sufficiently large, the highest payoff that can be achieved in a stationary-outcome, strongly symmetric equilibrium is bounded away from v^*, for all discount factors. We thus find circumstances under which asymmetric strategies are required to maximize symmetric payoffs.

We consider an equilibrium built around two prices, r and $r - \varepsilon$, with interest centering on small ε. Low cost firms use the price $r - \varepsilon$ to distinguish themselves from high-cost firms, ensuring that the firms produce efficiently. For small ε, the price $r - \varepsilon$ is a virtually costless announcement that one is low cost.

Consumers are indifferent between firms that set identical prices, and can condition on the public history to apportion themselves between the two firms in various ways. We excluded this possibility when examining the strongly symmetric equilibria of section 11.2.4, but it will play a role here.

We again ignore out-of-equilibrium prices, trusting the threat of a switch to the stage-game Nash equilibrium to ensure patient players will never set such prices. However, we can no longer hope to use only current incentives to enforce equilibrium prices. Because the prices r and $r - \varepsilon$ are arbitrarily close together, the threat of adverse future consequences will be necessary to ensure that neither firm finds it optimal to set price $r - \varepsilon$ even though it has drawn a high cost.

Proposition 11.2.1 *For any $\eta > 0$, there exists a $\bar{\delta} < 1$ such that for all $\delta \in (\bar{\delta}, 1)$, there exists a pure perfect public equilibrium with payoff at least $v_i^* - \eta$ for each player.*

Proof We exhibit an equilibrium with the desired property. Fix $\eta > 0$ with $\eta < \tfrac{1}{2}[v_i^* - \tfrac{1}{4}(\bar{\theta} - \underline{\theta})]$ and let ε be sufficiently small that choosing price r when high cost

11.2 ■ Repeated Adverse Selection

	Prices			
State	$r-\varepsilon, r-\varepsilon$	$r-\varepsilon, r$	$r, r-\varepsilon$	r, r
B	split	1	2	split
I	1	1	2	1
II	2	1	2	2

Figure 11.2.1 Market share regimes B, I, and II, each identifying how the market is split between the two firms, as a function of their prices. Allocations 1 and 2 grant all of the market to firm 1 and 2, respectively, and "split" divides the market equally between the two firms.

and $r-\varepsilon$ when low cost gives each firm a payoff at least $v_i^* - \eta$. Our candidate strategies for the firms then specify that a high-cost firm choose price r and a low-cost firm price $r-\varepsilon$, after any history featuring no other prices, and that any history featuring any other price prompts play of the stage-game Nash equilibrium. We also specify that if an out-of-equilibrium price has ever been set, consumers thereafter split equally between the two firms whenever the latter set identical prices.

Given our assumption about consumer behavior, the stage-game Nash equilibrium features the payoff profile $(\frac{1}{4}(\bar{\theta}-\underline{\theta}), \frac{1}{4}(\bar{\theta}-\underline{\theta}))$. As a result, there is a value $\delta_1 \in (0, 1)$ such that for any $\delta \in (\delta_1, 1)$, deviating to an out-of-equilibrium price after a history prescribing equilibrium prices is suboptimal. It then remains to ensure that firms have the incentives to choose the equilibrium prices prescribed for their cost levels.

The behavior of consumers in response to equilibrium prices is described by defining three market share "regimes," B, I, and II. Each regime specifies how consumers behave when the firms both set price r or both set price $r - \varepsilon$. (Consumers necessarily purchase from the low-price firm if one firm sets price r and the other price $r - \varepsilon$.) These regimes are shown in figure 11.2.1, where "split" indicates that the market is to be split equally, and otherwise the indicated firm takes the entire market.

The history of play determines the market share regime as shown in figure 11.2.2. Play begins in regime B. This regime treats the firms equally. Regime I rewards firm 1 and regime II rewards firm 2, where a reward is triggered by having chosen price r while the opponent chose price $r - \varepsilon$.

The prescribed actions always allocate the entire market to the low-cost producer, ensuring that the proposed equilibrium outcome is efficient. The three market share regimes differ in how the market is to be allocated when the two firms have the same cost level. Under the three regimes (for ε small), the payoffs shift along a frontier passing through the payoff profile v^*, with a slope of -1. Transitions between regimes thus correspond to transfers from one agent to the other, with no loss of efficiency.

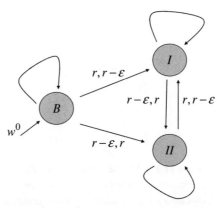

Figure 11.2.2 Equilibrium market-share regime transitions for the efficient equilibrium. Unlabeled transitions occur for the remaining equilibrium price combinations.

Verifying that the strategy profile is an equilibrium requires only showing that each firm prefers to "identify its cost level truthfully," in the sense that it prefers to set the appropriate price, in each regime. We examine the incentive constraints for the limiting case of $\varepsilon = 0$, establishing that they hold with strict inequality for sufficiently patient firms. They will continue to hold if ε is sufficiently small.

Let $\bar{V}$ be the value to firm 1 of regime I or equivalently the value of firm 2 of regime II. Also, let $\underline{V}$ be the value of firm 2 in regime I or firm 1 in regime II. To simplify the notation, let

$$r - \bar{\theta} = \beta$$
$$\text{and} \quad r - \underline{\theta} = \xi.$$

Then we have

$$\bar{V} = (1-\delta)\left(\tfrac{1}{2}\xi + \tfrac{1}{4}\beta\right) + \delta\left(\tfrac{3}{4}\bar{V} + \tfrac{1}{4}\underline{V}\right)$$

and

$$\underline{V} = (1-\delta)\tfrac{1}{4}\xi + \delta\left(\tfrac{3}{4}\underline{V} + \tfrac{1}{4}\bar{V}\right),$$

giving

$$\begin{bmatrix} 4-3\delta & -\delta \\ -\delta & 4-3\delta \end{bmatrix} \begin{bmatrix} \bar{V} \\ \underline{V} \end{bmatrix} = (1-\delta) \begin{bmatrix} 2\xi + \beta \\ \xi \end{bmatrix}.$$

Solving,

$$\bar{V} = \frac{(4-3\delta)(2\xi+\beta) + \delta\xi}{8(2-\delta)}$$

and

$$\underline{V} = \frac{(4-3\delta)\xi + \delta(2\xi+\beta)}{8(2-\delta)},$$

11.2 ■ Repeated Adverse Selection

so that

$$\bar{V} - \underline{V} = (1-\delta)\frac{\xi + \beta}{2(2-\delta)}. \qquad (11.2.7)$$

With this in hand, we turn to the incentive constraints. The requirement that a low-cost firm 1 be willing to set price $r - \varepsilon$ rather than r in regime I is

$$(1-\delta)\xi + \delta\left(\frac{1}{2}\underline{V} + \frac{1}{2}\bar{V}\right) \geq (1-\delta)\frac{1}{2}\xi + \delta\bar{V},$$

which we solve for

$$\xi \geq \frac{\delta}{1-\delta}(\bar{V} - \underline{V}).$$

The requirement that a high-cost firm 1 choose price r in regime I is:

$$(1-\delta)\frac{1}{2}\beta + \delta\bar{V} \geq (1-\delta)\beta + \delta\left(\frac{1}{2}\underline{V} + \frac{1}{2}\bar{V}\right),$$

giving

$$\frac{\delta}{1-\delta}(\bar{V} - \underline{V}) \geq \beta.$$

The requirement that a low-cost firm 2 set price $r - \varepsilon$ in regime I is

$$(1-\delta)\frac{1}{2}\xi + \delta\underline{V} \geq \delta\left(\frac{1}{2}\underline{V} + \frac{1}{2}\bar{V}\right),$$

giving

$$\xi \geq \frac{\delta}{1-\delta}(\bar{V} - \underline{V}).$$

The requirement that a high-cost firm 2 optimally choose price r in regime I is

$$\delta\left(\frac{1}{2}\underline{V} + \frac{1}{2}\bar{V}\right) \geq (1-\delta)\frac{1}{2}\beta + \delta\underline{V},$$

giving

$$\frac{\delta}{1-\delta}(\bar{V} - \underline{V}) \geq \beta.$$

Regime II yields equivalent incentive constraints. Our attention accordingly turns to regime B. Let V be the expected value of regime B, which is identical for the two firms. For a low-cost firm to optimally choose price $r - \varepsilon$, we have

$$(1-\delta)\frac{3}{4}\xi + \delta\left(\frac{1}{2}V + \frac{1}{2}V\right) \geq (1-\delta)\frac{1}{4}\xi + \delta\left(\frac{1}{2}\underline{V} + \frac{1}{2}\bar{V}\right),$$

which gives

$$\xi \geq \frac{\delta}{1-\delta}(\bar{V} - \underline{V}).$$

For a high-cost firm to optimally choose price r, we need

$$(1-\delta)\frac{1}{4}\beta + \delta\left(\frac{1}{2}\underline{V} + \frac{1}{2}\bar{V}\right) \geq (1-\delta)\frac{3}{4}\beta + \delta\left(\frac{1}{2}\bar{V} + \frac{1}{2}\underline{V}\right),$$

which we can solve for

$$\frac{\delta}{1-\delta}(\bar{V} - \underline{V}) \geq \beta.$$

Putting these results together, the incentive constraints are captured by

$$\xi \geq \frac{\delta}{1-\delta}(\bar{V} - \underline{V}) \geq \beta.$$

Using (11.2.7), this is

$$\xi \geq \frac{\delta(\xi + \beta)}{2(2-\delta)} \geq \beta.$$

In the limit as $\delta \to 1$, this inequality becomes

$$\xi \geq \frac{\xi + \beta}{2} \geq \beta,$$

which, because $\xi > \beta$, holds with the inequalities strict. There is then a sufficiently large $\underline{\delta} > \delta_1$ with the property that the on-equilibrium constraints hold for larger values of δ, yielding the desired result.

∎

Remark 11.2.2 In the efficient equilibrium, the choices of indifferent consumers provide the incentives that allow the firms to collude. It may seem unreasonable that consumers arrange their behavior to facilitate collusion. In our view, one cannot assess the plausibility of an equilibrium purely within the confines of the model. The selection of an equilibrium is not part of the analysis of the game; rather constructing the game and choosing the equilibrium are jointly part of the modeling process. The plausibility of the equilibrium must be evaluated in the context of the strategic interaction to be studied. In this respect, we note that "meet the competition" practices are both common and are often interpreted as a device for inducing consumer behavior that facilitates collusion.

◆

Remark 11.2.3 **Strong symmetry** The equilibrium constructed in this section features nonstationary outcomes but is strongly symmetric (definition 7.1.2), because all long-lived players are playing the same action (i.e., mapping from types to prices) after every history. As we have just seen, however, it is still possible to provide asymmetric incentives via the short-lived players, reinforcing our observation that strong symmetry is a less useful concept with short-lived players.

It is straightforward to construct a nonstrongly symmetric PPE in which identically pricing firms always evenly split the market but again come arbitrarily close to full efficiency. For ε small, the profile in regime I has the low-cost firm 1 pricing at $r - 3\varepsilon$, the high-cost firm 1 at $r - \varepsilon$, the low-cost firm 2 at $r - 2\varepsilon$, and the high-cost firm 2 at r (and similarly in regime II).

◆

11.3 Risk Sharing

This section examines consumers facing random income shocks. The model is similar to that of section 6.3, except that the income shocks are now privately observed. This will give rise to different incentive-compatibility considerations.[5]

There are two players, 1 and 2, and two equally likely states, H and L. Player 2 is endowed with 1 unit of the consumption good in each state. Player 1 is endowed with no consumption good in state L and 1 unit in state H. The state is observed only by player 1.[6]

Player 2 is risk neutral, with utility over consumption c given by $u_2(c) = c$. Player 1 is risk averse. Player 1 has a utility function u_1 that exhibits nonincreasing absolute risk aversion. This implies that for any positive $c_1, c_2, c_3,$ and c_4 and any $\lambda \in (0, 1)$

$$\lambda u_1(c_1) + (1-\lambda) u_1(c_2) \geq u_1(c_3)$$
$$\Rightarrow \lambda u_1(c_1 + c_4) + (1-\lambda) u_1(c_2 + c_4) \geq u_1(c_3 + c_4).$$

This assumption allows us to apply a convexity argument to a particular collection of constraints (compare with lemma 6.3.1).

In each period, player 1 has the opportunity to announce either $\hat{H}$ or $\hat{L}$, interpreted as a report about the state. If 1 announces $\hat{H}$, then 1 has the opportunity to transfer an amount of consumption $\tau_{\hat{H}} \geq 0$ to player 2. If 1 announces $\hat{L}$, then 2 has the opportunity to transfer an amount $\tau_{\hat{L}}$ to player 1.

The stage game has a unique Nash equilibrium outcome, in which $\tau_{\hat{H}} = \tau_{\hat{L}} = 0$ with utilities given by

$$\underline{v}_1 = \tfrac{1}{2} u_1(0) + \tfrac{1}{2} u_1(1)$$

and $\quad \underline{v}_2 = 1.$

These are the minmax values for the two players. Playing this Nash equilibrium after every history of the repeated game is a pure PPE of the latter. Because it entails minmax

---

5. The model examined here is a simplified version of Hertel (2004). Thomas and Worrall (1990) examine a risk-averse agent with a private income shock facing a risk-neutral insurer. Thomas and Worrall assume that the agents can commit to contracts, in the sense that participation constraints must be satisfied only at the beginning of the game, in terms of ex ante payoffs. The inability to completely smooth the risk-averse agent's consumption within a period prompts some risk to be optimally postponed until future periods. As a result, the risk-averse agent's equilibrium consumption shrinks to 0 as time proceeds and her continuation value becomes arbitrarily negative, both with probability 1, clearly eventually violating participation constraints in the continuation game. Extensions of this model are examined by Atkeson and Lucas (1992) (to a general equilibrium setting) and Atkeson and Lucas (1995) (to study unemployment insurance). Wang (1995) shows that if utility is bounded below, then (obviously) continuation utilities cannot become arbitrarily negative, and (more important) consumption does not shrink to 0 with probability 1, because then the accumulation of utility at the lower bound would preclude the creation of incentives. Ljungqvist and Sargent (2004, chapter 19) provide a useful discussion.
6. Unlike section 6.3, we have uncertainty in the aggregate endowment. We avoided such uncertainty in section 6.3 by taking the players' endowments to be perfectly (negatively) correlated. The essence of the problem here is that player 2 is uncertain about player 1's endowment, with the simplest case being that in which 2's own endowment provides no information about 1's. This requires either aggregate uncertainty or a continuum of agents.

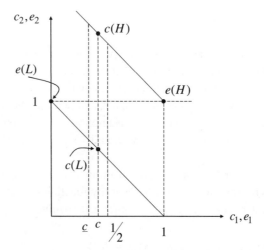

Figure 11.3.1 Endowments and full-insurance allocations. The endowments in state H and L are given by $e(H)$ and $e(L)$. Consumption vectors in states H and L, given c, are given by $c(H)$ and $c(L)$. Player 2 receives her minmax utility when $c = 1/2$, and player 1 receives his minmax utility when $c = \underline{c}$. Any $c \in (\underline{c}, 1/2)$ is a strictly individually rational, full-insurance consumption allocation.

payoffs for each player, it is the most severe punishment that can be imposed in the repeated game.

Because player 1 is risk averse, player 2 can insure player 1 against 1's income fluctuations. An *efficient allocation* in the stage game features complete insurance, or $1 - \tau_{\hat{H}} = \tau_{\hat{L}} \equiv c$ after every history, so that player 1's consumption does not vary across states. Setting $c = 1/2$ gives payoffs $(u_1(1/2), 1)$, with player 2 then receiving her minmax utility. Setting $c = \underline{c} < 1/2$, where $\underline{v}_1 = u_1(\underline{c})$, gives payoffs $(\underline{v}_1, 3/2 - \underline{c})$, with player 1 receiving his minmax utility. Any intermediate value of c gives an efficient, strictly individually rational allocation with player 2 preferring a lower value of c and player 1 a higher value. Figure 11.3.1 illustrates. We refer to such an outcome as one of perfect risk sharing.

A period t *ex post history* (for player 1) is a sequence of t realized player 1 endowments, announced endowments, and transfers, and the period t endowment realization. A period t *announcement history* (for player 2) is a sequence of t announcements and transfers, and the period t announcement. Observe that announcement histories are public. A strategy for player 1 maps from ex post histories into announcements and from ex post histories followed by an announcement of $\hat{H}$ into transfers. A strategy for player 2 maps from announcement histories ending in $\hat{L}$ into transfers.

This is again a game of repeated adverse selection. By proposition 9.5.1, we can come arbitrarily close to perfect risk sharing as an equilibrium outcome (in a finite approximation of this game) as the players become increasingly patient.

We are interested in efficient equilibria for arbitrary discount factors. Let $\tau_{\hat{H}}$ and $\tau_{\hat{L}}$ be defined as the *transfer* made after announcements $\hat{H}$ and $\hat{L}$, from player 1 to player 2 in the first case and in the opposite direction in the second. Efficient pure equilibria solve the optimization problem:

11.3 ■ Risk Sharing

$$\max_{\tau_{\hat{H}}, \tau_{\hat{L}}, \gamma(\hat{H}), \gamma(\hat{L})} \tfrac{1}{2}[(1-\delta)(1-\tau_{\hat{L}}) + \delta\gamma_2(\hat{L}) + (1-\delta)(1+\tau_{\hat{H}}) + \delta\gamma_2(\hat{H})]$$

such that

$$\tfrac{1}{2}[(1-\delta)u_1(\tau_{\hat{L}}) + \delta\gamma_1(\hat{L}) + (1-\delta)u_1(1-\tau_{\hat{H}}) + \delta\gamma_1(\hat{H})] \geq v_1,$$
$$(1-\delta)u_1(1-\tau_{\hat{H}}) + \delta\gamma_1(\hat{H}) \geq (1-\delta)u_1(1+\tau_{\hat{L}}) + \delta\gamma_1(\hat{L}),$$
$$(1-\delta)(1-\tau_{\hat{L}}) + \delta\gamma_2(\hat{L}) \geq 1,$$

and $\gamma(\hat{H}), \gamma(\hat{L}) \in \mathscr{E}^p(\delta)$.

We are thus maximizing player 2's payoff, with the first constraint imposing the requirement that player 1 receive at least v_1.

The next two inequalities are incentive constraints. The first indicates that player 1 should prefer to announce $\hat{H}$ and transfer $\tau_{\hat{H}}$ when the state is H, rather than announcing $\hat{L}$ to receive transfer $\tau_{\hat{L}}$. Player 1 faces a collection of additional incentive constraints, because he can announce $\hat{H}$ and then make any transfer from $[0, 1]$ (in state H). However, no restrictions are imposed by having the equilibrium strategies react to an announcement of H coupled with a transfer other than $\tau_{\hat{H}}$ by a permanent switch to the autarkic equilibrium: Player 1's most attractive alternative to announcing $\hat{H}$ and making the transfer $\tau_{\hat{H}}$, in state H, is to announce $\hat{L}$, because announcing $\hat{L}$ and earning the continuation payoff $\gamma_1(\hat{L})$ necessarily dominates an announcement of $\hat{H}$ followed by reversion to the autarkic equilibrium. Player 1 faces no incentive constraints in state L, where he is unable to make the transfer to player 2 that would be required in state H. Hence, the single player 1 constraint listed in the optimization program suffices for the optimality of player 1's actions.

The next constraint is the only nontrivial one for player 2—that she be willing to make transfer $\tau_{\hat{L}}$ in state L. The final constraint implies that continuation payoffs must themselves be equilibrium payoffs.

Let $G(v_1)$ be a function that describes, for each payoff v_1, the largest equilibrium payoff available to player 2, conditional on player 1 receiving at least v_1.

Lemma 11.3.1 *Let $\bar{v}_1$ be the largest equilibrium payoff possible for player 1. Then $G(v_1)$ is strictly concave on domain $[\underline{v}_1, \bar{v}_1]$.*

Proof It suffices to show that if v_1 and v'_1 are distinct equilibrium values for player 1 and $\lambda \in (0, 1)$, then there exists an equilibrium with payoff $\lambda v_1 + (1 - \lambda)v'_1$ to player 1 and a payoff exceeding $\lambda G(v_1) + (1 - \lambda)G(v'_1)$ to player 2. This establishes that G is strictly concave on $[\underline{v}_1, \bar{v}_1]$.

We construct the desired equilibrium. Let $\tilde{\mathscr{H}}$ denote the set of announcement histories. The equilibria yielding payoffs v_1 and v'_1 are described by two functions, $\tau : \tilde{\mathscr{H}} \to [0, 1]$ and $\tau' : \tilde{\mathscr{H}} \to [0, 1]$, each identifying the transfer, along the equilibrium paths of the equilibria giving payoffs v_1 and v'_1, after each announcement history in $\tilde{\mathscr{H}}$. If this is an announcement history ending in $\hat{H}$, then this is a transfer from player 1 to player 2. For an announcement history ending in $\hat{L}$, it is a transfer from player 2 to player 1. We use these functions to construct an equilibrium with the desired properties.

Throughout, we assume any out-of-equilibrium transfer prompts a reversion to the autarkic equilibrium. Specifying the remainder of an equilibrium then requires only specifying transfers after histories in $\mathcal{H}$. In particular, we can specify equilibrium transfers as a function of the announcement history (only), with the presumption that this is the appropriate transfer if there has been no previous deviation from prescribed play and that this transfer is irrelevant if play has already switched to the autarkic equilibrium.

We now construct a function τ'' characterizing the transfers in the equilibrium with the desired properties. For any history $\tilde{h} \in \mathcal{H}$, denote by $\tau''(\tilde{h})$ the transfer τ^* solving
$$u_1(\tau^*) = \lambda u_1(\tau(\tilde{h})) + (1 - \lambda) u_1(\tau'(\tilde{h}))$$
for $\tilde{h}$ ending in $\hat{L}$, and
$$u_1(1 - \tau^*) = \lambda u_1(1 - \tau(\tilde{h})) + (1 - \lambda) u_1(1 - \tau'(\tilde{h}))$$
for $\tilde{h}$ ending in $\hat{H}$. The implied continuation value γ_1'' clearly satisfies $\gamma_1''(\tilde{h}) = \lambda \gamma_1(\tilde{h}) + (1 - \lambda) \gamma_1'(\tilde{h})$, for all histories $\tilde{h}$, and so the allocation induced by the transfers $\tau''(\tilde{h})$ must give player 1 precisely utility $\lambda v_1 + (1 - \lambda) v_1'$. The concavity of player 1's utility function implies for all announcement histories ending in $\hat{L}$

$$\tau''(\tilde{h}) < \lambda \tau(\tilde{h}) + (1 - \lambda) \tau'(\tilde{h}) \tag{11.3.1}$$

whenever $\tau(\tilde{h}) \neq \tau'(\tilde{h})$, and for all announcement histories ending in $\hat{H}$

$$\tau''(\tilde{h}) > \lambda \tau(\tilde{h}) + (1 - \lambda) \tau'(\tilde{h}). \tag{11.3.2}$$

As a result, the allocation induced by the transfers $\tau''(\tilde{h})$ must give player 2 a strictly higher payoff than $\lambda G(v_1) + (1 - \lambda) G(v_1')$, as long as there is some history $\tilde{h}$ after which $\tau(\tilde{h}) \neq \tau'(\tilde{h})$. But there must exist such a history if $v_1 \neq v_1'$. This implies for every continuation history $\tilde{h}$, the continuation payoffs satisfy $\gamma_2''(\tilde{h}) \geq \lambda \gamma_2(\tilde{h}) + (1 - \lambda) \gamma_2'(\tilde{h})$, with a strict inequality holding in the initial period.

It remains to show that the transfers specified by $\tau''(\tilde{h})$ yield an equilibrium. This in turn requires that the two incentive constraints in the maximization problem characterizing efficient equilibria are satisfied, for any equilibrium history. Consider the constraint that player 2 be willing to transfer $\tau_{\hat{L}}$ to player 1 after announcement $\hat{L}$. Fix some history $\tilde{h}$ terminating in announcement $\hat{L}$. The player 2 incentive constraints for the equilibria supporting v and v' are (suppressing the history in our notation)

$$(1 - \delta)(1 - \tau_{\hat{L}}) + \delta \gamma_2(\hat{L}) \geq 1 \tag{11.3.3}$$
$$\text{and} \quad (1 - \delta)(1 - \tau_{\hat{L}}') + \delta \gamma_2'(\hat{L}) \geq 1. \tag{11.3.4}$$

Player 2's continuations under τ'' are larger than the average of the continuations under τ and τ', so (11.3.1)–(11.3.4) imply

$$(1 - \delta)(1 - \tau_{\hat{L}}'') + \delta \gamma_2''(\hat{L}) \geq 1,$$

11.3 ■ Risk Sharing

satisfying one of the incentive constraints for our putative equilibrium. We now consider player 1's constraint. We have

$$(1-\delta)u_1(1-\tau_{\hat{H}}) + \delta\gamma_1(\hat{H}) \geq (1-\delta)u_1(1+\tau_{\hat{L}}) + \delta\gamma_1(\hat{L}) \quad (11.3.5)$$

and

$$(1-\delta)u_1(1-\tau'_{\hat{H}}) + \delta\gamma'_1(\hat{H}) \geq (1-\delta)u_1(1+\tau'_{\hat{L}}) + \delta\gamma'_1(\hat{L}). \quad (11.3.6)$$

Our construction of τ'' gives

$$(1-\delta)u_1(1-\tau''_{\hat{H}}) + \delta\gamma''_1(\hat{H})$$
$$= \lambda[(1-\delta)u_1(1-\tau_{\hat{H}}) + \delta\gamma_1(\hat{H})] + (1-\lambda)[(1-\delta)u_1(1-\tau'_{\hat{H}}) + \delta\gamma'_1(\hat{H})]$$
$$\geq \lambda[(1-\delta)u_1(1+\tau_{\hat{L}}) + \delta\gamma_1(\hat{L})] + (1-\lambda)[(1-\delta)u_1(1+\tau'_{\hat{L}}) + \delta\gamma'_1(\hat{L})]$$
$$= \lambda[(1-\delta)u_1(1+\tau_{\hat{L}})] + (1-\lambda)[(1-\delta)u_1(1+\tau'_{\hat{L}})] + \delta\gamma''_1(\hat{L}), \quad (11.3.7)$$

where the first equality follows from our construction of τ'' and the subsequent inequality follows from (11.3.5)–(11.3.6). By the definition of τ'',

$$u_1(\tau''_{\hat{L}}) = \lambda u_1(\tau_{\hat{L}}) + (1-\lambda)u_1(\tau'_{\hat{L}}).$$

The nonincreasing absolute risk aversion of player 1 then implies

$$u_1(1+\tau''_{\hat{L}}) \leq \lambda u_1(1+\tau_{\hat{L}}) + (1-\lambda)u_1(1+\tau'_{\hat{L}}).$$

Inserting this in (11.3.7) gives

$$(1-\delta)u_1(1-\tau''_{\hat{H}}) + \delta\gamma''_1(\hat{H}) \geq (1-\delta)u_1(1+\tau''_{\hat{L}}) + \delta\gamma''_1(\hat{L}),$$

which is the needed incentive constraint for player 1.

∎

We can now characterize the equilibrium. In contrast to section 6.3, full insurance is unavailable for any discount factor:

Lemma 11.3.2 *In an efficient equilibrium, for any ex ante history h^t, $\tau_{\hat{L}}(h^t) < 1 - \tau_{\hat{H}}(h^t)$.*

The implication of this result is that player 1 always bears some risk. This in turn ensures that the folk theorem result for this model is truly an approximation result. As the discount factor approaches 1, we can come increasingly close to full insurance, but some residual risk inevitably remains.

Proof It is immediate that we cannot have $\tau_{\hat{L}}(h^t) > 1 - \tau_{\hat{H}}(h^t)$. If this were the case, we could instead replace $(\tau_{\hat{L}}, \tau_{\hat{H}})$ with $\tau'_{\hat{L}} = 1 - \tau_{\hat{H}}$ and $\tau'_{\hat{H}} = 1 - \tau_{\hat{L}}$ while leaving all other aspects of the strategy unchanged. Since both endowments are equally likely, the distribution of values in any period is unaffected, while causing both incentive constraints after history h^t to hold with strict inequality. We can then revise the equilibrium to offer more insurance to player 1 while offering a higher payoff to player 2 after h^t, without disrupting incentive constraints, contradicting the supposed efficiency of the equilibrium.

Suppose next that $\tau_{\hat{L}} = 1 - \tau_{\hat{H}}$, so that player 1 is perfectly insured. Then it must be the case that $\gamma_1(\hat{H}) > \gamma_1(\hat{L})$, because otherwise 1's incentive constraint for announcing state H fails. Now consider an adjustment in the supposedly efficient equilibrium that marginally decreases $\tau_{\hat{H}}$ and $\tau_{\hat{L}}$ at equal rates. This adjustment has only a second-order effect on player 1's current expected utility (because this effect is proportional to $(1 - \delta)(u'_1(\tau_{\hat{L}}) - u'_1(1 - \tau_{\hat{H}})) = 0$). However, it has a first-order effect on player 1's incentive constraint (because $(1 - \delta)(u'_1(1 + \tau_{\hat{L}}) + u'_1(1 - \tau_{\hat{H}})) > 0$). As a result, we could consider decreasing $\gamma_1(\hat{H})$ and increasing $\gamma_1(\hat{L})$ to preserve their expected value (and hence player 1's expected payoff in the equilibrium) until player 1's incentive constraint binds. From the strict concavity of G, however, this allows us to choose continuation equilibria that increase player 2's payoffs. Furthermore, this increase cannot disrupt any of player 2's incentive constraints in prior periods because it can only increase the left side of such incentive constraints. As a result, we have a contradiction to the putative efficiency of the equilibrium.

∎

In the terms introduced in section 6.3, we have just established that consumption will respond to current income more than is warranted by a full-insurance outcome. The intuition is straightforward. If player 1 faces no current consumption risk, then he must face future consumption risk to create the incentives for truthful current revelation. But this allocation of no risk to the current period and all risk to the future period is inefficient. There are gains to be had by smoothing his consumption, introducing some risk into the current period in response for reduced future risk. As the discount factor gets large, the risk in current consumption can be made quite small but is eliminated only in the limit as the discount factor becomes unity.

We also see history dependence in this model. The following is immediate.

Lemma 11.3.3 *After any announcement history at which $\tau_{\hat{H}}$ is positive, it must be that $\gamma_1(\hat{H}) > \gamma_1(\hat{L})$.*

Hence, favorable income realizations for player 1 give rise to more favorable continuation consumption streams. Turning this observation around, player 1's previous period(s) income is relevant for current consumption.

11.4 Principal-Agent Problems

11.4.1 Hidden Actions

Our point of departure is a canonical principal-agent problem with hidden actions. The agent (player 2) can choose an effort level $e \in [0, \bar{e}]$ that stochastically generates an output $y \in [0, \bar{y}]$. The cumulative distribution function governing the output y, given effort level e, is denoted by $F(y, e)$ with $F(0, 0) = 1$. The principal (player 1) and the agent have payoffs $u_1(y, w) = y - w$ and $u_2(e, w, y) = w - c(e)$, where w is a payment from player 1 to player 2 and $c(e)$ is an increasing, convex, continuously differentiable cost-of-effort function. We assume $F(y, e)$ first-order stochastically dominates $F(y, e')$ for $e > e'$ (higher effort levels are associated with higher outputs), F is twice continuously differentiable with density f, and F has the monotone likelihood ratio property ($f_e(y, e)/f(y, e)$ is strictly increasing in y).[7] Whenever convenient, we also assume $F(y, c^{-1}(x))$ is convex in x. This last condition is less easily interpreted but allows us to use the first-order approach to solving the principal-agent problem (Rogerson 1985).

The stage game proceeds as follows. Player 1 first offers a contract to player 2, consisting of a schedule $w(y)$ giving the payment from player 1 to 2 as a function of the observed output level. Player 2 either rejects, in which case both agents receive a payoff of 0, or accepts. In the latter case, player 2 then chooses an effort level, output is realized, and the appropriate payment made. Payments cannot be conditioned on the agent's effort level.

The efficient effort level e^* maximizes

$$\int_0^{\bar{y}} yf(y, e)dy - c(e).$$

We assume that this maximizer is positive, making this a nontrivial problem.

The unique subgame-perfect equilibrium of the stage game realizes the efficient effort level e^*, with

$$w(y) = y - \hat{w} \tag{11.4.1}$$

$$\text{and} \quad \hat{w} = \int_0^{\bar{y}} yf(y, e^*)dy - c(e^*). \tag{11.4.2}$$

This is easily recognized as the outcome in which the principal "sells the firm" to the agent, collecting from the agent a fee $\hat{w}$ equal to the expected surplus, given an efficient effort choice, and making the agent the residual claimant for any shortfall or excess in output. This causes the agent to be faced with just the incentives needed for efficiency.

The recommendation that the principal-agent problem be solved by selling the firm to the agent often seems unsatisfying. What might prevent such a sale? One commonly offered but often unmodeled explanation is that capital market imperfections preclude the agent's paying $\hat{w}$ to the principal. The remainder of this section explores two other common explanations.

11.4.2 Incomplete Contracts: The Stage Game

Suppose first that the output inevitably accrues to the principal and that there is nothing the principal can do to transfer this receipt to the agent. If the agent is to receive y, it must come via a transfer from the principal. Now suppose further that the parties are unable to contract on the output y. Such a constraint would obviously arise if y could not be observed. Instead, we assume that the principal and the agent both observe y,

---

7. This is equivalent to the formulation of the monotone likelihood ratio property invoked in section 11.1.

but y has an "I know it when I see it" nature that makes it prohibitively expensive to incorporate in a contract.[8]

The stage game begins with the principal proposing a fixed wage w to the agent, as well as an additional payment or bonus of $b(y)$. If the agent accepts the contract, the wage w is legally binding. However, this is the only provision of the principal-agent interaction that can be enforced. The noncontractability of y ensures that the bonus $b(y)$ is voluntary. Hence, once the output level has been realized, the principal has the option of making transfer $b(y)$ to the agent (if $b(y) > 0$) or not doing so, and similarly the agent has the option of making transfer $|b(y)|$ (if $b(y) < 0$) to the principal, or not doing so. Because the bonus is noncontractable, we ignore the announcement and consider a stage game where the principal decides on the size of the bonus after output is realized.

The subgame-perfect equilibrium of the stage game is trivial. Once the output has been produced, neither party will make transfers. As a result, the agent's remuneration will be given by w, and is independent of the realized output. The optimal effort level is then $e = 0$. Notice that the stage-game equilibrium gives each player his or her minmax payoff of 0.

11.4.3 Incomplete Contracts: The Repeated Game

We maintain the assumption that contracts cannot be conditioned on y, but now assume that the game is repeated. The result is sometimes called a "relational contract" (see, for example, Baker, Gibbons, and Murphy 2002; Levin 2003).

An action for the principal, in period t, consists of a transfer w^t and a bonus $b^t(y)$. The interpretation is that w^t is the fixed payment over which the principal and agent can contract, whereas $b^t(y)$ is a prescribed additional transfer from the principal to the agent (if $b^t(y) > 0$) or from agent to principal (if $b^t(y) < 0$). The players can write only short-term contracts, in the sense that the principal cannot renege on the period t wage w^t but cannot in period t make commitments to future wages.

The first observation is that if the players are sufficiently patient, we can again support the efficient effort level e^*. Let $w^*(y)$ be the remuneration scheme given by (11.4.1)–(11.4.2). On the equilibrium path, let the principal offer $w^t = w^*(0)$ and $b^t(y) = w^*(y) - w^t = w^*(y) - w^*(0)$. It is a best response for the agent to accept this contract and choose the efficient effort e^*. Let the equilibrium strategies call for the players to continue with this equilibrium path if the transfer $b^t(y)$ is made in each period t and otherwise revert to the minmax values of $(0, 0)$, enforced by the subsequent play of the stage-game Nash equilibrium.

By choosing $w^t = w^*(0)$, we ensure that only the principal is called on to make transfers. In particular, the strategy profile makes the agent indifferent between participating and not doing so. After each ex ante history, the agent's continuation value is the agent's minmax value of 0. We then cannot create incentives for the agent to make a transfer to the principal.

For the proposed strategies to be an equilibrium, it must be the case that the principal makes the transfer $b^t(y)$ when called on to do so. It is clearly most tempting to balk when $y = \bar{y}$. The incentive constraint to not do so is given by

---

8. Such "observable but not verifiable" assumptions are common but controversial in the literature on incomplete contracts. See Maskin and Tirole (1999).

$$(1 - \delta)(w^*(\bar{y}) - w^*(0)) \leq \delta \hat{w}.$$

Because $\hat{w} > 0$, this constraint is satisfied for sufficiently large δ. This result is another reflection of the ability to achieve efficient (or nearly efficient) outcomes in games of imperfect monitoring as the players become arbitrarily patient.

Player 1 extracts all of the surplus in this equilibrium. Unlike the complete-contracting stage game, where this is the only equilibrium division of the surplus, the repeated game has other efficient equilibria that divide the surplus between the two agents differently. These include equilibria in which the surplus is divided equally between the two agents as well as equilibria in which the principal earns a payoff of nearly 0, all enforced by punishments in which both agents receive continuation payoffs of 0. In addition, we can replace the inefficient punishments with continuation equilibria that are efficient but that feature a sufficiently low payoff for the transgressing player. This allows us to avoid renegotiation problems that might arise with punishments based on inefficiencies. Notice finally that we lose no generality in restricting attention to stationary-outcome efficient equilibria. Given an equilibrium in which continuation payoffs vary along the equilibrium path, we can rearrange the transfers between agents to make equilibrium continuation payoffs constant throughout the game.

What happens if the discount factor is too small to support this efficient outcome? Invoking the sufficient conditions to apply the first-order approach to the principal-agent problem, we can write the principal's problem in each period as

$$\max_{b(\cdot), e, w} \int_0^{\bar{y}} (\tilde{y} - (w + b(\tilde{y}))) f(\tilde{y}, e) d\tilde{y}$$

subject to

$$\int_0^{\bar{y}} (w + b(\tilde{y})) f_e(\tilde{y}, e) d\tilde{y} - c_e(e) = 0,$$

$$\int_0^{\bar{y}} (w + b(\tilde{y})) f(\tilde{y}, e) d\tilde{y} - c(e) \geq 0,$$

$$b(y) \geq 0, \quad \forall y \in [0, \bar{y}],$$

and

$$\delta \int_0^{\bar{y}} (\tilde{y} - (w + b(\tilde{y}))) f(\tilde{y}, e) d\tilde{y} - (1 - \delta) b(y) \geq 0, \quad \forall y \in [0, \bar{y}].$$

The first constraint is the first-order condition capturing an incentive constraint for the agent, ensuring that the agent finds optimal the effort level e involved in the solution. The next constraint ensures that the agent receives a nonnegative payoff. In equilibrium this constraint will bind. The following constraint, that $b(y) \geq 0$ for all $y \in [0, \bar{y}]$, ensures that the agent is never called on to make a transfer. This is an incentive constraint for the agent, who will never make a positive transfer in order to proceed with an equilibrium whose continuation value is 0. The final constraint is the principal's incentive constraint, ensuring that for every value of $y \in [0, \bar{y}]$, the principal prefers to make the accompanying transfer rather than abandon the equilibrium path.

Noting that the equilibrium will entail a zero payoff for the agent, we can rewrite this problem as

$$\max_{b(\cdot),e,w} \int_0^{\bar{y}} (\tilde{y} - c(e)) f(\tilde{y}, e) d\tilde{y}$$

subject to

$$\int_0^{\bar{y}} (w + b(\tilde{y})) f_e(\tilde{y}, e) d\tilde{y} - c_e(e) = 0, \qquad (11.4.3)$$

$$b(y) \geq 0, \quad \forall y \in [0, \bar{y}], \qquad (11.4.4)$$

and

$$\delta \int_0^{\bar{y}} (\tilde{y} - c(e)) f(\tilde{y}, e) d\tilde{y} - (1 - \delta) b(y) \geq 0, \quad \forall y \in [0, \bar{y}]. \qquad (11.4.5)$$

If $b(y)$ is interior (i.e., (11.4.4) and (11.4.5) do not bind), then the first-order condition $\lambda f_e(y, e) = 0$ must hold, where $\lambda > 0$ is the multiplier on the agent's incentive constraint (11.4.3). Hence, an interior value $b(y)$ requires $f_e(y, e) = 0$. If this equality is nongeneric, there will be *no* interior values $b(y)$ (up to a set of measure zero). The optimal bonus scheme thus has a bang-bang structure. If $f_e(y, e) < 0$, then $b(y)$ will be set at its minimum value of 0 (causing the agent's incentive constraint to bind). If $f_e(y, e) > 0$, then $b(y)$ will be set at a value $\bar{b}$ at which the principal's incentive constraint binds. The switching output level for the two bonus values is the output level at which the likelihood ratio $f_e(y, e)/f(y, e)$ switches from negative to positive.

In general, this contract will induce inefficiently little effort on the part of the worker. As the discount factor increases, so does the range of possible equilibrium transfers $[0, \bar{b}]$ and so does the principal's ability to create incentives for the agent. For sufficiently large discount factors, we achieve efficient outcomes.

11.4.4 Risk Aversion: The Stage Game

When the agent is risk averse, it is no longer efficient to sell her the firm. The efficient contract involves risk sharing between the risk-neutral principal and the risk-averse agent. Repetition may again make incentives available that cannot be created in a single interaction.

Simplifying the model makes the arguments more transparent. The stage game again involves two players, a principal and an agent. The agent can choose one of two effort levels, low (denoted by L) or high (denoted by H). Low effort is costless, whereas high effort incurs a cost of $c > 0$. There are two possible outcomes, success ($\bar{y}$) or failure ($\underline{y}$). The probability of a success in the event of low effort is q and the probability of a success in the event of high effort is $p > q$. The principal values a success at 1 and a failure at 0. The agent has an increasing and strictly concave utility function for money $\tilde{u}$, with $\tilde{u}(0) = 0$.

The stage game begins with the principal offering a contract to the agent, consisting of a payment $\bar{w}$ to be made in the event of a success and $\underline{w}$ in the event of a failure. The agent either accepts the contract or rejects, with both players receiving a payoff of 0 in the latter case. On accepting, the agent chooses an effort level and the outcome is

11.4 ■ Principal-Agent Problems

	High Effort	Low Effort
Success	$1-\bar{w}, \tilde{u}(\bar{w}) - c$	$1-\bar{w}, \tilde{u}(\bar{w})$
Failure	$-\underline{w}, \tilde{u}(\underline{w}) - c$	$-\underline{w}, \tilde{u}(\underline{w})$

Figure 11.4.1 Payoffs to the principal (first element) and agent (second element) as a function of the agent's effort level and realized outcome.

randomly drawn from the appropriate distribution. Payoffs are then received and the game ends. Payoffs, as a function of the agent's effort level and the outcome, are given in figure 11.4.1.

Because the agent is risk averse and the principal risk neutral, efficiency calls for the principal to insure the agent, and therefore $\bar{w} = \underline{w}$. Let $\hat{w}$ be the wage satisfying $\tilde{u}(\hat{w}) = c$. We assume that

$$p - \hat{w} > q.$$

This implies that any efficient allocation in the stage game calls for the agent to exert high effort (for potential payoff profile $(p - \hat{w}, 0)$) rather than low effort (payoff $(q, 0)$). There is a collection of efficient allocations for the stage game, differing in how the surplus is split between the principal and agent, with the payment w ranging from $\hat{w}$ to p, the former leaving no surplus to the agent and the latter no surplus to the principal.

These efficient allocations are unattainable in the stage game, because a fully insured agent faces no incentives and optimally chooses low effort. Instead, there are two candidates for subgame-perfect equilibria of the stage game, one involving high effort and one low effort. In one, $\underline{w} = 0$ and $\bar{w}$ satisfies $(p-q)\tilde{u}(\bar{w}) = c$, with the agent choosing high effort. Let $\tilde{w}$ be the resulting wage rate in the event of a success. The agent now bears some risk, and hence $p\tilde{w} > \hat{w}$. If the agent is not too risk averse, we will have $p - q > p\tilde{w}$ and a stage-game equilibrium in which the agent exerts high effort. If the agent is sufficiently risk averse, $p - q < p\tilde{w}$. The cost of inducing effort is then sufficiently large that the equilibrium calls for zero payments to the agent and low effort (though the agent still accepts the contract). We assume the former is the case and let $\tilde{v}$ be the resulting stage-game equilibrium payoff profile.

11.4.5 Risk Aversion: Review Strategies in the Repeated Game

Now suppose the game is repeated, with common discount factor δ. Suppose v^* is an efficient payoff profile.

It is an immediate implication of proposition 9.2.1 that if v^* is strictly individually rational, then as the players become increasingly patient, we can come arbitrarily close to v^* as an equilibrium payoff (in a finite approximation of the game). We examine here outcomes that can be achieved with *review strategies*, examined by Radner (1985), Rubinstein (1979b), and Rubinstein and Yaari (1983). These strategies have a simple and intuitive structure.[9] For large discount factors, there exist equilibria in review strategies that are approximately efficient.

Review strategies use the statistical idea that although one noisy signal may not be particularly informative, many signals may be. A single failure might then be ignored, whereas a string of signals with many failures may be more troubling. Incentives may then condition rewards or punishments on long histories of signals.

The ability to exploit strings of signals appears to be lost when working with self-generation techniques for studying equilibria. The recursive techniques involved in self-generation focus attention on how actions or signals in the current period affect future behavior and payoffs. However, this apparent memorylessness is possible only because the agents enter each period characterized by a state, consisting of a pair of target payoffs to be decomposed, that itself can be linked to past play in quite complicated ways. Any PPE in review strategies has a self-generation representation.

We simplify the analysis by examining an efficient profile v^* with $v_i^* > \tilde{v}_i$ for both players, so that v^* dominates the stage-game equilibrium profile $\tilde{v}$. Let w^* be the wage rate associated with payoff profile v^* in the stage game (i.e., $\tilde{u}(w^*) - c = v_2^*$). A review strategy proceeds as follows. For R periods, called the *review phase*, the wage w^* is offered and accepted. At the end of these R periods, the realized number of successes is compared to a *threshold* S. If the realized number of success equals or exceeds S, the agent is considered to have passed the review and a new review phase is started. If the number of successes falls short of S, the agent fails and a T period *punishment phase* begins, characterized by the stage-game subgame-perfect equilibrium in which $\bar{w} = \tilde{w}$ and $\underline{w} = 0$, with the agent accepting and exerting high effort, for payoffs $(\tilde{v}_1, \tilde{v}_2)$ (with $\tilde{v}_2 = 0$). At the end of the T periods, a new review process begins. Should the principal ever make a nonequilibrium wage offer, the game reverts to the permanent play of the stage-game subgame-perfect equilibrium.

It is clear that the review strategies exhibit equilibrium play during the punishment phase, in the form of a subgame-perfect equilibrium of the state game. It is also clear, given $v_i^* > \tilde{v}_i$, that the prospect of permanent reversion to the stage-game equilibrium suffices to ensure that a sufficiently patient principal will find the prescribed strategies optimal.

We have said nothing yet about how the agent responds to such a review strategy. Our task is to examine this behavior and to show that R, S, and T can be chosen so that the result is an equilibrium featuring a nearly efficient payoff.

To gain some intuition, suppose first that the agent must make a single choice of either high or low effort to be played in each of the R periods of the review phase, and suppose $T = \infty$ so that the punishment phase is absorbing. Suppose we let $S = ((p+q)R)/2$. Let $\phi(H, R)$ be the probability that the agent fails the review, given high effort, and $\phi(L, R)$ the probability that the agent fails, given low effort. Let $\bar{V}_2$ be the expected payoff to the agent at the beginning of a review phase (the expected payoff at the beginning of a punishment phase equals 0). Then the agent's payoff, at the beginning of the review phase, from choosing high effort is

$$(1 - \delta^R)(\tilde{u}(w^*) - c) + \delta^R(1 - \phi(H, R))\bar{V}_2$$

---

9. Review strategies not only provided some of the earliest positive results in public monitoring games, they now play a significant role in private monitoring games.

11.4 ■ Principal-Agent Problems

and from choosing low effort is

$$(1 - \delta^R)\tilde{u}(w^*) + \delta^R(1 - \phi(L, R))\bar{V}_2.$$

Hence, high effort will be optimal if

$$\bar{V}_2 = \frac{(1 - \delta^R)(\tilde{u}(w^*) - c)}{(1 - \delta^R(1 - \phi(H, R)))} \geq \frac{(1 - \delta^R)c}{\delta^R[\phi(L, R) - \phi(H, R)]}.$$

For sufficiently large R, $\phi(H, R)$ can be made arbitrarily close to 0 and $\phi(L, R) - \phi(H, R)$ to 1. In other words, by choosing R large, this looks like a game with perfect monitoring, and for large R the incentive constraint is satisfied for all large δ. Finally, by choosing δ^R sufficiently close to 1, and then choosing δ sufficiently close to 1 and R large, $\bar{V}_2$ can be made arbitrarily close to v_2^* with the principal's payoff approaching v_1^* in the process. Hence, review strategies can come arbitrarily close to attaining the efficient payoff v^*.

Review strategies are effective in this case because we have forced the agent to make a single effort choice at the beginning of the review phase that is binding throughout the review. In this case, the review strategies effectively convert the stage game of length 1, with imperfect monitoring, into a stage game of length R with very nearly perfect monitoring. It is then no surprise that we can achieve nearly efficient outcomes.

When the agent's choices are not constrained to be constant over the review phase, review strategies give rise to two sources of inefficiency. First, an agent who is in the middle of a review phase and who has had bad luck, obtaining a relatively large number of failures, may despair of passing the review (exceeding the threshold) and switch to low effort for the remainder of the phase. To make sure that this does not impose too large a cost, the principal must not set the threshold too high. However, an agent who has had good luck, obtaining a relatively large number of successes, is quite likely to pass the review and is tempted to coast on her past success by switching to low effort. To avoid this, the threshold must not be set too low.[10]

Fix $\xi \in (1/2, 1)$ and set

$$T = \lambda R$$
$$\text{and} \quad S = pR - R^\xi,$$

with R and $\lambda > 0$ chosen so that T and S are positive integers. The only restriction on ξ is that it lie in the interval $(1/2, 1)$. A review strategy requires a specification of both R and λ.

We now show that there exist equilibria, with the principal using review strategies with this structure, that give payoffs arbitrarily close to v^* when players are patient. We have seen that such payoffs could be obtained if we could ensure that the agent would always exert high effort during the review phase, but this is not possible. Nothing can prevent the agent from lapsing into low effort when either far ahead or far behind

---

10. If the agent only observes the R signals at the end of each review phase but chooses an effort in each period, the incentive constraints are effectively pooled, improving incentives (Abreu, Milgrom, and Pearce 1991).

the target for passing the review. However, by setting the threshold less than pR, the proportion of successes expected from always exerting high effort and then letting the review phase be very long, we can make it quite unlikely that the agent gets discouraged and abandons high effort. Moreover, as the review phase grows longer, we can set a smaller gap between pR and the proportion required to pass the review, while still constraining the losses from a discouraged agent. The agent will often get sufficiently far ahead of the target to coast into a favorable review with low effort, but allowing the target to approach pR ensures that the principal's losses from such behavior are small.

The key result is the following lemma.

Lemma 11.4.1 *There exists $\lambda > 0$ and $\eta > 0$ such that for any $\varepsilon \in (0, \eta)$, there exists R_ε such that for any $R \geq R_\varepsilon$, there exists $\delta_{(R,\varepsilon)} < 1$ such that for all $\delta \in (\delta_{(R,\varepsilon)}, 1)$, the agent's best response to the review strategy given by (λ, R_ε) gives the agent an expected utility at least $v_2^* - \varepsilon/4$ and a probability of failing the review of at most ε.*

Recall that for any fixed value of R, the principal's participation in the review strategy is optimal as long as the principal's expected payoff is sufficiently close to v_1^* and δ is sufficiently large. The lemma characterizes the result of any equilibrium behavior on the part of the agent. We use the lemma to show that if we first let $\delta \to 1$, then $R \to \infty$, and finally $\varepsilon \to 0$, the principal's expected payoff from the review strategy, coupled with any best response from the agent, approaches v_1^*. This ensures that if δ is sufficiently large, there exists an equilibrium with payoffs arbitrarily close to v^*.

The principal's expected equilibrium payoff satisfies

$$v_1 \geq (1-\varepsilon)[\delta^{R-pR+R^\xi}(1-\delta^{pR-R^\xi}) - (1-\delta^R)w^* + \delta^R v_1]$$
$$+ \varepsilon[(1-\delta^R)(-w^*) + \delta^R(1-\delta^{\lambda R})(p-\tilde{w}) + \delta^{(1+\lambda)R}v_1].$$

The first term on the right (multiplied by $1 - \varepsilon$) is a lower bound on the principal's payoff if the agent passes the review. In this case, there must have been at least $pR - R^\xi$ successes, which (in the worst case) occurred in the last $pR - R^\xi$ periods, and the payment w^* is made in every period. The second term (multiplied by ε) is a lower bound on the payoff in the event of a failure, which can at worst feature no successes at all.

Hence, v_1 is bounded below by

$$\frac{(1-\varepsilon)(\delta^{R-pR+R^\xi} - \delta^R - (1-\delta^R)w^*) + \varepsilon(\delta^R(1-\delta^{\lambda R})(p-\tilde{w}) - (1-\delta^R)w^*)}{1 - \delta^R(1-\varepsilon) - \delta^{(1+\lambda)R}\varepsilon}.$$

For fixed ε and R, this lower bound approaches (as δ approaches 1, applying l'Hôpital's rule)

$$\frac{(1-\varepsilon)(p - R^{\xi-1}) - w^* + \varepsilon\lambda(p-\tilde{w})}{1 + \lambda\varepsilon}.$$

As R gets large and then ε gets small, this approaches the efficient payoff $p - w^*$.

Our argument is then completed by proving the lemma.

Proof of Lemma 11.4.1 Fix $\varepsilon > 0$. If the principal plays a review strategy characterized by R, T $(= \lambda R)$ and S $(= pR - R^\xi)$, then the agent's equilibrium payoff, written as a function of δ, is given by

11.4 ■ Principal-Agent Problems

$$v_2(\delta) = (1-\delta)\sum_{t=1}^{R} \delta^{t-1} u_2^t + \phi[\delta^R(1-\delta^T)\tilde{v}_2 + \delta^{R+T} v_2(\delta)]$$

$$+ (1-\phi)\delta^R v_2(\delta)$$

$$= \frac{(1-\delta)\sum_{t=1}^{R} \delta^{t-1} u_2^t + \phi \delta^R (1-\delta^T)\tilde{v}_2}{1 - \phi \delta^{R+T} - (1-\phi)\delta^R}, \quad (11.4.6)$$

where ϕ is the probability of failing the review and u_2^t is the expected payoff received in period t during the review phase of the strategy profile.

One possibility open to the agent is to play the efficient action of high effort throughout the review phase. Using (11.4.6), this gives an expected payoff of

$$v_2^*(\delta) = \frac{(1-\delta^R)v_2^* + \phi^* \delta^R (1-\delta^T)\tilde{v}_2}{1 - \phi^* \delta^{R+T} - (1-\phi^*)\delta^R},$$

where ϕ^* is the probability of failing the review under this strategy.

Let $\kappa = p(1-p)$. Then κ is the variance of the outcome produced by high effort (taking a success to be 1 and a failure 0). It then follows from Chebychev's inequality that[11]

$$\phi^* \leq \frac{R\kappa}{R^{2\xi}} = \frac{\kappa}{R^{2\xi-1}}.$$

As a result, ϕ^* converges to 0 as R gets large. Using this and $T = \lambda R$, we can then find a value R' such that for any value $R \geq R'$ and for all δ we have[12]

$$v_2^*(\delta) > v_2^* - \tfrac{\varepsilon}{4}. \quad (11.4.7)$$

Hereafter restricting attention to values $R \geq R'$, this provides the first part of the lemma, ensuring that the agent can earn a payoff against the review strategy of at least $v_2^* - \varepsilon/4$.

It remains to show that under the agent's best response, the probability of failing the review is sufficiently small. The approach here is to calculate an upper bound on player 2's payoff, as a function of the failure probability ϕ. We then show that if this failure probability is too large, the upper bound on the agent's payoff must fall sufficiently short of v_2^* as to contradict the possibility result established by (11.4.7).

For an arbitrary strategy for the agent, denote by ϕ the implied failure probability and by e^t the expected effort in period t. The expected payoff to the agent

---

11. Let f be the realized number of failures. The probability of failing the review is the probability that f exceeds $R - (pR - R^\xi)$. The variance of f is $R\kappa$, and since the expected value of f is $(1-p)R$, ϕ^* is the probability that $f - (1-p)R$ exceeds R^ξ.
12. Uniformity in δ follows from

$$\lim_{\delta \to 1} v_2^*(\delta) = \frac{v_2^* + \phi^* \lambda \tilde{v}_2}{1 + \phi^* \lambda}$$

and $\quad \lim_{\phi^* \to 0} v_2^*(\delta) = v_2^*.$

in period t during the review phase under this strategy is

$$u_2^t = \tilde{u}(w^*) - ce^t = v_2^* + c(1 - e^t).$$

This allows us to rewrite (11.4.6) as

$$v_2(\delta) = \frac{(1-\delta^R)v_2^* + (1-\delta)c\sum_{t=1}^{R}\delta^{t-1}(1-e^t) + \phi\delta^R(1-\delta^T)\tilde{v}_2}{1 - \phi\delta^{R+T} - (1-\phi)\delta^R}. \quad (11.4.8)$$

The number of expected successes over the course of the review period, s^e, can be written as

$$s^e = \sum_{t=1}^{R}(pe^t + q(1-e^t))$$

$$= (p-q)\sum_{t=1}^{R}e^t + Rq,$$

and so, $\sum_{t=1}^{R} e^t = (s^e - Rq)/(p-q)$. Hence, because $1 - e^t \geq 0$,

$$\sum_{t=1}^{R}\delta^{t-1}(1-e^t) \leq \sum_{t=1}^{R}(1-e^t) = \frac{pR - s^e}{p - q}.$$

It takes $pR - R^\xi$ successes to pass the review, and the review is passed with probability $1 - \phi$, so the expected number of successes must satisfy $s^e \geq (1-\phi) \times (pR - R^\xi)$ or

$$pR - s^e \leq \phi pR + (1-\phi)R^\xi.$$

Combining these calculations with (11.4.8), we have

$$v_2(\delta) \leq \frac{(1-\delta^R)v_2^* + \phi\delta^R(1-\delta^T)\tilde{v}_2 + (1-\delta)K[\phi pR + (1-\phi)R^\xi]}{1 - \phi\delta^{R+T} - (1-\phi)\delta^R},$$

where $K = c/(p-q)$. Again applying l'Hôpital's rule and noting that $T = \lambda R$, we can fix R and let $\delta \to 1$ to find

$$v_2(1) \leq \frac{v_2^* + \phi\lambda\tilde{v}_2 + K[\phi p + (1-\phi)R^{\xi-1}]}{1 + \phi\lambda}.$$

Hence, for any value $R \geq R'$, we can find $\delta' \in (0, 1)$ such that for all larger values of δ,

$$|v_2(\delta) - v_2(1)| \leq \tfrac{\varepsilon}{4}. \quad (11.4.9)$$

We now let ϕ be the probability of failing the review under the agent's optimal strategy, suppose $\phi \geq \varepsilon$ and derive a contradiction. Fix a value η with $\eta/2 \in (0, v_2^* - \tilde{v}_2)$ (using the assumption that $v_2^* - \tilde{v}_2 > 0$). Choose λ so that

11.4 ■ Principal-Agent Problems

$$\lambda \geq \frac{1 + 2Kp}{2(v_2^* - \tilde{v}_2 - \eta/2)}.$$

Then, if $\varepsilon < \eta$, we have $\lambda(v_2^* - \tilde{v}_2 - \frac{\varepsilon}{2}) - Kp + KR^{\xi-1} > 0$ for all R. From our bound for $v_2(1)$, for $\varepsilon < \eta$, because the inequality

$$\frac{v_2^* + \phi\lambda\tilde{v}_2 + K[\phi p + (1-\phi)R^{\xi-1}]}{1 + \phi\lambda} \leq v_2^* - \frac{\varepsilon}{2}$$

is equivalent to the following lower bound on ϕ, we have

$$\phi \geq \frac{KR^{\xi-1} + \frac{\varepsilon}{2}}{\lambda(v_2^* - \tilde{v}_2 - \frac{\varepsilon}{2}) - Kp + KR^{\xi-1}} \quad \Rightarrow \quad v_2(1) \leq v_2^* - \frac{\varepsilon}{2}. \quad (11.4.10)$$

Now fix a value $R_\varepsilon \geq R'$ such that for any $R \geq R_\varepsilon$,

$$\frac{KR^{\xi-1} + \frac{\varepsilon}{2}}{\lambda(v_2^* - \tilde{v}_2 - \frac{\varepsilon}{2}) - Kp + KR^{\xi-1}} < \varepsilon.$$

Given any such value R, let $\delta_{(R,\varepsilon)}$ be a value $\delta \geq \delta'$ such that (11.4.9) holds for all $\delta \in (\delta_{(R,\varepsilon)}, 1)$. We now note that if $\phi \geq \varepsilon$, then (11.4.10) implies that $v_2(1) \leq v_2^* - \varepsilon/2$. Using (11.4.9), we then have $v_2(\delta) \leq v_2^* - \varepsilon/4$. But this upper limit on player 2's equilibrium payoffs contradicts the fact, established in (11.4.7), that a higher payoff is available from the strategy of always exerting high effort during the review phase.

■

We can compare this efficiency outcome to inefficiency results for two similar settings. Radner, Myerson, and Maskin (1986) examine a repeated partnership game with imperfect monitoring in which equilibrium payoffs are bounded away from efficiency, even as the discount factor gets large. Why are review strategies ineffective in this case? There is imperfect monitoring on both sides of the game in Radner, Myerson, and Maskin (1986). The review strategy argument uses the perfect observability of the principal's actions to ensure the principal's adherence to the review strategy. We then obtain the approximate efficiency result by using the payoff the agent receives when always choosing high effort and playing against the review strategy to impose a bound on the agent's equilibrium payoff. Connecting to our previous results, the two-sided imperfection in the monitoring also implies that the pairwise identifiability conditions from section 9.4 fail.

Section 8.3.2 showed that equilibria in games where some long-lived players are subject to binding moral hazard are necessarily inefficient. If the principal were short-lived, then the agent in this section would be subject to binding moral hazard: The agent's payoff is maximized by an action profile in which the principal makes payments to the agent, which is optimal for the principal only if the agent chooses high effort, which is not a best response for the agent. However, a long-lived principal need not play a stage-game best response in each period, and the game accordingly no longer exhibits binding moral hazard. We again see the importance of being able to use intertemporal incentives in inducing the principal to follow the review strategy.

Part III

Games with Private Monitoring

12 Private Monitoring

The next three chapters consider games with private monitoring. These games are the natural setting for many questions. For example, Stigler's (1964) story of secret price cutting is one of private monitoring. Stigler argued that collusion between firms is hard to sustain when other firms cannot observe discounts (deviations from collusion) that a firm may offer customers. If each firm's sales are themselves private, there is no public signal (as in Green and Porter 1984), and so the arguments of the previous chapters do not immediately apply. The construction of equilibria in general private monitoring games requires significantly different techniques.

A broad theme of the techniques we have surveyed is that players can be induced to not behave opportunistically (i.e., not play myopic best replies) through appropriate use of intertemporal incentives. Moreover, because it is important that the behavior be credible (sequentially rational), histories are used to coordinate the continuation play providing these intertemporal incentives. As we saw in section 7.1.3, a major motivation for the restriction to perfect public equilibria (PPE) in public monitoring games is the need to ensure that statements like "histories credibly coordinate continuation play" make sense. In contrast, it is not immediately clear what statements about coordinating continuation play mean in private monitoring games.

12.1 A Two-Period Example

We begin with a two-period game (inspired by Bhaskar and van Damme 2002), interpreting the second period as a proxy for the ability to implement different continuation values with future play in an infinitely repeated game. The first-period game is (yet again) the prisoners' dilemma of figure 1.2.1, reproduced on the left in figure 12.1.1. In the second period, the coordination game on the right is played. There is no discounting. All payoffs are received at the end of the second period, so first-period payoffs cannot convey information to the players.[1] We are interested in supporting the choice of E in the first period.

Consider first the case of *imperfect* public monitoring. Let $\rho(y \mid a)$ be the probability that public signal $y \in Y \equiv \{\underline{y}, \bar{y}\}$ is observed if action profile a is chosen, where the probability distribution is given by (7.2.1),

---

1. The formal development in section 12.2 assumes players receive an ex post payoff as a function of their action and signal, and so again payoffs cannot convey additional information.

	E	S
E	2, 2	−1, 3
S	3, −1	0, 0

	G	B
G	3, 3	0, 0
B	0, 0	1, 1

Figure 12.1.1 The first-period stage game is on the left with the second-period stage game on the right.

	$\underline{z}$	$\bar{z}$
$\underline{z}$	$(1-\eta)(1-\varepsilon)$	$\varepsilon/2$
$\bar{z}$	$\varepsilon/2$	$\eta(1-\varepsilon)$

Figure 12.1.2 An almost public monitoring distribution π, where $\eta = p$ if $a_1 a_2 = EE$, q if $a_1 a_2 = ES$ or SE, and r if $a_1 a_2 = SS$.

$$\rho(\bar{y} \mid a) = \begin{cases} p, & \text{if } a = EE, \\ q, & \text{if } a = ES \text{ or } SE, \\ r, & \text{if } a = SS, \end{cases} \tag{12.1.1}$$

where $0 < r < q < p < 1$. An obvious candidate for enforcing EE is a "trigger" profile specifying EE in the first period and GG in the second period if the public signal is $\bar{y}$, and BB in the second period if the signal is $\underline{y}$. Such a strategy profile is a PPE if and only if the first-period incentive constraint for cooperation is satisfied,

$$2 + 3p + (1 - p) \geq 3 + 3q + (1 - q), \tag{12.1.2}$$

that is, $2(p - q) \geq 1$. We assume $2(p - q) > 1$, noting for later use that (12.1.2) then holds strictly.

Our concern here is the case of private monitoring. Player i observes a *private* signal z_i drawn from the space $\{\underline{z}, \bar{z}\}$. For each $a \in A$, the probability that the private signal vector $(z_1, z_2) \in \{\underline{z}, \bar{z}\}^2$ is realized is given by $\pi(z_1 z_2 \mid a)$.

Private monitoring distributions that are highly correlated because they are close to some public monitoring distribution constitute one extreme. For example, interpreting $\underline{z}$ as $\underline{y}$ and $\bar{z}$ as $\bar{y}$, the monitoring distribution π given in figure 12.1.2, is ε-*close* to the ρ of (12.1.1) for any $p, q,$ and r. The probability that the two players each observe the *same* signal z under π is within ε of the probability that the corresponding signal y

12.1 ■ A Two-Period Example

EE	z_2	$\bar{z}_2$
z_1	ε^2	$(1-\varepsilon)\varepsilon$
$\bar{z}_1$	$(1-\varepsilon)\varepsilon$	$(1-\varepsilon)^2$

SE	z_2	$\bar{z}_2$
z_1	$(1-\varepsilon)\varepsilon$	ε^2
$\bar{z}_1$	$(1-\varepsilon)^2$	$(1-\varepsilon)\varepsilon$

Figure 12.1.3 A conditionally independent private monitoring distribution for two action profiles.

is realized under ρ. If ε is small, the two players' private signals will then be highly correlated, and so a player's signal will be very informative about the other player's signal. We also say such a distribution is *almost public*.

At the other extreme, monitoring is *conditionally independent* if

$$\pi(z_1 z_2 \mid a) = \pi_1(z_1 \mid a)\pi_2(z_2 \mid a), \tag{12.1.3}$$

where $\pi_i(z_i \mid a)$ is the marginal probability that player i observes z_i given the action profile a. An example of conditionally independent *almost perfect* monitoring for the prisoners' dilemma again has signal spaces $Z_1 = Z_2 = \{z, \bar{z}\}$, but with marginal distributions given by

$$\pi_i(z_i \mid a) = \begin{cases} 1 - \varepsilon, & \text{if } z_i = \bar{z} \text{ and } a_j = E, \text{ or} \\ & \quad z_i = z \text{ and } a_j = S, j \neq i, \\ \varepsilon, & \text{otherwise,} \end{cases} \tag{12.1.4}$$

and the joint distribution given by (12.1.3). For $\varepsilon < 1/2$, $\bar{z}$ is a signal that the other player is likely to have played E and z a signal that he is likely to have played S. The joint distributions for EE and SE under (12.1.4) are given in figure 12.1.3. As this figure makes clear, *given* the action profile, a player's signal provides no information whatsoever about the other player's signal. Perfect monitoring is the special (if trivial) case of $\varepsilon = 0$, and small ε corresponds to a particular form of almost perfect monitoring. The critical difference between the two types of private monitoring is that under almost public monitoring, the off-diagonal entries of the joint distribution are close to 0 relative to the diagonal elements (independent of the action profile), whereas this is not true with conditional independence. As ε becomes small, the limits of the almost public monitoring distributions and the conditionally independent almost perfect monitoring distributions are typically very different. In figure 12.1.2, the limit is the public monitoring distribution, and in figure 12.1.3, the limit is perfect monitoring.

12.1.1 Almost Public Monitoring

Almost public monitoring is the minimal perturbation of public monitoring in the direction of private monitoring. Histories that are public in public monitoring games (and so allow for the coordination of continuation play) are nearly so in almost public monitoring games.

Suppose the private monitoring is given by figure 12.1.2 and so is ε-close to the public monitoring of (12.1.1). Consider the private strategy profile induced in the private monitoring game by the trigger strategy, that is, both players play E in the first period, and in the second period, player i plays G after observing the private signal $\bar{z}$ and plays B after the private signal $\underline{z}$. It should not be surprising that this private profile is an equilibrium for ε sufficiently small when $2(p-q) > 1$. Intuitively, payoffs are continuous in ε, and so for small ε, the first-period incentives are close to those of the public monitoring game, which hold strictly. Moreover, for small ε, the probability that the two players observe the same value of the private signal is close to 1, and so the second-period incentive constraints are also satisfied (in the public monitoring game, second-period play after all histories is a strict Nash equilibrium).

We now provide the calculation underlying this intuition. The probability that player 1 assigns to player 2 observing the private signal $\underline{z}$ when he observed $\underline{z}$ is

$$Pr\{z_2 = \underline{z} \mid z_1 = \underline{z}\} = \frac{\pi(\underline{zz} \mid EE)}{\pi(\underline{zz} \mid EE) + \pi(\underline{z}\bar{z} \mid EE)} = \frac{1 - p - \varepsilon(1-p)}{1 - p - \varepsilon(1-2p)/2},$$

because player 1 knows that he chose E in the first period and he believes his partner also followed the profile's prescription. This fraction is larger than $3/4$ for ε sufficiently small. Consequently, if player 1 believes his partner will play B after privately observing $z_2 = \underline{z}$, then he finds B a best reply after privately observing $z_1 = \underline{z}$. A similar calculation shows that after observing $\bar{z}$, players are also sufficiently confident that their partner had observed the same signal. Summarizing, for ε small, the private profile is sequentially rational in the second period. Note that this would hold for any private monitoring distribution π for which the signals z_1 and z_2 were sufficiently correlated.

Turning to the first-period incentives for player 1 (player 2's are symmetric), exerting effort yields the expected payoff

$$2 + 3\pi(\bar{z}\bar{z} \mid EE) + 0 \times \pi(\bar{z}\underline{z} \mid EE) + 0 \times \pi(\underline{z}\bar{z} \mid EE) + \pi(\underline{zz} \mid EE),$$

and shirking yields the payoff (we now suppress the zero terms)

$$3 + 3\pi(\bar{z}\bar{z} \mid SE) + \pi(\underline{zz} \mid SE),$$

so deviating from the profile is not profitable if

$$1 \leq 3\{\pi(\bar{z}\bar{z} \mid EE) - \pi(\bar{z}\bar{z} \mid SE)\} + \{\pi(\underline{zz} \mid EE) - \pi(\underline{zz} \mid SE)\}$$
$$= 3\{p(1-\varepsilon) - q(1-\varepsilon)\} + \{(1-p)(1-\varepsilon) - (1-q)(1-\varepsilon)\}$$
$$= 2(p-q)(1-\varepsilon).$$

Because $2(p-q) > 1$ holds strictly, E is optimal in the first period for small ε. Note that these calculations relied on the closeness of π to ρ, and not just high correlation in the private signals. The closeness to the public monitoring distribution guaranteed that the intertemporal incentive constraints in the private monitoring game look sufficiently like those of the public monitoring game that similar conditions on parameters are sufficient for equilibrium.

As this discussion indicates, private monitoring that is sufficiently close to public monitoring does not introduce significantly new issues for *finitely* repeated games.

12.1 ■ A Two-Period Example

On the other hand, infinitely repeated games introduce significant complications. We provide an overview in section 12.3 and then devote chapter 13 to games with almost public monitoring.

12.1.2 Conditionally Independent Monitoring

Even in our two-period example, full-support conditionally independent private monitoring leads to a strikingly different conclusion. No matter how close to perfect monitoring π is, if π is a full-support conditionally independent distribution, then there is *no* pure strategy Nash equilibrium in which EE is played in the first period.

For simplicity, consider a profile similar to that analyzed for almost public monitoring: Player i plays E in the first period, G after $\bar{z}$, and B after $\underline{z}$. After playing E in the first period, player 1 knows that 2 will observe the signal $\bar{z}$ with probability $1 - \varepsilon$ and the signal $\underline{z}$ with probability ε, and so under the hypothesized profile, player 1 faces a second-period distribution of $(1 - \varepsilon) \circ G + \varepsilon \circ B$. For small ε, player 1 then plays G in the second period *independently* of the signal z_1 he observes. That is, the beliefs of player 1 about the play of player 2 in the second period (and so 1's best replies) are independent of the signals that 1 observes. This is the import of the assumption that the monitoring is conditionally independent and ensures that it is not sequentially rational for a player to respond to his private signals in the second period. But this implies that using pure strategies, intertemporal incentives cannot be provided to induce players to play E in the first period.[2]

The restriction to pure strategies (in particular in the first period) was critical in the analysis just presented. Under pure strategies, the observed signals conveyed *no* information about first-period behavior, which is assumed to match the equilibrium specification. Once there is randomization in the first period, the signals can convey information, so beliefs about continuation play may also depend on those realizations.

Suppose each player plays E in the first period with probability α and S with probability $1 - \alpha$. These mixtures can be viewed as inducing a joint distribution over a type space $T_1 \times T_2$, where a player's type identifies the action the player took and the signal the player receives. Hence, $T_i = A_i \times Z_i$ and the joint distribution is given in figure 12.1.4. The mixing in the first period induces correlation in the type space $T_1 \times T_2$. This is most easily seen by letting $\varepsilon \to 0$ (holding α constant), with the distribution over types converging to the distribution given in figure 12.1.5. The limit distribution has a block diagonal structure, so that $\Pr\{E\bar{z} \mid E\bar{z}\} \approx 1$ and $\Pr\{t_i \in \{E\underline{z}, S\bar{z}, S\underline{z}\} \mid t_j\} \approx 1$ for all $t_j \in \{E\underline{z}, S\bar{z}, S\underline{z}\}$ and ε small. However, it is not diagonal, as would be implied by almost public monitoring.

---

2. Bagwell (1995) makes a similar observation in the context of Stackelberg duopoly games.
    Intertemporal incentives may exist if, conditional on EE, the players are indifferent between G and B. In that event, the choice of G or B can depend on the realized signal, and so a player's deviation to S could lead to a disadvantageous change in the partner's behavior. An example is presented in Section 12.1.3. This possibility does not arise here. Each player can only be indifferent if after EE, he expects his partner to play B with probability $3/4$, giving a payoff of $3/4$. If $\varepsilon = 1/4$, the pure strategy for the partner of playing B after $\bar{z}$ and G after $\underline{z}$ makes the player indifferent between G and B (and so willing to follow the same strategy). Not only is this ruled out by small ε, but shirking results in the *more* advantageous distribution of $\varepsilon \circ B + (1 - \varepsilon) \circ G$.

	$E\bar{z}$	$E\underline{z}$	$S\bar{z}$	$S\underline{z}$
$E\bar{z}$	$\alpha^2(1-\varepsilon)^2$	$\alpha^2\varepsilon(1-\varepsilon)$	$\alpha(1-\alpha)\varepsilon(1-\varepsilon)$	$\alpha(1-\alpha)\varepsilon^2$
$E\underline{z}$	$\alpha^2\varepsilon(1-\varepsilon)$	$\alpha^2\varepsilon^2$	$\alpha(1-\alpha)(1-\varepsilon)^2$	$\alpha(1-\alpha)\varepsilon(1-\varepsilon)$
$S\bar{z}$	$\alpha(1-\alpha)\varepsilon(1-\varepsilon)$	$\alpha(1-\alpha)(1-\varepsilon)^2$	$(1-\alpha)^2\varepsilon^2$	$(1-\alpha)^2\varepsilon(1-\varepsilon)$
$S\underline{z}$	$\alpha(1-\alpha)\varepsilon^2$	$\alpha(1-\alpha)\varepsilon(1-\varepsilon)$	$(1-\alpha)^2\varepsilon(1-\varepsilon)$	$(1-\alpha)^2(1-\varepsilon)^2$

Figure 12.1.4 The joint distribution over first-period histories under the conditionally independent private monitoring of (12.1.4).

Consider now the strategy profile in which each player randomizes in the first period, with probability α on E, and in the second period plays

$$\sigma_i^2(a_i^1, z_i) = \begin{cases} G, & \text{if } a_i^1 = E \text{ and } z_i = \bar{z}, \\ B, & \text{otherwise.} \end{cases}$$

Hence, each player chooses G in the second period only if the player exerted effort in the first *and* received a signal suggesting the partner had also exerted effort. This will be an equilibrium if sufficient correlation is induced in the distribution over types (and the first-period incentive constraints are satisfied). This equilibrium is the "reduced form" of the equilibrium of the infinitely repeated prisoners' dilemma with private monitoring discussed in section 12.4.[3]

Let $T_i^B = \{E\underline{z}, S\bar{z}, S\underline{z}\}$, so that T_i^B is the set of player i types who choose B in the second period. Second period optimality requires

$$\Pr\{a_j^1 = E, z_j = \bar{z} \mid a_i^1 = E, z_i = \bar{z}\} = \frac{\alpha(1-\varepsilon)^2}{\alpha(1-\varepsilon) + (1-\alpha)\varepsilon} \geq \frac{1}{4}, \quad (12.1.5)$$

$$\Pr\{(a_j^1, z_j) \in T_j^B \mid a_i^1 = E, z_i = \underline{z}\} = \frac{\alpha\varepsilon^2 + (1-\alpha)(1-\varepsilon)}{\alpha\varepsilon + (1-\alpha)(1-\varepsilon)} \geq \frac{3}{4}, \quad (12.1.6)$$

$$\Pr\{(a_j^1, z_j) \in T_j^B \mid a_i^1 = S, z_i = \bar{z}\} = \frac{\alpha(1-\varepsilon)^2 + (1-\alpha)\varepsilon}{\alpha(1-\varepsilon) + (1-\alpha)\varepsilon} \geq \frac{3}{4}, \quad (12.1.7)$$

and

$$\Pr\{(a_j^1, z_j) \in T_j^B \mid a_i^1 = S, z_i = \underline{z}\} = \frac{\alpha\varepsilon(1-\varepsilon) + (1-\alpha)(1-\varepsilon)}{\alpha\varepsilon + (1-\alpha)(1-\varepsilon)} \geq \frac{3}{4}. \quad (12.1.8)$$

These inequalities all hold for sufficiently small ε as long as α is bounded away from 1.

---

3. There is also an equilibrium (for ε small) in which the players randomize with approximately equal probability on E and S in the first period, choosing B after $S\underline{z}$ and G otherwise. This is more efficient than the equilibrium discussed in the text, because GG occurs with higher probability.

12.1 ■ A Two-Period Example

	$E\bar{z}$	Ez	$S\bar{z}$	Sz
$E\bar{z}$	α^2	0	0	0
Ez	0	0	$\alpha(1-\alpha)$	0
$S\bar{z}$	0	$\alpha(1-\alpha)$	0	0
Sz	0	0	0	$(1-\alpha)^2$

Figure 12.1.5 The joint distribution over first-period histories, taking $\varepsilon \to 0$.

For the mixed strategy to be optimal in the first period, we must have the payoff from effort,

$$\alpha\{2 + 3(1-\varepsilon)^2 + \varepsilon^2\} + (1-\alpha)\{-1 + (1-\varepsilon)\},$$

equal the payoff from shirking,

$$\alpha\{3 + (1-\varepsilon)\} + (1-\alpha)\{0 + 1\}.$$

Solving for α gives

$$\alpha = \alpha(\varepsilon) = \frac{1+\varepsilon}{2(1-\varepsilon)^2 + 2\varepsilon^2}.$$

As $\varepsilon \to 0$, $\alpha(\varepsilon) \to 1/2$. Thus, there is an equilibrium with some first-period cooperation for sufficiently small ε.

Notice, however, that even as ε becomes small, S must be played with significant probability in the first period. As we have discussed, the randomization in the first period is required for behavior in the second period to depend on the realization of the private signals. This randomization does not require α close to $1/2$. Indeed, for any $\alpha < 1$, for ε sufficiently small, inequalities (12.1.5)–(12.1.8) will continue to hold. The action S receives a large equilibrium probability in the first period to lower the payoff difference between E and S (because S in the first period necessarily leads to B in the second), providing the necessary indifference. This suggests that if instead the expected continuation payoff after S could be increased, then it may be possible to construct an equilibrium in which E is played with very high probability in the first period.

Public correlation is the standard tool for achieving such goals (see also the discussion in sections 2.5.5 and 7.2.3). Suppose that in addition to their private signals, players also observe a public signal ω at the end of the first period, where ω is uniformly distributed on $[0, 1]$.[4] The profile we consider has each player again randomize in the first period, with probability α on E, and in the second period play

---

4. See remark 12.4.1, as well as Bhaskar and van Damme (2002) and Bhaskar and Obara (2002), for more on the role of public correlation.

$$\sigma_i^2(a_i^1, z_i, \omega) = \begin{cases} G, & \text{if } a_i^1 = E \text{ and } z_i = \bar{z}, \text{ or } \omega \leq \omega', \\ B, & \text{otherwise.} \end{cases}$$

Each player chooses G in the second period only if the player exerted effort in the first and received a signal suggesting the partner had also exerted effort, *or* the public signal had a realization less than ω'. In other words, the earlier strategy is modified so that on the event where B was played for sure, now player i randomizes between G and B, with probability $\phi \equiv \Pr\{\omega \leq \omega'\}$ on G. Moreover (and this is important), playing G or B is coordinated to the extent that both players' types are in $T_1^B \times T_2^B$. First, conditional on $\omega > \omega'$, the calculations leading to the inequalities (12.1.5)–(12.1.8) are unchanged, so they are all satisfied for ε small (given any $\alpha < 1$). Turning to the first period, the payoff from effort is now

$$\alpha\{2 + 3(1-\varepsilon)^2 + 6\varepsilon(1-\varepsilon)\phi + \varepsilon^2(3\phi + (1-\phi))\} \\ + (1-\alpha)\{-1 + (1-\varepsilon)(3\phi + (1-\phi)) + 3\varepsilon\phi\},$$

whereas the payoff from shirking is,

$$\alpha\{3 + (1-\varepsilon)(3\phi + (1-\phi)) + 3\varepsilon\phi\} + (1-\alpha)\{0 + (3\phi + (1-\phi))\}.$$

Equating the two and solving for α gives

$$\alpha \equiv \alpha(\varepsilon, \phi) = \frac{1 + \varepsilon - \varepsilon\phi}{2(1-\phi)(1 - 2\varepsilon + 2\varepsilon^2)},$$

and as $\varepsilon \to 0$,

$$\alpha(\varepsilon, \phi) \to \frac{1}{2(1-\phi)}.$$

This is a well-defined interior probability for $\phi < 1/2$, and can be made arbitrarily close to 1 by making ϕ arbitrarily close to $1/2$. Although we do not have efficiency in the second period, we do obtain EE with arbitrarily large probability in the first.

In concluding the discussion of this example, we emphasize that the order of quantifiers is crucial. For any $\alpha < 1$ we can construct an equilibrium (with a public correlating device) in which EE is played with probability at least α, as long as ε is sufficiently small. However, EE cannot be played with probability 1 in the first period, no matter how small is ε. Nontrivial updating requires α less than 1, and for any given $\alpha < 1$, (12.1.5)–(12.1.8) provide the relevant bounds on ε. We cannot fix ε and take α to 1.

Section 12.4 uses such *belief-based* techniques to construct nearly efficient equilibria in the prisoners' dilemma with (possibly conditionally independent) private monitoring.

12.1.3 Intertemporal Incentives from Second-Period Randomization

Here we describe an example (from Kandori 1991a) illustrating the role of conditioning behavior on histories when a player is indifferent between different actions.

12.1 ■ A Two-Period Example

	R_2	P_2
R_1	3, 3	0, 2
P_1	4, −2	−1, −1

Figure 12.1.6 The second-period stage game for section 12.1.3.

We build on the example of section 12.1.2. We keep the first-period stage game and conditionally independent private monitoring and replace the second period stage game by the game in figure 12.1.6. This stage game has a unique mixed strategy equilibrium in which each player randomizes with equal probability on each action. The payoff from this mixed equilibrium is $(3/2, 1/2)$.

The idea is to construct an equilibrium of the two-period game so that after EE (i.e., on the equilibrium path), second-period behavior unconditionally looks like $(\frac{1}{2} \circ R_1 + \frac{1}{2} \circ P_1, \frac{1}{2} \circ R_2 + \frac{1}{2} \circ P_2)$, but after a deviation to S, the deviating player faces a significantly increased probability of P_i.

The strategy for player i specifies E in the first period, and a probability of ξ_i on R_i after $\bar{z}$ and P_i for sure after $\underline{z}$. If player 1 plays E in the first period, then he expects R_2 to be played with probability $(1 - \varepsilon)\xi_2$, and setting this equal to $1/2$, gives

$$\xi_2 = \frac{1}{2(1 - \varepsilon)}.$$

This is a well-defined probability for all $\varepsilon \in [0, 1/2)$. The total payoff to player 1 from following his strategy when his partner also follows the strategy is $2 + 3/2 = 7/2$.

On the other hand, if player 1 deviates to S, then he expects R_2 with probability $\varepsilon \xi_2$ in the next period and so plays the best response R_1. The total payoff from this deviation is $3 + 3\xi_2\varepsilon$, which is less than $7/2$ for $\varepsilon < 1/4$, so the deviation is not profitable. A similar calculation holds for player 2.

The key idea is that both R_i and P_i are best replies for player i after both realizations of the private signal. Player i's choice between these best replies can then be arranged as a function of his signal to provide incentives for player j. A similar idea was used to construct equilibria in private strategies in the public monitoring two-period example in section 10.2 and its belief-free infinite-horizon version in section 10.4.[5] We will see that in infinitely repeated games, as in section 10.4, the relevant payoff ties can be endogenously obtained via appropriate mixing. We discuss an example in section 12.5 and study *belief-free equilibria* in chapter 14.

This equilibrium is fragile in that it cannot be purified (Harsanyi 1973), that is, it cannot be approximated by any equilibrium in nearby games of incomplete information. Each player is required to behave differently after different private histories, yet the player has identical beliefs about the behavior of his partner after these two

---

5. First-period mixing makes the informative public signals more informative in section 10.2, whereas here it is necessary if the private signals are to convey any information at all.

histories.[6] Suppose now that payoffs in the second period are perturbed: Each player receives a small private payoff shock distributed independently of the signal. As a result, the player is no longer indifferent (for some values of the payoff, he plays P_i and for others, he plays R_i—there will be a cutoff payoff type for which he is indifferent); because his beliefs about the opponent are independent of his private history, so almost surely is his behavior. For more details, see Bhaskar (1998) and Bhaskar and van Damme (2002).

12.2 Private Monitoring Games: Basic Structure

The infinitely repeated game with private monitoring is the infinite repetition of a stage game in which, at the end of the period, each player learns only the realized value of a private signal. There are n long-lived players, with a finite stage-game action set for player $i \in \{1, \ldots, n\}$ denoted A_i. At the end of each period each player i observes a private signal, denoted z_i, drawn from a finite set Z_i. The signal vector $z \equiv (z_1, \ldots, z_n) \in Z \equiv Z_1 \times \cdots \times Z_n$ occurs with probability $\pi(z \mid a)$ when the action profile $a \in A \equiv \prod_i A_i$ is chosen. Player i's marginal distribution over Z_i is denoted $\pi_i(\cdot \mid a)$. Player i does not receive any information other than z_i about the behavior of the other players. All players use the same discount factor, δ.

Because z_i is the only signal a player observes about opponents' play, we assume (as usual) that player i's ex post payoff after the realization (z, a) is given by $u_i^*(z_i, a_i)$. Ex ante stage game payoffs are then given by $u_i(a) \equiv \sum_z u_i^*(z_i, a_i) \pi(z \mid a)$. It will be convenient to index games by the monitoring technology (Z, π), fixing the set of players and action sets. In various examples, we will often fix ex ante stage-game payoffs and consider various monitoring distributions without explicitly mentioning the requisite adjustments to the payoffs u_i^*.

Definition 12.2.1 *A private monitoring game has ε-perfect monitoring if for each player i, there is a partition of Z_i, $\{Z_i(a)\}_{a \in A}$, such that for all action profiles $a \in A$,*

$$\sum_{z_i \in Z_i(a)} \pi_i(z_i \mid a) > 1 - \varepsilon.$$

When taking ε to 0, we often refer to the private monitoring games with ε-perfect monitoring as having *almost perfect monitoring*.

Remark 12.2.1 Private monitoring games include public (and so perfect) monitoring games as the special case where $Z_i = Z_j$ for all i and j, and the monitoring distribution satisfies $\sum_{z_i \in Z_i} \pi(z_i, \ldots, z_i \mid a) = 1$ for all $a \in A$, that is, all players always observe the same realization of the signal.
♦

A behavior strategy for player i in the private monitoring game is a function $\sigma_i : \mathcal{H}_i \to \Delta(A_i)$, where

$$\mathcal{H}_i \equiv \cup_{t=0}^{\infty} (A_i \times Z_i)^t$$

is the set of private histories for player i.

---

6. Clearly, if it is required that play in such circumstances is independent of the private signal, then only stage-game Nash equilibria can be played. See Matsushima (1991).

Definition 12.2.2 *A strategy is* action free *if, for all $h_i^t, \hat{h}_i^t \in \mathcal{H}_i$ satisfying $z_i^s = \hat{z}_i^s$ for all $s < t$,*

$$\sigma_i(h_i^t) = \sigma_i(\hat{h}_i^t).$$

At first glance, action-free strategies are similar to public strategies, and indeed they satisfy a similar property for a similar reason (and so we omit its proof):

Lemma 12.2.1 *Every pure strategy in a private monitoring game is realization equivalent to a pure action-free strategy. Every mixed strategy is realization equivalent to a mixture over action-free strategies.*

Remark 12.2.2 As for public strategies in public monitoring games (see section 10.1), behavior strategies realization equivalent to a mixed strategy will typically not be action-free.

♦

In public monitoring games, the public history is sufficient to describe continuation play when behavior is restricted to public strategies, and hence players have public best responses. In private monitoring games, not only are the private signals informative about the private history of the other players but player i's updated beliefs about player j's continuation play depend on i's actions. In other words, even if players $j \neq i$ are playing action-free strategies, it need not be true that i has a sequentially rational best reply in action-free strategies (the failure of sequential rationality occurs on histories reached by a deviation by player i and so does not contradict lemma 12.2.1).

If the marginal distributions of the private signals have full support, then the game has *no observable deviations*. Just as in full-support public monitoring games, no deviation by any player i is observed by the other players. (Note that full-support marginal distributions are consistent with full-support public monitoring.) Consequently, every Nash equilibrium outcome is the outcome of a profile satisfying stronger sequential rationality requirements, such as sequential equilibrium (proposition 12.2.1). It is not true, however, that Nash equilibria themselves need satisfy these requirements. In particular, Nash equilibria in action-free strategies need not be sequentially rational. We provide a detailed example in section 12.3.2. This is in contrast to full-support public monitoring games, where every Nash equilibrium in public strategies is a PPE.

Although sequential equilibrium was originally defined for finite extensive form games (Kreps and Wilson 1982b), the notion has a straightforward definition in our setting when all players' marginal distributions have full support. When the game has no observable deviations, given a strategy profile σ, after any private history h_i^t, player i's belief over the private histories of the other players is necessarily given by Bayes' rule (even if player i had deviated in the history h_i^t). Consequently, player i's belief over the continuation play of the other players after any private history h_i^t is determined.

Definition 12.2.3 *Suppose $\pi_i(z_i \mid a) > 0$ for all $z_i \in Z_i$ and all $a \in A$. A strategy profile σ is a* sequential equilibrium *of the repeated game with private monitoring if for all private histories h_i^t, $\sigma_i|_{h_i^t}$ is a best reply to $E[\sigma_{-i}|_{h_{-i}^t} \mid h_i^t]$, where $\sigma_{-i}|_{h_{-i}^t} = (\sigma_1|_{h_1^t}, \ldots, \sigma_{i-1}|_{h_{i-1}^t}, \sigma_{i+1}|_{h_{i+1}^t}, \ldots, \sigma_n|_{h_n^t})$.*

Because private monitoring games with $\pi_i(z_i \mid a) > 0$ for all i, $z_i \in Z_i$ and all $a \in A$ have no observable deviations, a Nash equilibrium σ fails to be sequential only if the behavior specified by σ_i (say) is suboptimal after a history h_i^t in which i has deviated (see proposition 12.3.2 for an example). But because the profile is Nash, replacing that part of σ_i with an optimal continuation strategy cannot destroy ex ante incentives. This suggests the following proposition. See Sekiguchi (1997, proposition 3) or Kandori and Matsushima (1998, appendix) for a proof.

Proposition 12.2.1 *Suppose $\pi_i(z_i \mid a) > 0$ for all i, $z_i \in Z_i$ and all $a \in A$. Every Nash equilibrium outcome is the outcome of a sequential equilibrium.*

Remark 12.2.3 **Automata** Every behavior strategy can be represented by an automaton with a set of states $\mathscr{W}_i$, an initial state w_i^0, a decision rule $f_i : \mathscr{W}_i \to \Delta(A_i)$ specifying a distribution over action choices for each state, and a transition function $\tau_i : \mathscr{W}_i \times A_i \times Z_i \to \mathscr{W}_i$. In the first period, player i chooses an action according to the distribution $f_i(w_i^0)$. At the end of period 0, the realized vector of actions, a^0, then generates a vector of private signals z^0 according to the distribution $\pi(\cdot \mid a^0)$, and player i observes the signal z_i^0. In period 1, player i chooses an action according to the distribution $f_i(w_i^1)$, where $w_i^1 = \tau_i(w_i^0, a_i^0 z_i^0)$, and so on.

In the representation of a pure action-free strategy, i's decision rule maps into A_i, and his transition function is $\tau_i : \mathscr{W}_i \times Z_i \to \mathscr{W}_i$. Any action-free strategy requires at most the countable set $\mathscr{W}_i = \cup_{t=0}^{\infty} Z_i^t$.

Any collection of pure action-free strategies can be represented by a set of states $\mathscr{W}_i$, a decision rule f_i, and a transition function τ_i (the initial state indexes the pure strategies). One class of action-free mixed strategies is described by $(\mathscr{W}_i, \mu_i, f_i, \tau_i)$, where μ_i is a probability distribution over the initial state w_i^0, and $\mathscr{W}_i$ is countable. Not all mixed strategies can be described in this way because the set of all pure strategies is uncountable (which would require $\mathscr{W}_i$ to be uncountable). An illustration of the automaton representation for private monitoring is provided later in figure 12.4.2. ◆

Remark 12.2.4 **Automata and beliefs** Using automata, we can describe player i's beliefs about the other players' continuation behavior as follows. Suppose σ_j is described by the automaton $(\mathscr{W}_j, w_j^0, f_j, \tau_j)$ for $j \neq i$. Then, player j's continuation play after h_j^t, $\sigma_j|_{h_j^t}$, is given by $(\mathscr{W}_j, w_j^t, f_j, \tau_j)$, where $w_j^t = \tau_j(w_j^0, h_j^t)$ is j's private state after the private history h_j^t (recall remark 2.3.2). Consequently, player i's beliefs over the continuation play of the other players can be described by a probability distribution β over $\prod_{j \neq i} \mathscr{W}_j$, the potential initial states of the continuation profile. ◆

In repeated games with private monitoring, calculating beliefs after all private histories can be a difficult (if not impossible) task. It is often easier to show that a profile is a Nash equilibrium and then appeal to proposition 12.2.1.

There are some cases in which it is possible to show that a profile is sequential, and in those cases the one-shot deviation principle is again useful. A *one-shot deviation* for player i from the strategy σ_i is a strategy $\hat{\sigma}_i \neq \sigma_i$ with the property that there exists a unique private history $\tilde{h}_i^t \in \mathscr{H}_i$ such that for all $h_i^s \neq \tilde{h}_i^t$,

$$\sigma_i(h_i^s) = \hat{\sigma}_i(h_i^s).$$

12.3 ■ Almost Public Monitoring

The proofs of the next two propositions are straightforward modifications of their perfect monitoring counterparts (propositions 2.2.1 and 2.4.1), and so are omitted.

Proposition 12.2.2 *The one-shot deviation principle* Suppose $\pi_i(z_i \mid a) > 0$ for all i, $z_i \in Z_i$ and all $a \in A$. A strategy profile σ is a sequential equilibrium if and only if there are no profitable one-shot deviations.

In private monitoring games, there is typically *no* nontrivial history that can serve to coordinate continuation play of different players. This is to be contrasted with perfect monitoring games (where any history can coordinate continuation play) and public monitoring games (where any public history can coordinate continuation play). As a consequence, sequential equilibria do not have a simple recursive formulation and so the statement of the private monitoring analog to proposition 2.4.1 is complicated. (A similar comment applies, of course, to sequential equilibria in private strategies of public monitoring games.) A simple example illustrating the use of proposition 12.2.3 is provided in section 12.3.1.

Proposition 12.2.3 Suppose $\pi_i(z_i \mid a) > 0$ for all i, $z_i \in Z_i$ and all $a \in A$. Suppose the strategy profile σ is described by the automata $\{(\mathcal{W}_j, w_j^0, f_j, \tau_j) : j = 1, \ldots, n\}$. Let $V_i(w_1, \ldots, w_n)$ be player i's average discounted value under the profile when player j is in initial state w_j. For all $h_i^t \in \mathcal{H}_i$, let $\beta_i(\cdot \mid h_i^t)$ be the beliefs of i over $\prod_{j \neq i} \mathcal{W}_j$ after h_i^t implied by the profile σ. The strategy profile σ is a sequential equilibrium if, and only if, for all $h_i^t \in \mathcal{H}_i$, $f_i(\tau(w_i^0, h_i^t))$ maximizes

$$\sum_{w_{-i}} \Big\{ (1-\delta) u_i(a_i, f_{-i}(w_{-i})) $$
$$+ \delta \sum_{a_{-i}} \sum_{z \in Z} V_i(\tau_i(w_i, a_i z_i), \tau_{-i}(w_{-i}, a_{-i} z_{-i})) \pi(z \mid a) \prod_{j \neq i} f_j(a_j \mid w_j) \Big\}$$
$$\times \beta(w_{-i} \mid h_i^t).$$

Remark 12.2.5 The introduction of public communication recovers a recursive structure in private monitoring games. See, for example, Compte (1998) and Kandori and Matsushima (1998).

◆

12.3 Almost Public Monitoring: Robustness in the Infinitely Repeated Prisoners' Dilemma

This section considers almost public monitoring (see section 12.1.1). We work with the prisoners' dilemma in the left panel of figure 12.1.1. The public signal space is $Y = \{\underline{y}, \bar{y}\}$ and the distribution ρ is given by (12.1.1) but only requiring $0 < q < p < 1$ and $0 < r < p$. Following Mailath and Morris (2002), we investigate the robustness of the forgiving PPE of section 7.2.2 and the grim trigger PPE of section 7.2.1 to the introduction of small degrees of private monitoring.

In the game with private monitoring, player i has two signals, $\underline{z}$ and $\bar{z}$, with distribution π, where (for $z_1 z_2 \in \{\underline{z}, \bar{z}\}^2$)

$$\pi(z_1 z_2 \mid EE) \equiv \pi^{EE}_{z_1 z_2}, \qquad \pi(z_1 z_2 \mid SS) \equiv \pi^{SS}_{z_1 z_2},$$

and

$$\pi(z_1 z_2 \mid ES) = \pi(z_1 z_2 \mid SE) \equiv \pi^{ES}_{z_1 z_2}.$$

We are interested in the case where π is ε-*close* to ρ:

$$|\pi(\bar{z}\bar{z} \mid a) - \rho(\bar{y} \mid a)| < \varepsilon, \quad \forall a \in A,$$
$$\text{and} \quad |\pi(\underline{z}\underline{z} \mid a) - \rho(\underline{y} \mid a)| < \varepsilon, \quad \forall a \in A.$$

Note that when π is ε-close to ρ, for $z_1 \neq z_2$, $\pi(z_1 z_2 \mid a) < 2\varepsilon$.

12.3.1 The Forgiving Profile

In this subsection, we assume $q = r$ in ρ and $\pi^{ES} = \pi^{SS}$.[7] Recall that the forgiving strategy profile of section 7.2.2 is described by the following automaton. There are two states, $\mathcal{W} = \{w_{EE}, w_{SS}\}$, initial state $w^0 = w_{EE}$, output function $f(w_{EE}) = EE$ and $f(w_{SS}) = SS$, and transition function

$$\tau(w, y) = \begin{cases} w_{EE}, & \text{if } y = \bar{y}, \\ w_{SS}, & \text{if } y = \underline{y}. \end{cases}$$

This is an equilibrium of the game with public monitoring if (7.2.12) and (7.2.13) are satisfied. Because $q = r$, (7.2.13) always holds strictly, and (7.2.12) becomes $\delta \geq [3(p-q)]^{-1}$. Moreover, the PPE is *strict* if this inequality is strict, that is,

$$\delta > \frac{1}{3(p-q)}. \tag{12.3.1}$$

Let $\underline{\delta}$ denote a lower bound for δ satisfying (12.3.1), and assume $\delta > \underline{\delta}$.

A notable feature of this profile is that the strategies have *bounded* (in fact, one-period) *recall*. The period t actions of the players depend only on the realization of the signal in the previous period. For this reason, we identify the state w_{EE} with the signal $\bar{y}$, and the state w_{SS} with $\underline{y}$. Consider now the same profile in the game with private monitoring: player i exerts effort in each period t if the player observed signal $\bar{z}$ in the previous period, and otherwise shirks.

If player 1 (say) observes $\bar{z}$, then for π sufficiently close to ρ, player 1 attaches a relatively high probability to the event that player 2 also observed $\bar{z}$. Because actions depend only on this last signal, player 1 thus attaches a relatively large probability (given the candidate equilibrium strategies) to player 2's playing E. If the incentive constraints in the PPE are strict, this should make E a best response for player 1.

---

7. These assumptions are inconsistent with the assumption that stage-game payoffs are expected ex post payoffs (see note 7 on page 238). These assumptions are made only to simplify calculations and the points hold in general. For our purposes here, it is enough to assume that players only observe their private signals but do not observe their ex ante payoffs.

12.3 ■ Almost Public Monitoring

Suppose player 1 assigns a probability β to his partner having also observed $\bar{z}$ after observing the private signal $\bar{z}$ (we provide an explicit expression for β shortly). We can write the incentive constraint for player 1 to follow the profile's specification of E as follows (recall $\pi^{ES} = \pi^{SS}$):

$$\beta\{(1-\delta)2 + \delta[\pi^{EE}_{\bar{z}\bar{z}}V_1^{\bar{z}\bar{z}} + \pi^{EE}_{\bar{z}z}V_1^{\bar{z}z} + \pi^{EE}_{z\bar{z}}V_1^{z\bar{z}} + \pi^{EE}_{zz}V_1^{zz}]\}$$
$$+ (1-\beta)\{(1-\delta)(-1) + \delta[\pi^{SS}_{\bar{z}\bar{z}}V_1^{\bar{z}\bar{z}} + \pi^{SS}_{\bar{z}z}V_1^{\bar{z}z} + \pi^{SS}_{z\bar{z}}V_1^{z\bar{z}} + \pi^{SS}_{zz}V_1^{zz}]\}$$
$$\geq \beta\{(1-\delta)3 + \delta[\pi^{SS}_{\bar{z}\bar{z}}V_1^{\bar{z}\bar{z}} + \pi^{SS}_{\bar{z}z}V_1^{\bar{z}z} + \pi^{SS}_{z\bar{z}}V_1^{z\bar{z}} + \pi^{SS}_{zz}V_1^{zz}]\}$$
$$+ (1-\beta)\{\delta[\pi^{SS}_{\bar{z}\bar{z}}V_1^{\bar{z}\bar{z}} + \pi^{SS}_{\bar{z}z}V_1^{\bar{z}z} + \pi^{SS}_{z\bar{z}}V_1^{z\bar{z}} + \pi^{SS}_{zz}V_1^{zz}]\},$$

where $V_1^{z_1 z_2}$ is the continuation to player 1 under the profile when player 1 is in state (has just observed) z_1 and player 2 is in state z_2. This expression simplifies to

$$\beta\delta\{(\pi^{EE}_{\bar{z}\bar{z}} - \pi^{SS}_{\bar{z}\bar{z}})V_1^{\bar{z}\bar{z}} + (\pi^{EE}_{\bar{z}z} - \pi^{SS}_{\bar{z}z})V_1^{\bar{z}z} + (\pi^{EE}_{z\bar{z}} - \pi^{SS}_{z\bar{z}})V_1^{z\bar{z}} + (\pi^{EE}_{zz} - \pi^{SS}_{zz})V_1^{zz}\}$$
$$\geq (1-\delta). \tag{12.3.2}$$

The continuation values satisfy:

$$V_1^{\bar{z}\bar{z}} = (1-\delta)2 + \delta\{\pi^{EE}_{\bar{z}\bar{z}}V_1^{\bar{z}\bar{z}} + \pi^{EE}_{\bar{z}z}V_1^{\bar{z}z} + \pi^{EE}_{z\bar{z}}V_1^{z\bar{z}} + \pi^{EE}_{zz}V_1^{zz}\},$$
$$V_1^{\bar{z}z} = -(1-\delta) + \delta\{\pi^{SS}_{\bar{z}\bar{z}}V_1^{\bar{z}\bar{z}} + \pi^{SS}_{\bar{z}z}V_1^{\bar{z}z} + \pi^{SS}_{z\bar{z}}V_1^{z\bar{z}} + \pi^{SS}_{zz}V_1^{zz}\},$$
$$V_1^{z\bar{z}} = (1-\delta)3 + \delta\{\pi^{SS}_{\bar{z}\bar{z}}V_1^{\bar{z}\bar{z}} + \pi^{SS}_{\bar{z}z}V_1^{\bar{z}z} + \pi^{SS}_{z\bar{z}}V_1^{z\bar{z}} + \pi^{SS}_{zz}V_1^{zz}\},$$

and

$$V_1^{zz} = \delta\{\pi^{SS}_{\bar{z}\bar{z}}V_1^{\bar{z}\bar{z}} + \pi^{SS}_{\bar{z}z}V_1^{\bar{z}z} + \pi^{SS}_{z\bar{z}}V_1^{z\bar{z}} + \pi^{SS}_{zz}V_1^{zz}\}.$$

Substituting V_1^{zz} into the second and third equations, and using the result to simplify the difference $V_1^{\bar{z}\bar{z}} - V_1^{zz}$ yields

$$V_1^{\bar{z}\bar{z}} - V_1^{zz} = \frac{(1-\delta)\{2 + \delta(\pi^{SS}_{\bar{z}z} - \pi^{EE}_{\bar{z}z}) + 3\delta(\pi^{EE}_{z\bar{z}} - \pi^{SS}_{z\bar{z}})\}}{1 - \delta(\pi^{EE}_{\bar{z}\bar{z}} - \pi^{SS}_{\bar{z}\bar{z}})}. \tag{12.3.3}$$

Suppose the private monitoring distribution is given by $\pi^{EE}_{\bar{z}\bar{z}} = p(1-2\varepsilon)$, $\pi^{SS}_{zz} = q(1-2\varepsilon)$, and $\pi^{SS}_{\bar{z}z} = \pi^{EE}_{\bar{z}z} = \pi^{SS}_{z\bar{z}} = \pi^{EE}_{z\bar{z}} = \varepsilon$ (this π is 2ε-close to ρ, see section 12.1). Then (12.3.2) simplifies to

$$\beta\delta(p-q)(1-2\varepsilon)(V_1^{\bar{z}\bar{z}} - V_1^{zz}) \geq (1-\delta),$$

and (12.3.3) simplifies to

$$V_1^{\bar{z}\bar{z}} - V_1^{zz} = \frac{2(1-\delta)}{1 - \delta(p-q)(1-2\varepsilon)},$$

so that the incentive constraint for effort following signal $\bar{z}$ is given by

$$(1 + 2\beta)(1 - 2\varepsilon) \geq \frac{1}{\delta(p-q)}.$$

A similar calculation for the incentive constraint describing behavior after $\underline{z}$ yields the inequality

$$\delta(p-q)(1-2\varepsilon)(3-2\zeta) \leq 1,$$

where $\zeta = \Pr(z_2 = \underline{z} \mid z_1 = \underline{z})$.

Finally, if the action a has been chosen in the previous period (using the assumption on the parametric form of π in the second line and $p > q$ in the third),

$$\beta = \Pr(z_2 = \bar{z} \mid z_1 = \bar{z}, a) = \frac{\Pr(z_2 = \bar{z}, z_1 = \bar{z} \mid a)}{\Pr(z_1 = \bar{z} \mid a)}$$

$$= \begin{cases} \frac{p(1-2\varepsilon)}{p(1-2\varepsilon)+\varepsilon}, & \text{if } a = EE, \\ \frac{q(1-2\varepsilon)}{q(1-2\varepsilon)+\varepsilon}, & \text{if } a \neq EE, \end{cases}$$

$$\geq \frac{q(1-2\varepsilon)}{q(1-2\varepsilon)+\varepsilon}.$$

Hence, β can be made uniformly close to 1 by choosing ε small (with a similar argument for ζ). Because $\underline{\delta}$ satisfies (12.3.1), there is an $\bar{\varepsilon}$ such that these incentive constraints hold, and applying proposition 12.2.3, we have a sequential equilibrium, for $\delta > \underline{\delta}$ and $\varepsilon < \bar{\varepsilon}$.

We study public profiles with bounded recall in more detail in section 13.3.

12.3.2 Grim Trigger

We now turn to the grim trigger profile studied in section 7.2.1: $\mathscr{W} = \{w_{EE}, w_{SS}\}$, $w^0 = w_{EE}$, $f_i(w_{EE}) = E$, $f_i(w_{SS}) = S$, and

$$\tau(w, y) = \begin{cases} w_{EE}, & \text{if } w = w_{EE} \text{ and } y = \bar{y}, \\ w_{SS}, & \text{otherwise.} \end{cases}$$

We assume grim trigger is a strict PPE of the public monitoring game (i.e., (7.2.4) holds).

In the game with private monitoring, the implied private profile has player i playing E in the first period, continuing to play E as long as $\bar{z}$ is observed and switching to permanent S after the first $\underline{z}$. We let w_E and w_S denote the states in the implied private automaton. Behavior in grim trigger potentially depends on signals that occur arbitrarily far in the past, so the situation is very different from that of the forgiving profile.

We say private monitoring has *full support* if $\pi(\mathbf{z} \mid a) > 0$ for all $\mathbf{z} \in \{\underline{z}, \bar{z}\}^2$ and $a \in A$. For games with full-support private monitoring, if $q > r$, the implied private profile is never a Nash equilibrium, even for π arbitrarily close to ρ.

Proposition 12.3.1
1. *If $q > r$, there exists $\underline{\varepsilon} > 0$ such that for all $\varepsilon \in (0, \underline{\varepsilon})$, the private profile implied by grim trigger is not a Nash equilibrium in the repeated prisoners' dilemma with full-support private monitoring ε-close to ρ.*
2. *If $q < r$,[8] there exists $\underline{\varepsilon} > 0$ such that for all $\varepsilon \in (0, \underline{\varepsilon})$, the private profile implied by grim trigger is a Nash equilibrium in the repeated prisoners' dilemma with full-support private monitoring ε-close to ρ.*

We begin with an initial observation. Because grim trigger is a strict PPE of the public monitoring game, if player 1 (say) is sufficiently confident that player 2's current

---

[8]. If $q = r$, the details of the private monitoring distribution determine whether the private profile is an equilibrium.

12.3 ■ Almost Public Monitoring

private state is w_E and the private monitoring is ε-close to ρ with ε sufficiently close to 0, then E is the only optimal choice for 1 in the current period. Similarly, if 1 is sufficiently confident that 2's current state is w_S and ε is sufficiently small, then S is the only optimal choice for 1 in the current period.[9] Therefore, if under grim trigger each player always assigns arbitrarily high probability to the other player being in the same state as himself, then grim trigger is a Nash equilibrium. This is an example of a more general phenomenon: Though in general there is no one-shot deviation principle for Nash equilibrium (recall the discussion just before example 2.2.3), profiles in almost public monitoring games induced by strict PPE inherit a modified version of the no one-shot deviation principle. See the proof of proposition 13.2.2.

We prove proposition 12.3.1 through two lemmas. The first lemma identifies the critical deviation that precludes grim trigger from being Nash when $q > r$.

Lemma 12.3.1 *Suppose $q > r$. There exists $\bar{\varepsilon} > 0$, such that if π is ε-close to ρ for $\varepsilon \in (0, \bar{\varepsilon})$, then it is not optimal for player 1 to play S following a sufficiently long history $(E\underline{z}, S\bar{z}, S\bar{z}, S\bar{z}, \ldots)$, as required by grim trigger.*

For π close to ρ, immediately following the signal $\underline{z}$, player 1 assigns a probability very close to 0 to player 2 being in the private state w_E (because with probability close to 1, player 2 also observed the signal $\underline{z}$). Thus, playing S in the subsequent period is optimal. However, because π has full support, player 1 is *not sure* that player 2 is in state w_S, and observing the signal $\bar{z}$ after playing S is an indication that player 2 had played E (recall that $\rho(\bar{z} \mid SE) = q > r = \rho(\bar{z} \mid SS)$). This makes player 1 *less* sure that 2 was in state w_S. Furthermore, if player 2 was in state w_E and observes $\bar{z}$, then 2 will still be in state w_E. Eventually, player 1 believes that player 2 is almost certainly in state w_E and so will have an incentive to exert effort.

Proof Suppose player 1 initially assigns prior probability η to player 2 being in state w_E. Write $\varphi^\pi(\eta \mid a_1 z_1)$ for the posterior probability that he assigns to player 2 being in state w_E one period later, if he chooses action a_1 and observes the private signal z_1, believing that his opponent is following grim trigger. Then,

$$\varphi^\pi(\eta \mid S\bar{z}) = \frac{\pi(\bar{z}\bar{z} \mid SE)\eta}{\{\pi(\bar{z}\bar{z} \mid SE) + \pi(\bar{z}\underline{z} \mid SE)\}\eta + \{\pi(\bar{z}\bar{z} \mid SS) + \pi(\bar{z}\underline{z} \mid SS)\}(1-\eta)}.$$

If π is ε-close to ρ (using $\pi(\bar{z}\bar{z} \mid a) + \pi(\bar{z}\underline{z} \mid a) = 1 - \pi(\underline{z}\underline{z} \mid a) - \pi(\underline{z}\bar{z} \mid a) < 1 - \pi(\underline{z}\underline{z} \mid a)$),

$$\varphi^\pi(\eta \mid S\bar{z}) > \frac{(q-\varepsilon)\eta}{(q+\varepsilon)\eta + (r+\varepsilon)(1-\eta)} \equiv \underline{\varphi}^\pi(\eta \mid S\bar{z}).$$

The function $\underline{\varphi}^\pi(\eta \mid S\bar{z})$ has a fixed point at $\eta_\varepsilon = (q - r - 2\varepsilon)/(q-r)$, with $\underline{\varphi}^\pi(\eta \mid S\bar{z}) \in (\eta, \eta_\varepsilon)$ if $\eta \in (0, \eta_\varepsilon)$ and $\underline{\varphi}^\pi(\eta \mid S\bar{z}) \in (\eta_\varepsilon, \eta)$ if $\eta \in (\eta_\varepsilon, 1]$ and with $\lim_{\varepsilon \to 0} \eta_\varepsilon = 1$. For any $\bar{\eta} \in (0, 1)$, we can choose ε sufficiently small that

---

9. Because of discounting, the continuation payoffs from playing E or S when 1 assigns probability 1 to player 2 being initially in a state can be made arbitrarily close to their public monitoring values by making ε sufficiently small. The strictness of the grim trigger PPE carries over to the private monitoring game. We can then deduce the strict optimality of E (respectively, S) when player 1 is sufficiently confident that player 2's current private state is w_E (resp., w_S). Lemma 13.2.3 verifies this intuition.

$\eta_\varepsilon > \bar{\eta}$. Then for any $\eta \in (0, \bar{\eta}]$, there exists a k such that after observing k signals $\bar{z}$, the posterior must exceed $\bar{\eta}$ (k goes to infinity as η becomes small). This implies that after a sufficiently long history ($E\underline{z}, S\bar{z}, S\bar{z}, S\bar{z}, \ldots$), player 1 eventually becomes very confident that player 2 is in fact in state w_E, and so (for sufficiently large $\bar{\eta}$) no longer has an incentive to play S. Because ($E\underline{z}, S\bar{z}, S\bar{z}, S\bar{z}, \ldots$) is a history that occurs with positive probability under grim trigger, grim trigger is not a Nash equilibrium.   ■

Lemma 12.3.2 *Suppose $q < r$. There exists $\bar{\varepsilon} > 0$, such that if π is ε-close to ρ for $\varepsilon \in (0, \bar{\varepsilon})$, then grim trigger is a Nash equilibrium of the game with full-support private monitoring.*

In this case, unlike the preceding, a signal of $\bar{z}$ reinforces the belief of player 1 (in state w_S) that the player 2 had also chosen S and so also is in state w_S. A signal of $\underline{z}$, on the other hand, now is a signal that player 2 chose E in the previous period (and so was in state w_E in the previous period). But for π close to ρ, player 1 also assigns high probability to 2 observing $\underline{z}$ and so transiting to w_S this period.

Proof For simplicity, we focus on player 1. Given any $\xi > 0$, we show that we can choose $\varepsilon > 0$ such that for every history *reached with positive probability* under grim trigger, if player 1 has observed $\underline{z}$ at least once, he assigns probability less than ξ to player 2 being in state w_E; if he has always observed $\bar{z}$, he assigns probability at least $1 - \xi$ to player 2 being in state w_E. By choosing ξ and ε sufficiently small, from the initial observation (given after the statement of proposition 12.3.1), grim trigger will then be a Nash equilibrium (and the induced outcome is a sequential equilibrium outcome).

Consider histories in which at least one $\underline{z}$ has been observed. The easy case is a history in which $\underline{z}$ was observed in the last period. Then, as we argued just before the proof, there exists ε' such that for $\varepsilon < \varepsilon'$, immediately following such a signal, player 1 assigns a probability of at least $1 - \xi$ that 2 also observed $\underline{z}$ (and so will be in state w_S).

We now turn to histories in which $\underline{z}$ has been observed and $\bar{z}$ was observed in the last period. An application of Bayes' rule shows that after observing $\underline{z}$, player 1 can attach probability at most $2\varepsilon'/(1 - p - \varepsilon') \equiv \bar{\eta}$ to player 2's being in state w_E. Intuitively, observing $\bar{z}$ after playing S is a further indication that player 2 is in state w_S (this requires $q < r$). Because

$$\varphi^\pi(\eta \mid S\bar{z}) < \frac{(q + \varepsilon)\eta}{(q - \varepsilon)\eta + (r - \varepsilon)(1 - \eta)},$$

there exists $\bar{\varepsilon} \in (0, \varepsilon')$ such that for all π that are ε-close to ρ for $\varepsilon < \bar{\varepsilon}$, $\varphi^\pi(\eta \mid S\bar{z}) < \eta$ for all $\eta \in (0, \bar{\eta})$ (set $\bar{\varepsilon} < (1 - \bar{\eta})(r - q)/2$). So, irrespective of the private signal a player observes, along every play path the player becomes increasingly confident that his opponent is in state w_S.

Finally, consider histories of the form ($E\bar{z}, E\bar{z}, E\bar{z}, E\bar{z}, \ldots$). At first glance, it would seem that after a sufficiently long history of $\bar{z}$, a player should believe that the partner has surely already observed at least *one* $\underline{z}$ and so i will wish to

12.3 ■ Almost Public Monitoring

preemptively switch to S. However, this ignores the role that recent $\bar{z}$ play in reassuring i that j had not earlier switched to S.

The argument for this case is quite similar to that used in proving lemma 12.3.1. The posterior on the relevant histories satisfies

$$\varphi^\pi(\eta \mid E\bar{z}) > \frac{(p-\varepsilon)\eta}{(p+\varepsilon)\eta + (q+\varepsilon)(1-\eta)} \equiv \underline{\varphi}^\pi(\eta \mid E\bar{z}). \quad (12.3.4)$$

The function $\underline{\varphi}^\pi(\eta \mid E\bar{z})$ has a fixed point at $\eta_\varepsilon = (p - q - 2\varepsilon)/(p - q)$, with $\underline{\varphi}^\pi(\eta \mid E\bar{z}) \in (\eta, \eta_\varepsilon)$ if $\eta \in (0, \eta_\varepsilon)$ and $\underline{\varphi}^\pi(\eta \mid E\bar{z}) \in (\eta_\varepsilon, \eta)$ if $\eta \in (\eta_\varepsilon, 1]$, and with $\lim_{\varepsilon \to 0} \eta_\varepsilon = 1$. The prior probability attached to player 2 being in state w_E is 1. The posterior $\varphi^\pi(\eta \mid E\bar{z})$ can fall as the history in question proceeds but can never fall below the fixed point η_ε. Choosing ε sufficiently small, and hence η_ε sufficiently close to 1, ensures that player 1 finds it optimal to play E.

■

Proposition 12.3.2 *There exists $\underline{\varepsilon} > 0$ such that for all $\varepsilon \in (0, \underline{\varepsilon})$, grim trigger is not a sequential equilibrium in the repeated prisoners' dilemma with full-support private monitoring ε-close to ρ.*

Proof For the case $q > r$, proposition 12.3.1 implies that grim trigger is not a Nash equilibrium, and so clearly cannot be a sequential equilibrium. For $q \leq r$, we now argue that it is not sequentially rational to follow grim trigger's specification of S after long private histories of the form $(E\underline{z}, E\bar{z}, E\bar{z}, E\bar{z}, \ldots)$. For π ε-close to ρ, player 1's posterior satisfies $\varphi^\pi(\eta \mid E\bar{z}) > \underline{\varphi}^\pi(\eta \mid E\bar{z})$, where $\underline{\varphi}^\pi(\eta \mid E\bar{z})$ is defined in (12.3.4). Hence, by the argument following (12.3.4), after a sufficiently long history $(E\underline{z}, E\bar{z}, E\bar{z}, E\bar{z}, \ldots)$, player 1 eventually becomes very confident that player 2 is in fact still in state w_E, and so 1 will find it profitable to deviate and play E. Because grim trigger specifies S after such a history, grim trigger is not a sequential equilibrium.

■

Propositions 12.3.1 and 12.3.2 illustrate an important feature of private monitoring games (mentioned in section 12.2). Even if opponents are playing action-free strategies, a player's sequentially rational best reply may necessarily not be action-free. Depending on player 1's actions, the same signal history $(\underline{z}, \bar{z}, \bar{z}, \bar{z}, \ldots)$ for player 1 implies very different beliefs over player 2's private state at the beginning of the next period. As we saw in the proof of lemma 12.3.2, for $q < r$ this signal history, when coupled with the actions specified by grim trigger, leads to beliefs that assign probability close to 1 that 2 will be in state w_S. On the other hand, if 1 deviates and always plays E, this signal history leads to beliefs that assign probability close to 1 that 2 will be in state w_E. That is, the interpretation of the signal is affected by the action of player 1 that generated that signal. Because $p > q$, when $a_1 = E$, $\bar{z}$ is a signal that $a_2 = E$, whereas if $q < r$, then $a_1 = S$ implies $\bar{z}$ is a signal that $a_2 = S$. In the former case, the continued play of S is not a best response, precluding the sequentiality of grim trigger. Nonetheless, if $q < r$, propositions 12.2.1 and 12.3.1 imply that the outcome produced by grim trigger is a sequential equilibrium outcome.

Remark **Forgiving grim triggers** As we discussed in section 7.2.1, as players become
12.3.1 patient, the payoffs from grim trigger converge to (0, 0). Section 7.7.1 studied grim triggers in the public monitoring repeated prisoners' dilemma that were more forgiving (i.e., profiles in which the specification of SS is absorbing, but the first realization of $\underline{z}$ need not trigger SS).[10] We analyze the implications of minimally private monitoring on two examples of such profiles in examples 13.4.2 and 13.4.3.

◆

12.4 Independent Monitoring: A Belief-Based Equilibrium for the Infinitely Repeated Prisoners' Dilemma

Building on the ideas of section 12.1.2, Sekiguchi (1997) constructs a belief-based equilibrium for some infinitely repeated prisoners' dilemmas that is arbitrarily efficient, providing the first example of efficiency in private monitoring repeated games.[11]

We again work with the prisoners' dilemma of figure 12.1.1. This game is infinitely repeated, and payoffs are discounted at rate δ. At the end of each period, each player observes a conditionally independent private signal about his partner's action in that period, as described in (12.1.3) and (12.1.4). Recall that such monitoring is ε-perfect (definition 12.2.1). Although some detailed calculations exploit the conditional independence of the private monitoring, the following result (and method of proof) actually covers any private monitoring structure sufficiently close to perfect monitoring (see Sekiguchi 1997 for details).

Proposition *For all $\zeta > 0$, there exists $\varepsilon > 0$ and $\underline{\delta} < 1$ such that for all $\delta \in (\underline{\delta}, 1)$, if the con-*
12.4.1 *ditionally independent private monitoring is ε-perfect, then there is a sequential equilibrium in which each player's average payoffs are at least $2 - \zeta$.*

The remainder of this section proves the proposition. We construct a Nash equilibrium, at which point proposition 12.2.1 delivers the result.[12]

The Nash equilibrium uses the following strategies as building blocks.[13] Let σ_i^S denote the strategy of always shirking,

$$\sigma_i^S(h_i^t) = S \ \forall h_i^t,$$

and σ_i^T denote the trigger strategy

$$\sigma_i^T(h_i^t) = \begin{cases} E, & \text{if } t = 0, \text{ or } z_i^s = \bar{z} \text{ for } 0 \leq s \leq t-1, \\ S, & \text{otherwise.} \end{cases}$$

---

10. This is the class of profiles studied by Compte (2002) for the conditionally independent private monitoring case.
11. Sekiguchi (1997) imposed a restriction on the payoffs of the stage game that is satisfied in our example—see note 16 on page 407.
12. See Bhaskar and Obara (2002) for an explicit description of the strategies in the sequential equilibrium.
13. The strategy σ_i^T is the action-free strategy realization equivalent to the grim-trigger strategy that Sekiguchi (1997) uses.

12.4 ■ Independent Monitoring

Clearly, (σ_1^S, σ_2^S) is a Nash equilibrium of the game with private monitoring, because history is ignored and the static Nash equilibrium of the stage game is played in each period.

On the other hand, (σ_1^T, σ_2^T) is not a Nash equilibrium of the private monitoring game. After observing $\underline{z}$ in the first period, player i is supposed to play S in every future period. But player i believes that player j played E and observed $\bar{z}$ with probability $1 - \varepsilon$, and so is still exerting effort with probability $1 - \varepsilon$.[14] By switching to S, player i will trigger shirking by j in the next period with probability $1 - \varepsilon$. By playing E, i maximizes the probability that player j continues to exert effort. For patient players and low ε, the deviation is profitable.

As in section 12.1.2, randomization will be used to allow behavior to depend on private histories. The equilibrium described in section 12.1.2 can be interpreted as the reduced form of such a mixed equilibrium, where G is σ_i^T and B is σ_i^S. In particular, each player will randomize between σ_i^T and σ_i^S. We first calculate the payoffs from different strategy profiles. The payoff to a player from the profile (σ_1^S, σ_2^S) is clearly 0. The payoff to player 1 from (σ_1^S, σ_2^T), v_1^{ST}, solves

$$v_1^{ST} = (1-\delta)3 + \delta\{(1-\varepsilon) \times 0 + \varepsilon v_1^{ST}\},$$

and so

$$v_1^{ST} = \frac{3(1-\delta)}{1-\varepsilon\delta}.$$

The payoff to 2 from (σ_1^S, σ_2^T), v_2^{ST}, solves

$$v_2^{ST} = (1-\delta)(-1) + \delta\varepsilon v_2^{ST},$$

and so

$$v_2^{ST} = \frac{-(1-\delta)}{1-\varepsilon\delta}.$$

Finally, the payoff to 1 (and 2) from (σ_1^T, σ_2^T), v_1^{TT}, solves

$$v_1^{TT} = (1-\delta)2 + \delta\{(1-\varepsilon)^2 v_1^{TT} + \varepsilon(1-\varepsilon)(v_1^{ST} + v_1^{TS})\},$$

and so

$$v_1^{TT} = \frac{2(1-\varepsilon^2\delta)(1-\delta)}{(1-\delta(1-\varepsilon)^2)(1-\varepsilon\delta)}.$$

For $\delta > 1/3$ and ε small, $v_1^{TT} > v_1^{ST}$, and we always have $v_1^{ST} > 0 > v_1^{TS}$. Hence the players are essentially playing the coordination game in figure 12.4.1 when they choose between σ_i^T and σ_i^S.[15] In addition to the two strict equilibria of this coordination game, (σ_1^S, σ_2^S) and (σ_1^T, σ_2^T), there is a mixed strategy equilibrium, with probability

---

14. Note that this is essentially the argument from the beginning of section 12.1.2 showing that there is no pure strategy equilibrium with EE in the first period of the two-period example.
15. This coordination game only captures first-period incentives. As we have argued, the pure profile (σ_1^T, σ_2^T) is not a Nash equilibrium of the repeated game, because some incentive constraints after the first period are violated.

Chapter 12 ■ Private Monitoring

	σ_2^T	σ_2^S
σ_1^T	v_1^{TT}, v_2^{TT}	v_1^{TS}, v_2^{TS}
σ_1^S	v_1^{ST}, v_2^{ST}	$0, 0$

Figure 12.4.1 The coordination game.

$$\xi(\varepsilon, \delta) = \frac{1 - \delta(1-\varepsilon)^2}{2\delta(1-2\varepsilon)} \tag{12.4.1}$$

on σ_i^T and $(1 - \xi(\varepsilon, \delta))$ on σ_i^S. Note that the randomization only occurs in period 1.

The following lemma is key in the construction. Unlike the final equilibrium, this asserts the existence of an approximately efficient equilibrium for a very limited range of discount factors.

Lemma 12.4.1 *For all $\xi' < 1$, there exist constants $\eta'' > \eta' > 0$ and $\varepsilon > 0$ such that for all $\delta \in (1/3 + \eta', 1/3 + \eta'')$, there is a Nash equilibrium of the private monitoring game in which player i plays σ_i^T with probability $\xi(\varepsilon, \delta) > \xi'$ and σ_i^S with probability $(1 - \xi(\varepsilon, \delta))$.*

Proof Player i's candidate equilibrium strategy is $\xi(\varepsilon, \delta) \circ \sigma_i^T + (1 - \xi(\varepsilon, \delta))\sigma_i^S$. The lower bound $\delta = 1/3$ is the critical discount factor at which a player is indifferent between σ_i^T and σ_i^S when the partner is playing σ_j^T for sure and $\varepsilon = 0$ (i.e., from (12.4.1), $\xi(0, 1/3) = 1$). Because ξ is a continuous function, for ε small and δ close to $1/3$, $\xi(\varepsilon, \delta)$ is close to 1 (in particular, larger than ξ'). Because ξ is decreasing in δ and increasing in ε, the lower bound η' ensures that ξ is bounded away from 1 for all ε small, so a bound on ε can be chosen independently of δ (the discussion at the end of section 12.1.2 on the order of limits applies here as well). That is, for η', η'', and $\tilde{\varepsilon}$ sufficiently small, if $\delta \in (1/3 + \eta', 1/3 + \eta'')$ then $\xi' < \xi(0, 1/3 + \eta'') < \xi(\varepsilon, \delta) < \xi(\tilde{\varepsilon}, 1/3 + \eta') < 1$ for all $0 < \varepsilon < \tilde{\varepsilon}$.

After the first period, it is best to think of the strategy of player i as being described by a two-state automaton (we suppress the player index because the description of the automaton is the same for both players), with state space $\{w_E, w_S\}$, output function $f : \{w_E, w_S\} \to \{E, S\}$ given by $f(w_a) = a$ for $a \in \{E, S\}$, and transition function

$$\tau(w, z) = \begin{cases} w_E, & \text{if } w = w_E \text{ and } z = \bar{z}, \\ w_S, & \text{otherwise.} \end{cases}$$

The strategy σ_i^T has initial state w_E, and σ_i^S has initial state w_S.

Under the profile of the proposition, the initial state of each player is randomly determined, with an initial probability of ξ on w_E. As each player accumulates his private histories, he updates his beliefs about the current *private* state of his

12.4 ■ Independent Monitoring

partner. We denote the probability that player i assigns to player j being in private state w_E in period t by ϕ_i^t. Thus $\phi_i^0 = \xi$.

The probability that player i assigns to player j being in state w_E last period, given a belief ϕ_i last period and after the signal $\bar{z}$ is

$$\frac{(1-\varepsilon)\phi_i}{(1-\varepsilon)\phi_i + \varepsilon(1-\phi_i)},$$

whereas after the signal $\underline{z}$,

$$\frac{\varepsilon\phi_i}{\varepsilon\phi_i + (1-\varepsilon)(1-\phi_i)}.$$

Player j is still in the state w_E this period if he was in that state last period and received the signal $\bar{z}$, which depends on the behavior of i last period, so i's belief about j's state this period is, if i observed $\bar{z}$ and chose E,

$$\varphi_i(\phi_i \mid E\bar{z}) = (1-\varepsilon)\frac{(1-\varepsilon)\phi_i}{(1-\varepsilon)\phi_i + \varepsilon(1-\phi_i)}, \tag{12.4.2}$$

and if i chose S,

$$\varphi_i(\phi_i \mid S\bar{z}) = \varepsilon\frac{(1-\varepsilon)\phi_i}{(1-\varepsilon)\phi_i + \varepsilon(1-\phi_i)}.$$

Before we check for sequential rationality, we show that for $\delta < 2/5$, after any private history that leads to player i having a belief satisfying $\phi_i^t \leq 1/2$, player i's best reply in that period is to play S.[16] The payoff to player i from playing S in the current period is bounded below by (because i can guarantee a continuation value of at least 0 by always playing S)

$$\underline{u}_i(S) \equiv \phi 3(1-\delta),$$

and the payoff from playing E is bounded above by (using the bounds of 3 if j is always playing E—even if i plays S in the future—and 0 from i detecting for sure that j is playing σ_j^S)

$$\bar{u}_i(E) \equiv \phi(2(1-\delta) + 3\delta) + (1-\phi)(-(1-\delta)).$$

If player i is sufficiently impatient ($\delta < 2/5$) and sufficiently pessimistic ($\phi \leq 1/2$), $\underline{u}_i(S) > \bar{u}_i(E)$. If players are patient, on the other hand, even a large probability that the partner is already in w_S may not be enough to ensure that the player shirks. One more observation before the player commits himself may be quite valuable, and the cost of inadvertently triggering the partner's transition from w_E to w_S too costly.

---

16. Though this upper bound on δ is larger than the lower bound for (σ_1^T, σ_2^T) to be an equilibrium of the game in figure 12.4.1 (1/3 for this parameterization of the prisoners' dilemma), this is not true for all prisoners' dilemmas. The argument presented here requires the upper bound exceed the lower bound. Bhaskar and Obara (2002) show that the restriction on the class of prisoners' dilemmas can be dropped.

There are four classes of histories that arise with positive probability under the profile, to check for sequential rationality when $\delta < 2/5$:

1. ($E\underline{z}$). Will player i, after starting with grim trigger, immediately switch to always shirk after observing $\underline{z}$ in the first period? After seeing $\underline{z}$, player i assigns probability of at least $1/2$ to j having played S in the initial period (and so being in private state w_S) if ε is taken to be less than $1 - \xi$ (from our discussion at the beginning of the proof, ξ is bounded away from 1 for ε small). In that case, by the above observation, i does indeed find it optimal to play S. Note, moreover, that by so playing, in the next period i believes j will observe $\underline{z}$ with high probability and so, even if j's private state had been w_E, he will transit to w_S.

2. ($E\bar{z}, E\bar{z}, \ldots, E\bar{z}$). If a player has been exerting effort for a long time and has always observed $\bar{z}$, will i continue to exert effort? The intuition for this case matches that encountered in proving lemma 12.3.2. From (12.4.2),

$$\phi' \equiv \frac{1 - 2\varepsilon - \varepsilon(1 - \varepsilon)}{1 - 2\varepsilon}$$

is a fixed point of $\varphi_i(\cdot \mid E\bar{z})$. For $0 < \phi_i < \phi'$ (if $\varepsilon < 1/2$), $\phi_i < \varphi_i(\phi_i \mid E\bar{z}) < \phi'$, and for $\phi' < \phi_i \leq 1$, $\phi' < \varphi_i(\phi_i \mid E\bar{z}) < \phi_i$. For $\xi(\varepsilon, \delta) > \phi'$, player i's beliefs fall over time in response to the positive signal due to the current imperfections in the monitoring. However, beliefs are bounded below by ϕ', which is close to 1 for ε small. Taking ε small enough that $\min\{\phi', \xi(\varepsilon, \delta)\}$ is sufficiently close to 1, E will be optimal after any history $(E\bar{z}, E\bar{z}, \ldots, E\bar{z})$.

3. ($E\bar{z}, E\bar{z}, \ldots, E\bar{z}, E\underline{z}$). Will a player shirk as soon as $\underline{z}$ is observed, after having observed $\bar{z}$? To clarify the nature of the argument, we first suppose player i observes $(E\bar{z}, E\underline{z}) = (E_i^0 \bar{z}_i^0, E_i^1 \underline{z}_i^1)$. There are two possibilities: Player j is still in w_E but i received an erroneous signal, and player j received an erroneous signal in the previous period (in which case j is now in w_S). Intuitively, these two events are of equal probability because they each involve one erroneous signal, and so i assigns probability approximately $1/2$ to j still being in state w_E. More formally, considering the most optimistic case, where player j is in state w_E in period 0,

$$\begin{aligned}
\Pr\bigl(w_j^2 = w_E \mid E_i^0 \bar{z}_i^0, E_i^1 \underline{z}_i^1\bigr) &= \Pr\bigl(\bar{z}_j^0 \bar{z}_j^1 \mid E_i^0 \bar{z}_i^0, E_i^1 \underline{z}_i^1\bigr) \\
&= \Pr\bigl(\bar{z}_j^0 \mid E_i^0 \bar{z}_i^0, E_i^1 \underline{z}_i^1; \bar{z}_j^1\bigr) \Pr\bigl(\bar{z}_j^1 \mid E_i^0 \bar{z}_i^0, E_i^1 \underline{z}_i^1\bigr) \\
&= \Pr\bigl(\bar{z}_j^0 \mid E_i^0, \underline{z}_i^1\bigr) \Pr\bigl(\bar{z}_j^1 \mid E_i^1\bigr) \\
&= \Pr\bigl(\bar{z}_j^0 \mid E_i^0, \underline{z}_i^1\bigr)(1 - \varepsilon),
\end{aligned}$$

where the penultimate equality follows from the conditional independence of the monitoring and the last from (12.1.4). Now,

$$\Pr\bigl(\bar{z}_j^0 \mid E_i^0, \underline{z}_i^1\bigr) = \frac{\Pr\bigl(\bar{z}_j^0, \underline{z}_i^1 \mid E_i^0\bigr)}{\Pr\bigl(\underline{z}_i^1 \mid E_i^0\bigr)}$$

12.4 ■ Independent Monitoring

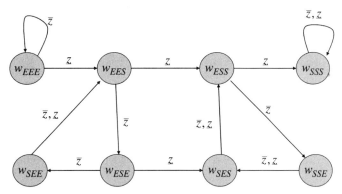

Figure 12.4.2 The automaton representation for the case $N = 3$. The decision rule is $f_i(w_{abc}) = a$. The initial distribution μ over $\mathcal{W} = \{w_{EEE}, w_{EES}, w_{ESE}, w_{SEE}, w_{ESS}, w_{SSE}, w_{SES}, w_{SSS}\}$ is given by $\mu(w_{abc}) = \xi^\ell (1-\xi)^{3-\ell}$, where ℓ is the number of times E appears in abc.

$$= \frac{\Pr(z_i^1 \mid E_i^0, \bar{z}_j^0)\Pr(\bar{z}_j^0 \mid E_i^0)}{\Pr(z_i^1 \mid E_i^0, \bar{z}_j^0)\Pr(\bar{z}_j^0 \mid E_i^0) + \Pr(z_i^1 \mid E_i^0, z_j^0)\Pr(z_j^0 \mid E_i^0)}$$

$$= \frac{\Pr(z_i^1 \mid E_j^1)\Pr(\bar{z}_j^0 \mid E_i^0)}{\Pr(z_i^1 \mid E_j^1)\Pr(\bar{z}_j^0 \mid E_i^0) + \Pr(z_i^1 \mid S_j^1)\Pr(z_j^0 \mid E_i^0)} = \frac{\varepsilon(1-\varepsilon)}{2\varepsilon(1-\varepsilon)} = \frac{1}{2},$$

where the antepenultimate equality follows from the structure of the profile. Thus i assigns probability less than $1/2$ to player j being in w_E after $(E\bar{z}, Ez)$, so i does indeed find it optimal to play S. For longer histories of the form $(E\bar{z}, E\bar{z}, \ldots, E\bar{z}, Ez)$, player i can only be more pessimistic about the current state of j (as an earlier transition to w_S may have occurred), and so i again finds it optimal to play S.

4. $(E\bar{z}, \ldots, E\bar{z}, Ez, S\bar{z}, \ldots, S\bar{z})$. Will a player continue to shirk, when the player is already in the private state w_S and the player observes the signal $\bar{z}$ (suggesting that in fact the partner is not in w_S)? The player still finds it optimal to shirk in the future, because he believes that his choice of shirk *this* period triggers a transition to the shirk private state by his partner with probability of at least $1 - \varepsilon$. ■

It remains to construct an equilibrium for large δ. Following Ellison (1994), the repeated game is divided into N distinct "games," with game k consisting of periods $k + tN$, where $t \in \mathbb{N}_0 \equiv \mathbb{N} \cup \{0\}$. This gives an effective discount rate of δ^N on each game. The profile consists of playing $\xi(\varepsilon, \delta^N) \circ \sigma_i^T + (1 - \xi(\varepsilon, \delta^N)) \circ \sigma_i^S$ on each of the N games. The automaton representation for this profile for $N = 3$ is illustrated in figure 12.4.2. Setting

$$\underline{\delta} \equiv \frac{1 + 3\eta'}{1 + 3\eta''},$$

for every $\delta > \underline{\delta}$, there is an N such that[17]

---

17. The desired value of N satisfies $\delta^N > 1/3 + \eta' > \delta^{(N+1)}$.

$$\frac{1}{3} + \eta' < \delta^N < \frac{1}{3} + \eta''.$$

By lemma 12.4.1, the result is a Nash equilibrium of the original game. Note that the lower bound $\underline{\delta}$ may be significantly larger than $1/3 + \eta''$.

Remark 12.4.1 **Public correlation** Fix $\xi' < 1$ and the implied η', η'', and ε from lemma 12.4.1. As Bhaskar and Obara (2002) emphasize, public correlation can lower the lower bound $\underline{\delta}$ on the discount factor for the purposes of constructing these equilibria. More precisely, for any $\delta \in (1/3 + \eta', 1/3 + \eta'')$, if players have access to a public correlating device, we can extend the equilibrium of lemma 12.4.1 to all $\delta' > \delta$ as follows. For any such δ', let $\theta = \delta/\delta' < 1$ and consider the profile that begins with $\xi(\varepsilon, \delta) \circ \sigma_i^T + (1 - \xi(\varepsilon, \delta)) \circ \sigma_i^S$, $i = 1, 2$, and, at the beginning of each period, conditions on a public randomization. With probability θ play continues under the existing specification, and with probability $1 - \theta$, each player begins again with $\xi(\varepsilon, \delta) \circ \sigma_i^T + (1 - \xi(\varepsilon, \delta)) \circ \sigma_i^S$. The effective discount rate is $\theta \delta' = \delta$, and so the profile is an equilibrium.

◆

12.5 A Belief-Free Example

The product choice game, which we analyzed in section 7.6, provides a convenient context to illustrate *belief-free equilibria* (previewed in section 12.1.3 and the subject of chapter 14). The stage game payoffs are reproduced in figure 12.5.1. Recall, from section 7.6.1, that for $\delta \geq 1/2$, the following profile is a subgame-perfect equilibrium of the repeated game with perfect monitoring: There are two states, $\mathscr{W} = \{w^L, w^H\}$, with initial state $w^0 = w^H$, output functions,

$$f_1(w) = \frac{1}{2} \circ H + \frac{1}{2} \circ L, \quad \text{and}$$

$$f_2(w) = \begin{cases} \alpha^h(H) \circ h + (1 - \alpha^h(H)) \circ \ell, & \text{if } w = w^H, \\ \alpha^h(L) \circ h + (1 - \alpha^h(L)) \circ \ell, & \text{if } w = w^L, \end{cases}$$

where $\alpha^h(H) = \alpha^h(L) + 1/(2\delta)$ and transitions,

$$\tau(w, a) = \begin{cases} w^H, & \text{if } a_1 = H, \\ w^L, & \text{if } a_1 = L. \end{cases}$$

	h	ℓ
H	2, 3	0, 2
L	3, 0	1, 1

Figure 12.5.1 The product choice game from section 7.6. Player 1 is long-lived, and player 2 is short-lived.

12.5 ■ A Belief-Free Example

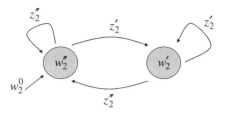

Figure 12.5.2 An automaton representation of player 2's strategy.

The profile is illustrated in figure 7.6.2, where the states are labeled γ^L and γ^H rather than w^L and w^H. Moreover, when $\alpha^h(H) = 1$, this profile yields the maximum equilibrium payoff of 2 for player 1.

Because player 1's behavior is independent of history, we can construct a similar equilibrium in the game with private monitoring. For our example, we consider conditionally independent private monitoring. Player i has private signal space $Z_i = \{z_i', z_i''\}$, with marginal distributions

$$\pi_1(z_1 \mid a) = \begin{cases} 1 - \varepsilon, & \text{if } z_1 = z_1' \text{ and } a_2 = \ell, \text{ or} \\ & z_1 = z_1'' \text{ and } a_2 = h, \\ \varepsilon, & \text{otherwise,} \end{cases}$$

and

$$\pi_2(z_2 \mid a) = \begin{cases} 1 - \varepsilon, & \text{if } z_2 = z_2' \text{ and } a_1 = L, \text{ or} \\ & z_2 = z_2'' \text{ and } a_1 = H, \\ \varepsilon, & \text{otherwise,} \end{cases}$$

and joint distribution $\pi(z \mid a) = \pi_1(z_1 \mid a)\pi_2(z_2 \mid a)$.

Consider the profile described by the two automata $(\mathscr{W}_i, w_i^0, f_i, \tau_i)$, $i = 1, 2$. Player 1's automaton is trivial: $\mathscr{W}_1 = \{w_1\}$, $w_1^0 = w_1$, $f_1(w_1) = \frac{1}{2} \circ H + \frac{1}{2} \circ L$, and $\tau_1(w, a_1 z_1) = w_1$. Player 2's automaton, illustrated in figure 12.5.2, has $\mathscr{W}_2 = \{w_2', w_2''\}$, $w_2^0 = w_2''$,

$$f_2(w_2) = \begin{cases} \alpha'' \circ h + (1 - \alpha'') \circ \ell, & \text{if } w_2 = w_2'', \\ \alpha' \circ h + (1 - \alpha') \circ \ell, & \text{if } w_2 = w_2', \end{cases}$$

and

$$\tau_2(w_2, a_2 z_2) = \begin{cases} w_2'', & \text{if } z_2 = z_2'', \\ w_2', & \text{if } z_2 = z_2'. \end{cases}$$

That is, as in the perfect monitoring profile, player 1 randomizes in every period with probability $1/2$ on H. If player 2 observes z_2'', player 2 randomizes, putting probability α'' on h. If player 2 observes z_2', then she randomizes, putting probability α' on h.

Note that as in the equilibrium of the perfect monitoring game, behavior in this profile is independent of player 2's earlier actions. Moreover, given player 1's behavior, player 2 is always indifferent between h and ℓ. It remains to verify that player 1 is indifferent between H and L, irrespective of the private history, h_1^t, that 1 observes.

A belief-based approach, like that of the previous sections, would require us to verify that player 1's behavior was optimal, given the beliefs that he has over $\mathcal{W}_2$ after each private history h_1^t (recall remark 12.2.4). The set of private histories increases without bound, so this is a daunting task.

The approach of *belief-free equilibria*, on the other hand, asks that for all private histories, $h_1^t \in \mathcal{H}_1$, player 1's behavior is a best reply to player 2's continuation strategy for *all* possible private histories $h_2^t \in \mathcal{H}_2$ that 2 could have observed (and not just against the expected continuation under the beliefs $\beta(\cdot \mid h_1^t) \in \Delta(\mathcal{W}_2)$). Equivalently, we must show that player 1 is indifferent between H and L regardless of player 2's current state. At the same time, player 1 will prefer player 2 to be in state w_2'', and the fact that playing L makes a transition to w_2' more likely lies behind player 1's indifference. Belief freeness, a stronger condition than simply requiring player 1 play a best response, is simpler to check and suffices to prove a partial folk theorem in some settings (see chapter 14).

Let $V_1^\varepsilon(a_1, w_2)$ be the value to player 1 from playing a_1 in the current period, when player 2's current private state is w_2, and future play is determined by the profile.[18] Then,

$$V_1^\varepsilon(H, w_2'') = (1-\delta)\alpha''2 + \delta(1-\varepsilon)\left\{\frac{1}{2}V_1^\varepsilon(H, w_2'') + \frac{1}{2}V_1^\varepsilon(L, w_2'')\right\} \quad (12.5.1)$$

$$+ \delta\varepsilon\left\{\frac{1}{2}V_1^\varepsilon(H, w_2') + \frac{1}{2}V_1^\varepsilon(L, w_2')\right\},$$

$$V_1^\varepsilon(L, w_2'') = (1-\delta)\{\alpha''3 + (1-\alpha'')\}$$

$$+ \delta\varepsilon\left\{\frac{1}{2}V_1^\varepsilon(H, w_2'') + \frac{1}{2}V_1^\varepsilon(L, w_2'')\right\} \quad (12.5.2)$$

$$+ \delta(1-\varepsilon)\left\{\frac{1}{2}V_1^\varepsilon(H, w_2') + \frac{1}{2}V_1^\varepsilon(L, w_2')\right\},$$

$$V_1^\varepsilon(H, w_2') = (1-\delta)\alpha'2 + \delta(1-\varepsilon)\left\{\frac{1}{2}V_1^\varepsilon(H, w_2'') + \frac{1}{2}V_1^\varepsilon(L, w_2'')\right\} \quad (12.5.3)$$

$$+ \delta\varepsilon\left\{\frac{1}{2}V_1^\varepsilon(H, w_2') + \frac{1}{2}V_1^\varepsilon(L, w_2')\right\},$$

and

$$V_1^\varepsilon(L, w_2') = (1-\delta)\{\alpha'3 + (1-\alpha')\}$$

$$+ \delta\varepsilon\left\{\frac{1}{2}V_1^\varepsilon(H, w_2'') + \frac{1}{2}V_1^\varepsilon(L, w_2'')\right\} \quad (12.5.4)$$

$$+ \delta(1-\varepsilon)\left\{\frac{1}{2}V_1^\varepsilon(H, w_2') + \frac{1}{2}V_1^\varepsilon(L, w_2')\right\}.$$

We are interested in randomizations by player 2 that have the property that

$$V_1^\varepsilon(H, w_2'') = V_1^\varepsilon(L, w_2'') \equiv V_1^\varepsilon(w_2'')$$

---

18. We emphasize, however, that apart from the initial period, player 1 does not know player 2's private state.

12.5 ■ A Belief-Free Example

and
$$V_1^\varepsilon(L, w_2') = V_1^\varepsilon(H, w_2') \equiv V_1^\varepsilon(w_2').$$

Equating the right sides of (12.5.1) and (12.5.2) gives
$$V_1^\varepsilon(w_2'') - V_1^\varepsilon(w_2') = \frac{(1-\delta)}{\delta(1-2\varepsilon)}.$$

Subtracting (12.5.3) from (12.5.1),
$$V_1^\varepsilon(w_2'') - V_1^\varepsilon(w_2') = 2(1-\delta)(\alpha'' - \alpha').$$

Solving these two equations,
$$\alpha'' - \alpha' = \frac{1}{2\delta(1-2\varepsilon)}. \tag{12.5.5}$$

Because player 1's myopic incentive to play L is independent of player 2's action, (12.5.4) is implied by (12.5.1)–(12.5.3).

Hence, for probabilities α' and α'' satisfying (12.5.5), we have constructed a belief-free equilibrium. Moreover, in this equilibrium, for ε small behavior is close to the equilibrium of the perfect monitoring game discussed at the beginning of this section, and for α'' close to 1 player 1's payoff is close to 2.

As in the mixed equilibria of the perfect monitoring game, player 1 is indifferent between L, the myopically dominant action, and H, due to the implied change in the distribution of continuation values. In particular, a play of H is rewarded in the next period by a probability $(1-\varepsilon)\alpha'' + \varepsilon\alpha'$ on h, rather than the lower probability $\varepsilon\alpha'' + (1-\varepsilon)\alpha'$.

Remark 12.5.1 Long-lived player 2 Player 1's history-independent behavior implies that the belief-free equilibrium constructed in this section remains a sequential equilibrium if player 2 is long-lived. ◆

13 Almost Public Monitoring Games

Although games with private monitoring have no public histories to coordinate continuation play, games with almost public monitoring have histories that are almost public. In this chapter, pursuing ideas introduced in section 12.3, we explore the extent to which it is possible to coordinate continuation play for such games.

We ask when an equilibrium strategy profile in a public monitoring game induces an equilibrium in a corresponding game whose private monitoring is almost public. After describing the sense in which private monitoring games can be almost public, we present two central results: It is always possible to coordinate continuation play by requiring behavior to have *bounded recall* (i.e., there is a bound L such that in any period, the last L signals are sufficient to determine behavior). Moreover, in games with general almost public private monitoring, this is essentially the only behavior that can coordinate continuation play. The structure of a public strategy profile is thus important in determining whether it remains an equilibrium in the private monitoring game. This chapter is accordingly focused more on the nature of equilibrium behavior than on equilibrium payoffs, though it does conclude with a folk theorem.

The benchmark public monitoring game is as described in chapter 7, with finite action and signal spaces. We use a ˜ to denote payoffs in the public monitoring game. Hence, player i's payoff after the realization (y, a) is given by $\tilde{u}_i^*(y, a_i)$. Stage game payoffs in the game with public monitoring are then given by $\tilde{u}_i(a) \equiv \sum_y \tilde{u}_i^*(y, a_i) \rho(y \mid a)$. We omit ˜ for payoffs in the private monitoring game.

In this chapter, by the phrase *public profile*, we will always mean a strategy profile for the public monitoring game that is itself public. Given a pure public profile (and the associated automaton), continuation play after any history is determined by the *public* state reached by that history. In games with private monitoring, by contrast, a sufficient statistic for continuation play after any history is the vector of current *private* states, one for each player (remark 12.2.4). A *private profile* is a strategy profile in the private monitoring game.

13.1 When Is Monitoring Almost Public?

When taking ε to 0 in the following definition, we often refer to private monitoring distributions ε-close (under some ξ) to the public monitoring distribution as being *almost public*.

Definition *The private monitoring distribution (Z, π) is ε-close under ξ to a full-support*
13.1.1 *public monitoring distribution (Y, ρ), where $\xi = (\xi_1, \ldots, \xi_n)$ is a vector of signal interpretations $\xi_i : Z_i \to Y$, if*

1. *for all $y \in Y$, $z_i \in \xi_i^{-1}(y)$, and all $a \in A$ with $\pi_i(z_i \mid a) > 0$,*

$$\pi(\{z_{-i} : \xi_j(z_j) = y \text{ for all } j \neq i\} \mid (a, z_i)) \geq 1 - \varepsilon,$$

and

2. *for each $a \in A$ and $y \in Y$,*

$$\left| \pi(\{z : \xi_i(z_i) = y \text{ for all } i\} \mid a) - \rho(y \mid a) \right| \leq \varepsilon.$$

The private monitoring distribution (Z, π) is ε-close to the full-support public monitoring distribution (Y, ρ) if it is ε-close under some signal interpretations ξ to (Y, ρ).

Under this definition, every private signal is interpretable as some public signal. The first condition guarantees that conditional on a player's private signal mapped into the public signal y, he assigns very high probability to *every* other player having also observed a private signal mapped into the *same* public signal y. The second condition guarantees that the probability that *all* players observe *some* private signal consistent with the public signal y is close to the probability of y under ρ. For sufficiently small ε, this condition requires that every player has at least one private signal mapped to each public signal.

Say the monitoring is *minimally private* if $Z_i = Y$ and ξ_i is the identity.[1] We denote minimally private monitoring distributions by (Y^n, π). In this case, the second condition implies the first for ε small and full-support public monitoring (lemma 13.1.1). The vector $(1, \ldots, 1)$ is denoted $\mathbf{1}$, whose dimension will be obvious from context. Thus, $\pi(y, \ldots, y \mid a)$ is written as $\pi(y\mathbf{1} \mid a)$.

Lemma *Fix a full-support public monitoring distribution (Y, ρ) and $\varepsilon > 0$. There exists*
13.1.1 *$\eta > 0$ such that for all distributions $\pi \in \Delta(Y^n)$, if $|\pi(y\mathbf{1} \mid a) - \rho(y \mid a)| < \eta$ for all $y \in Y$ and all $a \in A$, then the minimally private monitoring distribution (Y^n, π) is ε-close to (Y, ρ).*

Proof Fix the action profile $a \in A$. The probability that player i observes the private signal $z_i = y$ is $\sum_{\mathbf{y}_{-i}} \pi(y, \mathbf{y}_{-i} \mid a)$ and this probability is smaller than

$$(\rho(y \mid a) + \eta) + \sum_{\mathbf{y}_{-i} \neq y\mathbf{1}} \pi(y, \mathbf{y}_{-i} \mid a) < \rho(y \mid a) + \eta + \sum_{\tilde{y} \in Y} \sum_{\mathbf{y}_{-i} \neq \tilde{y}\mathbf{1}} \pi(\tilde{y}, \mathbf{y}_{-i} \mid a)$$

$$< \rho(y \mid a) + \eta + \left(1 - \sum_{\tilde{y} \in Y} \pi(\tilde{y}\mathbf{1} \mid a)\right)$$

$$< \rho(y \mid a) + \eta(1 + |Y|).$$

Thus, the probability player i assigns to the other players observing the same signal y, $\pi(y\mathbf{1} \mid a, z_i = y)$, is at least as large as $\pi(y\mathbf{1} \mid a)\{\rho(y \mid a) + \eta(1 + |Y|)\}^{-1} > (\rho(y \mid a) - \eta)\{\rho(y \mid a) + \eta(1 + |Y|)\}^{-1}$. Thus, by choosing

---

1. Mailath and Morris (2002) use the term *almost public monitoring* without qualification for minimally private monitoring.

13.1 ■ When Is Monitoring Almost Public?

$$\eta < \min_{a \in A, y \in Y} \frac{\varepsilon \rho(y \mid a)}{2 + |Y| - \varepsilon(1 + |Y|)}, \quad (13.1.1)$$

we have $\pi(y\mathbf{1} \mid a, z_i = y) > 1 - \varepsilon$ for all a.

■

Remark 13.1.1 **Full support** The assumption that the public monitoring distribution ρ in lemma 13.1.1 has full support is necessary for the conclusion that *all* minimally private monitoring distributions π satisfying $|\pi(y\mathbf{1} \mid a) - \rho(y \mid a)| < \eta$ for all $y \in Y$ and all $a \in A$, for some η, are ε-close. It is apparent from (13.1.1) that if some signal y has zero probability under an action profile a, then for any η, we can construct a π violating the first part of definition 13.1.1.

◆

Example 13.1.1 We now return to the prisoners' dilemma example of section 12.3, but with a richer set of private signals for player 1, $Z_1 = \{\underline{z}_1, \bar{z}'_1, \bar{z}''_1\}$. Player 2 still has two signals, $Z_2 = \{\underline{z}_2, \bar{z}_2\}$. The probability distribution of the signals is given in figure 13.1.1. This private monitoring distribution is $\sqrt{\varepsilon}$-close to the public monitoring distribution of (7.2.1) (reproduced in (12.1.1)) under the signal interpretations $\xi_1(\underline{z}_1) = \xi_2(\underline{z}_2) = \underline{y}$ and $\xi_2(\bar{z}_2) = \xi_1(\bar{z}'_1) = \xi_1(\bar{z}''_1) = \bar{y}$, as long as ε is sufficiently small (relative to $\min\{\zeta', \zeta - \zeta', 1 - \zeta\}$).

●

The condition of ε-closeness in definition 13.1.1 can be restated as follows. An event is *p-evident* if whenever it is true, everyone assigns probability at least p to it being true (Monderer and Samet 1989). The following lemma is a straightforward application of the definitions, so we omit the proof.

Lemma 13.1.2 *Suppose $\xi_i : Z_i \to Y$, $i = 1, \ldots, n$, is a collection of signal interpretations. The private monitoring distribution (Z, π) is ε-close under ξ to the public monitoring distribution (Y, ρ) if and only if for each public signal y, the set of private signal*

	$\underline{z}_2$	$\bar{z}_2$
$\underline{z}_1$	$(1-\zeta)(1-3\varepsilon)$	ε
$\bar{z}'_1$	ε	$\zeta'(1-3\varepsilon)$
$\bar{z}''_1$	ε	$(\zeta - \zeta')(1-3\varepsilon)$

Figure 13.1.1 The probability distribution of private signals for example 13.1.1. The distribution is given as a function of the action profile a, where $\zeta = p$ if $a = EE$, q if $a = ES$ or SE, and r if $a = SS$ (analogously, ζ' is given by p', q', or r' as a function of a).

profiles $\{z : \xi_i(z_i) = y \text{ for all } i\}$ is $(1 - \varepsilon)$-evident (conditional on any action profile) and has probability within ε of the probability of y (conditional on that action profile).

13.2 Nearby Games with Almost Public Monitoring

This section provides the tools for associating behavior in a public monitoring game with behavior in a *nearby* private monitoring game.

13.2.1 Payoffs

We say that public monitoring and private monitoring games are close if their signals and payoffs are close.

Definition 13.2.1 *A private monitoring game $(u^*, (Z, \pi))$ is ε-close under ξ to a public monitoring game $(\tilde{u}^*, (Y, \rho))$ if (Z, π) is ε-close under ξ to (Y, ρ) and $|\tilde{u}_i^*(\xi_i(z_i), a_i) - u_i^*(z_i, a_i)| < \varepsilon$ for all i, $a_i \in A_i$, and $z_i \in \xi_i^{-1}(Y)$.*

The ex ante stage game payoffs of any almost public monitoring game are close to the ex ante stage game payoffs of the benchmark public monitoring game.

Lemma 13.2.1 *For all $\eta > 0$, there is an $\varepsilon > 0$ such that if $(u^*, (Z, \pi))$ is ε-close to $(\tilde{u}^*, (Y, \rho))$, then for all i and all action profiles a,*

$$\left| \sum_{z_1, \ldots, z_n} u_i^*(z_i, a_i) \pi(z_1, \ldots, z_n \mid a) - \sum_y \tilde{u}_i^*(y, a_i) \rho(y \mid a) \right| < \eta.$$

Proof Suppose $(u^*, (Z, \pi))$ is ε-close to $(\tilde{u}^*, (Y, \rho))$ under $(\xi_1, \ldots, \xi_n)$. Then, for all a,

$$\left| \sum_{z_1, \ldots, z_n} u_i^*(z_i, a_i) \pi(z_1, \ldots, z_n \mid a) - \sum_y \tilde{u}_i^*(y, a_i) \rho(y \mid a) \right|$$

$$\leq \left| \sum_y \sum_{z_1 \in \xi_1^{-1}(y), \ldots, z_n \in \xi_n^{-1}(y)} u_i^*(z_i, a_i) \pi(z_1, \ldots, z_n \mid a) - \tilde{u}_i^*(y, a_i) \rho(y \mid a) \right|$$

$$+ |Y| \varepsilon \max_{z_i, a_i} |u_i^*(z_i, a_i)|$$

$$\leq \left| \sum_y \tilde{u}_i^*(y, a_i) \left\{ \sum_{z_1 \in \xi_1^{-1}(y), \ldots, z_n \in \xi_n^{-1}(y)} \pi(z_1, \ldots, z_n \mid a) - \rho(y \mid a) \right\} \right|$$

$$+ \varepsilon + |Y| \varepsilon \max_{z_i, a_i} |u_i^*(z_i, a_i)|$$

$$\leq 2|Y| \varepsilon \max_{z_i, a_i} |u_i^*(z_i, a_i)| + \varepsilon + \varepsilon^2 |Y|,$$

where the first inequality follows from $\sum_y \pi(\{z : \xi_i(z_i) = y \text{ for each } i\} \mid a) > 1 - \varepsilon |Y|$ (an implication of definition 13.1.1(2)); the second equality follows from $|\tilde{u}_i^*(y, a_i) - u_i^*(z_i, a_i)| < \varepsilon$ for all i, $a_i \in A_i$, and $z_i \in \xi_i^{-1}(y)$; and the third inequality follows from definition 13.1.1(2) and $\max_{y,a_i} |\tilde{u}_i^*(y, a_i)| \leq \max_{z_i,a_i} |u_i^*(z_i, a_i)| + \varepsilon$. The last term can clearly be made smaller than η by appropriate choice of ε.

∎

13.2.2 Continuation Values

Fix a public profile $(\mathcal{W}, w^0, f, \tau)$ of a full-support public monitoring game $(\tilde{u}^*, (Y, \rho))$ and an ε-close private monitoring game $(u^*, (Z, \pi))$, under ξ. The public profile *induces* a private profile in the private monitoring game in a natural way. Player i's strategy is described by the automaton $(\mathcal{W}, w^0, f_i, \tau_i)$, where $\tau_i(w_i, z_i) = \tau(w_i, \xi_i(z_i))$ for all $z_i \in Z_i$ and $w_i \in \mathcal{W}$. The set of states, initial state, and decision function are from the public profile. The transition function τ_i is well defined, because the signal interpretations all map into Y. If $\mathcal{W}$ is finite, each player can be viewed as following a finite state automaton.

This private strategy is action-free, and so i's state is well defined for private histories reached by i deviating, because the state is only a function of the private signals. We recursively calculate the states of player i as $w_i^1 = \tau(w^0, \xi_i(z_i^0)) = \tau_i(w^0, z_i^0)$, $w_i^2 = \tau_i(w_i^1, z_i^1)$, and so on. Thus, for any private history h_i^t, we write $w_i^t = \tau_i(w^0, h_i^t)$ (though, because the profile is action-free, i's past actions do not affect i's state transitions). When we can take the initial state as given, we write $w_i^t = \tau_i(w^0, h_i^t) = \tau_i(h_i^t)$. Though all players are in the same private state in the first period, because the signals are private, after the first period, different players may be in different private states. The *private profile* is the translation to the private monitoring game of the public profile (of the public monitoring game).

Consider now a strategy σ_i for i in a private monitoring game ε-close under ξ to a public monitoring game. The strategy is *measurable* with respect to the partition on Z_i induced by ξ_i if, for all $h_i^t, \hat{h}_i^t \in \mathcal{H}_i$, if $\xi_i(z_i^s) = \xi_i(\hat{z}_i^s)$ for all $s < t$, then $\sigma_i(h_i^t) = \sigma(\hat{h}_i^t)$. For brevity, we write σ_i is *measurable under* ξ_i. (With minimally private monitoring, all strategies are measurable under ξ_i, because ξ_i is the identity.) In this case, σ_i induces a strategy $\tilde{\sigma}_i$ in the game with public monitoring: For all histories $\tilde{h}_i^t \in (A_i \times Y)^{t-1}$, $\tilde{\sigma}_i(\tilde{h}_i^t) = \sigma_i(h_i^t)$, where $h_i^t \in (A_i \times Z_i)^{t-1}$ is the private history given by $a_i^s = \tilde{a}_i^s$ and $z_i^s \in Z_i$ is some private signal satisfying $\xi_i(z_i^s) = y_i^s$ (because σ_i is measurable under ξ_i, the particular choice of z_i^s is irrelevant).

Recalling remark 12.2.4, if player i believes that the other players are following a strategy induced by the public profile $(\mathcal{W}, w^0, f, \tau)$, a sufficient statistic of h_i^t for the purposes of evaluating continuation strategies is player i's private state and i's beliefs over the other players' private states, that is, (w_i^t, β_i^t), where $\beta_i^t \in \Delta(\mathcal{W}^{n-1})$.

Suppose $(u^*, (Z, \pi))$ is ε-close to $(\tilde{u}^*, (Y, \rho))$. Fix a public profile, represented by the automaton $(\mathcal{W}, w^0, f, \tau)$. Let $\tilde{U}_i(\tilde{\sigma}_i \mid w)$ be the expected value of the strategy $\tilde{\sigma}_i$ in the public monitoring game when players $j \neq i$ follow the automaton $(\mathcal{W}, w, f, \tau)$ (note that the initial state is now w, not w^0). Denote by $U_i(\sigma_i \mid h_i^t)$ the continuation value of the strategy σ_i in the private monitoring game, conditional on the private history h_i^t, when players $j \neq i$ follow the private strategies induced by $(\mathcal{W}, w^0, f, \tau)$.

If a player is sufficiently confident that all the other players are in the same private state w, then his payoff from any strategy measurable under ξ_i is close to the payoff from the comparable strategy in the nearby public monitoring game.

Lemma 13.2.2 *Fix a public profile $(\mathscr{W}, w^0, f, \tau)$ of the public monitoring game $(\tilde{u}^*, (Y, \rho))$ with discount factor δ. For all $\upsilon > 0$, there exists $\eta > 0$ and $\varepsilon > 0$, such that for all $(u^*, (Z, \pi))$ ε-close under ξ to $(\tilde{u}^*, (Y, \rho))$ and all i if the posterior belief implied by the induced private profile and the private history $\hat{h}_i^t$ satisfies $\beta_i(w\mathbf{1} \mid \hat{h}_i^t) > 1 - \eta$ for some $w \in \mathscr{W}$, then for any σ_i, with $\sigma_i|_{\hat{h}_i^t}$ measurable under ξ_i,*

$$\left| U_i(\sigma_i \mid \hat{h}_i^t) - \tilde{U}_i(\tilde{\sigma}_i \mid w) \right| < \upsilon, \tag{13.2.1}$$

where $\tilde{\sigma}_i$ is the strategy in the public monitoring game induced by $\sigma_i|_{\hat{h}_i^t}$.

Proof For fixed i and $w \in \mathscr{W}$, let $\tilde{\sigma}_{-i}^w$ denote the public profile for players $j \neq i$ described by the automaton $(\mathscr{W}, w, f, \tau)$, and σ_{-i}^w the induced private profile. Then, $U_i(\sigma_i \mid \hat{h}_i^t)$ is calculated assuming players $j \neq i$ have been following the profile induced by $(\mathscr{W}, w^0, f, \tau)$ from the initial period, and *not* that the continuation play of $j \neq i$ is σ_j^w, where $w = \tau(w^0, \hat{h}_i^t)$. The latter value, that is, the continuation value of σ_i after $\hat{h}_i^t$, conditional on the continuation play of $j \neq i$ being given by σ_j^w, is denoted $U_i(\sigma_i \mid \hat{h}_i^t, w\mathbf{1})$. We then have (using obvious notation),

$$U_i(\sigma_i \mid \hat{h}_i^t) = U_i(\sigma_i \mid \hat{h}_i^t, w\mathbf{1})\beta_i(w\mathbf{1} \mid \hat{h}_i^t) \\ + U_i(\sigma_i \mid \hat{h}_i^t, w_{-i} \neq w\mathbf{1})\{1 - \beta_i(w\mathbf{1} \mid \hat{h}_i^t)\}.$$

Payoffs are bounded, so there exists η sufficiently small (independent of i and σ_i), such that, if $\beta_i(w\mathbf{1} \mid \hat{h}_i^t) > 1 - \eta$, then

$$\left| U_i(\sigma_i \mid \hat{h}_i^t) - U_i(\sigma_i \mid \hat{h}_i^t, w\mathbf{1}) \right| < \frac{\upsilon}{3}. \tag{13.2.2}$$

For fixed $\delta < 1$, there is a T such that the continuation value from $T + 1$ on is less than $\upsilon/3$.

For any $\lambda > 0$, there exists $\varepsilon > 0$ (independent of i and σ_i), such that for any π ε-close to ρ, for all T-period sequences of public signals $(y^1, \ldots, y^T)$, the probability of the private monitoring event

$$\{(z^1, \ldots, z^T) \in Z^T : \xi_i(z_i^t) = y^t, \text{ for all } i \text{ and } t = 1, \ldots, T\}$$

under σ is within $\lambda > 0$ of the probability of $(y^1, \ldots, y^T)$ under $\tilde{\sigma}$. Moreover, by choosing λ sufficiently small, we can ensure that conditional on all players $j \neq i$ being in the same private state w in period t, the expected discounted value of the first T periods of play under σ is within $\upsilon/3$ of the expected discounted value of the first T periods of play under $\tilde{\sigma}$. Hence, from the determination of T, conditional on all players $j \neq i$ being in the same private state w in period t, the expected discounted value of play under σ is within $\upsilon/3$ of the expected discounted value of play under $\tilde{\sigma}$, that is,

$$\left| U_i(\sigma_i \mid \hat{h}_i^t, w\mathbf{1}) - \tilde{U}_i(\tilde{\sigma}_i \mid w) \right| < \frac{2\upsilon}{3},$$

which, with (13.2.2), implies (13.2.1). ∎

13.2.3 Best Responses

Consider a strict PPE of a public monitoring game, $(\mathscr{W}, w, f, \tau)$. Suppose player i is very confident that *all* the other players are in the same common private state w. Then, because the incentive constraints (as described by corollary 7.1.1) are strict as long as player i is sufficiently confident that the other players are in the same state w, $f_i(w)$ should still be optimal: Lemma 13.2.2 implies that the payoff implications of small perturbations in the continuation play of the opponents (reflecting both a player not being in the private state w and receiving future private signals inconsistent with those observed by i) are small (at least when i's continuation strategy is measurable under ξ_i). The next lemma formalizes the intuition of note 9 on page 401.

Lemma 13.2.3 *Suppose the public profile $(\mathscr{W}, w^0, f, \tau)$ is a strict PPE of the public monitoring game $(\tilde{u}^*, (Y, \rho))$ with discount factor δ. For all $w \in \mathscr{W}$, there exists $\eta > 0$ and $\varepsilon > 0$, such that for all $(u^*, (Z, \pi))$ ε-close under ξ to $(\tilde{u}^*, (Y, \rho))$, if $\beta_i(w\mathbf{1} \mid \hat{h}_i^t) > 1 - \eta$ for some $\hat{h}_i^t \in \mathscr{H}_i$, then for any σ_i, if the continuation strategy $\sigma_i|_{\hat{h}_i^t}$ is a best reply at $\hat{h}_i^t$ and is measurable under ξ_i, then $\sigma_i(\hat{h}_i^t) = f_i(w)$.*

Proof Fix a state $w' \in \mathscr{W}$ and private history $\hat{h}_i^t$ and strategy σ_i for player i. Suppose $\sigma_i(\hat{h}_i^t) \neq f_i(w')$ and $\sigma_i|_{\hat{h}_i^t}$ is measurable under ξ. Denote by $\tilde{\sigma}_i$ the public strategy induced by $\sigma_i|_{\hat{h}_i^t}$ and by $\tilde{\sigma}_i'$ the strategy for player i described by $(\mathscr{W}, w', f_i, \tau)$. Because the public profile is a strict PPE, the payoff loss from not choosing $f_i(w')$ in the current period, when the public profile is in state w', is at least $3\upsilon > 0$ for some υ (corollary 7.1.1), and so,

$$\tilde{U}_i(\tilde{\sigma}_i' \mid w') - \tilde{U}_i(\tilde{\sigma}_i \mid w') > 3\upsilon.$$

Let σ_i' denote the strategy in the private monitoring game induced by $\tilde{\sigma}_i'$, and let σ_i^* be the strategy given by

$$\sigma_i^*(h_i^s) = \begin{cases} \sigma_i(h_i^s), & \text{if } s < t, \\ \sigma_i'(\check{h}_i^{s-t}), & \text{if } s \geq t, \text{ where } h_i^s = (h_i^t, \check{h}_i^{s-t}). \end{cases}$$

In other words, for $s \geq t$, σ_i^* ignores the first t periods of the history and plays as if play began in period t with initial state w'. Applying lemma 13.2.2, for sufficiently small η and ε, if $\beta_i(w'\mathbf{1} \mid \hat{h}_i^t) > 1 - \eta$ and $(u^*, (Z, \pi))$ is ε-close under ξ to $(\tilde{u}^*, (Y, \rho))$, then (because $\tilde{\sigma}_i'$ is the public strategy induced by $\sigma_i^*|_{\hat{h}_i^t}$)

$$\begin{aligned} U_i(\sigma_i^* \mid \hat{h}_i^t) - U_i(\sigma_i \mid \hat{h}_i^t) &\geq U_i(\sigma_i^* \mid \hat{h}_i^t) - \tilde{U}_i(\tilde{\sigma}_i' \mid w') \\ &\quad + (\tilde{U}_i(\tilde{\sigma}_i' \mid w') - \tilde{U}_i(\tilde{\sigma}_i \mid w')) \\ &\quad + \tilde{U}_i(\tilde{\sigma}_i \mid w') - U_i(\sigma_i \mid \hat{h}_i^t) \\ &\geq \upsilon. \end{aligned}$$

Thus, $\sigma_i|_{\hat{h}_i^t}$ cannot be a best reply at $\hat{h}_i^t$. ∎

13.2.4 Equilibrium

Our bookkeeping lemmas, lemmas 13.2.2 and 13.2.3, immediately give us a simple and intuitive sufficient condition for when a strict PPE induces an equilibrium in nearby games with private monitoring: After all private histories $h_i^t \in \mathcal{H}_i$, player i must be sufficiently confident that all other players are in the same private state as himself, that is, in $\tau_i(h_i^t) = \tau_i(w^0, h_i^t)$. Because we need the bounds on η and ε to apply for all states $w \in \mathcal{W}$ of the automaton, we strengthen strict to uniformly strict (definition 7.1.4). Any strict PPE with finite $\mathcal{W}$ is uniformly strict.

Proposition 13.2.1 *Suppose the public profile $(\mathcal{W}, w^0, f, \tau)$ is a uniformly strict PPE of the public monitoring game $(\tilde{u}^*, (Y, \rho))$ with discount factor δ. For all $\kappa > 0$, there exists $\eta > 0$ and $\varepsilon > 0$, such that for all $(u^*, (Z, \pi))$ ε-close under ξ to $(\tilde{u}^*, (Y, \rho))$, if $\beta_i(\tau_i(h_i^t)\mathbf{1} \mid h_i^t) > 1 - \eta$ for all $h_i^t \in \mathcal{H}_i$, then the private profile is a sequential equilibrium of the game with private monitoring for the same δ, and the expected payoff in that equilibrium is within κ of the public equilibrium payoff.*

Proof From proposition 12.2.3, it is enough to show that no player has a profitable one-shot deviation. Because the private profile is measurable under ξ, this follows from lemma 13.2.3. The uniform strictness of the PPE allows us in lemma 13.2.3 to choose η and ε independently of $w \in \mathcal{W}$. The κ-closeness of equilibrium payoffs follows from lemma 13.2.2. ∎

A version of the one-shot deviation principle implies that private profiles induced by uniformly strict PPE are Nash equilibria.

Proposition 13.2.2 *Suppose the public profile $(\mathcal{W}, w^0, f, \tau)$ is a uniformly strict PPE of the public monitoring game $(\tilde{u}^*, (Y, \rho))$ with full support ρ and discount factor δ. For all $\kappa > 0$, there exists $\eta > 0$ and $\varepsilon > 0$, such that for all $(u^*, (Z, \pi))$ ε-close to $(\tilde{u}^*, (Y, \rho))$, if $\beta_i(\tau_i(h_i^t)\mathbf{1} \mid h_i^t) > 1 - \eta$ for all h_i^t consistent with the induced private profile, then the private profile is a Nash equilibrium of the game with private monitoring for the same δ, and the expected payoff in that equilibrium is within κ of the public equilibrium payoff. Moreover, if $\pi_i(z_i \mid a) > 0$ for all i, $z_i \in Z_i$, and $a \in A$, there is a realization-equivalent sequential equilibrium.*

Proof Let σ be the private profile induced by the public profile $\tilde{\sigma} \equiv (\mathcal{W}, w, f, \tau)$. Let $\sigma_i^* \neq \sigma_i$ be a deviation for player i, with $U_i(\sigma) \neq U_i(\sigma_i^*, \sigma_{-i})$. Then there exists some private history $h_i^t \in \mathcal{H}_i$ consistent with the strategy profile σ such that $\sigma_i^*(h_i^t) \neq \sigma_i(h_i^t)$. Let σ_i^{**} denote the strategy that agrees with σ_i^* at h_i^t and agrees with σ_i otherwise (σ_i^{**} is a one-shot deviation from σ_i). Observe that because $\tilde{\sigma}$ is a PPE, the continuation strategy implied by $\tilde{\sigma}_i^{**}$ is a best reply to the behavior of the other players in periods after h_i^t. Therefore,

$$\tilde{U}_i(\tilde{\sigma}_i^*|_{h_i^t} \mid \tau(h_i^t)) \leq \tilde{U}_i(\tilde{\sigma}_i^{**}|_{h_i^t} \mid \tau(h_i^t)).$$

Because the public profile is a uniformly strict PPE, there exists $\upsilon > 0$ (independent of h_i^t) such that

$$\tilde{U}_i(\tilde{\sigma}_i^{**}|_{h_i^t} \mid \tau(h_i^t)) < \tilde{U}_i(\tilde{\sigma}_i|_{h_i^t} \mid \tau(h_i^t)) - 3\upsilon.$$

Finally, we have (from lemma 13.2.2), for small η and ε,

$$U_i(\sigma_i^* \mid h_i^t) < \tilde{U}_i(\tilde{\sigma}_i^*|_{h_i^t} \mid \tau(h_i^t)) + \upsilon$$
$$\leq \tilde{U}_i(\tilde{\sigma}_i^{**}|_{h_i^t} \mid \tau(h_i^t)) + \upsilon$$

$$< \tilde{U}_i(\tilde{\sigma}_i|_{h_i^t} \mid \tau(h_i^t)) - 2\upsilon$$
$$< U_i(\sigma_i \mid h_i^t) - \upsilon,$$

so that σ_i^* is not a profitable deviation. Thus $(\mathscr{W}, w^0, f, \tau)$ induces a Nash equilibrium of the game with private monitoring.

Finally, when the marginals of π have full support we can apply proposition 12.2.1 to conclude that any Nash equilibrium outcome can be supported by a sequential equilibrium.

∎

In section 13.4, we provide conditions under which the private profile induced by a strict PPE is not a Nash equilibrium in nearby almost public monitoring games. In principle, this is a more difficult task than showing a profile is a Nash equilibrium. In the latter task, the hope is that the game with almost public monitoring is close enough that incentives carry-over to the private monitoring game essentially unchanged (and this is the content of propositions 13.2.1 and 13.2.2). In the former task, we have no such hope. However, proposition 12.3.1, on the failure of grim trigger to induce a Nash equilibrium, suggests that in some cases the game with almost public monitoring is close enough that incentives carry-over for deviations. And, indeed, lemma 13.2.3 immediately implies the following useful proposition.

Proposition 13.2.3 *Suppose the public profile $(\mathscr{W}, w^0, f, \tau)$ is a uniformly strict equilibrium of the public monitoring game $(\tilde{u}^*, (Y, \rho))$ for some δ. There exists $\eta > 0$ and $\varepsilon > 0$ such that for any private monitoring game $(u^*, (Z, \pi))$ ε-close to $(\tilde{u}^*, (Y, \rho))$, if there exists a player i, a positive probability private history for that player h_i^t, and a state w such that $f_i(w) \neq f_i(\tau_i(h_i^t))$ and $\beta_i(w\mathbf{1} \mid h_i^t) > 1 - \eta$, then the induced private profile is not a Nash equilibrium of the game with private monitoring for the same δ.*

13.3 Public Profiles with Bounded Recall

As we saw in proposition 12.3.1, a grim trigger PPE need not induce equilibria of almost public monitoring games because the public state in period t is determined in principle by the entire history h^t. Section 12.3.1, on the other hand, described how an equilibrium is induced in nearby games by a particular profile in which the last realized signal is sufficient to determine behavior. More generally, this section shows that equilibria in bounded recall strategies induce equilibria in almost public monitoring games.

Definition 13.3.1 *A public profile σ has L bounded recall if for all $h^t = (y^0, \ldots, y^{t-1})$ and $\hat{h}^t = (\hat{y}^0, \ldots, \hat{y}^{t-1})$, if $t \geq L$ and $y^s = \hat{y}^s$ for $s = t - L, \ldots, t - 1$, then*

$$\sigma(h^t) = \sigma(\hat{h}^t).$$

The following characterization of bounded recall is useful.

Lemma 13.3.1 *The public profile induced by the minimal automaton $(\mathscr{W}, w^0, f, \tau)$ has L bounded recall if and only if for all $w, w' \in \mathscr{W}$ reachable in the same period and for all $h \in Y^\infty$,*
$$\tau(w, h^L) = \tau(w', h^L).$$

Proof Suppose for all $w, w' \in \mathscr{W}$ reachable in the same period and for all $h \in Y^\infty$,
$$\tau(w, h^L) = \tau(w', h^L).$$

Then for all $w, w' \in \mathscr{W}$ reachable in the same period and for all $h \in Y^\infty$,
$$f(\tau(w, h^t)) = f(\tau(w', h^t)), \quad \forall t \geq L + 1.$$

If $w = \tau(w^0, y^1, \ldots, y^{t-L-1})$ and $w' = \tau(w^0, \hat{y}^1, \ldots, \hat{y}^{t-L-1})$, then for h^t and $\hat{h}^t$ as specified in definition 13.3.1,
$$\begin{aligned}\sigma(h^t) &= f(\tau(w, y^{t-L}, \ldots, y^{t-1})) \\ &= f(\tau(w', y^{t-L}, \ldots, y^{t-1})) \\ &= f(\tau(w', \hat{y}^{t-L}, \ldots, \hat{y}^{t-1})) = \sigma(\hat{h}^t),\end{aligned}$$

and thus σ has L bounded recall.

Suppose now the profile σ has L bounded recall. Let $(\mathscr{W}, w^0, f, \tau)$ be a minimal representation of σ. Suppose w and w' are two states reachable in the same period. Then there exists h^s and $\hat{h}^s$ such that $w = \tau(w^0, h^s)$ and $w' = \tau(w^0, \hat{h}^s)$. For all $h \in Y^\infty$, $h^s h^t$ and $\hat{h}^s h^t$ agree for the last t periods, and so if $t \geq L$, they agree for at least the last L periods, and so
$$\begin{aligned}f(\tau(w, h^t)) &= \sigma(h^s h^t) \\ &= \sigma(\hat{h}^s h^t) = f(\tau(w', h^t)).\end{aligned}$$

Minimality of the representing automaton then implies that for all $h \in Y^\infty$ and $w, w' \in \mathscr{W}$ reachable in the same period, $\tau(w, h^L) = \tau(w', h^L)$. ∎

Proposition 13.3.1 *Fix a public monitoring game $(\tilde{u}^*, (Y, \rho))$ with discount factor δ and a strict PPE $\tilde{\sigma}$, with bounded recall L. There exists $\varepsilon > 0$ such that for all private monitoring games $(u^*, (Z, \pi))$ ε-close under ξ to $(\tilde{u}^*, (Y, \rho))$, the induced private profile is a sequential equilibrium of the private-monitoring game with the same δ.*

Proof Fix a strict PPE with bounded recall, $(\mathscr{W}, w^0, f, \tau)$. Fix a private monitoring technology (Z, π) ε-close under ξ to (Y, ρ). Denote the private state reached after a private history h_i^t by $w_i(h_i^t)$, and note that by lemma 13.3.1, the private state depends only on t and the last L realizations of the private signals in h_i^t.

Because L is finite, the PPE is trivially uniformly strict. Moreover, for all $\eta > 0$, ε can be chosen such that all players assign at least probability $1 - \eta$ to all the other players observing sequences of L private signals that are consistent with their own sequence of L private signals, that is, for all $h_i^t \in \mathscr{H}_i$, $\beta_i(w_i(h_i^t)\mathbf{1} \mid h_i^t) >$

$1 - \eta$. Lemmas 13.2.2 and 13.2.3 then imply that for ε sufficiently small, the profile $(\hat{\sigma}_1, \ldots, \hat{\sigma}_n)$ is a sequential equilibrium where $\hat{\sigma}_i$ is the strategy

$$\hat{\sigma}_i^t(h_i^t) = f_i(w_i(h_i^t))$$

and $w_i(h_i^0) = w^0$.

∎

Remark 13.3.1 **Uninterpretable signals** Even if there is some small chance that players observe an uninterpretable signal, bounded recall PPE can be extended to sequential equilibria of nearby private monitoring games (Mailath and Morris 2006).

◆

Remark 13.3.2 **Scope of bounded recall strategies** A characterization of the set of PPE payoffs achievable using bounded recall strategies is still unavailable. Cole and Kocherlakota (2005) show that for some parameterizations of the prisoners' dilemma, the set of PPE payoffs achievable by bounded recall strongly symmetric profiles is degenerate, consisting of the singleton $\{u(SS)\}$, whereas the set of strongly symmetric PPE with unbounded recall is strictly larger.

◆

13.4 Failure of Coordination under Unbounded Recall

This section shows that we quite generally cannot coordinate behavior with unbounded recall strategies. We begin with three examples in the repeated prisoners' dilemma.

13.4.1 Examples

Example 13.4.1 From proposition 12.3.1, the implied private profile from grim trigger is a Nash equilibrium in minimally private almost public monitoring games only if $q < r$. We now argue that under the private monitoring of example 13.1.1, even if $q < r$, the implied profile is not a Nash equilibrium in some nearby games with almost public monitoring. Suppose $0 < r' < q' < q < r$. Under this parameter restriction and for ε sufficiently small, the signal $\bar{z}_1''$ after S is a signal that player 2 also played S (i.e., $\pi_1(\bar{z}_1'' \mid SS) = \varepsilon + (r - r')(1 - 3\varepsilon) > \varepsilon + (q - q')(1 - 3\varepsilon) = \pi_1(\bar{z}_1'' \mid SE)$). However, the signal $\bar{z}_1'$ after S is a signal that player 2 played E (i.e., $\pi_1(\bar{z}_1' \mid SS) < \pi_1(\bar{z}_1' \mid SE)$), and so there is a $\beta_\varepsilon < 1$ with $\lim_{\varepsilon \to 0} \beta_\varepsilon = 1$ so that, whenever $\Pr(w_2^t = w_{EE}) < \beta_\varepsilon$,[2]

$$\Pr(w_2^{t+1} = w_{EE} \mid \bar{z}_1', a_1 = S) = \frac{\Pr(w_2^{t+1} = w_{EE}, \bar{z}_1' \mid a_1 = S)}{\Pr(\bar{z}_1' \mid a_1 = S)}$$

---

[2]. Following the notational convention of this chapter, we label the states in players' private automata with their labels in the corresponding public automaton, in contrast to section 12.3.2.

$$= \frac{\Pr(\bar{z}_1' \bar{z}_2, w_2^t = w_{EE} \mid a_1 = S)}{\Pr(\bar{z}_1' \mid a_1 = S)}$$

$$= \frac{\pi(\bar{z}_1' \bar{z}_2 \mid SE) \Pr(w_2^t = w_{EE})}{\pi_1(\bar{z}_1' \mid SE) \Pr(w_2^t = w_{EE}) + \pi_1(\bar{z}_1' \mid SS) \Pr(w_2^t = w_{SS})}$$

$$= \frac{q'(1 - 3\varepsilon) \Pr(w_2^t = w_{EE})}{\varepsilon + (1 - 3\varepsilon)[q' \Pr(w_2^t = w_{EE}) + r' \Pr(w_2^t = w_{SS})]}$$

$$> \Pr(w_2^t = w_{EE}).$$

Hence a sufficiently long history of the form $(E\underline{z}_1, S\bar{z}_1', S\bar{z}_1', \ldots, S\bar{z}_1')$ will lead to a posterior for player 1 assigning a probability close to 1 (for sufficiently small ε) that player 2's private state is still w_{EE}, and so by proposition 13.2.3, the profile is not a Nash equilibrium.

●

Example 13.4.2 We return to the minimally private monitoring game of section 12.3 and consider the forgiving profile in figure 13.4.1. This profile is a strict PPE for sufficiently large δ if $3p - 2q > 1$ (that is, if (7.2.5) holds). As for grim trigger, this profile does not induce a Nash equilibrium in any nearby minimally-private monitoring game if $q > r$. However, the analysis for the $q < r$ case now breaks into two subcases, because isolated observations of $\underline{z}_i$ do not lead to a private state of $w_i = w_{SS}$. For histories in which player i observes either two consecutive or no $\underline{z}_i$, the arguments in the proof of lemma 12.3.2 show that player i has no profitable deviation. The remaining histories are those with isolated observations of $\underline{z}_i$. The critical history, because it contains the largest fraction of $\underline{z}_i$'s consistent with E, is $(E\underline{z}_i, E\bar{z}_i, E\underline{z}_i, E\bar{z}_i, \ldots, E\underline{z}_i, E\bar{z}_i)$, that is, alternating $\underline{z}_i$ and $\bar{z}_i$. If $p(1 - p) > q(1 - q)$, such a history indicates that the other player is still playing E, and conditional on both players being in one of w_{EE} or $\hat{w}_{EE}$, a player assigns very high probability to the other player being in the same state (the state being determined by the last signal). The profile thus induces a Nash equilibrium in nearby minimally-private monitoring games if $q < r$ and $p(1 - p) > q(1 - q)$. On the other hand, if $p(1 - p) < q(1 - q)$, histories of the form $(E\underline{z}_i, E\bar{z}_i, E\underline{z}_i, E\bar{z}_i, \ldots, E\underline{z}_i, E\bar{z}_i)$ indicate that the other player is playing S and so player i strictly prefers not to follow the profile after a sufficiently long such history.

●

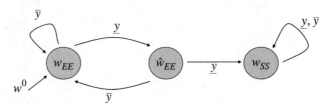

Figure 13.4.1 A forgiving profile in which two consecutive observations of $\underline{y}$ are needed to trigger permanent SS. As usual, the decision rule is given by $f(w_a) = a$.

13.4 ■ Failure of Coordination

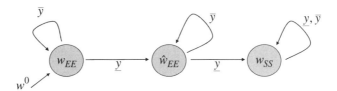

Figure 13.4.2 A forgiving profile in which two nonconsecutive observations of $\underline{y}$ trigger permanent SS.

Example 13.4.3 Our final forgiving profile in the context of the minimally private monitoring game of section 12.3 is displayed in figure 13.4.2. As for the profile from example 13.4.2, this profile is a strict PPE for sufficiently large δ if (7.2.5) holds. This PPE never induces a Nash equilibrium in any nearby minimally private monitoring game. Consider the player 1 private history $(E\bar{z}_1, E\bar{z}_1, \ldots, E\bar{z}_1)$ for T periods, followed by $E\underline{z}_1$. Under the profile in figure 13.4.2, player 1 is supposed to transit to the private state $\hat{w}_{EE}$ and play E. But for large T, it is significantly more likely that player 2 has observed $\underline{z}_2$ in exactly one of the first T periods than having observed $\bar{z}_2$ in every period. Consequently, for large T, player 1 assigns probability arbitrarily close to 1 that 2's private state is $w_2^T = \hat{w}_{EE}$.[3] The signal $\underline{z}_1$ then leads player 1 to assign a probability arbitrarily close to 1 that 2's private state is now $w_2^{T+1} = w_{SS}$, and so by lemma 13.2.3, 1 finds S uniquely optimal. ●

13.4.2 Incentives to Deviate

The examples in section 13.4.1 illustrate that the updating in almost public monitoring games can be very different than would be expected from the underlying public monitoring game. We now build on these examples to show that when the set of signals is sufficiently rich (in a sense to be defined), many profiles fail to induce equilibrium behavior in almost public monitoring games (proposition 13.4.1). This will occur whenever a player's beliefs can be manipulated through the selection of a private history so that the hypotheses of proposition 13.2.3 can be satisfied. In particular, we are interested in the weakest independent conditions on the private monitoring distributions and on the strategy profiles that would allow such manipulation. These conditions will need to rule out the drift in beliefs that disrupted the putative equilibrium of example 13.4.3, because such drift precludes the belief manipulation that lies at the heart of proposition 13.4.1 (and examples 13.4.1 and 13.4.2).

Fix a PPE of the public monitoring game and a nearby almost public monitoring game. The logic of example 13.4.1 runs as follows: Consider a player i in a private state $\hat{w}$ who assigns strictly positive (albeit small) probability to all the other players being in some other common private state $\bar{w} \neq \hat{w}$ (full-support private monitoring ensures that such an occurrence arises with positive probability). Let $\tilde{a} = (f_i(\hat{w}), f_{-i}(\bar{w}))$ be the action profile that results when i is in state $\hat{w}$ and all

---

3. This *drift* in beliefs can arise when players choose the same action in different states. Example 13.4.6 is a richer illustration.

	A	B	C
A	3,3	0,0	0,0
B	0,0	2,2	0,0
C	0,0	0,0	3,3

Figure 13.4.3 The stage game for examples 13.4.4–13.4.6.

the other players are in state $\bar{w}$. Suppose that if any other player is in a different private state $w \neq \bar{w}$, then the resulting action profile differs from $\tilde{a}$. Suppose, moreover, there is a signal y such that $\hat{w} = \tau(\hat{w}, y)$ and $\bar{w} = \tau(\bar{w}, y)$, that is, any player in the state $\hat{w}$ or $\bar{w}$ observing a private signal consistent with y stays in that private state (and so the profile cannot have bounded recall; see lemma 13.3.1). Suppose finally there is a private signal z_i for player i consistent with y that is more likely to have come from $\tilde{a}$ than *any* other action profile, that is, $z_i \in \xi_i^{-1}(y)$ and

$$\pi_i(z_i \mid \tilde{a}) > \pi_i(z_i \mid (f_i(\hat{w}), a'_{-i})) \ \forall a'_{-i} \neq f_{-i}(\bar{w}).$$

Then, after observing the private signal z_i, player i's posterior probability that all the other players are in $\bar{w}$ should increase. Moreover, because players in $\hat{w}$ and $\bar{w}$ do not change their private states, we can make player i's posterior probability that all the other players are in $\bar{w}$ as close to one as we like. If $f_i(\hat{w}) \neq f_i(\bar{w})$, an application of proposition 13.2.3 shows that the induced private profile is not an equilibrium.

13.4.3 Separating Profiles

The suppositions in the logic in section 13.4.2 can be weakened in two ways, culminating in proposition 13.4.1. The first weakening concerns the nature of the strategy profile. The logic assumed that there is a signal y such that $\hat{w} = \tau(\hat{w}, y)$ and $\bar{w} = \tau(\bar{w}, y)$. If there were only two states, $\hat{w}$ and $\bar{w}$, it would clearly be enough that there be a finite sequence of signals such that the automaton with initial state $\hat{w}$ (or $\bar{w}$) returns to $\hat{w}$ (or $\bar{w}$) under the sequence, as we now illustrate. We say that $\hat{w}$ and $\bar{w}$ *cycle* under the sequence of signals.

Example 13.4.4 The stage game is given in figure 13.4.3. In the public-monitoring game, there are two public signals, $\bar{y}$ and $\underline{y}$, with distribution ($0 < q < p < 1$)

$$\rho(\bar{y} \mid a) = \begin{cases} p, & \text{if } a_1 = a_2, \\ q, & \text{otherwise.} \end{cases} \quad (13.4.1)$$

The first profile we consider is illustrated in figure 13.4.4. This profile does not have bounded recall, and the states w_{AA} and w_{CC} cycle under both $\bar{y}$ and $\underline{y}$. The profile

13.4 ■ Failure of Coordination

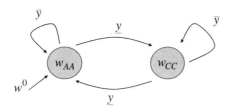

Figure 13.4.4 A simple cycling profile for the game from figure 13.4.3. As usual, the decision rule is given by $f(w_a) = a$.

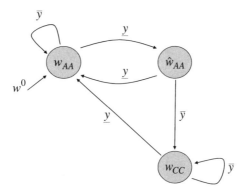

Figure 13.4.5 A more complicated cycling profile for the game from figure 13.4.3. As usual, the decision rule is given by $f(w_a) = a$.

is trivially a strict PPE, because the payoffs from AA and CC are identical. As $\underline{y}$ is an indication of miscoordination in behavior, the profile does not even induce a Nash equilibrium in nearby games with minimally private almost public monitoring: After a sufficiently long history of the form $(A\underline{y}_1, C\underline{y}_1, A\underline{y}_1, C\underline{y}_1, \ldots, A\underline{y}_1)$,[4] player 1 (say) is supposed to play C but assigns probability close to 1 that player 2 is in state w_{AA}.

●

Example 13.4.5 Consider now the profile in figure 13.4.5. This profile, a modification of the profile of figure 13.4.4, also does not have bounded recall. A sequence of private signals $\underline{y}\underline{y} \ldots \underline{y}$ cannot differentiate between w_{AA} and $\hat{w}_{AA}$, because behavior is identical in these two states. Under the sequence of signals $\bar{y}\bar{y}\ldots\bar{y}$, the states w_{AA} and w_{CC} do both cycle, and behavior is different in both states. However, under public monitoring (and so under minimally private monitoring as well), $\bar{y}$ is a signal that the players have coordinated. Suppose instead that the almost public monitoring is not minimally private. Similar to example 13.1.1, player 1's private signal space is $Z_1 = \{\underline{z}_1, \bar{z}'_1, \bar{z}''_1\}$, player 2's signal space is $Z_2 = \{\underline{z}_2, \bar{z}_2\}$, and the joint distribution is given by Figure 13.4.6.

As in example 13.4.1, when $q' > p'$, the signal $\bar{z}'_1$ is a signal to player 1 that player 2 had chosen a different action. It is then intuitive that if player 1 is in private

---

4. Because the monitoring is minimally private, $Z_i = Y$.

	$\underline{z}_2$	$\bar{z}_2$
$\underline{z}_1$	$(1-\zeta)(1-3\varepsilon)$	ε
$\bar{z}'_1$	ε	$\zeta'(1-3\varepsilon)$
$\bar{z}''_1$	ε	$(\zeta-\zeta')(1-3\varepsilon)$

Figure 13.4.6 The probability distribution of private signals for example 13.4.5. The distribution is given as a function of the action profile a, where $\zeta = p$ and $\zeta' = p'$ if $a_1 = a_2$, and q or q' otherwise.

state $w_1 = w_{AA}$ and observes $\bar{z}'_1$, he then assigns a higher posterior to player 2 being in private state $w_2 = w_{CC}$. After sufficiently many observations of $\bar{z}'_1$, player 1 should then assign arbitrarily high probability to $w_2 = w_{CC}$, at which point we can apply proposition 13.2.3. This intuition is correct, but potentially incomplete because of state $\hat{w}_{AA}$. In the presence of nontrivial private monitoring, player 1 cannot rule out the possibility that player 2 observes a signal (or sequence of signals) that leaves her in private state $w_2 = \hat{w}_{AA}$, a state that does not cycle under $\bar{y}$. However, it turns out that the logic of the intuition is saved by the observation that under $\bar{y}$, the profile takes $\hat{w}_{AA}$ into a state that does cycle (namely, w_{CC}). Consequently, in the long run, the possibility that player 2 is in state $\hat{w}_{AA}$ does not affect significantly 1's updating under histories $A\bar{z}'_1, A\bar{z}'_1, \ldots, A\bar{z}'_1$.

●

Summarizing the examples, to find a private history that both leaves a player in some private state $\hat{w}$ and after which that player assigns arbitrarily high probability to all the other players being in some other state $\bar{w} \neq \hat{w}$, it is enough to find a finite sequence of signals under which (1) $\bar{w}$ and at least one other distinct state cycle; (2) for any states that cycle, the actions chosen potentially allow the signals to reveal that the states are different from $\bar{w}$;[5] and (3) any states that don't cycle are taken into the cycling states by the finite sequence of signals.

Let $\mathscr{W}_t$ be the set of states reachable in period t, $\mathscr{W}_t \equiv \{w \in \mathscr{W} : w = \tau(w^0, y^0, y^1, \ldots, y^{t-1})$ for some $(y^0, y^1, \ldots, y^{t-1})\}$. Define $R(\tilde{w})$ as the set of states that are repeatedly reachable in the same period as $\tilde{w}$ (i.e., $R(\tilde{w}) = \{w \in \mathscr{W} : \{w, \tilde{w}\} \subset \mathscr{W}_t$ infinitely often$\}$).

Definition A *public profile is* cyclically separating *if there is a finite sequence of signals*
13.4.1 $\bar{y}^0, \ldots, \bar{y}^m$, *a collection of states* $\mathscr{W}_c$, *and a state* $\bar{w} \in \mathscr{W}_c$ *such that*

1. $\tau(w, \bar{y}^0, \ldots, \bar{y}^m) = w$ *for all* $w \in \mathscr{W}_c$;
2. $\tau(w, \bar{y}^0, \ldots, \bar{y}^m) \in \mathscr{W}_c$ *for all* $w \in R(\bar{w})$;

---

5. In example 13.4.2, taking $\bar{w} = w_{SS}$ and the cycle $\bar{y}\underline{y}$ shows that it not necessary that all cycling states be revealed to be distinct.

13.4 ■ Failure of Coordination

3. $\forall w \in \mathscr{W}_c \setminus \{\bar{w}\}$, $\forall i \, \exists k$, $0 \leq k \leq m$, such that

$$f_i(\tau(w, \bar{y}^0, \ldots, \bar{y}^k)) \neq f_i(\tau(\bar{w}, \bar{y}^0, \ldots, \bar{y}^k));$$

and

4. $|\mathscr{W}_c| \geq 2$.

Mailath and Morris (2006) show that this is, for finite public profiles, equivalent to the weaker property of *separating*. Given an outcome path $h^\infty \equiv (y^1, y^2, \ldots) \in Y^\infty$, let ${}^s h^\infty \equiv (y^s, y^{s+1}, \ldots) \in Y^\infty$ denote the outcome path from period s, so that $h^\infty = (h^s, {}^s h^\infty)$ and ${}^s h^{s+t} = (y^s, y^{s+1}, \ldots, y^{s+t-1})$.

Definition 13.4.2 *The public strategy profile is* separating *if there is some state $\tilde{w}$ and an outcome path $h^\infty \in Y^\infty$ such that there is another state $w \in R(\tilde{w})$ that satisfies $\tau(w, h^t) \neq \tau(\tilde{w}, h^t)$ for all t, and for all s and $w \in R(\tau(\tilde{w}, h^s))$, if $\tau(w, {}^s h^{s+t}) \neq \tau(\tilde{w}, h^{s+t})$ for all $t \geq 0$, then*

$$f_i(\tau(w, {}^s h^{s+t})) \neq f_i(\tau(\tilde{w}, h^{s+t})) \text{ infinitely often, for all } i.$$

Clearly, a separating profile cannot have bounded recall. Example 13.4.6 presents a PPE that neither has bounded recall nor is separating. Every other PPE in this book either has bounded recall or is separating. Example 13.4.6 also illustrates drift in beliefs and how drift is a distinct cause for the failure to induce Nash equilibrium in nearby games. Further discussion of separation can be found in Mailath and Morris (2006).

Example 13.4.6 The public profile is illustrated in figure 13.4.7. It is easily verified that this profile is again a strict PPE. This profile is not separating. Under any path in which $\bar{y}$ appears in any period before the last, all states transit to the same state. Under the remaining paths, only w_{AA} and $\hat{w}_{AA}$ appear. The definition of separation fails because play is the same at states w_{AA} and $\hat{w}_{AA}$.

The profile is also not robust to even minimally private monitoring, because of a drift in beliefs. To get a handle on the evolution of beliefs, we consider the minimally private distribution π obtained by the compound randomization where in the first stage a value of y is determined according to the ρ of (13.4.1), and then in the second stage, that value is reported to player i with probability $(1 - \varepsilon)$

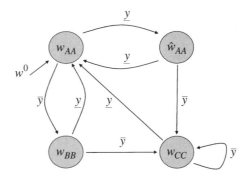

Figure 13.4.7 The strategy profile for example 13.4.6.

and the other value with probability ε. Conditional on the realization of the first stage, the second-stage randomizations are independent across players. For $\varepsilon = 0$, the monitoring is public.

We now consider the evolution of player 1's beliefs about the private state of player 2, as player 1 continually observes $\underline{y}_1$ and plays A. Let $\beta_t(w)$ be the belief that player 1 assigns to player 2 being in state w in the beginning of period t on such a history. The probability that player 1 observes $\underline{y}_1$ at the end of period t is

$$[(1-p)(1-\varepsilon) + p\varepsilon](\beta_t(w_{AA}) + \beta_t(\hat{w}_{AA}))$$
$$+ [(1-q)(1-\varepsilon) + q\varepsilon](\beta_t(w_{BB}) + \beta_t(w_{CC}))$$
$$= [(1-p)(1-\varepsilon) + p\varepsilon]\beta_t^{AA} + [(1-q)(1-\varepsilon) + q\varepsilon](1-\beta_t^{AA}) \equiv \xi(\beta_t^{AA}),$$

where $\beta_t^{AA} \equiv \beta_t(w_{AA}) + \beta_t(\hat{w}_{AA})$. Bayes' rule then implies

$$\beta_{t+1}(w_{AA}) = \xi(\beta_t^{AA})^{-1}\{[(1-p)(1-\varepsilon)^2 + p\varepsilon^2]\beta_t(\hat{w}_{AA})$$
$$+ [(1-q)(1-\varepsilon)^2 + q\varepsilon^2](1-\beta_t^{AA})\},$$
$$\beta_{t+1}(\hat{w}_{AA}) = \xi(\beta_t^{AA})^{-1}[(1-p)(1-\varepsilon)^2 + p\varepsilon^2]\beta_t(w_{AA}),$$
$$\beta_{t+1}(w_{BB}) = \xi(\beta_t^{AA})^{-1}\varepsilon(1-\varepsilon)\beta_t(w_{AA}),$$

and

$$\beta_{t+1}(w_{CC}) = \xi(\beta_t^{AA})^{-1}\varepsilon(1-\varepsilon)(1-\beta_t(w_{AA})).$$

These equations imply

$$\beta_{t+1}^{AA} = \xi(\beta_t^{AA})^{-1}\{[(1-p)(1-\varepsilon)^2 + p\varepsilon^2]\beta_t^{AA}$$
$$+ [(1-q)(1-\varepsilon)^2 + q\varepsilon^2](1-\beta_t^{AA})\}.$$

For ε small, the equation is a contraction, and so $\beta_t^{AA} \to \beta_\infty^{AA}$ as $t \to \infty$. This then implies that for small ε, β_∞^{AA} is close to 1 and so for large t, $\beta_t(w_{AA})$ is close to $1/2$. Given these beliefs, the first observation of $\bar{y}_1$ leads player 1 to assign approximately equal beliefs to player 2's private state being w_{BB} and w_{CC}, and so the best reply is C (irrespective of player 1's private state). ●

13.4.4 Rich Monitoring

We now address the second condition that appeared in section 13.4.2, the requirement that for some player i, for every action profile there is a player i private signal that is more likely to have come from that action profile than from any other. This is a much stronger condition than is needed. This section presents a weaker sufficient condition, that the monitoring be "rich." Because this condition is technical, some readers may prefer to proceed directly to section 13.4.5, substituting the stronger condition for richness.

Recall that $\tilde{a}$ in section 13.4.2 is the action profile that results when player i is in private state $\hat{w}$, whereas all the other players are in the common private state $\bar{w}$. It is not necessary that the *same* private signal z_i be more likely to have come from $\tilde{a}$ than

13.4 ■ Failure of Coordination

any other action profile. It is enough that for each action profile different from $\tilde{a}$, there is a private signal more likely to have come from $\tilde{a}$ than from that profile, as long as the signal does not disrupt the inferences about other states too badly. For example, suppose there are two players, with player 1's beliefs to be manipulated. In addition to state $\bar{w}$, player 2 could be in state $\hat{w}$ or w', both of which cycle under y. Suppose also $A_2 = \{\hat{a}_2, a'_2, \bar{a}_2\}$. We would like the odds ratio $\Pr(w_2 \neq \bar{w} \mid h_1^t)/\Pr(w_2 = \bar{w} \mid h_1^t)$ to converge to 0 as $t \to \infty$ for appropriate private histories. Let $\hat{a}_1 = f_1(\hat{w})$, $\bar{a}_2 = f_2(\bar{w})$, $\hat{a}_2 = f_2(\hat{w})$, and $a'_2 = f_2(w')$ (note that $\tilde{a} = (\hat{a}_1, \bar{a}_2)$). Suppose, moreover, there are two private signals z'_1 and z''_1 consistent with the same public signal, satisfying

$$\pi_1(z'_1 \mid \hat{a}_1, a'_2) > \pi_1(z'_1 \mid \tilde{a}) > \pi_1(z'_1 \mid \hat{a})$$

and

$$\pi_1(z''_1 \mid \hat{a}) > \pi_1(z''_1 \mid \tilde{a}) > \pi_1(z''_1 \mid \hat{a}_1, a'_2).$$

Then, after observing the private signal z'_1, we have

$$\frac{\Pr(w_2 = \hat{w} \mid h_1^t, z'_1)}{\Pr(w_2 = \bar{w} \mid h_1^t, z'_1)} = \frac{\pi_1(z'_1 \mid \hat{a})}{\pi_1(z'_1 \mid \tilde{a})} \frac{\Pr(w_2 = \hat{w} \mid h_1^t)}{\Pr(w_2 = \bar{w} \mid h_1^t)} < \frac{\Pr(w_2 = \hat{w} \mid h_1^t)}{\Pr(w_2 = \bar{w} \mid h_1^t)}$$

as desired, but $\Pr(w_2 = w' \mid h_1^t, z'_1)/\Pr(w_2 = \bar{w} \mid h_1^t, z'_1)$ increases.[6] On the other hand, the odds ratio $\Pr(w_2 = w' \mid h_1^t, z''_1)/\Pr(w_2 = \bar{w} \mid h_1^t, z''_1)$ falls after observing the private signal z''_1, while $\Pr(w_2 = \hat{w} \mid h_1^t, z''_1)/\Pr(w_2 = \bar{w} \mid h_1^t, z''_1)$ increases. However, it may be that the increases can be offset by appropriate decreases, so that, for example, z'_1 followed by two realizations of z''_1 results in a decrease in *both* odds ratios. If so, a sufficiently high number of realizations of $z'_1 z''_1 z''_1$ results in $\Pr(w_2 \neq \bar{w} \mid h_1^t)/\Pr(w_2 = \bar{w} \mid h_1^t)$ being close to 0.

In terms of the odds ratios, the sequence of signals $z'_1 z''_1 z''_1$ lowers both odds ratios if, and only if,

$$\frac{\pi_1(z'_1 \mid \hat{a})}{\pi_1(z'_1 \mid \tilde{a})} \left(\frac{\pi_1(z''_1 \mid \hat{a})}{\pi_1(z''_1 \mid \tilde{a})}\right)^2 < 1$$

and

$$\frac{\pi_1(z'_1 \mid \hat{a}_1, a'_2)}{\pi_1(z'_1 \mid \tilde{a})} \left(\frac{\pi_1(z''_1 \mid \hat{a}_1, a'_2)}{\pi_1(z''_1 \mid \tilde{a})}\right)^2 < 1.$$

A richness condition on private monitoring distributions captures this idea. For a private monitoring distribution, (Z, π), define $\gamma_{aa'_{-i}}(z_i) \equiv \ln \pi_i(z_i \mid a_i, a_{-i}) - \ln \pi_i(z_i \mid a_i, a'_{-i})$, and let $\gamma_a(z_i) = (\gamma_{aa'_{-i}}(z_i))_{a'_{-i} \in A_{-i}, a'_{-i} \neq a_{-i}}$ denote the vector in $\mathbb{R}^{|A_{-i}|-1}$ of the log odds ratios of the signal z_i associated with different action profiles. The last two displayed equations can then be written as $\frac{1}{3}\gamma_{\tilde{a}}(z'_1) + \frac{2}{3}\gamma_{\tilde{a}}(z''_1) > \mathbf{0}$, where $\mathbf{0}$ is the 2×1 zero vector.[7]

---

6. Repeated observations of the signal z'_1 will lead player 1 to assign high probability to player 2 being in private state w', suggesting that we should simply focus on w' instead of $\bar{w}$. This possibility would not arise if there were a fourth state that cycles under y and appropriate inequalities on the probabilities are satisfied.

7. The convex combination is strictly positive (rather than negative) because the definition of $\gamma_{aa'_{-i}}$ inverts the odds ratios from the displayed equations.

Definition 13.4.3 *A private monitoring distribution* (Z, π) *is* rich *for player i, if for all $y \in Y$ and $a \in A$, the convex hull of* $\{\gamma_a(z_i) : z_i \in f_i^{-1}(y) \text{ and } \pi_i(z_i \mid a_i, a'_{-i}) > 0 \text{ for all } a'_{-i} \in A_{-i}\}$ *has a nonempty intersection with* $\mathbb{R}_{++}^{|A_{-i}|-1}$.

Trivially, if for each action profile there is a private signal more likely to have come from that profile than any other profile, then the private monitoring distribution is rich. It will be useful to quantify the extent to which the conditions of definition 13.4.3 are satisfied. Because the space of signals and actions are finite, there are a finite number of constraints in definition 13.4.3, and so for any rich private monitoring distribution, the set of ζ over which the supremum is taken in the next definition is nonempty.[8]

Definition 13.4.4 *The* richness *of a private monitoring distribution (Z, π) rich for i is the supremum of all $\zeta > 0$ satisfying: for all $y \in Y$ and $a \in A$, the convex hull of the set of vectors* $\{\gamma_a(z_i) : z_i \in f_i^{-1}(y) \text{ and } \pi_i(z_i \mid a_i, a'_{-i}) \geq \zeta \text{ for all } a'_{-i} \in A_{-i}\}$ *has a nonempty intersection with* $\mathbb{R}_{\zeta}^{|A_{-i}|-1} \equiv \{x \in \mathbb{R}_{++}^{|A_{-i}|-1} : x_k \geq \zeta \text{ for } k = 1, \ldots, |A_{-i}| - 1\}$.

13.4.5 Coordination Failure

It remains to ensure that under private monitoring, players may transit to different states. It suffices to assume the following, weaker than full support condition:

Definition 13.4.5 *A private monitoring distribution (Z, π) that is ε-close to a public monitoring distribution (Y, ρ) has* essentially full support *if for all $(y_1, \ldots, y_n) \in Y^n$,*

$$\pi\{(z_1, \ldots, z_n) \in Z : \xi_i(z_i) = y_i\} > 0.$$

Mailath and Morris (2006) then prove the following result.

Proposition 13.4.1 *Fix a separating strict finite PPE of a full-support public monitoring game $(\tilde{u}^*, (Y, \rho))$. For all $\zeta > 0$, there exists $\varepsilon' > 0$ such that for all $\varepsilon < \varepsilon'$, if $(u, (Z, \pi))$ is a private monitoring game ε-close to $(\tilde{u}^*, (Y, \rho))$ with (Z, π) having richness at least ζ for some i and essentially full support, then the induced private profile is not a Nash equilibrium of the private monitoring game.*

We direct the reader to the original publication for the (long) proof. It is worth noting that the bound on ε is only a function of the richness of the private monitoring. It is *independent* of the probability that a disagreement in private states arises.

13.5 Patient Players

We now seek a folk theorem for almost public monitoring games. Proposition 13.3.1 showed that any *strict* (bounded recall) PPE of a public monitoring game induces a sequential equilibrium in any ε-close private monitoring game. The difficulty in

---

8. The bound ζ appears twice in the definition. Its first appearance ensures that for all $\zeta > 0$, there is uniform upper bound on the number of private signals satisfying $\pi_i(z_i \mid a_i, a'_{-i}) \geq \zeta$ in any private monitoring distribution with a richness of at least ζ.

13.5 ■ Patient Players

building a folk theorem on this result is that as players become patient and for any bounded recall strategy profile, values become equalized across states (see remark 7.2.1 for an example). This forces to 0 the value differences that support the construction in proposition 13.3.1, so that the implied bound on ε goes to 0 as δ goes to 1. At the same time, when players become increasing patient, the incentive to deviate is also disappearing. This section identifies conditions under which the balance between these forces allows strict equilibria in public monitoring games to carry over into equilibria or nearby private monitoring games for arbitrarily patient players.

13.5.1 Patient Strictness

A public profile is *finite* if the minimal representing automaton has a finite state space (see remark 7.1.3). We first rewrite the incentive constraints of the public monitoring game. Recall, from corollary 7.1.1, that $(\mathcal{W}, w^0, f, \tau)$ is a strict equilibrium if, for all i, $w \in \mathcal{W}$ and all $a_i \neq f_i(w)$,

$$\tilde{V}_i(w) > (1-\delta)u_i(a_i, f_{-i}(w)) + \delta \sum_y \tilde{V}_i(\tau(w,y))\rho(y \mid a_i, f_{-i}(w)),$$

where

$$\tilde{V}_i(w) = (1-\delta)u_i(f(w)) + \delta \sum_y \tilde{V}_i(\tau(w,y))\rho(y \mid f(w)). \quad (13.5.1)$$

The incentive constraints at w can be written more transparently, focusing on the transitions between states, as

$$\tilde{V}_i(w) > (1-\delta)u_i(a_i, f_{-i}(w)) + \delta \sum_{w'} \theta_{ww'}(a_i, f_{-i}(w))\tilde{V}_i(w'), \quad (13.5.2)$$

where $\theta_{ww'}(a)$ is the probability of transiting from state w to state w' under the action profile a,

$$\theta_{ww'}(a) = \sum_{\{y \in Y : \tau(w,y)=w'\}} \rho(y \mid a)$$

(the sum is 0 if $\tau(w,y) \neq w'$ for all $y \in Y$). Using (13.5.1) to substitute for $\tilde{V}_i(w)$ in (13.5.2) and rearranging yields (writing $\hat{\theta}_{ww'}$ for $\theta_{ww'}(f(w))$ and $\hat{\theta}_{ww'}(a_i)$ for $\theta_{ww'}(f_{-i}(w), a_i)$),

$$\delta \sum_{w'} (\hat{\theta}_{ww'} - \hat{\theta}_{ww'}(a_i))\tilde{V}_i(w') > (1-\delta)(u_i(a_i, f_{-i}(w)) - u_i(f(w))).$$

For any $\bar{w} \in \mathcal{W}$, because probabilities sum to 1, this is equivalent to

$$\delta \sum_{w'} (\hat{\theta}_{ww'} - \hat{\theta}_{ww'}(a_i))(\tilde{V}_i(w') - \tilde{V}_i(\bar{w}))$$
$$> (1-\delta)(u_i(a_i, f_{-i}(w)) - u_i(f(w))). \quad (13.5.3)$$

The property of *connectedness* now plays a critical role.[9]

---

9. Mailath and Morris (2002, section 3.2) discusses an example in which connectedness fails and players place significant probability on disagreement after long histories in nearby almost public monitoring games.

Definition 13.5.1 *A public profile is connected if for all $w, w' \in \mathscr{W}$, there exists $h^m \in Y^m$ for some m and $\bar{w} \in \mathscr{W}$ such that*

$$\tau(w, h^m) = \bar{w} = \tau(w', h^m).$$

Lemma 13.5.1 *For any connected finite public profile, there is a finite sequence of signals $h^m \in Y^m$ and a state $\bar{w}$ such that*

$$\tau(w, h^m) = \bar{w}, \quad \forall w \in \mathscr{W}.$$

Proof We prove this for $|\mathscr{W}| = 3$; the extension to an arbitrary finite number of states being a straightforward iteration. Fix w^1, w^2, and w^3. Let h^m be a sequence that satisfies $\tau(w^1, h^m) = \tau(w^2, h^m) \equiv w$. Because the profile is connected, there is a sequence of signals $\hat{h}^{m'}$ such that $\tau(w, \hat{h}^{m'}) = \tau(w', \hat{h}^{m'})$, where $w' \equiv \tau(w^3, h^m)$. The desired sequence of signals is the concatenation $h^m \hat{h}^{m'}$. ■

We need the following standard result (see, for example, Stokey and Lucas 1989, theorem 11.4). If $(\mathscr{Z}, R)$ is a finite-state Markov chain with state space $\mathscr{Z}$ and transition matrix R, then R^n is the matrix of n-step transition probabilities and $r_{ij}^{(n)}$ is the ijth element of R^n. For a vector $x \in \mathbb{R}^m$, define $\|x\|_\Sigma \equiv \sum_j |x_j|$.

Lemma 13.5.2 *Suppose $(\mathscr{Z}, R)$ is a finite-state Markov chain. Let $\eta_j^{(n)} = \min_i r_{ij}^{(n)}$ and $\eta^{(n)} = \sum_j \eta_j^{(n)}$. Suppose that there exists ℓ such that $\eta^{(\ell)} > 0$. Then, $(\mathscr{Z}, R)$ has a unique stationary distribution ϕ^* and, for all $\phi \in \Delta(\mathscr{Z})$,*

$$\|\phi R^{k\ell} - \phi^*\|_\Sigma \leq 2(1 - \eta^{(\ell)})^k.$$

If the profile is finite and connected and ρ has full support, the Markov chain on $\mathscr{W}$ implied by the profile is ergodic, and so has a unique stationary distribution. As a consequence, $\lim_{\delta \to 1} \tilde{V}_i(w)$ is independent of $w \in \mathscr{W}$, so simply taking $\delta \to 1$ in (13.5.3) yields $0 \geq 0$. The next lemma implies that we can instead divide by $(1 - \delta)$ and then evaluate the constraint.

Lemma 13.5.3 *Suppose the public monitoring has full support, and the public profile is finite and connected. The value of state $w \in \mathscr{W}$ for large δ, $\lim_{\delta \to 1} \tilde{V}_i(w)$, is independent of w. For any two states $w, \bar{w} \in \mathscr{W}$,*

$$\Delta_{w\bar{w}}\tilde{V}_i \equiv \lim_{\delta \to 1} \frac{(\tilde{V}_i(w) - \tilde{V}_i(\bar{w}))}{(1 - \delta)}$$

exists and is finite.

Proof Let Θ denote the matrix of transition probabilities on the finite state space $\mathscr{W}$ induced by the public profile. The ww'th element is $\theta_{ww'}(f(w)) = \hat{\theta}_{ww'}$. If $u_i(f) \in \mathbb{R}^\mathscr{W}$ and $\tilde{V}_i \in \mathbb{R}^\mathscr{W}$ are the vectors of stage payoffs and continuation values for player i associated with the states, then

$$\tilde{V}_i = (1 - \delta)u_i(f) + \delta \Theta \tilde{V}_i.$$

13.5 ■ Patient Players

Solving for $\tilde{V}_i$ yields

$$\tilde{V}_i = (1-\delta)(I_{\mathscr{W}} - \delta\Theta)^{-1} u_i(f)$$

$$= (1-\delta)\sum_{t=0}^{\infty}(\delta\Theta)^t u_i(f),$$

where $I_{\mathscr{W}}$ is the $|\mathscr{W}|$-dimensional identity matrix. Let e_w denote the wth standard basis vector. Then,

$$\tilde{V}_i(w) - \tilde{V}_i(\bar{w}) = (1-\delta)\sum_{t=0}^{\infty}(e_w - e_{\bar{w}})(\delta\Theta)^t u_i(f)$$

$$= (1-\delta)\sum_{t=0}^{\infty}\delta^t(e_w\Theta^t - e_{\bar{w}}\Theta^t)u_i(f).$$

Because the public profile is connected, for any two distributions on $\mathscr{W}$, ϕ and ϕ', $\|\phi\Theta^t - \phi'\Theta^t\|_{\Sigma} \to 0$ at an exponential rate (lemmas 13.5.1 and 13.5.2), so $\tilde{V}_i(w) - \tilde{V}_i(\bar{w}) \to 0$. Moreover, $\sum_{t=0}^{\infty}(e_w\Theta^t - e_{\bar{w}}\Theta^t)u_i(f(w))$ is absolutely convergent, so

$$\frac{(\tilde{V}_i(w) - \tilde{V}_i(\bar{w}))}{(1-\delta)}$$

has a finite limit as $\delta \to 1$.

∎

Hence, if a connected finite public profile is a strict equilibrium for discount factors arbitrarily close to 1,

$$\sum_{w'}(\hat{\theta}_{ww'} - \hat{\theta}_{ww'}(a_i))\Delta_{w'\bar{w}}\tilde{V}_i \geq (u_i(a_i, f_{-i}(w)) - u_i(f(w))).$$

Strengthening the weak inequality to a strict one gives a condition that implies (13.5.3) for δ large.

Definition 13.5.2 *A connected finite PPE is* patiently strict *if for some $\bar{w} \in \mathscr{W}$, all players i, states $w \in \mathscr{W}$, and actions $a_i \neq f_i(w)$,*

$$\sum_{w'}(\hat{\theta}_{ww'} - \hat{\theta}_{ww'}(a_i))\Delta_{w'\bar{w}}\tilde{V}_i > (u_i(a_i, f_{-i}(w)) - u_i(f(w))). \quad (13.5.4)$$

The value of the left side of (13.5.4) is independent of $\bar{w}$, because $\Delta_{w'\bar{w}}\tilde{V}_i + \Delta_{\bar{w}\tilde{w}}\tilde{V}_i = \Delta_{w'\tilde{w}}\tilde{V}_i$. The next lemma is obvious.

Lemma 13.5.4 *Suppose $(\tilde{u}^*, (Y, \rho))$ is a public monitoring game with full-support ρ. For any patiently strict connected finite public profile, there exists $\underline{\delta} < 1$ such that for all $\delta \in (\underline{\delta}, 1)$, the public profile is a strict PPE of the game with public monitoring.*

13.5.2 Equilibria in Nearby Games

This section proves the following proposition. It is worth remembering that every bounded recall public profile is both a connected finite public profile and induces posterior beliefs in nearby almost public monitoring games that assign uniformly large probability to agreement in private states. If monitoring is minimally private and we only impose the belief requirement after histories consistent with the strategy profile, the profile would still be a Nash equilibrium (and the induced outcome sequential) in the game with private monitoring (by proposition 13.2.2).

Proposition 13.5.1 *Suppose the public profile $(\mathscr{W}, w, f, \tau)$ is a connected patiently strict PPE of the public monitoring game $(\tilde{u}^*, (Y, \rho))$ with full-support ρ. For all $\kappa > 0$, there exist $\underline{\delta} < 1$, $\eta > 0$, and $\varepsilon > 0$ such that for all $(u^*, (Z, \pi))$ ε-close under ξ to $(\tilde{u}^*, (Y, \rho))$, if $\beta_i(\tau_i(h_i^t)\mathbf{1} \mid h_i^t) > 1 - \eta$ for all $h_i^t \in \mathcal{H}_i$, and $\delta \in (\underline{\delta}, 1)$, then the private profile is a sequential equilibrium of the game with private monitoring, and the payoff under the private profile is within κ of the payoff of the public profile.*

The finite public profile induces in the game with private monitoring a finite-state Markov chain $(\mathscr{Z}, Q^\pi)$, where $\mathscr{Z} \equiv \mathscr{W}^n$ and, for $\mathbf{w}, \mathbf{w}' \in \mathscr{Z}$,

$$q^\pi_{\mathbf{ww}'}(a) = \sum_{\{z_1 : \tau_1(w_1, z_1) = w_1'\}} \cdots \sum_{\{z_n : \tau_n(w_n, z_n) = w_n'\}} \pi(z \mid a)$$

(as for $\theta_{ww'}(a)$, the sum is 0 if for some i, $\tau(w_i, y) \neq w_i'$ for all $y \in Y$, because in that case, $\tau_i(w_i, z_i) \neq w_i'$ for all $z_i \in Z_i$). The value to player i at the vector of private states $\mathbf{w}$ is

$$V_i(\mathbf{w}) = (1 - \delta)u_i(f(\mathbf{w})) + \delta \sum_z \pi(z \mid f(\mathbf{w}))V_i(\tau_1(w_1, z_1), \ldots, \tau_n(w_n, z_n))$$

$$= (1 - \delta)u_i(f(\mathbf{w})) + \delta \sum_{\mathbf{w}'} q^\pi_{\mathbf{ww}'}(f(\mathbf{w}))V_i(\mathbf{w}')$$

$$= (1 - \delta)u_i(f(\mathbf{w})) + \delta \sum_{\mathbf{w}'} \hat{q}^\pi_{\mathbf{ww}'} V_i(\mathbf{w}'),$$

where $\hat{q}^\pi_{\mathbf{ww}'} \equiv q^\pi_{\mathbf{ww}'}(f(\mathbf{w}))$. We also define $\hat{q}^\pi_{\mathbf{ww}'}(a_i) \equiv q^\pi_{\mathbf{ww}'}(a_i, f_{-i}(\mathbf{w}))$.

Analogous to lemma 13.5.3, we have the following:

Lemma 13.5.5 *Suppose the public profile is finite and connected.*

1. *The value of state $\mathbf{w} \in \mathscr{Z}$ for large δ, $\lim_{\delta \to 1} V_i(\mathbf{w})$, is independent of $\mathbf{w}$. For any two vectors of private states $\mathbf{w}, \bar{\mathbf{w}} \in \mathscr{W}^n$,*

$$\Delta_{\mathbf{w}\bar{\mathbf{w}}} V_i \equiv \lim_{\delta \to 1} \frac{(V_i(\mathbf{w}) - V_i(\bar{\mathbf{w}}))}{(1 - \delta)}$$

exists and is finite,

2. *there exists $\varepsilon > 0$ such that for all (Z, π) ε-close to (Y, ρ), $\Delta_{\mathbf{w}\bar{\mathbf{w}}} V_i$ has an upper bound independent of π, and*

3. *for any $\zeta > 0$, there exists $\varepsilon > 0$ such that for all (Z, π) ε-close to (Y, ρ), and any two states $w, \bar{w} \in \mathscr{W}$, $|\Delta_{w\mathbf{1}, \bar{w}\mathbf{1}} V_i - \Delta_{w\bar{w}} \tilde{V}_i| < \zeta$.*

Proof The proof of the first assertion is identical to that of lemma 13.5.3, and also shows that

$$\Delta_{\mathbf{w}\bar{\mathbf{w}}} V_i = \sum_{t=0}^{\infty} (e_\mathbf{w}(Q^\pi)^t - e_{\bar{\mathbf{w}}}(Q^\pi)^t) u_i(f(\mathbf{w})). \tag{13.5.5}$$

13.5 ■ Patient Players

Because the public profile is finite and connected, for the purposes of applying lemma 13.5.2, we can take $\ell = m$, independent of π, where m is the length of the finite sequence of signals from lemma 13.5.1. Moreover, there exists $\varepsilon' > 0$ such that for all (Z, π) ε'-close to (Y, ρ),

$$\sum_{\mathbf{w}'} \min_{\mathbf{w}} q^{\pi,(m)}_{\mathbf{ww}'} > \frac{1}{2} \sum_{w'} \min_{w} \theta^{(m)}_{ww'} \equiv \eta^* > 0.$$

From lemma 13.5.2, for all (Z, π) ε'-close to (Y, ρ), $|e_{\mathbf{w}}(Q^\pi)^t - e_{\bar{\mathbf{w}}}(Q^\pi)^t| < 4(1-\eta^*)^t$ for all t, which, with (13.5.5), implies the second assertion.

From the proof of lemma 13.5.3,

$$\Delta_{w\bar{w}} \tilde{V}_i = \sum_{t=0}^{\infty} (e_w \Theta^t - e_{\bar{w}} \Theta^t) u_i(f(w)).$$

For a fixed $\zeta > 0$, there exists T such that

$$\left| \sum_{t=0}^{T} (e_w \Theta^t - e_{\bar{w}} \Theta^t) u_i(f(w)) - \Delta_{w\bar{w}} \tilde{V}_i \right| < \zeta/3 \tag{13.5.6}$$

and, for all (Z, π) ε'-close to (Y, ρ),

$$\left| \sum_{t=0}^{T} (e_{w\mathbf{1}}(Q^\pi)^t - e_{\bar{w}\mathbf{1}}(Q^\pi)^t) u_i(f(\mathbf{w})) - \Delta_{w\mathbf{1},\bar{w}\mathbf{1}} V_i \right| < \zeta/3. \tag{13.5.7}$$

Order the states in $(\mathcal{Z}, Q^\pi)$ so that the first $|\mathcal{W}|$ states are the states in which all players' private states are in agreement. Then, we can write the transition matrix as

$$Q^\pi = \begin{bmatrix} Q^\pi_{11} & Q^\pi_{12} \\ Q^\pi_{21} & Q^\pi_{22} \end{bmatrix},$$

and $[I_{\mathcal{W}} : 0] \mathbf{u}_i = \tilde{\mathbf{u}}_i$, where $\mathbf{u}_i$ is the $|\mathcal{Z}|$-vector with $\mathbf{w}$th element $u_i(f(\mathbf{w}))$ and $\tilde{\mathbf{u}}_i$ is the $|\mathcal{W}|$-vector with wth element $u_i(f(w))$. As π approaches ρ, $Q^\pi_{11} \to \Theta$, $Q^\pi_{12} \to 0$, and $Q^\pi_{22} \to 0$.

Now,

$$[(Q^\pi)^2]_{11} = (Q^\pi_{11})^2 + Q^\pi_{12} Q^\pi_{21}$$

and

$$[(Q^\pi)^2]_{12} = Q^\pi_{11} Q^\pi_{12} + Q^\pi_{12} Q^\pi_{22}.$$

In general,

$$[(Q^\pi)^t]_{11} = (Q^\pi_{11})^t + Q^\pi_{12}[(Q^\pi)^{t-1}]_{21}$$

and

$$[(Q^\pi)^t]_{12} = Q^\pi_{11}[(Q^\pi)^{t-1}]_{12} + Q^\pi_{12}[(Q^\pi)^{t-1}]_{22}.$$

Thus, for all t, $[(Q^\pi)^t]_{11} \to \Theta^t$ and $[(Q^\pi)^t]_{12} \to 0$, as π approaches ρ. Hence, there exists $\varepsilon'' > 0$, $\varepsilon'' \le \varepsilon'$, such that for all $t \le T$, if (Z, π) is ε''-close to (Y, ρ),

$$\left|\sum_{t=0}^{T}(e_w\Theta^t - e_{\bar{w}}\Theta^t)u_i(f(w)) - \sum_{t=0}^{T}(e_{w\mathbf{1}}(Q^\pi)^t - e_{\bar{w}\mathbf{1}}(Q^\pi)^t)u_i(f(\mathbf{w}))\right| < \zeta/3.$$

Combining this with (13.5.6) and (13.5.7) proves the third assertion. ∎

We now show that this lemma implies that an inequality similar to (13.5.4) holds.

Lemma 13.5.6 *If a connected finite public profile is patiently strict, then for any state $\bar{w}$ and for ε small, and (Z, π) ε-close to (Y, ρ), for all w,*

$$\sum_{\mathbf{w}'}(\hat{q}^\pi_{w\mathbf{1},\mathbf{w}'} - \hat{q}^\pi_{w\mathbf{1},\mathbf{w}'}(a_i))\Delta_{\mathbf{w}',\bar{w}\mathbf{1}}V_i > u_i(a_i, f_{-i}(w)) - u_i(f(w)). \quad (13.5.8)$$

Proof Let

$$\zeta = \frac{1}{2}\min_w\left\{\sum_{w''}(\hat{\theta}_{ww''} - \hat{\theta}_{ww''}(a_i))\Delta_{w''\bar{w}}\tilde{V}_i - [u_i(a_i, f_{-i}(w)) - u_i(f(w))]\right\}.$$

Because the public profile is finite and patiently strict, $\zeta > 0$ (recall the expression being minimized is independent of $\bar{w}$).

The left side of (13.5.8) is

$$\sum_{w''}(\hat{q}^\pi_{w\mathbf{1},w''\mathbf{1}} - \hat{q}^\pi_{w\mathbf{1},w''\mathbf{1}}(a_i))\Delta_{w''\mathbf{1},\bar{w}\mathbf{1}}V_i + \sum_{\substack{\mathbf{w}'\neq w''\mathbf{1},\\ w''\in\mathcal{W}}}(\hat{q}^\pi_{w\mathbf{1},\mathbf{w}'} - \hat{q}^\pi_{w\mathbf{1},\mathbf{w}'}(a_i))\Delta_{\mathbf{w}',\bar{w}\mathbf{1}}V_i$$

and, by lemma 13.5.5, there exists $\varepsilon > 0$ such that for (Z, π) ε-close to (Y, ρ),

$$\left|\sum_{\substack{\mathbf{w}'\neq w''\mathbf{1},\\ w''\in\mathcal{W}}}(\hat{q}^\pi_{w\mathbf{1},\mathbf{w}'} - \hat{q}^\pi_{w\mathbf{1},\mathbf{w}'}(a_i))\Delta_{\mathbf{w}',\bar{w}\mathbf{1}}V_i\right| < \zeta/2.$$

Moreover, again by lemma 13.5.5, by choosing ε sufficiently small,

$$\left|\sum_{w''}(\hat{\theta}_{ww''} - \hat{\theta}_{ww''}(a_i))\Delta_{w''\bar{w}}\tilde{V}_i - \sum_{w''}(\hat{q}^\pi_{w\mathbf{1},w''\mathbf{1}} - \hat{q}^\pi_{w\mathbf{1},w''\mathbf{1}}(a_i))\Delta_{w''\mathbf{1},\bar{w}\mathbf{1}}V_i\right| < \zeta/2,$$

so

$$\sum_{\mathbf{w}'}(\hat{q}^\pi_{w\mathbf{1},\mathbf{w}'} - \hat{q}^\pi_{w\mathbf{1},\mathbf{w}'}(a_i))\Delta_{\mathbf{w}',\bar{w}\mathbf{1}}V_i > \sum_{w''}(\hat{q}^\pi_{w\mathbf{1},w''\mathbf{1}} - \hat{q}^\pi_{w\mathbf{1},w''\mathbf{1}}(a_i))\Delta_{w''\mathbf{1},\bar{w}\mathbf{1}}V_i - \zeta/2$$

$$> \sum_{w''}(\hat{\theta}_{ww''} - \hat{\theta}_{ww''}(a_i))\Delta_{w''\bar{w}}\tilde{V}_i - \zeta$$

$$> u_i(a_i, f_{-i}(w)) - u_i(f(w)),$$

which is the desired inequality (13.5.8). ∎

The value player i assigns to being in state w, when she has beliefs β_i over the private states of her opponents, is

$$V_i(w; \beta_i) = \sum_{\mathbf{w}_{-i}} V_i(w, \mathbf{w}_{-i}) \beta_i(\mathbf{w}_{-i}),$$

and her incentive constraint in private state w is given by, for all $a_i \neq f_i(w)$,

$$V_i(w; \beta_i) \geq \sum_{\mathbf{w}_{-i}} \left\{ (1-\delta) u_i(a_i, f_{-i}(\mathbf{w})) + \delta \sum_{\mathbf{w}'} \hat{q}^\pi_{w\mathbf{w}_{-i}, \mathbf{w}'}(a_i) V_i(\mathbf{w}') \right\} \beta_i(\mathbf{w}_{-i}).$$

If β_i assigns probability close to 1 to the vector $w\mathbf{1}$, this inequality is implied by

$$V_i(w\mathbf{1}) > (1-\delta) u_i(a_i, f_{-i}(w)) + \delta \sum_{\mathbf{w}'} \hat{q}^\pi_{w\mathbf{1}, \mathbf{w}'}(a_i) V_i(\mathbf{w}').$$

Substituting for $V_i(w\mathbf{1})$ yields

$$\delta \sum_{\mathbf{w}'} (\hat{q}^\pi_{w\mathbf{1}, \mathbf{w}'} - \hat{q}^\pi_{w\mathbf{1}, \mathbf{w}'}(a_i)) V_i(\mathbf{w}') > (1-\delta)(u_i(a_i, f_{-i}(w)) - u_i(f(w))).$$

For any state $\bar{w} \in \mathscr{W}$, this is equivalent to

$$\delta \sum_{\mathbf{w}'} (\hat{q}^\pi_{w\mathbf{1}, \mathbf{w}'} - \hat{q}^\pi_{w\mathbf{1}, \mathbf{w}'}(a_i))(V_i(\mathbf{w}') - V_i(\bar{w}\mathbf{1}))$$
$$> (1-\delta)(u_i(a_i, f_{-i}(w)) - u_i(f(w))).$$

For large δ, this is implied by (13.5.8). If player i assigns a probability close to 1 to all her opponents being in the same private state as herself, the incentive constraint for i at that private state holds. Because there are only a finite number of incentive constraints, the bounds on δ and β_i are independent of $\bar{w} \in \mathscr{W}$.

Finally, it remains to argue that the payoff under the private profile can be made close to that of the public profile. Denote the stationary distribution of the Markov chain $(\mathscr{W}, \Theta)$ by μ^ρ and of the Markov chain $(\mathscr{Z}, Q^\pi)$ by μ^π. From lemma 13.5.3, $\lim_{\delta \to 1} \tilde{V}_i(w)$ is independent of $w \in \mathscr{W}$, and so is given by $\sum_w u_i(f(w)) \mu^\rho(w)$. Similarly, from lemma 13.5.5, $\lim_{\delta \to 1} V_i(\mathbf{w})$ is independent of $\mathbf{w} \in \mathscr{Z}$ and is given by $\sum_{\mathbf{w}} u_i(f(\mathbf{w})) \mu^\pi(\mathbf{w})$. The proposition then follows from the observation that because $Q^\pi_{11} \to \Theta$ as $|\pi - \rho| \to 0$ (recall that the first $|\mathscr{W}|$ states of $\mathscr{Z}$ are the states in which all players' private states are in agreement), $|\mu^\rho - [\,I_{\mathscr{W}} : 0\,] \mu^\pi| \to 0$.

13.6 A Folk Theorem

We now apply the results of the last section to prove a mutual minmax pure-action folk theorem for almost perfect almost public monitoring games.

Say that a public monitoring distribution (Y, ρ) is η-*perfect* if $Y = A$ and $\rho(a \mid a) > 1 - \eta$. It is easy to see that any private monitoring distribution (Z, π) ε-close to an η-perfect public monitoring distribution is $(\eta + \varepsilon)$-perfect, in the sense of definition 12.2.1.

Proposition 13.6.1 **A private monitoring folk theorem** *Suppose A is finite. Fix $\tilde{a} \in A$. If there exists $\underline{a} \in A$ satisfying*

$$\max_{a_i \in A_i} u_i(a_i, \underline{a}_{-i}) < u_i(\tilde{a}), \quad \text{for all } i, \quad (13.6.1)$$

then for all $\kappa < 1$, there exists $\bar{\delta} < 1$ and $\eta > 0$, such that for all η-perfect public monitoring distributions (Y, ρ), there exists $\varepsilon > 0$ such that for all private monitoring distributions (Z, π), ε-close to (Y, ρ), for all $\delta \in (\bar{\delta}, 1)$, there is a sequential equilibrium of the repeated game with private monitoring with payoffs within κ of $u(\tilde{a})$.

Proof We first consider the perfect monitoring game. Similar to the mutual minmaxing profile of proposition 3.3.1, the profile specifies $\tilde{a}$ on the equilibrium path of play, and any deviation results in L periods of $\underline{a}$. We cannot simply use the profile specified in the proof of proposition 3.3.1 because it does not have bounded recall. However, inequality (13.6.1) implies that for all i and $a_i \in A_i$, $(a_i, \underline{a}_{-i}) \neq \tilde{a}$, and for any weakly myopically profitable deviation by i from $\tilde{a}$ (i.e., $u_i(a_i, \tilde{a}_{-i}) \geq u_i(\tilde{a})$), $(a_i, \tilde{a}_{-i}) \neq \underline{a}$. Hence, any unilateral deviation from $\underline{a}$ or weakly myopically profitable deviation from $\tilde{a}$ is "detected," and so we can use a profile with L period memory in which $\tilde{a}$ is specified after L periods of $\tilde{a}$ or $\underline{a}$, and $\underline{a}$ otherwise. The bounded recall automaton has states $\mathscr{W} = \{w(i) : i = 0, \ldots, L\}$, initial state $w^0 = w(0)$, output function

$$f(w(\ell)) = \begin{cases} \tilde{a}, & \text{if } \ell = 0, \\ \underline{a}, & \text{if } \ell = 1, \ldots, L, \end{cases}$$

and transition rule

$$\tau(w(\ell), a) = \begin{cases} w(0), & \text{if } \ell = 0 \text{ or } \ell = L, \text{ and } a = \underline{a} \text{ or } \tilde{a}, \\ w(\ell+1), & \text{if } 0 < \ell < L \text{ and } a = \underline{a} \text{ or } \tilde{a}, \\ w(1), & \text{otherwise,} \end{cases}$$

where L satisfies

$$L \min_i (u_i(\tilde{a}) - u_i(\underline{a})) > M - \min_i u_i(\tilde{a}) \quad (13.6.2)$$

(as usual, $M = \max_{i,a} u_i(a)$).

The direct verification that this automaton describes a subgame-perfect equilibrium of the perfect monitoring game for large δ is identical to the proof of proposition 3.3.1. We instead show that the profile is patiently strict (implying it is a subgame-perfect equilibrium). Denote player i's value of being in state $w \in \mathscr{W}$ in the perfect monitoring game by $\hat{V}_i(w)$. Then,

$$\hat{V}_i(w(0)) - \hat{V}_i(w(1)) = u_i(\tilde{a}) - [(1-\delta^L)u_i(\underline{a}) + \delta^L u_i(\tilde{a})]$$
$$= (1-\delta^L)[u_i(\tilde{a}) - u_i(\underline{a})],$$

13.6 ■ A Folk Theorem

so
$$\Delta_{w(0)w(1)} \hat{V}_i \equiv \lim_{\delta \to 1} \frac{V_i(w(0)) - V_i(w(1))}{(1-\delta)} = L[u_i(\tilde{a}) - u_i(\underline{a})].$$

Consequently, using the observation that any weakly myopically profitable deviation at $w(0)$ takes the profile to $w(1)$, (13.5.4) is

$$\Delta_{w(0)w(1)} \hat{V}_i > u_i(a_i, \tilde{a}_{-i}) - u_i(\tilde{a}),$$

which, because $u_i(a_i, \tilde{a}_{-i}) \leq M$, is implied by (13.6.2). For any a_i satisfying $u_i(a_i, \tilde{a}_{-i}) < u_i(\tilde{a})$, the left side of (13.5.4) is at most 0 (because the deviation may be ignored), but the right side is strictly negative. Similarly,

$$\hat{V}_i(w(\ell)) - \hat{V}_i(w(1))$$
$$= (1 - \delta^{L+1-\ell})u_i(\underline{a}) + \delta^{L+1-\ell}u_i(\tilde{a}) - [(1 - \delta^L)u_i(\underline{a}) + \delta^L u_i(\tilde{a})]$$
$$= \delta^{L+1-\ell}(1 - \delta^{\ell-1})[u_i(\tilde{a}) - u_i(\underline{a})],$$

and so for all $\ell \geq 2$,

$$\Delta_{w(\ell)w(1)} \hat{V}_i \equiv \lim_{\delta \to 1} \frac{V_i(w(\ell)) - V_i(w(1))}{(1-\delta)} > u_i(\tilde{a}) - u_i(\underline{a}).$$

Now using the observation that any deviation takes the profile to $w(1)$, (13.5.4) is now

$$\Delta_{w(\ell)w(1)} \hat{V}_i > u_i(a_i, \underline{a}_{-i}) - u_i(\underline{a}),$$

which immediately holds because $u_i(\tilde{a}) > u_i(a_i, \underline{a}_{-i})$. Hence, the profile is patiently strict.

The profile in any η-perfect public monitoring game trivially has bounded recall and is connected. Thus, applying proposition 13.5.1 completes the proof once we have argued that the profile in any η-perfect public monitoring game is patiently strict for sufficiently small η. The profile has bounded recall, so the induced Markov chain on the set of states $\mathscr{W}$ is ergodic, and we can apply the reasoning of lemma 13.5.6 to show that the profile is patiently strict in the η-perfect public monitoring game. ■

Remark 13.6.1 There is no straightforward extension of the result to a folk theorem using player-specific punishments.[10] The difficulty is that a unilateral deviation by player i from player j's specific punishment may not be distinguishable from a unilateral deviation by player ℓ from player k's specific punishment. In such a case, it is not clear if there is a bounded recall version of the profile constructed in the proof of proposition 3.4.1. ◆

---

10. Mailath and Morris (2002) incorrectly claim that the profile described in the proof of their theorem 6.1 has bounded recall.

14 Belief-Free Equilibria in Private Monitoring Games

Chapter 13 and parts of chapter 12 focused on belief-based equilibria, in which players' beliefs about their opponents' histories play a central role. In this chapter, we study *belief-free* equilibria, for which beliefs are in some sense irrelevant. This is a tremendous computational advantage, because checking for equilibrium now does not involve calculating posterior beliefs over the private states of players. We have already seen an example of such an equilibrium in section 12.5. We do not require almost public monitoring in this chapter, but we often require almost perfect monitoring.

The descriptive relevance of belief-free equilibria is open to question because they typically require a significant amount of randomization. In particular, it is not clear if such equilibria can be purified (Harsanyi 1973).[1] However, belief-free equilibria have been the focus of study for only a few years, since their introduction by Piccione (2002) (simplified and extended by Ely and Välimäki 2002) and characterization by Ely, Hörner, and Olszewski (2005), and it is too early to evaluate them. Belief-free equilibria are important from a theoretical perspective because they play a key role in private equilibria in public monitoring games (section 10.4.2) and in the folk theorem for almost perfect private monitoring (remark 14.2.1).

14.1 Definition and Examples

The most natural definition of belief-free equilibria uses automata. Recall from remark 12.2.3 that every behavior strategy can be represented by an automaton, represented by a set of states $\mathscr{W}_i$, an initial state w_i^0, a decision rule $f_i : \mathscr{W}_i \to \Delta(A_i)$ specifying a distribution over action choices for each state, and a transition function $\tau_i : \mathscr{W}_i \times A_i \times Z_i \to \mathscr{W}_i$.[2] We begin with a restriction on the class of profiles to introduce the idea. For any history $h_i^t = (a_i^0 z_i^0, a_i^1 z_i^1, \ldots, a_i^{t-1} z_i^{t-1}) \in (A_i \times Z_i)^t$, the state reached in period t is denoted by $\tau_i(w_i^0, h_i^t)$. For example, $\tau_i(w_i^0, h_i^2) = \tau_i(\tau_i(w_i^0, a_i^0 z_i^0), a_i^1 z_i^1)$.

---

1. More specifically, it is not clear if in general a belief-free equilibrium can be approximated by any strict equilibrium in nearby games of incomplete information, where the incomplete information is generated by independently distributed (over time and players) payoff shocks. See also the discussion in section 12.1.3.
2. As illustrated in section 10.4.2, it is sometimes easier to allow for random state transitions as well (see also remark 7.1.2). Apart from obvious complications to notation, such representations cause no difficulty.

Definition 14.1.1 A strategy $(\mathscr{W}_i, w_i^0, f_i, \tau_i)$ is inclusive *if there exists T so that for every $t \geq T$ and every $w_i \in \mathscr{W}_i$, there exists a history $h_i^t = (a_i^0 z_i^0, a_i^1 z_i^1, \ldots, a_i^{t-1} z_i^{t-1})$ with $a_i^s \in \operatorname{supp} f_i(\tau(w_i^0, h_i^s))$ for all $s < t$, such that $w_i = \tau_i(w_i^0, h_i^t)$.*

Because the space of signals is finite, an inclusive strategy must have a finite state space. Under full-support private monitoring, when player i follows an inclusive strategy, after some period T, other players must assign strictly positive probability to each of player i's states in $\mathscr{W}$. This observation motivates the following definition.

Definition 14.1.2 *An inclusive strategy profile $\{(\mathscr{W}_i, w_i^0, f_i, \tau_i) : i = 1, \ldots, n\}$ is a belief-free equilibrium if for all $\mathbf{w} \in \prod_j \mathscr{W}_j$, and for all i, the strategy $(\mathscr{W}_i, w_i, f_i, \tau_i)$ is a best reply to the profile $\{(\mathscr{W}_j, w_j, f_j, \tau_j) : j \neq i\}$.*

The critical feature of this definition is that for all specifications of player i's initial state, the strategy induced by his automaton must be optimal against the induced strategies of the other players for all possible specifications of their initial states. As a result, the automaton is a best reply for *any* beliefs that i might have about the private states of the other players (and so about the private histories they may have observed). This discussion also implies the following lemma.

Lemma 14.1.1 *Suppose $\{(\mathscr{W}_i, w_i^0, f_i, \tau_i) : i = 1, \ldots, n\}$ is an inclusive belief-free equilibrium. Player i's value from any vector of initial states $\mathbf{w} \in \prod_j \mathscr{W}_j$, $V_i(\mathbf{w})$, is independent of w_i.*

Proof Because $(\mathscr{W}_i, w_i, f_i, \tau_i)$ is a best reply to $\{(\mathscr{W}_j, w_j, f_j, \tau_j) : j \neq i\}$ for all w_i, player i is indifferent over i's initial states, and so i's value is independent of w_i. ∎

Player i is thus indifferent over the various initial states in i's automaton, regardless of his beliefs about others' states. However, player i will in general *not* be indifferent over the initial states of his opponents, a fact crucial for the provision of nontrivial incentives in belief-free equilibria.

Remark 14.1.1 **Payoffs have a product structure** Suppose $\{(\mathscr{W}_i, w_i^0, f_i, \tau_i) : i = 1, \ldots, n\}$ is an inclusive belief-free equilibrium. Let $\mathscr{V}$ be the set of continuation payoffs induced by the equilibrium and let $\mathscr{V}_i$ be the projection of this set on player i's payoffs. Then, from lemma 14.1.1 $\mathscr{V} = \prod_{i=1}^n \mathscr{V}_i$. ♦

Belief-free equilibria in inclusive strategies trivially exist because the history-independent play of the same stage-game Nash equilibrium is belief-free.

Using ideas from section 13.4.3, we now extend the definition. Let $\mathscr{W}_{i,t}$ be the set of states reachable in period t, $\mathscr{W}_{i,t} \equiv \{w_i \in \mathscr{W}_i : w_i = \tau_i(w_i^0, h_i^t) \text{ for some } h_i^t = (a_i^0 z_i^0, a_i^1 z_i^1, \ldots, a_i^{t-1} z_i^{t-1}) \in \mathscr{H}_i^t, \ a_i^s \in \operatorname{supp} f_i(\tau(w_i^0, h_i^s)) \text{ for all } s < t\}$.

Definition 14.1.3 *A strategy profile $\{(\mathscr{W}_i, w_i^0, f_i, \tau_i) : i = 1, \ldots, n\}$ is a belief-free equilibrium if for all t and all $\mathbf{w} \in \prod_j \mathscr{W}_{j,t}$, and for all i, the strategy $(\mathscr{W}_i, w_i, f_i, \tau_i)$ is a best reply to the profile $\{(\mathscr{W}_j, w_j, f_j, \tau_j) : j \neq i\}$.*

We have not made any assumptions about the nature of the monitoring. In particular, the notion applies to games with perfect or public monitoring, where it is a more

14.1 ■ Definition and Examples

	E	S
E	2,2	−c,b
S	b,−c	0,0

Figure 14.1.1 The prisoners' dilemma from figure 2.5.1, where $b > 2$, $c > 0$, and $b - c < 4$.

stringent requirement than PPE (the equilibria described in section 12.5 are examples). Any belief-free equilibrium profile is sequential.

We illustrate the definition with Ely and Välimäki's (2002) example of a class of symmetric belief-free equilibria in the repeated prisoners' dilemma with almost perfect monitoring. Like the analysis of section 12.5, we first analyze symmetric belief-free equilibria of the perfect monitoring game. We use the payoffs of figure 2.5.1, which are reproduced in figure 14.1.1.

14.1.1 Repeated Prisoners' Dilemma with Perfect Monitoring

Ely and Välimäki (2002) construct a class of symmetric mixed-strategy equilibria with one-period memory. Players randomize in each period with player i placing probability $\alpha^{a_i a_j}$ on E after the previous-period action profile $a_i a_j$, where $j \neq i$. For $i = 1, 2$, the class of automata is $(\mathscr{W}_i, w_i^0, f_i, \tau_i)$, where

$$\mathscr{W}_i = \{w_i^{a_i a_j} : a_i a_j \in \{E, S\}^2\}, \quad w_i^0 \in \mathscr{W}_i, \tag{14.1.1}$$

$$f_i(w_i^{a_i a_j}) = \alpha^{a_i a_j} \circ E + (1 - \alpha^{a_i a_j}) \circ S, \tag{14.1.2}$$

and

$$\tau_i(w_i^{a_i a_j}, a_i' a_j') = w_i^{a_i' a_j'}. \tag{14.1.3}$$

The profile is constructed so that irrespective of the initial state of his opponent, the player is indifferent between E and S. The one-shot deviation principle then implies that the player's strategy is a best reply to the opponent's strategy induced by any initial state.

Symmetry allows us to focus on the incentives facing player 2 when verifying optimality. The requirement is then that for each $a_1 a_2$, player 2 is indifferent between playing E and S, when player 1 is in state $w_1^{a_1 a_2}$ and so is playing E with probability $\alpha^{a_1 a_2}$. This yields the following system, where $V_2^{a_1 a_2}$ is the value to player 2 when player 1 is in state $w_1^{a_1 a_2}$, and the first equality gives $V_2^{a_1 a_2}$ when E is chosen, whereas the second equality gives it when S is chosen, for all $a_1, a_2 \in \{E, S\}$:

$$V_2^{a_1 a_2} = (1 - \delta)(\alpha^{a_1 a_2} 2 + (1 - \alpha^{a_1 a_2})(-c)) + \delta \left\{ \alpha^{a_1 a_2} V_2^{EE} + (1 - \alpha^{a_1 a_2}) V_2^{SE} \right\} \tag{14.1.4}$$

$$= (1 - \delta)\alpha^{a_1 a_2} b + \delta \left\{ \alpha^{a_1 a_2} V_2^{ES} + (1 - \alpha^{a_1 a_2}) V_2^{SS} \right\}. \tag{14.1.5}$$

Subtracting (14.1.5) from (14.1.4) gives, again for all $a_1 a_2$,

$$\alpha^{a_1 a_2}\{(1-\delta)(2+c-b) + \delta[(V_2^{EE} - V_2^{ES}) - (V_2^{SE} - V_2^{SS})]\}$$
$$- (1-\delta)c + \delta(V_2^{SE} - V_2^{SS}) = 0.$$

At least two of the probabilities differ (otherwise, there are no intertemporal incentives and $\alpha^{a_1 a_2} = 0$ for all $a_1 a_2$), so the constant term and the coefficient of $\alpha^{a_1 a_2}$ are both 0, that is,

$$V_2^{SS} = V_2^{SE} - \frac{(1-\delta)c}{\delta} \tag{14.1.6}$$

and

$$V_2^{ES} = V_2^{EE} - \frac{(1-\delta)(b-c-2)}{\delta} + V_2^{SS} - V_2^{SE}$$
$$= V_2^{EE} - \frac{(1-\delta)(b-2)}{\delta}. \tag{14.1.7}$$

These two equations succinctly capture the trade-offs facing potentially randomizing players. Suppose player 2 knew her partner was going to shirk this period. The myopic incentive to also shirk is c, whereas the cost of shirking is that her continuation value falls from V_2^{SE} to V_2^{SS}, reflecting the change in 1's state. Equation (14.1.6) says that these two should exactly balance. Suppose instead player 2 knew her partner was going to exert effort this period. The myopic incentive to shirk is now $b - 2$, and the cost of shirking is now that her continuation value falls from V_2^{EE} to V_2^{ES}. This time, equation (14.1.7) says that these two should exactly balance. Notice that these two equations imply that a player's best replies are independent of the current realized behavior of the opponent.[3]

A symmetric profile described by the four probabilities $\{\alpha^{a_1 a_2} : a_1 a_2 \in \{E, S\}^2\}$ is a belief-free equilibrium when (14.1.4) and (14.1.5) are satisfied for the four action profiles $a_1 a_2 \in \{E, S\}^2$. At the risk of repeating the obvious, the four probabilities are to be determined, subject only to (14.1.6) and (14.1.7) (because the value functions are determined by the probabilities). This redundancy implies a two-dimensional indeterminacy in the solutions, and it is convenient to parameterize the solutions by $V_2^{EE} = \bar{v}$ and $V_2^{SE} = \underline{v}$ (recall that we are focusing on symmetric equilibria).

Solving (14.1.4) for $a_1 a_2 = EE$ gives

$$\alpha^{EE} = \frac{(1-\delta)c + \bar{v} - \delta\underline{v}}{(1-\delta)(2+c) + \delta(\bar{v} - \underline{v})}, \tag{14.1.8}$$

for $a_1 a_2 = SE$ gives

$$\alpha^{SE} = \frac{(1-\delta)c + \underline{v} - \delta\underline{v}}{(1-\delta)(2+c) + \delta(\bar{v} - \underline{v})}, \tag{14.1.9}$$

for $a_1 a_2 = ES$ (using (14.1.7)) gives

$$\alpha^{ES} = \frac{(1-\delta)(c - (b-2)/\delta) + \bar{v} - \delta\underline{v}}{(1-\delta)(2+c) + \delta(\bar{v} - \underline{v})}, \tag{14.1.10}$$

---

3. This is the starting point of Ely and Välimäki (2002), who work directly with the values to a player of having his opponent play E and S *this* period.

and finally, for $a_1a_2 = SS$ (using (14.1.6)) gives

$$\alpha^{SS} = \frac{(1-\delta)c(1-1/\delta) + \underline{v} - \delta\underline{v}}{(1-\delta)(2+c) + \delta(\bar{v}-\underline{v})}. \tag{14.1.11}$$

We have described an equilibrium if the expressions in (14.1.8)–(14.1.11) are probabilities. Before we provide conditions that guarantee this, we consider some special cases. For example, if we set $\bar{v} = 2$ (the highest symmetric feasible payoff), then $\alpha^{EE} = 1$. In other words, if we start at EE, the outcome path is *pure*: Under the profile, after EE, both players play E for sure. The outcome is supported by the specification that if player 2, for example, deviated and played S, in the next period, she will be punished by player 1, who now plays S in the next period with probability

$$1 - \alpha^{ES} = \frac{(1-\delta)(b-2)}{\delta((1-\delta)c + 2 - \delta\underline{v})} > 0.$$

A particularly simple example arises when $c = b - 2$, that is, the myopic incentives to shirk are independent of the partner's action. In this case, if $\bar{v} = \underline{v} = v$, the probabilities for a player are independent of his own action and given by

$$\alpha^{EE} = \alpha^{SE} = \frac{v+c}{2+c}$$

and

$$\alpha^{ES} = \alpha^{SS} = \frac{\delta v - c(1-\delta)}{\delta(2+c)}.$$

Proposition 14.1.1 *There is a two-dimensional manifold of symmetric mixed equilibria of the infinitely repeated perfect monitoring prisoners' dilemma: Suppose $0 < \underline{v} \leq \bar{v} \leq 2$ satisfy the inequalities*

$$(1-\delta)(b-2)/\delta + \delta\underline{v} \leq (1-\delta)c + \bar{v} \tag{14.1.12}$$

and

$$\underline{v} - \frac{(1-\delta)c}{\delta} \geq 0. \tag{14.1.13}$$

For each pair of initial states, the automata described by (14.1.1)–(14.1.3), with $\alpha^{a_1a_2}$ given by (14.1.8)–(14.1.11), are a belief-free equilibrium.

If (14.1.12) and (14.1.13) hold strictly, and $\bar{v} < 2$, then the probabilities from (14.1.8)–(14.1.11) are interior.

From our parameterization ($V_2^{EE} = \bar{v}$ and $V_2^{SE} = \underline{v}$) and (14.1.6)–(14.1.7):

Corollary 14.1.1 *Let $\underline{v}' = \underline{v} - (1-\delta)c/\delta$ and $\bar{v}' = \bar{v} - (1-\delta)(b-2)/\delta$. Every payoff vector in $\{\underline{v}', \underline{v}, \bar{v}', \bar{v}\}^2$ is the payoff of a belief-free equilibrium.*

The inequalities (14.1.12) and (14.1.13) are satisfied for any $0 < \underline{v} < \bar{v} \leq 2$, for δ sufficiently close to 1. Note that, setting $V_2^{EE} = \bar{v}$ and $V_2^{SE} = \underline{v}$, (14.1.13) is equivalent to $V_2^{SS} \geq 0$ (using (14.1.6)), whereas (given (14.1.13)) (14.1.12) is stronger than $V_2^{ES} \geq 0$ (using (14.1.7)). Also, using (14.1.5), the expressions for the probabilities can be written as, for all a_1a_2,

$$\alpha^{a_1 a_2} = \frac{V_2^{a_1 a_2} - \delta V_2^{SS}}{(1-\delta)b + \delta(V_2^{ES} - V_2^{SS})}.$$

Proof We need only verify that if $0 < \underline{v} \leq \bar{v} \leq 2$, (14.1.12) and (14.1.13) imply that the quantities described by (14.1.8)–(14.1.11) are probabilities because the belief-free nature of the profile is guaranteed by its construction. It is immediate that $\alpha^{ES} < \alpha^{EE}$ and $\alpha^{SS} < \alpha^{SE} \leq \alpha^{EE}$, so the only inequalities we need to verify are $\alpha^{ES}, \alpha^{SS} \geq 0$ and $\alpha^{EE} \leq 1$, with the inequalities strict when claimed. Observe first that the common denominator in (14.1.8)–(14.1.11) is strictly positive from $\underline{v} \leq \bar{v}$.

It is immediate that $\alpha^{EE} \leq 1$, because $\bar{v} \leq 2$, with the inequality strict when $\bar{v} < 2$. We also have $\alpha^{ES} \geq 0$ because

$$(1-\delta)(c - (b-2)/\delta) + \bar{v} - \delta \underline{v} \geq 0$$
$$\iff (1-\delta)c + \bar{v} \geq (1-\delta)(b-2)/\delta + \delta \underline{v},$$

which is (14.1.12), with strictness holding when claimed.

Finally, $\alpha^{SS} \geq 0$ is equivalent to (14.1.13), with strictness trivially holding when claimed.

∎

Remark 14.1.2 We restricted attention to symmetric profiles for tractability only. There is a four-dimensional manifold of belief-free equilibria, once we allow for asymmetries.

◆

Remark 14.1.3 **A partial folk theorem** This result is a partial folk theorem for the repeated prisoners' dilemma, in that any payoff in the square $(0, 2)^2$ can be achieved as an equilibrium payoff for sufficiently patient players. For any payoff vector $(v_1, v_2) \in (0, 2)^2$ with $v_1 \leq v_2$,[4] set

$$\underline{v} = v_1 \quad \text{and} \quad \bar{v} = v_2 + \frac{(1-\delta)(b-2)}{\delta}.$$

For δ sufficiently close to 1, the hypotheses of proposition 14.1.1 are satisfied, so there is a belief-free equilibrium with discounted average payoffs (v_1, v_2). In this profile, player 1's initial state is w_1^{ES} and 2's initial state is w_2^{SE}. Belief-free profiles of the type just constructed cannot achieve payoffs outside the square, because players must be indifferent between E and S in every period, and playing E in every period cannot give a payoff above 2. Ely and Välimäki (2002) extend this type of profile to obtain payoffs outside of the square.

As an illustration, consider again the case $c = b - 2$, and suppose δ is close to 1. The choices of $\bar{v} = 2$ and $\underline{v} = (1-\delta)c/\delta$ are the most extreme possible, given the restrictions in proposition 14.1.1. For these choices, $\alpha^{EE} = 1$ and $\alpha^{SS} = 0$, so both action profiles EE and SS are absorbing (the individual states w_i^{EE} and w_i^{SS} are not absorbing and the pairs of states (w_1^{EE}, w_2^{EE}) and (w_1^{SS}, w_2^{SS}) are).

---

4. Clearly, a symmetric argument applies for $v_1 > v_2$.

14.1 ■ Definition and Examples

period	1	2	6	11
EE	9.877×10^{-2}	0.1780	0.3665	0.4556
SE	1.235×10^{-2}	1.951×10^{-2}	2.283×10^{-2}	1.293×10^{-2}
ES	0.7901	0.6244	0.2442	7.593×10^{-2}
SS	9.877×10^{-2}	0.1780	0.3665	0.4556

Figure 14.1.2 The unconditional distribution over action profiles induced by the belief-free equilibrium with player 1 in initial state w_1^{EE} and player 2 in initial state w_2^{SS}. These calculations assume $b - 2 = c = 1$ and $\delta = 9/11$, so that $\alpha^{ES} = 8/9$ and $\alpha^{SE} = 1/9$.

Letting w_i^{EE} be the initial state for both players gives payoffs $(2, 2)$. It is also the case that the asymmetric profile $(0,2)$ is an equilibrium payoff with initial pure actions ES and continuations

$$\left(\frac{(1-\delta)c}{\delta}, 2 - \frac{(1-\delta)c}{\delta} \right),$$

with player 1 assigning E the relatively high probability

$$1 - \frac{(1-\delta)c}{2\delta}$$

and player 2 assigning E the relatively low probability

$$\frac{(1-\delta)c}{2\delta}.$$

In figure 14.1.2, we have described the unconditional evolution of play under this profile. The asymmetric payoff is generated by a process that has some persistence at ES and with equal probabilities switches to EE or SS forever. The potential transition to EE is needed to reward player 1 for having exerted effort while 2 shirked, and the threatened transition to SS (caused by 1 putting positive weight on S) is needed to provide incentives for 2 to put positive weight on E (facilitating a potential transition to EE).

◆

14.1.2 Repeated Prisoners' Dilemma with Private Monitoring

We now consider almost perfect private monitoring (we do *not* assume the monitoring is conditionally independent or almost public). For notational simplicity, we assume *symmetric* private monitoring, so that $Z_1 = Z_2$ and $\pi(z_1 z_2 \mid a_1 a_2) = \pi(z_2 z_1 \mid a_2 a_1)$.

Given an ε-perfect monitoring distribution (Z, π), let $Z_i(a)$ be the set of private signals satisfying $\pi_i(Z_i(a) \mid a) > 1 - \varepsilon$ (see definition 12.2.1).

For $i = 1, 2$, the class of automata is $(\mathscr{W}_i, w_i^0, f_i, \tau_i)$, where

$$\mathscr{W}_i = \{w_i^{a_i \hat{a}_j} : a_i \hat{a}_j \in \{E, S\} \times \{\hat{E}, \hat{S}\}\}, \quad w_i^0 \in \mathscr{W}, \tag{14.1.14}$$

$$f_i(w^{a_i \hat{a}_j}) = \alpha^{a_i \hat{a}_j} \circ E + (1 - \alpha^{a_i \hat{a}_j}) \circ S, \tag{14.1.15}$$

and

$$\tau_i(w_i^{a_i \hat{a}_j}, z_i, a_i') = \begin{cases} w_i^{a_i' \hat{E}}, & \text{if } z_i \in Z_i(a_i E), \\ w_i^{a_i' \hat{S}}, & \text{if } z_i \in Z_i(a_i S). \end{cases} \tag{14.1.16}$$

In other words, in each period, player i puts probability $\alpha^{a_i \hat{a}_j}$ on E, where a_i is i's last-period action choice, and $\hat{a}_j = \hat{E}$ ($\hat{S}$, respectively) if i's last-period private signal was in the set $Z_i(a_i, E)$ ($Z_i(a_i, S)$, respectively). The monitoring is ε-perfect, so $\hat{E}$ indicates that a signal from the $(1 - \varepsilon)$-probability event $Z_i(a_i, E)$ was observed.

As for the perfect monitoring case, symmetry allows us to focus on player 2's incentives. Let $V_2^{a_1 \hat{a}_2}(a_2)$ be the value (under the profile) to player 2 when player 1 is in the *private* state $w_1^{a_1 \hat{a}_2}$ and player 2 plays a_2. As in section 12.5, we look for probabilities, $\alpha^{a_1 \hat{a}_2}$, such that player 2 is indifferent between E and S at each of player 1's four private states, that is, $V_2^{w_1}(E) = V_2^{w_1}(S) \equiv V_2^{w_1}$ for all $w_1 \in \{E, S\} \times \{\hat{E}, \hat{S}\}$. If we are successful, the one-shot deviation principle implies that player 2's best replies will be independent of her beliefs about player 1's private state. Because the environment is symmetric, the resulting symmetric profile will be a belief-free equilibrium.

The equations implied by the indifferences are for all $w_1 \in \{E, S\} \times \{\hat{E}, \hat{S}\}$, where the first equality describes $V_2^{w_1}(E)$ while the second describes $V_2^{w_1}(S)$,

$$V_2^{w_1} = (1 - \delta)(\alpha^{w_1} 2 + (1 - \alpha^{w_1})(-c)) + \delta\{\alpha^{w_1}(\pi_1^{E\hat{E}} V_2^{E\hat{E}} + (1 - \pi_1^{E\hat{E}}) V_2^{E\hat{S}}) $$
$$+ (1 - \alpha^{w_1})(\pi_1^{S\hat{E}} V_2^{S\hat{E}} + (1 - \pi_1^{S\hat{E}}) V_2^{S\hat{S}})\} \tag{14.1.17}$$

$$= (1 - \delta)\alpha^{w_1} b + \delta\{\alpha^{w_1}(\pi_1^{E\hat{S}} V_2^{E\hat{S}} + (1 - \pi_1^{E\hat{S}}) V_2^{E\hat{E}}) $$
$$+ (1 - \alpha^{w_1})(\pi_1^{S\hat{S}} V_2^{S\hat{S}} + (1 - \pi_1^{S\hat{S}}) V_2^{S\hat{E}})\}, \tag{14.1.18}$$

where $\pi_1^{a_1 \hat{a}_2} = \pi_1(Z_1(a_1 a_2) \mid a_1 a_2)$ is the probability under a that player 1 observes a private signal in $Z_1(a)$ (recall the interpretation of $\hat{E}$ and $\hat{S}$ from before). Observe that these equations replicate (14.1.4) and (14.1.5) for perfect monitoring, because $\pi_1^{a_1 \hat{a}_2} = 1$ for all $a_1 \hat{a}_2$ when $\varepsilon = 0$.

Subtracting (14.1.18) from (14.1.17) gives a linear equation in α^{w_1}. As in the perfect monitoring case, the constant term and the coefficient of α^{w_1} must both equal 0 (otherwise there are no intertemporal incentives and $\alpha^{w_1} = 0$ for all w_1). The constant term is 0 if

$$V_2^{S\hat{S}} = V_2^{S\hat{E}} - \frac{(1-\delta)c}{\delta(\pi_1^{S\hat{E}} + \pi_1^{S\hat{S}} - 1)}, \tag{14.1.19}$$

and the coefficient equals 0 if

$$V_2^{E\hat{S}} = V_2^{E\hat{E}} - \frac{(1-\delta)(b-2-c) + \delta(\pi_1^{S\hat{E}} + \pi_1^{S\hat{S}} - 1)(V_2^{S\hat{E}} - V_2^{S\hat{S}})}{\delta(\pi_1^{E\hat{E}} + \pi_1^{E\hat{S}} - 1)}$$

$$= V_2^{E\hat{E}} - \frac{(1-\delta)(b-2)}{\delta(\pi_1^{E\hat{E}} + \pi_1^{E\hat{S}} - 1)}. \qquad (14.1.20)$$

Again following the approach from the perfect-monitoring case, we treat $V_2^{E\hat{E}}$ and $V_2^{S\hat{E}}$ as constants, use (14.1.19) and (14.1.20) to determine $V_2^{E\hat{S}}$ and $V_2^{S\hat{S}}$, and use equation (14.1.17) for w_1 to determine α^{w_1}.

Proposition 14.1.2 *Suppose $0 < \underline{v} \leq \bar{v} < 2$ satisfy the inequalities (14.1.12) and (14.1.13) strictly. There exists $\varepsilon > 0$ such that if the repeated prisoners' dilemma with symmetric private monitoring has ε-perfect monitoring, then there exist probabilities $\alpha^{a_i \hat{a}_j}$ such that for any initial states, the automata described by (14.1.14)–(14.1.16) are a belief-free equilibrium, with $V_i^{E\hat{E}} = \bar{v}$ and $V_i^{S\hat{E}} = \underline{v}$. Moreover, every payoff vector in $\{\underline{v}', \underline{v}, \bar{v}', \bar{v}\}^2$ is the payoff of a belief-free equilibrium, where $\underline{v}'$ is given by (14.1.19) and $\bar{v}'$ is given by (14.1.20).*

Proof Observe that we can make $\pi_1^{a_1 \hat{a}_2}$ arbitrarily close to 1 for all $a_1 \hat{a}_2$ by making ε sufficiently small. Let $V_2^{SS} = \underline{v} - (1-\delta)c/\delta$ and $V_2^{ES} = \bar{v} - (1-\delta)(b-2)/\delta$. Then, using $V_2^{E\hat{E}} = \bar{v}$ and $V_2^{S\hat{E}} = \underline{v}$, from (14.1.19) and (14.1.20), it is immediate that for $\pi_1^{a_1 \hat{a}_2}$ sufficiently close to 1 for all $a_1 \hat{a}_2$, $V_2^{S\hat{S}}$ can be made arbitrarily close to V_2^{SS} and $V_2^{E\hat{S}}$ can be made arbitrarily close to V_2^{ES}. This in turn implies that the solution α^{w_1} to (14.1.17) for $w_1 \in \{E, S\} \times \{\hat{E}, \hat{S}\}$ can be made arbitrarily close to the corresponding probability in (14.1.8)–(14.1.11) by choosing ε small enough. Moreover, because (14.1.12) and (14.1.13) are satisfied strictly and $\bar{v} < 2$, the corresponding probability is interior (proposition 14.1.1), and so $\alpha^{w_1} \in (0, 1)$. By construction, each player is indifferent between E and S, conditional on the other player's private state.

■

Remark 14.1.4 **A partial folk theorem for private monitoring** Just as for perfect monitoring (remark 14.1.3), this implies a partial folk theorem for the repeated prisoners' dilemma, in that any payoff in the square $(0, 2)^2$ can be achieved as an equilibrium payoff for sufficiently patient players. Ely and Välimäki (2002) again extend this type of profile to obtain payoffs outside of the square.

◆

14.2 Strong Self-Generation

For the case of two players, inclusive belief-free equilibria can be characterized in a manner similar to Abreu, Pearce, and Stacchetti's (1990) characterization of PPE in section 7.3.

Definition 14.2.1 *For any $\mathscr{V} \equiv \mathscr{V}_1 \times \mathscr{V}_2 \subset \mathbb{R}^2$ and $\hat{A} \equiv \hat{A}_1 \times \hat{A}_2 \subset A$, a mixed action $\alpha_j \in \Delta(\hat{A}_j)$ supports $\hat{A}_i$ on $\mathscr{V}_i$ if there exists a mapping $\gamma_i : A_j \times Z_j \to \mathscr{V}_i$, such that for all $a'_i \in A_i$,*

$$V_i^*(\alpha_j, \gamma_i) \equiv \min_{a_i \in \hat{A}_i} (1-\delta) u_i(a_i, \alpha_j) + \delta \sum_{z_j} \sum_{a_j} \gamma_i(a_j, z_j) \pi_j(z_j \mid a_i a_j) \alpha_j(a_j)$$

$$\geq (1-\delta) u_i(a'_i, \alpha_j) + \delta \sum_{z_j} \sum_{a_j} \gamma_i(a_j, z_j) \pi_j(z_j \mid a'_i a_j) \alpha_j(a_j). \quad (14.2.1)$$

Because inequality (14.2.1) is required to hold for all $a'_i \in A_i$, when α_j supports $\hat{A}_i$ (with j "promising" the continuation γ_i), player i is indifferent over all actions in $\hat{A}_i$, and weakly prefers any action in $\hat{A}_i$ to any other action. The critical feature of the notion of a supporting mixed action α_j is that the only constraint concerns player i, and the promised continuation γ_i is a function of player j's signal and action only.

There is an equivalent definition that emphasizes the connection with the notion of enforceability in definition 7.3.1.

Definition 14.2.2 *For any $\mathscr{V} \equiv \mathscr{V}_1 \times \mathscr{V}_2 \subset \mathbb{R}^2$ and $\hat{A} \equiv \hat{A}_1 \times \hat{A}_2 \subset A$, a mixed action profile $\alpha \in \prod_i \Delta(\hat{A}_i)$ is strongly enforceable on $(\mathscr{V}, \hat{A})$ if there exist mappings $\gamma_i : A_j \times Z_j \to \mathscr{V}_i$, $i = 1, 2$, such that for all i and $a'_i \in A_i$,*

$$(1-\delta) u_i(\alpha) + \delta \sum_{z_j} \sum_{a_j} \gamma_i(a_j, z_j) \pi_j(z_j \mid \alpha_i a_j) \alpha_j(a_j)$$

$$\geq (1-\delta) u_i(a'_i, \alpha_j) + \delta \sum_{z_j} \sum_{a_j} \gamma_i(a_j, z_j) \pi_j(z_j \mid a'_i a_j) \alpha_j(a_j),$$

with equality holding for all $a_i \in \hat{A}_i$.

The following is immediate.

Lemma 14.2.1 *The mixed action α_i supports $\hat{A}_j$ on $\mathscr{V}_j$ for $i, j = 1, 2$ and $j \neq i$ if and only if the profile $\alpha = (\alpha_1, \alpha_2)$ is strongly enforced on $(\mathscr{V}, \hat{A})$.*

For the case of public monitoring, weakening the restriction on the product structure of $\mathscr{V}$, taking $\hat{A}_i = A_i$, and restricting attention to γ_i independent of a_j, definition 14.2.2 agrees with the strong enforceability of Fudenberg, Levine, and Maskin (1994, definition 5.2). This latter notion strengthens our definition 7.3.1 by replacing the inequality in (7.3.1) with an equality. Lemma 9.2.2 effectively uses this version of strong enforceability to determine continuations for some versions of the folk theorem.

Definition 14.2.3 *A payoff v_i is strongly decomposable on $(\mathscr{V}_i, \hat{A})$, $\hat{A} = \hat{A}_1 \times \hat{A}_2$, if $v_i = V_i^*(\alpha_j, \gamma_i)$ for some $\alpha_j \in \Delta(\hat{A}_j)$ supporting $\hat{A}_i$ on $\mathscr{V}_i$ by γ_i. A set of payoffs $\mathscr{V}_i$ is strongly self-generating on $\hat{A}$ if every $v_i \in \mathscr{V}_i$ is strongly decomposable on $(\mathscr{V}_i, \hat{A})$. A set of payoff profiles $\mathscr{V} = \mathscr{V}_1 \times \mathscr{V}_2$ is strongly self-generating on $\hat{A}$ if $\mathscr{V}_i$ is strongly self-generating on $\hat{A}_i$ for $i = 1, 2$.*

We emphasize that a player i payoff v_i strongly decomposable on $(\mathscr{V}_i, \hat{A})$ is decomposed by player j's behavior and player j's promises of i's continuations. By construction, i is indifferent over every action in $\hat{A}_i$, so i's payoff is achieved through

14.2 ■ Strong Self-Generation

an appropriate combination of player j behavior and promises (with i's behavior being irrelevant). If $\mathscr{V}_i$ is strongly self-generating on $\hat{A}$, then these promises in turn are decomposed by further player j behavior and promises. We then have the following analogs to propositions 7.3.1 and 7.3.2.

Proposition 14.2.1 *If $\mathscr{V} \equiv \mathscr{V}_1 \times \mathscr{V}_2$ is strongly self-generating on $\hat{A} = \hat{A}_1 \times \hat{A}_2$, then it is a set of belief-free equilibrium payoffs.*

Proof The proof that $\mathscr{V}$ is a set of belief-free equilibrium payoffs is similar in spirit to that of proposition 7.3.1. For each i, denote the set of payoffs strongly decomposable on $(\mathscr{V}_i, \hat{A})$ by $\mathscr{B}_{\hat{A}}(\mathscr{V}_i)$, and (similar to definition 7.3.3), define the pair

$$\mathsf{Q}^*_j : \mathscr{B}_{\hat{A}}(\mathscr{V}_i) \to \Delta(\hat{A}_j) \text{ and } \mathsf{U}^*_i : \mathscr{B}_{\hat{A}}(\mathscr{V}_i) \to (\mathscr{V}_i)^{A_j \times Z_j},$$

so that $\mathsf{Q}^*_j(v_i)$ strongly enforces $\hat{A}_i$ on $\mathscr{V}_i$ using $\mathsf{U}^*_i(v_i)$, and $\mathsf{V}^*_i(\mathsf{Q}^*_j(v_i), \mathsf{U}^*_i(v_i)) = v_i$.

For player j, we now construct a collection of automata $\{(\mathscr{W}_j, w_j, f_j, \tau_j) : w_j \in \mathscr{W}_j\}$, where $\mathscr{W}_j = \mathscr{V}_i$ (i.e., the set of states is j's promises of i's continuations), the decision function is

$$f_j(v_i) = \mathsf{Q}^*_j(v_i)$$

for all $v_i \in \mathscr{V}_i$, and the transition function is

$$\tau_j(v_i, a_j z_j) = \mathsf{U}^*_i(v_i)(a_j, z_j)$$

for all $(a_j, z_j) \in A_j \times Z_j$. Because $\mathscr{V}$ is strongly self-generating on $\hat{A}$, we have $\mathscr{V}_i \subset \mathscr{B}_{\hat{A}}(\mathscr{V}_i)$ and the automata are well defined.

Player j's automata are designed to ensure that irrespective of i's action (as long as it is in $\hat{A}_i$), i's payoff is given by the initial state of j's automaton, which is by construction a value for player i. It is straightforward to verify from the definitions (using an argument similar to that of the proof of proposition 7.3.1) that the automata have the desired property, and that they are belief-free, for all choices of initial states. ■

Proposition 14.2.2 *If $\{(\mathscr{W}_i, w^0_i, f_i, \tau_i) : i = 1, 2\}$ is an inclusive belief-free equilibrium, then $\{(V_1(\mathbf{w}), V_2(\mathbf{w})) : \mathbf{w} \in \mathscr{W}_1 \times \mathscr{W}_2\}$ is strongly self-generating on some set $\hat{A}$, where $V_i(\mathbf{w})$ is player i's value from the pair of initial states $\mathbf{w} \in \mathscr{W}_1 \times \mathscr{W}_2$.*

Proof Let $\hat{A}_i = \cup_{w_i \in \mathscr{W}_i} \text{supp } f_i(w_i)$, and suppose $v_i = V_i(w_j)$ for some state w_j (recall from lemma 14.1.1 that V_i is independent of w_i). Then, because the profile is belief-free, $f_j(w_j)$ supports $\hat{A}_i$ on $\{V_i(w_j) : w_j \in \mathscr{W}_j\}$ using the continuations

$$\gamma_i(a_j, z_j) = V_i(\tau(w_j, a_j z_j)),$$

and so $\{V_i(w_j) : w_j \in \mathscr{W}_j\}$ is strongly self-generating on $\hat{A}_i$. ■

Remark 14.2.1 **Further characterization and folk theorems** In addition to developing the above characterization, Ely, Hörner, and Olszewski (2005) characterize the set of belief-free equilibrium payoffs as δ approaches 1. The prisoners' dilemma is quite special.

Though belief-free equilibria can support a large set of payoffs, in most games they are not sufficient to prove a folk theorem, even for vanishing noise. It is possible, on the other hand, to build on belief-free behavior to construct folk theorems. Matsushima (2004) extended the example of section 14.1.2 to the case of conditionally independent but very noisy signals, using review strategies (review strategies are discussed in section 11.4.5). Hörner and Olszewski (2005) prove a general folk theorem for almost perfect private monitoring using profiles that have some of the essential features of belief-free equilibria.

◆

Part IV

Reputations

15 Reputations with Short-Lived Players

The word *reputation* appears throughout our discussions of everyday interactions. Firms are said to have reputations for providing good service, professionals for working hard, people for being honest, newspapers for being unbiased, governments for being free from corruption, and so on. These reputation statements share two features. They establish links between past behavior and expectations of future behavior—one expects good service because good service has been provided in the past. In addition, they involve behavior that one might not expect in an isolated interaction—one is skeptical of a watch offered for sale by a stranger on a subway platform, but more confident of a special deal on a watch from a jeweler with whom one has regularly done business. Both characteristics make repeated games an ideal tool for studying reputations, and both suggest that reputations may be an important part of long-run relationships.

15.1 The Adverse Selection Approach to Reputations

There are two approaches to reputations in the repeated games literature. In the first one, an equilibrium of the repeated game is selected, involving actions along the equilibrium path that are not Nash equilibria of the stage game. As usual, incentives to choose these actions are created by attaching less favorable continuation paths to deviations. Players who choose the equilibrium actions are then interpreted as maintaining a reputation for doing so, with a punishment-triggering deviation interpreted as the loss of one's reputation. For example, players who exert effort in the repeated prisoners' dilemma are interpreted as maintaining a reputation for effort, whereas shirking destroys one's reputation. The firms in section 6.1 could be said to maintain (imperfect) reputations for collusion, the government in section 6.2 to maintain a reputation for not expropriating capital, and the consumers in section 6.3 to maintain a reputation for contributing to those with low incomes. Barro and Gordon (1983) and Canzoneri (1985) offer early examples of such reputations models, and Ljungqvist and Sargent (2004, Chapter 22) offer a more recent discussion. In this approach, the link between past behavior and expectations of future behavior is an equilibrium phenomenon, holding in some equilibria but not in others. The notion of reputation is used to interpret an equilibrium strategy profile, but otherwise adds nothing to the formal analysis.

The second or adverse selection approach to reputations begins with the assumption that a player is uncertain about key aspects of her opponent. For example,

	h	ℓ
H	2, 3	0, 2
L	3, 0	1, 1

Figure 15.1.1 The product-choice game.

player 2 may not know player 1's payoffs or may be uncertain about what constraints player 1 faces on his ability to choose various actions. This incomplete information is a device that introduces an intrinsic connection between past behavior and expectations of future behavior. Because incomplete information about players' characteristics can have dramatic effects on the *set* of equilibrium payoffs, reputations in this approach do not describe certain equilibria but place constraints on the set of possible equilibria.

In the course of introducing incomplete information, we have changed the game. Perhaps it is no surprise that we can get a different set of equilibrium outcomes when we examine a different game. Much of the interest in reputation models stems from the fact that seemingly quite small departures from perfect information about types can have large effects on the set of equilibrium payoffs.

Consider the product-choice game of figure 1.5.1, reproduced in figure 15.1.1, infinitely repeated with perfect monitoring. Player 1 is a long-lived player and player 2 is short-lived. Every payoff in the interval $[1, 2]$ is a subgame-perfect equilibrium for a sufficiently patient player 1 (section 7.6.1).

High payoffs for player 1 require player 1 to frequently play H, so that 2 will play her best response of h. Can player 1 develop a "reputation" for playing H by persistently doing so? This may be initially costly for player 1 if player 2 is not immediately convinced that 1 will play H and hence plays ℓ for some time, but the subsequent payoff could make this investment worthwhile for a sufficiently patient player 1.

It seems intuitive that if player 1 consistently chooses H, player 2 will eventually come to expect such play. However, nothing in the repeated game captures this intuition. Instead, repeated games have a recursive structure; the continuation game following any history is identical to the original game. No matter how many times player 1 has previously played H, the theory of complete information repeated games provides no reason for player 2 to believe that player 1 is more likely to play H now than at the beginning of the game.

The adverse selection approach to reputations allows player 2 to entertain the possibility that player 1 may be committed to playing H. Suppose player 2 thinks player 1 is most likely to be a normal player 1 but assigns some (small) probability $\hat{\mu} > 0$ to player 1 being a *commitment type* who always plays H. Even a tiny probability of a commitment type introduces a necessary relationship between past play of H and expectations of future play that can be magnified, over the course of repeated play, to have large effects.

Example Suppose the product-choice game is played twice, with player 1's payoffs added
15.1.1 over the two periods.[1] In the perfect monitoring game of complete information, $L\ell$
in both periods is the *unique* equilibrium outcome. Suppose now player 2 assigns
some (small) probability $\hat{\mu} > 0$ to player 1 being a commitment type who always
plays H and assigns complementary probability to 1 being as already described
(the *normal type*). The game still has perfect monitoring, so period 0 choices are
observed before period 1 choices are made. Consider the profile where the normal
type of player 1 plays L in both periods. Player 2 plays ℓ in period 0 (because $\hat{\mu}$
is small). In period 1, after observing H, player 2 concludes that she is facing the
commitment type of player 1 and best responds with h. On the other hand, after
observing L, player 2 concludes that she is facing the normal type of player 1
(who will play L in period 1) and best responds with ℓ. This profile is not an
equilibrium. By deviating and masquerading as the commitment type in period 0,
the normal player 1 sacrifices 1 in current payoff, but gains 2 in the next period.
The two period game does not have a pure strategy equilibrium. Example 15.3.1
describes a (pooling on H) pure strategy equilibrium of the infinite horizon game.

●

The reputation argument begins in section 15.3 with a basic implication of the link between current behavior and expectations of future behavior, introduced by the uncertainty about player 1's type. If a normal player 1 consistently plays like a commitment type, player 2 must eventually come to expect such behavior from player 1 and hence play a best response. The resulting payoff imposes a lower bound on player 1's equilibrium payoff. For a sufficiently patient normal player 1, this is as good as being the commitment type.

Much now hinges on the specification of the commitment type's behavior. If the commitment type happens to be a Stackelberg type (committed to the action most favorable to player 1), then player 1 has effectively been transformed into a "Stackelberg leader." However, the argument does not depend on there being *only* a single possible commitment type, nor on this commitment type having just the right behavior. The result holds for a general set of possible commitment types, with player 1 essentially choosing to develop a reputation for behaving as the most favorable type.

One interpretation of reputation results is that they provide a means of selecting among equilibria, providing a welcome antidote to the multiple equilibria reflected in the folk theorem. In our view, it is more useful to think of reputation results as an examination of the robustness of repeated-game results to variations in the specification of the game. In this sense, they provide an indication that a patient long-lived player facing short-lived opponents may have advantages beyond those captured by the standard complete information model. At the same time, the importance of commitment behavior directs attention to the need for a model of commitment types, a relatively neglected topic in the theory of reputations. We return to this issue in chapter 18.

Section 15.4 extends the argument to imperfect monitoring games. Here, reputations have more dramatic implications. In perfect monitoring games, a sufficiently rich

---

1. The adverse selection approach to reputations was first studied in finitely repeated games, where the effects can be particularly dramatic. See chapter 17.

set of commitment types ensures the normal player 1 a payoff arbitrarily close to his maximum equilibrium payoff in the complete information repeated game. Under imperfect monitoring, if the normal player 1 is subject to binding moral hazard, he may be assured a payoff in excess of *any* equilibrium payoff in the complete information game.

Section 15.5 shows that under general conditions in imperfect monitoring games, the incomplete information that is at the core of the adverse selection approach to reputations is a short-run phenomenon. Player 2 must eventually come to learn player 1's type and continuation play must converge to an equilibrium of the complete information game.

How do we reconcile this finding with the nontrivial bounds on ex ante payoffs, bounds that may push player 1 outside the set of equilibrium payoffs in the complete information game? There may well be a long period of time during which player 2 is uncertain of player 1's type and in which play does not resemble an equilibrium of the complete information game. The length of this period will depend on the discount factor, being longer for larger discount factors, and in general being long enough as to have a significant effect on player 1's payoffs. Eventually, however, such behavior must give way to a regime in which player 2 is (correctly) convinced of player 1's type.

We thus have an order of limits calculation. For any prior probability $\hat{\mu}$ that the long run player is the commitment type and for any $\varepsilon > 0$, there is a discount factor δ sufficiently large that player 1's expected payoff is ε-close to the commitment type payoff. This holds no matter how small is $\hat{\mu}$. As a result, it is tempting to think that even as the game is played and the posterior probability of the commitment type falls, we should be able to choose a period, think of it as the beginning of the game, and apply the standard reputation argument to conclude that uncertainty about player 1's type still has a significant effect. However, for any fixed δ and in any equilibrium, there is a time at which the posterior probability attached to the commitment type has dropped below the corresponding critical value of $\hat{\mu}$, becoming too small (relative to δ) for reputation effects to operate. We are then on the path to revealing player 1's type.

Which should command our interest, the ability of reputations to impose bounds on ex ante payoffs or the fact that such effects eventually disappear? These results reflect different views of a common model. Their relative importance depends on the context in which the model is applied rather than arguments that can be made within the model. We sometimes observe strategic interactions from a well-defined beginning, focusing attention on ex ante payoffs. We also often encounter ongoing interactions whose beginnings are difficult to identify, making long-run equilibrium properties a potentially useful guide to behavior. If one's primary interest is the long-lived player, then ex ante payoffs may again be paramount. One may instead take the view of a social planner who is concerned with the continuation payoffs of the long-run player and with the fate of all short-run players, even those in the distant future, directing attention to long-run properties. Finally, if one is interested in the steady state of a model with incomplete information, long-run properties are important.

We view the finding that reputations are temporary as an indication that a model of *long-run* reputations should incorporate some mechanism by which the uncertainty about types is continually replenished. For example, Holmström (1982), Cole, Dow, and English (1995), Mailath and Samuelson (2001), and Phelan (2001) assume that the type of the long-lived player is governed by a stochastic process rather than being determined once and for all at the beginning of the game. In such a situation, reputations can indeed have long-run implications. We return to this in chapter 18.

The reputation results in this chapter exploit the sharp asymmetry between players, with player 1 being long-lived and arbitrarily patient and player 2 being short-lived and hence myopic. In particular, short-lived player 2s allow us to move directly from the fact that player 2 believes 1 is playing like the commitment type to the conclusion that 2 plays a best response to the commitment type. There is no such direct link when player 2 is also a long-lived player, as we discuss in chapter 16, leading to considerably weaker reputation results.

15.2 Commitment Types

We consider the case of one long-lived player and one short-lived player, with the latter representing either a succession of players who live for one period or a continuum of small and anonymous infinitely lived players. The type of player 1 is unknown to player 2. A possible type of player 1 is denoted by $\xi \in \Xi$, where Ξ is a finite or countable set. Player 2's prior belief about 1's type is given by the distribution μ, with support Ξ.

We partition the set of types into *payoff types*, Ξ_1, and *commitment types*, $\Xi_2 \equiv \Xi \setminus \Xi_1$. Payoff types maximize the average discounted value of payoffs, which depend on their type and which may be nonstationary,

$$u_1 : A_1 \times A_2 \times \Xi_1 \times \mathbb{N}_0 \to \mathbb{R}.$$

Type $\xi_0 \in \Xi_1$ is the *normal type* of player 1, who happens to have a stationary payoff function, given by the stage game in the benchmark game of complete information,

$$u_1(a, \xi_0, t) = u_1(a) \quad \forall a \in A, \forall t \in \mathbb{N}_0.$$

It is standard to think of the prior probability $\mu(\xi_0)$ as being relatively large, so the games of incomplete information are a seemingly small departure from the underlying game of complete information, though there is no requirement that this be the case.

Commitment types (also called *action* or *behavioral types*) do not have payoffs and simply play a specified repeated game strategy. For any repeated game strategy from the complete information game, $\hat{\sigma}_1 : \mathcal{H}_1 \to \Delta(A_1)$, where $\mathcal{H}_1$ is the set of histories observed by player 1, denote by $\xi(\hat{\sigma}_1)$ the commitment type committed to the strategy $\hat{\sigma}_1$. In general, a commitment type of player 1 can be committed to any strategy in the repeated game. If the strategy in question plays the same (pure or mixed) stage-game action in every period, regardless of history, we refer to that type as a *simple commitment type*. For example, one simple commitment type in the product-choice game is a player who always exerts high effort. We let $\xi(a_1)$ denote the (simple commitment) type that plays the pure action a_1 in every period and $\xi(\alpha_1)$ denote the type that plays the mixed action α_1 in every period. Commitment types who randomize are important because they can imply a higher lower bound on player 1's payoff.

Other commitment types for player 1 are committed to more complicated sequences of actions. For example, in the repeated prisoners' dilemma, a type can play tit-for-tat, play E in every period up to and including t and then switch to S, or play E in prime-numbered periods and S otherwise.

Remark 15.2.1 **Payoff or commitment types** The distinction between payoff and commitment types is not clear-cut. For example, pure simple commitment types are easily modeled as payoff types. The type $\xi(a_1)$ for pure stage-game action a_1 can be interpreted as the payoff type for whom playing a_1 in every period is strictly dominant *in the repeated game* by specifying

$$u_1(a, \xi(a_1'), t) = \begin{cases} 1, & \text{if } a_1 = a_1', \\ 0, & \text{otherwise.} \end{cases}$$

The commitment type ξ' who plays a_1' in every period up to and including t, and then switches to a_1'', is the payoff type with payoffs

$$u_1(a, \xi', \tau) = \begin{cases} 1, & \text{if } a_1 = a_1' \text{ and } \tau \leq t, \\ & \text{or } a_1 = a_1'' \text{ and } \tau > t, \\ 0, & \text{otherwise.} \end{cases}$$

A payoff type for whom an action a_1 is a dominant action *in the stage game* is typically not equivalent to a commitment type who invariably plays the action a_1. To recast the latter as a payoff type, we need the constant play of the action a_1 to be a dominant strategy in the repeated game, a more demanding requirement. For example, shirking is a dominant action in the prisoners' dilemma but is not a dominant strategy in the repeated game.

It is possible to interpret, as Fudenberg and Levine (1992) do, mixed commitment types as payoff types as well. However, doing so requires an uncountable type space if the commitment type's strategy is to be strictly dominant in the repeated game, with associated technical complications.

The choice between payoff and commitment types is one of taste. The conceptual advantage of only having payoff types is that all types are expected utility maximizers. On the other hand, interest in reputation games stems from a belief that players may not be completely certain about the characteristics of other players. Requiring expected utility maximization may be less plausible than simply the belief that player 1 may be irrational or "crazy," and indeed this language has appeared in much of the literature. Alternatively, recognizing that the games with which we work are models of a more complicated strategic interaction, the uncertainty about a player's characteristic may include the possibility that the player models the strategic interaction quite differently, leading to payoffs that have no expected utility representation in the game in question. The player may then be completely rational but best represented as a commitment type.

15.2 ■ Commitment Types

In this book, we maintain the commitment type interpretation for $\xi(a_1)$, so that *by assumption*, a commitment type plays the specified strategy.

♦

Player 1's *pure-action Stackelberg payoff* is defined as

$$v_1^* = \sup_{a_1 \in A_1} \min_{\alpha_2 \in B(a_1)} u_1(a_1, \alpha_2), \qquad (15.2.1)$$

where $B(a_1)$ is the set of player 2 myopic best replies to a_1. If the supremum is achieved by some action a_1^*, that action is an associated Stackelberg action,

$$a_1^* \in \arg\max_{a_1 \in A_1} \min_{\alpha_2 \in B(a_1)} u_1(a_1, \alpha_2).$$

This is a pure action to which player 1 would commit, if he had the chance to do so (and hence the name Stackelberg action), given that such a commitment induces a best response from player 2. If there is more than one such action for player 1, we can choose one arbitrarily. The *(pure-action) Stackelberg type* of player 1 plays a_1^* and is denoted by $\xi(a_1^*) \equiv \xi^*$.

When player 2 is short-lived, any bound on player 1's ex ante payoffs that can be obtained using commitment types can be obtained using only simple commitment types. Chapter 16 shows that more complicated commitment types can be important when both players are long-lived.

Remark 15.2.2 **Mixed-action Stackelberg types** When considering perfect monitoring, we focus on simple commitment types who choose pure actions. As we have seen in section 2.7.2, it is advantageous in some games for a player to commit to a mixed action. In the product-choice game, for example, a commitment by player 1 to mixing between H and L, with slightly larger probability on H, induces player 2 to choose h and gives player 1 a larger payoff than a commitment to H. In effect, a player 1 who always plays H spends too much to induce response h from player 2.

Accordingly, define the mixed-action Stackelberg payoff as

$$\sup_{\alpha_1 \in \Delta(A_1)} \min_{\alpha_2 \in B(\alpha_1)} u_1(\alpha_1, \alpha_2), \qquad (15.2.2)$$

where $B(\alpha_1)$ is the set of player 2's best responses to α_1. Typically, the supremum is not achieved by any mixed action, so there is no mixed-action Stackelberg type. However, there are mixed commitment types that if player 2 is convinced she is facing such a type, will yield payoffs arbitrarily close to the mixed-action Stackelberg payoff.

In perfect-monitoring games, it is simpler to verify the lower bound on equilibrium payoffs implied by commitments to pure (rather than mixed) actions. In the former case, we need only analyze the updating of the short-lived players' beliefs on *one* path of informative actions, the path induced by the Stackelberg commitment type. In contrast, commitments to mixed actions require consideration of belief evolution on all histories that arise with positive probability. This consideration involves the same issues that arise when studying reputations in imperfect monitoring games. As we discuss in remarks 15.3.4 and 15.4.3, the reputation effects from mixed commitment types are qualitatively stronger than those from pure commitment types. We consider mixed commitment types together with imperfect monitoring in section 15.4.

◆

15.3 Perfect Monitoring Games

We begin by examining reputations in repeated games of perfect monitoring. We assume that the action set A_2 is finite, considering the case of an infinite A_2 in remark 15.3.6. Assumption 2.1.1 is otherwise maintained throughout. As usual, when any player has a continuum action space, we only consider behavior in which that player is playing a pure strategy (remark 2.1.1).

In section 2.7, we constructed equilibria with "high" payoffs for a long-lived player facing short-lived opponents. The basic reputation result is a lower bound on equilibrium payoffs for the normal long-lived player, in the game of incomplete information in which the short-lived player is uncertain about the characteristics of the long-lived player.[2]

The set of histories in the complete information game, $\mathcal{H}$, is the set of *public* histories in the incomplete information game and is also the set of player 2 histories in the incomplete information game. A history for player 1 in the incomplete information game is an element of $\Xi \times \mathcal{H}$, specifying player 1's type as well as the public history. A behavior strategy for player 1 in the incomplete information game is, using the notation on commitment types from section 15.2,

$$\sigma_1 : \mathcal{H} \times \Xi \to \Delta(A_1),$$

such that, for all commitment types $\xi(\hat{\sigma}_1) \in \Xi_2$,

$$\sigma_1(h^t, \xi(\hat{\sigma}_1)) = \hat{\sigma}_1(h^t) \qquad \forall h^t \in \mathcal{H}.$$

A behavior strategy for player 2 is, as in section 2.7, a map $\sigma_2 : \mathcal{H} \to \Delta(A_2)$.

Given a strategy profile σ, $U_1(\sigma, \xi)$ denotes the type ξ long-lived player's payoff in the repeated game. As is familiar, a Nash equilibrium is a collection of mutual best responses:

Definition 15.3.1 *A strategy profile $(\tilde{\sigma}_1, \tilde{\sigma}_2)$ is a* Nash equilibrium *of the reputation game with perfect monitoring if for all $\xi \in \Xi_1$, $\tilde{\sigma}_1$ maximizes $U_1(\sigma_1, \tilde{\sigma}_2, \xi)$ over player 1's repeated game strategies, and if for all t and all $h^t \in \mathcal{H}$ that have positive probability under $(\tilde{\sigma}_1, \tilde{\sigma}_2)$ and μ,*

$$E[u_2(\tilde{\sigma}_1(h^t, \xi), \tilde{\sigma}_2(h^t)) | h^t] = \max_{a_2 \in A_2} E[u_2(\tilde{\sigma}_1(h^t, \xi), a_2) | h^t].$$

Remark 15.3.1 **Existence of equilibrium** The existence of Nash equilibria when Ξ is finite follows by observing that every finite-horizon truncation of the game has a Nash

---

2. This result is first established in Fudenberg and Levine (1989), which also considers an uncountable Ξ.

15.3 ■ Perfect Monitoring

equilibrium and applying standard limiting arguments to obtain an equilibrium of the infinite horizon game (see, for example, Fudenberg and Levine 1983). When Ξ is countably infinite, existence is again an implication of Fudenberg and Levine (1983) if every finite-horizon truncation of the game has an ε-Nash equilibrium. To prove that the finite-horizon truncation of the game has an ε-Nash equilibrium, arbitrarily fix the behavior of all but a finite number of types of player 1 (because Ξ is countable, the set of types whose behavior is not fixed can be chosen so that its ex ante probability is close to 1). Then, in the finite-truncation game, all the short-lived players are maximizing while player 1 is ε-maximizing (because he is free to choose behavior for all but a small probability set of types).

If A_1 is a continuum, these arguments may not yield an equilibrium in pure strategies (contrasting with remark 2.1.1). In this chapter, we are concerned with lower bounds on equilibrium payoffs. The existence results assure us that we are bounding a nonempty set. Allowing for mixing by player 1 when A_1 is a continuum introduces some tedious details in the definition of equilibrium, but does not alter the nature of the bounds we calculate.

◆

Remark 15.3.2 **Sequential equilibrium** Because the lower bound on player 1's payoff applies to all Nash equilibria, we do not consider stronger equilibrium concepts.[3] The counterpart of a sequential equilibrium in this context is straightforward. Only player 1 has private information, and therefore sequential rationality for player 1 is immediate: After all histories, the continuation strategy of player 1 (of any type) should maximize his continuation payoffs. For player 2, after histories that have zero probability under the equilibrium and involve a deviation by player 1, we would simply require that her action be optimal, given some beliefs over Ξ, with subsequent player 2's updating the same beliefs when possible.

The consistency condition of sequential equilibrium has a powerful implication in the presence of commitment types. Should player 2 ever see an action that is not taken by a commitment type, then player 2 must thereafter attach probability zero to that commitment type, regardless of what she subsequently observes. This follows immediately from the fact that no perturbed strategies can generate such an outcome from the commitment type. The same is not the case with a payoff commitment type.[4]

◆

Example 15.3.1 We continue with the product-choice game (figure 15.1.1). The pure Stackelberg type of player 1 chooses H, with Stackelberg payoff 2. Suppose $\Xi = \{\xi_0, \xi^*, \xi(L)\}$. For $\delta \geq 1/2$, the grim trigger strategy profile of always playing Hh, with deviations punished by Nash reversion, is a subgame-perfect equilibrium of the complete information game. Consider the following adaptation of this profile in the incomplete information game:

---

3. The full strength of Nash equilibrium is not needed for the existence of reputation bounds. It essentially suffices that player 1 knows that the player 2s play a best response (see Battigali and Watson 1997).

4. Section 17.1 (see note 5 on page 552) presents an example where the stronger consistency implication of commitment types ensures that there is a unique sequential equilibrium outcome with a commitment type but not with an equivalent payoff type.

$$\sigma_1(h^t, \xi) = \begin{cases} H, & \text{if } \xi = \xi^*, \\ & \text{or } \xi = \xi_0 \text{ and } a^\tau = Hh \text{ for all } \tau < t, \\ L, & \text{otherwise,} \end{cases}$$

and

$$\sigma_2(h^t) = \begin{cases} h, & \text{if } a^\tau = Hh \text{ for all } \tau < t, \\ \ell, & \text{otherwise.} \end{cases}$$

In other words, player 2 and the normal type of player 1 follow the strategies from the Nash-reversion equilibrium in the complete information game, and the commitment types ξ^* and $\xi(L)$ play their actions.

This is a Nash equilibrium for $\delta \geq 1/2$ and $\mu(\xi(L)) < 1/2$. The restriction on $\mu(\xi(L))$ ensures that player 2 finds h optimal in period 0. Should player 2 ever observe L, then Bayes' rule causes her to place probability 1 on type $\xi(L)$ (if L is observed in the first period) or the normal type (if L is first played in a subsequent period), making her participation in Nash reversion optimal. The restriction on δ ensures that Nash reversion provides sufficient incentive to make H optimal for the normal player 1. After observing $a_1^0 = H$ in period 0, player 2 assigns zero probability to $\xi = \xi(L)$. However, the posterior probability that 2 assigns to the Stackelberg type does not converge to 1. In period 0, the prior probability is $\mu(\xi^*)$. After one observation of H, the posterior increases to $\mu(\xi^*)/[\mu(\xi^*) + \mu(\xi_0)]$, after which it is constant.[5]

Of more interest is the (im)possibility of a Nash equilibrium with a low payoff for the normal player 1. This contrasts with the game of complete information, where playing $L\ell$ in every period is a subgame-perfect equilibrium with a payoff of 1 to the normal player 1. It is an implication of proposition 15.3.1 that there is no Nash equilibrium of the incomplete information game with a payoff to the normal player 1 near 1. Here we argue that, if $\mu(\xi^*) < 1/3$ and $\mu(\xi(L)) < 1/3$, the normal player 1's payoff in any *pure strategy* Nash equilibrium is bounded below by 2δ and above by 2. The bounds on $\mu(\xi^*)$ and $\mu(\xi(L))$ imply that in any pure strategy Nash equilibrium outcome, the normal player 1 and player 2 choose either Hh or $L\ell$ in each period, and so 2 is the upper bound on 1's payoff.[6] Fix a pure strategy Nash equilibrium, and let t be the first period in which the normal player 1 chooses L. If $t = \infty$ (i.e., 1 never chooses L in equilibrium), then the normal player 1's payoff is 2. Suppose $t < \infty$. If the normal player 1 chooses H

---

5. We can complete player 2's beliefs to be consistent with sequentiality by stipulating that an observation of H in a history in which L has previously been observed causes her to place probability one on the normal type of player 1.
6. Player 2 must either expect L in the first period with probability $\mu^0(\xi(L)) + \mu^0(\xi_0)$ or expect H with probability $\mu^0(\xi^*) + \mu^0(\xi_0)$, with a best response of ℓ in the first case and h in the second. In subsequent periods, positive probability can be attached to only one commitment type. This probability must fall short of $1/2$ in the second period (because $\mu(\xi^*) < 1/3$ and $\mu(\xi(L)) < 1/3$) and, given that player 1 is normal, can never thereafter increase (because the equilibrium is pure), ensuring that player 2 always plays a best response to the normal player 1.

15.3 ■ Perfect Monitoring

	h	ℓ
H	3,3	1,2
L	2,0	0,1

	h	ℓ
H	3,3	3,2
L	2,0	0,1

Figure 15.3.1 A payoff type of player 1 for whom H is dominant in the stage game (left panel) and in the repeated game (right panel).

in period t and every subsequent period, then player 2 will choose h in period $t+1$ and every subsequent period, having concluded in period t that 1 is the Stackelberg type and then having received no evidence to the contrary. Hence, a lower bound on the normal player 1's payoff is

$$2(1-\delta^t) + 0 \times (1-\delta)\delta^t + 2\delta^{t+1}$$
$$= 2 - 2(1-\delta)\delta^t$$
$$\geq 2 - 2(1-\delta)$$
$$= 2\delta.$$

●

Example 15.3.2 **Payoff types** Continuing with the product-choice game (figure 15.1.1), we now consider the set of types $\Xi = \{\xi_0, \xi_1\}$, where ξ_1 is the payoff type with payoffs described in the left panel of figure 15.3.1. The lower bound from example 15.3.1 no longer holds, even though player 2 puts positive probability on player 1 being a type ξ_1 for whom the Stackelberg action is strictly dominant in the stage game. It is possible, even for δ arbitrarily close to 1, to construct sequential equilibria in which both types of player 1 receive a payoff arbitrarily close to their minmax values of 1. For example, first consider the profile in which in the absence of a deviation by player 1, both types of player 1 play L in even periods and H in odd periods, and player 2 plays ℓ in even periods and h in odd periods. Deviations by player 2 are ignored, and any deviation by player 1 results in player 2 concluding that player 1 is the normal type ξ_0 (and never subsequently revising her belief) and playing ℓ in every subsequent period. After any deviation by player 1, the normal type always plays L (while ξ_1 plays H), so the profile is sequential.[7] Profiles with lower payoffs can be constructed by increasing the frequency of $L\ell$ on the path of play (the result is still an equilibrium provided δ is large enough and the average payoff to player 1 of both types exceeds 1). Figure 15.3.1 also presents the payoffs for a payoff type who finds it a dominant strategy to play H after every history in the repeated game. This type is equivalent to the Stackelberg type.

●

---

7. This profile has the feature that the assumed belief for player 2 after a deviation to H in an even period is counterintuitive, because the deviation to ξ_1's most preferred action results in that type receiving zero probability. We do not discuss such *refinement* issues here. See Kreps and Wilson (1982a, p. 263) for further discussion.

15.3.1 Building a Reputation

Our first step toward the reputation result is to demonstrate that when player 2 assigns some probability to the simple type $\xi(a_1') \equiv \xi'$, if the normal player 1 persistently plays action a_1', then player 2 must eventually place high probability on that action being played. Of course, it may take a while to build such a reputation for playing a_1', and doing so may be quite costly in the meantime. However, this cost will be negligible if player 1 is sufficiently patient. If this action is the Stackelberg action a_1^*, when player 1 is sufficiently patient, the resulting lower bound on player 1's payoff is close to his Stackelberg payoff v_1^*.

Let $\Omega \equiv \Xi \times (A_1 \times A_2)^\infty$ be the space of outcomes. An outcome $\omega \in \Omega$ takes the form $\omega = (\xi, a_1^0 a_2^0, a_1^1 a_2^1, a_1^2 a_2^2, \ldots)$, specifying the type of player 1 and the actions chosen in each period. Associated with any outcome ω is the collection of period t public histories, one for each t, with $h^t = h^t(\omega) = (a_1^0(\omega)a_2^0(\omega), a_1^1(\omega)a_2^1(\omega), \ldots, a_1^{t-1}(\omega)a_2^{t-1}(\omega)) \in \mathscr{H}^t$.

A profile of strategies (σ_1, σ_2), along with the prior probability over types μ (with support Ξ), induces a probability measure on the set of outcomes Ω, denoted by $\mathbf{P} \in \Delta(\Omega)$. Denote by Ω' the event that the action a_1' is chosen in every period, that is,

$$\Omega' = \{\omega : a_1^t(\omega) = a_1' \; \forall t\} \subset \Omega = \Xi \times (A_1 \times A_2)^\infty.$$

The event Ω' contains a multitude of outcomes, differing in the type of player 1 and actions of player 2. For example, the action a_1' in every period is consistent with player 1 being the simple type $\xi(a_1')$, but also with player 1 being the normal type, as well as with a variety of other types and player 2 behavior.

Let q^t be the probability that the action a_1' is chosen in period t, conditional on the public history $h^t \in \mathscr{H}^t$, that is,

$$q^t \equiv \mathbf{P}(a_1^t = a_1' \mid h^t). \tag{15.3.1}$$

Note that q^t is a random variable, being a function of the form $q^t : \Omega \to [0, 1]$. Specifically, $q^t(\omega) = \mathbf{P}(a_1^t = a_1' \mid h^t(\omega))$. Because q^t depends on ω through $h^t(\omega)$, we will often write $q^t(h^t)$ rather than $q^t(\omega)$. Because q^t is conditioned on the public history, it provides a description of player 2's beliefs about player 1's play, after any history.

The normal player 1 receives a payoff of at least $\min_{a_2 \in B(a_1')} u_1(a_1', a_2)$ in any period t in which q^t is sufficiently large that player 2 chooses a best response to a_1', and player 1 in fact plays a_1'. The normal player 1 has the option of always playing a_1', so his payoff in any Nash equilibrium must be bounded below by the payoff generated by always playing a_1'. If there is a bound on the number of periods in which after always observing a_1', player 2's period t beliefs assign low probability to a_1', then there is a lower bound on the normal player 1's equilibrium payoff.

Hence, we are interested in the behavior of q^t on the set Ω'.

Example 15.3.3 q^t **may decrease on** Ω' Consider the product-choice game (figure 15.1.1) with $a_1' = H$ ($= a_1^*$, the Stackelberg action). Let $\tilde{\xi}_t$ describe a commitment type who plays H in every period $\tau < t$, and L thereafter, independently of history. In particular, $\tilde{\xi}_0$ is $\xi(L)$, the simple commitment type that plays L in every period,

15.3 ■ Perfect Monitoring

and $\tilde{\xi}_t$ is a nonsimple commitment type for $t \geq 1$. Let $\hat{\xi}$ denote the type that plays H in period 0 and plays H thereafter if and only if player 2 plays h in period 0; and otherwise plays L. The set of types is given by $\Xi = \{\xi_0, \xi^*, \hat{\xi}, \tilde{\xi}_0, \tilde{\xi}_1, \tilde{\xi}_2, \ldots\}$, with prior μ.

Consider first the strategy profile in which the normal type always plays H and player 2 always plays h. Recall that the set Ω' is the set of all outcomes in which player 1 always plays H. Then, $q^0 = 1 - \mu(\tilde{\xi}_0)$ because all types except $\tilde{\xi}_0$ play H in period 0. There are two period 1 histories consistent with Ω', Hh and $H\ell$, but $H\ell$ has zero probability under $\mathbf{P}$ (because 2 always plays h). Applying Bayes' rule,

$$q^1(Hh) = \frac{1 - \mu(\tilde{\xi}_0) - \mu(\tilde{\xi}_1)}{1 - \mu(\tilde{\xi}_0)}.$$

Because $q^0 < q^1(Hh)$ if and only if $\mu(\tilde{\xi}_0)(1 - \mu(\tilde{\xi}_0)) > \mu(\tilde{\xi}_1)$, q^t need not be monotonic on Ω' in t.

●

If the short-run players are playing pure strategies, then the conditional belief q^t is constant on a full-measure subset of Ω' (i.e., $q^t(\omega) = q^t(\omega')$ for all $\omega, \omega' \in \Omega^\dagger$ with $\mathbf{P}(\Omega^\dagger) = \mathbf{P}(\Omega')$), because there is then only one positive probability period t history h^t consistent with Ω'. If the short-lived players are randomizing, however, then h^t may be a nondegenerate random variable on Ω' and (because σ_1^t is a function of the short-lived players actions in earlier periods) so q^t may also be.

Example 15.3.4 q^t *can vary with* h^t *on* Ω' We continue with example 15.3.3. Consider now a profile in which the normal type of player 1 always plays H and player 2 plays h with probability $1/2$ and ℓ with probability $1/2$ in the first period, and then always plays h. Though our calculation of q^0 is unchanged, things are very different in period 1. Now, both period 1 histories consistent with Ω', Hh and $H\ell$, receive positive probability. So,

$$q^1(Hh) = \frac{1 - \mu(\tilde{\xi}_0) - \mu(\tilde{\xi}_1)}{1 - \mu(\tilde{\xi}_0)},$$

and

$$q^1(H\ell) = \frac{1 - \mu(\hat{\xi}) - \mu(\tilde{\xi}_0) - \mu(\tilde{\xi}_1)}{1 - \mu(\tilde{\xi}_0)}.$$

Consequently, for fixed t, q^t need not be constant as a function of h^t on a full-measure subset of Ω'.

●

Define $n_\zeta : \Omega \to \mathbb{N}_0 \cup \{\infty\}$ to be the number of random variables q^t ($t = 0, 1, \ldots$) for which $q^t \leq \zeta$. That is, for each $\omega \in \Omega$, $n_\zeta(\omega) = |\{t : q^t(\omega) \leq \zeta\}|$ is the number of terms in the sequence of conditional probabilities $\{q^t(\omega)\}_{t=1}^\infty$ that do not exceed ζ. Denote the event that player 1 is type ξ', $\{\xi'\} \times (A_1 \times A_2)^\infty$, by simply ξ'.

Lemma 15.3.1 *Fix $\zeta \in [0, 1)$. Suppose $\mu(\xi(a_1')) \in [\mu^\dagger, 1)$ for some $\mu^\dagger > 0$ and $a_1' \in A_1$. For any profile (σ_1, σ_2),*

$$\mathbf{P}\left\{ n_\zeta > \left. \frac{\ln \mu^\dagger}{\ln \zeta} \right| \Omega' \right\} = 0,$$

and for any outcome $\omega \in \Omega'$ such that all histories $\{h^t(\omega)\}_{t=0}^{\infty}$ have positive probability under $\mathbf{P}$, $\mathbf{P}(\xi(a_1') \mid h^t(\omega))$ is nondecreasing in t.

Thus, whenever player 2 observes ever longer strings of action a_1', eventually she must come to expect action a_1' to be played with high probability.

The restriction to histories $\{h^t(\omega)\}_{t=0}^{\infty}$ that have positive probability under $\mathbf{P}$ precludes outcomes ω that are impossible under strategy profile σ.

An important feature of this lemma is that the bound on n_ζ is *independent* of $\mathbf{P}$, allowing us to bound player 1's payoff in *any* equilibrium. Denote $\xi(a_1')$ by ξ'. This result does not assert $\mathbf{P}(\xi' \mid h^t) \to 1$ as $t \to \infty$, that is, that the posterior probability attached to the simple type ξ' converges to unity. Instead, it leaves open the possibility that player 1 is normal but plays like that simple type, as in the equilibrium in example 15.3.1.

The key idea behind the proof is the following. Suppose that under some history h^t, previous play is consistent with the simple type ($a_1^\tau = a_1'$ for all $\tau < t$) and the current expectation is that the action a_1' need *not* appear ($q^t < 1$). This can only happen if some probability is attached to the event that player 1 is not the simple type ξ' and will not play the action a_1'. Then, observing the action a_1' in period t results in a posterior that must put increased weight on ξ' and therefore (all else equal) must increase q^t in the future.

Proof Let $\Omega'' \equiv \{\omega : \mathbf{P}(h^t(\omega)) > 0, a_1^t(\omega) = a_1' \; \forall t\}$, that is, Ω'' is the set of outcomes ω such that all histories $h^t(\omega)$ have positive probability and a_1' is always played. Note that $\Omega'' \subset \Omega'$ and $\mathbf{P}(\Omega'') = \mathbf{P}(\Omega')$.

Step 1. Our first step is to show that, for $\omega \in \Omega''$,

$$\mathbf{P}(\xi' \mid h^t(\omega)) = \frac{\mathbf{P}(\xi' \mid h^{t-1}(\omega))}{q^{t-1}}.$$

This would be an immediate implication of Bayes' rule if only player 1's behavior were observed in period $t-1$. Establishing the result requires confirming that observing player 2's behavior as well does not confound the inference. Applying Bayes' rule, we have

$$\mathbf{P}(\xi' \mid h^t(\omega)) = \frac{\mathbf{P}(h^t(\omega) \mid \xi', h^{t-1}(\omega)) \mathbf{P}(\xi' \mid h^{t-1}(\omega))}{\mathbf{P}(h^t(\omega) \mid h^{t-1}(\omega))}. \qquad (15.3.2)$$

Reformulate the denominator by using the independence of any period t randomization of players 1 and 2 to obtain (suppressing ω),

$$\mathbf{P}(h^t \mid h^{t-1}) = \mathbf{P}(a_1^t a_2^t \mid h^{t-1}) = \mathbf{P}(a_1^t \mid h^{t-1}) \mathbf{P}(a_2^t \mid h^{t-1})$$
$$= \mathbf{P}(h^t(1) \mid h^{t-1}) \mathbf{P}(a_2^t \mid h^{t-1}),$$

15.3 ■ Perfect Monitoring

where $h^t(i)$ is the period t history of i's actions. Using the three observations that player 2's choice at t depends on 1's play only through h^{t-1}, $\omega \in \Omega''$ (so that $a_1^t = a_1'$), and $\mathbf{P}(\xi', h^{t-1}) > 0$,

$$\mathbf{P}(a_2^t \mid h^{t-1}) = \mathbf{P}(a_2^t \mid \xi', h^{t-1}) = \mathbf{P}(h^t(2) \mid \xi', h^{t-1}).$$

Turning to the numerator in (15.3.2), using the second observation again, $\mathbf{P}(a_1^t \mid \xi', h^{t-1}) = 1$, so

$$\mathbf{P}(h^t \mid \xi', h^{t-1}) = \mathbf{P}(h^t(2) \mid \xi', h^{t-1}).$$

Substituting these calculations into (15.3.2),

$$\begin{aligned}
\mathbf{P}(\xi' \mid h^t) &= \frac{\mathbf{P}(h^t(2) \mid \xi', h^{t-1}) \mathbf{P}(\xi' \mid h^{t-1})}{\mathbf{P}(h^t(1) \mid h^{t-1}) \mathbf{P}(h^t(2) \mid \xi', h^{t-1})} \\
&= \frac{\mathbf{P}(\xi' \mid h^{t-1})}{\mathbf{P}(h^t(1) \mid h^{t-1})} \\
&= \frac{\mathbf{P}(\xi' \mid h^{t-1})}{\mathbf{P}(a_1^t = a_1' \mid h^{t-1})} \\
&= \frac{\mathbf{P}(\xi' \mid h^{t-1})}{q^{t-1}}.
\end{aligned} \tag{15.3.3}$$

Because $q^t \leq 1$, $\mathbf{P}(\xi' \mid h^t)$ is nondecreasing.

Step 2. Next, because $\mu(\xi') \geq \mu^\dagger > 0$, we can use (15.3.3) to calculate that for all t,

$$\begin{aligned}
0 < \mu^\dagger &\leq \mathbf{P}(\xi' \mid \emptyset) \\
&= q^0 \mathbf{P}(\xi' \mid h^1) \\
&= q^0 q^1 \mathbf{P}(\xi' \mid h^2) \\
&\vdots \\
&= \left(\prod_{\tau=0}^{t-1} q^\tau \right) \mathbf{P}(\xi' \mid h^t),
\end{aligned}$$

and as $\mathbf{P}(\xi' \mid h^t) \leq 1$ for all t,

$$\prod_{\tau=0}^{t-1} q^\tau \geq \mu^\dagger.$$

Taking limits,

$$\prod_{\tau=0}^{\infty} q^\tau \geq \mu^\dagger.$$

That is, for all $\omega \in \Omega''$,

$$\prod_{\tau=0}^{\infty} q^\tau(\omega) \geq \mu^\dagger,$$

and so (using the observation at the beginning of the proof that $\mathbf{P}(\Omega'') = \mathbf{P}(\Omega')$),

$$\mathbf{P}\left\{\omega \in \Omega'' : \prod_{\tau=0}^{\infty} q^\tau(\omega) \geq \mu^\dagger\right\} = \mathbf{P}(\Omega'') = \mathbf{P}(\Omega').$$

Because $\mathbf{P}(\Omega') \geq \mathbf{P}(\xi') \geq \mu^\dagger > 0$ (where ω has again been suppressed),

$$\mathbf{P}\left\{\prod_{\tau=0}^{\infty} q^\tau \geq \mu^\dagger \,\Big|\, \Omega'\right\} = 1.$$

But

$$\prod_{\tau=0}^{\infty} q^\tau = \prod_{\{\tau : q^\tau \leq \zeta\}} q^\tau \prod_{\{\tau : q^\tau > \zeta\}} q^\tau < \prod_{\{\tau : q^\tau \leq \zeta\}} q^\tau \leq \zeta^{n_\zeta},$$

and so

$$\mathbf{P}\{\zeta^{n_\zeta} \geq \mu^\dagger | \Omega'\} = 1,$$

or

$$\mathbf{P}\{n_\zeta \ln \zeta \geq \ln \mu^\dagger | \Omega'\} = 1,$$

which gives the result

$$\mathbf{P}\{n_\zeta \ln \zeta < \ln \mu^\dagger | \Omega'\} = \mathbf{P}\left\{n_\zeta > \frac{\ln \mu^\dagger}{\ln \zeta} \,\Big|\, \Omega'\right\} = 0.$$

∎

15.3.2 The Reputation Bound

In the stage (or one-shot) game, player 1 can guarantee the payoff

$$v_1^*(a_1) \equiv \min_{a_2 \in B(a_1)} u_1(a_1, a_2) \qquad (15.3.4)$$

by committing to action a_1. We refer to $v_1^*(a_1)$ as the *one-shot bound* from a_1. Let $\underline{v}_1(\xi_0, \mu, \delta)$ be the infimum over the set of the normal player 1's payoff in any (pure or mixed) Nash equilibrium, given the distribution μ over types and discount factor δ. The basic reputation result establishes a lower bound on the equilibrium payoff of player 1.

Proposition *Suppose A_2 is finite and $\mu(\xi_0) > 0$. Suppose A_1' is a finite subset of A_1 with*
15.3.1 *$\mu(\xi(a_1)) > 0$ for all $a_1 \in A_1'$. Then there exists k such that*

$$\underline{v}_1(\xi_0, \mu, \delta) \geq \delta^k \max_{a_1 \in A_1'} v_1^*(a_1) + (1 - \delta^k) \min_{a \in A} u_1(a).$$

Proof Let a_1' be a best type in A_1', that is,

$$a_1' \in \arg\max_{a_1 \in A_1'} \min_{a_2 \in B(a_1)} u_1(a_1, a_2).$$

By hypothesis, the simple type $\xi' = \xi(a_1')$ is assigned positive probability by μ.

15.3 ■ Perfect Monitoring

Because A_2 is finite, there exists $\zeta \in (0, 1)$ such that if $\alpha_1(a_1') > \zeta$,

$$B(\alpha_1) \subset B(a_1').$$

In other words, as long as player 2 attaches sufficiently high probability to player 1's action a_1', player 2 will choose a best response to a_1'.

Fix a Nash equilibrium (σ_1, σ_2) and let **P** be the distribution on Ω induced by (σ_1, σ_2) and μ. Then, for all h^t such that $q^t(h^t) > \zeta$, $B(E(\sigma_1^t \mid h^t)) \subset B(a_1')$.[8] Letting $\hat{\mathcal{H}}^t \equiv \{h^t : q^t(h^t) > \zeta\}$, we have just argued that $\sigma_2^t(h^t) \in B(a_1')$ for all $h^t \in \hat{\mathcal{H}}^t$.

Set $k = \ln \mu(\xi')/ \ln \zeta$. From lemma 15.3.1, conditional on Ω' (i.e., conditional on 1 playing a_1' in every period), $q^t \leq \zeta$ for no more than k periods with **P** probability 1. Suppose now that the normal type plays according to the strategy "always play a_1'" (which may not be σ_1). This induces a probability measure **P**′ on Ω that generates the same distribution over public histories as does **P** conditional on ξ', that is, $\mathbf{P}'(C) = \mathbf{P}(C \mid \xi')$ for all $C \subset (A_1 \times A_2)^\infty$. Hence $q^t \leq \zeta$ for no more than k periods with **P**′ probability 1. The inequality for $\underline{v}_1(\xi_0, \mu, \delta)$ now follows from the observation that the normal type's equilibrium payoff must be no less than the payoff from this strategy, which is at least the payoff from receiving the worst possible payoff for the first k periods, after which a payoff of at least $\max_{a_1 \in A_1'} v_1^*(a_1)$ is received.

■

If the set of possible commitment types is sufficiently rich, the lower bound on the normal player 1's payoff is the Stackelberg payoff.

Corollary 15.3.1 *Suppose μ assigns positive probability to some sequence of simple types $\{\xi(a_1^k)\}_{k=0}^\infty$ with $\{a_1^k\}_k$ satisfying*

$$v_1^* = \lim_{k \to \infty} v_1^*(a_1^k).$$

For all $\varepsilon > 0$, there exists $\underline{\delta}' \in (0, 1)$ such that for all $\delta \in (\underline{\delta}', 1)$,

$$\underline{v}_1(\xi_0, \mu, \delta) \geq v_1^* - \varepsilon.$$

Proof Fix $\varepsilon > 0$ and choose a_1^k such that $v_1^*(a_1^k) > v_1^* - \varepsilon/2$. The result now follows from proposition 15.3.1, taking $A_1' = \{a_1^k\}$.

■

Remark 15.3.3 **Stackelberg bound** If there is a Stackelberg action, and the associated Stackelberg type has positive probability under μ, then the hypotheses of corollary 15.3.1 are trivially satisfied. In that case, the normal player 1 effectively builds a reputation for *playing* like the Stackelberg type, receiving a payoff (when patient) no less than the payoff v_1^*. Importantly, the normal player builds this reputation despite the fact that there are many other possible commitment types. However, this result tells us very little about player 1's equilibrium strategy. In particular, it does not imply that it is optimal for the normal player 1 to choose the Stackelberg action in each

---

8. Note that $\sigma_1^t : \mathcal{H}^t \times \Xi \to \Delta(A_1)$ and $q^t(h^t) = E(\sigma_1^t \mid h^t)(a_1') = \sum \sigma_1^t(h^t, \xi)(a_1')\mathbf{P}(\xi \mid h^t)$.

period, which is in general not the case. Section 15.5 discusses the consequences of this suboptimality.

♦

Remark 15.3.4 **Complete information games** Reputation effects from pure-action commitment types in perfect monitoring games yield a lower bound on equilibrium payoffs for player 1 that can be quite high. However, unlike mixed-action commitment types, or more generally imperfect monitoring games, they do not introduce the possibility of new payoffs. More precisely, for any pure action $a'_1 \in A_1$, there exists $\underline{\delta} \in (0, 1)$, such that for all $\delta \in (\underline{\delta}, 1)$ the complete information game has an equilibrium with player 1 payoffs at least $v_1^*(a'_1)$. This is immediate if there is a stage-game Nash equilibrium with player 1 payoffs at least $v_1^*(a'_1)$. If not, then Nash reversion can, for patient player 1, be used to support 1's choice of a'_1 in every period.

♦

Remark 15.3.5 **Diffuse beliefs** If A_1 is a continuum, then the hypotheses of corollary 15.3.1 are satisfied if A_1 has a countably dense subset $\{a_1^m\}_{m=1}^\infty$ with the property that $\mu(\xi(a_1^m)) > 0$ for all m.

♦

Remark 15.3.6 **Continuum short-lived player action set** The assumption that A_2 is finite in proposition 15.3.1 allowed us to conclude that if the probability attached to the action a'_1 exceeds some value ζ, then player 2 will play a best response to that action. The proposition is then proven by showing that under persistent play of a'_1, it takes at most k periods to push the probability of the action a'_1 past ζ.

Allowing A_2 to be a continuum forces only a slight weakening of this result. The continuity of player 2's payoff function ensures that as the probability attached to the action a'_1 approaches unity, the set of player 2's best responses converges to a subset of the set of best responses to that action. Although player 1 can never be certain that player 2 will choose an exact best response, he can be certain that player 2 will come arbitrarily close.

By the continuity in assumption 2.1.1, for all $\varepsilon > 0$, there is a value ζ such that if $\alpha_1(a'_1) > \zeta$, then $u_1(a'_1, a_2) > v_1^*(a'_1) - \varepsilon$ for any $a_2 \in B(\alpha_1)$. The proof of proposition 15.3.1 then gives:

Proposition 15.3.2 *Suppose A_2 is a continuum and $\mu(\xi_0) > 0$. Suppose A'_1 is a finite subset of A_1 with $\mu(\xi(a_1)) > 0$ for all $a_1 \in A'_1$. For all $\varepsilon > 0$, there exists k such that*

$$\underline{v}_1(\xi_0, \mu, \delta) \geq \delta^k (\max_{a'_1 \in A'_1} v_1^*(a'_1) - \varepsilon) + (1 - \delta^k) \min_{a \in A} u_1(a).$$

♦

Remark 15.3.7 **Multiple short-lived players** The arguments extend immediately to the case of multiple short-lived players, on appropriately reinterpreting the notion of a one-shot bound, which is now

$$v_1^*(a_1) = \min_{\alpha_{-1} \in B(a_1)} u_1(a_1, \alpha_{-1}),$$

where $B(a_1)$ is the set of Nash equilibria in the stage game played by the short-run players, given action a_1 from the long-lived player. The only minor qualification is that with many short-lived players, unlike one short-lived player with finite A_2, it is not true that $B(\alpha_1) \subset B(a_1')$ for α_1 sufficiently close to a_1'. However, as in remark 15.3.6, B is upper hemi-continuous, so a statement analogous to proposition 15.3.2 holds.

◆

Remark 15.3.8 **The role of discounting** The discount factor plays a somewhat different role in folk theorem and reputation arguments. In the case of the folk theorem, a higher discount factor has the effect of making future payoffs relatively more important. In the reputation argument, the discount factor plays a dual role. It makes future payoffs relatively more important, but it also discounts into insignificance the initial sequence of periods during which it may be costly for player 1 to mimic the commitment type.

To clarify these different roles for the discount factor, recall section 3.2.3's observation that future payoffs could attain sufficient weight to deter current deviations from equilibrium play in two ways. One is via a sufficiently large discount factor. The other is via stage-game payoffs that provide sufficiently small rewards for deviating from prescribed actions. We have seen that large discount factors also suffice to support reputation arguments. There is no reason to expect stage-game payoffs that provide relatively small incentives for deviating from an equilibrium to do likewise.

For example, section 3.2.3 discussed a version of the product-choice game parameterized by the cost of high effort. In such a game, the fear of future punishment can induce player 1 to consistently choose high effort without appealing to large discount factors, as long as the cost of high effort is sufficiently small. Will this also suffice to support a reputation result that imposes a lower bound on player 1's payoff? Here, another payoff difference is at issue, measuring not the temptation to deviate from high effort when player 2 chooses h, but the cost of choosing H when player 2 chooses ℓ. These in general are not the same, and we cannot be certain that the same stage-game payoffs that support folk theorem results will also support reputations. We return to this distinction in chapter 18.

◆

15.3.3 An Example: Time Consistency

Section 6.2 examined a model of a long-lived government facing a sequence of short-lived citizens. If the discount factor is high enough, the repeated game has an equilibrium in which actions $((R-1)/R, 0)$ are chosen in each period, yielding an equilibrium without expropriatory taxes. In such an equilibrium, the government receives its Stackelberg payoff of

$$v_1^* = \sup_{a_1 \in A_1} \min_{a_2 \in B(a_1)} u_1(a_1, a_2) = R + \gamma.$$

However, there are many other subgame-perfect equilibria, including one in which the actions $(1, 1)$ are taken in each period, giving each player their minmax payoff. Do we have any reason to regard some of these equilibrium payoffs as being more interesting or more likely than others?

If there is incomplete information about the government's type, then the government can build a reputation that ensures (with enough patience) a high payoff.[9] Let player 1's type be drawn at the beginning of the game from a distribution that attaches positive probability to the normal type and to each of a countable number of simple commitment types. Each commitment type is characterized by a tax rate and the set of such tax rates is dense in $[0, 1]$.

From corollary 15.3.1, we conclude that for any ε, there is a $\underline{\delta} < 1$ such that for all $\delta \in (\underline{\delta}, 1)$, the government's payoff in any Nash equilibrium is within ε of v_1^*. The government capitalizes on the uncertainty about its type to build a reputation for nonconfiscatory tax rates.

15.4 Imperfect Monitoring Games

We now examine reputations with imperfect monitoring. We again study games with one long-lived player (player 1) and one short-lived player (player 2), focusing on player 1's payoff. The stage game of the benchmark complete information game is the game with private monitoring described in section 12.2, with finite or continuum action spaces, A_i, and *finite* signal spaces, Z_i. This includes public monitoring as a special case (remark 12.2.1), and thus includes stage games with a nontrivial extensive form. Though the seminal study of reputations with imperfect monitoring (Fudenberg and Levine 1992) restricted attention to public monitoring games (see remark 15.4.2), this is unnecessary.

The distribution over private signals $z = (z_1, z_2)$ for each action profile a is denoted by $\pi(z \mid a)$, with π_i being player i's marginal distribution. As usual, the ex post payoffs of the normal type of player 1 and player 2, after the realization (z, a), are given by $u_i^*(z_i, a_i)$ $(i = 1, 2)$. If an action space is a continuum, we assume the appropriate analog of assumption 7.1.1, that is, $\pi : Z \times A \to [0, 1]$ is continuous, and u_i^* is continuous in all arguments and quasi-concave in a_i. Ex ante stage game payoffs are given by $u_i(a) \equiv \sum_z u_i^*(z_i, a_i) \pi(z \mid a)$.

The space of types Ξ is as described in section 15.2, with ξ_0 being the normal type. When A_1 is finite, we allow for simple commitment types that are committed to a mixed action.

The set of private histories for player 1 (excluding his type) is

$$\mathcal{H}_1 \equiv \cup_{t=0}^{\infty} (A_1 \times Z_1)^t,$$

and a behavior strategy for player 1 is, using the notation on commitment types from section 15.2,

$$\sigma_1 : \mathcal{H}_1 \times \Xi \to \Delta(A_1),$$

---

9. Celentani and Pesendorfer (1996) establish such a reputation result in the course of providing a more general treatment of reputations in dynamic games.

15.4 ■ Imperfect Monitoring

such that, for all $\xi(\hat{\sigma}_1) \in \Xi_2$,

$$\sigma_1(h_1^t, \xi(\hat{\sigma}_1)) = \hat{\sigma}_1(h_1^t) \qquad \forall h_1^t \in \mathcal{H}_1.$$

The set of private histories for the short-lived players is

$$\mathcal{H}_2 \equiv \cup_{t=0}^{\infty} (A_2 \times Z_2)^t,$$

and a behavior strategy for the short-lived players is

$$\sigma_2 : \mathcal{H}_2 \to \Delta(A_2). \tag{15.4.1}$$

We maintain throughout our convention of restricting attention to pure strategies when considering infinite action sets.

As before, given a strategy profile σ, $U_1(\sigma, \xi)$ denotes the type ξ long-lived player's payoff in the repeated game.

Definition 15.4.1 *A strategy profile $(\tilde{\sigma}_1, \tilde{\sigma}_2)$ is a* Nash equilibrium *of the reputation game with imperfect monitoring if for all $\xi \in \Xi_1$, $\tilde{\sigma}_1$ maximizes $U_1(\sigma_1, \tilde{\sigma}_2, \xi)$ over player 1's repeated game strategies, and if for all t and all $h_2^t \in \mathcal{H}_2$ that have positive probability under $(\tilde{\sigma}_1, \tilde{\sigma}_2)$ and μ,*

$$E[u_2(\tilde{\sigma}_1(h_1^t, \xi), \tilde{\sigma}_2(h_2^t)) \mid h_2^t] = \max_{a_2 \in A_2} E[u_2(\tilde{\sigma}_1(h_1^t, \xi), a_2) \mid h_2^t].$$

Remark 15.4.1 We have assumed in (15.4.1) that the short-lived player in period t observes all previous short-lived players' actions and signals. Although this is natural in games of perfect monitoring, the assumption requires discussion here.

When the game has truly private monitoring (that is, player 2 observes relevant signals that are not observed by player 1), it seems unnatural to distinguish between knowledge of past signals and past actions, and we accordingly require subsequent generations of player 2 to have access to all the information held by earlier generations of player 2. In many contexts, this assumption is unduly strong—a customer at a restaurant, for example, is unlikely to have better information about earlier customers' experience than the restaurant. In other contexts, the assumption is natural. Section 15.4.3 considers the case in which player 2 is a continuum of small and anonymous but long-lived players. The monitoring is truly private, but each player naturally knows her previous actions. ◆

Remark 15.4.2 **Public monitoring** The analysis requires only minor modifications when the signals are public and short-lived players do not observe the actions of previous short-lived players. We refer to the game with public signals and short-lived player actions *not* observed by subsequent short-lived players (i.e., $\mathcal{H}_2 \equiv \cup_{t=0}^{\infty} Y^t$) as the *canonical public monitoring* game, because it is often the most natural specification. This is the game studied in chapter 7 and in Fudenberg and Levine (1992), and is to be distinguished from the case when the signal is public and short-lived player actions are observed by subsequent short-lived players (a special case of the private monitoring game).

A special case of the canonical public-monitoring game has public short-lived player actions (and finite A_2, so as to preserve our assumption of a finite signal space). There is a space of signals Y_1 and a public-monitoring distribution ρ_1, so that the complete space of public signals is $Y = Y_1 \times A_2$ with probability distribution given by

$$\rho((y_1, a_2') \mid a) = \begin{cases} \rho_1(y_1 \mid a), & \text{if } a_2' = a_2, \\ 0, & \text{otherwise}. \end{cases}$$

♦

15.4.1 Stackelberg Payoffs

As in the case of perfect monitoring, the normal player 1 has an incentive to induce particular beliefs in the short-lived players in order to elicit beneficial best replies. However, because monitoring is imperfect, the best responses elicitable by a_1 are not simply those actions in $B(a_1)$.

We first consider the set of possible player 2 best responses when player 1 is almost certain to play some mixed action α_1. A (potentially mixed) action α_2 is an ε-*confirmed best response to* α_1 if there exists α_1' such that

$$\alpha_2(a_2) > 0 \Rightarrow a_2 \in \arg\max_{a_2'} u_2(\alpha_1', a_2')$$

and

$$\left| \pi_2(\cdot \mid \alpha_1, \alpha_2) - \pi_2(\cdot \mid \alpha_1', \alpha_2) \right| \leq \varepsilon.$$

Note that it is possible that a mixed action α_2 is an ε-confirmed best response to α_1, while at the same time no action in the support of α_2 is an ε-confirmed best response.[10] Denote the set of ε-confirmed best responses to α_1 by $B_\varepsilon(\alpha_1)$. Note that if there are different strategies α_1 and α_1' with $\pi_2(\cdot \mid \alpha_1, \alpha_2) = \pi_2(\cdot \mid \alpha_1', \alpha_2)$, then $B_0(\alpha_1)$ may contain strategies not in $B(\alpha_1)$, the set of best replies to α_1 (see example 15.4.2).

The private monitoring and the canonical public monitoring game differ in the information that short-lived players have about preceding short-lived player choices, leading to different constraints on optimal player 2 behavior (see note 11 on page 487).

For the private monitoring game, we define

$$B_\varepsilon^*(\hat{\alpha}_1) \equiv \{\alpha_2 : \text{supp}(\alpha_2) \subset B_\varepsilon(\hat{\alpha}_1)\}. \tag{15.4.2}$$

For the canonical public monitoring game, we define $B_\varepsilon^*(\hat{\alpha}_1) \equiv B_\varepsilon(\hat{\alpha}_1)$.

In this section, we prove that if player 2 assigns strictly positive probability to a simple type $\xi(\alpha_1')$, then a patient normal player 1's payoff in every Nash equilibrium can be (up to an $\varepsilon > 0$ approximation) no lower than $\underline{v}_1(\alpha_1')$, where

---

10. For example, in the product-choice game (figure 15.1.1), if the public monitoring is given by $Y = \{\underline{y}, \bar{y}\}$ with $\rho(\bar{y} \mid Hh) = \rho(\bar{y} \mid L\ell) = 1$ and 0 otherwise, then $\frac{1}{2} \circ h + \frac{1}{2} \circ \ell$ is a 0-confirmed best response to H, while ℓ is not. By adding an appropriate third action for player 2, we can ensure h is also not a 0-confirmed best response to H.

15.4 ■ Imperfect Monitoring

$$\underline{v}_1(\alpha_1') \equiv \min_{\alpha_2 \in B_0^*(\alpha_1')} u_1(\alpha_1', \alpha_2). \qquad (15.4.3)$$

Taking the supremum over α_1' yields the payoff

$$v_1^{**} \equiv \sup_{\alpha_1'} \min_{\alpha_2 \in B_0^*(\alpha_1')} u_1(\alpha_1', \alpha_2). \qquad (15.4.4)$$

Example 15.4.1 **Product-choice game with public monitoring** We return to the product-choice game, with the ex ante stage game payoffs in figure 15.1.1. Player 1's action is not public. As in section 7.6.2, there is a public signal with two possible values, $\underline{y}$ and $\bar{y}$, and distribution

$$\rho(\bar{y} \mid a) = \begin{cases} p, & \text{if } a_1 = H, \\ q, & \text{if } a_1 = L, \end{cases}$$

where $0 < q < p < 1$. Player 2's actions are public. Let $\hat{\alpha}_1$ denote 1's mixed action which randomizes equally between H and L. Then, for all $\varepsilon \geq 0$, $B_\varepsilon(\hat{\alpha}_1)$ contains every pure or mixed action for player 2, and hence we have $\min_{\alpha_2 \in B_0(\hat{\alpha}_1)} u_1(\hat{\alpha}_1, \alpha_2) = 1/2$. However, for any mixture α_1 under which H is more likely than L, for sufficiently small ε, $B_\varepsilon(\alpha_1) = \{h\}$. As a result, we have $v_1^{**} = 5/2$. This payoff is the mixed-action Stackelberg payoff (see (15.2.2)), and exceeds the upper bound on player 1's payoff in the corresponding public monitoring game of complete information, shown in section 7.6.2 to be

$$2 - \frac{1-p}{p-q} < 2.$$

●

Remark 15.4.3 **New possibilities even for perfect monitoring** We observed in remark 15.3.4 that reputation effects from pure-action commitment types in perfect monitoring games cannot introduce new payoff possibilities. Taking $\hat{\alpha}_1 = H$ in example 15.4.1 shows that pure commitment types in imperfect-monitoring games *can* introduce new possibilities in terms of equilibrium payoffs. Similarly, mixed-action commitment types in perfect monitoring games can introduce new possibilities. A game with perfect monitoring is a special case of a game with imperfect monitoring, where the set of signals Z is the set of pure-action profiles A, and $\pi(z \mid a) = 1$ if and only if $z = a$. Consequently, for perfect monitoring games, $B_0^* = B_0 = B$ and v_1^{**} is the mixed-action Stackelberg payoff, (15.2.2). This section thus extends the reputation result for perfect monitoring games of section 15.3 to mixed commitment types. In the process, we obtain a stronger bound on payoffs, as the mixed-action Stackelberg payoff can exceed the pure-action Stackelberg payoff. The pure and mixed Stackelberg payoffs for the product-choice game are given by:

$$v_1^* = \max_{a_1} \min_{\alpha_2 \in B(a_1)} u_1(a_1, \alpha_2) = 2$$

and

$$v_1^{**} = \sup_{\alpha_1} \min_{\alpha_2 \in B(\alpha_1)} u_1(\alpha_1, \alpha_2) = 2\tfrac{1}{2}.$$

The lower bound on player 1's payoff can be strictly higher than what player 1 could achieve in the perfect monitoring game of complete information. In section 2.7.2, examining games of complete information in which a long-lived player 1 faces short-lived opponents, we introduced the upper bound $\bar{v}_1$ (cf. (2.7.2)) on player 1's payoff. In the product-choice game,

$$\bar{v}_1 = \max_{\alpha \in \mathbf{B}} \min_{a_1 \in \mathrm{supp}(\alpha_1)} u_1(a_1, \alpha_{-1}) = 2.$$

◆

The bound $\underline{v}_1(\alpha_1')$ differs from $\min_{\alpha_2 \in B(\alpha_1')} u_1(\alpha_1', \alpha_2)$ in allowing player 2's action to be a minimizer from the set $B_0^*(\alpha_1')$ rather than $B(\alpha_1')$. In general, this difference can yield a bound that is even lower than the pure-action Stackelberg payoff.

Example 15.4.2 **The purchase game** The stage game in this example has a nontrivial dynamic structure, with the short-lived customer first deciding between "buy" (b) and "don't buy" (d), and then after b, the long-lived firm deciding on the level of effort, high (H) or low (L). The extensive form is given in figure 15.4.1, with the normal form in figure 15.4.2. The action profile Ld is the unique pure Nash equilibrium of the stage game, and $(0, 0)$ is the unique stage-game Nash equilibrium payoff.

Now let the game be infinitely repeated. In each period, the terminal node reached in that period's extensive-form stage game is observed. Because no information about player 1 is revealed if player 2 chooses d, this is a game of imperfect monitoring. Noting that $B(H) = \{b\}$, we have $\max_{a_1 \in A_1} \min_{a_2 \in B(a_1)} u_1(a_1, a_2) = 1$, and the pure-action Stackelberg payoff is achieved by the action H. The pure-action Stackelberg type ξ^* plays action H.

Now consider the repeated game of incomplete information, with player 1's type drawn from the set $\Xi = \{\xi_0, \xi^*\}$. Does this suffice to ensure player 1 a payoff

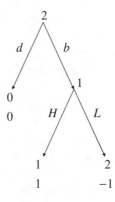

Figure 15.4.1 The purchase game. The game has three public signals, corresponding to the three terminal nodes.

15.4 ■ Imperfect Monitoring

	d	b
H	0,0	1,1
L	0,0	2,−1

Figure 15.4.2 The normal form for the purchase game.

near 1? Notice that $|\pi_2(\cdot \mid Hd) - \pi_2(\cdot \mid Ld)| = 0$, and hence b and d are both 0-confirmed best responses to H (as, indeed, is d to any $\alpha_1 \in \Delta(A_1)$). As a result, $v_1^{**} = 0$.

This surprisingly low value for v_1^{**} is not simply a matter of having calculated a loose lower bound. If $\delta > 1/2$ and $\mu(\xi^*) < 1/2$, there is a sequential equilibrium with payoff $(0, 0)$.

As a first attempt at constructing such an equilibrium, suppose that after any history, all short-lived players choose d, and the normal type of player 1 chooses L. Given μ, choosing b causes each short-lived player to face a lottery that assigns more than probability $1/2$ to L, making d a best response. Because all short-lived players are choosing d, the normal long-lived player is indifferent over all actions, so L is a best reply. We thus have a Nash equilibrium with payoffs $(0, 0)$. However, this strategy profile fails a minimal sequential rationality requirement. Suppose a short-lived player chooses b. Will the normal type actually choose L? By doing so, he reveals himself as the normal type (and the profile then specifies permanent Ld). If the normal player 1 instead chooses H, future short-lived players will believe they face the commitment type ξ^* and so will choose b as long as H continues to be chosen. When $\delta > 1/2$, this deviation is profitable for the normal player 1, ensuring that the profile is not a sequential equilibrium.

This issue is addressed by the strategy profile illustrated in figure 15.4.3. The equilibrium path features the constant play of Ld and payoffs $(0, 0)$. No short-lived player purchases because the first short-lived player to do so elicits low effort. The normal type has no incentive to choose H when the first (out-of-equilibrium) customer purchases, because once such a choice of b has been made, the normal player 1 receives his "Stackelberg" payoff of 1 in the continuation game, regardless of what choice he makes in the current period.

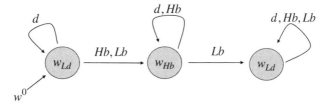

Figure 15.4.3 The behavior of player 2 and the normal type of player 1 in an equilibrium of the purchase game. The subscripts in the labels of the states identify the actions taken in those states.

Not only are reputation effects weak in this type of "outside option" game, but the presence of "bad" types can impose severe upper bounds on equilibrium payoffs (see sections 18.6.2–18.6.6).

●

The difficulty in the purchase game is that if player 2 chooses d, her signals reveal no information about player 1's action. Hence, if d is a best response to anything, then it is in $B_0(\alpha_1)$ for every α_1, allowing in particular $d \in B_0(H)$. A necessary condition, then, for $B(\alpha_1) = B_0(\alpha_1)$ is that there be no such uninformative actions. That is, no two actions for player 1 should generate the same distribution of signals, for any action of player 2.

Assumption 15.4.1 *For all $a_2 \in A_2$, the collection of probability distributions $\{\pi_2(\cdot \mid (a_1, a_2)) : a_1 \in A_1\}$ is linearly independent.*

Equivalently, applying definition 9.2.1 to the distribution π_2, this is the requirement that every profile $a \in A$ have individual full rank for player 1. The purchase game fails this result for the action d. In section 15.5, this assumption plays a key role in ensuring that player 2 can eventually identify player 1's strategy in a repeated game of incomplete information.

For the canonical public-monitoring game, a stronger version of assumption 15.4.1 is needed.

Assumption 15.4.2 *For all $\alpha_2 \in \Delta(A_2)$, the collection of probability distributions $\{\pi_2(\cdot \mid (a_1, \alpha_2)) : a_1 \in A_1\}$ is linearly independent.*

We immediately have the following.

Lemma 15.4.1 *For the private-monitoring game, if assumption 15.4.1 holds, $B(\alpha_1) = B_0^*(\alpha_1)$ and v_1^{**} equals the mixed-action Stackelberg payoff.*

For the canonical public-monitoring game, if assumption 15.4.2 holds, $B(\alpha_1) = B_0^*(\alpha_1) = B_0(\alpha_1)$ and v_1^{**} equals the mixed-action Stackelberg payoff.

15.4.2 The Reputation Bound

Recall that $\underline{v}_1(\xi_0, \mu, \delta)$ is the infimum over the set of the normal player 1's payoff in any (pure or mixed) Nash equilibrium in the repeated private monitoring or canonical public monitoring game, given the distribution μ over types and the discount factor δ. Our goal is to prove the following result (which implies corollary 15.4.1, the reputation lower bound):

Proposition 15.4.1 *Let $\hat{\xi}$ denote the simple commitment type that always plays $\hat{\alpha}_1 \in \Delta(A_1)$ (if A_1 is finite) or $\hat{\alpha}_1 \in A_1$ (A_1 infinite). Suppose $\mu(\xi_0), \mu(\hat{\xi}) > 0$. In the private monitoring or canonical public monitoring game, for every $\varepsilon > 0$, there is a value K such that for all δ,*

$$\underline{v}_1(\xi_0, \mu, \delta) \geq (1-\varepsilon)\delta^K \inf_{\alpha_2 \in B_\varepsilon^*(\hat{\alpha}_1)} u_1(\hat{\alpha}_1, \alpha_2) + (1 - (1-\varepsilon)\delta^K) \min_{a \in A} u_1(a). \tag{15.4.5}$$

15.4 ■ Imperfect Monitoring

Corollary 15.4.1 *Suppose μ assigns positive probability to some sequence of simple types $\{\xi(\alpha_1^k)\}_{k=1}^\infty$ with each α_1^k in $\Delta(A_1)$ (if A_1 is finite) or A_1 (if A_1 is a continuum) satisfying*

$$v_1^{**} = \lim_{k \to \infty} \min_{\alpha_2 \in B_0^*(\alpha_1^k)} u_1(\alpha_1^k, \alpha_2).$$

For all $\varepsilon' > 0$, there exists $\underline{\delta} < 1$ such that for all $\delta \in (\underline{\delta}, 1)$,

$$\underline{v}_1(\xi_0, \mu, \delta) \geq v_1^{**} - \varepsilon'.$$

As for perfect monitoring games, the normal player 1 again effectively builds a reputation for *playing* like a simple type, and this occurs despite the presence of many other possible commitment types (a similar remark to remark 15.3.5 holds here). The value K in proposition 15.4.1 depends on the prior distribution μ only through the probability $\mu(\hat{\xi})$.

Not surprisingly, the key ingredient in the proof of proposition 15.4.1 is an imperfect monitoring version of lemma 15.3.1. Following section 15.3, let $\Omega \equiv \Xi \times (A_1 \times Z_1 \times A_2 \times Z_2)^\infty$ denote the space of outcomes. A strategy profile σ and the prior distribution over types, μ, induces a probability measure $\mathbf{P} \in \Delta(\Omega)$. Suppose $\hat{\mathbf{P}}$ and $\tilde{\mathbf{P}}$ are the probability measures induced on Ω by $\mathbf{P}$, conditioning on the type $\hat{\xi}$ and the set of types $\tilde{\Xi} \equiv \Xi \setminus \{\hat{\xi}\}$, respectively. Hence, $\mathbf{P} = \hat{\mu}\hat{\mathbf{P}} + (1 - \hat{\mu})\tilde{\mathbf{P}}$, where $\hat{\mu} = \mu(\hat{\xi})$. Denote the event that player 1 is type $\hat{\xi}$, $\{\hat{\xi}\} \times (A_1 \times Z_1 \times A_2 \times Z_2)^\infty$, by simply $\{\hat{\xi}\}$, so that $\hat{\mu} = \mathbf{P}(\{\hat{\xi}\})$ and $\hat{\mathbf{P}}(C) = \mathbf{P}(C \mid \{\hat{\xi}\})$ for all $C \subset \Omega$. Because $\tilde{\mathbf{P}}$ and $\hat{\mathbf{P}}$ are absolutely continuous with respect to $\mathbf{P}$, any statement that holds $\mathbf{P}$-almost surely also holds $\tilde{\mathbf{P}}$- and $\hat{\mathbf{P}}$-almost surely.

Player 2's posterior belief in period t that player 1 is the commitment type $\hat{\xi}$ is the $\mathscr{G}_2^t$-measurable random variable $\mathbf{P}(\{\hat{\xi}\} \mid \mathscr{G}_2^t) : \Omega \to [0, 1]$, where $\mathscr{G}_2^t$ is the σ-algebra generated by $\mathscr{H}_2^t$. We let $\hat{\mu}^t$ denote the period t posterior, so $\hat{\mu}^0 = \hat{\mu}$.

It is helpful to collect some basic facts concerning Bayesian updating. Given a measurable space $(\Omega, \mathscr{G})$, a sequence of σ-algebras $\{\mathscr{G}^t\}_t$ is called a *filtration* if it is increasing, that is, $\mathscr{G}^t \subset \mathscr{G}^{t+1} \subset \mathscr{G}$. A sequence of random variables $\{X^t\}_t$ is *adapted* to a filtration $\{\mathscr{G}^t\}_t$ if X^t is $\mathscr{G}^t$-measurable for each t, and it is a *martingale* under a probability measure $\mathbf{P}$ if, in addition, $E[|X^t|] < \infty$ and

$$E[X^t \mid \mathscr{G}^{t-1}] = X^{t-1}, \quad \mathbf{P}\text{-almost surely.} \tag{15.4.6}$$

We often emphasize the role of the measure $\mathbf{P}$ in this statement (while leaving other details to be inferred from the context) by saying simply that $\hat{\mu}^t$ is a $\mathbf{P}$-martingale. The sequence is a *submartingale* (resp., *supermartingale*) if (15.4.6) is replaced with $E[X^t \mid \mathscr{G}^{t-1}] \geq X^{t-1}$ (resp., $E[X^t \mid \mathscr{G}^{t-1}] \leq X^{t-1}$).

Much of the usefulness of these concepts stems from posterior beliefs being a bounded martingale and so converging (the martingale convergence theorem). This ensures that although we may not be sure what player 2 eventually believes about player 1, we can be certain that player 2's beliefs do converge to something.

Lemma 15.4.2 *Fix a strategy profile σ.*

1. *The posterior belief $\{\hat{\mu}^t\}_t$ is a bounded martingale adapted to the filtration $\{\mathscr{G}_2^t\}_{t=0}^{\infty}$ under the measure $\mathbf{P}$. It therefore converges $\mathbf{P}$-almost surely (and hence $\tilde{\mathbf{P}}$- and $\hat{\mathbf{P}}$-almost surely) to a random variable μ^{∞} defined on Ω.*
2. *The odds ratio $\{\hat{\mu}^t/(1-\hat{\mu}^t)\}_t$ is a $\tilde{\mathbf{P}}$-martingale.*
3. *The posterior belief $\{\hat{\mu}^t\}_t$ is a $\tilde{\mathbf{P}}$-supermartingale and a $\hat{\mathbf{P}}$-submartingale with respect to the filtration $\{\mathscr{G}_2^t\}_{t=0}^{\infty}$.*

The first statement delivers the convergence of player 2's beliefs. Notice that these beliefs are a submartingale with respect to $\hat{\mathbf{P}}$. Therefore, conditional on player 1 playing as does the commitment type, player 2's posterior that 1 is the commitment type (in expectation) rises (leading to the counterpart under imperfect monitoring of the second statement in lemma 15.3.1).

Proof 1. Because $\hat{\mu}^t = E[\chi_{\{\hat{\xi}\}} \mid \mathscr{G}_2^t]$ and $\mathscr{G}_2^t \subset \mathscr{G}_2^{t+1}$,

$$E[\hat{\mu}^{t+1} \mid \mathscr{G}_2^t] = E\big[E[\chi_{\{\hat{\xi}\}} \mid \mathscr{G}_2^{t+1}]\big|\mathscr{G}_2^t\big] = E[\chi_{\{\hat{\xi}\}} \mid \mathscr{G}_2^t] = \hat{\mu}^t.$$

The convergence of $\hat{\mu}^t$ $\mathbf{P}$-a.s. then follows from the martingale convergence theorem (Billingsley 1979, theorem 35.4). Because $\hat{\mathbf{P}}$ and $\tilde{\mathbf{P}}$ are both absolutely continuous with respect to $\mathbf{P}$, we thus have convergence $\hat{\mathbf{P}}$-a.s. and $\tilde{\mathbf{P}}$-a.s.

2. Suppose the $(t+1)$ period history consisting of h_2^t followed by $a_2^t z_2^t$ has positive probability under $\tilde{\mathbf{P}}$ and so under $\mathbf{P}$. Then Bayes' rule gives

$$\frac{\hat{\mu}^{t+1}}{1-\hat{\mu}^{t+1}} = \frac{\hat{\mu}^t \hat{\mathbf{P}}(a_2^t z_2^t \mid h_2^t)(\mathbf{P}(a_2^t z_2^t \mid h_2^t))^{-1}}{(1-\hat{\mu}^t)\tilde{\mathbf{P}}(a_2^t z_2^t \mid h_2^t)(\mathbf{P}(a_2^t z_2^t \mid h_2^t))^{-1}} = \frac{\hat{\mu}^t \hat{\mathbf{P}}(a_2^t z_2^t \mid h_2^t)}{(1-\hat{\mu}^t)\tilde{\mathbf{P}}(a_2^t z_2^t \mid h_2^t)}.$$

We then have

$$\tilde{E}\left[\frac{\hat{\mu}^{t+1}}{1-\hat{\mu}^{t+1}}\bigg| h_2^t\right] = \frac{\hat{\mu}^t}{1-\hat{\mu}^t} \tilde{E}\left[\frac{\hat{\mathbf{P}}(a_2^t z_2^t \mid h_2^t)}{\tilde{\mathbf{P}}(a_2^t z_2^t \mid h_2^t)}\bigg| h_2^t\right]$$

$$= \frac{\hat{\mu}^t}{1-\hat{\mu}^t} \sum_{a_2 \in A_2, z_2 \in Z_2} \hat{\mathbf{P}}(a_2^t z_2^t \mid h_2^t)$$

$$= \frac{\hat{\mu}^t}{1-\hat{\mu}^t}.$$

The ratio $\hat{\mu}^t/(1-\hat{\mu}^t)$ is thus a $\tilde{\mathbf{P}}$-martingale.

3. Applying Jensen's inequality to the convex function that maps $\hat{\mu}^t/(1-\hat{\mu}^t)$ into $-\hat{\mu}^t$, shows that $\hat{\mu}^t$ is a $\tilde{\mathbf{P}}$-supermartingale and hence a $\hat{\mathbf{P}}$-submartingale. ∎

Given the filtration $\{\mathscr{G}^t\}_t$, a *one-step ahead probability* or *prediction* under $\mathbf{P}$ is $\mathbf{P}(A \mid \mathscr{G}^t)$ for some $\mathscr{G}^{t+1}$-measurable event A. For the private monitoring game, $\{\mathscr{G}^t\}_t$ will be the filtration implied by the histories $\mathscr{H}_2^t = \cup_t (A_2 \times Z_2)^t$ and the realized period t action of player 2: $\mathscr{G}^t$ is the σ-algebra describing the period t information and action of player 2. For the canonical public monitoring game, $\{\mathscr{G}^t\}_t$ will be the filtration implied by the histories $\mathscr{H}_2^t = \cup_t Y^t$; the realized player 2 period t action is now excluded from $\mathscr{G}^t$ because player 2 actions are not part of the player 2 filtration.

15.4 ■ Imperfect Monitoring

The relevant $\mathscr{G}^{t+1}$-measurable events are the realizations of the signal observed by player 2 in period t. The imperfect monitoring version of lemma 15.3.1 asserts that under $\hat{\mathbf{P}}$, there is a small probability that player 2 makes very different one-step ahead predictions about her signals under $\mathbf{P}$ and under $\hat{\mathbf{P}}$ too many times.

Lemma 15.4.3 *Let $(\Omega, \mathscr{G})$ be a Borel measurable space. For all $\varepsilon, \psi > 0$ and $\mu^{\dagger} \in (0, 1]$, there exists a positive integer K such that for all $\hat{\mu} \in [\mu^{\dagger}, 1]$, for every $\mathbf{P}, \hat{\mathbf{P}}$, and $\tilde{\mathbf{P}}$ probability measures on $(\Omega, \mathscr{G})$ with $\mathbf{P} = \hat{\mu}\hat{\mathbf{P}} + (1-\hat{\mu})\tilde{\mathbf{P}}$, and for every filtration $\{\mathscr{G}^t\}_{t\geq 1}$, $\mathscr{G}^t \subset \mathscr{G}$,*

$$\hat{\mathbf{P}}(|\{t \geq 1 : d^t(\mathbf{P}, \hat{\mathbf{P}}) \geq \psi\}| \geq K) \leq \varepsilon, \tag{15.4.7}$$

where

$$d^t(\mathbf{P}, \hat{\mathbf{P}}) \equiv \sup_{A \in \mathscr{G}^{t+1}} \left| \mathbf{P}(A \mid \mathscr{G}^t) - \hat{\mathbf{P}}(A \mid \mathscr{G}^t) \right|.$$

The distance d^t can be used to capture a notion of merging (see Kalai and Lehrer 1994). Remarkably, the bound K (as in lemma 15.3.1) holds for all measures $\hat{\mathbf{P}}$ and $\tilde{\mathbf{P}}$ and all mixing probabilities $\hat{\mu} \in [\mu^{\dagger}, 1)$. This allows us to obtain corollary 15.4.1 for all Nash equilibria. We defer the proof of lemma 15.4.3, an elegant merging argument due to Sorin (1999), to the end of this section.

Proof of Proposition 15.4.1 We present the argument for the private monitoring game (the argument for the canonical public monitoring game being an obvious modification). Fix $\varepsilon > 0$. Let $\{\mathscr{G}^t\}_t$ be the filtration with $\mathscr{G}^t$ the σ-algebra generated by $\mathscr{H}_2^t$, the period t information, *and* A_2, the period t action of player 2. Denote by z_2^t a period t realization of player 2's private signal; thus, $\{z_2^t\} \in \mathscr{G}^{t+1}$. Letting $a_2^t(\omega)$ be the period t action in outcome ω, for $(h_2^t, a_2^t) = (h_2^t(\omega), a_2^t(\omega))$, we write $\mathbf{P}(z_2^t \mid h_2^t, a_2^t)$ for $\mathbf{P}(z_2^t \mid \mathscr{G}^t)(\omega)$, that is, the realization of the conditional probability evaluated at an outcome ω with player 2 history h_2^t and action a_2^t. From the linearity of π_2 in α_1, we have

$$\mathbf{P}(z_2^t \mid h_2^t, a_2^t) = \pi_2(z_2^t \mid E[\sigma_1(h_1^t, \xi) \mid h_2^t], a_2^t),$$

and

$$\hat{\mathbf{P}}(z_2^t \mid h_2^t, a_2^t) = \pi_2(z_2^t \mid \hat{\alpha}_1, a_2^t).$$

Hence, because any $a_2^t \in \operatorname{supp} \sigma_2(h_2^t)$ maximizes $E[u_2(\sigma_1(h_1^t, \xi), a_2) \mid h_2^t] = u_2(E[\sigma_1(h_1^t, \xi) \mid h_2^t], a_2)$, if $d^t(\mathbf{P}, \hat{\mathbf{P}}) < \psi$, then a_2^t is a ψ-confirmed best response to $\hat{\alpha}_1$.[11]

---

11. For the canonical public monitoring game, it is also true that each $a_2^t \in \operatorname{supp} \sigma_2(h^t)$ maximizes $u_2(E[\sigma_1(h_1^t, \xi) \mid h^t], a_2)$, and that if $|\rho(y^t \mid E[\sigma_1(h_1^t, \xi) \mid h^t], a_2^t) - \rho(y^t \mid \hat{\alpha}_1, a_2^t)| < \psi$, then a_2^t is a ψ-confirmed best response to $\hat{\alpha}_1$. However, we cannot apply lemma 15.4.3 to bound the number of periods in which a_2^t is not a ψ-confirmed best response to $\hat{\alpha}_1$, because the lemma requires the conditioning in the sequence of predictions to be based on a filtration. The argument for the canonical public monitoring game proceeds by observing that if $|\rho(y^t \mid E[\sigma_1(h_1^t, \xi) \mid h^t], \sigma_2(h^t)) - \rho(y^t \mid \hat{\alpha}_1, \sigma_2(h^t))| < \psi$ (that is, the difference in one-step predictions given $\sigma_2(h^t)$ rather than the realized action is less than ψ), then $\sigma_2(h^t)$ is a ψ-confirmed best response to $\hat{\alpha}_1$, and then applying lemma 15.4.3.

Now applying lemma 15.4.3 (with $\psi = \varepsilon$), there is a K (independent of $\mathbf{P}$ and $\hat{\mathbf{P}}$) such that with $\hat{\mathbf{P}}$-probability at least $1 - \varepsilon$, in all but K periods

$$d^t(\mathbf{P}, \hat{\mathbf{P}}) < \varepsilon.$$

Therefore, with $\hat{\mathbf{P}}$-probability $1 - \varepsilon$, in all but K exceptional periods, we have

$$\sigma_2^t(h^t) \in B_\varepsilon^*(\hat{\alpha}_1).$$

A lower bound for $\underline{v}_1(\xi_0, \mu, \delta)$ can now be calculated from the deviation by the normal type to the simple strategy of $\hat{\alpha}_1$ in every period, ignoring history. This deviation induces a probability measure on Ω that generates the same distribution over player 2 histories as does $\hat{\mathbf{P}}$, and so player 1's expected payoff from the deviation is at least

$$(1 - \varepsilon)\left[(1 - \delta^K)\min_{a \in A} u_1(a) + \delta^K \inf_{\alpha_2 \in B_\varepsilon^*(\hat{\alpha}_1)} u_1(\hat{\alpha}_1, \alpha_2)\right] + \varepsilon \min_{a \in A} u_1(a),$$

which equals the right side of (15.4.5).

∎

Proof of Corollary 15.4.1 Fix $\varepsilon' > 0$, and recall that $M = \max_a |u_1(a)|$. Let ξ^k denote the simple commitment type that always plays α_1^k. There is an α_1^k satisfying

$$\inf_{\alpha_2 \in B_0^*(\alpha_1^k)} u_1(\alpha_1^k, \alpha_2) > v_1^{**} - \varepsilon'/6.$$

Because $B_\varepsilon^*(\alpha_1^k)$ is upper hemicontinuous in ε, there exists $\bar{\varepsilon} > 0$ such that for $\varepsilon \in (0, \bar{\varepsilon})$

$$\inf_{\alpha_2 \in B_\varepsilon^*(\alpha_1^k)} u_1(\alpha_1^k, \alpha_2) > v_1^{**} - \varepsilon'/3.$$

Applying proposition 15.4.1 with $\hat{\xi} = \xi^k$, for all $\varepsilon \in (0, \bar{\varepsilon})$, there exists K such that

$$\underline{v}_1(\xi_0, \mu, \delta) \geq (1 - \varepsilon)\delta^K(v_1^{**} - \varepsilon'/3) - (1 - (1 - \varepsilon)\delta^K)M. \quad (15.4.8)$$

Choose $\varepsilon \leq \min\{\varepsilon'/(6M), \bar{\varepsilon}\}$ and $\underline{\delta}$ to satisfy

$$\underline{\delta}^K > 1 - \frac{\varepsilon'}{6M(1 - \varepsilon)}.$$

Then, for all $\delta \in (\underline{\delta}, 1)$, $1 - (1 - \varepsilon)\delta^K = \varepsilon + (1 - \varepsilon)(1 - \delta^K) < \varepsilon'/(3M)$. Moreover, because $v_1^{**} < M$, $(1 - \varepsilon)\delta^K v_1^{**} > v_1^{**} - \varepsilon'/3$. Substituting into (15.4.8),

$$\underline{v}_1(\xi_0, \mu, \delta) \geq v_1^{**} - \varepsilon'.$$

∎

Fudenberg and Levine (1992) also establish an upper bound on the set of player 1 Nash equilibrium payoffs.

15.4 ■ Imperfect Monitoring

Remark 15.4.4 For finite A_1, if $\{\alpha_1^k\}_k$ is a countably dense subset of $\Delta(A_1)$ and $\mu(\xi(\alpha_1^k)) > 0$ for all k, then we can improve the bound in corollary 15.4.1 because we have effectively assumed full support over stage game mixed actions. Consequently, no player 2 can play a weakly dominated action, and so we could redefine the set of ε-confirmed best replies to exclude weakly dominated actions. ◆

The remainder of this section proves lemma 15.4.3. We begin with three preliminary lemmas.

Lemma 15.4.4 *Suppose X^t is a martingale under $\mathbf{P}$ adapted to $\{\mathscr{G}^t\}_t$, with $0 \leq X^t \leq 1$ for all t. For all $\eta > 0$ and all $K \geq 1$,*

$$\mathbf{P}(|\{t \geq 1 : V^{1,t} \geq \eta\}| \geq K) \leq \frac{1}{K\eta^2}, \quad (15.4.9)$$

where $V^{1,t} = E[|X^{t+1} - X^t| | \mathscr{G}^t]$.

Proof For each $m \geq 1$,

$$E\left[\sum_{t=1}^m (X^{t+1} - X^t)^2\right] = E\left[\sum_{t=1}^m (X^{t+1})^2 - 2X^{t+1}X^t + (X^t)^2\right]$$
$$= E\left[\sum_{t=1}^m (X^{t+1})^2 - 2E[X^{t+1}X^t \mid \mathscr{G}^t] + (X^t)^2\right]$$

(using X^t measurable with respect to $\mathscr{G}^t$)

$$= E\left[\sum_{t=1}^m (X^{t+1})^2 - 2E[X^{t+1} \mid \mathscr{G}^t]X^t + (X^t)^2\right]$$

(using the martingale property, $E[X^{t+1} \mid \mathscr{G}^t] = X^t$)

$$= E\left[\sum_{t=1}^m (X^{t+1})^2 - (X^t)^2\right]$$
$$= E[(X^{m+1})^2 - (X^1)^2] \leq 1. \quad (15.4.10)$$

Because the summands in the following are nonnegative, we can apply the law of iterated expectations and Lebesgue's Monotone Convergence theorem, and use (15.4.10) to conclude

$$E\left[\sum_{t=1}^\infty E\left[(X^{t+1} - X^t)^2 \mid \mathscr{G}^t\right]\right] = \lim_{m \to \infty} E\left[\sum_{t=1}^m (X^{t+1} - X^t)^2\right] \leq 1.$$
$$(15.4.11)$$

Let χ_C denote the indicator function, $\chi_C(\omega) = 1$ if $\omega \in C$ and 0 otherwise, and let $V^{2,t} = E[(X^{t+1} - X^t)^2 | \mathscr{G}^t]$. Then, for all $\eta > 0$ (using (15.4.11) for the first inequality),

$$1 \geq E\left[\sum_{t=1}^\infty V^{2,t}\right] \geq E\left[\sum_{t=1}^\infty V^{2,t} \chi_{\{V^{2,t} \geq \eta^2\}}\right] \geq \eta^2 E\left[\sum_{t=1}^\infty \chi_{\{V^{2,t} \geq \eta^2\}}\right].$$

Because, for $\omega \in \Omega$, $\sum_{t=1}^{\infty} \chi_{\{V^{2,t}(\omega) \geq \eta^2\}}$ is the number of terms in the sequence $\{V^{2,t}(\omega)\}_t$ that weakly exceed η^2, we have, for all $\eta > 0$

$$E[|\{t : V^{2,t} \geq \eta^2\}|] \leq \eta^{-2}.$$

Applying Jensen's inequality to $V^{2,t}$ yields $V^{2,t} = E[(X^{t+1} - X^t)^2 | \mathcal{G}^t] \geq (E[|X^{t+1} - X^t| | \mathcal{G}^t])^2 = (V^{1,t})^2$, so that, for all $\eta > 0$,

$$E[|\{t : V^{1,t} \geq \eta\}|] \leq \eta^{-2}.$$

Finally, for all $\eta > 0$ and K,

$$E[|\{t : V^{1,t} \geq \eta\}|] \geq E[|\{t : V^{1,t} \geq \eta\}| \chi_{\{|\{t:V^{1,t} \geq \eta\}| \geq K\}}]$$
$$\geq K E[\chi_{\{|\{t:V^{1,t} \geq \eta\}| \geq K\}}] = K \mathbf{P}(|\{t : V^{1,t} \geq \eta\}| \geq K)$$

implying (15.4.9).

■

Let $\hat{\Omega} \subset \Omega$ be the event that the stochastic process follows the measure $\hat{\mathbf{P}}$ from the statement of lemma 15.4.3, so that $\hat{\mu} = \mathbf{P}(\hat{\Omega})$. In our reputation context, $\hat{\Omega}$ is the event $\{\hat{\xi}\} \times (A_1 \times Z_1 \times A_2 \times Z_2)^{\infty}$.

Lemma 15.4.5 Let $\phi^t = \mathbf{P}(\hat{\Omega} | \mathcal{G}^t)$. Then,

$$E[|\phi^{t+1} - \phi^t| | \mathcal{G}^t] \geq \phi^t d^t(\mathbf{P}, \hat{\mathbf{P}}). \qquad (15.4.12)$$

Proof For all $t \geq 1$ and $C \in \mathcal{G}^{t+1}$,

$$E[|\phi^{t+1} - \phi^t| | \mathcal{G}^t] \geq E[|\phi^{t+1} - \phi^t| \chi_C | \mathcal{G}^t]$$
$$\geq |E[\phi^{t+1} \chi_C | \mathcal{G}^t] - E[\phi^t \chi_C | \mathcal{G}^t]|$$
$$= |E[\phi^{t+1} \chi_C | \mathcal{G}^t] - \phi^t E[\chi_C | \mathcal{G}^t]|$$
$$= |E[\phi^{t+1} \chi_C | \mathcal{G}^t] - \phi^t \mathbf{P}(C | \mathcal{G}^t)|.$$

We also have

$$E[\phi^{t+1} \chi_C | \mathcal{G}^t] = E[E[\chi_{\hat{\Omega}} | \mathcal{G}^{t+1}] \chi_C | \mathcal{G}^t]$$
$$= E[E[\chi_{C \cap \hat{\Omega}} | \mathcal{G}^{t+1}] | \mathcal{G}^t]$$
$$= \mathbf{P}(C \cap \hat{\Omega} | \mathcal{G}^t)$$
$$= \mathbf{P}(\hat{\Omega} | \mathcal{G}^t) \mathbf{P}(C | \mathcal{G}^t, \hat{\Omega})$$
$$= \phi^t \hat{\mathbf{P}}(C | \mathcal{G}^t).$$

So $E[|\phi^{t+1} - \phi^t| | \mathcal{G}^t] \geq |\phi^t \hat{\mathbf{P}}(C | \mathcal{G}^t) - \phi^t \mathbf{P}(C | \mathcal{G}^t)|$, and taking the supremum over $C \in \mathcal{G}^{t+1}$ implies (15.4.12).

■

15.4 ■ Imperfect Monitoring

Because the posterior beliefs $\{\phi^t\}_t$ are a martingale under $\mathbf{P}$ adapted to $\{\mathscr{G}^t\}_t$, lemmas 15.4.4 and 15.4.5 imply that for all $\eta > 0$ and all K,

$$\mathbf{P}(|\{t : \phi^t d^t(\mathbf{P}, \hat{\mathbf{P}}) \geq \eta\}| \geq K) \leq \frac{1}{K\eta^2}. \tag{15.4.13}$$

In other words, the unconditional probability that in many periods, both ϕ^t is large and the one-step ahead predictions under $\mathbf{P}$ and $\hat{\mathbf{P}}$ are very different, is small. Intuitively, if the one-step ahead predictions under $\mathbf{P}$ and $\hat{\mathbf{P}}$ are very different in some period, then the signals in that period are informative and the posterior belief will reflect that. In particular, when the one-step predictions are different, the uninformed agents must assign significant probability to the outcome not being in $\hat{\Omega}$, and so with significant probability, $\phi^t \to 0$. On the other hand, the only way $\phi^t \not\to 0$ is for the one-step ahead predictions to be close.

Lemma 15.4.6 *For all $\gamma \in (0, 1]$,*

$$\hat{\mathbf{P}}(\cup_{t \geq 1}\{\phi^t \leq \gamma\hat{\mu}\}) \leq \gamma \mathbf{P}(\cup_{t \geq 1}\{\phi^t \leq \gamma\hat{\mu}\}).$$

Proof Let $C^0 = \Omega$ and for all $t \geq 1$, $C^t = \{\phi^m > \gamma\hat{\mu}, \forall m \leq t\}$. Then,

$$\begin{aligned}
\gamma\hat{\mu}\mathbf{P}(C^{t-1} \cap \{\phi^t \leq \gamma\hat{\mu}\}) &= E\left[\gamma\hat{\mu}\chi_{C^{t-1} \cap \{\phi^t \leq \gamma\hat{\mu}\}}\right] \\
&\geq E\left[\phi^t \chi_{C^{t-1} \cap \{\phi^t \leq \gamma\hat{\mu}\}}\right] \\
&= E\left[\mathbf{P}(\hat{\Omega} \mid \mathscr{G}^t)\chi_{C^{t-1} \cap \{\phi^t \leq \gamma\hat{\mu}\}}\right] \\
&= E\left[E\left[\chi_{\hat{\Omega} \cap C^{t-1} \cap \{\phi^t \leq \gamma\hat{\mu}\}} \mid \mathscr{G}^t\right]\right] \\
&= \mathbf{P}(\hat{\Omega} \cap C^{t-1} \cap \{\phi^t \leq \gamma\hat{\mu}\}) \\
&= \hat{\mu}\hat{\mathbf{P}}(C^{t-1} \cap \{\phi^t \leq \gamma\hat{\mu}\}),
\end{aligned}$$

so that

$$\hat{\mathbf{P}}(C^{t-1} \cap \{\phi^t \leq \gamma\hat{\mu}\}) \leq \gamma\mathbf{P}(C^{t-1} \cap \{\phi^t \leq \gamma\hat{\mu}\}).$$

The collection of sets $\{C^{t-1} \cap \{\phi^t \leq \gamma\hat{\mu}\}\}$ are pairwise disjoint and

$$\bigcup_{t \geq 1}\{C^{t-1} \cap \{\phi^t \leq \gamma\hat{\mu}\}\} = \bigcup_{t \geq 1}\{\phi^t \leq \gamma\hat{\mu}\}.$$

Consequently,

$$\begin{aligned}
\hat{\mathbf{P}}(\cup_{t \geq 1}\{\phi^t \leq \gamma\hat{\mu}\}) &= \hat{\mathbf{P}}(\cup_{t \geq 1}\{C^{t-1} \cap \{\phi^t \leq \gamma\hat{\mu}\}\}) \\
&= \sum_{t \geq 1} \hat{\mathbf{P}}(C^{t-1} \cap \{\phi^t \leq \gamma\hat{\mu}\}) \\
&\leq \gamma \sum_{t \geq 1} \mathbf{P}(C^{t-1} \cap \{\phi^t \leq \gamma\hat{\mu}\}) \\
&= \gamma\mathbf{P}(\cup_{t \geq 1}\{\phi^t \leq \gamma\hat{\mu}\}).
\end{aligned}$$

■

Proof of Lemma 15.4.3 Fix $\varepsilon > 0$ and $\psi > 0$. Then, for $K \geq 1$ and $\gamma > 0$,

$$\hat{\mathbf{P}}(|\{t \geq 1 : d^t(\mathbf{P}, \hat{\mathbf{P}}) \geq \psi\}| \geq K)$$
$$= \hat{\mathbf{P}}(|\{t \geq 1 : \phi^t d^t(\mathbf{P}, \hat{\mathbf{P}}) \geq \phi^t \psi\}| \geq K)$$
$$= \hat{\mathbf{P}}(\{|\{t \geq 1 : \phi^t d^t(\mathbf{P}, \hat{\mathbf{P}}) \geq \phi^t \psi\}| \geq K\} \cap \{\phi^t \geq \gamma \hat{\mu}, \forall t\})$$
$$\quad + \hat{\mathbf{P}}(\{|\{t \geq 1 : \phi^t d^t(\mathbf{P}, \hat{\mathbf{P}}) \geq \phi^t \psi\}| \geq K\} \cap \{\phi^t < \gamma \hat{\mu}, \text{ some } t\})$$
$$\leq \hat{\mathbf{P}}(|\{t \geq 1 : \phi^t d^t(\mathbf{P}, \hat{\mathbf{P}}) \geq \gamma \hat{\mu} \psi\}| \geq K)$$
$$\quad + \hat{\mathbf{P}}(\{|\{t \geq 1 : \phi^t d^t(\mathbf{P}, \hat{\mathbf{P}}) \geq \phi^t \psi\}| \geq K\} \cap \{\phi^t < \gamma \hat{\mu}, \text{ some } t\})$$
$$\leq \hat{\mathbf{P}}(|\{t \geq 1 : \phi^t d^t(\mathbf{P}, \hat{\mathbf{P}}) \geq \gamma \hat{\mu} \psi\}| \geq K)$$
$$\quad + \hat{\mathbf{P}}(\{\phi^t < \gamma \hat{\mu}, \text{ some } t\}).$$

Because $\mathbf{P} = \hat{\mu}\hat{\mathbf{P}} + (1 - \hat{\mu})\tilde{\mathbf{P}} \geq \hat{\mu}\hat{\mathbf{P}}, \hat{\mathbf{P}} \leq \mathbf{P}/\hat{\mu}$. Hence, (15.4.13) and lemma 15.4.6 imply

$$\hat{\mathbf{P}}(|\{t \geq 1 : d^t(\mathbf{P}, \hat{\mathbf{P}}) \geq \psi\}| \geq K) \leq \frac{1}{K\hat{\mu}(\gamma\hat{\mu}\psi)^2} + \gamma$$
$$\leq \frac{1}{K(\hat{\mu})^3(\gamma\psi)^2} + \gamma.$$

Choosing $\gamma < \varepsilon/2$ and $K > 2/[\varepsilon(\hat{\mu})^3(\gamma\psi)^2]$ then gives (15.4.7).

∎

15.4.3 Small Players with Idiosyncratic Signals

We have remarked in sections 2.7 and 7.8 that the short-lived player is often naturally interpreted as a continuum of small and anonymous long-lived players. We now illustrate how such an interpretation is consistent with the discussion in remark 15.4.1. Consider a large player 1 facing a continuum of small and anonymous long-lived opponents in the role of player 2. The actions of individual small players are private (to ensure anonymity, see remark 2.7.1). The large player observes a private signal $z_1 \in Z_1$ of the aggregate behavior of the small players (this signal may simply be the distribution over small player actions).[12] The set of period t histories for player 1 is, as usual, $\mathcal{H}_1^t = (A_1 \times Z_1)^t$. Player 1's ex post payoff is $u_1^*(z_1, a_1)$.

Each small player receives a private signal from the finite set Z_2. There are now two possibilities. If the private signal is common across all small players, then when we restrict the small players to identical pure equilibrium strategies, the model is formally equivalent to that of section 15.4.

The second possibility is that different small players observe different realizations of the private signal, that is, the signals are *idiosyncratic*. In each period t with signal distribution $\pi_2(\cdot \mid \alpha_1^t)$, precisely $\pi_2(z_2 \mid \alpha_1^t)$ of the player 2's who are characterized by the private history $h_2^\tau \in \mathcal{H}_2$ receive signal z_2, for each signal $z_2 \in Z_2$ and history $h_2^\tau \in \mathcal{H}_2$. For example, if there are signals $\{\underline{z}, \bar{z}\}$ with $\pi_2(\bar{z} \mid \alpha_1^0) = 1/2$, then in the

---

12. Because z_1 is private, no small player learns anything about the private signals of other small players.

first period half the population of small players observe $\underline{z}$ and half observe $\bar{z}$. Then if $\pi_2(\bar{z} \mid \alpha_1^1) = 1/3$ in the next period, a third of those small players who observed $\underline{z}$ in the first period see $\bar{z}$ in the second, as do a third of those who observed $\bar{z}$ in the first. We continue in this fashion for subsequent histories. In each period t, the probability that each individual player 2 observes signal z_2 is given by $\pi_2(z_2 \mid \alpha_1^t)$. Hence, although there is no aggregate uncertainty, each individual player 2 faces a random signal whose probabilities match the aggregate proportions in which each signal appears (for further discussion, see remark 18.1.3).

Each small player has ex post payoff $u_2^*(z_2, a_2)$. As long as the behavior of the population of small players is measurable, in each period this behavior induces a vector of population proportions choosing the various actions in A_2 that we can denote by α_2 and that is formally equivalent to a mixed action drawn from $\Delta(A_2)$.

Suppose now that the normal long-lived player chooses the action $\hat{\alpha}_1$ in every period. Then lemma 15.4.3 ensures that for any $\varepsilon > 0$, there is a K such that in all but K periods, at least $1 - \varepsilon$ of the short-lived players will be choosing pure actions from $B_\varepsilon^*(\hat{\alpha}_1)$, and hence player 1 faces aggregate player 2 behavior that places probability at least $1 - \varepsilon$ on a (possibly mixed) action from $B_\varepsilon^*(\hat{\alpha}_1)$. The following is then immediate.

Proposition 15.4.2 *Suppose there is a continuum of small and anonymous players in the role of player 2, each receiving idiosyncratic signals. Let $\hat{\xi}$ denote the simple commitment type that always plays $\hat{\alpha}_1 \in \Delta(A_1)$ (if A_1 is finite) or $\hat{\alpha}_1 \in A_1$ (A_1 infinite). Suppose $\xi_0, \hat{\xi} \in \Xi$. For every $\varepsilon > 0$, there is a value K such that for all δ,*

$$\underline{v}_1(\xi_0, \mu, \delta) \geq (1-\varepsilon)\delta^K \inf_{\alpha_2 \in B_\varepsilon^*(\hat{\alpha}_1)} u_1(\hat{\alpha}_1, \alpha_2) + (1-(1-\varepsilon)\delta^K) \min_{a \in A} u_1(a).$$

15.5 Temporary Reputations

This section, drawing on Cripps, Mailath, and Samuelson (2004a,b), shows that the incomplete information that is at the core of the adverse selection approach to reputations is in the presence of imperfect monitoring a short-run phenomenon. Under fairly general conditions, player 2 must eventually learn player 1's type, with play converging to an equilibrium of the complete information game defined by player 1's type.[13]

Our argument first shows that either player 2 eventually learns player 1's type, or player 2 must come to expect the different types of player 1 to play identically. We have encountered similar reasoning in lemma 15.4.3. The first outcome immediately yields our desired conclusion. In the second case, as player 2 comes to expect commitment-type behavior from player 1, player 2 will play a best response.

Under perfect monitoring, there are often pooling equilibria in which the normal and commitment type of player 1 behave identically on the equilibrium path (as in example 15.3.1). Deviations on the part of the normal player 1 are deterred by the

---

13. Benabou and Laroque (1992) study an example in which the uninformed players respond continuously to their beliefs (see section 18.1.4). They show that the informed player eventually reveals his type in any Markov perfect equilibrium (remark 5.6.1).

prospect of the resulting punishment. Under imperfect monitoring, such pooling equilibria do not exist. The normal and commitment types may play identically for a long period of time, but the normal type always eventually has an incentive to cheat at least a little on the commitment strategy, contradicting player 2's belief that player 1 will exhibit commitment behavior. Player 2 must then eventually learn player 1's type.

We follow the lead of section 15.4 in presenting the argument for games of private monitoring, though the analysis covers imperfect public monitoring as a special case.[14]

15.5.1 Asymptotic Beliefs

Our setting is the incomplete information private monitoring game of section 15.4 with finite action sets (essentially identical results hold for the canonical public monitoring game, see remark 15.5.1). We assume full-support marginal private monitoring distributions—each signal is observed with positive probability under every action profile:

Assumption 15.5.1 *For all $i = 1, 2$, $a \in A$, $z_i \in Z_i$, $\pi_i(z_i \mid a) > 0$.*

Because this assumption does not require $\pi(z \mid a) > 0$ for all z and a, public monitoring is a special case. This assumption implies that Bayes' rule determines the beliefs of player 2 about the type of player 1 after all histories.

We assume assumption 15.4.1 and a similar condition for player 1.

Assumption 15.5.2 *For all $a_1 \in A_1$, the collection of probability distributions $\{\pi_1(\cdot, (a_1, a_2)) : a_2 \in A_2\}$ is linearly independent.*

These two assumptions ensure that with sufficient observations, player i can correctly identify from the frequencies of the signals any fixed stage-game action of player j. Assumption 15.4.1 implies (from lemma 15.4.1 and corollary 15.4.1) that if the Stackelberg type has prior positive probability, the normal player 1 can force player 2 to best respond to the Stackelberg action. In the current context, assumption 15.4.1 implies that nonetheless player 2 eventually either learns player 1's type or learns that player 1 is behaving like the commitment type, and assumption 15.5.2 implies that player 1 eventually learns player 2 is playing a best response to this commitment type, two key steps in the proof.

We focus on the case of one simple commitment type for player 1, so $\Xi = \{\xi_0, \hat{\xi}\}$, where $\hat{\xi} = \xi(\hat{\alpha}_1)$ for some $\hat{\alpha}_1 \in \Delta(A_1)$. The analysis is extended to many commitment types in Cripps, Mailath, and Samuelson (2004a, section 6.1). It is convenient to denote a strategy for player 1 as a pair of functions $\tilde{\sigma}_1$ and $\hat{\sigma}_1$ (so $\hat{\sigma}_1(h_1^t) = \hat{\alpha}_1$ for all $h_1^t \in \mathcal{H}_1$), the former for the normal type and the latter for the commitment type. It will sometimes be convenient to write a strategy σ_i as a sequence $(\sigma_i^0, \sigma_i^1, \ldots)$ of

---

14. If monitoring is perfect and the commitment type plays a mixed strategy, the game effectively has imperfect monitoring (as Fudenberg and Levine 1992 observe). For example, in the perfect monitoring version of the product-choice game, if the commitment type randomizes with probability 3/4 on H, then the realized action choice is a noisy signal of the commitment type. Proposition 15.5.1 immediately applies to the perfect monitoring case, as long as the commitment type plays a mixed strategy with full support.

15.5 ■ Temporary Reputations

functions $\sigma_i^t : \mathcal{H}_i^t \to \Delta(A_i)$. We let $\{\mathcal{G}_i^t\}_{t=0}^\infty$ denote the filtration on Ω generated by player i's histories, and $\mathcal{G}^\infty$ the σ-algebra generated by $\cup_{t=0}^\infty \mathcal{G}_i^t$.

Recall that $\mathbf{P} \in \Delta(\Omega)$ is the unconditional probability measure induced by the prior μ, and the strategy profile $(\hat{\sigma}_1, \tilde{\sigma}_1, \sigma_2)$, whereas $\hat{\mathbf{P}}$ is the measure induced by conditioning on $\hat{\xi}$. Because $\{\xi_0\} = \Xi \setminus \{\hat{\xi}\}$, $\tilde{\mathbf{P}}$ (from lemma 15.4.3) is the measure induced by conditioning on ξ_0. That is, $\hat{\mathbf{P}}$ is induced by the strategy profile $\hat{\sigma} = (\hat{\sigma}_1, \sigma_2)$ and $\tilde{\mathbf{P}}$ by $\tilde{\sigma} = (\tilde{\sigma}_1, \sigma_2)$, describing how play evolves when player 1 is the commitment and normal type, respectively. We denote expectations taken with respect to the measure $\mathbf{P}$ by $E[\cdot]$. We also use $\tilde{E}[\cdot]$ and $\hat{E}[\cdot]$ to denote expectations taken with respect to $\tilde{\mathbf{P}}$ and $\hat{\mathbf{P}}$. The expression $\tilde{E}[\sigma_i^t \mid \mathcal{G}_i^s]$ is thus the standard conditional expectation of the player i's period t strategy, viewed as a $\mathcal{G}_i^s$-measurable random variable on Ω. We also write $\tilde{E}[\cdot \mid h_i^t]$ for the expectation conditional on having observed the history h_i^t.

The action of the commitment type satisfies the following assumption.

Assumption 15.5.3 *Player 2 has a unique stage-game best response to $\hat{\alpha}_1$ (denoted by $\hat{a}_2$), and $\hat{\alpha} \equiv (\hat{\alpha}_1, \hat{a}_2)$ is not a stage-game Nash equilibrium.*

Let $\hat{\sigma}_2$ denote the strategy of playing the unique best response $\hat{a}_2$ to $\hat{\alpha}_1$ in each period independently of history. Because $\hat{\alpha}$ is not a stage-game Nash equilibrium, $(\hat{\sigma}_1, \hat{\sigma}_2)$ is not a Nash equilibrium of the complete information infinite horizon game.

Example 15.5.1 This assumption requires a unique best response to $\hat{\alpha}_1$. For example, every action for player 2 is a best response to player 1's mixture $\frac{1}{2} \circ H + \frac{1}{2} \circ L$ in the product-choice game. Section 7.6.2 exploited this indifference to construct an equilibrium in which (the normal) player 1 plays $\frac{1}{2} \circ H + \frac{1}{2} \circ L$ after every history. This will still be an equilibrium in the game of incomplete information in which the commitment type plays $\frac{1}{2} \circ H + \frac{1}{2} \circ L$, with the identical play of the normal and commitment types ensuring that player 2 never learns player 1's type. In contrast, player 2 has a unique best response to any other mixture on the part of player 1. Thus, if the commitment type is committed to any mixed action other than $\frac{1}{2} \circ H + \frac{1}{2} \circ L$, player 2 will eventually learn player 1's type.

●

Proposition 15.5.1 *Suppose the monitoring distribution π satisfies assumptions 15.4.1, 15.5.1, and 15.5.2; action spaces are finite; and the commitment action $\hat{\alpha}_1$ satisfies assumption 15.5.3. In any Nash equilibrium of the game with incomplete information,*

$$\hat{\mu}^t \equiv \mathbf{P}(\{\hat{\xi}\} \mid \mathcal{G}_2^t) \to 0, \qquad \tilde{\mathbf{P}}\text{-a.s.}$$

The intuition is straightforward: Suppose there is a Nash equilibrium of the incomplete information game in which both the normal and the commitment type receive positive probability in the limit (on a positive probability set of histories). On this set of histories, player 2 cannot distinguish between signals generated by the two types (otherwise player 2 could ascertain which type she is facing), and hence must believe that the normal and commitment types are playing the same strategies on average. But then player 2 must play a best response to this strategy and thus to the commitment type. Because the commitment type's behavior is not a best response for the normal type (to this player 2 behavior), player 1 must eventually find it optimal to *not* play the commitment-type strategy, contradicting player 2's beliefs.

Remark **Canonical public monitoring** Reputations are also temporary in the canonical
15.5.1 public monitoring game. The proof is essentially identical, once Assumption
15.4.1 is replaced by Assumption 15.4.2.
♦

We first note some implications of this result, deferring its proof to section 15.6.

15.5.2 Uniformly Disappearing Reputations

Proposition 15.5.1 leaves open the possibility that for any period T, there may be equilibria in which uncertainty about player 1's type survives beyond T, even though such uncertainty asymptotically disappears in any equilibrium. We show here that this possibility cannot arise. The existence of a sequence of Nash equilibria with uncertainty about player 1's type persisting beyond period $T \to \infty$ would imply the (contradictory) existence of a limiting Nash equilibrium in which uncertainty about player 1's type persists.

Proposition *Suppose the monitoring distribution π satisfies assumptions 15.4.1, 15.5.1,*
15.5.2 *and 15.5.2; action spaces are finite; and the commitment action $\hat{a}_1$ satisfies assumption 15.5.3. For all $\varepsilon > 0$, there exists T such that for any Nash equilibrium of the game with incomplete information,*

$$\tilde{\mathbf{P}}(\hat{\mu}^t < \varepsilon, \forall t > T) > 1 - \varepsilon.$$

Proof Suppose not. Then there exists $\varepsilon > 0$ such that for all T, there is a Nash equilibrium $\sigma_{(T)}$ such that

$$\tilde{\mathbf{P}}_{(T)}(\hat{\mu}^t_{(T)} < \varepsilon, \forall t > T) \leq 1 - \varepsilon,$$

where $\tilde{\mathbf{P}}_{(T)}$ is the measure induced by the normal type under $\sigma_{(T)}$, and $\hat{\mu}^t_{(T)}$ is the posterior in period t under $\sigma_{(T)}$.

Because the space of strategy profiles is sequentially compact in the product topology, there is a sequence $\{T_{(k)}\}$ such that the subsequence $\{\sigma_{(T_k)}\}$ converges to a limit $\sigma_{(\infty)}$, with corresponding measure $\tilde{\mathbf{P}}_{(\infty)}$ on Ω induced by the normal type. We denote $\sigma_{(T_k)}$ by $\sigma_{(k)}$, and so

$$\tilde{\mathbf{P}}_{(k)}\left(\hat{\mu}^t_{(k)} < \varepsilon, \forall t > T_{(k)}\right) \leq 1 - \varepsilon,$$

that is,

$$\tilde{\mathbf{P}}_{(k)}\left(\hat{\mu}^t_{(k)} \geq \varepsilon \text{ for some } t > T_{(k)}\right) \geq \varepsilon.$$

Because each $\sigma_{(k)}$ is a Nash equilibrium, $\hat{\mu}^t_{(k)} \to 0$ $\tilde{\mathbf{P}}_{(k)}$-a.s. (proposition 15.5.1), and so there exists $K_{(k)}$ such that

$$\tilde{\mathbf{P}}_{(k)}\left(\hat{\mu}^t_{(k)} < \varepsilon, \forall t \geq K_{(k)}\right) \geq 1 - \tfrac{\varepsilon}{2}.$$

Consequently, for all k,

$$\tilde{\mathbf{P}}_{(k)}\left(\hat{\mu}^t_{(k)} \geq \varepsilon, \text{ for some } t, T_{(k)} < t < K_{(k)}\right) \geq \tfrac{\varepsilon}{2}.$$

15.5 ■ Temporary Reputations

Let $\tau_{(k)}$ denote the stopping time

$$\tau_{(k)} = \min\{t > T_{(k)} : \hat{\mu}^t_{(k)} \geq \varepsilon\},$$

and $X^t_{(k)}$ the associated stopped process,

$$X^t_{(k)} = \begin{cases} \hat{\mu}^t_{(k)}, & \text{if } t < \tau_{(k)}, \\ \varepsilon, & \text{if } t \geq \tau_{(k)}. \end{cases}$$

Note that $X^t_{(k)}$ is a supermartingale under $\tilde{\mathbf{P}}_{(k)}$ and that for $t < T_{(k)}$, $X^t_{(k)} = \hat{\mu}^t_{(k)}$. Observe that for all k and $t \geq K_{(k)}$,

$$\tilde{E}_{(k)} X^t_{(k)} \geq \varepsilon \tilde{\mathbf{P}}_{(k)}(\tau_{(k)} \leq t) \geq \tfrac{\varepsilon^2}{2},$$

where $\tilde{E}_{(k)}$ denotes expectation with respect to $\tilde{\mathbf{P}}_{(k)}$.

Because $\sigma_{(\infty)}$ is a Nash equilibrium, $\hat{\mu}^t_{(\infty)} \to 0$ $\tilde{\mathbf{P}}_{(\infty)}$-a.s. (appealing to proposition 15.5.1 again), and so there exists a date s such that

$$\tilde{\mathbf{P}}_{(\infty)}\left(\hat{\mu}^s_{(\infty)} < \tfrac{\varepsilon^2}{12}\right) > 1 - \tfrac{\varepsilon^2}{12}.$$

Then,

$$\tilde{E}_{(\infty)} \hat{\mu}^s_{(\infty)} \leq \tfrac{\varepsilon^2}{12}\left(1 - \tfrac{\varepsilon^2}{12}\right) + \tfrac{\varepsilon^2}{12} < \tfrac{\varepsilon^2}{6},$$

where $\tilde{E}_{(\infty)}$ denotes expectation with respect to $\tilde{\mathbf{P}}_{(\infty)}$. Because $\sigma_{(k)} \to \sigma_{(\infty)}$ in the product topology and there are only a finite number of player 1 and 2 s-length or shorter histories, there is a k' with $T_{(k')} > s$ such that for all $k \geq k'$,

$$\tilde{E}_{(k)} \hat{\mu}^s_{(k)} < \frac{\varepsilon^2}{3}.$$

But because $T_{(k')} > s$, $X^s_{(k)} = \hat{\mu}^s_{(k)}$ for $k \geq k'$ and so for any $t \geq K_{(k)}$,

$$\frac{\varepsilon^2}{3} > \tilde{E}_{(k)} \hat{\mu}^s_{(k)} = \tilde{E}_{(k)} X^s_{(k)}$$
$$\geq \tilde{E}_{(k)} X^t_{(k)} \geq \frac{\varepsilon^2}{2}, \tag{15.5.1}$$

which is a contradiction.

■

15.5.3 Asymptotic Equilibrium Play

Given proposition 15.5.1, we should expect continuation play to converge to an equilibrium of the complete information game. The results here are strongest if the monitoring technology is such that player 1 knows player 2's belief. We accordingly work temporarily with the special class of canonical public monitoring games described in remark 15.4.2: Player 2's action is public, and there is a public signal y_1

drawn from a finite set Y_1. For notational simplicity we drop the subscript 1, so the public signal is a pair $ya_2 \in Y \times A_2$. Assumption 15.5.1 now requires the distribution ρ satisfy $\rho(y \mid a) > 0$ for all $y \in Y$ and $a \in A$ (where $\rho(\cdot \mid a) \in \Delta(Y)$ is the distribution denoted by ρ_1 in remark 15.4.2). Because player 2's actions are public, assumption 15.5.2 no longer plays a role, whereas assumption 15.4.1 now becomes that for all $a_2 \in A_2$, the collection of probability distributions $\{\rho(\cdot \mid (a_1, a_2)) : a_1 \in A_1\}$ is linearly independent.

We use the term *continuation game* for the game with initial period in period t, ignoring the period t histories. We use the notation $t' = 0, 1, 2, \ldots$ for a period of play in a continuation game (which may be the original game) and t for the time elapsed prior to the start of the continuation game. A pure strategy for player 1, s_1, is a map $s_1 : \mathcal{H}_1 \to A_1$, so $s_1 \in A_1^{\mathcal{H}_1} \equiv S_1$, and similarly $s_2 \in A_2^{\mathcal{H}} \equiv S_2$. The spaces S_1 and S_2 are countable products of finite sets. We equip S_1 and S_2 with the σ-algebras generated by the cylinder sets, denoted by $\mathscr{S}_1$ and $\mathscr{S}_2$. Note that S_i is the set of player i pure strategies in the original complete information game, as well as in any continuation game. Player 1's and the period t' player 2's payoffs in the (infinitely repeated) continuation game, as a function of the pure strategy profile s, are given by

$$U_1(s_1, s_2) \equiv E\left[(1-\delta) \sum_{t'=0}^{\infty} \delta^{t'} u_1(a_1^{t'}, a_2^{t'})\right]$$

and $\quad u_2^{t'}(s_1, s_2) \equiv E[u_2(a_1^{t'}, a_2^{t'})].$

These expectations are taken over the action pairs $(a_1^{t'}, a_2^{t'})$.

For $\ell = 1, 2$, let $\mathscr{D}_\ell$ denote the space of probability measures λ_ℓ on $(S_\ell, \mathscr{S}_\ell)$. We say a sequence of measures $\lambda_1^{(n)} \in \mathscr{D}_1$ *converges* to $\lambda_1 \in \mathscr{D}_1$ if, for each τ, we have

$$\lambda_1^{(n)}\big|_{A_1^{(A_1 \times A_2 \times Y)^\tau}} \to \lambda_1\big|_{A_1^{(A_1 \times A_2 \times Y)^\tau}} \qquad (15.5.2)$$

and a sequence of measures $\lambda_2^{(n)} \in \mathscr{D}_2$ *converges* to $\lambda_2 \in \mathscr{D}_2$ if for each τ, we have

$$\lambda_2^{(n)}\big|_{A_2^{(A_2 \times Y)^\tau}} \to \lambda_2\big|_{A_2^{(A_2 \times Y)^\tau}}. \qquad (15.5.3)$$

Moreover, each $\mathscr{D}_\ell$ is sequentially compact in the topology of this convergence. Payoffs for players 1 and 2 are extended to $\mathscr{D} = \mathscr{D}_1 \times \mathscr{D}_2$ in the obvious way.

Fix an equilibrium of the incomplete information game. If the normal type of player 1 observes a private history $h_1^t \in \mathcal{H}_1^t$, his strategy $\tilde{\sigma}_1$ specifies a behavior strategy in the continuation game. This behavior strategy is realization equivalent to a mixed strategy $\tilde{\lambda}_1^{h_1^t} \in \mathscr{D}_1$ for the continuation game. We let $\tilde{\lambda}_1^{h^t}$ denote the expected value of $\tilde{\lambda}_1^{h_1^t}$, conditional on the public history h^t. From the point of view of player 2, who observes only the public history, $\tilde{\lambda}_1^{h^t}$ is the strategy of the normal player 1 following history h^t. We let $\lambda_2^{h^t} \in \mathscr{D}_2$ denote player 2's mixed strategy in the continuation game.

The limit of every convergent subsequence of $(\tilde{\lambda}_1^{h^t}, \lambda_2^{h^t})$ is a Nash equilibrium (with the sequential compactness of $\mathscr{D}$ ensuring that such subsequences exist).

15.5 ■ Temporary Reputations

Proposition 15.5.3 *Suppose the monitoring distribution π satisfies assumptions 15.4.1 and 15.5.1, action spaces are finite, and the commitment action $\hat{\alpha}_1$ satisfies assumption 15.5.3. For any Nash equilibrium of the incomplete information game and for $\tilde{\mathbf{P}}$-almost all sequences of histories $\{h^t\}_t$, every cluster point of the sequence of continuation profiles $\{(\tilde{\lambda}_1^{h^t}, \lambda_2^{h^t})\}_t$ is a Nash equilibrium of the complete information game with normal player 1.*

Proof At the given equilibrium, the normal type is playing in an optimal way from time t onward given his (private) information. Thus, for each history h_1^t, associated public history h^t, and strategy $s_1' \in S_1$,

$$E^{(\tilde{\lambda}_1^{h_1^t}, \lambda_2^{h^t})}[U_1(s_1, s_2)] \geq E^{\lambda_2^{h^t}}[U_1(s_1', s_2)].$$

The superscripts on the expectation operator are the measures on (s_1, s_2) involved in calculating the expectation. Moreover, for the associated public history h^t and any strategy $s_1' \in S_1$,

$$E^{(\tilde{\lambda}_1^{h^t}, \lambda_2^{h^t})}[U_1(s_1, s_2)] \geq E^{\lambda_2^{h^t}}[U_1(s_1', s_2)]. \tag{15.5.4}$$

Player 2 is also playing optimally from time t onward given the public information, which implies that for all $s_2' \in S_2$, all $h^{t'}$ and all $t' > 0$,

$$E^{(\hat{\mu}^t \hat{\lambda}_1^{h^t} + (1-\hat{\mu}^t)\tilde{\lambda}_1^{h^t}, \lambda_2^{h^t})}[u_2^{t'}(s_1, s_2)] \geq E^{\hat{\mu}^t \hat{\lambda}_1^{h^t} + (1-\hat{\mu}^t)\tilde{\lambda}_1^{h^t}}[u_2^{t'}(s_1, s_2')], \tag{15.5.5}$$

where $\hat{\lambda}_1^{h^t}$ is the play of the commitment type. Because player 2 is a short-run player, this inequality is undiscounted and holds for all t'.

From proposition 15.5.1, $\hat{\mu}^t \to 0$ $\tilde{P}$-a.s. Suppose $\{h^t\}_t$ is a sequence of public histories with $\hat{\mu}^t \to 0$, and suppose $\{(\tilde{\lambda}_1^{h^t}, \lambda_2^{h^t})\}_{t=1}^{\infty} \to (\tilde{\lambda}_1^*, \lambda_2^*)$ on this sequence. We need to show that $(\tilde{\lambda}_1^*, \lambda_2^*)$ satisfies (15.5.4) and (15.5.5) (the latter for all $t' > 0$). It suffices for the result that the expectations $E^{(\lambda_1, \lambda_2)}[U_1(s_1, s_2)]$ and $E^{(\lambda_1, \lambda_2)}[u_2(s_1, s_2)]$ are continuous in (λ_1, λ_2). Continuity is immediate from the continuity of u_2 for player 2, for each time t'. For player 1, continuity is an implication of discounting. For any $\varepsilon > 0$, one can find a $\gamma > 0$ and $\tau > 0$ such that if $|\lambda_1^{(n)}|_{A_1^{(A_1 \times A_2 \times Y)^\tau}} - \lambda_1^*|_{A_1^{(A_1 \times A_2 \times Y)^\tau}}| < \gamma$ and $|\lambda_2^{(n)}|_{A_2^{(A_2 \times Y)^\tau}} - \lambda_2^*|_{A_2^{(A_2 \times Y)^\tau}}| < \gamma$, then $|E^{(\lambda_1^{(n)}, \lambda_2^{(n)})}[U_1(s_1, s_2)] - E^{(\lambda_1^*, \lambda_2^*)}[U_1(s_1, s_2)]| < \varepsilon$. It is then clear that $E^{(\lambda_1, \lambda_2)}[U_1(s_1, s_2)]$ is continuous, given the convergence notion given by (15.5.2) and (15.5.3).

■

Example 15.5.2 Recall that in the product-choice game, the unique player 2 best response to the H is to play h, and Hh is not a stage-game Nash equilibrium. Proposition 15.4.1 ensures that the normal player 1's expected value in the repeated game of incomplete information with an H-commitment type is arbitrarily close to 2 when player 1 is very patient. In particular, if the normal player 1 plays H in every period, then player 2 will at least eventually play her best response of h. If the normal player 1 persisted in mimicking the commitment type by playing H in each period, this behavior would persist indefinitely. It is the feasibility of such a strategy that lies at the heart of the reputation bounds on expected payoffs. However, this strategy is not optimal. Instead, player 1 does even better by attaching some probability to L, occasionally reaping the rewards of his reputation by earning a stage-game payoff even larger than 2. The result of such equilibrium behavior, however, is that player 2 must eventually learn player 1's type. The continuation payoff is then bounded below 2 (see example 15.4.1).

●

Remark 15.5.2 **A partial converse** Cripps, Mailath, and Samuelson (2004a, theorem 3) provide a partial converse to proposition 15.5.3, identifying a class of equilibria of the complete information game to which (under a continuity hypothesis) equilibrium play of the incomplete information game can converge.

◆

Remark 15.5.3 **Private beliefs** Cripps, Mailath, and Samuelson (2004b) examine asymptotic equilibrium play when player 1 does not know player 2's beliefs (as arises under the private-monitoring case of Section 15.5.1). When 2's beliefs are not public, Cripps, Mailath, and Samuelson (2004b, theorem 5) show that play must eventually be a correlated equilibrium of the complete information game, with the players' private histories providing the correlating device. Correlated equilibria in repeated games remain relatively unexplored. For example, it is not known whether a limiting correlated equilibrium of the incomplete information game is a public randomization over Nash equilibria of the complete information game. Nor is it known whether proposition 8.3.1—imperfect monitoring breeds inefficiency in binding moral hazard games like the product-choice game—extends to correlated equilibria.

◆

15.6 Temporary Reputations: The Proof of Proposition 15.5.1

15.6.1 Player 2's Posterior Beliefs

The first step is to show that either player 2's expectation (given her history) of the strategy played by the normal type is in the limit identical to the strategy played by the commitment type, or player 2's posterior probability that player 1 is the commitment type converges to 0 (given that player 1 is indeed normal). This is a merging argument. If the distributions generating player 2's signals are different for the normal and commitment types, then these signals provide information that player 2 will use in updating her posterior beliefs about the type she faces. This (converging, by lemma 15.4.2) belief can converge to an interior probability only if the distributions generating the signals are asymptotically uninformative, which requires (by assumption 15.4.1) that they be identical.

Lemma 15.6.1 *Suppose the monitoring distribution π satisfies assumptions 15.4.1 and 15.5.1, and the action spaces are finite. Then in any Nash equilibrium,*

$$\lim_{t \to \infty} \hat{\mu}^t (1 - \hat{\mu}^t) \| \hat{\alpha}_1 - \tilde{E}[\tilde{\sigma}_1^t \mid \mathscr{G}_2^t] \| = 0, \quad \textbf{P}\text{-}a.s. \quad (15.6.1)$$

15.6 ■ Proof

Proof **Step 1.** Equation (15.4.13) implies,

$$\mathbf{P}(\hat{\mu}^t d^t(\mathbf{P}, \hat{\mathbf{P}}) \geq \eta \text{ infinitely often}) = 0, \qquad \forall \eta > 0 \qquad (15.6.2)$$

because $\hat{\mu} = \mathbf{P}(\hat{\Omega} \mid \mathscr{G}_2^t) = \phi^t$, where d^t is defined in lemma 15.4.3. Equation (15.6.2) is equivalent to

$$\hat{\mu}^t d^t(\mathbf{P}, \hat{\mathbf{P}}) \to 0, \qquad \mathbf{P}\text{-a.s.}$$

Because $\{z_2^t\} \in \mathscr{G}_2^{t+1}$, this implies, for all $z_2 \in Z_2$,

$$\hat{\mu}^t \left| \pi_2(z_2 \mid \hat{\alpha}_1, a_2^t) - \pi_2(z_2 \mid E[\sigma_1(h_1^t, \xi) \mid \mathscr{G}_2^t], a_2^t) \right| \to 0, \qquad \mathbf{P}\text{-a.s.}, \qquad (15.6.3)$$

where a_2^t is any $\mathscr{G}_2^t$-measurable mapping from Ω into A_2 such that $a_2^t(\omega)$ has positive probability under $\sigma_2(h_2^t(\omega))$. Then because $E[\sigma_1(h_1^t, \xi) \mid \mathscr{G}_2^t] = \hat{\mu}^t \hat{\alpha}_1 + (1 - \hat{\mu}^t) \tilde{E}[\tilde{\sigma}_1(h_1^t) \mid \mathscr{G}_2^t]$, we have

$$\pi_2(z_2 \mid E[\sigma_1(h_1^t, \xi) \mid \mathscr{G}_2^t], a_2^t)$$
$$= \hat{\mu}^t \pi_2(z_2 \mid \hat{\alpha}_1, a_2^t) + (1 - \hat{\mu}^t) \pi_2(z_2 \mid \tilde{E}[\tilde{\sigma}_1(h_1^t) \mid \mathscr{G}_2^t], a_2^t).$$

Substituting into (15.6.3) and rearranging,

$$\hat{\mu}^t(1 - \hat{\mu}^t) \left| \pi_2(z_2 \mid \hat{\alpha}_1, a_2^t) - \pi_2(z_2 \mid \tilde{E}[\tilde{\sigma}_1(h_1^t) \mid \mathscr{G}_2^t], a_2^t) \right| \to 0, \qquad \mathbf{P}\text{-a.s.},$$

that is, for all $z_2 \in Z_2$,

$$\hat{\mu}^t(1 - \hat{\mu}^t) \left| \sum_{a_1 \in A_1} \pi_2(z_2 \mid a_1, a_2^t)(\hat{\alpha}_1(a_1) - \tilde{E}[\tilde{\sigma}_1^{a_1}(h_1^t) \mid \mathscr{G}_2^t]) \right| \to 0, \qquad \mathbf{P}\text{-a.s.} \qquad (15.6.4)$$

Therefore, if both types are given positive probability in the limit then the frequency that any signal is observed is identical under the two types. (See Cripps, Mailath, and Samuelson (2004a, p. 419) for a direct, but longer, proof that only uses Bayes' rule.)

Step 2. We now show that (15.6.4) implies (15.6.1). Let $\Pi_{a_2^t}$ be a $|Z_2| \times |A_1|$ matrix whose z_2^{th} row, for each signal $z_2 \in Z_2$, contains the terms $\pi_2(z_2 \mid a_1, a_2^t)$ for $a_1 \in A_1$. Then as (15.6.4) holds for all z_2 (and Z_2 is finite), (15.6.4) can be restated as

$$\hat{\mu}^t(1 - \hat{\mu}^t) \left\| \Pi_{a_2^t}(\hat{\alpha}_1 - \tilde{E}[\tilde{\sigma}_1^t \mid \mathscr{G}_2^t]) \right\| \to 0, \qquad \mathbf{P}\text{-a.s.} \qquad (15.6.5)$$

By assumption 15.4.1, the matrices $\Pi_{a_2^t}$ have $|A_1|$ linearly independent columns for all a_2^t, so $x = 0$ is the unique solution to $\Pi_{a_2^t} x = 0$ in $\mathbb{R}^{|A_1|}$. In addition, there exists a strictly positive constant $b = \inf_{a_2 \in A_2, x \neq 0} \|\Pi_{a_2} x\| / \|x\|$. Hence $\|\Pi_{a_2} x\| \geq b \|x\|$ for all $x \in \mathbb{R}^{|A_1|}$ and all $a_2 \in A_2$. From (15.6.5), we then get

$$\hat{\mu}^t(1-\hat{\mu}^t)\|\Pi_{a_2^t}(\hat{\alpha}_1 - \tilde{E}[\tilde{\sigma}_1^t \mid \mathscr{G}_2^t])\|$$
$$\geq \hat{\mu}^t(1-\hat{\mu}^t)b\|\hat{\alpha}_1 - \tilde{E}[\tilde{\sigma}_1^t \mid \mathscr{G}_2^t]\| \to 0, \qquad \textbf{P}\text{-a.s.,}$$

which implies (15.6.1). ∎

Condition (15.6.1) says that either player 2's best prediction of the normal type's behavior converges to the commitment type's behavior (that is, $\|\hat{\alpha}_1 - \tilde{E}[\tilde{\sigma}_1^t \mid \mathscr{G}_2^t]\| \to 0$ or the type is revealed (that is, $\hat{\mu}^\infty(1-\hat{\mu}^\infty) = 0$, **P**-a.s.). However, $\hat{\mu}^\infty < 1$ $\tilde{\textbf{P}}$-a.s., and hence (15.6.1) implies a simple corollary:[15]

Corollary 15.6.1 *At any equilibrium of a game with monitoring distribution π satisfying assumptions 15.4.1 and 15.5.1, and finite action spaces,*

$$\lim_{t\to\infty} \hat{\mu}^t \|\hat{\alpha}_1 - \tilde{E}[\tilde{\sigma}_1^t \mid \mathscr{G}_2^t]\| = 0, \qquad \tilde{\textbf{P}}\text{-a.s.}$$

15.6.2 Player 2's Beliefs about Her Future Behavior

We now examine the consequences of the existence of a $\tilde{\textbf{P}}$-positive measure subset of Ω on which player 1's type is not learned, that is, on which $\lim_{t\to\infty} \hat{\mu}^t(\omega) > 0$. The normal and the commitment types eventually play the same strategy on these states (lemma 15.6.1). Consequently, on a positive probability subset of these states, player 2 eventually attaches high probability to the event that in all future periods he will play a best response to the commitment type. Before stating this formally, we prove an intermediate result, which may seem intuitive but requires some care. The argument is based on Hart (1985, lemma 4.24).

Lemma 15.6.2 *Suppose $\{X^t\}_t$ is a bounded sequence of random variables on $(\Omega, \mathscr{G}, \textbf{P})$ with $X^t \to 0$ **P**-a.s. Then for any filtration $\{\mathscr{G}^t\}_t$, $E[X^t \mid \mathscr{G}^t] \to 0$ **P**-a.s.*

Proof Let $\mathbb{X}^t = \sup_{t' \geq t} |X^{t'}|$. Then the sequence $\{\mathbb{X}^t\}_t$ is a nonincreasing sequence of random variables converging to 0 **P**-a.s. By definition, we have $E[\mathbb{X}^{t+1} \mid \mathscr{G}^t] \leq E[\mathbb{X}^t \mid \mathscr{G}^t]$ **P**-a.s., and hence $\{E[\mathbb{X}^t \mid \mathscr{G}^t]\}_t$ is a bounded supermartingale adapted to the filtration $\{\mathscr{G}^t\}_t$. From the martingale convergence theorem (Billingsley 1979, theorem 35.4), there exists a limit $\mathbb{X}^\infty$ with $E[\mathbb{X}^t \mid \mathscr{G}^t] \to \mathbb{X}^\infty$ **P**-a.s. But because $E[E[\mathbb{X}^t \mid \mathscr{G}^t]] = E[\mathbb{X}^t] \to 0$, we have $E[\mathbb{X}^\infty] = 0$. Because $\mathbb{X}^\infty \geq 0$ **P**-a.s., we must have $\mathbb{X}^\infty = 0$ **P**-a.s. Then noting that $-E[\mathbb{X}^t \mid \mathscr{G}^t] \leq E[X^t \mid \mathscr{G}^t] \leq E[\mathbb{X}^t \mid \mathscr{G}^t]$, we conclude that $E[X^t \mid \mathscr{G}^t] \to 0$ **P**-almost surely. ∎

We are now in a position to state and prove the result of this section. The event that player 2 plays a best response to the commitment strategy in all periods $s \geq t$ is

$$G^t = \{\omega : \sigma_2^s(h_2^s(\omega)) = \hat{a}_2, \forall s \geq t\}.$$

---

[15]. Because the odds ratio $\hat{\mu}^t/(1-\hat{\mu}^t)$ is a $\tilde{\textbf{P}}$-martingale (lemma 15.4.2), $\hat{\mu}^0/(1-\hat{\mu}^0) = \tilde{E}[\hat{\mu}^t/(1-\hat{\mu}^t)]$ for all t. The left side of this equality is finite, so $\lim \hat{\mu}^t < 1$ $\tilde{\textbf{P}}$-a.s.

15.6 ■ Proof

Lemma 15.6.3 *Suppose the monitoring distribution π satisfies assumptions 15.4.1 and 15.5.1, the action spaces are finite, and there is a Nash equilibrium in which player 2 does not necessarily learn player 1's type, that is, $\tilde{\mathbf{P}}(A) > 0$, where $A \equiv \{\hat{\mu}^t \nrightarrow 0\}$. There exists $\eta > 0$ and $F \subset A$, with $\tilde{\mathbf{P}}(F) > 0$, such that for any $\theta > 0$, there exists T for which on F,*

$$\hat{\mu}^t > \eta, \qquad \forall t \geq T,$$

and

$$\tilde{\mathbf{P}}(G^t \mid \mathscr{G}_2^t) > 1 - \theta, \qquad \forall t \geq T. \tag{15.6.6}$$

Proof *Step 1.* Because $\tilde{\mathbf{P}}(A) > 0$ and $\hat{\mu}^t$ converges a.s., there exists $\psi > 0$ and $\eta > 0$ such that $\tilde{\mathbf{P}}(D) > 2\psi$, where $D \equiv \{\omega : \lim_{t \to \infty} \hat{\mu}^t(\omega) > 2\eta\}$. The random variables $\|\hat{\alpha}_1 - \tilde{E}[\tilde{\sigma}_1^t | \mathscr{G}_2^t]\|$ tend $\tilde{\mathbf{P}}$-a.s. to 0 on D (by corollary 15.6.1). Consequently, the random variables $X^t \equiv \sup_{s \geq t} \|\hat{\alpha}_1 - \tilde{E}[\tilde{\sigma}_1^s | \mathscr{G}_2^s]\|$ also converge $\tilde{\mathbf{P}}$-a.s. to 0 on D. This in turn implies, applying lemma 15.6.2, that on D the expected value of X^t, conditional on player 2's information, converges to 0, that is, $\tilde{E}[\chi_D X^t | \mathscr{G}_2^t]$ converges a.s. to 0, where χ_D is the indicator for the event D. Now define $A^t \equiv \{\omega : \tilde{E}[\chi_D | \mathscr{G}_2^t](\omega) > 1/2\}$. The $\mathscr{G}_2^t$-measurable event A^t approximates D (because player 2 knows her own beliefs, the random variables $d^t \equiv |\chi_D - \chi_{A^t}|$ converge $\tilde{\mathbf{P}}$-a.s. to 0). Hence

$$\chi_D \tilde{E}[X^t \mid \mathscr{G}_2^t] \leq \chi_{A^t} \tilde{E}[X^t \mid \mathscr{G}_2^t] + d^t$$
$$= \tilde{E}[\chi_{A^t} X^t \mid \mathscr{G}_2^t] + d^t$$
$$\leq \tilde{E}[\chi_D X^t \mid \mathscr{G}_2^t] + \tilde{E}[d^t \mid \mathscr{G}_2^t] + d^t,$$

where the first and third lines use $X^t \leq 1$ and the second uses the measurability of A^t with respect to $\mathscr{G}_2^t$. All the terms on the last line converge $\tilde{\mathbf{P}}$-a.s. to 0, and so $\tilde{E}[X^t | \mathscr{G}_2^t] \to 0$ $\tilde{\mathbf{P}}$-a.s. on the set D. Egorov's theorem (Chung 1974, p. 74) then implies that there exists $F \subset D$ such that $\tilde{\mathbf{P}}(F) > 0$ on which the convergence of $\hat{\mu}^t$ and $\tilde{E}[X^t | \mathscr{G}_2^t]$ is uniform.

Step 2. From the upper hemicontinuity of the best response correspondence, there exists $\phi > 0$ such that for any history h_1^s and any action $\alpha_1 \in \Delta(A_1)$ satisfying $\|\alpha_1 - \hat{\alpha}_1\| \leq \phi$, a best response to α_1 is also a best response to $\hat{\alpha}_1$ and so necessarily equals $\hat{\alpha}_2$. The uniform convergence of $\tilde{E}[X^t | \mathscr{G}_2^t]$ on F implies that for any $\theta > 0$, there exists a time T such that on F for all $t \geq T$ and $\hat{\mu}^t > \eta$ and

$$\tilde{E}\left[\sup_{s \geq t} \|\hat{\alpha}_1 - \tilde{E}[\tilde{\sigma}_1^s | \mathscr{G}_2^s]\| \,\bigg|\, \mathscr{G}_2^t\right] < \theta \phi.$$

As $\tilde{E}[X^t | \mathscr{G}_2^t] < \theta \phi$ for all $t \geq T$ on F and $X^t \geq 0$, $\tilde{\mathbf{P}}(\{X^t > \phi\} | \mathscr{G}_2^t) < \theta$ for all $t \geq T$ on F, implying (15.6.6). ∎

15.6.3 Player 1's Beliefs about Player 2's Future Behavior

Our next step is to show that with positive probability, player 1 eventually expects player 2 to play a best response to the commitment type for the remainder of the game.

The potential difficulty in proving this result is that player 1 does not know player 2's private history and hence 2's beliefs.

We first show that it cannot be too disadvantageous for too long, in terms of predicting 2's behavior, for player 1 to not know player 2's history. Though player 2's private history h_2^t is typically of use to player 1 in predicting 2's period s behavior for $s > t$, this usefulness vanishes as $s \to \infty$. If period s behavior is eventually (as s becomes large) independent of h_2^t, then clearly h_2^t is eventually of no use in predicting that behavior. Suppose then that h_2^t is essential to predicting player 2's behavior in all periods $s > t$. Then, player 1 continues to receive information about this history from subsequent observations, reducing the value of having h_2^t explicitly revealed. As time passes player 1 will learn whether h_2^t actually occurred from his own observations, again reducing the value of independently knowing h_2^t.

Denote by $\beta(\mathscr{A}, \mathscr{B})$ the smallest σ-algebra containing the σ-algebras $\mathscr{A}$ and $\mathscr{B}$. Thus, $\beta(\mathscr{G}_1^s, \mathscr{G}_2^t)$ is the σ-algebra describing what player 1's information at time s would be if he were to learn the private history of player 2 at time t.

Lemma 15.6.4 *Suppose the monitoring distribution π satisfies assumptions 15.5.1 and 15.5.2, and the action spaces are finite. For any $t > 0$ and $\tau \geq 0$,*

$$\lim_{s \to \infty} \| \tilde{E}[\sigma_2^{s+\tau} | \beta(\mathscr{G}_1^s, \mathscr{G}_2^t)] - \tilde{E}[\sigma_2^{s+\tau} | \mathscr{G}_1^s] \| = 0, \qquad \tilde{\mathbf{P}}\text{-a.s.}$$

Proof *Step 1.* We first prove the result for $\tau = 0$. Suppose $K \subset \mathscr{H}_2^t$ is a set of t-period player 2 histories. We also denote by K the corresponding event (i.e., subset of Ω). By Bayes' rule and the finiteness of the action and signal spaces, we can write the conditional probability of the event K given the observation by player 1 of $h_1^{s+1} = (h_1^s, a_1^s z_1^s)$ as

$$\tilde{\mathbf{P}}[K | h_1^{s+1}] = \tilde{\mathbf{P}}[K | h_1^s, a_1^s z_1^s]$$
$$= \frac{\tilde{\mathbf{P}}[K | h_1^s] \tilde{\mathbf{P}}[a_1^s z_1^s | K, h_1^s]}{\tilde{\mathbf{P}}[a_1^s z_1^s | h_1^s]}$$
$$= \frac{\tilde{\mathbf{P}}[K | h_1^s] \sum_{a_2} \pi_1(z_1^s | a_1^s a_2) \tilde{E}[\sigma_2^{a_2}(h_2^s) | K, h_1^s]}{\sum_{a_2} \pi_1(z_1^s | a_1^s a_2) \tilde{E}[\sigma_2^{a_2}(h_2^s) | h_1^s]},$$

where $\sigma_2^{a_2}(h_2^s)$ is the probability assigned to the action a_2 by σ_2 after the history h_2^s and the last equality uses $\tilde{\mathbf{P}}[a_1^s | K, h_1^s] = \tilde{\mathbf{P}}[a_1^s | h_1^s]$.

Subtract $\tilde{\mathbf{P}}[K | h_1^s]$ from both sides to obtain

$$\tilde{\mathbf{P}}[K | h_1^{s+1}] - \tilde{\mathbf{P}}[K | h_1^s]$$
$$= \frac{\tilde{\mathbf{P}}[K | h_1^s] \sum_{a_2} \pi_1(z_1^s | a_1^s a_2)(\tilde{E}[\sigma_2^{a_2}(h_2^s) | K, h_1^s] - \tilde{E}[\sigma_2^{a_2}(h_2^s) | h_1^s])}{\sum_{a_2} \pi_1(z_1^s | a_1^s a_2) \tilde{E}[\sigma_2^{a_2}(h_2^s) | h_1^s]}.$$

The term $\sum_{a_2} \pi_1(z_1^s | a_1^s a_2) \tilde{E}[\sigma_2^{a_2}(h_2^s) | h_1^s]$ is player 1's conditional probability of observing the period s signal z_1^s given he takes action a_1^s and hence is strictly positive and less than 1. Thus,

15.6 ■ Proof

$$\left|\tilde{\mathbf{P}}[K|h_1^{s+1}] - \tilde{\mathbf{P}}[K|h_1^s]\right|$$
$$\geq \tilde{\mathbf{P}}[K|h_1^s] \left|\sum_{a_2} \pi_1(z_1^s \mid a_1^s a_2)(\tilde{E}[\sigma_2^{a_2}(h_2^s)|K, h_1^s] - \tilde{E}[\sigma_2^{a_2}(h_2^s)|h_1^s])\right|.$$

Because the sequence of random variables $\{\tilde{\mathbf{P}}[K|\mathscr{G}_1^s]\}_s$ is a $\tilde{\mathbf{P}}$-martingale adapted to $\{\mathscr{G}_1^s\}_s$, it converges $\tilde{\mathbf{P}}$-a.s. to a nonnegative limit $\tilde{\mathbf{P}}[K|\mathscr{G}_1^\infty]$ as $s \to \infty$. Consequently, the left side of this inequality converges $\tilde{\mathbf{P}}$-a.s. to 0. The signals generated by player 2's actions satisfy assumption 15.5.2, so an argument identical to that of step 2 of the proof of lemma 15.6.1 establishes that $\tilde{\mathbf{P}}$-almost everywhere on K,

$$\lim_{s \to \infty} \tilde{\mathbf{P}}[K|\mathscr{G}_1^s] \left\|\tilde{E}[\sigma_2^s \mid \beta(\mathscr{G}_1^s, K)] - \tilde{E}[\sigma_2^s \mid \mathscr{G}_1^s]\right\| = 0,$$

where $\beta(\mathscr{A}, B)$ is the smallest σ-algebra containing both the σ-algebra $\mathscr{A}$ and the event B. Moreover, $\tilde{\mathbf{P}}[K|\mathscr{G}_1^\infty](\omega) > 0$ for $\tilde{\mathbf{P}}$-almost all $\omega \in K$. Thus, $\tilde{\mathbf{P}}$-almost everywhere on K,

$$\lim_{s \to \infty} \left\|\tilde{E}[\sigma_2^s|\beta(\mathscr{G}_1^s, K)] - \tilde{E}[\sigma_2^s|\mathscr{G}_1^s]\right\| = 0.$$

Because this holds for all $K \in \mathscr{G}_2^t$,

$$\lim_{s \to \infty} \left\|\tilde{E}[\sigma_2^s|\beta(\mathscr{G}_1^s, \mathscr{G}_2^t)] - \tilde{E}[\sigma_2^s|\mathscr{G}_1^s]\right\| = 0, \qquad \tilde{\mathbf{P}}\text{-a.s.},$$

giving the result for $\tau = 0$.

Step 2. The proof for $\tau \geq 1$ follows by induction. We have

$$\Pr[K \mid h_1^{s+\tau+1}] = \Pr[K \mid h_1^s, a_1^s z_1^s, \ldots, a_1^{s+\tau} z_1^{s+\tau}]$$
$$= \frac{\Pr[K \mid h_1^s]\Pr[a_1^s z_1^s, \ldots, a_1^{s+\tau} z_1^{s+\tau} \mid K, h_1^s]}{\Pr[a_1^s z_1^s, \ldots, a_1^{s+\tau} z_1^{s+\tau} \mid h_1^s]}$$
$$= \frac{\Pr[K \mid h_1^s] \prod_{\ell=s}^{s+\tau} \sum_{a_2} \pi_1(z_1^\ell \mid a_1^\ell a_2) \tilde{E}[\sigma_2^{a_2}(h_2^\ell) \mid K, h_1^s]}{\prod_{\ell=s}^{s+\tau} \sum_{a_2} \pi_1(z_1^\ell \mid a_1^\ell a_2) \tilde{E}[\sigma_2^{a_2}(h_2^\ell) \mid h_1^s]},$$

where $h_1^{\ell+1} = (h_1^\ell, a_1^\ell z_1^\ell)$. Therefore,

$$\left|\Pr[K \mid h_1^{s+\tau+1}] - \Pr[K \mid h_1^s]\right|$$
$$\geq \Pr[K \mid h_1^s] \left|\prod_{\ell=s}^{s+\tau} \sum_{a_2} \pi_1(z_1^\ell \mid a_1^\ell a_2) \tilde{E}[\sigma_2^{a_2}(h_2^\ell) \mid K, h_1^s]\right.$$
$$\left. - \prod_{\ell=s}^{s+\tau} \sum_{a_2} \pi_1(z_1^\ell \mid a_1^\ell a_2) \tilde{E}[\sigma_2^{a_2}(h_2^\ell) \mid h_1^s]\right|.$$

The left side of this inequality converges to 0 $\tilde{\mathbf{P}}$-a.s., and hence so does the right side. Moreover, applying the triangle inequality and rearranging, we find that the right side is larger than

$$\Pr[K \mid h_1^s] \left| \prod_{\ell=s}^{s+\tau-1} \sum_{a_2} \pi_1(z_1^\ell \mid a_1^\ell a_2) \tilde{E}[\sigma_2^{a_2}(h_2^\ell) \mid h_1^s] \right|$$

$$\times \left| \sum_{a_2} \pi_1(z_1^{s+\tau} \mid a_1^{s+\tau} a_2) \tilde{E}[\sigma_2^{a_2}(h_2^{s+\tau}) \mid K, h_1^s] \right.$$

$$\left. - \sum_{a_2} \pi_1(z_1^{s+\tau} \mid a_1^{s+\tau} a_2) \tilde{E}[\sigma_2^{a_2}(h_2^{s+\tau}) \mid h_1^s] \right|$$

$$- \Pr[K \mid h_1^s] \left| \prod_{\ell=s}^{s+\tau-1} \sum_{a_2} \pi_1(z_1^\ell \mid a_1^\ell a_2) \tilde{E}[\sigma_2^{a_2}(h_2^\ell) \mid K, h_1^s] \right.$$

$$\left. - \prod_{\ell=s}^{s+\tau-1} \sum_{a_2} \pi_1(z_1^\ell \mid a_1^\ell a_2) \tilde{E}[\sigma_2^{a_2}(h_2^\ell) \mid h_1^s] \right|$$

$$\times \left| \sum_{a_2} \pi_1(z_1^{s+\tau} \mid a_1^{s+\tau} a_2) \tilde{E}[\sigma_2^{a_2}(h_2^{s+\tau}) \mid K, h_1^s] \right|.$$

From the induction hypothesis that $\|\tilde{E}[\sigma_2^\ell \mid \beta(\mathcal{G}_1^s, \mathcal{G}_2^t)] - \tilde{E}[\sigma_2^\ell \mid \mathcal{G}_1^s]\|$ converges to 0 $\tilde{\mathbf{P}}$-a.s. for every $\ell \in \{s, \ldots, s + \tau - 1\}$, the negative term also converges to 0 $\tilde{\mathbf{P}}$-a.s. But then the first term also converges to 0, and, as before, the result holds for $\ell = s + \tau$.

∎

Now we apply lemma 15.6.4 to a particular piece of information player 2 could have at time t. By lemma 15.6.3, with positive probability, we reach a time t at which player 2 assigns high probability to the event that all her future behavior is a best reply to the commitment type. Intuitively, lemma 15.6.4 implies that player 1 will eventually have expectations about player 2 that match those he would have if he knew player 2 expected (in time t) to always play a best response to the commitment type.

This step is motivated by the observation that if player 1 eventually expects player 2 to always play a best response to the commitment type, then the normal type of player 1 will choose to deviate from the behavior of the commitment type (which is not a best response to player 2's best response to the commitment type). At this point, we appear to have a contradiction between player 2's belief on the event F (from lemma 15.6.3) that the normal and commitment types are playing identically and player 1's behavior on the event $F^\dagger$ (the event where player 1 expects player 2 to always play a best response to the commitment type, identified in the next lemma). This contradiction would be immediate if $F^\dagger$ was both a subset of F and measurable for player 2. Unfortunately we have no reason to expect either. However, the next lemma shows that $F^\dagger$ is in fact close to a $\mathcal{G}_2^s$-measurable set on which player 2's beliefs that player 1 is the commitment type do not converge to 0. In this case we will (eventually) have a contradiction. On all such histories, the normal and commitment types are playing identically. However, nearly everywhere on a relatively large subset of these states, the normal player 1 is deviating from the commitment strategy in an identifiable way.

15.6 ■ Proof

The proof of the next lemma is somewhat technical, and some readers may accordingly prefer to skip to the completion of the proof of proposition 15.5.1 in section 15.6.4.

Lemma 15.6.5 *Suppose the monitoring distribution π satisfies assumptions 15.4.1, 15.5.1, and 15.5.2, and action spaces are finite. Suppose there is a Nash equilibrium in which player 2 does not necessarily learn player 1's type, that is, $\tilde{\mathbf{P}}(\{\hat{\mu}^t \nrightarrow 0\}) > 0$. Let $\eta > 0$ be the constant and F the positive probability event identified in lemma 15.6.3. For any $\nu > 0$ and number of periods $\tau > 0$, there exists an event $F^\dagger$ and a time $T(\nu, \tau)$ such that for all $s > T(\nu, \tau)$ there exists $C^{\dagger s} \in \mathscr{G}_2^s$ with:*

$$\hat{\mu}^s > \eta \text{ on } C^{\dagger s}, \tag{15.6.7}$$

$$F^\dagger \cup F \subset C^{\dagger s}, \tag{15.6.8}$$

$$\tilde{P}(F^\dagger) > \tilde{P}(C^{\dagger s}) - \nu \tilde{P}(F), \tag{15.6.9}$$

and for any $s' \in \{s, s+1, \ldots, s+\tau\}$, on $F^\dagger$,

$$\tilde{E}[\sigma_2^{s', \hat{a}_2} \mid \mathscr{G}_1^s] > 1 - \nu, \quad \tilde{\mathbf{P}}\text{-a.s.,} \tag{15.6.10}$$

where $\sigma_2^{s', \hat{a}_2}$ is the probability player 2 assigns to the action $\hat{a}_2$ in period s'.

Proof Fix $\nu \in (0, 1)$ and a number of periods $\tau > 0$. Fix $\theta < (\frac{1}{4}\nu\tilde{P}(F))^2$, and let T denote the critical period identified in lemma 15.6.3 for this value of θ.

Player 1's minimum estimated probability on $\hat{a}_2$ over periods $s, \ldots, s+\tau$ can be written as $f^s \equiv \min_{s \leq s' \leq s+\tau} \tilde{E}[\sigma_2^{s', \hat{a}_2} \mid \mathscr{G}_1^s]$. Notice that $f^s > 1 - \nu$ is a sufficient condition for inequality (15.6.10).

Step 1. We first find a lower bound for f^s. For any $t \leq s$, the triangle inequality implies

$$1 \geq f^s \geq \min_{s \leq s' \leq s+\tau} \tilde{E}[\sigma_2^{s', \hat{a}_2} \mid \beta(\mathscr{G}_1^s, \mathscr{G}_2^t)] - k^{t,s},$$

where $k^{t,s} \equiv \max_{s \leq s' \leq s+\tau} |\tilde{E}[\sigma_2^{s', \hat{a}_2} \mid \beta(\mathscr{G}_1^s, \mathscr{G}_2^t)] - \tilde{E}[\sigma_2^{s', \hat{a}_2} \mid \mathscr{G}_1^s]|$ for $t \leq s$. By lemma 15.6.4, $\lim_{s \to \infty} k^{t,s} = 0$ $\tilde{\mathbf{P}}$-a.s.

As $\sigma_2^{s', \hat{a}_2} \leq 1$ and is equal to 1 on G^t, the above implies

$$f^s \geq \tilde{P}(G^t \mid \beta(\mathscr{G}_1^s, \mathscr{G}_2^t)) - k^{t,s}.$$

Moreover, the sequence of random variables $\{\tilde{P}(G^t | \beta(\mathscr{G}_1^s, \mathscr{G}_2^t))\}_s$ is a martingale with respect to the filtration $\{\mathscr{G}_1^s\}_s$, and so converges a.s. to a limit, $g^t \equiv \tilde{P}(G^t | \beta(\mathscr{G}_1^\infty, \mathscr{G}_2^t))$. Hence

$$1 \geq f^s \geq g^t - k^{t,s} - \ell^{t,s}, \tag{15.6.11}$$

where $\ell^{t,s} \equiv |g^t - \tilde{P}(G^t | \beta(\mathscr{G}_1^s, \mathscr{G}_2^t))|$ and $\lim_{s \to \infty} \ell^{t,s} = 0$ $\tilde{\mathbf{P}}$-a.s.

Step 2. We now determine the sets $C^{\dagger s}$ and a set that we will use to later determine $F^\dagger$. For any $t \geq T$, define
$$K^t \equiv \{\omega : \tilde{\mathbf{P}}(G^t \mid \mathscr{G}_2^t) > 1 - \theta, \; \hat{\mu}^t > \eta\} \in \mathscr{G}_2^t.$$

Let $F^{t,s}$ denote the event $\cap_{\tau=t}^s K^\tau$ and set $F^t \equiv \cap_{\tau=t}^\infty K^\tau$. Note that $\liminf K^t \equiv \cup_{t=T}^\infty \cap_{\tau=t}^\infty K^\tau = \cup_{t=T}^\infty F^t$. By lemma 15.6.3, $F \subset K^t$ for all $t \geq T$, so $F \subset F^{t,s}$, $F \subset F^t$, and $F \subset \liminf K^t$.

Define $N^t \equiv \{\omega : g^t \geq 1 - \sqrt{\theta}\}$. Set $C^{\dagger s} \equiv F^{T,s} \in \mathscr{G}_2^s$ and define an intermediate set F^* by $F^* \equiv F^T \cap N^T$. Because $C^{\dagger s} \subset K^s$, (15.6.7) holds. In addition, $F^* \cup F \subset C^{\dagger s}$, and hence (15.6.8) holds with F^* in the role of $F^\dagger$. By definition,
$$\tilde{\mathbf{P}}(C^{\dagger s}) - \tilde{\mathbf{P}}(F^*) = \tilde{\mathbf{P}}(C^{\dagger s} \cap (F^T \cap N^T)^c) = \tilde{\mathbf{P}}((C^{\dagger s} \cap (F^T)^c) \cup (C^{\dagger s} \cap (N^T)^c)),$$
where we use S^c to denote the complement of a set S. By our choice of $C^{\dagger s}$, the event $C^{\dagger s} \cap (N^T)^c$ is a subset of the event $K^T \cap (N^T)^c$. Thus, we have the bound
$$\tilde{\mathbf{P}}(C^{\dagger s}) - \tilde{\mathbf{P}}(F^*) \leq \tilde{\mathbf{P}}(C^{\dagger s} \cap (F^T)^c) + \tilde{\mathbf{P}}(K^T \cap (N^T)^c). \quad (15.6.12)$$

We now find upper bounds for the two terms on the right side of (15.6.12). First notice that $\tilde{\mathbf{P}}(C^{\dagger s} \cap (F^T)^c) = \tilde{\mathbf{P}}(F^{T,s}) - \tilde{\mathbf{P}}(F^T)$. Because $\lim_{s \to \infty} \tilde{\mathbf{P}}(F^{T,s}) = \tilde{\mathbf{P}}(F^T)$, there exists $T' \geq T$ such that
$$\tilde{\mathbf{P}}(C^{\dagger s} \cap (F^T)^c) < \sqrt{\theta} \quad \text{for all } s \geq T'. \quad (15.6.13)$$

Also, as $\tilde{\mathbf{P}}(G^t \mid K^t) > 1 - \theta$ and $K^t \in \mathscr{G}_2^t$, the properties of iterated expectations imply that $1 - \theta < \tilde{\mathbf{P}}(G^t \mid K^t) = \tilde{E}[g^t \mid K^t]$. Because $g^t \leq 1$, we have
$$1 - \theta < \tilde{E}[g^t \mid K^t] \leq (1 - \sqrt{\theta})\tilde{\mathbf{P}}((N^t)^c \mid K^t) + \tilde{\mathbf{P}}(N^t \mid K^t)$$
$$= 1 - \sqrt{\theta}\tilde{\mathbf{P}}((N^t)^c \mid K^t).$$

The extremes of the inequality imply that $\tilde{\mathbf{P}}((N^t)^c \mid K^t) < \sqrt{\theta}$. Thus, taking $t = T$ we get
$$\tilde{\mathbf{P}}(K^T \cap (N^T)^c) < \sqrt{\theta}. \quad (15.6.14)$$

Using (15.6.13) and (15.6.14) in (15.6.12), $\tilde{\mathbf{P}}(C^{\dagger s}) - \tilde{\mathbf{P}}(F^*) < 2\sqrt{\theta}$ for all $s \geq T'$. Given $F \subset C^{\dagger s}$, the bound on θ, and $\nu < 1$, it follows that
$$\tilde{\mathbf{P}}(F^*) > \tilde{\mathbf{P}}(F) - 2\sqrt{\theta} > \tfrac{1}{2}\tilde{\mathbf{P}}(F) > 0.$$

Step 3. Finally, we combine the first two steps to obtain $F^\dagger$. As $\tilde{\mathbf{P}}(F^*) > 0$ and $k^{T,s} + \ell^{T,s}$ converges a.s. to 0, by Egorov's theorem, there exists $F^\dagger \subset F^*$ such that $\tilde{\mathbf{P}}(F^* \setminus F^\dagger) < \sqrt{\theta}$ and a time $T'' > T$ such that $k^{T,s} + \ell^{T,s} < \sqrt{\theta}$ on $F^\dagger$ for all $s \geq T''$. Because $F^\dagger \cup F \subset F^* \cup F \subset C^{\dagger s}$, (15.6.8) holds. Let $T(\nu, \tau) \equiv \max\{T'', T'\}$. Also, $g^T \geq 1 - \sqrt{\theta}$ on $F^\dagger$, because $F^\dagger \subset N^T$. Hence on $F^\dagger$, by (15.6.11), $f^s > 1 - 2\sqrt{\theta}$ for all $s > T(\nu, \tau)$. This and the bound on θ imply (15.6.10). Moreover, as $\tilde{\mathbf{P}}(F^* \setminus F^\dagger) < \sqrt{\theta}$ and $\tilde{\mathbf{P}}(C^{\dagger s}) - \tilde{\mathbf{P}}(F^*) < 2\sqrt{\theta}$, (15.6.9) holds for all $s > T(\nu, \tau)$. ∎

15.6.4 Proof of Proposition 15.5.1

We have shown that when player 2 does not necessarily learn player 1's type, there exists a set $F^\dagger$ on which (15.6.10) holds and $F^\dagger \subset C^{\dagger s} \in \mathscr{G}_2^s$. In broad brushstrokes, the remaining argument proving proposition 15.5.1 is as follows. First, we conclude that on $F^\dagger$, the normal type will not be playing the commitment strategy. To be precise, on $F^\dagger$ there will exist a stage-game action played by $\hat{\alpha}_1$ but not by the normal type. This will bias player 2's expectation of the normal type's actions away from the commitment strategy on $C^{\dagger s}$, because there is little probability weight on $C^{\dagger s} \setminus F^\dagger$. We then get a contradiction because the fact that $\hat{\mu}^s > \eta$ on $C^{\dagger s}$ implies player 2 must believe the commitment type's strategy and the normal type's average strategy are the same on $C^{\dagger s}$.

Suppose, en route to the contradiction, that there is a Nash equilibrium in which player 2 does not necessarily learn player 1's type. Then $\tilde{\mathbf{P}}(\hat{\mu}^t \nrightarrow 0) > 0$. Let $\hat{\alpha}_1 \equiv \min_{a_1 \in A_1}\{\hat{\alpha}_1(a_1) : \hat{\alpha}_1(a_1) > 0\}$, that is, $\hat{\alpha}_1$ is the smallest nonzero probability attached to an action under the commitment strategy $\hat{\alpha}_1$.

Because $(\hat{\alpha}_1, \hat{\alpha}_2)$ is not a Nash equilibrium, there exists $\gamma > 0$, $a_1' \in A_1$ with $\hat{\alpha}_1(a_1') > 0$ and $\bar{\nu} > 0$ such that

$$\gamma < \min_{\|\alpha_2 - \hat{\alpha}_2\| \leq \bar{\nu}} \left(\max_{a_1 \in A_1} u_1(a_1, \alpha_2) - u_1(a_1', \alpha_2) \right).$$

Finally, for a given discount factor $\delta_1 < 1$ there exists a τ sufficiently large such that the loss of γ for one period is larger than any feasible potential gain deferred by τ periods: $(1 - \delta_1)\gamma > \delta_1^\tau (M - m)$.

Fix the event F from lemma 15.6.3. For $\nu < \min\{\bar{\nu}, \hat{\alpha}_1/2\}$ and τ fixed as above, let $F^\dagger$ and, for $s > T(\nu, \tau)$, $C^{\dagger s}$ be the events described in lemma 15.6.5. Now consider the normal type of player 1 in period $s > T(\nu, \tau)$ at some state in $F^\dagger$. By (15.6.10), he expects player 2 to play within $\nu < \bar{\nu}$ of $\hat{\alpha}_2$ for the next τ periods. Playing the action a_1' is conditionally dominated in period s, because the most he can get from playing a_1' in period s is worse than playing a best response to $\hat{\alpha}_2$ for τ periods and then being minmaxed. Thus, on $F^\dagger$ the normal type plays action a_1' with probability zero: $\sigma_1^{s,a_1'} = 0$.

Now we calculate a lower bound on the difference between player 2's beliefs about the normal type's probability of playing action a_1' in period s, $\tilde{E}[\sigma_1^{s,a_1'} \mid \mathscr{G}_2^s]$, and the probability the commitment type plays action a_1' on the set of states $C^{\dagger s}$:

$$\tilde{E}\big[|\hat{\alpha}_1(a_1') - \tilde{E}[\sigma_1^{s,a_1'} \mid \mathscr{G}_2^s]|\chi_{C^{\dagger s}}\big] \geq \tilde{E}\big[(\hat{\alpha}_1(a_1') - \tilde{E}[\sigma_1^{s,a_1'} \mid \mathscr{G}_2^s])\chi_{C^{\dagger s}}\big]$$
$$\geq \hat{\alpha}_1 \tilde{\mathbf{P}}(C^{\dagger s}) - \tilde{E}[\sigma_1^{s,a_1'} \chi_{C^{\dagger s}}]$$
$$\geq \hat{\alpha}_1 \tilde{\mathbf{P}}(C^{\dagger s}) - (\tilde{\mathbf{P}}(C^{\dagger s}) - \tilde{\mathbf{P}}(F^\dagger))$$
$$\geq \hat{\alpha}_1 \tilde{\mathbf{P}}(C^{\dagger s}) - \nu \tilde{\mathbf{P}}(F)$$
$$\geq \tfrac{1}{2}\hat{\alpha}_1 \tilde{\mathbf{P}}(F). \qquad (15.6.15)$$

The first inequality follows from removing the absolute values. The second inequality applies $\hat{\alpha}_1(a_1') \geq \hat{\alpha}_1$, uses the $\mathscr{G}_2^s$-measurability of $C^{\dagger s}$ and applies the properties of conditional expectations. The third applies the fact that $\sigma_1^{s,a_1'} = 0$ on $F^\dagger$ and $\sigma_1^{s,a_1'} \leq 1$.

The fourth inequality applies (15.6.9) in lemma 15.6.5. The fifth inequality follows from $\nu < \hat{\underline{\alpha}}_1/2$ and $F \subset C^{\dagger s}$ (by (15.6.8)).

From corollary 15.6.1, $\hat{\mu}^s \|\hat{\alpha}_1 - \tilde{E}(\tilde{\sigma}_1^s \mid \mathscr{G}_2^s)\| \to 0$ $\tilde{\mathbf{P}}$-a.s. It follows that

$$\hat{\mu}^s |\hat{\alpha}_1(a_1') - \tilde{E}(\tilde{\sigma}_1^{s,a_1'} \mid \mathscr{G}_2^s)|\chi_{C^{\dagger s}} \to 0, \qquad \tilde{\mathbf{P}}\text{-a.s.}$$

But by lemma 15.6.5, $\hat{\mu}^s > \eta$ on the set $C^{\dagger s}$, and so

$$|\hat{\alpha}_1(a_1') - \tilde{E}(\tilde{\sigma}_1^{s,a_1'} \mid \mathscr{G}_2^s)|\chi_{C^{\dagger s}} \to 0, \qquad \tilde{\mathbf{P}}\text{-a.s.}$$

This concludes the proof of proposition 15.5.1, because we now have a contradiction with $\tilde{\mathbf{P}}(F) > 0$ (from lemma 15.6.3) and (15.6.15), which holds for all $s > T(\nu, \tau)$.

16 Reputations with Long-Lived Players

In this chapter, we show that the introduction of nontrivial intertemporal incentives for the uninformed player qualifies and complicates the analysis of chapter 15. For example, when we consider only simple Stackelberg types, the Stackelberg payoff may not bound equilibrium payoffs. The situation is further complicated by the possibility of nonsimple commitment types (i.e., types that follow nonstationary strategies).

16.1 The Basic Issue

Consider applying the logic from chapter 15 to obtain a Stackelberg reputation bound when both players are long-lived and player 1's characteristics are unknown under perfect monitoring. The first step is to demonstrate that if the normal player 1 persistently plays the Stackelberg action and there exists a type committed to that action, then player 2 must eventually attach high probability to the event that the Stackelberg action is played in the future. This argument is simply lemma 15.3.1, which depends only on the properties of Bayesian belief revision, independently of whether the person holding the beliefs is a long-lived or short-lived player.

When player 2 is short-lived, the next step is to note that if she expects the Stackelberg action, then she will play a best response to this action. If player 2 is instead a long-lived player, she may have an incentive to play something other than a best response to the Stackelberg type.

The key step when working with two long-lived players is thus to establish conditions under which as player 2 becomes increasing convinced that the Stackelberg action will appear, player 2 must eventually play a best response to that action. One might begin such an argument by observing that as long as player 2 discounts, any losses from not playing a current best response must be recouped within a finite length of time. But if player 2 is "very" convinced that the Stackelberg action will be played not only now but for sufficiently many periods to come, there will be no opportunity to accumulate subsequent gains, and hence she might just as well play a stage-game best response.

Once we have player 2 best responding to the Stackelberg action, the remainder of the argument proceeds as in the case of a short-lived player 2. The normal player 1 must eventually receive very nearly the Stackelberg payoff in each period of the repeated game. By making player 1 sufficiently patient (*relative* to player 2, so we are in

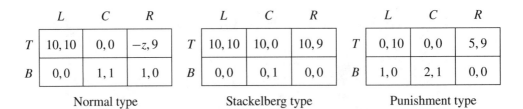

Figure 16.1.1 Payoffs for three types of player 1, where $z \in (0, 8)$, with player 2's payoffs also shown in each case. The Stackelberg type has a dominant repeated game strategy of always playing T and hence is equivalent to a commitment type.

an environment of differing discount factors), we can ensure that this consideration dominates player 1's payoffs, putting a lower bound on the latter. Hence, the obvious handling of discount factors is to fix player 2's discount factor δ_2, and to consider the limit as player 1 becomes patient, that is, δ_1 approaching 1. (Section 16.6 examines the case of equal discounting and discusses the extent to which some asymmetry is important in establishing reputation results.)

This intuition misses the following possibility. Player 2 may be choosing something other than a best response to the Stackelberg action out of fear that a current best response may trigger a disastrous future punishment. This punishment would not appear if player 2 faced the Stackelberg type, but player 2 can be made confident only that she faces the Stackelberg *action*, not the Stackelberg type. The fact that the punishment lies off the equilibrium path makes it difficult to assuage player 2's fear of such punishments. Short-lived players in the same situation are similarly uncertain about the future ramifications of best responding, but being short-lived, this uncertainty does not affect their behavior.

Example 16.1.1 **Failure of reputation effects** The game is shown in figure 16.1.1. The normal type of player 1 is joined by two other types, a Stackelberg type and a "punishment" type. We represent the punishment type here as a payoff type to make it clear that there is nothing perverse about this type's behavior. Having done this, we also show the Stackelberg type as a payoff type. The prior distribution puts probability 0.8 on the normal type and probability 0.1 on each of the others.

The repeated game has perfect monitoring. We are interested in the strategy profile described as follows:

Normal player 1: Play T after any history except one in which player 1 has played B in the past, in which case play B.
Stackelberg player 1: Play T after every history.
Punishment type of player 1: Play T initially and after any equilibrium history, otherwise play B.
Player 2, in equilibrium: Alternate between one period of L and one of R. After any history in which player 1 has ever played B, attach probability one to the punishment type and play C.

16.1 ■ The Basic Issue

Player 2, out of equilibrium: After any history featuring a first deviation by
player 2, play L. If 1 plays T next period, attach probability zero to the punishment type, make no more belief revisions, and play L forever. If 1 plays B next period, attach probability one to the punishment type, make no more belief revisions, and play C forever. Otherwise, make no belief revisions.

The equilibrium path alternates between TL and TR.

To show that this is an equilibrium, we first argue that player 1's behavior is optimal. Incentives for the Stackelberg type are trivial, because it is a dominant strategy in the repeated game to play T in every period. Along the equilibrium path, the normal and punishment types of player 1 earn payoffs that approach (as $\delta_1 \to 1$) $(10-z)/2$ and $5/2$ respectively, and deviations lead to continuation payoffs of 1 and 2, respectively. Deviations from the equilibrium path are thus not optimal for player 1, if sufficiently patient.

Should play ever leave the equilibrium path as a result of player 1's having chosen B, subsequent play constitutes a stage-game Nash equilibrium for player 2 and the normal and punishment types of player 1. Should play leave the equilibrium path because player 2 has deviated, then the punishment player 1 has the choice of playing T in the next period, which earns a subsequent payoff of 1 (from playing B against a player 2 who stubbornly persists in paying L, on the strength of the belief that player 1 is certainly normal or Stackelberg). Alternatively, the punishment player 1 can play B, leading to the subsequent play of BC and a payoff of 2. The punishment type is thus playing optimally off the equilibrium path. The normal player 1 earns a payoff of 10 after a deviation by player 2 and thus is playing optimally off the equilibrium path.

Along the equilibrium path, player 2 learns nothing about player 1. If player 2 deviates from the equilibrium, she has a chance to screen the types of player 1, earning a continuation payoff of 10 against the normal or Stackelberg type and a continuation payoff of at most 1 against the punishment type. The resulting expected payoff is at most 9.1, falling short of the equilibrium payoff of almost 9.5 (for a patient player 2). This completes the argument that these strategies are an equilibrium under the conditions on discount factors that appear in the reputation result, namely, that we fix δ_2 (allowed to be sufficiently large) and then let δ_1 approach 1.

By increasing the absolute value of z in figure 16.1.1, though perhaps requiring more patience for player 1, we obtain an equilibrium in which the normal player 1's payoff is arbitrarily close to his pure minmax payoff of 1. It is thus apparent that reputation considerations can be quite ineffective.

In equilibrium, player 2 *is* convinced that she will face the Stackelberg action in every period. However, she does not play a best response out of fear that doing so has adverse future consequences, a fear made real by the punishment type. Celentani, Fudenberg, Levine, and Pesendorfer (1996, section 5) describe an example with similar features but involving only a normal and Stackelberg type of player 1, using the future play of the normal player 1 to punish player 2 for choosing a best response to the Stackelberg action when she is not supposed to. Section 16.6.1, building on Cripps and Thomas (1997), also presents a game with common interests and with only normal and Stackelberg types for player 1, in which equilibria exist with the normal player-1's payoffs arbitrarily close to 1's minmax level.[1]

This chapter describes several positive reputation results for two long-lived players. As one would expect from example 16.1.1, these results are built around sufficient conditions for the uninformed player to best respond to the Stackelberg action.

In section 16.2, there is a simple commitment type who minmaxes player 2. Because no punishment can ever push player 2 below her minmax payoff, when she believes she almost certainly faces the commitment action, she plays a best response.

This allows us to establish a strong lower bound on player 1's payoff if there is an action that both minmaxes player 2 *and* provides a high payoff to player 1 (when 2 plays a best response). In games of *conflicting interests*, in which player 1's pure Stackelberg action minmaxes player 2, this brings player 1 arbitrarily close to his Stackelberg payoff.

Section 16.3 pursues this logic to show that for any commitment action, player 2 must at least eventually play a response that gives her at least her minmax payoff (though this need not be a stage-game best response for player 2). Hence, a lower bound on player 1's payoff is given by maximizing over simple commitment types the minimum payoff 1 receives when 2's behavior is limited to responses that provide at least her minmax payoff.

Section 16.4 pursues a different tack. The difficulty described in example 16.1.1 is that player 1 can bring player 2 to expect the behavior of the commitment type on the equilibrium path but cannot control 2's beliefs about behavior following nonequilibrium histories. In a game with full-support public monitoring, there are no nonequilibrium (public) histories. Here, with a sufficiently rich set of commitment types, player 1 can be assured of at least his Stackelberg payoff. Indeed, player 1 can often be assured of an even higher payoff in the presence of commitment types who play nonstationary strategies.

Section 16.5 explores another possibility for capitalizing on more complicated commitment types. Again, the difficulty in example 16.1.1 is player 2's fear that playing a best response to the commitment type may trigger a punishment. Section 16.5 examines a particular class of commitment types who punish player 2 for *not* playing a best response. Once again, the result is a high lower bound on player 2's payoff, given the appropriate commitment type.

As for short-lived player 2s, these results constitute a marked contrast between reputation effects under perfect and imperfect monitoring. Most notably, the sufficient conditions on commitment types required for an effective player 1 bound are milder under full-support imperfect monitoring.

---

1. The example in section 16.6.1 requires *equal* discount factors, whereas the current example and Celentani, Fudenberg, Levine, and Pesendorfer (1996, section 5) are consistent with fixing player 2's discount factor and allowing $\delta_1 \to 1$. The example in 16.6.1 allows player 1's equilibrium payoff to be pushed arbitrarily close to his minmax payoff without adjusting stage-game payoffs (though requiring more patience and a smaller probability of the commitment type). Celentani, Fudenberg, Levine, and Pesendorfer (1996) achieve payoffs arbitrarily close to minmax payoffs for fixed discount factors and prior distribution.

16.2 Perfect Monitoring and Minmax-Action Reputations

The lesson in these results is that the case of two long-lived players, even with one arbitrarily more patient than the other, is qualitatively different than that of one long-lived and one short-lived player. Player 1 has the option of leading a long-lived player 2 to expect commitment behavior, but this no longer suffices to ensure a best response from player 2, no matter how firm the belief. Bounds on player 1's payoffs then depend on some special structure of the game (such as conflicting interests), imperfect monitoring, or special commitment types. Section 16.6 shows that reputation results are even more elusive when the two long-lived players are comparably patient. Finally, under imperfect monitoring, uncertainty about player 1's type must again be temporary (section 16.7).

16.2 Perfect Monitoring and Minmax-Action Reputations

We consider a perfect monitoring repeated game with two long-lived players, 1 and 2, with finite action sets. We depart from our practice in the remainder of this book and allow the two long-lived players to have different discount factors, with player i's discount factor given by δ_i.

Player 1's type is determined by a probability distribution μ with a finite or countable support Ξ. The characteristics of player 2 are known. The support of μ contains the normal type of player 1, ξ_0, and a collection of commitment types (see section 15.2).

The set of public histories is $\mathscr{H} = \cup_{t=0}^{\infty}(A_1 \times A_2)^t$. A behavior strategy for player 1 is $\sigma_1 : \mathscr{H} \times \Xi \to \Delta(A_1)$, and for player 2 is $\sigma_2 : \mathscr{H} \to \Delta(A_2)$. A Nash equilibrium strategy profile $\sigma = (\sigma_1, \sigma_2)$ and the prior distribution μ induce a measure $\mathbf{P}$ over the set of outcomes $\Omega \equiv \Xi \times (A_1 \times A_2)^{\infty}$.

16.2.1 Minmax-Action Types and Conflicting Interests

Our first positive result follows Schmidt (1993b) in focusing on a commitment type that minmaxes player 2. Recall that $\underline{v}_2$ is player 2's mixed-action minmax utility and that $v_1^*(a_1)$ is the one-shot bound on player 1's payoffs when he commits to the action a_1 (see (15.3.4)). In this section, we show that if there is a pure action a_1' that mixed-action minmaxes player 2 (i.e., $\underline{v}_2 = \max_{a_2} u_2(a_1', a_2)$) and the prior μ attaches positive probability to the simple type $\xi(a_1')$ (who invariably plays the pure action a_1'), then a sufficiently patient normal player 1 earns a payoff arbitrarily close to $v_1^*(a_1')$. If there are multiple pure actions that minmax player 2, then the relevant payoff is the maximum of $v_1^*(a_1')$ over all the minmax actions whose corresponding simple type is assigned positive probability by 2's prior.

Definition 16.2.1 *The stage game has* conflicting interests *if a pure Stackelberg action a_1^* mixed-action minmaxes player 2.*

The highest reputation bound is obtained when the game has conflicting interests, because the reputation bound is then $v_1^*(a_1^*) = v_1^*$, player 1's Stackelberg payoff. Let $\underline{v}_1(\xi_0, \mu, \delta_1, \delta_2)$ be the infimum, over the set of Nash equilibria, of the normal player 1's payoffs.

Proposition *Suppose $\mu(\xi(a_1')) > 0$ for some pure action a_1' that mixed-action minmaxes*
16.2.1 *player 2. There exists a value k, independent of δ_1 (but depending on δ_2), such that*

$$\underline{v}_1(\xi_0, \mu, \delta_1, \delta_2) \geq \delta_1^k v_1^*(a_1') + (1 - \delta_1^k) \min_a u_1(a).$$

If player 1 always plays the action a_1', then only in some k periods can player 2 play anything other than a best response to a_1'. The value of k depends on δ_2, but not δ_1. Hence, by making player 1 quite patient, we can ensure that these k periods have a small effect on player 1's payoffs.

Corollary *Suppose $\mu(\xi(a_1')) > 0$ for some pure action a_1' that mixed-action minmaxes*
16.2.1 *player 2. For any $\varepsilon > 0$, there exists a $\underline{\delta}_1 \in (0, 1)$ such that for all $\delta_1 \in (\underline{\delta}_1, 1)$*

$$\underline{v}_1(\xi_0, \mu, \delta_1, \delta_2) > v_1^*(a_1') - \varepsilon.$$

The proof of the proposition requires an intermediate result that we state separately for later use. Let Ω' be the set of outcomes at which player 1 plays some action a_1' in every period. For any history $h^t \in \mathcal{H}^t$, let $E[U_2(\sigma|_{h^t}) \mid h^t]$ be player 2's expected continuation payoff, conditional on the history h^t.[2] For any history h^t that arises with positive probability given Ω' (i.e., $\mathbf{P}\{\omega \in \Omega' : h^t = h^t(\omega)\} > 0$), let $E[U_2(\sigma|_{h^t}) \mid \Omega']$ be 2's expected continuation payoff, conditional on the history h^t and Ω'.[3]

Lemma *Fix $\delta_2 \in (0, 1)$, $\eta > 0$, and an action $a_1' \in A_1$. There exists L and $\varepsilon \in (0, 1]$ such*
16.2.1 *that for all Nash equilibria σ, pure strategies $\tilde{\sigma}_2$ satisfying $\tilde{\sigma}_2(\bar{h}^t) \in \operatorname{supp} \sigma_2(\bar{h}^t)$ for all $\bar{h}^t \in \mathcal{H}$, and histories $h^t \in \mathcal{H}$ with positive probability under Ω', if*

$$E[U_2((\sigma_1, \tilde{\sigma}_2)|_{h^t}) \mid \Omega'] \leq \underline{v}_2 - \eta, \qquad (16.2.1)$$

then there is a period τ, $t \leq \tau \leq t + L$, such that if player 1 has always played a_1' and player 2 follows $\tilde{\sigma}_2$, then player 2's posterior probability of player 1's action being a_1' in period τ is less than $1 - \varepsilon$.

Intuitively, if player 2's equilibrium strategy gives player 2 a payoff below her minmax payoff, conditional on a_1' always being played, then it must be that player 2 does not expect a_1' to always be played.

Proof Fix a history h^t that occurs with positive probability under Ω' and suppose (16.2.1) holds. We assume that in each of the next $L + 1$ periods, if player 1 has always played a_1' and player 2 follows $\tilde{\sigma}_2$, then player 2's posterior probability of player 1's action being a_1' in that period is at least $1 - \varepsilon$, and derive a contradiction for sufficiently large L and small ε. Given such beliefs, an upper bound on player 2's period t expected continuation payoff is given by (where $u_2^{t+\ell}$ is player 2's payoff under the strategy profile $\tilde{\sigma} \equiv (\sigma_1, \tilde{\sigma}_2)$ in period $t + \ell$ when player 1 plays a_1' in period $t + \ell$),

---

2. The history h^t appears twice in the notation, first in determining the continuation strategy, and second to determine 2's beliefs over the type of player 1.
3. Conditional on Ω', the history is only needed to determine the continuation strategy of player 2.

16.2 ■ Minmax-Action Reputations

$$E[U_2(\tilde{\sigma}|_{h^t}) \mid h^t] \leq (1-\delta_2)[(1-\varepsilon)u_2^t + \varepsilon M]$$
$$+ (1-\delta_2)\delta_2[(1-\varepsilon)^2 u_2^{t+1} + (1-(1-\varepsilon)^2)M]$$
$$+ (1-\delta_2)\delta_2^2[(1-\varepsilon)^3 u_2^{t+2} + (1-(1-\varepsilon)^3)M]$$
$$\ddots$$
$$+ (1-\delta_2)\delta_2^L[(1-\varepsilon)^{L+1} u_2^{t+L} + (1-(1-\varepsilon)^{L+1})M]$$
$$+ \delta_2^{L+1} M$$

$$= (1-\delta_2)[(1-\varepsilon)u_2^t + \varepsilon M]$$
$$+ (1-\delta_2)\delta_2(1-\varepsilon)^2 \sum_{\ell=0}^{L-1} \delta_2^\ell (1-\varepsilon)^\ell u_2^{t+\ell+1}$$
$$+ (1-\delta_2)\delta_2 \left[\frac{1-\delta_2^L}{1-\delta_2} - \frac{(1-\varepsilon)^2(1-\delta_2^L(1-\varepsilon)^L)}{1-\delta_2(1-\varepsilon)} \right] M$$
$$+ \delta_2^{L+1} M.$$

As L gets large and ε approaches 0, this upper bound approaches $(1-\delta_2) \times \sum_{\ell=0}^{\infty} \delta_2^\ell u_2^{t+\ell} = E[U_2(\tilde{\sigma}|_{h^t}) \mid \Omega']$, that is, it approaches player 2's expected continuation payoff conditioning on the event Ω'. Hence, we can find an L sufficiently large and ε sufficiently small that

$$E[U_2(\tilde{\sigma}|_{h^t}) \mid h^t] \leq E[U_2(\tilde{\sigma}|_{h^t}) \mid \Omega'] + \frac{\eta}{2} < \underline{v}_2.$$

But then player 2's continuation value $E[U_2(\tilde{\sigma}|_{h^t}) \mid h^t]$ falls short of her minmax payoff, a contradiction.

∎

If a_1' minmaxes player 2 and she fails to play a best response to a_1' in some period, then her continuation payoff conditional on Ω' must be less than her minmax payoff. The following is then an immediate application of lemma 16.2.1.

Corollary 16.2.2 *Fix $\delta_2 \in (0, 1)$ and $a_1' \in A_1$, where a_1' mixed-action minmaxes player 2. There exists L and $\varepsilon > 0$ such that for all Nash equilibria (σ_1, σ_2), pure strategies $\tilde{\sigma}_2$ satisfying $\tilde{\sigma}_2(\bar{h}^t) \in \operatorname{supp} \sigma_2(\bar{h}^t)$ for all $\bar{h}^t \in \mathcal{H}$, and histories h^t with positive probability under Ω', if*

$$\tilde{\sigma}_2(h^t) \notin B(a_1'),$$

then there is a period τ, $t \leq \tau \leq t + L$, such that if player 1 has always played a_1' and player 2 follows $\tilde{\sigma}_2$, then player 2's posterior probability of player 1's action being a_1' in period τ is less than $1 - \varepsilon$.

Proof of Proposition 16.2.1 Fix a Nash equilibrium and let L and ε be the corresponding values from corollary 16.2.2. Let

$$k = L \frac{\ln \mu(\xi(a_1'))}{\ln(1-\varepsilon)}.$$

Let Ω' be the set of outcomes in which a_1' is always played, and Ω^k the set of outcomes in which player 2 chooses an action $a_2^t \notin B(a_1')$ in more than k periods. From corollary 16.2.2, for any period t in which $a_2^t \notin B(a_1')$, there must be a period $\tau \in \{t, \ldots, t+L\}$ in which $q^\tau < (1-\varepsilon)$ (where q^t is 2's posterior probability that the action a_1' is chosen in period t, i.e., (15.3.1)). Hence, for every L periods in which $a_2^t \notin B(a_1')$, there must be at least one period in which $q^\tau < (1-\varepsilon)$. If player 2 chooses an action $a_2^t \notin B(a_1')$ in more than k periods, then there must be more than $\ln \mu(\xi(a_1'))/\ln(1-\varepsilon)$ periods in which $q^t < (1-\varepsilon)$. From lemma 15.3.1, $\mathbf{P}(\Omega' \cap \Omega^k) = 0$, and we have the claimed upper bound.

∎

One might suspect that a game in which a long-lived player 1 faces a succession of short-lived players is the limiting case of a game with two long-lived players, as player 2 becomes quite impatient. However, as we have seen, the *analysis* for two long-lived players is quite different, irrespective of the degree of impatience of player 2.

It might be advantageous for the normal player 1 to mimic a mixed rather than pure commitment type, giving rise to a game of effectively imperfect monitoring (see section 16.4). Example 16.2.1 explains why types committed to more complicated pure strategies are not helpful in the current context, and sections 16.4 and 16.5 explore the potential advantages of such types.

16.2.2 Examples

Example 16.2.1 The prisoners' dilemma (figure 16.2.1) is a game of conflicting interests. Player 1's Stackelberg action is S. Player 2's unique best response of S yields a payoff of $(0, 0)$, giving player 2 her minmax level. Proposition 16.2.1 establishes conditions under which the normal player 1 must earn nearly his Stackelberg payoff, though this is no improvement on the observation that player 1's payoff must be weakly individually rational.

The normal player 1 could benefit from committing in the prisoners' dilemma to a strategy that is not simple. Suppose, for example, that Ξ includes a type committed to tit-for-tat. A long-lived opponent, once reasonably convinced that tit-for-tat was being played (by either the normal or commitment type of player 1), would respond by exerting effort. Unfortunately, tit-for-tat sacrifices the conflicting interests property, because a best-responding player 2 no longer earns her minmax payoff.

	E	S			h	ℓ			In	Out
E	2, 2	−1, 3		H	2, 3	0, 2		A	2, 2	5, 0
S	3, −1	0, 0		L	3, 0	1, 1		F	−1, −1	5, 0

Figure 16.2.1 Prisoners' dilemma, product-choice game, and a normal form version of the chain store game.

16.2 ■ Minmax-Action Reputations

In the absence of coincidental payoff ties, when the Stackelberg action a_1^* minmaxes player 2, player 2's best response to other pure player 1 actions must give her more than the minmax level. A repeated game strategy that allows player 1 a higher payoff (when player 2 plays a best response) than does the repeated play of a_1^* must then sometimes put player 2 in the position of expecting a continuation payoff that exceeds her minmax payoff. But then we have the prospect of histories in which player 2 is close to certain that she faces committment behavior but earns more than her minmax payoff, disrupting the method of proof used to establish proposition 16.2.1. As a result, an appeal to more general commitment types pushes us outside the techniques used to establish the lower bound in proposition 16.2.1 on the normal player 1's payoff.

●

Example 16.2.2 The product-choice game is not a game of conflicting interests. The Stackelberg action H prompts a best response of h that earns player 2 a payoff of 3, above her minmax payoff of 1. In contrast to conflicting interests games, the normal player 1 and player 2 both fare better when 1 chooses the Stackelberg action (and 2 best responds) than in the stage-game Nash equilibrium. This coincidence of interests precludes using the reasoning behind proposition 16.2.1.

●

Example 16.2.3 The chain store game is a game of conflicting interests, in which (unlike the prisoners' dilemma) the reputation result has some impact. The Stackelberg action prompts a best response of Out, producing the minmax payoff of 0 for player 2. The lower bound on the normal player 1's payoff established in corollary 16.2.1 is then (up to some $\varepsilon > 0$) close to his Stackelberg payoff of 5, the highest player 1 payoff in the game.

●

Example 16.2.4 Consider a variant of the ultimatum game. Two players must allocate a surplus of size one between them. The actions available are $A_1 = A_2 = \{0, 1/n, 2/n, \ldots, n/n\}$, where an action denotes the amount of the surplus allocated to player 2. Player 1 announces an amount from A_1 to be offered to player 2. Player 2 simultaneously announces a demand from A_2 that we interpret as the smallest demand player 2 is willing to accept. If player 1's offer equals or exceeds player 2's demand, 2 receives player 1's offer and 1 retains the remainder. Otherwise, both players receive a payoff of 0.

This game is not a game of conflicting interests. Player 2 has two best responses to the Stackelberg action $a_1^* = 1/n$, which are to announce either 0 or $1/n$, each giving player 2 a payoff that exceeds her minmax payoff of 0.

The bound in proposition 16.2.1 is based on player 1 playing the minmax action $\hat{a}_1^2 = 0$ and player 2 choosing her best response that is least favorable for player 1. One best response for player 2 to an offer of 0, is to demand more than 0, producing a payoff of 0 for player 1. A commitment to offer 0 thus places a rather unhelpful bound of 0 on player 1's payoff.

If we removed the offer (demand) 0 from the game, so that $A_1 = A_2 = \{1/n, 2/n, \ldots, n/n\}$, then the Stackelberg action would again be $1/n$, with player 2's

	L	C	R
T	3, 2	0, 1	0, 1
B	0, −1	2, 0	0, −1

Figure 16.2.2 A game without conflicting interests.

unique best response of $1/n$ now generating her minmax utility. The game thus satisfies conflicting interests, and corollary 16.2.1 indicates that a patient normal player 1 is assured of virtually all the surplus.

Example 16.2.5 Consider the game shown in figure 16.2.2. The Nash equilibria of the stage game are TL, BC, and a mixed equilibrium $\left(\frac{1}{2} \circ T + \frac{1}{2} \circ B, \frac{2}{5} \circ L + \frac{3}{5} \circ C\right)$. Player 1 minmaxes player 2 by playing B, for a minmax value for player 2 of 0. The Stackelberg action for player 1 is T, against which player 2's best response is L, for payoffs of $(3, 2)$. This is accordingly not a game of conflicting interests, and we cannot be sure that it would be helpful for player 1 to commit to T. However, from corollary 16.2.1, if the set of possible player 1 types includes a type committed to B (perhaps as well as a type committed to T), then (up to some $\varepsilon > 0$) the normal player 1 must earn a payoff no less than 2.

16.2.3 Two-Sided Incomplete Information

Standard reputation models allow uncertainty about player 1's type but not about player 2's. When player 2 is a short-lived player, this does not seem a particularly unnatural extension of the asymmetry between players. The asymmetry is somewhat more apparent when player 2 is also a long-lived player.

Suppose that there is uncertainty about the types of both (long-lived) players 1 and 2. Each player's type is drawn from a countable set before the game begins and made known to that player only. Let $\lambda^0 > 0$ be the probability that player 2 is normal. As in the proof of proposition 16.2.1, player 1 has the option of playing a_1' in every period, and if $\mu(\xi(a_1')) > 0$, there is an upper bound on the number of periods that the normal type of player 2 can play anything other than a best response. The other types of player 2 may do quite different things, but the prior probability distribution provides a bound on how important these types can be in player 1's payoff. We thus immediately have the following:

Proposition 16.2.2 *Suppose $\mu(\xi(a_1')) > 0$ for some action a_1' minmaxing player 2. There exists a constant k, independent of player 1's discount factor, such that the normal player 1's payoff in any Nash equilibrium of the repeated game is at least*

$$\lambda^0 \delta_1^k v_1^*(a_1') + (1 - \lambda^0 \delta_1^k) \min_a u_1(a).$$

16.3 Weaker Reputations for Any Action

We thus see that the crucial ingredient in player 1's ability to establish a reputation is not that incomplete information be one-sided. Instead, it is important that player 2 assign some probability to the simple commitment type $\xi(a_1')$ and player 1 be arbitrarily patient. The first feature ensures that player 1 can induce player 2 to eventually play a best response to a_1'. The second feature ensures that the "eventually" in this statement is soon enough to be of value.

16.3 Weaker Reputations for Any Action

If player 2 puts positive prior probability on $\xi(a_1')$, when facing a steady stream of a_1', player 2 must eventually come to expect a_1'. Short-lived player 2s thus must eventually best respond to a_1', implying the one-shot bound $v_1^*(a_1')$. With a long-lived player 2, if a_1' minmaxes player 2, there is a bound on the number of times player 2 can take an action other than the best response to a_1', and so again the one-shot bound $v_1^*(a_1')$ is applicable. If the action a_1' does not minmax player 2, we can no longer bound the number of periods in which player 2 is not best responding. We can, however, bound the number of times player 2 can expect a continuation payoff (rather than current-period payoff) less than her minmax value. The implied player 1 payoff bound is

$$v_1^\dagger(a_1') \equiv \min_{\alpha_2 \in \mathscr{D}(a_1')} u_1(a_1', \alpha_2), \tag{16.3.1}$$

where

$$\mathscr{D}(a_1') = \{\alpha_2 \in \Delta(A_2) \mid u_2(a_1', \alpha_2) \geq \underline{v}_2\}$$

is the set of player-2 actions that in conjunction with a_1', imply at least her minmax utility.

Cripps, Schmidt, and Thomas (1996) prove the following result.

Proposition 16.3.1 *Fix $\delta_2 \in [0, 1)$ and $a_1' \in A_1$ with $\mu(\xi(a_1')) > 0$. For any $\varepsilon > 0$, there exists a $\underline{\delta}_1 < 1$ such that for all $\delta_1 \in (\underline{\delta}_1, 1)$,*

$$\underline{v}_1(\xi_0, \mu, \delta_1, \delta_2) \geq v_1^\dagger(a_1') - \varepsilon.$$

If a_1' minmaxes player 2, $v_1^\dagger(a_1') = v_1^*(a_1')$, the one-shot bound defined in (15.3.4); otherwise $v_1^\dagger(a_1')$ may be strictly smaller. Moreover, it need not be the case that the Stackelberg action maximizes $v_1^\dagger(a_1)$. However, this bound holds for all actions, not just those minmaxing player 2.

Proof Fix $\varepsilon > 0$. Because for all $\alpha_2 \in \mathscr{D}(a_1')$, we have $u_1(a_1', \alpha_2) \geq v_1^\dagger(a_1')$, there exists an $\eta > 0$ such that

$$u_2(a_1', \alpha_2) \geq \underline{v}_2 - \eta \implies u_1(a_1', \alpha_2) \geq v_1^\dagger(a_1') - \tfrac{\varepsilon}{2}. \tag{16.3.2}$$

The set of stage-game payoff profiles consistent with player 1 choosing the action a_1' is

$$\mathscr{F}(a_1') \equiv \{u(a_1', \alpha_2) \mid \alpha_2 \in \Delta(A_2)\}.$$

Because this set is convex and each of its elements satisfies (16.3.2), all of its convex combinations also satisfy (16.3.2).

Let Ω' denote the set of outcomes at which player 1 plays a_1' in every period and let $E[U_2(\sigma|_{h^t}) \mid \Omega']$ denote player 2's expected period t continuation payoff, conditional on the history h^t and Ω' (see note 3 on page 516). We can write this payoff as

$$E[U_2(\sigma|_{h^t}) \mid \Omega'] = (1-\delta_2) E\left[\sum_{\tau=t}^{\infty} \delta_2^{\tau-t} u_2(a_1', \sigma_2(h^\tau)) \mid \Omega'\right].$$

Hence, $E[U_2(\sigma|_{h^t}) \mid \Omega']$ is a convex combination of player 2 payoffs from stage-game payoff profiles in $\mathscr{F}(a_1')$. Denoting player 1's expected period t continuation payoff, conditional on the history h^t and Ω', *and* using player 2's discount factor, by $E[U_1(\sigma|_{h^t}, \delta_2) \mid \Omega']$, the vector $(E[U_1(\sigma|_{h^t}, \delta_2) \mid \Omega'], E[U_2(\sigma|_{h^t}) \mid \Omega'])$ is a convex combination of terms in $\mathscr{F}(a_1')$. We have noted that any such combination satisfies (16.3.2), or

$$E[U_2(\sigma|_{h^t}) \mid \Omega'] \geq \underline{v}_2 - \eta \Rightarrow E[U_1(\sigma|_{h^t}, \delta_2) \mid \Omega'] \geq v_1^\dagger(a_1') - \tfrac{\varepsilon}{2}. \quad (16.3.3)$$

We thus have a lower bound on player 1's payoff, when calculated at player 2's discount factor, under the assumption that player 2's expected continuation value *against the commitment type* is within η of her minmax value. We now convert this into a statement involving player 1's discount factor and player 2's equilibrium expected continuation value (which may not be her continuation value against the commitment type).

We first note that

$$E[U_1(\sigma|_{h^t}, \delta_2) \mid \Omega'] = E[(1-\delta_2) u_1(a_1', \sigma_2(h^t)) + \delta_2 U_1(\sigma|_{(h^t, a_1' a_2^t)}, \delta_2) \mid \Omega']$$

or

$$E[(1-\delta_2) u_1(a_1', \sigma_2(h^t)) \mid \Omega'] = E[U_1(\sigma|_{h^t}, \delta_2) \mid \Omega']$$
$$- \delta_2 E[U_1(\sigma|_{(h^t, a_1' a_2^t)}, \delta_2) \mid \Omega'].$$

We can then calculate

$$E[U_1(\sigma, \delta_1) \mid \Omega']$$
$$= (1-\delta_1) \sum_{t=0}^{\infty} \delta_1^t E[u_1(a_1', a_2^t) \mid \Omega']$$
$$= E\left[\sum_{t=0}^{\infty} \delta_1^t \frac{(1-\delta_1)}{(1-\delta_2)} \{E[U_1(\sigma|_{h^t}, \delta_2) \mid \Omega'] - \delta_2 E[U_1(\sigma|_{h^{t+1}}, \delta_2) \mid \Omega']\} \bigg| \Omega'\right]$$
$$= \frac{(1-\delta_1)}{(1-\delta_2)} \bigg\{ E[U_1(\sigma, \delta_2) \mid \Omega']$$
$$+ E\left[\sum_{t=0}^{\infty} E[\delta_1^t (\delta_1 - \delta_2) U_1(\sigma|_{h^{t+1}}, \delta_2) \mid \Omega'] \bigg| \Omega'\right] \bigg\}, \quad (16.3.4)$$

where the outer expectation in the third line is taken over h^t and the final equality is obtained by pulling $E[U_1(\sigma, \delta_2) \mid \Omega'] = E[U_1(\sigma|_\varnothing, \delta_2) \mid \Omega']$ out of the sum and regrouping terms. Let

16.3 ■ Weaker Reputations for Any Action

$$k = L \frac{\ln \mu(\xi(a_1'))}{\ln(1-\varepsilon)},$$

where the value of L is from lemma 16.2.1 (ε and η were determined at the beginning of this proof). Condition (16.3.3) and lemmas 16.2.1 and 15.3.1 then imply that the expectation of $E[U_1(\sigma|_{h^{t+1}}, \delta_2) \mid \Omega']$ can fall short of $v_1^* - \varepsilon/2$ at most Nk times. We thus have the bound

$$\sum_{t=0}^{\infty} E[\delta_1^t(\delta_1 - \delta_2)U_1(\sigma|_{h^{t+1}}, \delta_2) \mid \Omega']$$

$$\geq \frac{(\delta_1 - \delta_2)}{(1-\delta_1)} \left(v_1^\dagger(a_1') - \frac{\varepsilon}{2}\right) - (\delta_1 - \delta_2)(v_1^\dagger(a_1') - \min_a u_1(a))k. \quad (16.3.5)$$

Inserting this limit in (16.3.4) and taking the limit as $\delta_1 \to 1$,

$$\lim_{\delta_1 \to 1} E[U_1(\sigma, \delta_1) \mid \Omega'] \geq v_1^\dagger(a_1') - \frac{\varepsilon}{2}.$$

We then need only choose a value $\underline{\delta}_1$ such that the left side of this expression is within $\varepsilon/2$ of its limit to conclude, as desired, that for all $\delta_1 \in (\underline{\delta}_1, 1)$,

$$E[U_1(\sigma, \delta_1) \mid \Omega'] \geq v_1^\dagger(a_1') - \varepsilon.$$
∎

If μ attaches positive probability to all simple pure commitment types, then a normal player 1 is assured a payoff of nearly

$$\max_{a_1 \in A_1} v_1^\dagger(a_1).$$

Example 16.3.1 The stage game is the battle of the sexes game in figure 16.3.1. Because the mixed-action minmax utility for player 2 is $3/4$, there is no pure action minmaxing player 2, and proposition 16.2.1 cannot be used to bound player 1's payoffs. The Stackelberg action is T, and the set of responses to T in which player 2 receives at least her minmax utility is the set of actions that place at least probability $3/4$ on R. Hence, if 2 assigns positive probability to 1 being the Stackelberg type, the lower bound on 1's payoff is $9/4$.
●

Remark 16.3.1 **Two-sided incomplete information** Once again, proposition 16.3.1 has exploited only the properties of player 2's best response behavior, in any Nash equilibrium,

	L	R
T	0, 0	3, 1
B	1, 3	0, 0

Figure 16.3.1 Battle of the sexes game.

to an outcome in which player 1 invariably plays a_1'. Like proposition 16.2.1, it accordingly generalizes to uncertainty about player 2's type.

◆

16.4 Imperfect Public Monitoring

In example 16.1.1, although there is some chance that player 1 is a Stackelberg type, there is an equilibrium in which the normal player 1's payoff falls well below the Stackelberg payoff. The difficulty is that player 2 frequently plays an action that is not her best response to 1's Stackelberg action. Why does player 2 do so? She fears that playing the best response when she is not supposed to will push the game off the equilibrium path into a continuation phase where she is punished. The normal and Stackelberg types of player 1 would not impose such a punishment, but there is another "punishment" type who will. Along the equilibrium path player 2 has no opportunity to discern whether she is facing the normal type or punishment type. This poses no difficulties in games of conflicting interests, because player 1's payoff along the equilibrium path is sufficiently low that she has nothing to fear in being pushed off the path.

In games of imperfect public monitoring, the sharp distinction between being on and off the equilibrium path disappears. Player 2 may then have ample opportunity to become well acquainted with player 1's behavior, including any punishment possibilities.[4] Indeed, the arguments in this section show that under full-support public monitoring, with the set of types of example 16.1.1, a sufficiently patient normal player 1 can be assured of a payoff arbitrarily close to 10.

The imperfect public monitoring game has the structure described in section 7.1 for two long-lived players and finite action sets A_1 and A_2. The public monitoring distribution ρ has full support, that is, for all $y \in Y$ and $a \in A_1 \times A_2$,

$$\rho(y \mid a) > 0.$$

We also assume a slight strengthening of the public monitoring analog of assumption 15.4.1. For any mixed action $\alpha_2 \in \Delta(A_2)$,

$$\rho(\cdot \mid (\alpha_1, \alpha_2)) = \rho(\cdot \mid (\alpha_1', \alpha_2)) \Rightarrow \alpha_1 = \alpha_1'. \tag{16.4.1}$$

It is important that player 2's actions be imperfectly monitored by player 1, so that a sufficiently wide range of player 1 behavior occurs in equilibrium. It is also important that player 2 be able to update her beliefs about the type of player 1, in response to the behavior she observes. Full-support public monitoring with (16.4.1) satisfies both desiderata. We do not require the signals to be informative about player 2's behavior.

---

4. Our analysis essentially follows Celentani, Fudenberg, Levine, and Pesendorfer (1996). Aoyagi (1996) presents a similar analysis, with trembles instead of imperfect monitoring blurring the distinction between play on and off the equilibrium path, and with player 1 infinitely patient while player 2 discounts.

16.4 ■ Imperfect Public Monitoring

As usual, player i's discount factor is denoted δ_i, player 1's type is drawn according to a prior μ with a finite or countable support Ξ, and the support of μ contains the normal type ξ_0.

The set of histories for player 1 is $\mathcal{H}_1 = \cup_{t=0}^{\infty}(A_1 \times Y)^t$. A behavior strategy for player 1 is $\sigma_1 : \mathcal{H}_1 \times \Xi \to \Delta(A_1)$. The set of histories for player 2 is $\mathcal{H}_2 = \cup_{t=0}^{\infty}(A_2 \times Y)^t$, and a behavior strategy for 2 is $\sigma_2 : \mathcal{H}_2 \to \Delta(A_2)$. A Nash equilibrium strategy profile $\sigma = (\sigma_1, \sigma_2)$ and the prior distribution μ induce a measure $\mathbf{P}$ over the set of outcomes $\Omega \equiv \Xi \times (A_1 \times A_2 \times Y)^{\infty}$.

We now take seriously the prospect that player 1 may be committed to a mixed strategy or to a strategy that is not simple. In the prisoners' dilemma, for example, player 1 may prefer to play tit-for-tat rather than either always exerting effort or always shirking. We can thus no longer define player 1's payoff target in terms of the stage game. However, we would like to define it in terms of some finite game. Hence, let $G^N(\delta_2)$ be the complete information finitely repeated game that plays the complete information stage game N times, retaining the discount factor δ_2 for player 2. That is, in $G^N(\delta_2)$, player 1's payoff is[5]

$$\frac{1}{N} \sum_{t=0}^{N-1} u_1(a^t) \qquad (16.4.2)$$

and player 2's payoff is

$$\frac{1-\delta_2}{1-\delta_2^N} \sum_{t=0}^{N-1} \delta_2^t u_2(a^t). \qquad (16.4.3)$$

We denote player 1's expected payoff from a strategy profile σ^N in $G^N(\delta_2)$ using (16.4.2) by $U_1^N(\sigma^N)$. The set of player 2's best replies to a strategy σ_1^N in $G^N(\delta_2)$ is denoted by $B^N(\sigma_1^N; \delta_2)$.

For every strategy σ_1^N in the finitely repeated game of length N, there is a corresponding strategy in the infinitely repeated game that plays σ_1^N in the first N periods, then wipes the history clean, starts over, plays it again, and so on. We also let σ_1^N denote this strategy in the infinitely repeated game, or in any finitely repeated game whose length is an integer multiple of N, trusting to context to make the appropriate interpretation clear. The commitment type playing strategy σ_1^N is denoted by $\xi(\sigma_1^N)$, and we say a strategy σ_1^N is in Ξ if $\xi(\sigma_1^N) \in \Xi$. Thus, for any behavior strategy for player 1, σ_1, we have $\sigma_1(\xi(\sigma_1^N), h_1^t) = \sigma_1^N(h_1^t)$ for all $h_1^t \in \mathcal{H}_1$. We assume that in addition to the normal type, Ξ contains at least one such type $\xi(\sigma_1^N)$.

The target for player 1's payoff is the largest payoff that can be obtained by the strategies in Ξ, the support of μ, each evaluated in the corresponding finitely repeated game for the case in which player 1 is arbitrarily patient. Define the set of player 1 payoffs,

$$\mathcal{V}_1(\delta_2, \Xi) \equiv \{v_1 : \forall \varepsilon > 0, \exists N, \xi(\sigma_1^N) \in \Xi$$
$$\text{s.t. } \forall \sigma_2^N \in B^N(\sigma_1^N; \delta_2), \; U_1^N(\sigma_1^N, \sigma_2^N) \geq v_1 - \varepsilon\}, \qquad (16.4.4)$$

---

5. Because we are concerned with the payoff target of an arbitrarily patient player 1, it is convenient to work with average payoffs.

and set

$$v_1^{\ddagger}(\delta_2, \Xi) = \sup \mathcal{V}_1(\delta_2, \Xi).$$

Remark 16.4.1 **Simple commitment types** If the simple commitment type $\xi(\alpha_1)$ is contained in Ξ, then $v_1^{\ddagger}(\delta_2, \Xi) \geq v_1^*(\alpha_1)$. If Ξ contains only simple commitment types corresponding to a countably dense subset of A_1, then $v_1^{\ddagger}(\delta_2, \Xi) = v_1^{**}$. ◆

Remark 16.4.2 **Payoff bounds** The bound $v_1^{\ddagger}$ may be much higher than v_1^{**}. Consider, for example, the public monitoring prisoners' dilemma of section 7.2. Let Ξ contain a type committed to the player 1 component of the forgiving strategy of section 7.2.2. If

$$3\delta_2(p-q) \geq 1$$

and q is not too much larger than r, it is a best response to the forgiving strategy for player 2 to also exert effort after $\bar{y}$ and shirk after $\underline{y}$ in every period, except the last, where she shirks after both signals. Suppose $p = 1 - r^2$. Player 1's payoff according to (16.4.2), under the forgiving strategy (with 2's best response) then approaches 2 as $r \to 0$ and $N \to \infty$ (see remark 7.2.1). Celentani, Fudenberg, Levine, and Pesendorfer (1996, theorem 2) establish conditions under which $\lim_{\delta_2 \to 1} v_1^{\ddagger}(\delta_2, \Xi) = \max\{v_1 : v \in \mathscr{F}^{\dagger p}\}$, that is, as player 2 becomes increasingly patient, the bound $v_1^{\ddagger}(\delta_2, \Xi)$ approaches the largest feasible payoff for player 1 consistent with player 2's pure-minmax payoff. ◆

Recall that $\underline{v}_1(\xi_0, \mu, \delta_1, \delta_2)$ denotes the infimum, over the set of Nash equilibria, of the normal player 1's payoffs.

Proposition 16.4.1 *Suppose (16.4.1) holds. For any $\eta > 0$ and δ_2, there is a $\underline{\delta}_1 < 1$ such that for all $\delta_1 \in (\underline{\delta}_1, 1)$,*

$$\underline{v}_1(\xi_0, \mu, \delta_1, \delta_2) \geq v_1^{\ddagger}(\delta_2, \Xi) - \eta.$$

We first collect some preliminary results to be used in the proof.

Lemma 16.4.1 *For any $\eta > 0$ and $\delta_2 > 0$, there exists N', δ_1', ε', and a strategy $\sigma_1^{N'}$ for $G^{N'}(\delta_2)$, with $\xi(\sigma_1^{N'}) \in \Xi$, such that if player 2 plays an ε'-best response to $\sigma_1^{N'}$ in $G^{N'}(\delta_2)$, then player 1's δ_1-discounted payoff in $G^{N'}(\delta_2)$ is at least $v_1^{\ddagger}(\delta_2, \Xi) - \eta/2$.*

Proof Fix $\eta > 0$. Because $v_1^{\ddagger}(\delta_2, \Xi)$ is the supremum of $\mathcal{V}_1(\delta_2, \Xi)$, we can find an N' sufficiently large and a strategy $\sigma_1^{N'}$ with $\xi(\sigma_1^{N'}) \in \Xi$ for player 1 such that for every best response $\sigma_2^{N'}$ for player 2 in $G^{N'}(\delta_2)$,

$$U_1^{N'}(\sigma_1^{N'}, \sigma_2^{N'}) \geq v_1^{\ddagger}(\delta_2, \Xi) - \frac{\eta}{6}.$$

Fix this N' and $\sigma_1^{N'}$. Because the stage game is finite and hence has bounded payoffs, we can find a value δ_1' such that for all $\delta_1 \in (\delta_1', 1)$ and for any sequence

16.4 ■ Imperfect Public Monitoring

of N' action profiles $(a^0, a^1, \ldots, a^{N'-1})$, the average payoff to player 1 is within $\eta/6$ of the δ_1-discounted payoff, that is,

$$\left| \frac{1}{N'} \sum_{t=0}^{N'-1} u_1(a^t) - \frac{1-\delta_1}{1-\delta_1^{N'}} \sum_{t=0}^{N'-1} \delta_1^t u_1(a^t) \right| < \frac{\eta}{6}.$$

Fix this value of δ_1'. Finally, note that the ε-best response correspondence is upper hemicontinuous in ε. Hence, there exists an ε' such that if player 2 plays an ε'-best response $\tilde{\sigma}_2^{N'}$ to $\sigma_1^{N'}$ in $G^{N'}(\delta_2)$, then the δ_1-discounted payoff to player 1 from the profile $(\sigma_1^{N'}, \tilde{\sigma}_2^{N'})$ and from the profile $(\sigma_1^{N'}, \sigma_2^{N'})$ are $\eta/6$-close, for some best response $\sigma_2^{N'}$. Combining these three yields the result.
∎

Let σ^N be a strategy profile in $G^N(\delta_2)$ and let $\rho^N(\cdot \mid \sigma^N)$ be the induced distribution over public histories in Y^N, in game $G^N(\delta_2)$. Because Y^N is finite, $\rho^N(\cdot \mid \sigma^N)$ is an element of $\mathbb{R}^{|Y|^N}$.

The following lemma requires full-support monitoring. In a perfect monitoring, finitely repeated prisoners' dilemma, for example, player 2's strategy of playing grim trigger until the final period and then defecting is a best response to tit-for-tat but not to a strategy that exerts effort after every history. Nonetheless, the strategy profiles generate identical signal distributions, causing the counterpart of lemma 16.4.2 to fail in that context. The result holds with full-support public monitoring because there is then no indistinguishable out-of-equilibrium behavior that can affect whether player 2 is playing a best response.

Lemma 16.4.2 *For every $\varepsilon > 0$, $N \in \mathbb{N}$, and σ_1^N, there exists a $\gamma > 0$ such that for all $\tilde{\sigma}_1^N$, if $|\rho^N(\cdot \mid (\sigma_1^N, \sigma_2^N)) - \rho^N(\cdot \mid (\tilde{\sigma}_1^N, \sigma_2^N))| < \gamma$ for σ_2^N an ε-best response to σ_1^N in $G^N(\delta_2)$, then σ_2^N is a 2ε-best response to $\tilde{\sigma}_1^N$.*

Proof Fix a strategy profile (σ_1^N, σ_2^N) in $G^N(\delta_2)$. Because payoffs are given by (7.1.1), it suffices to show that if $\{\sigma_{1,(n)}^N\}_{n=1}^\infty$ is a sequence under which $\rho^N(\cdot \mid (\sigma_{1,(n)}^N, \sigma_2^N))$ converges to $\rho^N(\cdot \mid (\sigma_1^N, \sigma_2^N))$, then $\sigma_{1,(n)}^N$ converges to σ_1^N, where we do not distinguish between strategies that are realization equivalent.[6] It suffices for this result to show that $\rho^N(\cdot \mid (\cdot, \sigma_2^N))$ is a one-to-one mapping from the set of player 1 strategies in game $G^N(\delta_2)$ to the set of distributions over Y^N. Because the mapping is also continuous, it has a continuous inverse, which gives the desired conclusion.

We accordingly seek a contradiction by assuming that $\rho^N(\cdot \mid (\sigma_1^N, \sigma_2^N)) = \rho^N(\cdot \mid (\tilde{\sigma}_1^N, \sigma_2^N))$ but $\sigma_1^N \neq \tilde{\sigma}_1^N$. If the latter is to be the case, there must be some period $t \leq N-1$ and a period t history $h_1^t = (a_1^0 y^0, a_1^1 y^1, \ldots, a_1^{t-1} y^{t-1})$ with $a_1^\tau \in \operatorname{supp} \sigma_1^\tau(h_1^\tau) = \operatorname{supp} \tilde{\sigma}_1^\tau(h_1^\tau)$ such that

$$\tilde{\sigma}_1^N(h_1^t) \neq \sigma_1^N(h_1^t). \tag{16.4.5}$$

---

6. Given full-support public monitoring, distinct public strategies cannot be realization equivalent. Two strategies σ_1^N and $\tilde{\sigma}_1^N$ can only be realization equivalent if they differ only at information sets that both strategies preclude from being reached.

Let $h_2^t = (a_2^0 y^0, a_2^1 y^1, \ldots, a_2^{t-1} y^{t-1})$ for some $a_2^\tau \in \text{supp}\,\sigma_2^\tau(h_2^\tau)$. It follows from (16.4.1) and (16.4.5) that

$$\rho(\cdot \mid (\tilde{\sigma}_1^N(h_1^t), \sigma_2^N(h_2^t))) \neq \rho(\cdot \mid (\sigma_1^N(h_1^t), \sigma_2^N(h_2^t))).$$

Since ρ has full support, this inequality then contradicts $\rho^N(\cdot \mid (\sigma_1^N, \sigma_2^N)) = \rho^N(\cdot \mid (\tilde{\sigma}_1^N, \sigma_2^N))$. ∎

Remark 16.4.3 **Observed player 1 actions** The displayed inequality $\rho(\cdot \mid (\tilde{\sigma}_1^N(h_1^t), \sigma_2^N(h_2^t))) \neq \rho(\cdot \mid (\sigma_1^N(h_1^t), \sigma_2^N(h_2^t)))$ contradicts $\rho^N(\cdot \mid (\sigma_1^N, \sigma_2^N)) = \rho^N(\cdot \mid (\tilde{\sigma}_1^N, \sigma_2^N))$ even when player 1's actions are perfectly monitored, as long as player 2's actions are monitored with full support signals. ◆

We use finitely repeated games of the form $G^N(\delta_2)$ to formulate our payoff target, so it will be helpful to divide the infinitely repeated game into blocks of length N. We refer to the kth block of periods of length N (i.e., periods Nk to $N(k+1) - 1$) as "block" $G^{N,k}$ (note that the initial block is the 0th block).

Though the incomplete information infinitely repeated game has a fixed prior $\mu \in \Delta(\Xi)$, the proof also considers the posteriors $\mu' \in \Delta(\Xi)$, and it is convenient to therefore consider different priors, μ'. Given a strategy profile σ and a prior μ', let $\rho_{\sigma,\mu'}^N(\cdot \mid h_2^{Nk})$ be player 2's "one-block" ahead prediction of the distribution over signals in block $G^{N,k}$, for any private history $h_2^{Nk} \in \mathcal{H}_2^{Nk}$.

Let $\mathbf{P}^{(\sigma_1^N, \sigma_2)}$ be the probability measure over outcomes Ω implied by σ_2 and conditioning on the event that player 1's type is $\xi(\sigma_1^N) \in \Xi$. Because the measure assigns probability one to the event $\{\xi = \xi(\sigma_1^N)\}$, it does not depend on the prior μ'.

Lemma 16.4.3 Fix $\lambda, \mu^\dagger \in (0, 1)$ and $\gamma > 0$, integer N, and a strategy σ_1^N. There exists an integer L such that for any pair of strategies σ_1 and σ_2, and all $\mu' \in \Delta(\Xi)$ with $\mu'(\xi(\sigma_1^N)) \geq \mu^\dagger$,

$$\mathbf{P}^{(\sigma_1^N, \sigma_2)}(|\{k \geq 0 : |\rho_{(\sigma_1^N, \sigma_2),\mu'}^N(\cdot \mid h_2^{Nk}) - \rho_{(\sigma_1, \sigma_2),\mu'}^N(\cdot \mid h_2^{Nk})| \geq \gamma\}| \leq L) \geq 1 - \lambda.$$

Proof This is an immediate application of lemma 15.4.3, indexing time by k and taking $\mathcal{G}^k$ to be the σ-algebra generated by $\mathcal{H}_2^{Nk}$, the player 2 histories after block $G^{N,k}$. ∎

Proof of Proposition 16.4.1 Fix a Nash equilibrium σ of the infinitely repeated game of incomplete information. Recall that M and m denote the maximum and minimum stage-game payoffs.

Fix a value $\eta > 0$ with $v_1^\ddagger(\delta_2, \Xi) - \eta > m$.[7] Then, from lemma 16.4.1, there is a game length N', a strategy $\sigma_1^{N'}$ in $G^{N'}(\delta_2)$ with $\mu(\xi(\sigma_1^{N'})) > 0$, a lower bound on player 1's discount factor $\delta_1' < 1$, and an $\varepsilon' > 0$, so that if player 2 plays an ε'-best response to $\sigma_1^{N'}$ in $G^{N'}(\delta_2)$ and $\delta_1 \in (\delta_1', 1)$, player 1's δ_1-discounted payoff in $G^{N'}(\delta_2)$ is at least $v_1^\ddagger(\delta_2, \Xi) - \eta/2$. The strategy $\sigma_1^{N'}$ is our building

---

7. If $v_1^\ddagger(\delta_2, \Xi) - \eta \leq m$, then the result is immediate.

block for providing player 1 with a payoff within η of $v_1^{\ddagger}(\delta_2, \Xi)$. Recall that $\sigma_1^{N'}$ also denotes a strategy in G^{∞} for the normal type of player 1 and a commitment type in Ξ.

Fix a value κ so that

$$\delta_2^{\kappa N'}(M - m) < \frac{\varepsilon'(1 - \delta_2)}{2}. \tag{16.4.6}$$

We will use this to bound the impact of variations in payoffs that occur after $\kappa N'$ periods on player 2's payoffs, and so obtain an ε'-best response.

Set $\varepsilon = \varepsilon'(1 - \delta_2)/[2(1 - \delta_2^N)]$ and $N = \kappa N'$. Denote by σ_1^N the strategy in the game $G^N(\delta_2)$ that plays $\sigma_1^{N'}$ by restarting $\kappa - 1$ times. For later reference, note that the strategies $\sigma_1^{N'}$ and σ_1^N are identical when viewed as infinitely repeated game strategies, and the commitment type $\xi(\sigma_1^{N'})$ plays the strategy σ_1^N. Consequently, the measures induced by $\mathbf{P}^{(\sigma_1^{N'}, \sigma_2)}$ and $\mathbf{P}^{(\sigma_1^N, \sigma_2)}$ on $(A_1 \times A_2 \times Y)^{\infty}$ are also identical. Fix γ at the value from lemma 16.4.2 for these values of ε, N, and the strategy σ_1^N.

Let $\underline{\mu}(\sigma_1^{N'}, N, \mu)$ be the infimum of the posterior probability assigned to type $\xi(\sigma_1^{N'})$ in the first N periods of play in the repeated game, given the prior distribution μ, where the infimum is taken over all possible histories that could be observed in the first N periods and over strategy profiles σ. Because $\rho(y \mid a)$ is positive for all y and a and $\mu(\xi(\sigma_1^{N'})) > 0$, $\underline{\mu}(\sigma_1^{N'}, N, \mu) > 0$.

Fix a value of λ to satisfy

$$(1 - (1 - \lambda)^{\kappa})(M - m) < \frac{\eta}{4}. \tag{16.4.7}$$

For $\mu^{\dagger} = \underline{\mu}(\sigma_1^{N'}, N, \mu)$ and the values N, γ, λ, and strategy σ_1^N already determined, lemma 16.4.3 yields a value of L with the property stated in that lemma. We can thus conclude that with $\mathbf{P}^{(\sigma_1^{N'}, \sigma_2)}$-probability at least $(1 - \lambda)$, there are at most L "exceptional" blocks $G^{N,k}$ in which the distribution over public histories induced by $(\sigma_1^{N'}, \sigma_2)$ differs from that induced by the Nash equilibrium strategy σ by more than γ.

Consider a positive probability history h_2^{Nk} that reaches an unexceptional block. By (16.4.6), player 2's continuation strategy $\sigma_2|_{h_2^{Nk}}$ must be at least an ε-best response in the block $G^{N,k}$, because otherwise it could not be a best response in the infinitely repeated game. From lemma 16.4.2, this ensures that player 2 is playing a 2ε-best response to $\sigma_1^{N'}$ in $G^{N,k}$. But then player 2 must be playing at least a $2\varepsilon(< \varepsilon')$-best response to $\sigma_1^{N'}$ in the game $G^{N'}$ comprising the first N' periods of the block $G^{N,k}$, and hence by lemma 16.4.1 player 1 must receive a δ_1-discounted payoff at least $v_1^{\ddagger}(\delta_2, \Xi) - \eta/2$ during these periods.

We have thus divided the infinitely repeated game into blocks of length N and have shown that with probability at least $1 - \lambda$, in the first N' periods of all but at most L exceptional blocks, player 1 earns a payoff at least $v_1^{\ddagger}(\delta_2, \Xi) - \eta/2$. This is helpful but applies only to the first N' periods of each block of length N, and hence covers at most a κ^{-1}-proportion of the periods in the game.

Now suppose that period N' has been reached with a player 2 private history $h_2^{N'}$. We can view the continuation as a repeated game of incomplete information, called the N' game. The state space is $\Omega' = \Xi \times (A_1 \times A_2 \times Y)^\infty$ (a copy of Ω). The prior distribution over Ξ is given by player 2's posterior beliefs conditional on her private history $h_2^{N'}$, denoted μ'. The important observation here is that the posterior probability assigned to $\xi(\sigma_1^{N'})$ is at least $\underline{\mu}(\sigma_1^{N'}, N, \mu) = \mu^\dagger$, a fact we use presently to apply lemma 16.4.3. Player 2's continuation strategy is given by $\sigma_2' = \sigma_2|_{h_2^{N'}}$. Finally, player 2 also has beliefs over the previous private actions of player 1, and so has a belief over player 1 continuation strategies $\sigma_1|_{h_1^{N'}}$. This belief is equivalent to a mixed strategy, which has a realization equivalent behavior strategy σ_1', giving player 1's strategy in the N' game.

We are interested in the case in which player 1 plays in the N' game as the commitment type $\sigma_1^{N'}$. Let $\mathbf{P}_{N'}^{(\sigma_1^{N'}, \sigma_2)}$ denote the marginal of $\mathbf{P}^{(\sigma_1^{N'}, \sigma_2)}(\cdot \mid h_2^{N'})$ on Ξ and the set of player 1 and 2 histories beginning in period N' (where $\mathbf{P}^{(\sigma_1^{N'}, \sigma_2)}$ is the measure over outcomes in the original game, given that player 1 plays as $\sigma_1^{N'}$). Then $\mathbf{P}_{N'}^{(\sigma_1^{N'}, \sigma_2)}$ is the measure over outcomes in the N' game. We can now apply lemma 16.4.3 to the game starting in period N', concluding that with $\mathbf{P}_{N'}^{(\sigma_1^{N'}, \sigma_2)}$-probability at least $(1 - \lambda)$, there are at most L exceptional blocks in this game. These blocks now run from periods $Nk + N'$ to $N(k+1) + N' - 1$ in the original game. As this result holds for every player 2 private history, we can conclude that in the original repeated game of incomplete information, with $\mathbf{P}^{(\sigma_1^{N'}, \sigma_2)}$-probability at least $(1 - \lambda)^2$, there are at most $2L$ exceptional blocks.

We can argue as before for the unexceptional blocks that have appeared in the application of lemma 16.4.3 to the N' game, concluding that in the *second* N' periods of each of the blocks of length N of the original game, player 1 earns an average payoff at least $v_1^\ddagger(\delta_2, \Xi) - \eta/2$.

We now do the same for the game beginning at period $2N'$, and so on, through period $(\kappa - 1)N'$. We conclude that of the blocks of length N' into which the infinitely repeated game has now been divided, with probability at least $(1 - \lambda)^\kappa$, player 1 has an average payoff at least $v_1^\ddagger(\delta_2, \Xi) - \eta/2$ in all but the possibly κL exceptional cases (L exceptional cases for each of the κ infinitely repeated games we have examined, beginning at periods $0, N, \ldots, N(\kappa - 1)$). Assuming the worst case—namely, that player 1's payoff is m in the exceptional cases and that the latter all occur at the beginning of the game—player 1's payoff in the repeated game is at least

$$(1 - \lambda)^\kappa \left[(1 - \delta_1^{\kappa LN})m + \delta_1^{\kappa LN}\left(v_1^\ddagger(\delta_2, \Xi) - \frac{\eta}{2}\right)\right] + (1 - (1 - \lambda)^\kappa)m$$

$$= (1 - \delta_1^{\kappa LN})m + \delta_1^{\kappa LN}\left(v_1^\ddagger(\delta_2, \Xi) - \frac{\eta}{2}\right) - \delta_1^{\kappa LN}(1 - (1 - \lambda)^\kappa)$$

$$\times \left[v_1^\ddagger(\delta_2, \Xi) - \frac{\eta}{2} - m\right]$$

$$\geq (1 - \delta_1^{\kappa LN})m + \delta_1^{\kappa LN}\left(v_1^\ddagger(\delta_2, \Xi) - \frac{\eta}{2}\right) - \delta_1^{\kappa LN}\frac{\eta}{4}$$

$$= (1 - \delta_1^{\kappa LN})m + \delta_1^{\kappa LN}\left(v_1^\ddagger(\delta_2, \Xi) - \frac{3\eta}{4}\right),$$

where the inequality uses (16.4.7). Letting δ_1 get large now gives the result. ∎

16.5 Commitment Types Who Punish

In example 16.1.1, player 2 does not play a best response to the Stackelberg type because she fears that doing so will trigger an adverse reaction from the punishment type. The key virtue of imperfect monitoring in section 16.4 lies in ensuring that no such features of player 1's strategy remain hidden from player 2. Evans and Thomas (1997) achieve a similar result, with a similar bound on the normal player 1's payoff, in a game of perfect monitoring. The key to their result is the presence of a commitment type who punishes player 2 for *not* behaving appropriately.

Fix an action $a_1' \in A_1$, with best response a_2' for player 2 for which $u_2(a_1', a_2') > \underline{v}_2^p$, so that 2's best response gives her more than her pure minmax payoff. Let $\hat{a}_1^2$ be the action for player 1 that (pure-action) minmaxes player 2. Consider a commitment type for player 1 who plays as follows. Begin in phase 0. In general, phase k consists of k periods of $\hat{a}_1^2$, followed by the play of a_1'. If player 2 plays anything other than a_2' in periods $n+1, \ldots$ of phase n, the strategy switches to phase $n+1$.

The commitment type thus punishes player 2, in strings of ever-longer punishments, for not playing a_2'. Let $\hat{\sigma}_1$ denote this strategy. We assume throughout that $\hat{\sigma}_1 \in \Xi$ and let $\hat{\mu}^0$ be the prior probability attached to this commitment type. Let $\hat{\Omega}$ be the set of outcomes in which player 1 plays as $\hat{\sigma}_1$.

The first step in assessing why such a strategy might be useful is the following:

Lemma 16.5.1 *Fix an integer $K > 0$ and $\eta > 0$. Then there exists an integer $T(K, \eta, \hat{\mu}^0)$ such that for any pure strategy σ_2 and any $\omega \in \hat{\Omega}$, , there are no more than $T(K, \eta, \hat{\mu}^0)$ periods t in which 2 attaches probability no greater than $1 - \eta$ to the event that player 1 plays as $\hat{\sigma}_1$ in periods $t, \ldots, t + K$, given that 2 plays as σ_2.*

Proof Fix $K, \eta, \hat{\mu}^0$, and a pure strategy σ_2. Let q^t be the probability that player 2 attaches to the event that player 1 plays $\hat{\sigma}_1$ in periods $t, \ldots, t + K$, given that 2 plays as σ_2. Then q^t is a random variable on Ω. Let $\hat{\mu}^t$ be the period t posterior probability that player 1 is committed to $\hat{\sigma}_1$. Fix an outcome $\omega \in \hat{\Omega}$ consistent with σ_2.[8] Suppose $\{t_\ell\}_{\ell=0}^L$ is a sequence of periods in which $q^{t_\ell} \leq 1 - \eta$, where $t_{\ell+1} \geq t_\ell + K + 1$ and the value of L (possibly infinite) remains to be investigated. By Bayes' rule,

$$\hat{\mu}^{t_\ell + K + 1} \geq \frac{\hat{\mu}^{t_\ell}}{1 - \eta}.$$

Because $\hat{\mu}^t$ is nondecreasing in t on $\hat{\Omega}$, we then have

$$\hat{\mu}^{t_L} \geq \frac{\hat{\mu}^0}{(1-\eta)^L},$$

ensuring, because $\hat{\mu}^{t_L} \leq 1$, that

$$L \leq \frac{\ln \hat{\mu}^0}{\ln(1-\eta)}$$

and hence $T(K, \eta, \hat{\mu}^0) \leq L(K+1)$. ∎

---

8. A similar argument, presented in more detail, appears in the proof of lemma 15.3.1.

The bound $T(K, \eta, \hat{\mu}^0)$ in this result is independent of the players' discount factors and the strategy σ_2.

Proposition 16.5.1 *Fix $\varepsilon > 0$. Let Ξ contain $\hat{\sigma}_1$, for some action profile a' with $u_2(a') > \underline{v}_2^p$. Then there exists a $\underline{\delta}_2 < 1$ such that for all $\delta_2 \in (\underline{\delta}_2, 1)$, there exists a $\underline{\delta}_1$ such that for all $\delta_1 \in (\underline{\delta}_1, 1)$,*

$$\underline{v}(\xi_0, \mu, \delta_1, \delta_2) \geq u_1(a') - \varepsilon.$$

Proof It suffices to show, for any Nash equilibrium σ^* of the repeated game with player 2 assigning positive probability to pure strategy σ_2, that there is a finite upper bound on the number of times player 2 triggers a punishment when playing σ_2, on the subset of $\hat{\Omega}$ consistent with σ_2. This ensures that if player 1 plays as $\hat{\sigma}_1$, player 2 will choose a_2' in all but (a bounded number of) finitely many periods, at which point the payoff bound holds for a sufficiently patient player 1.

Fix the bound $\underline{\delta}_2 < 1$ such that for all $\delta_2 \in [\underline{\delta}_2, 1)$,

$$(1 - \delta_2)M + \delta_2 \underline{v}_2^p < \hat{\mu}^0 u_2(a') + (1 - \hat{\mu}^0)((1 - \delta_2)m + \delta_2 \underline{v}_2^p).$$

Now fix $\delta_2 \in (\underline{\delta}_2, 1)$, and then fix $\eta > 0$ sufficiently small and an integer L sufficiently large that

$$\eta M + (1 - \eta)[(1 - \delta_2)M + (\delta_2 - \delta_2^L)\underline{v}_2^p + \delta_2^L M]$$
$$< \hat{\mu}^0 u_2(a') + (1 - \hat{\mu}^0)[(1 - \delta_2)m + \delta_2 \underline{v}_2^p].$$

Notice that this inequality then also holds for all larger values of L. If player 2 attaches probability at least $1 - \eta$ to player 1 playing as $\hat{\sigma}_1$ in periods $t, \ldots, t + L$ (given σ_2), if 2 has triggered $L - 1$ previous punishments, and if strategy $\hat{\sigma}_1$ plays a_1' in the current period, then the left side of this expression is an upper bound on the payoff player 2 can receive if 2 does not play a_2' in period t, and the right side is a lower bound on the payoff she receives from doing so. The inequality then implies that player 2 will not trigger the punishment. From lemma 16.5.1, player 2 can thus trigger at most $L - 1 + T(L, \eta, \hat{\mu}^0)$ punishments. ∎

Remark 16.5.1 **Stackelberg payoff** If the action a_1' on which $\hat{\sigma}_1$ is based is player 1's Stackelberg action, or if Ξ includes a variety of such types, based on a sufficiently rich set of actions, then proposition 16.5.1 gives us player 1's Stackelberg payoff as an approximate lower bound on his equilibrium payoff in the game of incomplete information. The only restriction is that his Stackelberg payoff be consistent with player 2 earning more than her pure-strategy minmax.

Evans and Thomas (1997) show that this phenomenon is more general. Somewhat more complicated commitment types can be constructed in which the commitment type $\hat{\sigma}_1$ plays a sequence of actions during its nonpunishment periods, rather than simply playing a fixed action a_1, and punishes player 2 for not playing an appropriate sequence in response. Using such constructions, Evans and Thomas (1997, theorem 1) establish that $\lim_{\delta_2 \to 1} \lim_{\delta_1 \to 1} \underline{v}_1(\xi_0, \mu, \delta_2, \delta_1) = \max\{v_1 : v \in \mathscr{F}^{\dagger p}\}$.[9] The idea behind the argument is to construct a commitment type

---

9. The same payoff bound is obtained in proposition 16.4.1 when Ξ is sufficiently rich (remark 16.4.2).

consisting of a phase in which payoffs at least $(\max\{v_1 : v \in \mathscr{F}^{\dagger p}\} - \varepsilon, \underline{v}_2^p + \varepsilon)$ are received, as long as player 2 behaves "appropriately," with inappropriate behavior triggering ever-longer punishment phases. Conditional on seeing behavior consistent with the commitment type, a sufficiently patient player 2 must then eventually find it optimal to play the appropriate response to the commitment type. Making player 1 arbitrarily patient then gives the inner limit.

◆

The payoff bound for the normal player 1 is established by showing that player 2 will risk only finitely many punishments from the commitment type of player 1. Then why build a commitment type with arbitrarily severe punishment capabilities? If we fix both the prior probability of the commitment type and player 2's discount factor, then we could work with a finitely specified commitment type. If we seek a single type that works for all nonzero prior probabilities and player 2 discount factors, as is commonly the case in establishing reputation results, then we require the infinite specification.

Like the other reputation results we have presented, this one requires that the uncertainty about player 1 contains the appropriate commitment types. However, the types in this case are more complicated than the commitment types that appear in many reputation models, particularly the simple types that suffice with short-lived player 2s.[10] In this case, the commitment type not only repeatedly plays the action that brings player 1 the desired payoff but also consistently punishes player 2 for not fulfilling her role in producing that payoff. We can thus think of the commitment as involving behavior both along the path of a proposed outcome and on paths following deviations. Work on reputations would be well served by a better-developed model of which commitment types are likely to be contained in Ξ.

16.6 Equal Discount Factors

The reputation results established so far in this chapter allow player 2 to be arbitrarily patient, but then hold player 2's discount factor δ_2 fixed and make player 1 yet arbitrarily more patient by taking $\delta_1 \to 1$. What if we eliminate this asymmetry by requiring the players to share a common discount factor?[11]

Example 16.1.1 has shown that if the set Ξ of player 1 types includes an appropriate punishment type, the normal player 1 may receive an equilibrium payoff arbitrarily close to his minmax payoff, even when Ξ includes the pure Stackelberg type. Though presented in the context of player 1 being arbitrarily more patient than player 2, none

---

10. As Evans and Thomas (1997) note, a commitment type with punishments of arbitrary length cannot be implemented by a finite automaton. Evans and Thomas (2001), working with two infinitely patient long-lived players, argue that commitment strategies capable of imposing arbitrarily severe punishments are necessary if reputation arguments are to be effective in restricting attention to efficient payoff profiles.
11. The symmetric case of *perfectly* patient players on both sides is studied by Cripps and Thomas (1995), Shalev (1994), and Israeli (1999). Cripps and Thomas (1995) show that when there is some prior probability that player 1 (only) may be a Stackelberg type, the normal player 1's payoff in any Nash equilibrium must be at least the bound established by Cripps, Schmidt, and Thomas (1996) for discounted games with player 1 arbitrarily more patient than player 2.

	L	R
T	1,1	0,0
B	0,0	0,0

Figure 16.6.1 A common interest game.

of the arguments in example 16.1.1 required asymmetric patience. It is then clear that some assumption like conflicting interests will be necessary if we hope to obtain a strong reputation-induced payoff bound with symmetric patience. In fact, given that *asymmetric* patience played a central role in proposition 16.2.1, it will be no surprise that we require even more stringent conditions with symmetric patience. We illustrate the possibilities with five examples.

16.6.1 Example 1: Common Interests

This section presents an example, following Cripps and Thomas (1997), that retains much of the structure of example 16.1.1 but in which player 1 is restricted to being either normal or Stackelberg. Player 2's deviations must now be punished only by the normal and Stackeberg types, which will call for more complication in constructing equilibria. We show that if the two players have identical discount factors, then there exist equilibria with normal player 1 payoffs arbitrarily close to his minmax payoff, if the commitment type is not too likely and the players are sufficiently patient.[12] The possibility of the Stackelberg type does not place an effective lower bound on normal player 1's equilibrium payoff.

We consider the game shown in figure 16.6.1. Player 1 may be a normal type, with probability $1 - \hat{\mu}^0$, or a Stackelberg type committed to playing T, with probability $\hat{\mu}^0$. Players 1 and 2 obviously must receive the same equilibrium payoffs. We show that for any $\gamma > 0$, there exists $\hat{\mu}(\gamma) > 0$ and $\underline{\delta}(\gamma) < 1$ such that if $\hat{\mu}^0 < \hat{\mu}(\gamma)$ and $\delta_1 = \delta_2 \in (\underline{\delta}(\gamma), 1)$, then there exists an equilibrium in which the normal type of player 1 receives a payoff less than γ.

To construct the equilibrium, we begin with period N, whose value is to be determined (and depends on δ). In each period $t \in \{0, \ldots, N-1\}$, we say that we are in phase one of the strategy profile if every previous period has produced either the action profile TL or TR. In phase one, player 2 plays R. The normal player 1 mixes between T and B, placing probability ϕ^t on T in period t. If this mixture produces the outcome T, then we continue to the next period, still in phase one. If this mixture produces B in period t, then player 1 is revealed to be normal. Players 1 and 2 then condition their

---

12. The fact that the two players have the same discount factor is important in this example. Celentani, Fudenberg, Levine, and Pesendorfer (1996, section 5) present an example of a game that similarly fails conflicting interests and involves only normal and Stackelberg types for player 1. In equilibrium, the normal player 1 receives his pure minmax payoff (which is less than his Stackelberg payoff), for a fixed probability of the commitment type and sufficiently (symmetric or asymmetric) patient players.

16.6 ■ Equal Discount Factors

continuation behavior on the outcome of a public correlating device, choosing with probability δ^{N-t-1} to play *TL* in every subsequent period and with the complementary probability to play *BR* in every subsequent period. Both continuation outcomes are subgame-perfect equilibria of the continuation game. Should period N be reached while still in phase one, *TL* is then played in every period, regardless of player 1's type or of previous play. Once again, this is an equilibrium of the continuation game. Should player 2 play *L* and player 1 play *B* in phase one, then 1 is again revealed to be normal and thereafter *BR* is played.

If each of the N possible periods of phase one produces the action profile *TR*, then player 1's realized payoff is δ^N. Moreover, player 1 is mixing between *T* and *B* in each of these periods (in phase 1), and hence must be indifferent over all of the outcomes that could occur in the first N periods. This ensures that player 1's *equilibrium* payoff must be δ^N. Because our goal is to show that equilibria exist in which player 1 receives a small payoff, much will hinge on evaluating δ^N.

We first verify that player 1 is indifferent between *T* and *B* when in phase one. A play of *T* in period $N-1$ gives an expected payoff of δ (because *TL* is then played in every subsequent period), as does a play of *B* (because it leads with probability $\delta^{N-(N-1)-1} = 1$ to the play of *TL* in every subsequent period). Working backward, the expected payoff from playing either *T* or *B* in any preceding period t is δ^{N-t}. Player 1's equilibrium mixture is thus optimal.

Ensuring that we have an equilibrium then requires only that player 2 prefers to choose *R* in each of the first N periods in phase one. The payoff from *R* in period t is δ^{N-t}. A play of *L* brings a payoff of 0 if 1 plays *B* and $(1-\delta) + \delta^{N-t}$ if 1 plays *T*. Player 2's choice of *R* is optimal if

$$[\hat{\mu}^t + (1-\hat{\mu}^t)\phi^t](1 - \delta + \delta^{N-t}) \leq \delta^{N-t}. \quad (16.6.1)$$

We use this equation to define ϕ^t for $t = 1, \ldots, N-1$ (considering period 0 later) as the solution to

$$\hat{\mu}^t + (1-\hat{\mu}^t)\phi^t = \frac{\delta^{N-t}}{1 - \delta + \delta^{N-t}}, \quad (16.6.2)$$

making player 2 indifferent in every period.

This ensures that player 2's strategy is optimal (with period 0 yet to be considered). It remains to identify the probabilities $\hat{\mu}^t$ for $t \in \{0, \ldots, N-1\}$, verify that the values of ϕ^t we have specified are contained within the unit interval, and then to examine δ^N. We begin by identifying the probabilities $\hat{\mu}^0, \ldots, \hat{\mu}^{N-1}$ in phase one, in the process ensuring that they are less than 1 and are increasing in t. The former property ensures that $\phi^t \geq 0$, whereas the latter ensures that $\phi^t \leq 1$ because, from Bayes' rule,

$$\hat{\mu}^{t+1} = \frac{\hat{\mu}^t}{\hat{\mu}^t + (1-\hat{\mu}^t)\phi^t}. \quad (16.6.3)$$

We then set $\hat{\mu}^N = 1$ and then work backward. From (16.6.2) and (16.6.3), we have

$$\hat{\mu}^{N-1} = \delta \quad \text{and} \quad \phi^{N-1} = 0.$$

More generally, successively performing this calculation gives

$$\hat{\mu}^t = \prod_{\tau=t}^{N-1} \frac{\delta^{N-\tau}}{1-\delta+\delta^{N-\tau}} \tag{16.6.4}$$

for $t \in \{1, \ldots, N-1\}$. This ensures that the $\hat{\mu}^t$ are increasing and all less than unity. The value of N is now set to anchor this sequence of posterior probabilities at the prior $\hat{\mu}^0$. Choose N such that the expression on the right side of (16.6.4) for $t=0$ is not smaller than the prior, that is, choose N such that

$$\prod_{\tau=0}^{N-1} \frac{\delta^{N-\tau}}{1-\delta+\delta^{N-\tau}} \geq \hat{\mu}^0. \tag{16.6.5}$$

Having fixed N, ϕ^0 is determined by the requirement that the posterior value $\hat{\mu}^1$, following a period 0 outcome of T, equals the value determined by (16.6.4), that is, ϕ^0 solves

$$\hat{\mu}^1 = \frac{\hat{\mu}^0}{\hat{\mu}^0 + (1-\hat{\mu}^0)\phi^0}. \tag{16.6.6}$$

Our choice of N ensures

$$[\hat{\mu}^0 + (1-\hat{\mu}^0)\phi^0] = \frac{\hat{\mu}^0}{\hat{\mu}^1} \leq \frac{\delta^N}{1-\delta+\delta^N},$$

and hence that player 2 finds R optimal in period 0 (see (16.6.1)).

We now relate the size of N, δ, and an upper bound on $\hat{\mu}^0$. Letting $N(\delta)$ denote the largest integer satisfying (16.6.5) and taking $N = N(\delta)$ in the construction yields an equilibrium, for all δ. Moreover, $N(\delta)+1$ then fails (16.6.5), that is,

$$\hat{\mu}^0 > \prod_{\tau=0}^{N(\delta)} \frac{\delta^{N(\delta)+1-\tau}}{1-\delta+\delta^{N(\delta)+1-\tau}}.$$

Taking logs, using $\ln x \geq 1 - 1/x$, and reorganizing the summation,

$$\ln \hat{\mu}^0 > \sum_{\tau=1}^{N(\delta)+1} \left(1 - \frac{1-\delta+\delta^\tau}{\delta^\tau}\right)$$

$$= -\frac{1-\delta}{\delta} \sum_{\tau=0}^{N(\delta)} \frac{1}{\delta^\tau}$$

$$= -\frac{1-\delta}{\delta} \times \frac{1-(\frac{1}{\delta})^{N(\delta)+1}}{1-\frac{1}{\delta}}$$

$$= 1 - \frac{1}{\delta^{N(\delta)+1}}.$$

Rearranging the inequality given by the extreme terms, we have

$$\delta^{N(\delta)} < \frac{1}{\delta(1-\ln\hat{\mu}^0)}.$$

Hence, choosing $\hat{\mu}(\gamma)$ and $\underline{\delta}(\gamma)$ so that

$$\frac{1}{\underline{\delta}(\gamma)(1 - \ln \hat{\mu}(\gamma))} < \gamma$$

ensures that, if $\hat{\mu}^0 < \hat{\mu}(\gamma)$ and $\delta \in (\underline{\delta}(\gamma), 1)$, then player 1's equilibrium payoff is at most γ.

16.6.2 Example 2: Conflicting Interests

It is surprising that reputation effects need not appear in the game shown in figure 16.6.1, where the players both stand to gain from 1's reputation. Nonetheless, our experience with asymmetrically patient players (sections 16.2–16.3) suggests that we should expect reputations to be effective only under strong conditions. We now show that conflicting interests alone does not suffice.

The stage game is shown in figure 16.6.2. The minmax playoffs for the two players are $(0, 0)$. The stage game has a unique component of Nash equilibria in which player 2 chooses R and player 1 mixes with probability greater than or equal to $1/2$ on B. Player 1's (pure) Stackelberg action is T. Player 2's best response to T gives 2 her minmax payoff of 0, and hence this is a game of conflicting interests.

Player 1 is the Stackelberg type with probability $\hat{\mu}^0$ and is otherwise normal. As in the previous example, for all $\gamma > 0$, there exists $\hat{\mu}(\gamma) > 0$ and $\underline{\delta}(\gamma) < 1$ such that if $\hat{\mu}^0 < \hat{\mu}(\gamma)$ and $\delta_1 = \delta_2 \in (\underline{\delta}(\gamma), 1)$, then there exists an equilibrium in which the normal type of player 1 receives a payoff less than γ.

We again work backward from a period N to be determined. Play during periods $0, \ldots, N - 1$ is said to be in phase one if the history consists only of realizations TL or TR. Should play reach period N while in phase one, TL is thereafter played forever (with subsequent player 1 deviations punished by switching to perpetual play of BR and player 2 deviations ignored).

In phase one, player 2 chooses R and the normal player 1 mixes (so that in all, except perhaps the initial period, 2 is indifferent between L and R), with probability ϕ^t on T. The outcome BR prompts a public correlation with probability δ^{N-t-1} attached to a continuation equilibrium with payoffs $(1, 1/2)$—the normal player 1 plays B and 2 mixes equally between L and R along the equilibrium path, with deviations punished by a temporary phase of mutual minmaxing, followed by a return to the equilibrium

	L	R
T	1,0	0,−1
B	2,0	0,1

Figure 16.6.2 A game of conflicting interests, in which player 1's Stackelberg action T and player 2's best response combine to minmax player 2.

	L	R
T	$\delta^{N-1-t}(1,0)$	$\delta^{N-1-t}(1,0)$
B	$(1,0)$	$\delta^{N-1-t}\left(1,\tfrac{1}{2}\right)+(1-\delta^{N-1-t})(0,1)$

Figure 16.6.3 Period $t+1$ continuation payoffs for $t<N$, if in phase one in period t (only outcomes TL and TR in previous periods), as a function of the period t action profile, assuming (for $t<N-1$) player 2 indifference between L and R in subsequent periods.

path—and the remaining probability to the perpetual play of the stage-game Nash equilibrium BR. An outcome of BL causes a switch to perpetual play of TL (with subsequent player 1 deviations punished by switching to perpetual play of BR).

Figure 16.6.3 shows the discounted continuation payoffs for each period t outcome, for periods $t<N$, if in phase one. For sufficiently large (common) discount factors, we obviously have equilibrium continuation play after any history not in phase one. In addition, any mixtures by the normal player 1 are best responses in phase 1, because T and B in period t both bring the expected payoff δ^{N-t}. We must then consider the conditions for R to be a best response for player 2 throughout phase one. Choosing L gives player 2 a current and future payoff of 0, and R gives a payoff of (from figure 16.6.3)

$$[\hat{\mu}^t+(1-\hat{\mu}^t)\phi^t]((1-\delta)(-1))$$
$$+(1-[\hat{\mu}^t+(1-\hat{\mu}^t)\phi^t])((1-\delta)+\delta^{N-t}\tfrac{1}{2}+(\delta-\delta^{N-t})).$$

Indifference in periods $t=1,\ldots,N-1$ requires

$$\hat{\mu}^t+(1-\hat{\mu}^t)\phi^t=\frac{2-\delta^{N-t}}{4-2\delta-\delta^{N-t}}.$$

Setting $\hat{\mu}^N=1$, for $t=1,\ldots,N-1$, we can then solve for

$$\hat{\mu}^t=\prod_{\tau=t}^{N-1}\frac{2-\delta^{N-\tau}}{4-2\delta-\delta^{N-\tau}}. \tag{16.6.7}$$

As in the previous example, for fixed δ, the value of N is chosen to satisfy

$$\hat{\mu}^0\leq\prod_{\tau=0}^{N-1}\frac{2-\delta^{N-\tau}}{4-2\delta-\delta^{N-\tau}}, \tag{16.6.8}$$

and the initial randomization by the normal player 1 is determined by the requirement that the posterior value $\hat{\mu}^1$ after a realization of T is equal to the expression in (16.6.7). Given this choice of N and randomization ϕ^0, player 2 finds R optimal in the initial period.

16.6 ■ Equal Discount Factors

We now argue that for sufficiently small prior probability on the Stackelberg type and sufficiently patient players, there is an equilibrium with low payoffs. Let $N(\delta)$ be the largest integer satisfying (16.6.8) and note that setting $N = N(\delta)$ yields an equilibrium, for all δ. Moreover, because $N(\delta) + 1$ fails (16.6.8), we have

$$\hat{\mu}^0 > \prod_{\tau=0}^{N(\delta)} \frac{2 - \delta^{N(\delta)+1-\tau}}{4 - 2\delta - \delta^{N(\delta)+1-\tau}},$$

Taking logs and using $\ln x \geq 1 - 1/x$,

$$\ln \hat{\mu}^0 > \sum_{\tau=0}^{N(\delta)} \left(1 - \frac{4 - 2\delta - \delta^{N(\delta)+1-\tau}}{2 - \delta^{N(\delta)+1-\tau}}\right)$$

$$= -2(1-\delta) \sum_{\tau=1}^{N(\delta)+1} \frac{1}{2 - \delta^{\tau}}$$

$$> -2(N(\delta)+1)(1-\delta).$$

Rearranging the extreme terms, we have

$$N(\delta) > -\frac{\ln \hat{\mu}^0}{2(1-\delta)} - 1.$$

Hence, there is an equilibrium in which the normal type has a log payoff $N(\delta) \ln \delta$ satisfying

$$N(\delta) \ln \delta < -\left(\frac{\ln \hat{\mu}^0}{2(1-\delta)} + 1\right) \ln \delta.$$

Because (using l'Hôpital's rule)

$$\lim_{\delta \to 1} -\left(\frac{\ln \hat{\mu}^0}{2(1-\delta)} + 1\right) \ln \delta = \frac{\ln \hat{\mu}^0}{2},$$

for any $\varepsilon > 0$ and sufficiently large δ,

$$\delta^{N(\delta)} < \sqrt{\hat{\mu}^0} + \varepsilon.$$

For any $\gamma > 0$, we can then take $\hat{\mu}^0(\gamma) = (\gamma - \varepsilon)^2$, at which point there exists a $\underline{\delta}(\gamma)$ such that for all $\delta \in (\underline{\delta}(\gamma), 1)$, the normal player 1's payoff is less than γ.

We can then hope for reputation effects to take hold only for games satisfying stronger conditions than conflicting interests.[13] The next two examples illustrate.

---

13. Cripps and Thomas (2003) study games of incomplete information with payoff types with positive probability attached to a (payoff) Stackelberg type. They show that as the common discount factor of player 1 and 2 approaches unity, there exist equilibria in which player 1's payoff falls short of the bound established in Cripps, Schmidt, and Thomas (1996). Because Cripps and Thomas (1995) showed that equilibria do respect this bound in the limiting case of perfectly patient players, we thus have a discontinuity.

	L	R
T	2, 1	0, 0
B	0, 0	−1, 2

Figure 16.6.4 A strictly dominant action game. Player 1's Stackelberg action, T, is a dominant action in the stage game for the normal type and player 2's best response gives 1 the largest feasible stage-game payoff.

16.6.3 Example 3: Strictly Dominant Action Games

Suppose player 1's Stackelberg action is a strictly dominant action for the normal type in the stage game, and player 2's best response to this action produces the highest stage-game payoff available to player 1 in the set $\mathscr{F}^p$. Figure 16.6.4 presents an example. We say that such a game is a *strictly dominant action game*. Notice that player 2's best response to 1's Stackelberg action need not minmax player 2, and hence conflicting interests are not required. Chan (2000) shows that if such a game is perturbed to add a single possible commitment type for player 1, in the form of a type who always plays the Stackelberg action, then the normal player 1 receives the Stackelberg payoff v_1^* in any sequential equilibrium. This result holds regardless of the discount factors of the two agents. We prove this result here for the game shown in figure 16.6.4.

Low payoffs could be achieved for player 1 in sections 16.6.1–16.6.2, despite the possibility of being a Stackelberg type, because the normal player 1 could be induced to reveal his type. In figure 16.6.4, because the commitment action is strictly dominant for the normal player 1 and provides player 1 with the largest possible payoff when player 2 chooses a best response, one cannot force the normal player 1 to reveal his type. As a result, player 1 can build a reputation for playing like the Stackelberg type that imposes a lower bound on 1's payoff.

Let $\sigma = (\sigma_1, \sigma_2)$ be a candidate equilibrium. Our first task is to show that the normal player 1 always chooses T after every history in this equilibrium.[14] Suppose this is not the case, and thus that there exists a period t and a history h^t at which $\sigma_1|_{h^t}(B) > 0$. We construct a recursive argument that leads to a contradiction.

To begin, let t_0 be the first period in which there is a history h^{t_0} with $\sigma_1|_{h^{t_0}}(B) > 0$. If this is to be optimal, it must be the case that σ attaches positive probability to the event that if player 1 chooses T in period t_0 and every subsequent period, then a future period t and history h^t, consistent with h^{t_0} and subsequent choices of T by player 1, appears under which player 2 chooses R. Otherwise, playing T in period t_0 and subsequent periods ensures the continuation payoff 2 for player 1, and so a payoff superior to any that can be obtained by playing B in period t_0.

Call the first such future period t and the corresponding history h^t. Because $\sigma_2|_{h^t}(R) > 0$, there must exist $\varepsilon > 0$, and a time $t' \geq t$ and history $h^{t'}$ (consistent

---

14. Here we see that the result is established for sequential rather than Nash equilibria.

with h^t) under which 1 plays T in periods $t, \ldots, t'$ and in period t', $\sigma_1|_{h^{t'}}(B) > \varepsilon$. If not, and therefore player 1 chooses T with probability larger than $1 - \varepsilon$ after every continuation history featuring the persistent play of T, then as ε approaches 0 a lower bound on 2's continuation payoff when playing L in the current period approaches 1, whereas an upper bound on the payoff from playing R approaches δ, contradicting the optimality of R.

Set $t_1 = t'$ and $h^{t_1} = h^{t'}$. Continuing in this fashion, we obtain a sequence of periods $\{t_n\}_{n=0}^\infty$ and histories $\{h^{t_n}\}_{n=0}^\infty$ with each h^{t_n} an initial segment of $h^{t_{n+1}}$ and with the properties that (i) player 1's realized action is T in every period of each history h^{t_n}, (ii) $\sigma_1|_{h^{t_n}}$ puts probability bounded away from 0 on B after infinitely many of the histories h^{t_n}, and (iii) player 2 puts positive probability on R after infinitely many such histories h^{t_n}. This is a contradiction. In particular, the posterior probability that player 2 attaches to the event that 1 is the Stackelberg type must rise to 1 along this sequence of histories, ensuring that a best response cannot put positive probability on R infinitely often.[15]

We conclude that player 1 must choose T after every history. The best response is then for 2 to choose L after every history, yielding the result. Notice that we have used no assumptions about player 1's patience in this argument, and it holds even if player 1 is less patient than player 2.

16.6.4 Example 4: Strictly Conflicting Interests

This example, following Cripps, Dekel, and Pesendorfer (2004), illustrates a second class of games in which a reputation result can be achieved with two long-lived players and equal patience. A game of strictly conflicting interests is a game of conflicting interests in which the combination of player 1's Stackelberg action and each of player 2's best responses yields the highest stage-game payoff possible for player 1, or $\max_{v \in \mathscr{F}^p} v_1$, and the (mixed) minmax playoff $\underline{v}_2$ to player 2, and in which every other payoff profile $v' \in \mathscr{F}^p$ with $v'_1 = \max_{v \in \mathscr{F}^p} v_1$ also gives $v'_2 = \underline{v}_2$.

The low equilibrium payoffs obtained in sections 16.6.1–16.6.2 appear because player 1 can provide incentives for player 2 to *not* play a best response to the Stackelberg action while also causing the posterior probability of player 1's type to change very slowly. This is done by relying heavily on continuation payoffs to create incentives (again, an impossibility with short-lived player 2s), allowing the normal and Stackelberg types to play very similarly and hence reveal little information. Strictly conflicting interests preclude such a possibility. Player 2 can be given incentives to not best respond to the Stackelberg action only if the normal and Stackelberg types play quite differently in the current period, ensuring rapid belief revision. This allows the normal player 1, by mimicking the Stackelberg type, to enforce a payoff bound near the Stackelberg payoff. We illustrate this argument by showing that the type of equilibrium constructed in sections 16.6.1 and 16.6.2 cannot be used to give the normal player 1 a payoff near his minmax level.

---

15. Notice we use the fact that the only commitment type is the pure Stackelberg *type*, because the argument is that player 2 attach high probability to the Stackelberg type rather than Stackelberg action.

	L	R
T	2, 0	0, −1
B	1, 0	0, 1

Figure 16.6.5 A game of strictly conflicting interests.

	L	R
T	$\delta^{N-1-t}(2,0)$	$\delta^{N-1-t}(2,0)$
B	$\approx(1,0)$	$\delta^{N-1-t}(2,0) + (1-\delta^{N-1-t})(0,1)$

Figure 16.6.6 Period $t+1$ continuation payoffs for $t < N$, if period t is in phase one.

The stage game is shown in figure 16.6.5. The Stackelberg action is T. Player 2's best response gives her a payoff of 0, the minmax level.

During the first N periods of an equilibrium patterned after sections 16.6.1 and 16.6.2, the game is in phase one if only outcomes TL and TR have been observed. Reaching period N in phase one produces perpetual play of TL. An outcome of BL reveals that player 1 is the normal type, at which point we continue with an equilibrium of the complete information game that we can choose to give a payoff arbitrarily close to $(1, 0)$ (as δ gets large). A play of BR induces a public mixture between continuing with permanent play of TL (probability δ^{N-t-1}) and BR to support the normal player 1's indifference.

The period $t+1$ continuation payoffs under this strategy, for $t < N$, in phase one and as a function of the current action profile, are given in figure 16.6.6. With what probability must the normal player 1 choose B in period $t < N$ to support player 2's mixture? Making the approximation that the continuation payoff following BL is literally $(1, 0)$ allows us to calculate a lower bound on this probability (by making L less attractive to player 2). Player 2's indifference requires

$$0 = (1-\psi^t)(1-\delta)(-1) + \psi^t[(1-\delta) + \delta - \delta^{N-t}],$$

where $\psi^t = (1-\hat{\mu}^t)(1-\phi^t)$ is the product of the posterior probability $(1-\hat{\mu}^t)$ that player 1 is normal and the probability $(1-\phi^t)$ that the normal player 1 chooses B. In period $N-1$, for example, we must have $\psi^{N-1} = 1/2$. The probability attached to B drops off slowly as we move backward through previous periods. As a result, the posterior probability attached to player 1's being the Stackelberg type increases rapidly, and N is relatively small. This ensures that δ^N approaches unity as does δ.

To see what lies behind the result, compare the conflicting interests game of section 16.6.2 with the strictly conflicting interests game of this section. In each case, we are interested in the possibility of a relatively low equilibrium payoff for player 1.

A low equilibrium payoff for player 1 requires that player 2 choose R (or more generally, that player 2 choose an action that is not a best response to player 1's commitment type). If the incentives to play R can be created with a player 1 strategy that prompts player 2 to revise her beliefs about player 1 only very slowly, then the possibility arises of a low equilibrium payoff for player 1. In particular, the obvious route to a high player 1 equilibrium payoff, persistently playing T while player 2's expectation that she faces such behavior increases to the point that she plays L, now takes ineffectively long. This is the case in section 16.6.2. Suppose instead that the only way to create incentives for player 2 to play R ensures that player 2 will revise her beliefs about player 1 quite rapidly. Then playing as the commitment type promises a relatively high payoff for player 1, leading to a large lower bound on player 1's equilibrium payoff.

In both figures 16.6.2 and 16.6.5, player 2 is willing to play R only if either player 1 attaches a sufficiently large probability to B or playing R gives rise to a sufficiently lucrative continuation payoff (or some combination of both). The continuation payoffs listed in figure 16.6.3 show that in the game of section 16.6.2, a relatively large continuation payoff for player 2 could be attached to play of BR. This is possible because the game does not exhibit *strictly* conflicting interests, and hence player 1's Stackelberg payoff is not his largest stage-game payoff. A relatively large payoff for player 1 can then be attached to BR, maintaining 1's incentives to play B while also providing a relatively lucrative payoff for player 2. This ensures that incentives for 2 to play R can be created while attaching only a small probability to 1 playing B. But then very little information is revealed about player 1.[16] In figure 16.6.5, the strictly conflicting interests property ensures that if player 1 is to be provided the relatively high continuation payoff required to induce play of B, then 2 must receive a low continuation payoff. This in turn ensures that the incentives for 2 to choose R must come primarily from player 1's current play of B. But if this is the case, B must be played with high probability, ensuring that player 2's posteriors about player 1's type evolve rapidly. There is then little cost to player 1 of always choosing T, ensuring a payoff close to the Stackelberg payoff. Player 1's payoff in any equilibrium must then be near the Stackelberg payoff.

Cripps, Dekel, and Pesendorfer (2004) show that this result depends neither on the particular game we have examined nor the candidate equilibrium. For any game of strictly conflicting interests and any $\varepsilon > 0$, there is a $\underline{\delta}$ and μ such that if $\delta > \underline{\delta}$ and the total probability of all commitment types other than the Stackelberg type is less than μ, then player 1's equilibrium payoff must be within ε of his pure Stackelberg payoff.

---

16. With a short-lived player 2, or even with two differentially patient long-lived players, there is no issue of how rapidly player 2 learns about player 1. We simply make player 1 sufficiently patient that however long it takes player 2 to play a best response to the commitment type, it is soon enough.

	E	S
E	3,3	0,0
S	0,0	1,1

Figure 16.6.7 Game of common interest.

16.6.5 Bounded Recall

A game features common interests if the stage game contains a payoff profile that strictly dominates any other distinct payoff profile. The game examined in section 16.6.1 is an example. Aumann and Sorin (1989) examine repeated games of common interest played by two long-lived and equally patient players. There is uncertainty about both players' types. The set Ξ_i of player i's types includes every pure simple type, and includes only types committed to pure strategies with bounded recall. Aumann and Sorin (1989) show that for every $\gamma > 0$, there is a discount factor $\underline{\delta}(\gamma)$ and probability μ such that for any discount factor in $(\underline{\delta}(\gamma), 1)$ and any prior distribution μ^0 over Ξ that places probability at least $1 - \mu$ on the normal type, every pure-strategy equilibrium of the repeated game must give payoffs within γ of the efficient payoff profile.

Example 16.6.1 The game given in figure 16.6.1 is a game of common interest. In Aumann and Sorin's (1989) setting, the normal players must receive a payoff near 1. We illustrate the reasoning behind their result with the variation of this game shown in figure 16.6.7, modified so as to have two strict equilibria and labeled symmetrically to conserve on notation.

Let the set Ξ include the normal type and a type committed to each of the one-period recall strategies. We specify a commitment type by a strategy with a triple, xyz, where $x, y, z \in \{E, S\}$ and where x specifies the first action taken by the strategy, y specifies the action if the opponent played E in the previous period, and z specifies the action taken if the opponent played S in the previous period.[17] Figure 16.6.8 shows the eight one-period recall strategies, with the payoffs they obtain against one another, evaluated in the limit as discount factors approach unity. Notice that these include a type committed to S (SSS) and one committed to E. Let the prior distribution attach probability $\hat{\mu}$ to each of these eight types, with $1 - 8\hat{\mu}$ probability of a normal type.

Now consider a pure-strategy equilibrium in which the normal types of players 1 and 2 choose strategies σ_1 and σ_2. The bounded recall of the commitment types ensures that for any commitment type ξ that produces a different outcome path against σ_1 than does σ_2, σ_1 must earn close (arbitrarily close to, as δ approaches 1) the highest possible payoff against ξ. Intuitively, on observing an

---

17. To complete the strategy specification, assume that one's own past deviations are ignored when choosing continuation play.

16.6 ■ Equal Discount Factors

	EEE	EES	ESE	ESS	SEE	SES	SSE	SSS
EEE	3, 3	3, 3	0, 0	0, 0	3, 3	3, 3	0, 0	0, 0
EES	3, 3	3, 3	1, 1	1, 1	3, 3	0, 0	1, 1	1, 1
ESE	0, 0	1, 1	2, 2	0, 0	0, 0	1, 1	0, 0	0, 0
ESS	0, 0	1, 1	0, 0	1, 1	0, 0	1, 1	0, 0	1, 1
SEE	3, 3	3, 3	0, 0	0, 0	3, 3	3, 3	0, 0	0, 0
SES	3, 3	0, 0	1, 1	1, 1	3, 3	1, 1	1, 1	1, 1
SSE	0, 0	1, 1	0, 0	0, 0	0, 0	1, 1	2, 2	0, 0
SSS	0, 0	1, 1	0, 0	1, 1	0, 0	1, 1	0, 0	1, 1

Figure 16.6.8 Payoffs from commitment types with one-period recall.

action not played by σ_2, the strategy σ_1 can experiment sufficiently to identify the opponent's commitment type and then play a best response to that type. Given the limited recall of that type's strategy, the experimentation has no long-run impact on payoffs.[18] Notice that we use the restriction to pure strategies here. Mixtures play a crucial role in obtaining the low equilibrium payoffs in sections 16.6.1 and 16.6.2.

These considerations leave two possibilities for the equilibrium strategy σ_2. One is that σ_1 and σ_2 combine to produce a path of persistent mutual effort. In this case, the normal player equilibrium payoff becomes arbitrarily close to $(3, 3)$ as $\hat{\mu}$ gets small and δ gets large. The other possibility is that σ_1 and σ_2 do not produce persistent mutual effort. In this case, by playing commitment strategy EEE instead of σ_2, player 2 can ensure that 2's strategy and σ_1 produce an outcome path that eventually features persistent mutual effort and hence a payoff that again approaches $(3, 3)$ as $\hat{\mu}$ gets small and δ gets large. Thus the equilibrium payoff must be close to $(3, 3)$.

It is important for this result that set of types be sufficiently rich. If the only commitment types are SSS and SES, there is an equilibrium of the repeated game in which the normal types invariably shirk. Second, bounded recall among the commitment types is also important. Suppose the eight commitment types of figure 16.6.8 are joined by another commitment type who plays E in odd periods and S in even ones, as long as the opponent has always done so, and plays S after every other history. Consider a candidate equilibrium in which the normal players

---

18. Grim trigger does not have bounded recall. If player 1 assigns positive probability to 2 being a grim trigger type, a play of S can lead to permanent S, making experimentation potentially too costly.

choose E in odd periods and S in even ones. In the absence of this additional type, player 1 would have an incentive to continually play E, causing player 2 to play a best response to what appears to be an EEE player. Doing so in the presence of our newest type raises the risk of perpetual shirking, making this strategy less attractive. The restriction to bounded-recall commitment types excludes such possibilities.

●

16.6.6 Reputations and Bargaining

Reputations have played a role in work generalizing Rubinstein's (1982) model of alternating offers bargaining to incomplete information, raising issues related to those that appear in repeated games. Chatterjee and Samuelson (1987, 1988) begin with a model of bargaining between a buyer and seller, each of whom may be either *hard* or *soft*. The gains from trade are largest with two soft opponents, smaller but positive if one is soft and one hard, and nonexistent if both are hard. Chatterjee and Samuelson (1987) limit the offers available to the agents to model the bargaining as a war of attrition. Individual rationality limits hard agents to a single demand that they make after every history, whereas soft agents face a choice between mimicking the hard agents or making a concession. The first concession ends the game with an agreement.

The game generically has a unique sequential equilibrium in which soft agents continually mix between conceding and mimicking their hard types. As play proceeds without a concession, each agent revises upward his posterior that his opponent actually *is* hard. Eventually, these posteriors become sufficiently high that it is no longer worthwhile for soft agents to continue, and the game either ends with a soft-agent concession or the realization that both agents are hard and that there are no gains from trade. The ex ante expected division of the surplus depends on the prior distribution of types, with a soft agent's expected value increasing as the agent is more likely to be hard and the opponent more likely to be soft.

The hard agents in this bargaining model are the counterpart of action commitment types in repeated games. The soft agents are normal types whose payoffs are bounded by the payoffs they would receive if they mimicked hard types. The gains from doing so are a result of the belief revision that such behavior elicits from the opponent. The key difference is that a concession ends the bargaining game. Thus, there are histories that induce unique continuation values, eliminating the ability to use future play in creating the rewards and punishments that lie behind the multiplicity of equilibria in repeated games.

The equilibria in Chatterjee and Samuelson (1987) depend heavily on the structure of the model, most notably the specification of possible types and the restriction on actions that makes the bargaining a concession game. Abreu and Gul (2000) show that these equilibrium properties can be derived as necessarily characterizing the equilibria of sufficiently patient players in a more general game. Abreu and Gul (2000) again work with two-sided uncertainty about types,[19] but allow a richer set of commitment

---

19. Schmidt (1993a) examines finite-horizon bargaining games with one-sided incomplete information about types.

types and an unrestricted set of offers. This opens the possibility that play may look nothing like a concession game. However, if players can make offers sufficiently rapidly, equilibria of the game have a concession game structure. Normal types choose a commitment type to mimic and do so until a normal type makes a concession that essentially ends the game.

These results rest on two important insights. First, once the type of each player is known, the players face a version of Rubinstein (1982) complete information bargaining game featuring a unique equilibrium outcome (with immediate agreement). We thus again have histories after which future rewards and punishments cannot be used to create incentives. Second, by focusing on short time periods, "Coase conjecture" arguments (Gul, Sonnenschein, and Wilson 1986) can be invoked to again achieve immediate agreement in games of one-sided incomplete information. Because the first concession (i.e., the first deviation from mimicking a commitment type) reveals that the conceding player is normal and yields one-sided incomplete information, we again have a concession game.

16.7 Temporary Reputations

The results of section 15.5 extend to games with two long-lived players. Examining games with uncertainty about player 1's type, we present the conditions under which a long-lived player 2 must eventually almost surely learn the type of player 1. The model (and notation) is as in section 15.5, except that player 2 is now a long-lived player.

As we saw in section 16.4, when player 2 is long-lived, nonsimple Stackelberg types may give rise to higher lower bounds on player 1's payoff than do simple types. We accordingly do not restrict attention to simple commitment types.

When dealing with commitment types that are not simple, we must impose a noncredibility analogue of assumption 15.5.3 directly on the infinitely repeated game of complete information. We assume the commitment type plays a repeated-game strategy $\hat{\sigma}_1$ with the properties that (i) player 2's best response $\hat{\sigma}_2$ is unique on the equilibrium path and (ii) there exists a finite time T^o such that for every $t > T^o$, a normal player 1 would almost surely want to deviate from $\hat{\sigma}_1$, given player 2's best response. That is, there is a period t continuation strategy for player 1 that strictly increases her utility. A strategy $\hat{\sigma}_1$ satisfying these criteria at least eventually loses its credibility and hence is said to have "no long-run credibility."

Player 2's set of best responses to strategy σ_1 in the game of complete information is given by:

$$B(\sigma_1) \equiv \{\sigma_2 : U_2(\sigma_1, \sigma_2) \geq U_2(\sigma_1, \sigma_2') \; \forall \sigma_2'\}.$$

Let U_i^t be player i's period t continuation value, viewed as a random variable defined on the set of outcomes Ω.

Definition 16.7.1 *The strategy $\hat{\sigma}_1$ has* no long-run credibility *if there exists T^o and $\varepsilon^o > 0$ such that for every $t \geq T^o$,*

1. $\hat{\sigma}_2 \in B(\hat{\sigma}_1)$ implies that with $\mathbf{P}^{(\hat{\sigma}_1,\hat{\sigma}_2)}$-probability one, $\hat{\sigma}_2^t$ is pure and

$$E^{\hat{\sigma}}[U_2^t \mid \mathcal{G}_2^t] > E^{(\hat{\sigma}_1,\sigma_2')}[U_2^t \mid \mathcal{G}_2^t] + \varepsilon^o, \qquad (16.7.1)$$

for all σ_2' attaching probability zero to the action played by $\hat{\sigma}_2(h_2^t)$ after $\mathbf{P}^{(\hat{\sigma}_1,\hat{\sigma}_2)}$-almost all $h_2^t \in \mathcal{H}_2^t$, and

2. there exists $\tilde{\sigma}_1$ such that for $\hat{\sigma}_2 \in B(\hat{\sigma}_1)$, $\mathbf{P}^{(\hat{\sigma}_1,\hat{\sigma}_2)}$-almost surely,

$$E^{(\tilde{\sigma}_1,\hat{\sigma}_2)}[U_1^t \mid \mathcal{G}_1^t] > E^{\hat{\sigma}}[U_1^t \mid \mathcal{G}_1^t] + \varepsilon^o.$$

This definition captures the two main features of assumption 15.5.3, a unique best response and absence of equilibrium, in a dynamic setting. In particular, the stage-game action of any simple strategy satisfying definition 16.7.1 satisfies assumption 15.5.3. In assuming the best response is unique, we need to avoid the possibility that there are multiple best responses to the commitment action "in the limit" (as t gets large). We do so by imposing a uniformity condition in definition 16.7.1(1), that inferior responses reduce payoffs by at least ε^o. The condition on the absence of equilibrium in definition 16.7.1(2) similarly ensures that for all large t, player 1 can strictly improve on the commitment action. Again it is necessary to impose uniformity to avoid the possibility of an equilibrium in the limit.

Proposition 16.7.1 *Suppose the monitoring distribution π satisfies assumptions 15.4.1, 15.5.1, and 15.5.2 and the commitment type's strategy $\hat{\sigma}_1$ is public and has no long-run credibility. Then in any Nash equilibrium of the game with incomplete information, $\hat{\mu}^t \to 0$ $\tilde{\mathbf{P}}$-almost surely.*

We require the commitment strategy $\hat{\sigma}_1$ to be public so that player 1 can usefully anticipate player 2's best response to $\hat{\sigma}_1$, once 1 is convinced 2 plays a best response to $\hat{\sigma}_1$. This is automatically the case in section 15.5, where the commitment type is simple. The combination of publicness and no long-run credibility still leave a vast array of behavior as possible commitment strategies.

The proof of proposition 16.7.1 (Cripps, Mailath, and Samuelson 2004b, section 4) follows that of proposition 15.5.1, accounting for two complications. First, an additional argument is required to show that a *long-lived* player 2 best responds to the commitment type once convinced she is almost certainly facing the commitment strategy. Second, we can ensure only that for some fixed period t, the normal type's contradictory deviation from the commitment strategy occurs within in some finite number of periods, $t, \ldots, t + L$, exacerbating the measurability problems confronted in the proof of proposition 15.5.1.

17 Finitely Repeated Games

The concept of a reputation was introduced into economics in response to the observation that certain finitely repeated games have counterintuitive subgame-perfect equilibria. Selten (1978) used the chain store game, shown in figure 17.0.1 (and previously encountered in figures 5.4.1 and 16.2.1), to dramatize this issue. The only subgame-perfect equilibrium of the extensive-form stage game involves player 2 choosing *In* and player 1 choosing *A*. Building on this uniqueness, a backward induction argument shows that the only subgame-perfect equilibrium in the finitely repeated chain store game, no matter how long, features entry and acquiescence in every period. However, one's intuition is that if player 2 enters in the first period, player 1 might fight, implicitly making a statement of the form "I know fighting is costing me now, but it will be well worth it if subsequent player 2s are convinced by this display that I will fight them as well, and hence decide not to enter." Similar reasoning arises in the finitely repeated prisoners' dilemma. The unique Nash equilibrium, featuring persistent shirking, contrasts with the intuition that players are likely to exert effort in early periods.

With Rosenthal (1981) as a precursor, Kreps, Milgrom, Roberts, and Wilson (1982), Kreps and Wilson (1982a), and Milgrom and Roberts (1982) addressed this "chain store paradox" with a model centered around incomplete information about player 1's type.[1] With some (possibly quite small) probability, player 1 is a commitment type who always fights entry. If the finitely repeated game is long enough, this allows an equilibrium in which the player 2 chooses *Out* virtually all of the time. Notice that player 1's payoff under incomplete information lies outside the set of equilibrium payoffs in the repeated game with complete information. We have seen a similar possibility in public monitoring infinitely repeated games in section 15.4.

A similar argument allows equilibrium effort to appear in the finitely repeated prisoners' dilemma. Continuing in this vein, Fudenberg and Maskin (1986, section 5) present a folk theorem for finitely repeated games of incomplete information about players' types.

This chapter presents three examples of reputation equilibria in finitely repeated games, examining the chain store game, the perfect monitoring prisoners' dilemma, and the imperfect monitoring product-choice game.

---

1. See Aumann (2000) for a related investigation. Masso (1996) shows that entry can occur in a subgame-perfect equilibrium of the chain store game with complete information if the player 2s are uncertain as to their place in the line of potential player 2s, and if they cannot observe perfectly the previous history of play. The idea behind this result is that no player 2 can ever know that she is the last one, ensuring that the backward induction argument never gets started.

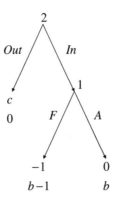

Figure 17.0.1 Chain store game. Player 2 can choose either to stay out of a market (*Out*) or enter (*In*), at which point player 1 can choose to fight (*F*) or acquiesce (*A*). Player 1's payoffs appear first at each terminal node. We assume $c > 1$ and $b \in (0, 1)$.

These stage games have unique equilibria. Section 4.4 explains why equilibria of repeated games whose stage games have unique equilibria are *not* representative of equilibria in finitely repeated games whose stage games have multiple equilibria. Finitely repeated games with multiple stage-game equilibria, when repeated sufficiently often, look much like infinitely repeated games.

Section 15.5 showed that uncertainty about a player 1's type cannot persist forever in an infinitely repeated game of imperfect monitoring. Similarly, types must also become known in sufficiently long finitely repeated games (see Cripps, Mailath, and Samuelson 2004b, section 3.4). In contrast, sections 17.1 and 17.2 construct equilibria for the finitely repeated chain store game and prisoners' dilemma game in which no information is revealed about player 1's type until the very end of the game, no matter how long the game. We see again that perfect and imperfect monitoring have quite different implications for reputation building.[2]

17.1 The Chain Store Game

The chain store stage game is shown in figure 17.0.1. The game is played in T periods, with a long-lived player 1 facing a new player 2 in each period. Player 1 does not discount. In each period, both players observe all of the previous choices. These

---

2. Jackson and Kalai (1999) take an alternative approach to arguing that reputations must be temporary in a finitely repeated game. They consider a sequence of finitely repeated games, set in two possible environments. In the pure environment, there are only normal player 1s. In a mixed environment, at the beginning of each finitely repeated game, player 1 is drawn to be a commitment type with probability $\hat{\mu}$ and is otherwise normal. Over the course of the repeated finitely repeated games, player 2s draw inferences about the probability that the environment is mixed. Jackson and Kalai (1999) show that if the environment is indeed normal, then player 2's posterior that the environment is normal approaches 1 almost surely. This in turn implies that the probability attached to a commitment type of player 1 eventually falls so low as to eliminate any reputation effects in the finitely repeated game. Conlon (2003) argues that the rate at which reputations disappear in Jackson and Kalai's (1999) model can be very slow, and hence reputation effects can persist for a very long time.

choices will not be very informative about player 1's behavior unless player 2 has chosen *In* in the past.

The calculations are easier to follow if we number periods in reverse, so that 1 is the last period and T the first, with period t following period $t+1$. With prior probability $\hat{\mu}^T$, player 1 is a commitment type who chooses the Stackelberg action F. We assume $\hat{\mu}^T$ is sufficiently small that player 2 would prefer to enter, given probability $1 - \hat{\mu}^T$ of a normal player 1, if certain that the normal player 1 would acquiesce. The posterior probability attached to 1 being the commitment type in period t is given by $\hat{\mu}^t$.

We present a sequential equilibrium for the finitely repeated game. We begin with the evolution of player 2's posterior beliefs about player 1. Let a^t be the realized profile of actions in period t. Then[3]

$$\hat{\mu}^t = \begin{cases} \hat{\mu}^{t+1}, & \text{if } a^{t+1} = (Out, \cdot), \\ \max\{b^t, \hat{\mu}^{t+1}\}, & \text{if } a^{t+1} = (In, F) \text{ and } \hat{\mu}^{t+1} > 0, \\ 0, & \text{if } a^{t+1} = (In, A) \text{ or } \hat{\mu}^{t+1} = 0. \end{cases}$$

Player 2's behavior in period t is given by

$$\alpha_2^t(In) = \begin{cases} 0, & \text{if } \hat{\mu}^t > b^t, \\ 1 - \frac{1}{c}, & \text{if } \hat{\mu}^t = b^t, \\ 1, & \text{if } \hat{\mu}^t < b^t. \end{cases}$$

Finally, the commitment player 1 always fights entry. The normal player 1 fights in period t, should player 2 enter, as follows:

$$\alpha_1^t(F) = \begin{cases} 0, & \text{if } t = 1, \\ 1, & \text{if } t > 1 \text{ and } \hat{\mu}^t \geq b^{t-1}, \\ \frac{(1-b^{t-1})\hat{\mu}^t}{(1-\hat{\mu}^t)b^{t-1}}, & \text{if } t > 1 \text{ and } \hat{\mu}^t < b^{t-1}. \end{cases}$$

Throughout the game, the critical event is the observation of A from player 1. Once this happens, the posterior probability that player 1 is the commitment type drops to 0 and remains there, with each subsequent period featuring (In, A). Describing the equilibrium is thus a matter of identifying play following histories in which an A has not been observed. Play begins in an initial phase in which the posterior probability of a commitment type, $\hat{\mu}^t$, is unrevised from the prior $\hat{\mu}^T$ and exceeds b^{t-1}. During this phase, no entry occurs, and the posterior proceeds to the next period unchanged at $\hat{\mu}^{t+1} = \hat{\mu}^T$. Should entry occur, the normal player 1 fights with probability one and the posterior again remains unchanged. As the game length T increases, the length of this initial period similarly increases, becoming an arbitrarily large fraction of total play as T gets large. Hence, for very long finitely repeated games, entry is deterred in virtually all periods.

This first phase ends with a pair of transition periods. The first of these periods is the last period t in which $\hat{\mu}^T > b^t$. In this period, player 2 remains out and the posterior

---

3. Notice that b^t denotes the parameter b raised to the power t, whereas a^t and $\hat{\mu}^t$ have their usual meanings of the period t action profile and posterior belief.

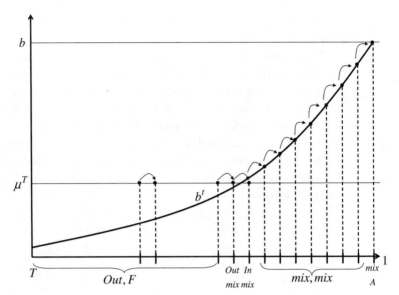

Figure 17.1.1 Illustration of equilibrium strategies for the chain store game. Player 2's action is listed first in each case. The figure traces the evolution of the posterior belief $\hat{\mu}^t$, given that each nontrivial player 2 mixture produces the realization In and player 1's mixture produces the realization F after any such entry. All other belief updating is degenerate.

probability remains at $\hat{\mu}^T$.[4] In the out-of-equilibrium event that player 2 enters in this period, however, the normal player 1 would mix between F and A, with the posterior (in the event of F) increasing from $\hat{\mu}^t$ to $\hat{\mu}^{t-1} = b^{t-1}$. Equilibrium play would then continue as prescribed. In the second transition period, say, t', player 2 enters with probability one, and the normal player 1 mixes his response to bring the posterior from $\hat{\mu}^T$ up to $\hat{\mu}^{t'-1} = b^{t'-1}$.

This transition period is followed by a phase in which both player 1 and player 2 mix (see figure 17.1.1). If player 2's mixture produces entry and player 1 fights, then the posterior probability climbs along the function b^t, with $\hat{\mu}^t = b^t$ in each period. If a period t occurs in which the realization of player 2's mixture is to stay out, then there is no revision in the posterior probability, with $\hat{\mu}^{t-1}$ then lagging behind b^{t-1}. There then follows a period in which player 2 enters with certainty and the normal player 1 mixes in response, restoring the equality $\hat{\mu}^{t-2} = b^{t-2}$ (assuming player 1's realization is to fight). The game culminates in a final period in which player 1 is thought to be the commitment type with probability b, player 2 mixes, and the normal player 1 acquiesces.

Why would we expect the equilibrium to take this form?[5] It is clear that if the game is long and the commitment type rare, then the normal type of player 1 must fight entry in early periods. If not, a single instance of F would push the posterior on

---

4. We assume there is no value of t for which $\hat{\mu}^T = \hat{\mu}^t = b^t$.
5. Because there is no t such that $b^t = \hat{\mu}^T$, this is the unique sequential-equilibrium outcome (Kreps and Wilson 1982a). Kreps and Wilson (1982a) work with a payoff rather than action commitment type of player 1, allowing additional sequential equilibrium outcomes (see remark 15.3.2).

17.1 ■ The Chain Store Game

the commitment type to 1, ensuring that there is no subsequent entry and thus offering a bonus well worth the cost of fighting. In the final period, in contrast, the normal player 1 will surely acquiesce and the commitment type will fight. The task is now to specify play during a transition period that connects these two pieces.

Actions during these transition periods will in general have to be mixed. If not, then we have an initial phase in which player 2s stay out and player 1 fights (if entry occurs), followed by a final phase in which player 2s enter and the normal player 1 acquiesces. Consider the final period of the first phase. Given the putative equilibrium strategies, no information is revealed, and no adjustment in future play is prompted by fighting entry. But given that this is the case, a normal player 1 would not fight, disrupting the equilibrium. If we attempt to rescue the equilibrium by assuming that only the commitment player 1 fights in this period, then we have the difficulty that player 2 would find it optimal to enter (given that the posterior of a commitment player 1 remains fixed at its prior value of $\hat{\mu}^T$). Therefore, the transition phase must involve mixed actions, creating the dependence between current actions and future play that sustains incentives to fight just before the transition phase.

The equilibrium strategies are constructed to preserve the indifferences required for these mixtures. First, player 2 is willing to mix in the final period, knowing that the normal player 1 acquiesces and the commitment player 1 fights, only if $b(1 - \hat{\mu}^1) + (b - 1)\hat{\mu}^1 = 0$, leading to the requirement that the terminal posterior probability must equal b. From this beginning, we determine player 2's mixture in each period to make the normal player 1 indifferent between acquiescing and fighting in the previous period, and jointly choose a sequence of posterior probabilities that player 1 is the commitment type and a sequence of mixed strategies for the normal player 1 so that player 2 is indifferent between entering and staying out in each period. We work backward in this way until the required mixtures hit pure strategies, giving the first phase in which entry does not occur and the normal player 1 would fight if it did.

We now verify that the proposed strategies constitute an equilibrium. Begin with the calculation of posterior probabilities. The nontrivial cases are those in which entry has occurred in period t, the equilibrium calls for player 1 to mix, and player 1's realized action is to fight. Then we have

$$\hat{\mu}^{t-1} = \frac{\hat{\mu}^t}{\hat{\mu}^t + (1 - \hat{\mu}^t)\frac{(1-b^{t-1})\hat{\mu}^t}{(1-\hat{\mu}^t)b^{t-1}}}$$

$$= b^{t-1},$$

matching the equilibrium specification.

We next consider the optimality of player 2s' behavior. The only nontrivial case here is a period in which $\hat{\mu}^t = b^t$ and hence player 2 mixes between entering and not doing so. The latter gives a payoff of 0, so that optimality requires

$$b\left[(1-\hat{\mu}^t)\left(1 - \frac{(1-b^{t-1})\hat{\mu}^t}{(1-\hat{\mu}^t)b^{t-1}}\right)\right] + (b-1)\left[\hat{\mu}^t + (1-\hat{\mu}^t)\frac{(1-b^{t-1})\hat{\mu}^t}{(1-\hat{\mu}^t)b^{t-1}}\right] = 0,$$

where the first term in brackets on the right side is the probability that player 1 acquiesces, and the second the probability that player 1 fights. A straightforward simplification of the right side, using $\hat{\mu}^t = b^t$, verifies the equality.

It remains to verify the optimality of player 1's strategy. Again, the nontrivial case is one in which player 1 mixes. It is a straightforward calculation that player 1 is indifferent in period 2 if called on to randomize. In earlier periods, acquiescing gives a (repeated-game) payoff of 0. Fighting gives a payoff of $-1 + (1/c)c = 0$.[6] It is then clear player 1 is indifferent between fighting and acquiescing, giving an equilibrium.

Finally, we describe the length of the second two phases of the equilibrium, in which the possibility of entry appears. The first period in which the player 2s potentially enter corresponds to the largest integer smaller than the (noninteger) value of t solving

$$\hat{\mu}^T = b^t.$$

Hence the smaller the prior probability of a commitment player 1, the longer the mixing phase of the equilibrium. Viewing this phase as a transition from the prior probability of $\hat{\mu}^T$ to the terminal posterior of b^t, with the rate at which the posterior increases set by the requirement that each player's mixed actions are calculated to preserve indifference on the part of the other player, this is expected. Rearranging this expression, we can say that the length of the final two phases is on the order of $-\ln \hat{\mu}^T$. As a result, entry will be deterred in most periods of long games.

17.2 The Prisoners' Dilemma

This section presents an equilibrium for the T period prisoners' dilemma. We assume that both players are long-lived with no discounting. In addition, there is uncertainty about the types of both players. We present a sequential equilibrium in which for sufficiently long games, the outcome features mutual effort in most periods.[7]

It will help in interpreting various steps of the derivation to represent the prisoners' dilemma as in figure 17.2.1. It then simplifies the calculations to assume that $d = b - c$, as is the case with the prisoners' dilemma of figure 1.2.1 with which we commonly work. We will note as we proceed where the argument depends upon this assumption. Define

$$\phi = \frac{c}{b} < 1.$$

We again number the periods so that the final period is period 1 and the first period is T.

---

6. The first term (-1) is the current cost of fighting. The second is calculated by noting that if player 2 enters in the next period (probability $1 - 1/c$), then arguing recursively on the period, the normal player 1 will be indifferent between fighting and acquiescing, with the latter producing a payoff of 0; whereas if player 2 stays out (probability $1/c$), then the normal player 1 reaps a payoff of c, but the absence of belief revision ensures that player 2 enters in the next period, with player 1's mixture again ensuring an expected payoff of 0.

7. The equilibrium we examine is taken from Conlon (2003). Kreps, Milgrom, Roberts, and Wilson (1982) examine prisoners' dilemmas of arbitrary (finite) length in which one player is known to be normal and the other may be either normal or a commitment type who plays tit-for-tat. They show that there is an upper bound K on the number of stages in which defection can occur with positive probability, in any sequential equilibrium of the finitely repeated game of any length.

17.2 ■ The Prisoners' Dilemma

	E	S
E	d,d	$-c,b$
S	$b,-c$	$0,0$

	E	S
E	$2,2$	$-1,3$
S	$3,-1$	$0,0$

Figure 17.2.1 Prisoners' dilemma, where we assume $d = b - c$ (left). Our common example of a prisoners' dilemma, from figure 1.2.1 and reproduced here on the right, satisfies this assumption. As we have seen (e.g., sections 2.2 and 2.5.6), this case has some special and convenient properties.

Each player is a normal type with probability $1 - \hat{\mu}^T$ and is a commitment type who plays grim trigger with probability $\hat{\mu}^T$. The players' types are independently drawn. Using grim trigger rather than (say) tit-for-tat as the commitment type simplifies the calculations by making it easy to keep track of the commitment type's action once a player has shirked. Each player observes only his own type.

We construct a strongly symmetric sequential equilibria. Define

q^t = probability player i exerts effort in period t,

$\hat{\mu}^t$ = period t probability that player i is the commitment type,

and r^t = probability a normal player i exerts effort in period t.

We will be interested in these probabilities only after histories in which no shirking has yet occurred. Given the symmetry of the prior distribution from which types are drawn and our focus on strongly symmetric equilibria, these probabilities will be identical for the two players after such a history, obviating the need for player subscripts. These three probabilities are linked by the identity

$$q^t = (1 - \hat{\mu}^t)r^t + \hat{\mu}^t,$$

allowing us to solve for

$$r^t = \frac{q^t - \hat{\mu}^t}{1 - \hat{\mu}^t}. \tag{17.2.1}$$

Once shirking occurs, both players shirk for the remainder of the game. In particular, suppose player 1 shirks in period t and that this is the first incidence of shirking. Given that commitment types play grim trigger, this reveals that player 1 is normal. If player 2 is the commitment type, then player 2 will shirk for the remainder of the game. If player 2 is normal, then exerting effort in period $t - 1$ will reveal that player 2 is indeed normal, moving us to a continuation game in which both players are known to be normal and in which the continuation equilibrium calls for subsequent shirking. As a result, player 2 has nothing to gain from exerting effort in period $t - 1$, and hence must shirk. A similar argument shows that the players must also shirk in each subsequent period.

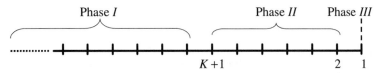

Figure 17.2.2 Equilibrium phases for the equilibrium constructed for the finitely repeated prisoner's dilemma. In any period $t > K + 1$, both players exert effort (given no previous shirking). Periods $K + 1$ to 2 are a transition phase during which normal players mix between effort and shirking (given no previous shirking). In period 1 (the final period), normal players shirk and commitment players exert effort (again, if there has been no previous shirking). Any instance of shirking produces shirking from both players in all subsequent periods.

To specify an equilibrium, it then suffices to consider histories in which there has been no shirking. We identify the posterior probabilities q^t and $\hat{\mu}^t$ for each $t = 1, \ldots, T$, counting on (17.2.1) to fill in the actions of the normal players. We then verify that beliefs satisfy Bayes' rule and that behavior is optimal.

We begin by choosing an even $K \in \mathbb{N}_0$ so that (recall $\phi < 1$, so that either $\hat{\mu}^T \geq \phi$, or for some k, $\phi^{k+1} \leq \hat{\mu}^T < \phi^k$)

$$\phi \leq \phi^{-\frac{K}{2}} \hat{\mu}^T < 1. \tag{17.2.2}$$

We examine an equilibrium in which both players exert effort during the first $T - K - 1$ periods of play. The next K periods are a transition phase during which normal players mix between effort and shirking, as long as there has been no previous shirking. The first incidence of shirking prompts all players to shirk thereafter. Should period 1 be reached with no shirking having occurred, normal players shirk and commitment players exert effort. Figure 17.2.2 illustrates.

To see the role played by K, suppose first that $\hat{\mu}^T > \phi$. In this case, commitment players are relatively plentiful. We can then take $K = 0$, meaning that there is no intermediate phase of behavior and every player exerts effort until the final period, with normal players then shirking. To verify that this behavior is optimal, consider the choice facing a normal player in the penultimate period (period 2). No information has yet been revealed about the probability that the opponent is a commitment type, which accordingly stands at its prior value of $\hat{\mu}^T$. The equilibrium action of effort gives a current-period payoff of d plus an expected payoff in the final period of $\hat{\mu}^T b$, obtained by shirking while a commitment opponent (present with probability $\hat{\mu}^T$) exerts effort. Shirking in period 2 gives a payoff of b, with a final-period payoff of 0 (given mutual shirking in the final period). Using the equality $d = b - c$, the proposed behavior is optimal if, as we have assumed, $\hat{\mu}^T > \phi$. Hence, the prior probability of a commitment opponent is sufficiently large that normal players are willing to exert effort to induce effort from commitment opponents, keeping alive until the final period the chance to fleece commitment opponents by shirking.

Suppose instead that $\hat{\mu}^T < \phi$. It now cannot be an equilibrium for all players to exert effort until the final period. A normal player facing such a supposed equilibrium

17.2 ■ The Prisoners' Dilemma

would instead shirk in the penultimate period, deeming it better to earn a payoff of b now and sacrifice final-period payoffs than to earn only d now in return for the unlikely event that the opponent is committed and hence exerts effort in the final period. Instead, an intermediate phase of play is required to connect the initial phase of unbroken effort with the final phase in which only commitment players (if any) exert effort. During this intermediate phase, the mixed strategies of normal players ensure that if effort is observed, the posterior probability attached to a commitment opponent increases. The requirement is that this posterior probability reach ϕ by period 1 to sustain normal player effort in period 2. The number of periods required to make this transition is K.

We continue the specification of an equilibrium with the unconditional probability of player 1 effort, given by (where these are well-defined probabilities by (17.2.2))

$$q^1 = q^3 = \cdots = q^{K+1} = \phi^{-\frac{K}{2}} \hat{\mu}^T,$$

$$q^2 = q^4 = \cdots = q^K = \phi^{\frac{K}{2}+1} \frac{1}{\hat{\mu}^T},$$

and

$$q^t = 1 \quad \text{if } t > K+1.$$

Effort is thus chosen until the final $K+1$ periods are reached. Within these periods, the probability of effort is constant across even-numbered periods, as it is across odd-numbered periods. Notice that if we take the product of the probability of effort across any two periods t and $t+1$ for $t \leq K+1$, we have

$$q^t q^{t+1} = \phi, \tag{17.2.3}$$

a relationship that simplifies calculations.

The equilibrium is completed by specifying the posterior probability of a commitment opponent as

$$\hat{\mu}^t = \begin{cases} \hat{\mu}^T, & \text{if } t \geq K+1, \\ \dfrac{\hat{\mu}^T}{q^{K+1} q^K \cdots q^{t+1}}, & \text{if } t \leq K. \end{cases} \tag{17.2.4}$$

Again, we see that the probability of a commitment opponent does not move until period $K+1$, after which it climbs steadily.

We begin the verification of equilibrium by ensuring that Bayes' rule is satisfied. As long as there has been no previous shirking, a commitment player exerts effort with probability one. Bayes' rule then implies that the probability of a commitment player in period t, given effort in period $t+1$, is given by the probability of a commitment player in period $t+1$ (times the unitary probability of effort) divided by the probability of effort in period $t+1$, or

$$\hat{\mu}^t = \frac{\hat{\mu}^{t+1}}{q^{t+1}}.$$

This immediately leads to (17.2.4).

It remains only to check the optimality of the proposed behavior of the normal types. We begin with period 1. The relevant observation here is that we have

$$\hat{\mu}^1 = \frac{\hat{\mu}^T}{q^{K+1}\cdots q^2} = \frac{\hat{\mu}^T}{\phi^{\frac{K}{2}}} = q^1,$$

where the central equality is obtained by applying (17.2.3) to the $K/2$ adjacent products that appear in the denominator. Hence, in the final period, the probability of effort equals the probability of a commitment opponent. As a result, normal players are shirking with probability one. The specified behavior in the final period is optimal, because shirking must be a best response for normal players in this period.

Now consider a period $t \in \{2, \ldots, K+1\}$. In each such period, normal players mix between effort and shirking. The payoff from S given by

$$bq^t,$$

and the payoff from E is

$$(d + bq^{t-1})q^t - c(1 - q^t).$$

To obtain these expressions, we argue recursively, noting that they trivially hold and are equal for $t = 2$. Suppose they hold for $t - 1$, and consider t. Shirking gives an immediate payoff of b if the opponent exerts effort, for an expected payoff of bq^t, followed by the zero payoffs of mutual shirking. Effort gives an immediate payoff of $-c$ if the opponent shirks, followed by a future of mutual shirking and zero payoffs, for an expectation of $-c(1 - q^t)$. If the opponent instead exerts effort (which occurs with probability q^t), the player in question receives d in the current period, followed by a continuation payoff that can be evaluated at the expected payoff bq^{t-1} of shirking next period (because the player will either be indifferent between effort and shirking next period, or strictly prefer to shirk when the next period is the final period).

Using the relationship $d = b - c$, the condition for indifference becomes

$$q^t q^{t-1} = \phi.$$

This ensures that we have optimality for any $t \leq K + 1$. For larger values of t, the prescription that normal types exert effort is optimal if[8]

$$q^t q^{t-1} \geq \phi.$$

---

8. We are underestimating the value of effort here with the presumption that the player shirks in the following period.

Period	q^t	$\hat{\mu}^t$	r^t
10	1	1/50 (=.02)	1
9	1	1/50 (=.02)	1
8	1	1/50 (=.02)	1
7	27/50	1/50 (=.02)	26/49 (=.53)
6	50/81	1/27 (=.04)	47/78 (=.60)
5	27/50	3/50 (=.06)	24/47 (=.51)
4	50/81	1/9 (=.11)	41/72 (=.57)
3	27/50	9/50 (=.18)	18/41 (=.44)
2	50/81	1/3 (=.33)	23/54 (=.43)
1	27/50	27/50 (=.54)	0

Figure 17.2.3 Example equilibrium for finitely repeated prisoners' dilemma. Conditional on not having observed shirking in previous rounds, q^t is the probability of effort in period t, $\hat{\mu}^t$ is the probability a player is a commitment type, and r^t is the probability that a normal type exerts effort.

For values of $t \geq K + 3$, $q^t q^{t-1} = 1 > \phi$. For the case of $t = K + 2$, this is $q^{K+2} q^{K+1} = q^{K+1} \geq q^{K+1} q^K = \phi$, giving the result.

To illustrate this equilibrium, we take an example from Conlon (2003). Let the payoffs of the prisoners' dilemma be as given in the right panel of figure 17.2.1 and let $\hat{\mu}^T = 1/50$. We have $c/b = \phi = 1/3$, and a value of $K = 6$ satisfies (17.2.2). This ensures that the ending phase of the game will consist of the final period and a transitional phase of six periods. The path of play is given in figure 17.2.3 for a game of length 10. Should the game be longer, additional periods would simply be added to the initial phase in which effort occurs with probability one.

As expected, the posterior probability of a commitment type increases from its prior of 1/50 to a posterior of 27/50, the latter being high enough to support effort from normal types in period 2, even though only commitment types will exert effort in the last period. A curious feature of the equilibrium is the see-saw nature of the probability of effort in the transitional periods. This is an implication of the relationship that $q^t q^{t-1} = \phi = 1/3$ in these periods. This requires either that the probability of effort alternate between high and low values in the final phases or that it be constant. Suppose we attempted to construct an equilibrium in which the probability of effort was

Period	q^t	$\hat{\mu}^t$
8	$1/\sqrt{3}$	$1/81$
7	$1/\sqrt{3}$	$1/(27\sqrt{3})$
6	$1/\sqrt{3}$	$1/27$
5	$1/\sqrt{3}$	$1/(9\sqrt{3})$
4	$1/\sqrt{3}$	$1/9$
3	$1/\sqrt{3}$	$1/(3\sqrt{3})$
2	$1/\sqrt{3}$	$1/3$
1	$1/\sqrt{3}$	$1/\sqrt{3}$

Figure 17.2.4 Outcome for candidate equilibrium in which $q^t q^{t-1}$ is a constant in the transition phase.

constant in the final periods. That probability would then have to be $1/\sqrt{3}$ to satisfy $q^t q^{t-1} = \phi = 1/3$, and the requirement that $q^1 = \hat{\mu}^1$ would fix the terminal posterior of a commitment type at the same value. We can then work backward to characterize the equilibrium path, with the results reported in figure 17.2.4. The difficulty is now apparent. For these strategies to be an equilibrium, the descending sequence of posterior probabilities $\hat{\mu}^t$, as we move backward from the final period, must hit the prior probability of $1/50$. This terminal condition will typically fail, as it does here, ensuring that an equilibrium with a constant probability of effort over the final phase does not exist.

17.3 The Product-Choice Game

We present here an example of a finitely repeated game with imperfect public monitoring. The presence of imperfect monitoring significantly complicates the calculations. We accordingly limit the example to two periods and return to our convention that the initial period is period 0.

Player 1 is long-lived and player 2 short-lived. Player 1 discounts, but because the game lasts only two periods, we ignore the normalization $(1 - \delta)$ on payoffs.

The stage game is the product-choice game, shown again in figure 17.3.1. The set of public signals is given by $Y = \{\underline{y}, \bar{y}\}$. The signal depends only on player 1's action

17.3 ■ The Product-Choice Game

	h	ℓ
H	2, 3	0, 2
L	3, 0	1, 1

Figure 17.3.1 The product-choice game.

a_1 according to the distribution

$$\rho(\bar{y} \mid a_1 a_2) = \begin{cases} p, & \text{if } a_1 = H, \\ q, & \text{if } a_1 = L, \end{cases}$$

where $0 < q < p < 1$. Player 2's actions are public.

We assume that there are only two types, the normal type (ξ_0) and the pure Stackelberg type ($\xi(H)$, who plays H in every period). We let the prior probability that player 1 is the Stackelberg type be denoted by $\hat{\mu}^0$, with $\hat{\mu}^1$ denoting the corresponding probability in period 1.

We suppose player 1 is relatively patient, that is,

$$\delta \geq 1/(2p - 2q).$$

We see shortly that this assumption is required for the equilibrium we construct.

Two calculations will be useful. Suppose that player 1 is thought to be the Stackelberg type with probability $\hat{\mu}^0$ and that the normal type of player 1 is expected to choose H with probability γ. Then the updated posteriors that player 1 is the Stackelberg type, following signals $\bar{y}$ and $\underline{y}$, are given by

$$\varphi(\hat{\mu}^0 \mid \bar{y}) = \frac{p\hat{\mu}^0}{p\hat{\mu}^0 + (1 - \hat{\mu}^0)[\gamma p + (1 - \gamma)q]}$$

and

$$\varphi(\hat{\mu}^0 \mid \underline{y}) = \frac{(1 - p)\hat{\mu}^0}{(1 - p)\hat{\mu}^0 + (1 - \hat{\mu}^0)[\gamma(1 - p) + (1 - \gamma)(1 - q)]}.$$

If $\gamma = 1$, the Stackelberg and normal types play identically, and signals reveal no information:

$$\varphi(\hat{\mu}^0 \mid \underline{y}) = \hat{\mu}^0 = \varphi(\hat{\mu}^0 \mid \bar{y}).$$

For any $\gamma < 1$, the Stackelberg type is more likely to generate signal $\bar{y}$, and hence

$$\varphi(\hat{\mu}^0 \mid \underline{y}) < \hat{\mu}^0 < \varphi(\hat{\mu}^0 \mid \bar{y}).$$

17.3.1 The Last Period

Suppose that the last period, period 1, has been reached via a history h^1 and with a posterior probability $\hat{\mu}^1$ that player 1 is the Stackelberg type. Player 1's behavior in the second period is

$$\sigma_1(h^1, \xi(H)) = H,$$
$$\text{and} \quad \sigma_1(h^1, \xi_0) = L.$$

The Stackelberg type chooses H, and the normal type chooses the strictly dominant action L.

Player 2 thus faces a lottery of $\hat{\mu}^1 \circ H + (1 - \hat{\mu}^1) \circ L$, to which player 2's best response is

$$\sigma_2(h^1) \begin{cases} = h, & \text{if } \hat{\mu}^1 > \frac{1}{2}, \\ \in [0, 1], & \text{if } \hat{\mu}^1 = \frac{1}{2}, \\ = \ell, & \text{if } \hat{\mu}^1 < \frac{1}{2}. \end{cases}$$

The value to the normal player 1 as a function of the posterior is

$$V_1(\hat{\mu}^1) \begin{cases} = 3, & \text{if } \hat{\mu}^1 > \frac{1}{2}, \\ \in [1, 3], & \text{if } \hat{\mu}^1 = \frac{1}{2}, \\ = 1, & \text{if } \hat{\mu}^1 < \frac{1}{2}. \end{cases}$$

Hence, a posterior of $1/2$ is critical for player 1. The second-period implications of first-period behavior depend only on whether the result is a posterior above, equal to, or below $1/2$, with higher posteriors being more valuable for the normal player.

17.3.2 The First Period, Player 1

First period behavior depends on the prior probability that player 1 is the Stackelberg type. The normal player 1 can only pool with $\xi(H)$ when $\hat{\mu}^0 = 1/2$. For other priors, the lack of belief revision under pooling implies that player 2's behavior in the last period is independent of the realized signal, precluding pooling.

Three prior probabilities will be critical points for player 1's behavior, given by

$$0 < \hat{\mu}' < \tfrac{1}{2} < \hat{\mu}'' < 1.$$

The prior $\hat{\mu}''$ is determined by the requirement that if the normal type of player 1 is expected to play H with probability zero in period 0 and signal y is observed, then the posterior belief is $1/2$, that is,

$$\varphi(\hat{\mu}'' \mid y) = \tfrac{1}{2},$$

or

$$\frac{(1 - p)\hat{\mu}''}{(1 - p)\hat{\mu}'' + (1 - \hat{\mu}'')(1 - q)} = \frac{1}{2},$$

and so

$$\hat{\mu}'' = \frac{1 - q}{2 - p - q} > \frac{1}{2}.$$

17.3 ■ The Product-Choice Game

First-period behavior for $\hat{\mu}^0 > \hat{\mu}''$ is unambiguous. Even a negative signal leads to the most valuable posterior probability, exceeding $1/2$. As a result, the normal player 1 will choose L in the first period.

Analogously, the prior $\hat{\mu}'$ is determined by the requirement that if the normal player 1 is expected to choose H with probability zero in the first period and the favorable signal $\bar{y}$ is observed, then the posterior probability attached to the Stackelberg type is $1/2$, that is,

$$\varphi(\hat{\mu}' \mid \bar{y}) = \tfrac{1}{2},$$

or

$$\frac{p\hat{\mu}'}{p\hat{\mu}' + (1-\hat{\mu}')q} = \frac{1}{2},$$

and so

$$\hat{\mu}' = \frac{q}{p+q} < \frac{1}{2}.$$

If $\hat{\mu}^0 < \hat{\mu}'$, then there is no signal that can move the second-period posterior probability above $1/2$, no matter what behavior is expected of player 1 in the first period. As a result, the normal player 1 has no incentive to deviate from choosing L and will choose L in the first period when $\hat{\mu} < \hat{\mu}'$.

We see here a potential nonmonotonicity in player 1's behavior. We can create incentives for the normal player 1 to play H in the first period only for intermediate priors, with L being optimal for both quite optimistic or quite pessimistic priors. A choice of H is an investment in player 2's future (second-period) behavior. Such an investment is worthwhile only for intermediate priors, which are relatively sensitive to new information. In contrast, extreme priors are sufficiently insensitive to new information that they are not worth trying to influence, either because they are safely high or irretrievably low. This difficulty in creating incentives for extreme priors will play an important role in chapter 18.

For priors $\hat{\mu}^0 \in (\hat{\mu}', \hat{\mu}'')$, posterior beliefs may land on either side of $1/2$, depending on the first-period signal. The normal player 1 now faces a nontrivial intertemporal trade-off. Our assumption that player 1 is relatively patient then becomes important. Let γ^0 be the probability attached to H in period 0. For priors $\hat{\mu}^0 \in (\hat{\mu}', 1/2)$ [$\hat{\mu}^0 \in (1/2, \hat{\mu}'')$, respectively], $\gamma^0 \in (0, 1)$—if $\gamma^0 = 0$, the update after $\bar{y}$ [$\underline{y}$] leads to h [ℓ] in the last period for sure, making a deviation to H profitable.

We proceed in three steps.

Step 1. Consider first $\hat{\mu}^0 = 1/2$. We argue that in this case, equilibrium requires $\gamma^0 = 1$, and thus $\hat{\mu}^1 = \hat{\mu}^0 = 1/2$ regardless of which signal is received. Suppose $0 \leq \gamma^0 < 1$, so that $\varphi(\hat{\mu}^0 \mid \underline{y}) < 1/2 < \varphi(\hat{\mu}^0 \mid \bar{y})$. Let θ be the probability that player 2 chooses h in the first period. Then the payoff from H in period $t = 0$ is

$$2\theta + \delta p3 + \delta(1-p) = 2\theta + \delta + 2\delta p,$$

and the payoff from L is

$$3\theta + (1-\theta) + \delta q3 + \delta(1-q) = 2\theta + 1 + \delta + 2\delta q.$$

Consequently, the payoff from H is strictly larger than from L,

$$2\theta + \delta + 2\delta p - (2\theta + 1 + \delta + 2\delta q) = -1 + 2\delta(p-q) > 0,$$

and so the normal player 1 is not willing to place any probability on L, a contradiction. Therefore, we must have $\gamma^0 = 1$.

This player 1 behavior requires that player 2 in the second period play differently after $\bar{y}$ and $\underline{y}$. To see this, let $\bar{\beta}$ and $\underline{\beta}$ be the probabilities placed on h by player 2 in the second period, after observing signals $\bar{y}$ and $\underline{y}$ in the first period (and given $\hat{\mu}^1 = 1/2$). Then for H to be a best response for player 1 in the first period, we require

$$1 \leq \delta\{p(\bar{\beta}3 + (1-\bar{\beta})) + (1-p)(\underline{\beta}3 + (1-\underline{\beta})) \\ - (q(\bar{\beta}3 + (1-\bar{\beta})) + (1-q)(\underline{\beta}3 + (1-\underline{\beta})))\} \\ = 2\delta(p-q)(\bar{\beta} - \underline{\beta}),$$

which holds if and only if

$$\bar{\beta} \geq \underline{\beta} + \frac{1}{2\delta(p-q)}.$$

The equilibrium allows any second-period play consistent with this inequality. If we evaluate this equality at $\delta = 1/(2(p-q))$, we get $\bar{\beta} = 1$ and $\underline{\beta} = 0$.[9]

Step 2. Consider $\hat{\mu}^0 \in (\hat{\mu}', 1/2)$. In this case, the posterior after $\underline{y}$ must be less than $1/2$, causing player 2 to choose ℓ in the second period. The only opportunity for making player 2 indifferent in the second period, and thus for designing player 2's second-period behavior to support indifference on the part of the normal player 1 in the first period, is for player 2's posterior after $\bar{y}$ to be $1/2$. This requires

$$\frac{p\hat{\mu}^0}{p\hat{\mu}^0 + (1-\hat{\mu}^0)[\gamma^0 p + (1-\gamma^0)q]} = \frac{1}{2},$$

which we solve for

$$\gamma^0 = \frac{p\hat{\mu}^0 + \hat{\mu}^0 q - q}{(p-q)(1-\hat{\mu}^0)}.$$

Note that $\gamma^0 = 0$ if $\hat{\mu}^0 = \hat{\mu}'$ and $\gamma^0 = 1$ if $\hat{\mu}^0 = 1/2$.

To support player 1's first-period indifference, player 2 must randomize in the second period. Because player 2 cannot randomize after receiving signal $\underline{y}$ (the posterior after $\underline{y}$ is less than $1/2$ and hence $\underline{\beta} = 0$), we must have

$$\bar{\beta} = \frac{1}{2\delta(p-q)}$$

and $\underline{\beta} = 0$.

Step 3. Consider a prior probability $\hat{\mu}^0 \in (1/2, \hat{\mu}'')$. In this case, the posterior probability after signal $\bar{y}$ exceeds $1/2$, inducing player 2 to choose h in period 2. The only possibility for player 2 to be indifferent in the second period is for the posterior after

---

9. The restriction $\delta \geq 1/(2(p-q))$ implies the inequality can be satisfied by $\underline{\beta}, \bar{\beta} \in [0, 1]$.

17.3 ■ The Product-Choice Game

Prior	Player 1	Player 2	
$\hat{\mu}^0$	γ^0	$\bar{\beta}$	$\underline{\beta}$
$[0, \hat{\mu}']$	0	0	0
$[\hat{\mu}', \frac{1}{2}]$	$\frac{p\hat{\mu}^0 + \hat{\mu}^0 q - q}{(p-q)(1-\hat{\mu}^0)}$	$\frac{1}{2\delta(p-q)}$	0
$\frac{1}{2}$	1	$\left[\frac{1}{2\delta(p-q)}, 1\right]$	$\bar{\beta} - \frac{1}{2\delta(p-q)}$
$[\frac{1}{2}, \hat{\mu}'']$	$\frac{1 - 2\hat{\mu}^0 + \hat{\mu}^0 p + \hat{\mu}^0 q - q}{(p-q)(1-\hat{\mu}^0)}$	1	$1 - \frac{1}{2\delta(p-q)}$
$[\hat{\mu}'', 1]$	0	1	1

Figure 17.3.2 The normal player 1's first period probability of playing H (γ^0) and player 2's second period probability of h after signal $\bar{y}$ ($\bar{\beta}$) and after signal $\underline{y}$ ($\underline{\beta}$), as a function of the prior probability $\hat{\mu}^0$ of the Stackelberg type.

signal $\underline{y}$ to equal 1/2. Hence, the normal player 1 randomizes in the first period so that player 2's posterior after $\underline{y}$ is 1/2:

$$\frac{(1-p)\hat{\mu}^0}{(1-p)\hat{\mu}^0 + (1-\hat{\mu}^0)[\gamma^0(1-p) + (1-\gamma^0)(1-q)]} = \frac{1}{2}.$$

Solving, we have

$$\gamma^0 = \frac{1 - 2\hat{\mu}^0 + \hat{\mu}^0 p + \hat{\mu}^0 q - q}{(p-q)(1-\hat{\mu}^0)}.$$

Note that $\gamma^0 = 0$ if $\hat{\mu}^0 = \hat{\mu}''$ and $\gamma^0 = 1$ if $\hat{\mu}^0 = 1/2$.

To support this player 1 behavior, player 2 must again randomize in the second period. Analogously to step 2, we then have

$$\bar{\beta} = 1 \quad \text{and} \quad \underline{\beta} = 1 - \frac{1}{2\delta(p-q)}.$$

Summary. The normal player 1's first-period behavior and player 2's second-period behavior is summarized in figure 17.3.2.

17.3.3 The First Period, Player 2

In the first period, player 2 is thus facing a probability on H of $\hat{\mu}^0 + (1-\hat{\mu}^0)\gamma^0(\hat{\mu}^0)$. Let $\hat{\mu}^*$ be the critical value of $\hat{\mu}^0$ that will make player 2 indifferent between first-period actions, so that $\hat{\mu}^*$ satisfies

$$\hat{\mu}^* + (1 - \hat{\mu}^*)\gamma^0(\hat{\mu}^*) = \tfrac{1}{2}.$$

Because commitment player 1 types always choose H and normal types mix, we will have $\hat{\mu}^* < 1/2$, allowing us to substitute for $\gamma^0(\hat{\mu}^*)$ (see figure 17.3.2) to obtain

$$\hat{\mu}^* + (1 - \hat{\mu}^*)\frac{p\hat{\mu}^* + \hat{\mu}^* q - q}{(p - q)(1 - \hat{\mu}^*)} = \frac{1}{2}$$

and solve for

$$\hat{\mu}^* = \frac{p + q}{4p},$$

with

$$\hat{\mu}' < \hat{\mu}^* < \tfrac{1}{2} < \hat{\mu}''.$$

Hence, $\hat{\mu}^0 < \hat{\mu}^*$ implies that the probability of H falls short of $1/2$, whereas $\hat{\mu}^0 > \hat{\mu}^*$ implies that the probability of H exceeds $1/2$. The first-period player 2 is indifferent between h and ℓ if and only if $\hat{\mu}^0 = \hat{\mu}^*$, playing h if $\hat{\mu}^0 > \hat{\mu}^*$ and playing ℓ if $\hat{\mu}^0 < \hat{\mu}^*$. When $\hat{\mu}^0 = \hat{\mu}^*$, any mixture on player 2's part is compatible with equilibrium.

Remark 17.3.1 **Markov equilibria** Player 2's second-period mixed action in this equilibrium, when her posterior probability of the Stackelberg firm is $1/2$, depends on the prior probability of the Stackelberg firm and the signal by which this posterior was reached. A prior greater than $1/2$ combined with signal $\underline{y}$ elicits a different probability on h than does a prior smaller than $1/2$ combined with signal $\bar{y}$. Such an equilibrium is not Markov.[10]

The equilibria we have constructed for both the chain store and prisoners' dilemma games are Markov equilibria. Kreps and Wilson (1982a, p. 265) note that if the parameters in the chain store game in figure 17.0.1 are altered so that $c < 1$, then the chain store game no longer has a Markov equilibrium, with player 2's behavior near the end of the game depending on the posterior probability that player 1 is the commitment type and on the history of play in the last k periods, where k is the smallest integer for which $kc > 1$. If $k > 1$, the equilibrium is not a weak Markov equilibrium (see remark 5.6.1). ◆

---

10. Because different priors can be viewed as describing different games, and the posterior of $1/2$ in the last period only arises after one signal realization, *given* a prior not equal to $1/2$, it would be consistent with the formal definition (but not its spirit) to view the equilibrium as Markov.

18 Modeling Reputations

This chapter explores models of reputations that move beyond the framework developed in chapters 15–17. Our motivation is fourfold.

First, existing models do not readily capture the full spectrum of issues encompassed in the popular use of the word *reputation*. It is common to think of reputations as assets—things of value that require costly investments to build and maintain, that can be enhanced or allowed to deteriorate, that gradually run down if neglected, and that can be bought and sold. We would like a model that captures this richness.

The repeated games of adverse selection examined in chapters 15–17 may well have equilibria capturing many of these features. The focus on payoffs that is characteristic of the reputation literature leaves the nature of the attendant equilibria largely unexplored, but the analysis leaves hints as to the structure of these equilibria. The argument that player 2 must eventually come to expect the commitment action if player 1 invariably plays it (sections 15.3.1 and 15.4.2) is suggestive of a reputation-building phase. The results of section 15.5 and 16.7 suggest that ultimately reputations are optimally depleted. As intriguing as these results are, however, these models do not provide the explicit links between the structure of the interaction and equilibrium behavior that would be especially useful in studying reputations.

Second, in a similar spirit, we seek not only a characterization of equilibrium payoffs but also equilibrium behavior. We typically proceed in this chapter by constructing equilibria, feeling free to limit attention to equilibria whose features we find particularly interesting. This is in keeping with our view that much is to be learned by focusing on the behavioral implications of the theory of repeated games.

Third, reputations in standard models are built by mimicking behavior to which one would like to be committed. We refer to these as pooling reputations, because the payoff bounds arise out of pooling one's actions with those of the commitment type. In contrast, this chapter focuses on separating reputations, in which players strive to distinguish themselves from types for whom they would like to not be mistaken. Stackelberg types may not always be conveniently available. Consumers may approach the market not in terms of finding a firm who necessarily provides good service, but of avoiding the one who is incapable of doing so. The normal firm may then find that there are effectively no Stackelberg types with whom to pool but that providing good service is essential in distinguishing himself from inept types.

The reputation results presented in chapter 15 allow player 1 to "choose" his most preferred type from a possibly countably infinite set of possible types, and hence depend on neither the presence of Stackelberg types nor the absence of other types. The presence of inept types, even if quite likely, has no effect on the argument. Separating reputations thus appear to require not only inept types but *only* inept types. Why pursue a model built on such a strong restriction?

In our view, this question must be addressed not in the context of the model but in the setting for which the model is intended. The question of which types should be included in the model is one of how the agents view their environment. We think the possibility of a Stackelberg commitment type is sometimes quite relevant, but we are also interested in situations in which consumers' evaluations of the market are dominated by the possibility of inept types.

Fourth, many equilibria in repeated games require what often appears to be an implausible degree of coordination among the players. We will work in this chapter with models deliberately designed to limit such coordination. This in turn will provide a natural setting for reputations based on separation.

We regard the ability to use histories to coordinate future play as one of the fundamental hallmarks of repeated games. At the same time, we prefer some caution when embedding the possibilities for such coordination in a model. Indeed, as we explained at the beginning of chapter 15, we could build a theory of reputations on pure coordination with no need for incomplete information. Instead, even after using incomplete information to create an intrinsic link between current play and future expectations, we are concerned that reputation models may be leaning too heavily on coordination. Once again, we return to the view that work on repeated games would be well served by a better understanding of the structure of the equilibria involved.

18.1 An Alternative Model of Reputations

18.1.1 Modeling Reputations

We begin with a variant of the product-choice game (see figure 1.5.1), and much of this chapter will have the flavor of an extended example built around this game. As is typically the case, our first step is to add incomplete information about player 1's type. The normal player 1 has the option of choosing high or low effort. We add a single commitment type of player 1, $\xi(L)$, who necessarily chooses low effort. For example, effectively exerting high effort may require complementary skills or inputs that this type lacks. In keeping with this interpretation, we will frequently refer to the commitment type of player 1 as inept. We will also often refer to player 1 as a firm and player 2 as a consumer.

We are interested in equilibria in which the normal type of player 1 exerts high effort. We already have payoff bounds for the normal player 1 in the presence of (for example) a Stackelbeg commitment type. Thus, whether the motivation is to pool with a good type or separate from a bad type, it seems the effect is that player 1 exerts high effort. How do the two models differ? Our separating model always features an equilibrium in which both the inept and normal type of player 1 exert low effort, with player 2 never drawing any inferences about player 1 and with both players receiving their minmax payoffs. Rather than imposing a lower bound on player 1's payoff, the incomplete information raises the *possibility* that player 1 can achieve a higher payoff.

18.1 ■ An Alternative Model

If we are simply concerned with the possibility of high payoffs for player 1, why is this not simply another folk theorem exercise? Why does incomplete information play a role? We view player 2 as a continuum of small and anonymous players (see sections 2.7 and 7.8) and in addition assume that the various player 2s are idiosyncratic, in the sense that different consumers receive different realizations from an imperfect monitoring technology (see section 15.4.3), each observing only their own signal. This idiosyncrasy disrupts the coordination that typically plays a central role in creating intertemporal incentives. A consumer who has just received a bad meal from a restaurant has no way of knowing whether this is simply an unlucky draw from a restaurant who continues to exert high effort or whether it is a signal that the restaurant is shirking. By itself, this inference problem is not particularly problematic. In a standard public monitoring game, bad signals trigger punishments even though players know they are *not* an indication of shirking (in equilibrium). However, for this behavior to be consistent with equilibrium, it is important that there be coordination in the punishment, not only among the small anonymous players but also with player 1. The idiosyncratic signals rob the players of this coordination possibility.

Each short-lived player's signals provide information about player 1's type. As in pooling models of reputations, this introduces a link between current and future behavior. In particular, the firm may now find it optimal to exert high effort because doing so increases the consumer posterior that the firm is normal rather than inept, in turn allowing the firm to reap the rewards of high effort. Eventually, however, the consumers' posteriors will come arbitrarily close to attaching probability one to the firm's being normal. At this point, further experience has virtually no effect on consumer posteriors and hence, given a belief that normal firms exert high effort, on their actions. But then the firm has an irresistible incentive to deviate to low effort, unraveling the putative equilibrium. Increased patience might allow this unraveling to be postponed, but it cannot be avoided.

To obtain an equilibrium with consistent high effort, consumers' posteriors about the firm must be bounded away from certainty. This will be the case, for example, if consumers have bounded memory, using only some finite number of their most recent observations in drawing inferences about the firm's type. Overwhelming evidence that the firm is normal could then never be amassed.

We adopt a different approach here, assuming that in every period there is some possibility that the firm is replaced by a new firm whose type is randomly drawn from a prior distribution over types. Consumers understand the possibility of such replacements but cannot observe them. Intuitively, the possibility of changing types plays a role whenever one's response to a disappointing experience with a firm known for good outcomes is not simply "I've been unlucky" but also "I wonder if something has changed." This again ensures that consumers can never be too certain about the firm, and thus that the firm always faces incentives to choose high effort.

Remark 18.1.1 **Interpreting replacements** A firm's type, normal or inept, can be interpreted as reflecting a variety of factors, including the composition and skills of its workforce, technology and capital stock, access to appropriate materials, management style and organization, and workplace culture. A replacement may literally be a change in ownership of the firm but may also be a change in some key characteristic.

Throughout this chapter, we take a replacement to be a departure of the firm's current owner. This is a convenience for the bulk of the chapter, but is necessary for the analysis in section 18.7.2.

◆

Remark 18.1.2 **The role of patience** By introducing the prospect that a firm's characteristics or even identity are constantly subject to revision, we place an upper bound on the effective discount factor, no matter how patient the firm happens to be. As a result, appealing to the limit as δ gets arbitrarily close to 1 no longer creates arbitrarily effective incentives. Instead, we follow the suggestion of section 3.2.3 and remark 15.3.8 in focusing on cases in which the cost of high effort is relatively small.

◆

18.1.2 The Market

We consider a single long-lived player, or *the firm*, facing a continuum of small and anonymous players (*consumers*), indexed by $i \in [0, 1]$. In each period t, the firm chooses an effort level $a_1^t \in \{L, H\}$. Each consumer is long-lived and observes an idiosyncratic realization of a signal (see section 15.4.3). The signal has two possible values, $\bar{z}$ (good) and $\underline{z}$ (bad), with marginal distribution[1]

$$\pi_i(\bar{z} \mid a) = \begin{cases} \rho_H & \text{if } a_1 = H, \\ \rho_L, & \text{if } a_1 = L, \end{cases}$$

where

$$0 < \rho_L < \rho_H < 1.$$

In each period t and for each group of consumers having experienced a common history of signals, a proportion $\rho_{a_1^t}$ of this group receives the good signal. A deterministic sequence of effort choices thus yields a deterministic sequence of distributions of consumers' histories.

Remark 18.1.3 **Idiosyncratic signals** There are well-known technical complications in modeling a continuum of independent random variables (see, for example, Al-Najjar 1995). In our case, independence is unnecessary. The critical feature of the model is that if the firm is expected to exert effort level a_1^t in period t, then each consumer assigns a probability of $\rho_{a_1^t}$ to the event that she receives a good signal and believes that a fraction $\rho_{a_1^t}$ of consumers will receive the good signal. This can be achieved as follows. Fix the first-period effort choice a_1. Let ω be the realization of a uniform random variable on $[0, 1]$. Then consumers in $\left([\omega - 1, \omega + \rho_{a_1} - 1] \cup [\omega, \omega + \rho_{a_1}]\right) \cap [0, 1]$ receive the good signal and consumers in the complement the bad signal. The subpopulation who received a good signal in the first period can be viewed in a natural way as a population distributed on $I_{\bar{z}}$, an interval of

---

1. Using ρ_H and ρ_L (rather that π_H and π_L) to denote the private-signal distribution facilitates the transition to the subsequent model with public signals.

18.1 ■ An Alternative Model

length ρ_{a_1}, whereas the subpopulation who received a bad outcome in the first period can be viewed as a population distributed on $I_{\underline{z}}$, an interval of length $1 - \rho_{a_1}$. Second-period signals are determined from second-period effort and the realizations of a uniformly distributed random variable on $I_{\bar{z}}$ and one distributed on $I_{\underline{z}}$. These two random variables are independent of each other and of the realization of ω. This construction has an obvious recursion, yields the desired pattern of signal realizations, and only requires a countable number of independent random variables (one for each possible finite history of utility realizations).

◆

The aggregate distribution of the signals received by consumers in any period is perfectly informative about the firms's effort choice in that period—consumers need only observe the fraction of good signals to infer effort. However, a consumer observes neither the aggregate distribution nor the signal of any other consumer.

The (normal) firm's stage-game payoff is the difference between its revenue and its costs. Low effort is costless, but high effort requires a cost of c.

Rather than explicitly modeling the interaction between firm and consumers as a noncooperative game, we specify the firm's revenues as a function of consumer expectations about effort (we followed a similar modeling strategy in example 5.5.3). A consumer receives payoff 1 from signal $\bar{z}$ and 0 from $\underline{z}$. Consumer expectations are given by a distribution function F, with $F(x)$ being the proportion of consumers who expect the firm to exert high effort with probability less than or equal to x. We denote by $\mathfrak{F}$ the set of possible distribution functions on $[0, 1]$ describing consumer expectations. The firm's revenue, as a function of $F \in \mathfrak{F}$, is denoted by $p : \mathfrak{F} \to \mathbb{R}$. We assume p is strictly increasing, so that higher expectations of high quality lead to higher revenue,

$$F' \succ F \Rightarrow p(F') > p(F),$$

where $\succ$ is strict first-order stochastic dominance. We also assume that $p(F^n) \to p(F)$ for all sequences $\{F^n\}$ converging weakly to F.

An obvious example of a market interaction with these properties is perfect price discrimination, in which each consumer buys one unit of the good in each period and the firm charges each consumer her reservation price. We let $p(1)$ and $p(0)$ denote the revenue of the firm in the special cases in which every consumer expects high effort with probability 1 and 0, respectively.

We assume that

$$\rho_H - \rho_L > c,$$

ensuring that high effort is the efficient choice. We further assume that

$$p(1) - p(0) > c,$$

making H the pure Stackelberg action for the firm.[2]

---

2. The condition $\rho_H - \rho_L > c$ suffices for H to be the firm's Stackelberg action if the firm is a perfect price discriminator, but not if the consumers retain some surplus.

In the repeated game, the normal firm maximizes the discounted sum of expected profits, with discount factor δ. There are two types of firm, normal and inept. An inept firm can only choose low effort.

Before play begins, nature determines the original type of the firm, choosing normal with probability μ_0^0 (with the subscript distinguishing the normal type and the superscript the prior expectation) and inept with probability $1 - \mu_0^0$. The firm learns its type, but consumers do not. Moreover, in each subsequent period, there is a probability λ that the firm is replaced, with probability μ_0^0 of the new firm being normal. We view λ as the probability that an exogenous change in personal circumstances, such as reaching retirement age, causes an existing owner to leave the market, to be replaced by a new owner.

Remark 18.1.4 **Replacements** Given our interpretation of a replacement as the departure of the current firm, the firm's effective discount factor is $\delta(1 - \lambda)$ and the firm is concerned only with payoffs conditional on not being replaced. If we interpreted a replacement as a change in characteristic of a continuing firm (see remark 18.1.1), the appropriate discount factor would be δ and the firm's expected payoff would include flow payoffs received (perhaps as a different type) after having been "replaced." Because the firm cannot affect the replacement probability, the two formulations yield qualitatively similar results. ◆

Consumers cannot observe whether a replacement has occurred. For example, the ownership of a restaurant might change without changing the restaurant's name and without consumers being aware of the change. Section 18.2 explains why the possibility of changing type is crucial to the results.

At the beginning of period t, each consumer i is characterized by her posterior probability that the firm is normal and her posterior probability that the firm will exert high effort, denoted υ_i^t. If the firm is normal, it makes its (unobserved) effort choice. The firm receives revenues that depend on the distribution F^t of consumers' beliefs about the firm's effort, but not on the firm's type or action in that period. Consumers observe their own signals and update beliefs about the type of firm. Finally, with probability λ, the firm is replaced.

For consumer $i \in [0, 1]$, a period t history is a t-tuple of signals, $h_i^t \in \{\underline{z}, \bar{z}\}^t \equiv \mathcal{H}_2^t$, describing the payoffs consumer i has received in periods 0 through $t - 1$. The set of all consumer histories is $\mathcal{H}_2 = \cup_{t \geq 0} \mathcal{H}_2^t$. A *belief function* for consumer i is a function $\upsilon_i : \mathcal{H}_2 \to [0, 1]$, where $\upsilon_i(h_i^t)$ is the probability consumer i assigns to the firm exerting high effort in period t, given history h_i^t. Because every history of signals has positive probability under any sequence of effort choices of the firm, it is necessarily the case that as long as consumers use Bayes' rule and start with a common prior, two consumers observing the same sequence of signals have the same beliefs about the firm's behavior. In particular, $\upsilon_i(h_2^t) = \upsilon_j(h_2^t)$ for all $h_2^t \in \mathcal{H}_2$ and all $i, j \in [0, 1]$. We accordingly describe consumers' beliefs by a single function $\upsilon : \mathcal{H}_2 \to [0, 1]$. Given a sequence of realized effort choices by the firm, h_1^t, there is an induced probability measure on $\mathcal{H}_2^t$, denoted $\psi^t(\cdot \mid h_1^t)$. Then, given υ and h_1^t, $F_{\upsilon, h_1^t}(x) = \psi^t(\{h_2^t : \upsilon(h_2^t) \leq x\} \mid h_1^t)$ and the revenue in period t after the history h_1^t is given by $p(F_{\upsilon, h_1^t})$.

18.1 ■ An Alternative Model

Recalling that an effort level deterministically induces a distribution of signals, we can take a period t history h_1^t for the firm to be the t-tuple of realized effort choices, $h_1^t \in \{L, H\}^t \equiv \mathcal{H}_1^t$, describing the choices made in periods 0 through $t-1$. The set of all possible firm histories is $\mathcal{H}_1 = \cup_{t \geq 0} \mathcal{H}_1^t$. A pure strategy for a normal firm is a strategy, $\sigma_1 : \mathcal{H}_1 \to \{L, H\}$, giving the effort choice after observing history h_1^t.

If $\lambda > 0$, then with probability one, there will be an infinite number of replacement events, infinitely many of which will introduce new normal firms into the game. This description of histories ignores such replacement events. Restricting attention to firm histories in $\mathcal{H}_1$ implies that a new normal firm, entering after the effort history h_1^t, behaves in the same way as an existing normal firm after the same history. Although this restriction may rule out some equilibria, any equilibrium under this assumption will again be an equilibrium without it. We sometimes refer to a strategy σ_1 as the normal firm's (or normal type's) strategy, although it describes the behavior of all new normal firms as well.

The pair (σ_1, υ) will be an equilibrium if $\sigma_1(h_1^t)$ is maximizing for normal firms after every effort history $h_1^t \in \mathcal{H}_1$, and consumers' beliefs about effort choice, υ, are (correctly) determined by Bayes' rule. A precise general definition of equilibrium requires tedious notation. The difficulty is that a mixed strategy, or a pure strategy in which the firm sometimes takes high effort and there are replacements that might be either normal or inept, gives rise to a random sequence of effort levels.[3] Because the firm's strategy may call for different effort choices after different effort histories, a consumer must then use her outcome history to form a posterior over the firm's effort histories, yielding an intricate updating process. In particular, the posterior probability that a consumer assigns to the firm being normal is *not* necessarily a sufficient statistic for her history of outcomes. In the equilibria we examine, however, the normal firm takes the same equilibrium action after any realized effort-level history, implying that a consumer's posterior belief concerning the firm's competency *is* a sufficient statistic for her outcome history.

18.1.3 Reputation with Replacements

We examine a pure-strategy equilibrium in which the normal firm always chooses high effort. Let $\varphi(\mu_0 \mid z)$ denote the posterior probability that the firm is normal, after the consumer has received a single signal, $z \in \{\underline{z}, \bar{z}\}$, given a prior probability of μ_0 that the firm is normal and that the normal firm chooses high effort:

$$\varphi(\mu_0 \mid \bar{z}) = (1 - \lambda) \frac{\rho_H \mu_0}{\rho_H \mu_0 + \rho_L (1 - \mu_0)} + \lambda \mu_0^0$$

and

$$\varphi(\mu_0 \mid \underline{z}) = (1 - \lambda) \frac{(1 - \rho_H) \mu_0}{(1 - \rho_H) \mu_0 + (1 - \rho_L)(1 - \mu_0)} + \lambda \mu_0^0.$$

These equations incorporate the assumption that a normal firm always chooses high effort and the possibility of replacement. For any history h_2^t, let $\varphi(\mu_0^0 \mid h_2^t)$ denote the posterior belief of a consumer who had observed $h_2^t \in \mathcal{H}_2$.

---

3. Even if $\sigma_1(h_1^t) = H$ for all $h_1^t \in \mathcal{H}_1$, the induced effort path will switch between high and low effort whenever a replacement changes the type of the firm.

Definition 18.1.1 *A profile* (σ_1, υ) *is a* **high-effort equilibrium** *if*

1. $\sigma_1(h_1^t) = H$ *is maximizing for the normal firm, for all* $h_1^t \in \mathscr{H}_1$, *given* υ, *and*
2. $\upsilon(h_2^t) = \varphi(\mu_0^0 \mid h_2^t)$ *for all* $h_2^t \in \mathscr{H}_2$.

Because the only off-the-equilibrium-path information sets are those of the normal firm, a high-effort equilibrium is trivially sequential.

Remark 18.1.5 **Low-effort equilibrium** The pair (σ_1, υ) with $\sigma_1(h_1^t) = L$ for all h_1^t and $\upsilon(h_2^t) = 0$ for all h_2^t is the *low-effort profile*. Because consumers never expect the normal firm to exert effort under this profile, the signals are uninformative, and the normal firm has no incentive to exert effort. The low-effort profile is thus a sequential equilibrium for all costs of effort and all discount factors. ◆

Proposition 18.1.1 *Suppose* $\lambda \in (0, 1)$. *There exists* $\bar{c} > 0$ *such that for all* $0 \leq c < \bar{c}$, *there exists a high-effort equilibrium.*

Proof Let μ_0' solve $\varphi(\mu_0 \mid \underline{z}) = \mu_0$, and let μ_0'' solve $\varphi(\mu_0 \mid \bar{z}) = \mu_0$. Suppose the normal firm always exerts high effort. Then, $\varphi(\mu_0 \mid z) \in [\lambda \mu_0^0, 1 - \lambda + \lambda \mu_0^0]$ for all $\mu_0 \in [0, 1]$ and $z \in \{\underline{z}, \bar{z}\}$. Moreover, $\lambda \mu_0^0 < \mu_0' < \mu_0^0 < \mu_0'' < 1 - \lambda + \lambda \mu_0^0$ and $\varphi(\mu_0 \mid z) \in [\mu_0', \mu_0'']$ for all $\mu_0 \in [\mu_0', \mu_0'']$ and all $z \in \{\underline{z}, \bar{z}\}$. For $z \in \{\underline{z}, \bar{z}\}$, $\varphi^{-1}(\mu_0 \mid z)$ denotes the inverse of μ_0 under $\varphi(\cdot \mid z)$; for $\mu_0 < \min_{\check{\mu}_0} \varphi(\check{\mu}_0 \mid z)$, set $\varphi^{-1}(\mu_0 \mid z) = 0$ and for $\mu_0 > \max_{\check{\mu}_0} \varphi(\check{\mu}_0 \mid z)$, set $\varphi^{-1}(\mu_0 \mid z) = 1$. As $\varphi^{-1}(\mu_0 \mid \underline{z}) - \varphi^{-1}(\mu_0 \mid \bar{z}) > 0$ for all $\mu_0 \in (\lambda \mu_0^0, 1 - \lambda + \lambda \mu_0^0)$, there is a constant $\beta > 0$ such that $\varphi^{-1}(\mu_0 \mid \underline{z}) - \varphi^{-1}(\mu_0 \mid \bar{z}) \geq \beta$ for all $\mu_0 \in [\mu_0', \mu_0'']$.

Let G_{a_1} denote the distribution over posteriors that the firm is normal in period $t+1$ (suppressing time superscripts) that results from a choice of effort a_1, given the distribution of consumer posteriors, G. Then we have $G_{a_1}(\mu_0) = \rho_{a_1} G(\varphi^{-1}(\mu_0 \mid \bar{z})) + (1 - \rho_{a_1}) G(\varphi^{-1}(\mu_0 \mid \underline{z}))$, so that

$$G_L(\mu_0) - G_H(\mu_0) = (\rho_H - \rho_L)(G(\varphi^{-1}(\mu_0 \mid \underline{z})) - G(\varphi^{-1}(\mu_0 \mid \bar{z}))) \geq 0.$$

Observe that the average consumer posterior under G is $\int \mu_0 dG(\mu_0) = 1 - \int G(\mu_0) d\mu_0$. Choose ε satisfying $0 < \varepsilon < \min\{\beta, \mu_0' - \varphi^{-1}(\mu_0' \mid \bar{z}), 1 - \mu_0''\}$. Then, because $\varepsilon < \beta$,

$$\int_0^1 G(\varphi^{-1}(\mu_0 \mid \underline{z})) - G(\varphi^{-1}(\mu_0 \mid \bar{z})) \, d\mu_0$$
$$\geq \int_{\mu_0'}^{\mu_0''} G(\varphi^{-1}(\mu_0 \mid \bar{z}) + \varepsilon) - G(\varphi^{-1}(\mu_0 \mid \bar{z})) \, d\mu_0.$$

Let K be the largest integer k for which $\varphi^{-1}(\mu_0' \mid \bar{z}) + k\varepsilon \leq \mu_0''$. By construction $\varphi^{-1}(\mu_0' \mid \bar{z}) + (K+1)\varepsilon < 1$. Construct an increasing sequence $\{\mu_0'^{(k)}\}_{k=0}^{K+1}$ by setting $\mu_0'^{(k)} = \varphi(\varphi^{-1}(\mu_0' \mid \bar{z}) + k\varepsilon \mid \bar{z})$; note that $\mu_0'^{(0)} = \mu_0'$. For $k = 0, \ldots, K$, if $f^{(k)} : [\mu_0'^{(0)}, \mu_0'^{(1)}] \to [\mu_0'^{(k)}, \mu_0'^{(k+1)}]$ is defined by the function $f^{(k)}(\mu_0) = \varphi(\varphi^{-1}(\mu_0 \mid \bar{z}) + k\varepsilon \mid \bar{z})$, then $f^{(k)}(\mu_0'^{(0)}) = \mu_0'^{(k)}$ and $f^{(k)}(\mu_0'^{(1)}) = \mu_0'^{(k+1)}$. As $f^{(k)}$ is continuous, it is onto. Also, $f^{(k+1)}(\mu_0) = \varphi(\varphi^{-1}(f^{(k)}(\mu_0) \mid \bar{z}) + \varepsilon \mid \bar{z})$,

18.1 ■ An Alternative Model

and because $\varphi(\cdot \mid \bar{z})$ is concave, $df^{(k+1)}(\mu_0)/d\mu_0 \leq df^{(k)}(\mu_0)/d\mu_0$. Then for $k = 0, \ldots, K$,

$$\int_{\mu_0^{\prime(k)}}^{\mu_0^{\prime(k+1)}} G(\varphi^{-1}(\mu_0 \mid \bar{z}) + \varepsilon) \, d\mu_0$$

$$= \int_{\mu_0^{\prime(0)}}^{\mu_0^{\prime(1)}} G(\varphi^{-1}(f^{(k)}(\mu_0) \mid \bar{z}) + \varepsilon) \frac{df^{(k)}(\mu_0)}{d\mu_0} \, d\mu_0$$

$$= \int_{\mu_0^{\prime(0)}}^{\mu_0^{\prime(1)}} G(\varphi^{-1}(f^{(k+1)}(\mu_0) \mid \bar{z})) \frac{df^{(k)}(\mu_0)}{d\mu_0} \, d\mu_0$$

$$\geq \int_{\mu_0^{\prime(0)}}^{\mu_0^{\prime(1)}} G(\varphi^{-1}(f^{(k+1)}(\mu_0) \mid \bar{z})) \frac{df^{(k+1)}(\mu_0)}{d\mu_0} \, d\mu_0.$$

Then, because $\mu_0^{\prime(0)} = \mu_0'$, $\mu_0^{\prime(K+1)} > \mu_0''$, and the support of G is a subset of $[\mu_0', \mu_0'']$,

$$\int_{\mu_0'}^{\mu_0''} G(\varphi^{-1}(\mu_0 \mid \bar{z}) + \varepsilon) - G(\varphi^{-1}(\mu_0 \mid \bar{z})) \, d\mu_0$$

$$= \sum_{k=0}^{K} \int_{\mu_0^{\prime(k)}}^{\mu_0^{\prime(k+1)}} G(\varphi^{-1}(\mu_0 \mid \bar{z}) + \varepsilon) - G(\varphi^{-1}(\mu_0 \mid \bar{z})) \, d\mu_0$$

$$\geq \int_{\mu_0^{\prime(0)}}^{\mu_0^{\prime(1)}} \sum_{k=0}^{K-1} G(\varphi^{-1}(f^{(k+1)}(\mu_0) \mid \bar{z})) \frac{df^{(k+1)}(\mu_0)}{d\mu_0}$$

$$- \sum_{k=0}^{K} G(\varphi^{-1}(f^{(k)}(\mu_0) \mid \bar{z})) \frac{df^{(k)}(\mu_0)}{d\mu_0} \, d\mu_0$$

$$+ \int_{\mu_0^{\prime(K)}}^{\mu_0^{\prime(K+1)}} G(\varphi^{-1}(\mu_0 \mid \bar{z}) + \varepsilon) \, d\mu_0$$

$$= \int_{\mu_0^{\prime(K)}}^{\mu_0^{\prime(K+1)}} G(\varphi^{-1}(\mu_0 \mid \bar{z}) + \varepsilon) \, d\mu_0 - \int_{\mu_0^{\prime(0)}}^{\mu_0^{\prime(1)}} G(\varphi^{-1}(\mu_0 \mid \bar{z})) \, d\mu_0$$

$$\geq G(\varphi^{-1}(\mu_0^{\prime(K)} \mid \bar{z}) + \varepsilon)(\mu_0^{\prime(K+1)} - \mu_0^{\prime(K)}) - G(\varphi^{-1}(\mu_0^{\prime(1)} \mid \bar{z}))(\mu_0^{\prime(1)} - \mu_0^{\prime(0)})$$

$$\geq \mu_0^{\prime(K+1)} - \mu_0^{\prime(K)},$$

where the last inequality is implied by $\varphi^{-1}(\mu_0^{\prime(1)} \mid \bar{z}) \leq \mu_0'$ and $\varphi^{-1}(\mu_0^{\prime(K)} \mid \bar{z}) + \varepsilon \geq \mu_0''$. Thus, $\int_0^1 [G_L(\mu_0) - G_H(\mu_0)] \, d\mu_0$ is bounded away from 0 (the bound depends only on ε and not on the period t distribution G or on t), and (because the normal firm is always expected to be choosing high effort) so is $\int_0^1 F_L(\mu_0) - F_H(\mu_0) \, d\mu_0$, where F_{a_1} is the distribution in period $t+1$ of consumer expectations of the probability that the firm exerts high effort. This in turn implies that there is a constant Δ (independent of t) such that period $t+1$ revenues after high effort in period t exceed those after low effort in period t by at least Δ. Thus a sufficient condition for an equilibrium with high effort is that the discounted value of this difference exceeds the cost, $c < \delta(1 - \lambda)\Delta$. ■

It is expected that a high-effort equilibrium exists only if the cost of high effort is not too large. The upper bound for c being "not too large" is in turn nonzero only if there is a positive probability of an inept replacement ($\lambda(1 - \mu_0^0) > 0$). The value functions induced by high and low effort approach each other as the posterior probability μ_0^t assigned to the normal type approaches 1, because the values diverge only through the effect of current outcomes on future expectations, and current outcomes have very little effect on future expectations when consumers are currently quite sure of the firm's type. The prospect of inept replacements bounds the posterior μ_0^t away from 1, ensuring that high effort is always (before cost) more valuable than low effort, and thus that high effort will be optimal for sufficiently small $c > 0$.

18.1.4 How Different Is It?

The model of section 18.1.3 exhibits five features, the first four of which are not commonly found in reputation models.

1. Player 1's type is not fixed immutably at the beginning of the game but evolves according to a Markov process.
2. The commitment type is a "bad" type, from whom player 1 would like to separate.
3. The collection of small and anonymous uninformed players receive idiosyncratic signals.
4. The small and anonymous players respond continuously to changes in their beliefs.
5. The signals received by the uninformed players do not depend on their actions. Given the interpretation of player 1 as a firm and player 2 as consumers, this suggests that consumers never stop purchasing from the firm, no matter how pessimistic they might be concerning the firm's type or effort level.

The next five sections investigate the role of each feature in the reputation result.

18.2 The Role of Replacements

Insight into the role of replacements is most effectively provided by examining the special case of the model in which there are no replacements, that is, $\lambda = 0$.

Suppose first there are no replacements *and* $\mu_0^0 = 1$, so that the original and only firm is known to be normal. It is then immediate that the only pure-strategy equilibrium of the repeated game calls for the firm to always exert low effort. A consumer who receives the (idiosyncratic) signal $\underline{z}$ when the firm is supposed to exert high effort will assume the firm exerted high effort but the consumer received an unlucky draw from the monitoring distribution. Given that the signals are idiosyncratic, bad signals will then prompt no punishments from consumers, creating an irresistible incentive for the firm to exert low effort and destroying the equilibrium.

The logic of the complete information case also holds with incomplete information ($\mu_0^0 < 1$), in the absence of replacements, though the argument is more involved.

18.2 ■ The Role of Replacements

Suppose the normal firm is following a pure strategy. The posterior probability that the firm is normal, given a prior probability of μ_0 and the signal $z \in \{\underline{z}, \bar{z}\}$ is (where $\alpha_1 \in \{0, 1\}$ is the probability of H)

$$\varphi(\mu_0 \mid \bar{z}) = \frac{[\alpha_1 \rho_H + (1 - \alpha_1) \rho_L] \mu_0}{[\alpha_1 \rho_H + (1 - \alpha_1) \rho_L] \mu_0 + \rho_L (1 - \mu_0)} \quad (18.2.1)$$

and

$$\varphi(\mu_0 \mid \underline{z}) = \frac{[\alpha_1 (1 - \rho_H) + (1 - \alpha_1)(1 - \rho_L)] \mu_0}{[\alpha_1 (1 - \rho_H) + (1 - \alpha_1)(1 - \rho_L)] \mu_0 + (1 - \rho_L)(1 - \mu_0)}. \quad (18.2.2)$$

Extending the notation in an obvious manner, we write $\varphi(\mu_0^0 \mid h_2^t)$ for the update from a prior μ_0^0 after the history h_2^t. If a consumer believes the normal firm is following the pure strategy σ_1, she attaches probability $\varphi(\mu_0^0 \mid h_2^t) \alpha_1^t(\sigma_1)$ to the firm exerting high effort, after observing history h_2^t.

Proposition *If there are no replacements ($\lambda = 0$), there is a unique pure-strategy sequential*
18.2.1 *equilibrium, and in this equilibrium the normal player 1 exerts low effort in every period.*

Proof Suppose (σ_1, υ) is an equilibrium and σ_1 is a pure strategy. Suppose σ_1 calls for the normal firm to sometimes exert high effort in equilibrium. Because there are no replacements, the firm's history evolves deterministically (recall that an effort level maps deterministically into a distribution of consumer signals), and hence σ_1 determines the periods in which the firm exerts high effort.

It is immediate that σ_1 must call for high effort infinitely often. If this were not the case, there is a final period t^* in which the firm exerts high effort. The revenue in every subsequent period would then be $p(0)$, independently of the outcome in period t^*, ensuring that high effort is suboptimal in period t^*.

Hence, let $T_n(H) \equiv \{t \geq n : a_1^t(\sigma_1) = H\}$ be those periods larger than n in which high effort is exerted. Similarly, let $T_n(L) \equiv \{t \geq n : a_1^t(\sigma_1) = L\} = \{n, n+1, \ldots\} \setminus T_n(H)$. Then, for $t \in T_n(L)$, all consumers expect the firm to choose low effort with probability one, that is,

$$p(F^t) = p(0),$$

where F^t is the distribution of consumer expectations over player 1's effort level in period t.

For any period $t \in T_n(H)$, we have, for any history h_2^t,

$$\upsilon(h_2^t) = \varphi(\mu_0^0 \mid h_2^t),$$

and hence the probability distribution function F^t of consumers' period t probabilities of high effort is given by the distribution function of consumers' posterior beliefs that the firm is normal, G^t. If $t \in T_n(L)$, then $G^{t+1} = G^t$. Moreover, because $T_n(H) \neq \emptyset$ for all n, conditional on the firm being normal, $G^t(x) \to 0$ for all $x < 1$ as $t \to \infty$, that is, consumers eventually become convinced they are facing the normal firm. Thus, for all $\varepsilon > 0$, there exists $t(\varepsilon)$ such that for all $t \geq t(\varepsilon)$, $G^t(1-\varepsilon) < \varepsilon$. That is, at least a fraction $1-\varepsilon$ of consumers have observed a private history h_2^t that yields an update $\varphi(\mu_0^0 \mid h_2^t) > 1-\varepsilon$. Observe that for all $\eta > 0$ and $k \in \mathbb{N}$, there exists $\varepsilon(\eta, k) > 0$ such that $\mu_0 > 1 - \varepsilon(\eta, k) \Rightarrow \varphi(\mu_0 \mid \underline{z}^{(k)}) > 1 - \eta$, where $\underline{z}^{(k)}$ is the history of k consecutive realizations of the bad signal $\underline{z}$, which in turn implies that for any k-period history h_2^k, $\varphi(\mu_0 \mid h_2^k) > 1 - \eta$. Let G^η be the distribution function given by $G^\eta(x) = \eta$ for $x < 1 - \eta$ and $G^\eta(x) = 1$ for $x \geq 1 - \eta$. Then, for all $t' \geq t(\varepsilon(\eta, k))$ and all $t \in \{t', \ldots, t'+k\}$, G^t first order stochastically dominates G^η. Thus the continuation payoff from deviating in a period $t' \in T_{t(\varepsilon(\eta,k))}(H)$ is at least (omitting the normalization $(1-\delta)$)

$$p(G^\eta) + \sum_{\substack{t \in T_{t'}(H) \\ t'+1 \leq t \leq t'+k}} \delta^{t-t'}(p(G^\eta) - c) + \sum_{\substack{t \in T_{t'}(L) \\ t'+1 \leq t \leq t'+k}} \delta^{t-t'} p(0) + \frac{\delta^{k+1}}{(1-\delta)}(p(0) - c),$$

for any $k > 0$. Because $p(G^\eta) \to p(1)$ as $\eta \to 0$, by choosing k large and η small, this lower bound can be made arbitrarily close to

$$p(1) + \sum_{\substack{t \in T_{t'}(H), \\ t \geq t'+1}} \delta^{t-t'}(p(1) - c) + \sum_{\substack{t \in T_{t'}(L), \\ t \geq t'+1}} \delta^{t-t'} p(0).$$

Because the continuation payoff from following σ_1 is no more than

$$\sum_{t \in T_{t'}(H)} \delta^{t-t'}(p(1) - c) + \sum_{t \in T_{t'}(L)} \delta^{t-t'} p(0),$$

the normal firm has a profitable deviation. Thus, there is no equilibrium in pure strategies with the normal firm ever choosing high effort.

∎

The possibility of an inept type potentially provides the firm with an incentive to exert high effort, because a consumer who receives signal $\underline{z}$ "punishes" the firm by increasing the probability with which the consumer thinks the firm is inept. The difficulty is that a firm who builds a reputation is too successful at building the reputation. Eventually, almost all consumers become almost certain that the firm is normal, in the sense that the posterior probability attached to a normal firm gets arbitrarily close to 1 for an arbitrarily large subset of consumers. The incentive to exert high effort arises only out of the desire to affect consumers' beliefs about the firm. As the posterior probability of a normal firm approaches unity, the effect of signal $\underline{z}$ or $\bar{z}$ on this belief becomes smaller. At some point, the current signal will have such a small effect on the current belief that the cost c of high effort overwhelms the very small difference in beliefs caused by signal $\bar{z}$ rather than $\underline{z}$, and the normal firm will then find it optimal to revert to low effort. Consumers and the firm can foresee that this will happen, however, causing the equilibrium to unravel. The only pure-strategy equilibrium calls for only low effort to be exerted.

Remark 18.2.1 **Mixed strategies** The complete information version of this model shares many features with the model analyzed in section 12.5. In section 12.5 (and in a variant

18.2 ■ The Role of Replacements

of the current model discussed in section 18.4.1), the uninformed players choose from a discrete set, so mixed strategy equilibria can be constructed (using similar techniques to belief-free equilibria). It is not known if an analogous approach works in the current context, nor whether belief-free equilibria exist in incomplete information games (see the discussion at the end of section 18.3.1).

◆

Remark 18.2.2 **Symmetric information** A similar role for replacements was described by Holmström (1982) in the context of a signal-jamming model of managerial employment. The wage of the manager in his model is higher if the market posterior over the manager's type is higher, even if the manager chooses no effort. In contrast, the revenue of a firm in the model of section 18.1 is higher for higher posteriors only if consumers also believe that the normal firm is choosing high effort. Because the market directly values managerial talent, Holmström's manager always has an incentive to increase effort in an attempt to enhance the market estimation of his talent. In contrast to proposition 18.2.1, an equilibrium then exists (without replacements) in which the manager chooses effort levels that are higher than the myopic optimum. In agreement with the spirit of proposition 18.2.1, however, this overexertion disappears over time, as the market's posterior concerning the manager's type approaches 1. Holmström (1982) is one of the first publications to use changing types to obtain a sustained reputation effect.

Neither the market *nor the manager* knows the talent of the manager in Holmström's (1982) model. The manager's evaluation of the profitability of effort then reflects only market beliefs. In contrast, the normal firms in section 18.1 are more optimistic about the evolution of posterior beliefs than are consumers. However, the underlying mechanism generating incentives is the same. When the firms in section 18.1 do not know their types, being symmetrically uninformed with consumers, the existence of a high-effort equilibrium is possible with replacements and impossible without.

◆

Propositions 18.1.1 and 18.2.1 combine to provide the seemingly paradoxical result that it can be good news for the firm to have consumers constantly fearing that the firm might "go bad." The purpose of a reputation is to convince consumers that the firm is normal and will exert high effort. As we have just seen, the problem with maintaining a reputation in the absence of replacements is that the firm essentially succeeds in convincing consumers it is normal. If replacements continually introduce the possibility that the firm has turned inept, then there is an upper bound, short of unity, on the posterior μ_0^t, so the difference in posteriors after different signals is bounded away from 0. The incentive to exert high effort in order to convince consumers that the firm is still normal then always remains.

Remark 18.2.3 **Competition and beliefs** Replacements are not the only mechanism by which incentives for high effort can be sustained. Section 18.4.6 discusses competition in a model where consumers can leave a firm (and so impose a significant cost) after *any* reduction in beliefs.

◆

18.3 Good Types and Bad Types

The next item on our agenda is to explore the differing roles of good and bad commitment types in reputation results. We do this in the context of the product-choice game, reproduced in figure 18.3.1, played by a long-lived player 1 facing a single short-lived player 2. This is the standard long-lived/short-lived player setting considered in sections 2.7 and 7.6 and examples 15.3.1 and 15.4.1.

18.3.1 Bad Types

Suppose first that the set of possible types Ξ is given by $\{\xi_0, \xi(L)\}$, so that player 1 may be either normal or committed to action L. The lower bound on player 1's payoff, established in proposition 15.3.1, in this case tells us only that in the repeated game of perfect monitoring, player 1 must earn at least his minmax payoff of 1.

A tighter bound is not available, and the possibility of an inept type in this setting has no effect on the set of payoff possibilities for player 1. For example, the pooling profile in which the normal type (as well as the inept type) always plays L and player 2 always plays ℓ is trivially an equilibrium. The introduction of a rare inept type also does not alter the conclusion of proposition 3.6.1. Any payoff in the interval $(1, 2]$ is also an equilibrium payoff for a sufficiently patient player 1 in the game of incomplete information.

Proposition 18.3.1 *For any $\varepsilon > 0$ and $v_1 \in [1, 2]$, there is a discount factor $\underline{\delta}$ such that for any $\delta \in (\underline{\delta}, 1)$ there exists a subgame-perfect equilibrium with payoff to the normal player 1 within ε of v_1.*

Proof Fix $v_1 \in [1, 2]$ and $\varepsilon > 0$. In the game of complete information with normal player 1, there exists an equilibrium σ with payoff v_1 for player 1 (proposition 3.6.1). Let strategies in the incomplete information game specify H for the normal player 1 and L for the commitment type in the first period (with player 2 playing a best response), with any history featuring L in the first period followed by permanent play of $L\ell$ and any history featuring H by strategy profile σ. For sufficiently large δ, this is an equilibrium of the game of incomplete information with a payoff for the normal player 1 contained in $[\delta v_1, 2(1 - \delta) + \delta v_1]$, and hence within ε of v_1. ∎

The difficulty in using the possibility of the inept type to impose bounds on player 1's payoff is that there exists a low equilibrium payoff for the normal player 1 in the

	h	ℓ
H	2, 3	0, 2
L	3, 0	1, 1

Figure 18.3.1 Product-choice game.

game of *incomplete* information. This can be used as a punishment if player 1 does not choose an action revealing his type. An equilibrium strategy profile can thus be constructed that creates incentives for player 1 to reveal his type and then proceeds as in a game of complete information.

Suppose that the monitoring is imperfect and public (as in section 7.6.2). Once again, the lower bound established by the appropriate reputation argument, proposition 15.4.1, has no force, implying only that player 1 must earn at least his minmax payoff. Less is known about the structure of equilibria in this case. For example, consider the belief-free equilibrium constructed for the complete information product-choice game of section 7.6.2. In that equilibrium, player 1 plays $\frac{1}{2} \circ H + \frac{1}{2} \circ L$ in each period, ensuring player 2's indifference, with player 2 mixing as a function of her signal to ensure the optimality of player 1's mixture. In an attempt to reproduce this in the game of incomplete information, suppose the normal player 1 chooses L with probability

$$\frac{1 - 2\mu^t(\xi(L))}{2 - 2\mu^t(\xi(L))},$$

where $\mu^t(\xi(L))$ is the period t posterior that player 1 is type $\xi(L)$. As long as $\mu^t(\xi(L)) < 1/2$, this probability is well defined and ensures that player 2 faces the expected action $\frac{1}{2} \circ H + \frac{1}{2} \circ L$. However, each signal y pushes upward the posterior belief that player 1 is type $\xi(L)$. As this belief increases, the probability the normal type attaches to L decreases, exacerbating the belief revision. With probability 1, the posterior that player 1 is the type $\xi(L)$ will be pushed above $1/2$, at which point the equilibrium collapses.[4]

The structure of equilibria in repeated games of imperfect monitoring and incomplete information remains relatively unexplored. Section 17.3 constructs an equilibrium for the twice-played product-choice game that illustrates the complexities.

18.3.2 Good Types

Now consider the infinitely repeated product-choice game with the set of types Ξ given by $\{\xi_0, \xi(H)\}$, so that player 1 is either normal or committed to the pure Stackelberg action H. With either perfect or imperfect monitoring, the result is straightforward. Propositions 15.3.1 and 15.4.1 ensure that a sufficiently patient normal player 1 earns a payoff arbitrarily close to 2. For example, there is an equilibrium in the perfect monitoring game in which the normal player 1 plays H in every period, supported by the threat that any deviation to L prompts the perpetual play of $L\ell$.[5]

We now add replacements to the perfect monitoring version of the model. In each period, player 1 continues to the next period with probability $1 - \lambda > 1/2$, and is replaced by a new player 1 with probability λ. The type of the new player 1 is $\xi(H)$

---

4. The integrity of the construction of the belief-free equilibrium can be preserved using replacements, as long as the probability of inept types is sufficiently low and the replacement rate is sufficiently high that the posterior on $\xi(L)$ can never exceed $1/2$.
5. Player 1 might consider playing H once this punishment has been triggered, in hopes of resuscitating player 2's belief that 1 is the commitment type, but the consistency condition of sequential equilibrium precludes success (remark 15.3.2).

with probability $\mu^0(\xi(H)) \equiv \hat{\mu}^0$, the prior probability of the commitment type, with complementary probability on the normal type.

A profile in which the normal player 1 always plays H, supported by permanent reversion to $L\ell$ once L is observed, is no longer a sequential equilibrium, when player 1 is sufficiently patient and replacements sufficiently unlikely. Suppose that under this profile, the punishment is triggered. At this point, the first play of H by player 1 causes player 2 to believe that player 1 has been replaced by a commitment type. Given that replacements are not too likely, 2's best response is to play h. But then the normal type of player 1, if sufficiently patient, will play H, vitiating the optimality of the punishment. Hence we cannot simply transfer the no-replacement equilibrium profile to the game with replacements.

We construct an equilibrium for the case in which player 1 is sufficiently patient. In each period t in which $\hat{\mu}^t$ is less than or equal to $1/2$, the normal player 1 attaches the probability

$$\alpha_1(\hat{\mu}^t) \equiv \frac{1 - 2\hat{\mu}^t}{2 - 2\hat{\mu}^t}$$

to H. This implies that in such a period, player 2 faces the mixed action $\frac{1}{2} \circ H + \frac{1}{2} \circ L$, because

$$\hat{\mu}^t + (1 - \hat{\mu}^t)\frac{1 - 2\hat{\mu}^t}{2 - 2\hat{\mu}^t} = \frac{1}{2}, \quad (18.3.1)$$

and thus is indifferent between h and ℓ. When $\hat{\mu}^t \leq 1/2$, player 2 mixes, putting probability $\alpha_2(\hat{\mu}^t)$ (to be calculated) on h. For values $\hat{\mu}^t > 1/2$, player 2 plays h for sure and the normal player 1 chooses L. Notice that the actions of the consumer and the normal player 1 depend only on the posterior probability that player 1 is the Stackelberg type, giving us a profile in Markov strategies (section 5.6).

It remains only to determine the mixtures chosen by player 2, which are designed so that the normal player 1 is behaving optimally. Let $\varphi(\hat{\mu} \mid H)$ be the posterior probability attached to player 1 being $\xi(H)$, given the prior $\hat{\mu}$ and an observation of H. If the normal type chooses H with probability $\alpha_1(\hat{\mu})$, we have

$$\varphi(\hat{\mu} \mid H) = (1 - \lambda)\frac{\hat{\mu}}{\hat{\mu} + (1 - \hat{\mu})\alpha_1(\hat{\mu})} + \lambda\hat{\mu}^0 = 2(1 - \lambda)\hat{\mu} + \lambda\hat{\mu}^0, \quad (18.3.2)$$

using (18.3.1) for the second equality, and the corresponding calculation for L is

$$\varphi(\hat{\mu} \mid L) = \lambda\hat{\mu}^0.$$

For each value $\hat{\mu} \in [\lambda\hat{\mu}^0, \varphi(\lambda\hat{\mu}^0 \mid H))$, let $\hat{\mu}^{(0)}, \hat{\mu}^{(1)}, \ldots, \hat{\mu}^{(N)}$ be the sequence of posterior probabilities (with dependence on $\hat{\mu}$ suppressed in the notation) satisfying $\hat{\mu}^{(i)} = \varphi(\hat{\mu}^{(i-1)} \mid H)$ for $(i = 1, \ldots, N)$, with $\hat{\mu}^{(N)}$ being the first such probability to equal or exceed $1/2$. We associate one such sequence with every value $\hat{\mu} \in [\lambda\hat{\mu}^0, \varphi(\lambda\hat{\mu}^0 \mid H))$, giving an uncountable collection of finite sequences.

The function $\varphi(\hat{\mu} \mid H)$ defined in (18.3.2) is linear in $\hat{\mu}$, with positive intercept and slope exceeding 1. This ensures that no two of the sequences we have constructed

18.3 ■ Good Types and Bad Types

have terms in common. In addition, the union of the sequences' terms is a superset of $[\lambda\hat{\mu}^0, 1/2]$.

We now construct player 2's strategy. For each sequence of posteriors we have just constructed, we attach a player 2 action to each posterior in the sequence. Fix such a sequence $\{\hat{\mu}^{(0)}, \ldots, \hat{\mu}^{(N)}\}$. Let $V_0^{(i)}$ be the value to the *normal* player 1 of continuation play, beginning at posterior $\hat{\mu}^{(i)}$. Player 2 must randomize so that the normal type of player 1 is indifferent between L and H, and hence to satisfy, for $k = 0, \ldots, N-1$,

$$V_0^{(k)} = (2\alpha_2(\hat{\mu}^{(k)}) + 1)(1 - \delta(1-\lambda)) + \delta(1-\lambda)V_0^{(0)} \qquad (18.3.3)$$

$$= 2\alpha_2(\hat{\mu}^{(k)})(1 - \delta(1-\lambda)) + \delta(1-\lambda)V_0^{(k+1)}, \qquad (18.3.4)$$

where we normalize by $(1 - \delta(1-\lambda))$. Finally,

$$V_0^{(N)} = 3(1 - \delta(1-\lambda)) + \delta(1-\lambda)V_0^{(0)}, \qquad (18.3.5)$$

because the normal player 1 chooses L with certainty for $\hat{\mu}^t > 1/2$.

Solving (18.3.3) for $k = 0$ gives

$$V_0^{(0)} = 2\alpha_2(\hat{\mu}^{(0)}) + 1. \qquad (18.3.6)$$

Solving the equality of the right side of (18.3.3) with (18.3.4) yields, for $k = 0, \ldots, N-1$,

$$V_0^{(k+1)} = V_0^{(0)} + \frac{1 - \delta(1-\lambda)}{\delta(1-\lambda)}. \qquad (18.3.7)$$

These two equations tell us a great deal about the equilibrium. The value $V_0^{(0)}$ is relatively low, whereas the remaining values $V_0^{(1)} = V_0^{(2)} = \cdots = V_0^{(N)} \equiv \bar{V}_0$ are higher and equal to one another. This in turn implies (from (18.3.3)) that $\alpha_2(\hat{\mu}^{(0)})$ is relatively low, and the remaining probabilities $\alpha_2(\hat{\mu}^{(1)}) = \cdots = \alpha_2(\hat{\mu}^{(N-1)}) \equiv \bar{\alpha}_2$ are higher and equal to one another. These properties reflect the special structure of the product-choice game, most notably the fact that the stage-game payoff gain to player 1 from playing L rather than H is independent of player 2's action.

It remains to calculate $\alpha_2(\hat{\mu}^{(0)})$ and $\bar{\alpha}_2$, confirm that both are probabilities, and confirm that a normal player 1 facing posterior $\hat{\mu}^{(N)}$ prefers to play L. All other aspects of player 1's strategy are optimal because he is indifferent between H and L at every other posterior belief.

Rearrange (18.3.5) and (18.3.7) to give

$$\begin{bmatrix} 1 & -\delta(1-\lambda) \\ 1 & -1 \end{bmatrix} \begin{bmatrix} \bar{V}_0 \\ V_0^{(0)} \end{bmatrix} = \begin{bmatrix} 3(1 - \delta(1-\lambda)) \\ \frac{1-\delta(1-\lambda)}{\delta(1-\lambda)} \end{bmatrix},$$

and then solve for

$$\bar{V}_0 = 2 \qquad (18.3.8)$$

and

$$V_0^{(0)} = \frac{3\delta(1-\lambda) - 1}{\delta(1-\lambda)}.$$

Inserting the second of these in (18.3.6), we calculate

$$\alpha_2(\hat{\mu}^{(0)}) = \frac{2\delta(1-\lambda)-1}{2\delta(1-\lambda)},$$

which falls in the interval $(0, 1)$ if and only if $\delta(1-\lambda) > 1/2$, which becomes the "sufficient patience" condition for this equilibrium. From (18.3.4) for any $k > 0$, we have (because $V_0^{(k)} = \bar{V}_0$),

$$\bar{V}_0 = 2\bar{\alpha}_2(1-\delta(1-\lambda)) + \delta(1-\lambda)\bar{V}_0,$$

so that

$$\bar{V}_0 = 2\bar{\alpha}_2,$$

and so (from (18.3.8)) $\bar{\alpha}_2 = 1$.

The final step is to verify that player 1 prefers L when the posterior is $\hat{\mu}^{(N)}$, at the cost of temporarily removing any possibility in player 2's mind that he is type $\xi(H)$ and triggering a temporary reduction in the probability with which consumers play h. This is not obvious, because choosing H leaves the posterior unchanged and preserves a regime in which consumers choose h with probability one. However, a simple argument suffices to show that L is optimal for any $\hat{\mu}^{(N)} \geq 1/2$, on noting that the equilibrium value of any such posterior equals $V_0^{(N)}$. The required optimality condition is

$$3(1-\delta(1-\lambda)) + \delta(1-\lambda)V_0^{(0)} \geq 2(1-\delta(1-\lambda)) + \delta(1-\lambda)V_0^{(N)}.$$

This is implied by the fact that mixing is optimal for posterior $\hat{\mu}^{(N-1)}$, or

$$(2\alpha_2(\hat{\mu}^{(N-1)}) + 1)(1-\delta(1-\lambda)) + \delta(1-\lambda)V_0^{(0)}$$
$$= 2\alpha_2(\hat{\mu}^{(N-1)})(1-\delta(1-\lambda)) + \delta(1-\lambda)V_0^{(N)}.$$

Phelan (2001) constructs a similar equilibrium in a game that does not have the equal-gain property of the product-choice game, that player 1's payoff increment from playing H is independent of player 2's action. The equilibrium then has a richer dynamic structure including mixing on the part of player 2 for any posterior less than $1/2$.

The breakdown of player 1's reputation in this model is unpredictable, in the sense that the probability of a reputation-ending exertion of low effort is the same regardless of the current posterior that player 1 is good. A normal player 1 who has labored long and hard to build his reputation is just as likely to spend it as is one just starting. Once the reputation has been spent, it can be rebuilt, but only gradually, as the posterior probability of a good type gets pushed upward once more.

18.4 Reputations with Common Consumers

If we interpret short-lived players as a continuum of long-lived but anonymous players and assume that they receive common signals and play pure strategies, then the model is formally equivalent to the case of a single short-lived player (see section 15.4.3).

In contrast, our analysis rests heavily on the assumption that the anonymous players receive idiosyncratic signals. For example, section 18.2 notes that in the absence of replacements, consumers who receive bad signals do not punish the firm because they know (in a high-effort profile) that the firm exerted high effort. If the consumers receive common signals, there is no difficulty in using bad signals to trigger punishments, even if consumers are confident the firm exerted high effort. This type of dependence on public signals underlies the typical intertemporal incentive in public monitoring games. The difficulty with idiosyncratic signals is that a bad signal brings a consumer no information about what other consumers have seen, preventing the consumer coordination that is essential for effective incentives.

This section explores the role of the idiosyncratic consumer signals with which we worked in section 18.1 by examining the corresponding model with common consumer signals. We are especially interested in exploring the extent to which the idiosyncrasy of signals (like private monitoring in part III), by obstructing the ability to coordinate continuation play, imposes constraints on the ability to construct equilibria.

18.4.1 Belief-Free Equilibria with Idiosyncratic Consumers

Because idiosyncratic signals and private monitoring appear to make the coordination of continuation play difficult in a similar manner, it is natural to consider the possibility of constructing belief-free equilibria in a model with small and anonymous players who receive idiosyncratic signals. As an illustration, consider a version of the private monitoring product-choice game analyzed in section 12.5. The stage game is again figure 18.3.1, with the set of player 2 private signals $\{\underline{z}, \bar{z}\}$ and the marginal signal distribution given by

$$\pi_2(\bar{z} \mid a) = \begin{cases} p, & \text{if } a_1 = H, \\ q, & \text{if } a_1 = L, \end{cases}$$

From the analysis in sections 7.6.2 and 12.5, there is an equilibrium in which player 1 plays $\frac{1}{2} \circ H + \frac{1}{2} \circ L$ in every period, and player 2 chooses h with probability $\bar{\alpha}_2$ after signal $\bar{z}$ and probability α_2 after signal $\underline{z}$, where $\bar{\alpha}_2 = \alpha_2 + 1/(2\delta(p-q))$ (hence requiring $2\delta(p-q) \geq 1$). There is a continuum of such equilibria, defined by the requirements that $\alpha_2, \bar{\alpha}_2 \in [0, 1]$. Note that these equilibria do not require that player 1 receive *any* signal about the play of player 2 or about player 2's signals.

Now suppose there is a continuum of small and anonymous consumers who receive idiosyncratic signals with the above marginal. Consider the profile in which player 1 (the firm) randomizes uniformly over H and L in every period and each consumer follows a personal version of the equilibrium strategy, playing h with probability $\bar{\alpha}_2$ after she receives signal $\bar{z}$ and probability α_2 after signal $\underline{z}$. Because the original profile is a belief-free equilibrium (chapter 14), it is immediate that this profile is an equilibrium of the idiosyncratic consumer game. The firm's mixture ensures that the consumers are playing optimally, whereas the firm's continuation value is independent of history, ensuring that the firm's actions are optimal.

There is no obvious way to extend the connection between games with private and idiosyncratic monitoring beyond belief-free equilibria. Indeed, it is not clear what it means to transfer a profile that is not belief-free from a game of private monitoring and short-lived player 2 to the corresponding game with idiosyncratic player 2 signals, because it is not clear how the player 1's behavior is to be specified as a function of the distribution of signals about player 2. Moreover, it is not clear how to apply belief-free techniques to the (complete information version of the) model of section 18.1, where consumers do not have a discrete choice.

18.4.2 Common Consumers

We now investigate the extent to which a model with common signals can be used to shed light on the model with idiosyncratic signals. Because the model with idiosyncratic signals is difficult to analyze beyond the straightforward results in section 18.1, we are especially interested in the extent to which the more tractable case of common signals is a good substitute.

We retain the model of section 18.1, except that in each period, either all consumers receive a common good outcome $\bar{y}$ (with probability ρ_{a_1}, given effort level a_1 from the firm) or all receive a common bad outcome $\underline{y}$ (with probability $1 - \rho_{a_1}$). We refer to this as the case of common consumers.

A deterministic sequence of effort levels from the firm now induces a stochastic rather than deterministic sequence of signals, in the aggregate as well as from the point of view of the consumer. We restrict attention to public strategy profiles. Hence after any history, every consumer holds the same expectation of high effort. It is then natural to assume the pricing function from section 18.1 sets the price equal to consumers' expected payoff. If the firm is thought to be normal with probability μ_0, and if the normal firm is thought to choose high effort with probability α, then the price will be

$$p(\mu_0 \alpha) = \mu_0 \alpha \rho_H + (1 - \mu_0 \alpha) \rho_L.$$

It is immediate that there exist equilibria in which the normal firm often exerts high effort. Let the normal firm initially exert high effort and continue to do so as long as signal $\bar{y}$ is realized, with the period t price given by $\mu_0^t \rho_H + (1 - \mu_0^t) \rho_L$, where μ_0^t is the period t posterior probability of a normal firm. Let signal $\underline{y}$ prompt $L \geq 1$ periods of low effort and price ρ_L. These strategies will be an equilibrium as long as the cost c is sufficiently small. Allowing consumers to receive a common signal thus has an important impact, restoring high-effort equilibria without an appeal to either incomplete information or replacements.

The equilibria we have just sketched, with punishments triggered by the signal $\underline{y}$, involve precisely the coordinated behavior that is impossible with idiosyncratic signals. To bring the common consumer model more in line with the idiosyncratic consumer model, we now rule out such coordination by restricting attention to Markov strategies. Section 5.5.2 discusses Markov strategies in complete information games. The extension of these ideas to games of incomplete information is in general not trivial, but in the current context μ_0^t is the natural Markov state. In other words, we require that play of all players be identical after two histories that lead to the same posterior belief of consumers that the firm is normal.

A *Markov strategy* for the normal firm can be written as a mapping

$$\alpha : [0, 1] \to [0, 1],$$

18.4 ■ Common Consumers

where $\alpha(\mu_0)$ is the probability of choosing action H when the consumer's posterior probability of a normal firm is μ_0. The consumers' beliefs are updated according to

$$\varphi(\mu_0 \mid \bar{y}) = (1-\lambda)\frac{[\rho_H \alpha(\mu_0) + \rho_L(1-\alpha(\mu_0))]\mu_0}{[\rho_H \alpha(\mu_0) + \rho_L(1-\alpha(\mu_0))]\mu_0 + \rho_L(1-\mu_0)} + \lambda \mu_0^0$$

and

$$\varphi(\mu_0 \mid \underline{y})$$
$$= \frac{(1-\lambda)[(1-\rho_H)\alpha(\mu_0) + (1-\rho_L)(1-\alpha(\mu_0))]\mu_0}{[(1-\rho_H)\alpha(\mu_0) + (1-\rho_L)(1-\alpha(\mu_0))]\mu_0 + (1-\rho_L)(1-\mu_0)} + \lambda \mu_0^0.$$

Given a public history $h^t \in \mathcal{H}^t \equiv \{\underline{y}, \bar{y}\}^t$, the resulting consumer posterior is denoted by $\varphi(\mu_0^0 \mid h^t)$, where μ_0^0 is the prior in period 0. Given a Markov strategy α, and the updating, let $\sigma_1^\alpha : \mathcal{H} \to [0,1]$ denote the implied strategy, $\sigma_1^\alpha(h^t) = \alpha(\varphi(\mu_0^0 \mid h^t))$.

Definition 18.4.1 *The strategy α is a* Markov equilibrium *if σ_1^α is maximizing for the normal firm.*

Because every public history has positive probability, every Markov equilibrium is sequential.

18.4.3 Reputations

The following gives us a reputation result analogous to proposition 18.1.1 of the idiosyncratic consumer case.

Proposition 18.4.1 *Suppose $\lambda \in (0,1)$. Then there exists $\bar{c} > 0$ such that for all $0 \leq c < \bar{c}$, there exists a Markov equilibrium in which the normal firm always exerts high effort.*

Proof We simplify the notation in this proof by shortening μ_0 to simply μ. Suppose the normal firm always exerts high effort. Then given a posterior probability μ that the firm is normal, the firm's revenue is given by $p(\mu) = \mu\rho_H + (1-\mu)\rho_L$. Let $\mu_y = \varphi(\mu \mid y)$ and $\mu_{xy} = \varphi(\mu \mid xy)$ for $x, y \in \{\underline{y}, \bar{y}\}$. Then $\mu_{\bar{y}\bar{y}} > \mu_{\bar{y}} > \mu > \mu_{\underline{y}} > \mu_{\underline{y}\underline{y}}$ and $\mu_{\bar{y}y} > \mu_{\underline{y}y}$ for $y \in \{\underline{y}, \bar{y}\}$. The value function of the normal firm is given by

$$V_0(\mu) = (1-\delta(1-\lambda))(p(\mu)-c) + \delta(1-\lambda)[\rho_H V_0(\mu_{\bar{y}}) + (1-\rho_H)V_0(\mu_{\underline{y}})].$$

The function V_0 is monotonic in μ. In particular, let $G^t(\mu^t; \mu, t')$ be the distribution of consumer posteriors μ^t at time $t > t'$ given that normal firms exert effort and given period t' posterior μ. If $\mu > \mu'$, then $G^t(\mu^t; \mu, t')$ first-order stochastically dominates $G^t(\mu^t; \mu', t')$ for all $t > t'$. The same is then true for the distribution of revenues, which suffices for the monotonicity of V_0.

The payoff from exerting low effort and thereafter adhering to the equilibrium strategy is

$$V_0(\mu; L) \equiv (1-\delta(1-\lambda))p(\mu) + \delta(1-\lambda)[\rho_L V_0(\mu_{\bar{y}}) + (1-\rho_L)V_0(\mu_{\underline{y}})].$$

Thus, $V_0(\mu) - V_0(\mu; L)$ is given by

$$-c(1 - \delta(1 - \lambda)) + \delta(1 - \lambda)(1 - \delta(1 - \lambda))(\rho_H - \rho_L)(p(\mu_{\bar{y}}) - p(\mu_{\underline{y}}))$$
$$+ \delta^2(1 - \lambda)^2(\rho_H - \rho_L)\{\rho_H[V_0(\mu_{\bar{y}\bar{y}}) - V_0(\mu_{\underline{y}\bar{y}})] + (1 - \rho_H)[V_0(\mu_{\bar{y}\underline{y}}) - V_0(\mu_{\underline{y}\underline{y}})]\}$$
$$\geq (1 - \delta(1 - \lambda))\{-c + \delta(1 - \lambda)(\rho_H - \rho_L)[p(\mu_{\bar{y}}) - p(\mu_{\underline{y}})]\}, \quad (18.4.1)$$

where the monotonicity of V_0 in μ yields the inequality.

From an application of the one-shot deviation principle, it is an equilibrium for the normal firm to always exert high effort, with the implied consumer beliefs, if and only if $V_0(\mu) - V_0(\mu; L) \geq 0$ for all feasible μ. From (18.4.1), a sufficient condition for this inequality is

$$p(\mu_{\bar{y}}) - p(\mu_{\underline{y}}) \geq \frac{c}{\delta(1 - \lambda)(\rho_H - \rho_L)}. \quad (18.4.2)$$

The set of feasible posteriors is contained in the interval $[\lambda\mu^0, 1 - \lambda(1 - \mu^0)]$, where μ^0 is the prior probability of a normal firm. In addition, the minimum of $p(\mu_{\bar{y}}) - p(\mu_{\underline{y}}) = p(\varphi(\mu \mid \bar{y})) - p(\varphi(\mu \mid \underline{y}))$ over $\mu \in [\lambda\mu^0, 1 - \lambda(1 - \mu^0)]$ is strictly positive because p and φ are continuous. We can thus find a value of c sufficiently small that (18.4.2) holds for all $\mu \in [\lambda\mu^0, 1 - \lambda(1 - \mu^0)]$. ∎

As for idiosyncratic consumers, in equilibrium, the difference between the value of choosing high effort and the value of choosing low effort must exceed the cost of high effort. However, the value functions corresponding to high and low effort approach each other as $\mu_0 \to 1$, because the values diverge only through the effect of current outcomes on future posteriors and current outcomes have very little effect on future posteriors when consumers are currently quite sure of the firm's type. The smaller the probability of an inept replacement, the closer the posterior expectation of a normal firm can approach unity. Replacements ensure that μ_0 can never reach unity, and thus there is always a wedge between the high-effort and low-effort value functions. As long as the cost of the former is sufficiently small, high effort will be an equilibrium.

18.4.4 Replacements

Once again, the possibility of replacements is important in this result. In the absence of replacements, consumers eventually become so convinced the firm is normal (i.e., the posterior μ_0 becomes so high), that subsequent evidence can only shake this belief very slowly. Once this happens, the incentive to choose high effort disappears. If replacements continually introduce the possibility that the firm has become inept, then the firm cannot be "too successful" at convincing consumers it is normal, and so there is an equilibrium in which the normal firm always exerts high effort.

We illustrate the importance of replacements by characterizing equilibrium behavior in the model without replacements. We begin with a particularly simple case, characterized by a restriction on the parameters ρ_H and ρ_L.

Proposition 18.4.2 *Suppose there are no replacements ($\lambda = 0$) and there is some $m \in \mathbb{N}$ such that $(1 - \rho_H)\rho_H^m = (1 - \rho_L)\rho_L^m$. Then there is a unique Markov equilibrium in pure*

18.4 ■ Common Consumers

strategies, and in this equilibrium the normal firm exerts low effort with probability one, that is, $\alpha(\mu_0) = 0$ for all μ_0.

One obvious circumstance in which $(1 - \rho_H)\rho_H^m = (1 - \rho_L)\rho_L^m$ for some m is the symmetric case in which $\rho_H = 1 - \rho_L$. In this case, the argument is particularly straightforward. For any posterior belief μ_0 with $\alpha(\mu_0) = \alpha(\varphi(\mu_0 \mid \bar{y})) = 1$ we have $\varphi(\mu_0 \mid \bar{y}y) = \mu_0$. But then no punishment can be triggered after $\bar{y}y$, so there are no incentives to exert effort at belief $\varphi(\mu_0 \mid \bar{y})$.

Proof The strategy $\alpha(\mu_0) = 0$ is clearly a Markov equilibrium. We need to argue that this is the only pure-strategy Markov equilibrium. Fix a pure-strategy Markov equilibrium α, and let $\upsilon(\mu_0) = \alpha(\mu_0)\mu_0$ be the probability of high effort under α. Letting $V_0(\mu_0)$ denote the value function of normal firm, we have

$$V_0(\mu_0) = (1 - \delta)(p(\upsilon(\mu_0)) - \alpha(\mu_0)c)$$
$$+ \delta\alpha(\mu_0)(\rho_H - \rho_L)(V_0(\varphi(\mu_0 \mid \bar{y})) - V_0(\varphi(\mu_0 \mid \underline{y})))$$
$$+ \delta(\rho_L V_0(\varphi(\mu_0 \mid \bar{y})) + (1 - \rho_L)V_0(\varphi(\mu_0 \mid \underline{y}))). \quad (18.4.3)$$

Because $p(\upsilon(\mu_0)) \geq p(0)$, a lower bound on V_0 is $p(0)$, the payoff obtained by setting $\alpha(\mu_0) = 0$ for all μ_0 and hence incurring no cost. Denoting by $V_0(\mu_0; L)$ the value of a one-period deviation to choosing low effort and then reverting to the equilibrium strategy of α, we have

$$V_0(\mu_0; L) = (1 - \delta)p(\upsilon(\mu_0)) + \delta\rho_L V_0(\varphi(\mu_0 \mid \bar{y}))$$
$$+ \delta(1 - \rho_L)V_0(\varphi(\mu_0 \mid \underline{y})), \quad (18.4.4)$$

so that from (18.4.3) and (18.4.4),

$$V_0(\mu_0) - V_0(\mu_0; L) = -\alpha(\mu_0)(1 - \delta)c$$
$$+ \alpha(\mu_0)\delta(\rho_H - \rho_L)\{V_0(\varphi(\mu_0 \mid \bar{y})) - V_0(\varphi(\mu_0 \mid \underline{y}))\}.$$

We now suppose there exists some μ_0 with $\alpha(\mu_0) = 1$ and seek a contradiction. Choosing H for sure at μ_0 is only optimal if

$$\delta(\rho_H - \rho_L)\{V_0(\varphi(\mu_0 \mid \bar{y})) - V_0(\varphi(\mu_0 \mid \underline{y}))\} \geq (1 - \delta)c. \quad (18.4.5)$$

If $\mu_0 \in \{0, 1\}$, then $\varphi(\mu_0 \mid \bar{y}) = \varphi(\mu_0 \mid \underline{y})$, ensuring $\alpha(\mu_0) = 1$ is suboptimal. Hence, let $\mu_0 \in (0, 1)$ and define a sequence $\{\mu_0^{(k)}\}_{k=0}^{\infty}$ by $\mu_0^{(k+1)} = \varphi(\mu_0^{(k)} \mid \bar{y})$. First observe that the firm must choose effort as long as $\bar{y}$ only has been realized. If not, there is some k for which $\alpha(\mu_0^{(k)}) = 0$. Then $\varphi(\mu_0^{(k)} \mid \underline{y}) = \varphi(\mu_0^{(k)} \mid \bar{y}) = \mu_0^{(k)}$, so that $V_0(\varphi(\mu_0^{(k)} \mid \bar{y})) = V_0(\varphi(\mu_0^{(k)} \mid \underline{y})) = p(0)$ and so $\alpha(\mu_0^{(k-1)}) = 0$. A recursion then implies $\alpha(\mu_0) = 0$, a contradiction.

Now,

$$\varphi(\mu_0^{(k)} \mid \underline{y}) = \frac{(1 - \rho_H)\rho_H^k \mu_0}{(1 - \rho_H)\rho_H^k \mu_0 + (1 - \rho_L)\rho_L^k(1 - \mu_0)}, \quad (18.4.6)$$

so that, for $k \geq m$ and using $(1 - \rho_H)\rho_H^m = (1 - \rho_L)\rho_L^m$,

$$\varphi(\mu_0^{(k)} \mid \underline{y}) = \mu_0^{(k-m)}.$$

Using this equality in (18.4.5), for $\ell \in \mathbb{N}$ we have

$$V_0(\mu_0^{(\ell(m+1))}) \geq V_0(\mu_0) + \frac{(1-\delta)\ell c}{\delta(\rho_H - \rho_L)},$$

which gives

$$\lim_{\ell \to \infty} V_0(\mu_0^{(\ell(m+1))}) = \infty.$$

But this is impossible, because V_0 is bounded above by $(p(1) - c)$. It must then not be the case that $\alpha(\mu_0) = 1$, yielding the result. ∎

18.4.5 Continuity at the Boundary and Markov Equilibria

Propositions 18.4.1 and 18.4.2 give us a preliminary indication that the model with idiosyncratic consumers is much like the model with common consumers, once we restrict attention to Markov equilibria in the latter. Each has an equilibrium in which the normal firm exerts high effort, when there are replacements, and high-effort equilibria are problematic in the absence of replacements. However, the condition $(1 - \rho_H)\rho_H^m = (1 - \rho_L)\rho_L^m$ in proposition 18.4.2 is rather specific. A claim that the common consumer model appropriately captures aspects of the idiosyncratic model should not rely on such a condition.

Proposition 18.4.3 *Suppose there are no replacements ($\lambda = 0$) and $(1 - \rho_H)\rho_H^m \neq (1 - \rho_L)\rho_L^m$ for all $m \in \mathbb{N}$.*

1. *For sufficiently small $c > 0$ and large $\delta < 1$, there exists a Markov equilibrium (α, υ) in which $\alpha(\mu_0^0) = 1$. If high effort is ever exerted in a pure-strategy Markov equilibrium outcome, then $\alpha(\mu_0^0) = 1$.*
2. *In any pure-strategy Markov equilibrium in which $\alpha(\mu_0^0) > 0$, we have*

$$\limsup\nolimits_{\mu_0 \to 1} \alpha(\mu_0) = 1$$
$$\text{and} \quad \liminf\nolimits_{\mu_0 \to 1} \alpha(\mu_0) = 0. \tag{18.4.7}$$

3. *There exist positive integers κ, T, and a number $\zeta \in (0, 1)$, such that in any pure-strategy Markov equilibrium the probability (conditional on the firm being normal) of high effort being exerted in any period $t > T$ is less than $\kappa \zeta^t$.*

As long as the specific condition of proposition 18.4.2 is not satisfied, there exist pure-strategy Markov equilibria in which the normal firm sometimes exerts high effort. Once low effort is exerted in a pure-strategy Markov equilibrium, it must be exerted in every subsequent period. In particular, the posterior μ_0 cannot change after a period in which low effort is exerted as part of a pure strategy, so that the posterior thereafter remains fixed and the firm is locked into low effort. As a result, if high effort is ever

18.4 ■ Common Consumers

exerted, it must be in period 0. The possibility of high effort appears only for a finite time period, after which the firm is doomed to low effort.

Proof *Statement 1.* Suppose $\mu_0 \in (0, 1)$. From (18.4.6), $\varphi(\mu_0 \mid \bar{y}^{(m)}y) = \mu_0$ if and only if $(1 - \rho_H)\rho_H^m = (1 - \rho_L)\rho_L^m$, and so $\varphi(\mu_0 \mid \bar{y}^{(m)}\underline{y}) \neq \mu_0$ for all m. Let

$$\Phi = \left\{ \mu_0 : \mu_0 = \frac{\rho_H^m \mu_0^0}{\rho_H^m \mu_0^0 + \rho_L^m (1 - \mu_0^0)} \text{ for some } m \in \mathbb{N}_0 \right\}$$

be the set of posterior probabilities that are attached to histories in which only $\bar{y}$ has been observed (including the null history), assuming the normal firm has always chosen high effort. Then assuming the normal firm has always chosen high effort, for all m, $\varphi(\mu_0 \mid \bar{y}^{(m)}\underline{y}) \notin \Phi$. Consider now the pure strategy $\alpha(\mu_0) = 1$ for $\mu_0 \in \Phi$ and $\alpha(\mu_0) = 0$ otherwise.

The normal firm's value function for $\mu_0 \in \Phi$ is

$$V_0(\mu_0) = (1 - \delta)(p(\mu_0) - c) + \delta\{\rho_H V_0(\varphi(\mu_0 \mid \bar{y})) + (1 - \rho_H)V_0(\varphi(\mu_0 \mid \underline{y}))\},$$
(18.4.8)

whereas for $\mu_0 \notin \Phi$, because $\alpha(\mu_0) = 0$,

$$V_0(\mu_0) = p(0).$$

Because

$$V_0(\varphi(\mu_0 \mid \bar{y})) - V_0(\mu_0)$$
$$= (1 - \delta) \sum_{m=1}^{\infty} (\delta\rho_H)^{m-1}(p(\varphi(\mu_0 \mid \bar{y}^{(m)})) - p(\varphi(\mu_0 \mid \bar{y}^{(m-1)}))),$$

V_0 is increasing in μ_0 on Φ. Evaluating (18.4.8) at $\varphi(\mu_0 \mid \bar{y})$, we have (because $\varphi(\mu_0 \mid \bar{y}\underline{y}) \notin \Phi$),

$$V_0(\varphi(\mu_0 \mid \bar{y})) \geq \frac{(1 - \delta)(p(\varphi(\mu_0 \mid \bar{y})) - c)}{1 - \delta\rho_H} + \frac{\delta(1 - \rho_H)p(0)}{1 - \delta\rho_H}.$$

The normal firm's payoff from deviating to L at $\mu_0 \in \Phi$ is

$$V_0(\mu_0; L) = (1 - \delta)p(\mu_0) + \delta\{\rho_L V_0(\varphi(\mu_0 \mid \bar{y})) + (1 - \rho_L)p(0)\}.$$

The deviation is not profitable if

$$(1 - \delta)c \leq \delta(\rho_H - \rho_L)\{V_0(\varphi(\mu_0 \mid \bar{y})) - p(0)\},$$

which certainly holds if

$$c \leq \delta(\rho_H - \rho_L) \left\{ \frac{(p(\varphi(\mu_0 \mid \bar{y})) - c)}{(1 - \delta\rho_H)} + \frac{\delta(1 - \rho_H)p(0)}{(1 - \delta\rho_H)(1 - \delta)} - \frac{p(0)}{(1 - \delta)} \right\}.$$

Simplifying yields

$$c \leq \frac{\delta(\rho_H - \rho_L)(p(\varphi(\mu_0 \mid \bar{y})) - p(0))}{(1 - \delta\rho_L)}.$$

Thus because the right side is increasing in $\mu_0 \in \Phi$, a sufficient condition for (α, υ) to be an equilibrium is

$$c \le \frac{\delta(\rho_H - \rho_L)(p(\varphi(\mu_0^0 \mid \bar{y})) - p(0))}{(1 - \delta \rho_L)}.$$

Statement 2. Consider a pure-strategy Markov equilibrium in which high effort is sometimes exerted. Though the equilibrium constructed in proving the first statement assumed that the switch from high to low effort is necessarily triggered by the first realization of $\underline{y}$, this may not hold in general. Hence let $\mathcal{Y}$ be the collection of finite strings of signals defined by the property that if a string of length t is in $\mathcal{Y}$, then the history of realized outcomes described by this string would generate posterior expectations that in turn induce high effort in each of periods 0 through t but induce low effort in period $t + 1$. First, note that no string in $\mathcal{Y}$ can end with signal $\bar{y}$. Second, $\mathcal{Y} \ne \varnothing$. (If $\mathcal{Y} = \varnothing$, then every sequence of realized utility outcomes results in high-effort choices. But then the normal firm has no incentive to choose high effort.) This in turn ensures that for any $m \in \mathbb{N}$ there is a history in $\mathcal{Y}$ whose initial segment has m signals $\bar{y}$, which ensures that $\limsup_{\mu_0 \to 1} \alpha(\mu_0) = 1$.

Now let an m-history be a history in $\mathcal{Y}$ terminating with m $\underline{y}$ signals. We claim that there exists a number M such that $m < M$ for all m-histories in $\mathcal{Y}$. Fix m sufficiently large that

$$(1 - \delta) \sum_{t=0}^{m-1} \delta^t \rho_H + \delta^m \rho_L > (1 - \delta) \sum_{t=0}^{\infty} \delta^t (\rho_H - c).$$

If no such M exists, we can find histories in $\mathcal{Y}$ that include m realizations of $\underline{y}$, which implies a contradiction: The left side is then a lower bound on the continuation payoff from never exerting effort after reaching a point at which m signals $\underline{y}$ would not move the history outside $\mathcal{Y}$, and the right side is an upper bound on the continuation payoff under the equilibrium strategy (which requires high effort along histories in $\mathcal{Y}$).

This then implies $\liminf_{\mu_0 \to 1} \alpha(\mu_0) = 0$ (by considering histories consisting of n observations of $\bar{y}$, followed exclusively by observations of $\underline{y}$, for arbitrarily large n), completing the proof of the second statement.

Statement 3. Let $m(n)$ be the largest natural number with the property that there is a string of length n in $\mathcal{Y}$ whose $m(n)$ final elements are $\underline{y}$. From the proof of the second statement, there is a smallest number M with $m(n) \le M$ for all $n > 0$. An upper bound on the probability that the normal firm chooses high effort in period t in any pure-strategy equilibrium is given by the probability p_t^* that the normal firm chooses high effort in period t when following the profile of always choosing high effort until M consecutive $\underline{y}$ signals occur, after which low effort is always chosen. The outcome under this strategy can be described by a Markov chain with state space $\{0, 1, \ldots, M\}$. From state $s \in \{0, 1, \ldots, M-1\}$, the process transits to state 0 with probability ρ_H and to state $s + 1$ with probability $1 - \rho_H$ (s is the length of the current sequence of $\underline{y}$ outcomes). The process starts in

18.4 ■ Common Consumers

state 0, and state M is absorbing. Note that $1 - p_t^*$ is the probability the Markov chain reaches M by period t. Because M is the only absorbing state of the finite state Markov process, its stationary distribution assigns probability one to M. Because the sequence of distributions on the state space under the Markov transitions converge exponentially to the stationary distribution, there are positive integers κ, T, and a number $\zeta \in (0, 1)$ such that for any $t > T$ the probability that the process is not in state M is at most $\kappa \zeta^t$.

■

Proposition 18.4.3 describes equilibria in which the firm exerts high effort and the consumers expect high effort, as long as there have been only realizations of signal $\bar{y}$. The first realization of y relegates the firm to low effort thereafter. Though this equilibrium does not sound Markovian, the parameter restriction that $(1 - \rho_H)\rho_H^m \neq (1 - \rho_L)\rho_L^m$ ensures that any history consisting of a sequence of $\bar{y}$ followed by a single $\underline{y}$ must yield a posterior distinct from that yielded by any sequence of only $\bar{y}$ realizations. The posterior probabilities can then be used as a code identifying the signal $\underline{y}$, and any such signal can trigger a punishment while still using Markov strategies. Consumers can thus use their posterior to attach (infinite) punishments to bad signals.

Remark 18.4.1 The ability to code information about previous play into posteriors persists, even if there exists m for which $(1 - \rho_H)\rho_H^m = (1 - \rho_L)\rho_L^m$, if we allow mixed strategies. Mailath and Samuelson (2001, proposition 2) show:

Proposition 18.4.4 *Suppose $\lambda = 0$ and $\rho_H = 1 - \rho_L \equiv 1 - \rho$.*

1. *A mixed-strategy Markov perfect equilibrium with $\alpha(\mu_0) > 0$ for some μ_0 exists if*

$$\rho + c(1 - \delta\rho)/\{\delta(1 - 2\rho)\} < 1.$$

2. *In any Markov perfect equilibrium with $\alpha(\mu_0) > 0$ for some μ_0,*

$$\limsup_{\mu_0 \to 1} \alpha(\mu_0) > \liminf_{\mu_0 \to 1} \alpha(\mu_0). \tag{18.4.9}$$

◆

The discontinuity reflected in (18.4.7) and (18.4.9) is required to use the current state as an indicator of whether the most recent realized utility has been good or bad. This in turn is necessary to maintain incentives for a normal firm to produce high quality when the posterior probability μ_0 is high. Realized outcomes have a very small effect on beliefs in this case, but we need realized outcomes to have a large effect on prices if the firm is to have an incentive to produce high quality. This can only be accomplished if small changes in posterior beliefs can cause large changes in expected utility, which in turn requires the discontinuity.

We are thus inclined to view the model with common consumers and Markov strategies *continuous at the boundaries* as a natural analog of the idiosyncratic consumer case. In both cases, equilibria exist with normal firms always exerting high effort when there are replacements, and cannot exist without replacements. More important, our interest in idiosyncratic consumers arose out of a desire to limit the degree of coordination in imposing punishments. The discontinuous behavior required to support high effort in a common consumer Markov equilibrium strikes us as going beyond the spirit of Markov equilibria and as entailing the same implausible degree of coordination on the part of consumers that prompted us to impose a Markov restriction on behavior.

18.4.6 Competitive Markets

When consumers are idiosyncratic, small changes in consumer beliefs necessarily have small changes in behavior, forcing an appeal to replacements to ensure that changes in beliefs never become too small. In a richer model, even small changes in consumer behavior may have a large impact on the firm, obviating the need to impose a lower bound on belief revision. For example, when there are competing firms, a small change in consumers' beliefs about a firm may lead them to change to a very similar firm, with the resulting loss in customers being a significant cost on the firm in question. Analyzing this possibility with idiosyncratic consumers is formidable, so this section discusses a model of competitive firms with common consumers (following Hörner 2002). The model illustrates both the possibility that competition can allow firms to sustain reputations for high effort as well as the fragility of such constructions. This model also endogenizes the prior probability of the inept type.

At date t, there is a population of firms who may be inept or normal. As in section 18.4.2, normal firms can exert either high (H) or low effort (L), and inept firms can only choose L. In each period, all customers of a firm either receive a common good outcome $\bar{y}$ (with probability ρ_{a_1}, given effort level a_1 from that firm) or all receive a common bad outcome $\underline{y}$ (probability $1 - \rho_{a_1}$). For a normal firm, L is costless, whereas H incurs a cost of $c > 0$.

There is a continuum of consumers, of mass $I > 0$, with types indexed by i uniformly distributed on $[0, I]$. We take I large enough to ensure that the market has an interior equilibrium, being more precise shortly as to just what this means. A consumer receives a payoff of 1 from the good signal $\bar{y}$ and payoff 0 from the bad signal $\underline{y}$. A consumer of type i (or simply "consumer i") who purchases at price p from a firm choosing H with probability υ receives a payoff of

$$\upsilon \rho_H + (1 - \upsilon)\rho_L - p - i.$$

The term $\upsilon \rho_H + (1 - \upsilon)\rho_L - p$ is the surplus consumer i receives from participating in this market, and the consumer's type is the opportunity cost of participation.

We solve for the *steady state* (N_0, N_L) of the market with entry and exit, where N_0 is the mass of active normal firms in steady state, and N_L the mass of active inept firms. In each period, a mass N_0^0 of normal firms enter the market and mass N_L^0 inept firms enter the market. These quantities will be determined by the free-entry condition that the value of entering the market just compensate the firm for its opportunity cost of participating.

Because the total mass of active firms in the market is $N = N_0 + N_L$, in equilibrium the active firms sell to mass N of consumers. The opportunity cost of the marginal consumer in a market serving N consumers is N. Consequently, each firm must produce a surplus in the form of the difference between its expected probability of signal $\bar{y}$ and its price, equal to N. In equilibrium, different firms will set different prices, reflecting their different market experiences and hence the different probabilities with which they are expected to produce signal $\bar{y}$.

18.4 ■ Common Consumers

Consumers observe the histories of signals produced by each firm. On first producing signal y, a firm is abandoned by consumers and leaves the market.[6] The composition of the firms in the market is described by a pair of sequences $\{N_0^\tau\}_{\tau=0}^\infty$ and $\{N_L^\tau\}_{\tau=0}^\infty$, where N_0^τ is the mass of firms in the market who are normal and who have been in the market for τ periods, during which they have exhibited τ straight realizations of $\bar{y}$, and N_L^τ is the analogous term for inept firms. Thus, we have $N_0^t = \rho_H N_0^{t-1}$, $N_L^t = \rho_L N_L^{t-1}$, and so

$$N_0 \equiv \sum_{\tau=0}^\infty N_0^\tau = \frac{N_0^0}{1-\rho_H}$$

$$\text{and} \quad N_L \equiv \sum_{\tau=0}^\infty N_L^\tau = \frac{N_L^0}{1-\rho_L}.$$

The posterior probability that a firm is a normal firm after t consecutive realizations of $\bar{y}$, μ_0^t, is given by

$$\mu_0^t = \frac{N_0^t}{N_0^t + N_L^t}$$

$$= \frac{\rho_H^t N_0^0}{\rho_H^t N_0^0 + \rho_L^t N_L^0}$$

$$= \left(1 + \left(\frac{\rho_L}{\rho_H}\right)^t \frac{N_L^0}{N_0^0}\right)^{-1}.$$

Market clearing requires that consumers be indifferent over the different active firms. This implies particular pricing on the part of active firms. Letting p^t be the price commanded by a firm after t consecutive realizations of $\bar{y}$, we thus have

$$p^t = \mu_0^t(\rho_H - \rho_L) + \rho_L - N.$$

To make such pricing an equilibrium, and to make it an equilibrium for a firm who has produced a failure to leave the market, we assume that the out-of-equilibrium event of a lower price or continued participation in the market after a failure gives rise to the consumer expectation that the firm in question is *certainly* inept. As long as $N > \rho_L$, this ensures a negative price (and hence the optimality of exit) for the firm.

The expected payoff to a firm in this market is determined once N_0^0 and N_L^0, the measures of normal and inept entrants in each period, are fixed. Let $U(N_0^0, N_L^0, \xi_0)$ and $U(N_0^0, N_L^0, \xi(L))$ be the payoffs of the normal and inept firm in the resulting configuration. The model is closed by the requirement that entrants earn zero profits, or

$$U(N_0^0, N_L^0, \xi_0) = (1-\delta)K_0$$
$$\text{and} \quad U(N_0^0, N_L^0, \xi(L)) = (1-\delta)K_L,$$

where K_0 and K_L are the entry costs of a normal and inept firm.

---

6. We return to the optimality of such behavior shortly. Equilibria with less extreme behavior remain unexplored.

We have

$$U(N_0^0, N_L^0, \xi_0) = (1-\delta) \sum_{\tau=0}^{\infty} \rho_H^\tau \delta^\tau \left[\left(1 + \left(\frac{\rho_L}{\rho_H}\right)^\tau \frac{N_L^0}{N_0^0}\right)^{-1} (\rho_H - \rho_L) + \rho_L - N - c \right]$$

and

$$U(N_0^0, N_L^0, \xi(L)) = (1-\delta) \sum_{\tau=0}^{\infty} \rho_L^\tau \delta^\tau \left[\left(1 + \left(\frac{\rho_L}{\rho_H}\right)^\tau \frac{N_L^0}{N_0^0}\right)^{-1} (\rho_H - \rho_L) + \rho_L - N \right],$$

allowing us to write the entry conditions as

$$\sum_{\tau=0}^{\infty} \rho_H^\tau \delta^\tau \left[\left(1 + \left(\frac{\rho_L}{\rho_H}\right)^\tau \frac{N_L^0}{N_0^0}\right)^{-1} (\rho_H - \rho_L) \right] = K_0 + \frac{N - \rho_L + c}{1 - \rho_H \delta} \quad (18.4.10)$$

and

$$\sum_{\tau=0}^{\infty} \rho_L^\tau \delta^\tau \left[\left(1 + \left(\frac{\rho_L}{\rho_H}\right)^\tau \frac{N_L^0}{N_0^0}\right)^{-1} (\rho_H - \rho_L) \right] = K_L + \frac{N - \rho_L}{1 - \rho_L \delta}. \quad (18.4.11)$$

It remains to determine the steady-state values of N_0 and N_L, or equivalently, the steady-state values of $N = N_0 + N_L$ and N_L^0/N_0^0. Fix a value of K_L, and for each N associate the value $N_L^0/N_0^0 = g(N)$ that satisfies (18.4.11). Then g is strictly decreasing, with ρ_L being the minimum value of N consistent with equilibrium (to ensure the departure of firms who have had a failure) and $\bar{N} = g^{-1}(0)$ the maximum value N consistent with (18.4.11). We can now fix a value $N \in (\rho_L, \bar{N})$ and use (18.4.10) to identify the locus of pairs (K_0, c) that satisfy the normal firm's entry condition.

If K_0 and c are too small, no inept firms will enter the market (in equilibrium). Normal firms will nonetheless exert effort, with a failure prompting consumers to abandon the firm despite the absence of any revision in beliefs. This is clearly not a Markov equilibrium. The equilibrium with inept firms is Markov, given the assumption that consumers expect a firm continuing in the market after a failure to be certainly inept.

Completing the equilibrium requires showing that the normal firm finds it optimal to exert high effort. Because a failure leads to a continuation payoff of 0, and a success leads to a continuation payoff of at least K_0, this will be the case as long as $c \in (0, \bar{c})$ is chosen sufficiently small. It suffices to increase K_0 while decreasing c to preserve (18.4.10).

18.5 Discrete Choices

In the model of section 18.1, consumer responses to changes in their beliefs are continuous. In this section, we consider the impact of discrete choices by the consumers. We work with the version of the product-choice game shown in figure 18.5.1, now

18.5 ■ Discrete Choices

	h	ℓ
H	3−c, 3	1−c, 2
L	3, 0	1, 1

Figure 18.5.1 Version of the product-choice game, parameterized by the cost of high effort c ($= 1$ in our usual formulation).

formulated to make explicit the cost c of high effort. As usual, high effort produces signal $\bar{y}$ with probability ρ_H, low effort produces $\bar{y}$ with probability $\rho_L < \rho_H$. Normal firms can exert either high or low effort, and inept firms inevitably exert low effort. We again assume that the firm is replaced in each period with probability λ, with the replacement being normal with probability $\mu_0^0 \in (0, 1)$. As will be clear from the proof, the incentives now come from the possibility that, when beliefs are in the ℓ-regime (i.e., $\mu_0 < 1/2$), a larger probability on $\bar{y}$ increases the probability of leaving the ℓ-regime by some finite time, and when beliefs are in the h-regime (i.e., $\mu_0 > 1/2$), a larger probability on $\underline{y}$ increases the probability of entering the ℓ-regime by some finite time.

Proposition 18.5.1 *Given $\mu_0^0 \in (0, 1)$, for sufficiently small λ and c, there is a pure-strategy Markov equilibrium in which the normal firm always exerts high effort, and consumers choose h if and only if $\mu_0 \geq 1/2$.*

Proof Given the specified behavior for the consumers, we need only verify that high effort is optimal for the firm. We begin by noting that the posterior updating rules are given by

$$\varphi(\mu_0 \mid \bar{y}) = (1 - \lambda)\frac{\rho_H \mu_0}{\rho_H \mu_0 + \rho_L(1 - \mu_0)} + \lambda \mu_0^0$$

and

$$\varphi(\mu_0 \mid \underline{y}) = (1 - \lambda)\frac{(1 - \rho_H)\mu_0}{(1 - \rho_H)\mu_0 + (1 - \rho_L)(1 - \mu_0)} + \lambda \mu_0^0.$$

Let μ_0' solve $\varphi(\mu_0 \mid \underline{y}) = \mu_0$ and μ_0'' solve $\varphi(\mu_0 \mid \bar{y}) = \mu_0$. There exists $\bar{\lambda}$ such that for all $\lambda \in (0, \bar{\lambda})$, we have $\mu_0' < 1/2 < \mu_0''$.

Let $V_0(\mu_0)$ be the expected continuation payoff to the normal firm under the proposed equilibrium, given that the current posterior is μ_0.

The payoff to the normal firm from high effort, given posterior μ_0, is given by

$$(1 - \delta(1 - \lambda))(f(\mu_0) - c) + \delta(1 - \lambda)[\rho_H V_0(\varphi(\mu_0 \mid \bar{y})) + (1 - \rho_H)V_0(\varphi(\mu_0 \mid \underline{y}))],$$

and the payoff from low effort is

$$(1 - \delta(1 - \lambda))f(\mu_0) + \delta(1 - \lambda)[\rho_L V_0(\varphi(\mu_0 \mid \bar{y})) + (1 - \rho_L)V_0(\varphi(\mu \mid \underline{y}))],$$

where $f(\mu_0) = 3$ if $\mu_0 \geq 1/2$ and equals 1 otherwise. Optimality then requires, for any c,

$$V_0(\varphi(\mu_0 \mid \bar{y})) - V_0(\varphi(\mu_0 \mid \underline{y})) \geq \frac{(1 - \delta(1 - \lambda))c}{(\rho_H - \rho_L)\delta(1 - \lambda)}.$$

It thus suffices to show that $V_0(\varphi(\mu_0 \mid \bar{y})) - V_0(\varphi(\mu_0 \mid \underline{y}))$ is uniformly (in μ_0) bounded away from 0. The value function is

$$V_0(\mu_0) = (1 - \delta(1 - \lambda))(f(\mu_0) - c)$$
$$+ \delta(1 - \lambda)[\rho_H V_0(\varphi(\mu_0 \mid \bar{y})) + (1 - \rho_H)V_0(\varphi(\mu_0 \mid \underline{y}))]. \quad (18.5.1)$$

Because the distribution of posteriors in every period is increasing (using first-order stochastic dominance) in μ_0 and f is weakly increasing, V_0 is also weakly increasing.

Suppose first that $\mu_0 \geq 1/2$, so $f(\mu_0) = 3$. It suffices to show that $V_0(\mu_0) - V_0(\varphi(\mu_0 \mid \underline{y}))$ is bounded away from 0. Using (18.5.1) and $V_0(\varphi(\mu_0 \mid \bar{y})) \geq V_0(\mu_0)$, we have

$$V_0(\mu_0) - V_0(\varphi(\mu_0 \mid \underline{y})) \geq \frac{1 - \delta(1 - \lambda)}{1 - \delta(1 - \lambda)\rho_H}(3 - c)$$
$$+ \left(\frac{\delta(1 - \lambda)(1 - \rho_H)}{1 - \delta(1 - \lambda)\rho_H} - 1\right) V_0(\varphi(\mu_0 \mid \underline{y}))$$
$$= \frac{1 - \delta(1 - \lambda)}{1 - \delta(1 - \lambda)\rho_H}(3 - c - V_0(\varphi(\mu_0 \mid \underline{y}))).$$

We now argue that for $\lambda \in (0, \bar{\lambda})$, there exists $\eta > 0$ such that for all $\mu_0 \geq 1/2$, we have $V_0(\varphi(\mu_0 \mid \underline{y})) + \eta < 3 - c$. The value $3 - c$ requires that player 2 play h in every future period. Because of replacements, there is an upper bound $(1 - \lambda) + \lambda\mu_0^0 < 1$ on the posterior, so there is a period t and a probability $\theta > 0$ such that under the profile, with probability at least θ, beginning at the belief $\varphi(\mu_0 \mid \underline{y})$, the posterior falls below $1/2$ and player 2 plays ℓ in period t. Consequently, $V_0(\varphi(\mu_0 \mid \underline{y})) < 3 - c - 2\delta^t(1 - \delta)\theta$.

Suppose now that $\mu_0 < 1/2$, so $f(\mu_0) = 1$. It now suffices to show that $V_0(\varphi(\mu_0 \mid \bar{y})) - V_0(\mu_0)$ is bounded away from 0. From (18.5.1) and $V_0(\mu_0) \geq V_0(\varphi(\mu_0 \mid \underline{y}))$, we have

$$V_0(\varphi(\mu_0 \mid \bar{y})) - V_0(\mu_0) \geq \left(1 - \frac{\delta(1 - \lambda)\rho_H}{1 - \delta(1 - \lambda)(1 - \rho_H)}\right) V_0(\varphi(\mu_0 \mid \bar{y}))$$
$$- \frac{1 - \delta(1 - \lambda)}{1 - \delta(1 - \lambda)(1 - \rho_H)}(1 - c)$$
$$= \frac{1 - \delta(1 - \lambda)}{1 - \delta(1 - \lambda)(1 - \rho_H)}(V_0(\varphi(\mu_0 \mid \bar{y})) - (1 - c)).$$

We argue (as before) that for $\lambda \in (0, \bar{\lambda})$, there exists $\eta' > 0$ such that for all $\mu_0 < 1/2$, we have $1 - c + \eta' < V_0(\varphi(\mu_0 \mid \bar{y}))$. The value $1 - c$ results from player 2 playing ℓ in every future period. Because of replacements, there is a lower bound $\lambda \mu_0^0 > 0$ on the posterior, so there is a period t' and a probability $\theta' > 0$ such that under the profile with probability at least θ', beginning at the belief $\varphi(\mu_0 \mid \bar{y})$, the posterior is above $1/2$ and player 2 plays h in period t'. Consequently, $V_0(\varphi(\mu_0 \mid \bar{y})) > 1 - c + 2\delta^{t'}(1 - \delta)\theta'$.

∎

18.6 Lost Consumers

The consumers in Section 18.1 continue to purchase from the firm, no matter how discouraging the signals produced by the firm. We examine here the implications of allowing consumers an outside option that induces sufficiently pessimistic consumers to abandon the firm, and so not observe any signals. The ability to sustain a reputation hinges crucially on whether the behavior of a firm on the brink of losing its consumers makes the firm's product more (section 18.6.1) or less (sections 18.6.2–18.6.6) valuable to consumers.

18.6.1 The Purchase Game

We begin with a version of the purchase game of figure 15.4.1, shown in figure 18.6.1. The firm receives 2 if consumers buy its product and must pay a cost c if exerting high effort. If the consumer buys (chooses b), high effort produces a good outcome with probability ρ_H and low effort a good outcome with probability $\rho_L < \rho_H$. The consumer values a good signal at $(2 - \rho_H - \rho_L)/(\rho_H - \rho_L)$ and a bad signal at $-(\rho_H + \rho_L)/(\rho_H - \rho_L)$. If the consumer does not buy (chooses d), then no signal is observed (just as in example 15.4.2).

Normal firms can exert either high or low effort, and inept firms inevitably exert low effort. We again assume that the firm is replaced in each period with probability λ, with the replacement being normal with probability μ_0^0.

The essential message of the previous sections continues to hold in the presence of the outside option.

Proposition 18.6.1 *Fix $\mu_0^0 \in (0, 1)$.*

1. *If there are no replacements ($\lambda = 0$), in every pure-strategy Markov equilibrium continuous at 1, the firm always chooses low effort.*

	d	b
H	0, 0	2−c, 1
L	0, 0	2, −1

Figure 18.6.1 Version of the purchase game, parameterized by the cost c of high effort ($= 1$ in figure 15.4.2).

2. If $\lambda > 0$, for sufficiently small λ and c, there is a pure-strategy Markov equilibrium in which the normal firm exerts high effort and consumers buy for all $\mu_0 \geq 1/2$, and the firm exerts low effort and consumers do not buy for $\mu_0 < 1/2$.

Proof 1. The proof essentially replicates the proof of proposition 18.4.3(2).
2. The proof is essentially that of proposition 18.5.1 for the case $\mu_0 \geq 1/2$. ∎

If the consumers are pessimistic, that is, $\mu_0^0 < 1/2$, then even with replacements the outcome path involves the consumers never buying. On the other hand, if consumers are optimistic, $\mu_0^0 > 1/2$, then the outcome path begins with consumers buying. They continue to do so until a sufficient string of bad signals pushes the posterior below $1/2$, at which point they desist. However, this no-trade outcome is not absorbing. The possibility of replacements causes the consumers' posteriors to drift upward toward μ_0^0, until they exceed $1/2$, when once again the consumers purchase.

18.6.2 Bad Reputations: The Stage Game

Once the firm's posterior probability falls below $1/2$ in section 18.6.1, consumers no longer buy from the firm. This effectively suspends the relationship until the consumer is sufficiently confident the firm has been replaced by a normal type. The incentives for the normal firm to exert effort arise out of his desire to avoid falling into this no-trade zone. This section, drawing on Ely and Välimäki (2003), presents a model in which the firms' efforts to avoid a no-trade region destroy the incentives needed for a nontrivial equilibrium.

There are two players, referred to as the firm (player 1) and the customer (player 2). We think of the customer as hiring the firm to perform a service, with the appropriate nature of the service depending on a diagnosis that only the firm can perform. For example, the firm may be a doctor who must determine whether the patient needs to take two aspirin tablets daily or needs a heart transplant. The firm may be a computer support service that must determine whether the customer needs to reformat her hard disk or needs a new computer.

The interaction is modeled as a repeated game with random states. There are two states of the world, θ_H and θ_L. In the former, the customer requires a high level of service, denoted by H, in the latter a low level, denoted by L. The two states are equally likely.

The stage game is an extensive-form game. The state is first drawn by nature and revealed to the firm but not the customer. The customer then decides whether to hire the firm. If the firm is hired, he chooses the level of service to provide.

Figure 18.6.2 identifies the payoffs attached to each terminal node of the extensive form game. The firm and the customer thus have identical payoffs. Both prefer that high service be provided when necessary and that low service be provided when appropriate. If the firm is not hired, then both players receive 0. This model thus differs from many expert-provider models, in which the firm has an incentive to provide high service regardless of what is needed.

A pure stage-game action for the customer is a choice to either hire or not hire the firm. A mixed stage-game action for the firm is (α_H, α_L), where α_H is the probability

18.6 ■ Lost Consumers

	Hire	Not hire
H	u, u	$0, 0$
L	$-w, -w$	$0, 0$

State θ_H

	Hire	Not hire
H	$-w, -w$	$0, 0$
L	u, u	$0, 0$

State θ_L

Figure 18.6.2 Payoffs for each terminal node of the extensive-form stage game, as a function of the state (θ_H or θ_L), the customer's decision of whether to hire the firm, and (if hired) the firm's choice of service. We assume $w > u > 0$.

of action H in state θ_H and α_L the probability of action L in state θ_L. We thus identify the firm's strategy with the probability of providing the appropriate service in each of the two states. The action $(1, 1)$ is the outcome in which the appropriate service is always provided, and $(1, 0)$ corresponds to always providing high service and $(0, 0)$ to always providing the wrong service.

It is straightforward to verify that the stage game presents no incentive problems, working backward from the observation that the only sequentially rational action for the firm is $(1, 1)$:

Proposition 18.6.2 *The stage game has a unique sequential equilibrium, and in this equilibrium the firm chooses $(1, 1)$.*

18.6.3 The Repeated Game

Suppose now that the (extensive-form) stage game is repeated. The firm is a long-run player who discounts at rate δ. The customer is a short-run player.

Each period features the arrival of a new customer. Nature then draws the state and reveals its realization to the firm (only). These draws are independent across periods, with the two states equally likely in each case. The customer decides whether to hire the firm, and the firm then chooses a level of service. At the end of the period, a public signal from the set $Y \equiv \{X, H, L\}$ is observed, indicating either that the firm was not hired (X) or was hired and provided either high (H) or low (L) service. Short-lived players thus learn nothing about the firm's stage-game strategy when the firm is not hired and never learn anything about previous states.

A period t public history is denoted by $h^t \in Y^t$. As one would expect, there is an equilibrium of the repeated game in which the firm is always hired and provides the appropriate service. However, as one would also expect, arguments familiar from the various folk theorems allow us to show that there is another equilibrium in which player 1 receives his minmax payoff.

Proposition 18.6.3 *The repeated game has a sequential equilibrium in which the firm is always hired and chooses $(\alpha_H, \alpha_L) = (1, 1)$, earning a payoff of u. However, for sufficiently large values of $\delta < 1$, there is also a sequential equilibrium in which the firm is never hired, for a payoff of 0.*

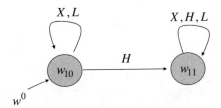

Figure 18.6.3 Firm's strategy for an equilibrium of the repeated complete information game in which the firm is never hired. The firm chooses action (α_H, α_L) in state $w_{\alpha_H \alpha_L}$.

Proof It is immediate that it is an equilibrium for the customer to always hire the firm and the firm to choose $(1, 1)$.

To construct an equilibrium in which the firm is optimally never hired, let the firm's strategy be described by a two-state automaton.[7] Letting $\mathscr{W} \equiv \{w_{10}, w_{11}\}$ be the set of states with initial state w_{10}, with $f(w_{\alpha_H \alpha_L}) = (\alpha_H, \alpha_L)$ and with the transition rule

$$\tau(w, y) = \begin{cases} w_{10}, & \text{if } w = w_{10} \text{ and } y = X \text{ or } L, \\ w_{11}, & \text{if } (w, y) = (w_{10}, H) \text{ or } w = w_{11}. \end{cases}$$

Figure 18.6.3 illustrates. In the first state of this automaton, the firm always provides high service. All is well once the firm's automaton reaches its second state, at which point the firm always provides the appropriate service. In addition, the firm's strategy is public, so the customers know which state the firm is in. The difficulty is that reaching the second state requires that the firm be hired when in automaton state w_{10}. Here the firm provides the appropriate service in game state θ_H but not θ_L, for a negative expected value. No short-lived player wants to be the first to hire the firm, nor the first to hire the firm after a history in which H has never been observed. The customers' strategy is thus to hire the firm if and only if the history includes at least one observation of H, leading to an outcome in which the firm is never hired.

The only aspect of these strategies that is not obviously equilibrium behavior is the prescription that the firm choose H if hired, when in automaton state w_{10} and the state of nature is θ_L. Doing so trades a current loss against the ability to generate the signal H and thus make a transition to state w_{11}. The payoff to the firm from following this strategy is

$$(1 - \delta)(-w) + \delta u,$$

and the payoff from choosing L is

$$(1 - \delta)u.$$

The proposed strategies are thus an equilibrium if

$$(1 - \delta)(-w) + \delta u \geq (1 - \delta)u.$$

---

7. The equilibrium in which the firm is never hired has a similar structure to the low-payoff equilibrium of the purchase game in example 15.4.2.

18.6 ■ Lost Consumers

or, equivalently,

$$\delta \geq \frac{u+w}{2u+w}.$$

We thus have an equilibrium if the firm is sufficiently patient.

■

18.6.4 Incomplete Information

We now introduce incomplete information. The result is a bound on the firm's payoff, but it is now an upper bound that consigns the firm to a surprisingly low payoff.

With probability $\mu_0^0 > 0$, the firm is normal. With complementary probability $\hat{\mu}^0 > 0$, the firm is "bad" and follows a strategy of choosing from $\{H, L\}$ independently and identically in each period, regardless of the period's state. The bad firm chooses H with probability γ, that is, follows the strategy $(\gamma, 1 - \gamma)$, where

$$\gamma > \frac{w+3u}{2w+2u} > 1/2. \tag{18.6.1}$$

As we shall see, this ensures that the bad firm behaves sufficiently differently from any relevant normal firm behavior that, should the firm be hired, an observation of H increases the posterior probability that the firm is bad.

Let $\bar{v}_1(\xi_0, \mu, \delta)$ be the supremum of the (discounted average) Nash equilibrium payoffs for the normal firm. The result is that patient firms receive a payoff arbitrarily close to their minmax payoff.

Proposition 18.6.4
$$\lim_{\delta \to 1} \bar{v}_1(\xi_0, \mu, \delta) = 0.$$

Remark 18.6.1 **Overproviding firms** The intuition behind this result is nicely illustrated by considering the special case on which Ely and Välimäki (2003) focus, namely, that in which the bad firm is committed to the strategy $(1, 0)$ and hence always provides high service. This is in keeping with a popular impression of unscrupulous firms as always providing excess service.

In this case, a single choice of L reveals the firm to be normal. Let us further assume that if the firm is believed to be normal with probability one, the firm is thereafter always hired and hence earns a continuation payoff of u.[8] It then seems as if it should be straightforward for the normal firm to establish a reputation. If nothing else, the firm could simply reveal his type by once playing L, thereafter enjoying the fruits of being known to be normal.

A moment's reflection reveals that things are not so clear. Suppose, for example, that the normal firm's strategy is to play L the first time hired, regardless of the state, thereby convincing the customers that the firm is normal and enjoying continuation payoff u. This strategy ensures that the firm is never hired.

Proceeding more carefully, fix an equilibrium strategy profile and let $\hat{\mu}^\dagger$ be the supremum of the set of probabilities of a bad firm for which the firm is

---
8. Proposition 18.6.3 shows that this is an assumption rather than a necessary feature of the equilibrium. Ely and Välimäki (2003) motivate this as a renegotiation-proofness requirement.

(in equilibrium) hired with positive probability. We show that $\hat{\mu}^\dagger > 0$ is a contradiction. If the firm is ever to be hired, there must be a significant chance that he chooses L (in state θ_L), because otherwise his value to the consumer is negative. Then for any posterior probability $\hat{\mu}'$ sufficiently close to $\hat{\mu}^\dagger$ at which the firm is hired, an observation of H must push the posterior of a bad firm past $\hat{\mu}^\dagger$, ensuring that the firm is never again hired. But then no sufficiently patient normal firm, facing a posterior probability $\hat{\mu}'$, would ever choose H in state θ_H. Doing so gives a payoff of $(1 - \delta)u$ (a current payoff of u, followed by a posterior above $\hat{\mu}^\dagger$ and hence a continuation payoff of 0), whereas choosing L reveals the firm to be normal and hence gives a higher (for large δ) payoff of $-(1 - \delta)w + \delta u$. The normal firm thus cannot be induced to choose H at posterior $\hat{\mu}'$. But this now ensures that the firm will not be hired for any such posterior, giving us a contradiction to the assumption that $\hat{\mu}^\dagger$ is the supremum of the posterior probabilities for which the firm is hired. We thus have:

Proposition 18.6.5 *Let $\gamma = 1$ and assume that in any period in which the firm is believed to be normal with probability 1, the firm is hired. Then there is a unique Nash equilibrium outcome in which the firm is never hired.*

The difficulty facing the normal firm is that an unlucky sequence of (θ_H) states may push the posterior probability that the firm is bad disastrously high. At this point, the normal firm will choose L in both states in a desperate attempt to stave off a career-ending bad reputation. Unfortunately, customers will anticipate this and not hire the firm, ending his career even earlier. The normal firm might attempt to forestall this premature end by playing L (in state θ_H) somewhat earlier, but the same reasoning unravels the firm's incentives back to the initial appearance of state θ_H. We can thus never construct incentives for the firm to choose H in state θ_H, and the firm is never hired.

Ely and Välimäki (2003) note that if we relax the assumption that the firm is always hired when believed to be normal with probability 1, then there are equilibria in which the firm is sometimes hired. However, proposition 18.6.4 implies that in any such equilibrium, the firm's payoff must converge to 0 as he becomes increasingly patient. ◆

Proof of Proposition 18.6.4 A short-lived player only hires the firm if

$$0 \leq \frac{1}{2}[\hat{\alpha}_H u - (1 - \hat{\alpha}_H)w] + \frac{1}{2}[\hat{\alpha}_L u - (1 - \hat{\alpha}_L)w],$$

where $\hat{\alpha}_H = \hat{\mu}\gamma + (1 - \hat{\mu})\alpha_H$ and $\hat{\alpha}_L = \hat{\mu}(1 - \gamma) + (1 - \hat{\mu})\alpha_L$ are the probabilities the short-lived player assigns to the action H being taken in state θ_H and the action L being taken in state θ_L, and $\hat{\mu}$ is the probability the short-lived player assigns the firm being bad. This inequality can be rewritten as

$$\alpha_L + \alpha_H \geq \frac{2w - (u + w)\hat{\mu}}{(1 - \hat{\mu})(u + w)}. \tag{18.6.2}$$

18.6 ■ Lost Consumers

Because the left side is no larger than 2, this inequality requires

$$2 \geq \frac{2w - (u+w)\hat{\mu}}{(1-\hat{\mu})(u+w)},$$

which holds if and only if

$$\hat{\mu} \leq \frac{2u}{(u+w)} \equiv \hat{\mu}^*.$$

Moreover,

$$\frac{2w - (u+w)\hat{\mu}}{(1-\hat{\mu})(u+w)} > 1,$$

because

$$\frac{2w - (u+w)\hat{\mu}}{(1-\hat{\mu})(u+w)} - 1 = \frac{w-u}{(1-\hat{\mu})(u+w)} > 0.$$

Hence, if the firm is hired in any period, then it must be that $\hat{\mu} \leq \hat{\mu}^*$ and (taking a bound that holds for all $\hat{\mu}$)

$$\alpha_L, \alpha_H \geq \alpha^* \equiv (w-u)/(w+u). \tag{18.6.3}$$

We now assume that the firm is hired at some point and use these necessary conditions to derive a contradiction. Consider a Nash equilibrium in which the firm is hired in period t, after some history h^t that causes the customer's period t posterior belief that the firm is bad to be $\hat{\mu}$. If the customer observes H, then her updated posterior belief is

$$\varphi(\hat{\mu} \mid H) = \frac{\hat{\mu}\gamma}{\hat{\mu}\gamma + (1-\hat{\mu})[\frac{1}{2}\alpha_H + \frac{1}{2}(1-\alpha_L)]}.$$

A lower bound on this probability is obtained by assuming that the normal firm is as much like the bad firm as possible. However, setting $\alpha_H = \gamma$ and $\alpha_L = 1 - \gamma$ violates (18.6.3), given (18.6.1). The lower bound, consistent with (18.6.3), is obtained by setting $\alpha_H = 1$ and setting $\alpha_L = \alpha^*$, and so

$$\varphi(\hat{\mu} \mid H) \geq \frac{\hat{\mu}\gamma}{\hat{\mu}\gamma + (1-\hat{\mu})[\frac{1}{2} + \frac{1}{2}(1-\alpha^*)]}$$

$$= \frac{\hat{\mu}\gamma}{\hat{\mu}\gamma + (1-\hat{\mu})[\frac{w+3u}{2w+2u}]}$$

$$\equiv \Phi(\hat{\mu}).$$

This defines a function $\Phi : [0, 1] \to [0, 1]$ with $\Phi' > 0$, where $\Phi(\hat{\mu})$ is the smallest possible posterior probability of a bad type that can emerge from a prior probability of $\hat{\mu}$ and an observation of H. Because $\gamma > (w + 3u)/(2w + 2u)$, we have $\Phi(\hat{\mu}) > \hat{\mu}$ for all $\hat{\mu} \in (0, 1)$. Hence, an observation of H pushes upward the posterior probability that the firm is bad. This reflects the fact that the normal firm must sometimes choose L, if he is to be hired, and the bad firm is (by construction) more likely to choose H.

Define $\hat{\mu}_{(1)} = \hat{\mu}^*$ and $\hat{\mu}_{(m)} = \Phi^{-1}(\hat{\mu}_{(m-1)})$ for $m > 1$. A string of the form $\{\hat{\mu}_{(m)}, \hat{\mu}_{(m-1)}, \ldots, \hat{\mu}_{(2)}, \hat{\mu}_{(1)}\}$ is thus a string of (increasing) posterior probabilities that starts at $\hat{\mu}_{(m)}$, with succeeding posteriors providing a lower bound on the updated probability that could appear if the firm is (optimally) hired and H observed, and that culminates with $\hat{\mu}_{(1)}$.

Let $V_0(\hat{\mu}, \delta)$ be the supremum over equilibrium payoffs for the normal firm, when thought to be bad with probability $\hat{\mu}$ and given discount factor δ. Note that for $\hat{\mu} > \hat{\mu}_{(1)}$, no customer hires the firm, and hence $V(\hat{\mu}, \delta) = 0$ for all δ. Moreover, $\hat{\mu}_{(m)} \to 0$ as $m \to \infty$. In other words, no matter how low the initial prior, a long enough string of H's will be sufficiently negative news that the posterior is larger than $\hat{\mu}^*$, and no further customer hires the firm.

Define $V_{(m)}(\delta)$ recursively as follows: $V_{(1)}(\delta) = 0$, and for $m \geq 2$,

$$V_{(m)}(\delta) = \frac{(1-\delta)(3u+w)}{2} + \delta V_{(m-1)}(\delta)$$
$$= \frac{(1-\delta^{m-1})(3u+w)}{2}.$$

Note that $V_{(m)}(\delta) \nearrow (3u+w)/2$ as $m \to \infty$. However, its more important feature is that for any m, $\lim_{\delta \to 1} V_{(m)}(\delta) = 0$. The proof of proposition 18.6.4 is then completed by the following lemma. ∎

Lemma 18.6.1 *For all prior beliefs $\hat{\mu} > \hat{\mu}_{(m)}$, $V_0(\hat{\mu}, \delta) \leq V_{(m)}(\delta)$.*

Proof The proof is by induction. If $\hat{\mu} > \hat{\mu}_{(1)} = \hat{\mu}^*$, the firm is never hired and we have $V_0(\hat{\mu}, \delta) = 0 = V_{(1)}(\delta)$.

Suppose for all prior beliefs $\hat{\mu} > \hat{\mu}_{(m-1)}$, $V_0(\hat{\mu}, \delta) \leq V_{(m-1)}(\delta)$. Suppose $\hat{\mu} \in (\hat{\mu}_{(m)}, \hat{\mu}_{(m-1)}]$ and the normal firm is hired in the first period. Then it must be the case that $\alpha_L, \alpha_H > \alpha^*$, and hence the normal firm puts positive probability on the appropriate level of service in each state. But then $(1, 1)$ is also a best reply for the normal firm, so

$$V_0(\hat{\mu}, \delta) \leq (1-\delta)u + \frac{\delta}{2}\{V_0(\varphi(\hat{\mu} \mid H), \delta) + V_0(\varphi(\hat{\mu} \mid L), \delta)\}. \quad (18.6.4)$$

We now seek upper bounds for the terms on the right side. Because $\hat{\mu} > \hat{\mu}_{(m)}$, $\varphi(\hat{\mu} \mid H) > \hat{\mu}_{(m-1)}$ and

$$V_0(\varphi(\hat{\mu} \mid H), \delta) \leq V_{(m-1)}(\delta). \quad (18.6.5)$$

It remains to bound $V_0(\varphi(\hat{\mu} \mid L), \delta)$.[9] When the realized state is θ_H, the normal firm finds it optimal to choose H, and so the following incentive constraint must hold:

---

9. If we assumed $\gamma = 1$, then $\varphi(\hat{\mu} \mid L) = 0$ for all $\hat{\mu} \in (0, 1)$, and so $V_0(\varphi(\hat{\mu} \mid L), \delta) \leq u$. However, this bound is not sufficiently tight for our purpose, because using this bound gives

$$V_0(\hat{\mu}, \delta) \leq (1-\delta)u + \frac{\delta}{2}\{V_{(m-1)}(\delta) + u\},$$

which converges to $u/2$ as $\delta \to 1$.

$$(1-\delta)u + \delta V_0(\varphi(\hat{\mu}\mid H),\delta) \geq (1-\delta)(-w) + \delta V_0(\varphi(\hat{\mu}\mid L),\delta),$$

which we rearrange to give

$$V_0(\varphi(\hat{\mu}\mid L),\delta) \leq \frac{(1-\delta)(u+w)}{\delta} + V_0(\varphi(\hat{\mu}\mid H),\delta). \qquad (18.6.6)$$

Inserting (18.6.5) and (18.6.6) in (18.6.4), we find

$$\begin{aligned}V_0(\hat{\mu},\delta) &\leq (1-\delta)u + \frac{\delta}{2}\left[V_{(m-1)}(\delta) + \frac{(1-\delta)(u+w)}{\delta} + V_{(m-1)}(\delta)\right]\\ &= \frac{(1-\delta)(3u+w)}{2} + \delta V_{(m-1)}(\delta)\\ &= V_{(m)}(\delta),\end{aligned}$$

the desired result. ∎

This result does not imply that the firm is never hired in equilibrium, only that quite patient firms are very seldom hired, seldom enough that the firm's payoff converges to 0 as the firm becomes arbitrarily patient.

The key intuition in the argument is that for $[0,\hat{\mu}^*]$ to be a set of posteriors at which the customer is willing to hire the firm, the firm must be willing to choose H in state θ_H when faced with such a prior. But because H pushes the firm toward the "zero-payoff set" $(\hat{\mu}^*,1]$, this must in turn imply that the expected payoff of remaining in the "hire" set $[0,\hat{\mu}^*]$ is not too high. The lemma exploits this implication to show that the positive payoffs earned in the hire set cannot be large enough and last long enough to overwhelm the eventual zero-payoff future.

18.6.5 Good Firms

Ely, Fudenberg, and Levine (2002) explore the possibility of allowing a richer set of commitment types. In the basic model, the firm can be either normal or bad. What if we also allowed a Stackelberg type, committed to the strategy $(\alpha_H,\alpha_L) = (1,1)$? In pursuing this question, assume that bad types are characterized by $\gamma = 1$, so they always choose H.

Suppose first that there are only Stackelberg and bad types. Only the customers make nontrivial decisions in this case. Letting η be the probability of the Stackelberg type, the consumers will enter if and only if $\eta u + (1-\eta)(u-w)/2 \geq 0$, and hence if and only if

$$\eta \geq \frac{w-u}{u+w} = \eta^*.$$

We thus have the situation shown in figure 18.6.4. If the prior probability of a bad type is more than $1-\eta^*$, with the firm otherwise being Stackelberg, then the consumer will never hire the firm. It must then be the case that for any configuration in which the bad type has prior probability at least $1-\eta^*$, consumers will never enter, because

608 Chapter 18 ■ Modeling Reputations

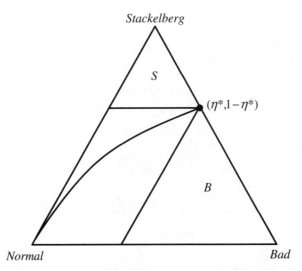

Figure 18.6.4 Space of priors for bad reputation game with Stackelberg, normal, and bad firms. The point $(\eta^*, 1 - \eta^*)$ identifies a prior with probability η^* on the Stackelberg type and $1 - \eta^*$ on the bad type.

normal firms can make things no more attractive than Stackelberg firms. The region of such priors is labeled B in figure 18.6.4. Any prior in this region induces a unique equilibrium in which consumers never hire the firm, who earns payoff of 0.

Consumers will always hire the firm if the Stackelberg type has probability at least η^*, because normal types can make things no worse than bad types. This is the region S in figure 18.6.4. If the prior falls in this region, there is an equilibrium in which a sufficiently patient normal player receives a payoff arbitrarily close to u. In particular, let the normal player choose $(0, 1)$ in the first period, thus always choosing L. The probability of the Stackelberg type is sufficiently high that consumers will still hire the firm. Conditional on the firm being normal, the posterior probability attached to the bad firm thus drops to 0 after the first period. There is then a continuation equilibrium in which the normal and Stackelberg types both choose $(1, 1)$, for payoff u. This gives an overall payoff close to u for patient normal types.

What happens in the remaining region? Ely, Fudenberg, and Levine (2002) show that if the prior probability is below the curve that divides the remaining region of priors, then the normal type's payoff in any Nash equilibrium is bounded above by a value that converges to 0 as the normal type becomes arbitrarily patient. In addition, this curve is asymptotic to the left side of the simplex at the lower left corner. This implies that even with Stackelberg types, the bad reputation result holds for virtually all prior distributions that put sufficient weight on the normal type.

18.6.6 Captive Consumers

A seemingly critical element of the bad reputation model is that there are histories after which no information about the type of firm is revealed. This feature lies behind the contrast between this model and the temporary reputation models of sections 15.5 and 16.7.

18.6 ■ Lost Consumers

The complete absence of additional information about the firm may be an extreme assumption. Suppose instead that there is always some probability that more information becomes available. Will this reverse the result? We pursue this possibility in this section, under the presumption that $\gamma = 1$, so that a single observation of L reveals the firm's type. This appears to provide the best case for such additional information making a difference.

We model this by assuming that in each period, the (short-run) consumer is a normal type, as already modeled, with probability $1 - \varepsilon$. However, with probability ε, the consumer is a captive who comes to the firm regardless of the firm's history. Every time a captive appears, more is learned about the firm. Much now hinges on the relative sizes of the discount factor δ and the likelihood ε of a captive consumer.

It is a straightforward modification of our previous proof that "very small" ε do not alter the conclusion.

Proposition 18.6.6 Let $\bar{v}_1(\xi_0, \hat{\mu}, \delta, \varepsilon)$ be the supremum of the (discounted average) Nash-equilibrium payoffs for the normal firm. Then,

$$\lim_{\delta \to 1} \lim_{\varepsilon \to 0} \bar{v}_1(\xi_0, \hat{\mu}, \delta, \varepsilon) = 0.$$

As ε gets arbitrarily small, the continuation values in the perturbed game approach those of the unperturbed game. Because our arguments in the latter exploited only strict inequalities, they carry over for sufficiently small ε.

Higher payoffs can be achieved if the firm is patient enough that it is not too costly to wait for a (perhaps very rare) captive consumer.

Proposition 18.6.7 For all $\varepsilon > 0$,

$$\lim_{\delta \to 1} \bar{v}_1(\xi_0, \hat{\mu}, \delta, \varepsilon) = u.$$

Proof Suppose $\varepsilon > 2u/(3u + w)$. We show that for sufficiently large $\delta < 1$, it is an equilibrium for the firm to always perform the appropriate repair, that is, $(\alpha_H, \alpha_L) = (1, 1)$, and for normal consumers to hire the firm if and only if the posterior $\hat{\mu} \leq 2u/(w + u)$. To verify the latter, note that substituting $(\alpha_H, \alpha_L) = (1, 1)$ in (18.6.2), the consumer's behavior is optimal. The firm's behavior is also optimal after any history featuring a previous play of L (revealing the firm to be normal) or in which the current state is θ_L.

Suppose the firm has been hired after a history featuring only signals X and H, with state θ_H. Then the normal firm's value of choosing L is given by

$$V_0(L) = (1 - \delta)(-w) + \delta u,$$

coupling the current loss of w with the future benefit of being known to be normal and earning u. The value of playing H is bounded by

$$V_0(H) = (1 - \delta)u + \varepsilon\delta\left(\tfrac{1}{2}u + \tfrac{1}{2}V_0(H)\right) + (1 - \varepsilon)\varepsilon\delta^2\left(\tfrac{1}{2}u + \tfrac{1}{2}V_0(H)\right) \\ + (1 - \varepsilon)^2\varepsilon\delta^3\left(\tfrac{1}{2}u + \tfrac{1}{2}V_0(H)\right) + \cdots,$$

The expression $V_0(H)$ gives the value to the normal firm of always performing the appropriate repair when normal consumers only hire the firm after the first observation of L (and so the bound is tight for $\hat{\mu} > 2u/(w+u)$). Solving,

$$V_0(H) = \frac{(2(1-\delta)(1-(1-\varepsilon)\delta) + \varepsilon\delta)u}{2(1-(1-\varepsilon)\delta) - \varepsilon\delta}.$$

As $V_0(L) = V_0(H) = u$ when $\delta = 1$, a sufficient condition for $V_0(H) > V_0(L)$ when δ is large is that the derivative of $V_0(H) - V_0(L)$ with respect to δ evaluated at $\delta = 1$ be negative. But this is an implication of the bound on ε.

The same argument shows that if $0 < \varepsilon \leq 2u/(3u+w)$, the profile in which the normal firm chooses L on first hiring (irrespective of the state) and thereafter always performs the appropriate repair, and normal consumers only hire the firm after the first observation of L, is an equilibrium for large δ with value approaching u.

∎

18.7 Markets for Reputations

This section examines markets for reputations. It is far from obvious what it means to buy or sell a reputation. We consider cases in which the value of an object depends on the reputation of its previous owner, so that buying the object is essentially buying the reputation. For this reputation to be valuable, it must have some effect on the payoff of the object's new owner. We ensure this by assuming that the sale of the object is not observed by others in the game, as with the replacements of section 18.1.

18.7.1 Reputations Have Value

We take a first look at this possibility in the context of a model patterned after that of Tadelis (1999, 2002). The model is a two-period snapshot of an overlapping generations economy.[10] At the beginning of period 0, there is a unit mass of "old" firms, half of whom are good and half of whom are normal. There is also a unit mass of young firms, again with half being good and half being normal. Each firm is distinguished by a name.

Good firms provide a service that is successful with probability $\rho > 1/2$ and a failure otherwise. A normal firm faces a choice in each period. It can exert high effort, at cost $c > 0$, in which case its output is a success with probability ρ and a failure with probability $1 - \rho$. Alternatively, it can exert low effort at no cost, in which case its output is surely a failure.

A success has a value of 1 to a consumer, and a failure has value 0. Firms sell their outputs to consumers who cannot distinguish good firms from normal ones and cannot condition their price on whether the service will be a success or failure, and who are sufficiently numerous to bid prices for the firms' products up to their expected value.

At the end of period 0, all old firms disappear, to be replaced by a generation of new firms. Firms that were previously new become old firms. These continuing firms have the option of retaining their name or abandoning it to either invent a new one or buy a name, at the (possibly zero) market price, from either a departing or continuing

---

10. Tadelis (2003) generalizes the model to a continuum of types and an infinite horizon.

18.7 ■ Markets for Reputations

firm. Each new firm has the option of either inventing a new name for themselves or buying a name from a continuing or departing firm. Firms in the market for two periods maximize the sum of the payoffs in the two periods.

Consumers in the second period observe the name of each firm. For each name that is not new, consumers observe whether that name was associated with a success or failure in the previous period. However, they do not observe whether the name is owned by a continuing firm or a new firm.

We thus have three markets, the first-period and second-period markets for the firms' services and an interperiod market for names. An equilibrium is a specification of effort levels and a six-tuple of prices (including a price for first-period services, a price for names bearing a previous success and names bearing a previous failure in the interperiod market for names, and a price in the second-period market for the services of a firm bearing a new name, an old name with a first-period success, and an old name with a first-period failure) with the properties that firms' effort choices are optimal and the prices clear the relevant markets.

The name the firm bears has no effect on the quality of good it supplies. There is thus a sense in which names are intrinsically worthless. However, the interperiod market for names must be active in any equilibrium.

Proposition 18.7.1 *Names are traded in any equilibrium.*

Proof If no firm adopts the name of a departing firm, then normal old firms optimally exert low effort in period 0. If normal young firms exert low effort, then all names persisting in the second-period market after a first-period success are owned by good firms, whose service price is thus ρ. In contrast, a new name belongs to a (continuing normal or new) firm who is good with probability 1/3 and who commands price $\rho/3$. If (some) normal young firms optimally exert high effort, the service price commanded by a firm with a success-bearing name must exceed that of a firm with a new name. In either case, selling the name of a successful old firm to a new firm allows mutual gains, a contradiction. ■

We now present a class of equilibria. Every continuing firm experiencing a failure in period 0 abandons its name, every young firm whose service was successful retains its name, and no names carrying a failure are purchased. (Other equilibria do exist.) We accordingly do not specify the interperiod price of the name of a failing firm, nor the second-period price of a firm whose name carries a first-period failure.

In the initial period, all firms command the same price, to be determined. In the second period, consumers potentially perceive two types of firm in the market, bearing continuing names whose service was successful in period 0 and new names. Let these command prices p_s and p_n. Note that a continuing name need not be a continuing firm.

It is clear that normal firms will exert no effort in period 1. We seek an equilibrium in which they exert effort in period 0. Let x_0 be the proportion of old normal firms who exert effort in period 0, and x_1 the proportion of young normal firms who exert effort in period 0. Then the price in the period 0 output market is given by

$$\rho\left(\tfrac{1}{2} + \tfrac{1}{4}(x_0 + x_1)\right).$$

Type of name	Type of firm	Measure
s	Old, good	$\frac{1}{2}\rho$
s	Old, normal	$\frac{1}{2}\rho x_1$
s	Good	$\beta\theta$
s	Normal	$\beta(1-\theta)$
n	Good	$\frac{1}{2}(1-\rho) + \frac{1}{2} - \beta\theta$
n	Normal	$\frac{1}{2}(1-\rho x_1) + \frac{1}{2} - \beta(1-\theta)$

Figure 18.7.1 Composition of firms in the period 1 market. An s name is a name present in the market and experiencing a success in the first period and an n name is new in the market.

Consider the interperiod market for names. There are $\rho(1+x_0)/2$ names in the market whose period 0 service was successful. Denote by $\beta \leq \rho(1+x_0)/2$ the measure of such names that are purchased and by θ the proportion of these names purchased by good firms. Figure 18.7.1 shows the composition of firms in the second period.

The period 1 output market features two prices, p_s and p_n, being the prices commanded by a name whose past service was successful and by a new name. Because only good firms exert high effort in the second period, each of these prices is ρ times the probability that the firm is good, where this probability is conditioned on the information contained in the firm's name. These probabilities are readily calculated from figure 18.7.1, using Bayes' rule, giving prices

$$p_s = \rho \left(\frac{\frac{1}{2}\rho + \beta\theta}{\frac{1}{2}\rho + \beta\theta + \frac{1}{2}\rho x_1 + \beta(1-\theta)} \right)$$

and

$$p_n = \rho \left(\frac{\frac{1}{2}(1-\rho) + \frac{1}{2} - \beta\theta}{\frac{1}{2}(1-\rho) + \frac{1}{2} - \beta\theta + \frac{1}{2}(1-\rho x_1) + \frac{1}{2} - \beta(1-\theta)} \right).$$

The interperiod market price of a name is given by $p_s - p_n$.

Now consider the incentives for normal firms to exert effort in period 0. An old firm who exerts low effort finishes the period with a worthless name, whereas high effort gives a probability ρ that his name can be sold. High effort is thus optimal if $\rho(p_s - p_n) \geq c$. A young firm who exerts low effort produces a failure, adopts a new name, and receives price p_n in the second-period output market. High effort gives a probability ρ that a price p_s can be commanded in period 1. High effort is again optimal if $\rho(p_s - p_n) \geq c$. Firms of different vintages thus face the same incentives.

18.7 ■ Markets for Reputations

We seek an equilibrium in which normal firms exert high effort in period 0 and hence in which $x_0 = x_1 = 1$. In such a configuration, there are ρ successful names in the names market, and we examine an equilibrium in which all are sold, or $\beta = \rho$. This implies the period 1 prices

$$p_s = \rho \left(\frac{1 + 2\theta}{4} \right)$$

and

$$p_n = \rho \left(\frac{2 - \rho - 2\rho\theta}{4 - 4\rho} \right).$$

The price of a name is then

$$p_s - p_n = \rho \left(\frac{2\theta - 1}{4 - 4\rho} \right).$$

For names to command a positive price, we must then have more than half of the names purchased by good firms ($\theta > 1/2$). We have a continuum of potential equilibria. If $\rho \leq 2/3$, the price of a name increases from 0 to $\rho/(4 - 4\rho)$ as θ increases from $1/2$ to its maximum of 1. If $\rho \geq 2/3$, the price of a name increases from 0 to $1/2$ as θ increases from $1/2$ to its maximum of $(2 - \rho)/(2\rho)$.[11] Equilibrium requires that the price of a name suffice to compensate a normal firm for high effort in the first period, or $\rho(p_s - p_n) \geq c$. Putting these results together, there will exist equilibria in which high effort is expended as long as c is sufficiently small.

The market for names plays an important role in creating the incentives for normal firms to exert high effort. Suppose that there was no market for names. It is then clear that old normal firms will not exert effort in period 0, because there is no prospect of a future reward from doing so. Similarly, it cannot be an equilibrium for all young normal firms to exert effort. If they do, the first-period outcome is uninformative as to the firm's type, and names bearing a success will command the same price in the second period as new names. For sufficiently small c, there is an equilibrium in which some but not all zero-year-old normal firms exert high effort, with the proportion doing so set so that $\rho(p_s - p_n) = c$.

18.7.2 Buying Reputations

The model in section 18.7.1 yields a continuum of equilibria, differing in the characteristics of the firms who buy reputations. Are reputations more likely to be purchased by good firms or bad firms? We endeavor here to gain some insight into this question.

Our point of departure is the model of section 18.4.2. We assume $\rho_H = 1 - \rho_L$ and set $\rho_L \equiv \rho$ (unlike section 18.7.1). To use this as a model in which reputations are traded, we allow an entrant's characteristics to be endogenously determined.

We interpret the model as one in which an entrepreneur owns a resource that is essential for a firm to operate in the market, whether it be a name, location, patent,

---

11. These calculations follow from noting that there are ρ names bearing a success in the market from departing firms who achieved a success, and $(1 - \rho)/2 + 1/2$ good buyers, including old firms who produced a failure and new firms.

skill, or some other input. The entrepreneur sells the right to use this resource to a firm, for the duration of the firm's lifetime. On the death of this firm, the entrepreneur offers the resource for sale again. It is now important to the analysis that replacements entail the departure of the current firm, so that the new firm's type can be determined by the resulting market.

Whenever the resource is sold, there is a large number of potential new firms who are inept. We normalize the opportunity cost of potential inept firms to 0. Normal firms are scarce. The existence of a potential normal firm, and the opportunity cost of such a firm, is randomly determined in each period, independently and identically across periods. With probability $\nu + \kappa D(d_0)$, there is a potential normal firm whose normalized opportunity cost of participating in the market is less than or equal to $d_0 \geq 0$. We assume $\nu \in (0, 1)$, $\kappa \geq 0$, $\nu + \kappa \leq 1$, and D is a strictly increasing, continuously differentiable cumulative distribution function on $[0, 1/\delta]$ with $D(0) = 0$. Hence, ν is the probability that there is a normal firm with opportunity cost zero. With probability $\kappa D(d_0)$, there is a normal firm whose opportunity cost exceeds 0 but not d_0. With probability $1 - \nu - \kappa$, there is no potential normal firm.

We assume there is at most one normal firm. When the current firm exits, the right to the resource is sold by a sealed-bid, second-price auction. The second-price auction is convenient because it ensures that the right to the resource is sold to the firm with the highest net valuation. Coupled with our assumption that there is at most one normal firm and at least two inept firms among the potential entrants, this allows us to easily identify equilibrium prices and the circumstances under which entrants are likely to be either normal or inept.

If we set $\kappa = 0$, the model is formally identical to that of exogenous replacements, with ν being the probability that a replacement is normal. In particular, with probability ν, there is a normal firm who will win the second-price auction, giving a normal replacement.

Let V_0 (V_I) denote the value function for the normal (inept) firm. These value functions are functions of the common consumer beliefs μ_0 that the firm is normal. The *price* of a resource currently characterized by belief μ_0 is the net (of opportunity cost) value of the resource to the second highest bidder, who will be an inept firm with zero opportunity cost and value $V_I(\mu_0)$, giving a price of $V_I(\mu_0)$. A normal firm with opportunity cost d_0 thus buys the name if

$$V_0(\mu_0) \geq d_0 + V_I(\mu_0). \tag{18.7.1}$$

One of the inept firms buys the name if the inequality is reversed and strict. The probability that the replacement is normal is then $\nu + \kappa D(V_0(\mu_0) - V_I(\mu_0))$, which depends on the consumers' posterior μ_0.

We seek an equilibrium in which normal firms always exert high effort. In such an equilibrium, posterior beliefs of the consumers are given by

$$\varphi(\mu_0 \mid \bar{y}) = (1 - \lambda) \frac{(1 - \rho)\mu_0}{(1 - \rho)\mu_0 + \rho(1 - \mu_0)} + \lambda \nu \\ + \lambda \kappa D(V_0(\varphi(\mu_0 \mid \bar{y})) - V_I(\varphi(\mu_0 \mid \bar{y}))), \tag{18.7.2}$$

18.7 ■ Markets for Reputations

and

$$\varphi(\mu_0 \mid \underline{y}) = (1-\lambda)\frac{\rho\mu_0}{\rho\mu_0 + (1-\rho)(1-\mu_0)} + \lambda\nu$$
$$+ \lambda\kappa D(V_0(\varphi(\mu_0 \mid \underline{y})) - V_I(\varphi(\mu_0 \mid \underline{y}))). \quad (18.7.3)$$

The belief functions $\varphi(\mu_0 \mid \bar{y})$ and $\varphi(\mu_0 \mid \underline{y})$ enter both sides of (18.7.2) and (18.7.3). This reflects the fact that beliefs depend on the likelihood that entrants are normal or inept firms, which in turn depends on beliefs. The beliefs of the consumers are then a fixed point of (18.7.2) and (18.7.3).

Given that normal firms exert high effort, a posterior probability μ_0 that the firm is normal gives a price of

$$p(\mu_0) = (1-2\rho)\mu_0 + \rho.$$

The value function of the inept firm is then

$$V_I(\mu_0) = (1 - \delta(1-\lambda))((1-2\rho)\mu_0 + \rho)$$
$$+ \delta(1-\lambda)\{\rho V_I(\varphi(\mu_0 \mid \bar{y})) + (1-\rho)V_I(\varphi(\mu_0 \mid \underline{y}))\}, \quad (18.7.4)$$

and the value function of the normal firm is

$$V_0(\mu_0) = (1 - \delta(1-\lambda))((1-2\rho)\mu_0 + \rho - c)$$
$$+ \delta(1-\lambda)\{(1-\rho)V_0(\varphi(\mu_0 \mid \bar{y})) + \rho V_0(\varphi(\mu_0 \mid \underline{y}))\}. \quad (18.7.5)$$

From (18.7.5), it is sufficient for normal firms to always optimally exert high effort if, for all possible posteriors μ_0, a one shot deviation to low effort is not profitable, that is,

$$\delta(1-\lambda)(1-2\rho)\{V_0(\varphi(\mu_0 \mid \bar{y})) - V_0(\varphi(\mu_0 \mid \underline{y}))\} \geq c(1 - \delta(1-\lambda)). \quad (18.7.6)$$

Moreover, as in the proof of proposition 18.4.1, (18.7.6) is also sufficient. The triple (α, p, φ) is then a *reputation equilibrium* if the normal firm chooses high effort in every state ($\alpha(\mu_0) = 1$ for all μ_0), the expectation updating rules and the value functions of the firms satisfy (18.7.2)–(18.7.5), and the normal firm is maximizing at every μ_0.

From Mailath and Samuelson (2001, proposition 3) we have:

Proposition 18.7.2 Suppose $\nu > 0$, $\lambda > 0$, $\delta(1-\lambda) < \rho(1-\rho)/(1 - 3\rho + 3\rho^2)$, and D' is bounded. Then there exists $\kappa^* > 0$ and $c^* > 0$ such that a reputation equilibrium exists for all $\kappa \in [0, \kappa^*]$ and $c \in [0, c^*]$.

The difficulty in establishing the existence of a reputation equilibrium arises from the linkage between the posterior updating rules and the firms' value functions. In the case of exogenous replacements, the updating rules are defined independently of the value functions. We could accordingly first calculate posterior beliefs, use these calculations to obtain value functions, and then confirm that the proposed strategies are optimal given the value functions. With endogenous replacements, we require a fixed-point argument to establish the existence of mutually consistent updating rules and value functions. After concluding that consistent value functions and updating rules exist, familiar arguments establish that as long as c and κ are not too large, the proposed strategies are optimal.

We again require that c be sufficiently small that the potential future gains of maintaining a reputation can exceed the current cost. The requirements that $\nu > 0$ and $\lambda > 0$ ensure that there exist μ'_0 and μ''_0, with $0 < \mu'_0 < \mu''_0 < 1$, for any allowable values of κ, such that for any φ satisfying (18.7.2) and (18.7.3), $\varphi(\mu_0 \mid y) \in [\mu'_0, \mu''_0]$ for all $\mu_0 \in [0, 1]$ and $y \in \{\underline{y}, \bar{y}\}$. As in the case of proposition 18.4.1, this bounding of posterior probabilities away from the ends of the unit interval is necessary to preserve the incentive for normal firms to exert high effort.

The inequality restriction on δ, λ, and ρ in proposition 18.7.2 ensures that the one-period discounted "average" derivative of the no-replacement updating rule is strictly less than 1. Coupled with the requirements that D' is bounded and that κ is not too large, this ensures that the value functions have uniformly bounded derivatives. This in turn allows us to construct a compact set of potential value functions to which a fixed point argument can be applied to yield consistent belief updating rules and value functions. Taking κ to be small also ensures that the type of an entering firm is not too sensitive to the difference $V_0(\mu_0) - V_I(\mu_0)$. Otherwise, the possibility arises that for some values of μ_0, consumers and potential entrants might coordinate on an equilibrium in which entrants are likely to be normal, because the value of a normal firm is high, because consumers expect entrants to be normal. For other values of μ_0, entrants may be unlikely to be normal, because the value is low, because consumers expect inept entrants. This allows us to introduce sharp variations in the value function $V_0(\mu_0)$, potentially destroying the convention that higher reputations are good, which lies at the heart of a reputation equilibrium.

We now turn our attention to the market for reputations. In particular, which posteriors are most likely to attract normal firms as replacements, and which are most likely to attract inept firms? A normal firm is more likely to enter as the difference $V_0(\mu_0) - V_I(\mu_0)$ increases.

Proposition 18.7.3 *Suppose a reputation equilibrium exists for all $\kappa < \kappa^*$. For any $\xi > 0$, there is a $\kappa^\dagger \leq \kappa^*$ such that for any $\kappa < \kappa^\dagger$, $V_0(\mu_0) - V_I(\mu_0)$ is strictly increasing for $\mu_0 < 1/2 - \xi$ and strictly decreasing for $\mu_0 > 1/2 + \xi$.*

Replacements are more likely to be normal firms for intermediate values of μ_0 and less likely to be normal firms for extreme values of μ_0.[12] Thus, firms with low reputations are relatively likely to be replaced by inept firms. Good firms find it too expensive to build up the reputation of such a name. On the other hand, firms with very good reputations are also relatively likely to be replaced by inept firms. These names are attractive to normal firms, who would prefer to inherit a good reputation to having to build up a reputation, and who would maintain the existing good reputation. However, these names are even more attractive to inept entrants, who will enjoy the fruits of running down the existing high reputation (recall that if consumers believe that the firm is almost certainly normal, then bad outcomes do not change consumer beliefs by a large amount).

---

12. The function $V_0(\mu_0) - V_I(\mu_0)$ has its maximum near $1/2$ because of the symmetry assumption $\Pr\{\bar{y} \mid H\} = \Pr\{\underline{y} \mid L\}$. This result holds without the symmetry assumption, but with the maximum possibly no longer near $1/2$.

18.7 ■ Markets for Reputations

Replacements are more likely to be normal firms for intermediate reputations. These are attractive to normal firms because less expenditure is required to build a reputation than is the case when the existing firm has a low reputation. At the same time, these reputations are less attractive than higher reputations to inept entrants, because the intermediate reputation offers a smaller stock that can be profitably depleted.

As a result, we expect reputations to exhibit two features. There will be churning: High reputations will be depleted and intermediate reputations will be enhanced. Low reputations are likely to remain low.

Proof Consider first the case of exogenous entry, $\kappa = 0$. The corresponding value functions $V_0(\mu_0)$ and $V_I(\mu_0)$ can be written as

$$V_0(\mu_0) = \rho - c + (1 - \delta(1-\lambda))(1-2\rho)\mu_0$$
$$+ (1-\delta(1-\lambda))(1-2\rho) \sum_{t=1}^{\infty} \delta^t(1-\lambda)^t \sum_{h^t \in \{\underline{y},\bar{y}\}^t} \varphi(\mu_0 \mid h^t) \Pr(h^t \mid H),$$

and

$$V_I(\mu_0) = \rho + (1 - \delta(1-\lambda))(1-2\rho)\mu_0$$
$$+ (1-\delta(1-\lambda))(1-2\rho) \sum_{t=1}^{\infty} \delta^t(1-\lambda)^t \sum_{h^t \in \{\underline{y},\bar{y}\}^t} \varphi(\mu_0 \mid h^t) \Pr(h^t \mid L),$$

where $\Pr(h^t \mid L)$ is the probability of realizing the sequence of outcomes h^t given that the firm chooses low effort in every period. Combining these expressions,

$$V_0(\mu_0) - V_I(\mu_0)$$
$$= (1-\delta(1-\lambda))(1-2\rho) \sum_{t=1}^{\infty} \left\{ \sum_{h^t} \delta^t(1-\lambda)^t \Pr(h^t \mid H) \varphi(\mu_0 \mid h^t) \right.$$
$$\left. - \sum_{h^t} \delta^t(1-\lambda)^t \Pr(h^t \mid L) \varphi(\mu_0 \mid h^t) \right\} + k,$$

(18.7.7)

where k is independent of μ_0. The set of histories $\{\underline{y}, \bar{y}\}^t$ can be partitioned into sets of "mirror images," $\{h^t, \hat{h}^t\}$, where h^t specifies $\bar{y}$ in period $\tau \le t$ if and only if $\hat{h}^t$ specifies $\underline{y}$ in period $\tau \le t$. It suffices to show that

$$\beta(\mu_0) \equiv \varphi(\mu_0 \mid h^t)\Pr(h^t \mid H) + \varphi(\mu_0 \mid \hat{h}^t)\Pr(\hat{h}^t \mid H)$$
$$- \varphi(\mu_0 \mid h^t)\Pr(h^t \mid L) - \varphi(\mu_0 \mid \hat{h}^t)\Pr(\hat{h}^t \mid L)$$

is concave and maximized at $\mu_0 = 1/2$, because (18.7.7) is a weighted sum of such terms. Now notice that

$$\Pr(h^t \mid H) = \Pr(\hat{h}^t \mid L) \equiv x$$

and $$\Pr(\hat{h}^t \mid H) = \Pr(h^t \mid L) \equiv y,$$

which implies

$$\varphi(\mu_0 \mid h^t) = (1-\lambda)^t \frac{x\mu_0}{x\mu_0 + y(1-\mu_0)} + (1-(1-\lambda)^t)\gamma$$

and

$$\varphi(\mu_0 \mid \hat{h}^t) = (1-\lambda)^t \frac{y\mu_0}{y\mu_0 + x(1-\mu_0)} + (1-(1-\lambda)^t)\gamma,$$

where γ does not depend on μ_0. Letting $x\mu_0 + y(1-\mu_0) \equiv Z_x$ and $y\mu_0 + x(1-\mu_0) \equiv Z_y$, we can then calculate (where β' and β'' denote first and second derivatives)

$$\beta' = (1-\lambda)^t \left[\frac{x^2 y}{Z_x^2} + \frac{xy^2}{Z_y^2} - \frac{xy^2}{Z_x^2} - \frac{x^2 y}{Z_y^2} \right],$$

which equals 0 when $\mu_0 = 1/2$. We can then calculate

$$\beta'' = -2xy(1-\lambda)^t \left[\frac{(x-y)^2}{Z_x^3} + \frac{(y-x)^2}{Z_y^3} \right] \leq 0,$$

with the inequalities strict whenever h^t specifies an unequal number of good and bad outcomes, so that $V_0 - V_I$ is strictly concave and maximized at $\mu_0 = 1/2$.

Moreover, because $d\{V_0(\mu_0) - V_I(\mu_0)\}/d\mu_0$ is strictly decreasing, it is bounded away from 0 from below for $\mu_0 \leq 1/2 - \xi$ and it is bounded away from 0 from above for $\mu_0 \geq 1/2 + \xi$. The extension to κ small but nonzero is then an immediate implication of the sequential compactness of the spaces of updating rules and value functions. ∎

Remark 18.7.1 Section 18.7 illustrates the possibilities for putting reputation models to work. Much remains to be done in terms of exploring alternative models and their uses, with an eye toward better capturing the idea of reputation. ◆

Bibliography

Abreu, D. (1986): "Extremal Equilibria of Oligopolistic Supergames," *Journal of Economic Theory*, 39(1), 191–225.

——— (1988): "On the Theory of Infinitely Repeated Games with Discounting," *Econometrica*, 56(2), 383–396.

Abreu, D., P. Dutta, and L. Smith (1994): "The Folk Theorem for Repeated Games: A NEU Condition," *Econometrica*, 62(4), 939–948.

Abreu, D., and F. Gul (2000): "Bargaining and Reputation," *Econometrica*, 68(1), 85–117.

Abreu, D., P. Milgrom, and D. Pearce (1991): "Information and Timing in Repeated Partnerships," *Econometrica*, 59(6), 1713–1733.

Abreu, D., D. Pearce, and E. Stacchetti (1986): "Optimal Cartel Equilibria with Imperfect Monitoring," *Journal of Economic Theory*, 39(1), 251–269.

——— (1990): "Toward a Theory of Discounted Repeated Games with Imperfect Monitoring," *Econometrica*, 58(5), 1041–1063.

——— (1993): "Renegotiation and Symmetry in Repeated Games," *Journal of Economic Theory*, 60(2), 217–240.

Abreu, D., and A. Rubinstein (1988): "The Structure of Nash Equilibrium in Repeated Games with Finite Automata," *Econometrica*, 56(6), 1259–1281.

Ahn, I., and M. Suominen (2001): "Word-of-Mouth Communication and Community Enforcement," *International Economic Review*, 42(2), 399–415.

Al-Najjar, N. I. (1995): "Decomposition and Characterization of Risk with a Continuum of Random Variables," *Econometrica*, 63(5), 1195–1224.

Al-Najjar, N. I., and R. Smorodinsky (2001): "Large Nonanonymous Repeated Games," *Games and Economic Behavior*, 37(1), 26–39.

Aliprantis, C. D., and K. C. Border (1999): *Infinite Dimensional Analysis: A Hitchhiker's Guide*, 2nd edition. Springer-Verlag, Berlin.

Aoyagi, M. (1996): "Reputation and Dynamic Stackelberg Leadership in Infinitely Repeated Games," *Journal of Economic Theory*, 71(2), 378.

Athey, S., and K. Bagwell (2001): "Optimal Collusion with Private Information," *RAND Journal of Economics*, 32(3), 428–465.

Athey, S., K. Bagwell, and C. Sanchirico (2004): "Collusion and Price Rigidity," *Review of Economic Studies*, 71(2), 317–349.

Atkeson, A., and R. E. Lucas Jr. (1992): "On Efficient Distribution with Private Information," *Review of Economic Studies*, 59(3), 427–453.

——— (1995): "Efficiency and Equality in a Simple Model of Efficient Unemployment Insurance," *Journal of Economic Theory*, 66(1), 64–88.

Aumann, R. J. (1965): "Integrals of Set-Valued Functions," *Journal of Mathematical Analysis and Applications*, 12(1), 1–12.

Aumann, R. J. (1974): "Subjectivity and Correlation in Randomized Strategies," *Journal of Mathematical Economics*, 1(1), 67–96.

—— (1990): "Nash Equilibria Are Not Self-Enforcing," in *Economic Decision Making: Games, Econometrics and Optimisation: Essays in Honor of Jacques Dreze*, ed. by J. J. Gabszewics, J.-F. Richard, and L. Wolsey, pp. 201–206. Elsevier Science Publishers, Amsterdam.

—— (2000): "Irrationality in Game Theory," in *Collected Papers of Robert J. Aumann, Volume 1*, pp. 621–634. MIT Press, Cambridge, MA.

Aumann, R. J., M. B. Maschler, and R. E. Stearns (1968): "Repeated Games of Incomplete Information: An Approach to the Non-Zero Sum Case," reprinted in *Repeated Games with Incomplete Information*, by R. J. Aumann and M. B. Maschler (1995), MIT Press, Cambridge, MA.

Aumann, R. J., and L. S. Shapley (1976): "Long-Term Competition—A Game Theoretic Analysis," reprinted in *Essays in Game Theory in Honor of Michael Maschler*, ed. by N. Megiddo (1994), pp. 1–15, Springer-Verlag, New York.

Aumann, R. J., and S. Sorin (1989): "Cooperation and Bounded Recall," *Games and Economic Behavior*, 1(1), 5–39.

Ausubel, L. M., P. Cramton, and R. J. Deneckere (2002): "Bargaining with Incomplete Information," in *Handbook of Game Theory, Volume 3*, ed. by R. J. Aumann and S. Hart, pp. 1897–1945. North Holland, New York.

Ausubel, L. M., and R. J. Deneckere (1989): "A Direct Mechanism Characterization of Sequential Bargaining with One-Sided Incomplete Information," *Journal of Economic Theory*, 48(1), 18–46.

Axelrod, R. M. (1984): *The Evolution of Cooperation*. Basic Books, New York.

Bagwell, K. (1995): "Commitment and Observability in Games," *Games and Economic Behavior*, 8(2), 271–280.

Bagwell, K., and R. W. Staiger (1997): "Collusion over the Business Cycle," *RAND Journal of Economics*, 28(1), 82–106.

Baker, G., R. Gibbons, and K. J. Murphy (2002): "Relational Contracts and the Theory of the Firm," *Quarterly Journal of Economics*, 117(1), 39–84.

Barro, R. J., and D. B. Gordon (1983): "Rules, Discretion and Reputation in a Model of Monetary Policy," *Journal of Monetary Economics*, 12(1), 101–122.

Battigalli, P., and J. Watson (1997): "On "Reputation" Refinements with Heterogeneous Beliefs," *Econometrica*, 65(2), 369–374.

Benabou, R., and G. Laroque (1992): "Using Privileged Information to Manipulate Markets: Insiders, Gurus, and Credibility," *Quarterly Journal of Economics*, 107(3), 921–958.

Benoit, J., and V. Krishna (1985): "Finitely Repeated Games," *Econometrica*, 53(4), 905–922.

—— (1993): "Renegotiation in Finitely Repeated Games," *Econometrica*, 61(2), 303–323.

Bernheim, B. D., and A. Dasgupta (1995): "Repeated Games with Asymptotically Finite Horizons," *Journal of Economic Theory*, 67(1), 129–152.

Bernheim, B. D., B. Peleg, and M. D. Whinston (1987): "Coalition-Proof Nash Equilibria I: Concepts," *Journal of Economic Theory*, 42(1), 1–12.

Bernheim, B. D., and D. Ray (1989): "Collective Dynamic Consistency in Repeated Games," *Games and Economic Behavior*, 1(4), 295–326.

Bernheim, B. D., and M. D. Whinston (1987): "Coalition-Proof Nash Equilibria II: Applications," *Journal of Economic Theory*, 42(1), 13–29.

—— (1990): "Multimarket Contact and Collusive Behavior," *RAND Journal of Economics*, 21(1), 1–26.

Bhaskar, V. (1998): "Informational Constraints and the Overlapping Generations Model: Folk and Anti-Folk Theorems," *Review of Economic Studies*, 65(1), 135–149.

Bhaskar, V., and I. Obara (2002): "Belief-Based Equilibria in the Repeated Prisoners' Dilemma with Private Monitoring," *Journal of Economic Theory*, 102(1), 40–69.

Bhaskar, V., and E. van Damme (2002): "Moral Hazard and Private Monitoring," *Journal of Economic Theory*, 102(1), 16–39.

Billingsley, P. (1979): *Probability and Measure*. Wiley, New York.

Binmore, K. (1994): *Game Theory and the Social Contract Volume I: Playing Fair*. MIT Press, Cambridge, MA.

——— (1998): *Game Theory and the Social Contract Volume II: Just Playing*. MIT Press, Cambridge, MA.

Blackwell, D. (1951): "The Comparison of Experiments," in *Proceedings of the Second Berkeley Symposium on Mathematical Statistics and Probability, 1950*, pp. 93–102. Berkeley, University of California Press.

Bond, E. W., and L. Samuelson (1984): "Durable Good Monopolies with Rational Expectations and Replacement Sales," *RAND Journal of Economics*, 15(3), 336–345.

——— (1987): "The Coase Conjecture Need Not Hold for Durable Good Monopolies with Depreciation," *Economics Letters*, 24, 93–97.

Börgers, T. (1989): "Perfect Equilibrium Histories of Finite and Infinite Horizon Games," *Journal of Economic Theory*, 47(1), 218–227.

Canzoneri, M. B. (1985): "Monetary Policy Games and the Role of Private Information," *American Economic Review*, 75(5), 1056–1070.

Carlton, D. W., and J. M. Perloff (1992): *Modern Industrial Organization*. Scott, Foresman, London.

Carmichael, H. L., and W. B. MacLeod (1997): "Gift Giving and the Evolution of Cooperation," *International Economic Review*, 38(3), 485–509.

Celentani, M., D. Fudenberg, D. K. Levine, and W. Pesendorfer (1996): "Maintaining a Reputation against a Long-Lived Opponent," *Econometrica*, 64(3), 691–704.

Celentani, M., and W. Pesendorfer (1996): "Reputation in Dynamic Games," *Journal of Economic Theory*, 70(1), 109–132.

Chan, J. (2000): "On the Non-Existence of Reputation Effects in Two-Person Infinitely-Repeated Games," Johns Hopkins University.

Chari, V. V., and P. J. Kehoe (1990): "Sustainable Plans," *Journal of Political Economy*, 98(4), 783–802.

——— (1993a): "Sustainable Plans and Debt," *Journal of Economic Theory*, 61(2), 230–261.

——— (1993b): "Sustainable Plans and Mutual Default," *Review of Economic Studies*, 60(1), 175–195.

Chatterjee, K., and L. Samuelson (1987): "Bargaining with Two-Sided Incomplete Information: An Infinite Horizon Model with Alternating Offers," *Review of Economic Studies*, 54(2), 175–192.

——— (1988): "Bargaining with Two-Sided Incomplete Information: The Unrestricted Offers Case," *Operations Research*, 36(4), 605–638.

Chung, K. L. (1974): *A Course in Probability Theory*. Academic Press, New York.

Clemhout, S., and H. Y. Wan Jr. (1994): "Differential Games—Ecomonic Applications," in *Handbook of Game Theory, Volume 2*, ed. by R. J. Aumann and S. Hart, pp. 801–825. North Holland, New York.

Coate, S., and G. C. Loury (1993): "Will Affirmative-Action Policies Eliminate Negative Stereotypes?" *American Economic Review*, 83(5), 1220–1240.

Cole, H. L., J. Dow, and W. B. English (1995): "Default, Settlement, and Signalling: Lending Resumption in a Reputational Model of Sovereign Debt," *International Economic Review*, 36(2), 365–385.

Cole, H. L., and N. R. Kocherlakota (2005): "Finite Memory and Imperfect Monitoring," *Games and Economic Behavior*, 53(1), 59–72.

Compte, O. (1998): "Communication in Repeated Games with Imperfect Private Monitoring," *Econometrica*, 66(3), 597–626.

——— (2002): "On Failing to Cooperate When Monitoring Is Private," *Journal of Economic Theory*, 102(1), 151–188.

Conlon, J. (2003): "Hope Springs Eternal: Learning and the Stability of Cooperation in Short Horizon Repeated Games," *Journal of Economic Theory*, 112(1), 35–65.

Cooper, R., and A. John (1988): "Coordinating Coordination Failures in Keynesian Models," *Quarterly Journal of Economics*, 103(3), 441–463.

Coslett, S. R., and L. Lee (1985): "Serial Correlation in Latent Discrete Variable Models," *Journal of Econometrics*, 27(1), 79–97.

Crawford, V. P., and J. Sobel (1982): "Strategic Information Transmission," *Econometrica*, 50(6), 1431–1351.

Cripps, M. W., E. Dekel, and W. Pesendorfer (2004): "Reputation with Equal Discounting in Repeated Games with Strictly Conflicting Interests," *Journal of Economic Theory*, 121(2), 259–272.

Cripps, M. W., G. J. Mailath, and L. Samuelson (2004a): "Imperfect Monitoring and Impermanent Reputations," *Econometrica*, 72(2), 407–432.

——— (2004b): "Disappearing Private Reputations," *Journal of Economic Theory*, forthcoming.

Cripps, M. W., K. M. Schmidt, and J. P. Thomas (1996): "Reputation in Perturbed Repeated Games," *Journal of Economic Theory*, 69(2), 387–410.

Cripps, M. W., and J. P. Thomas (1995): "Reputation and Commitment in Two-Person Repeated Games without Discounting," *Econometrica*, 63(6), 1401–1419.

——— (1997): "Reputation and Perfection in Repeated Common Interest Games," *Games and Economic Behavior*, 18(2), 141–158.

——— (2003): "Some Asymptotic Results in Discounted Repeated Games of One-Sided Incomplete Information," *Mathematics of Operations Research*, 28(3), 433–462.

Diamond, D. W. (1989): "Reputation Acquisition in Debt Markets," *Journal of Political Economy*, 97(4), 828–862.

Diamond, D. W., and P. H. Dybvig (1983): "Bank Runs, Deposit Insurance and Liquidity," *Journal of Political Economy*, 91(3), 401–419.

Dutta, P. (1995): "A Folk Theorem for Stochastic Games," *Journal of Economic Theory*, 66(1), 1–32.

Ellickson, R. C. (1991): *Order without Law: How Neighbors Settle Disputes*. Harvard University Press, Cambridge, MA.

Ellison, G. (1994): "Cooperation in the Prisoner's Dilemma with Anonymous Random Matching," *Review of Economic Studies*, 61(3), 567–588.

Ely, J. C., D. Fudenberg, and D. K. Levine (2002): "When Is Reputation Bad?" Northwestern University, Harvard University, and University of California at Los Angeles.

Ely, J. C., J. Hörner, and W. Olszewski (2005): "Belief-Free Equilibria in Repeated Games," *Econometrica*, 73(2), 377–416.

Ely, J. C., and J. Välimäki (2002): "A Robust Folk Theorem for the Prisoner's Dilemma," *Journal of Economic Theory*, 102(1), 84–105.

——— (2003): "Bad Reputation," *Quarterly Journal of Economics*, 118(3), 785–814.

Evans, R., and E. Maskin (1989): "Efficient Renegotiation-Proof Equilibria in Repeated Games," *Games and Economic Behavior*, 1(4), 361–369.

Evans, R., and J. P. Thomas (1997): "Reputation and Experimentation in Repeated Games with Two Long-Run Players," *Econometrica*, 65(5), 1153–1173.

——— (2001): "Cooperation and Punishment," *Econometrica*, 69(4), 1061–1075.

Farrell, J., and E. Maskin (1989): "Renegotiation in Repeated Games," *Games and Economic Behavior*, 1(4), 327–360.

Forges, F. (1992): "Repeated Games of Incomplete Information: Non-Zero-Sum," in *Handbook of Game Theory, Volume 1*, ed. by R. J. Aumann and S. Hart, pp. 155–177. North Holland, New York.

Friedman, A. (1994): "Differential Games," in *Handbook of Game Theory, Volume 2*, ed. by R. J. Aumann and S. Hart, pp. 781–799. North Holland, New York.

Friedman, J. W. (1971): "A Noncooperative Equilibrium for Supergames," *Review of Economic Studies*, 38(1), 1–12.

——— (1985): "Cooperative Equilibria in Finite Horizon Non-Cooperative Supergames," *Journal of Economic Theory*, 52(2), 390–398.

Fudenberg, D., D. Kreps, and E. Maskin (1990): "Repeated Games with Long-Run and Short-Run Players," *Review of Economic Studies*, 57(4), 555–574.

Fudenberg, D., and D. K. Levine (1983): "Subgame-Perfect Equilibria of Finite- and Infinite-Horizon Games," *Journal of Economic Theory*, 31(2), 251–268.

——— (1986): "Limit Games and Limit Equilibria," *Journal of Economic Theory*, 38(2), 261–279.

——— (1989): "Reputation and Equilibrium Selection in Games with a Patient Player," *Econometrica*, 57(4), 759–778.

——— (1992): "Maintaining a Reputation When Strategies Are Imperfectly Observed," *Review of Economic Studies*, 59(3), 561–579.

——— (1994): "Efficiency and Observability with Long-Run and Short-Run Players," *Journal of Economic Theory*, 62(1), 103–135.

Fudenberg, D., D. K. Levine, and E. Maskin (1994): "The Folk Theorem with Imperfect Public Information," *Econometrica*, 62(5), 997–1039.

Fudenberg, D., D. K. Levine, and W. Pesendorfer (1998): "When Are Non-Anonymous Players Negligible?" *Journal of Economic Theory*, 79(1), 46–71.

Fudenberg, D., D. K. Levine, and J. Tirole (1985): "Infinite-Horizon Models of Bargaining with One-Sided Incomplete Information," in *Game-Theoretic Models of Bargaining*, ed. by A. E. Roth, pp. 73–98. Cambridge University Press, New York.

Fudenberg, D., and E. Maskin (1986): "The Folk Theorem in Repeated Games with Discounting or with Incomplete Information," *Econometrica*, 54(3), 533–554.

——— (1991): "On the Dispensability of Public Randomization in Discounted Repeated Games," *Journal of Economic Theory*, 53(2), 428–438.

Fudenberg, D., and J. Tirole (1991): *Game Theory*. MIT Press, Cambridge, MA.

Gale, D. (1960): *The Theory of Linear Economic Models*. McGraw-Hill, New York.

Ghosh, P., and D. Ray (1996): "Cooperation in Community Interaction without Information Flows," *Review of Economic Studies*, 63(3), 491–519.

Glicksberg, I. L. (1952): "A Further Generalization of the Kakutani Fixed Point Theorem, with Application to Nash Equilibrium Points," *Proceedings of the American Mathematical Society*, 3(1), 170–174.

Gossner, O. (1995): "The Folk Theorem for Finitely Repeated Games with Mixed Strategies," *International Journal of Game Theory*, 24(1), 95–107.

Green, E. J. (1980): "Noncooperative Price Taking in Large Dynamic Markets," *Journal of Economic Theory*, 22(2), 155–182.

Green, E. J., and R. H. Porter (1984): "Noncooperative Collusion under Imperfect Price Formation," *Econometrica*, 52(1), 87–100.

Greif, A. (1997): "Microtheory and Recent Developments in the Study of Economic Institutions through Economic History," in *Advances in Economics and Econometrics: Theory and Applications, Seventh World Congress*, ed. by D. M. Kreps and K. F. Wallis, pp. 79–113. Cambridge University Press, Cambridge.

——— (2005): *Institutions and the Path to the Modern Economy: Lessons from Medieval Trade*. Cambridge University Press, Cambridge.

Greif, A., P. Milgrom, and B. R. Weingast (1994): "Coordination, Commitment, and Enforcement: The Case of the Merchant Guild," *Journal of Political Economy*, 102(4), 745–776.

Gul, F., H. Sonnenschein, and R. Wilson (1986): "Foundations of Dynamic Monopoly and the Coase Conjecture," *Journal of Economic Theory*, 39(1), 155–190.

Haltiwanger, J., and J. E. Harrington Jr. (1991): "The Impact of Cyclical Demand Movements on Collusive Behavior," *RAND Journal of Economics*, 22(1), 89–106.

Hamermesh, D. S. (2004): *Economics Is Everywhere*. McGraw-Hill/Irwin, New York.

Harrington, J. E. Jr. (1995): "Cooperation in a One-Shot Prisoners' Dilemma," *Games and Economic Behavior*, 8(2), 364–377.

Harris, C. (1985): "A Characterization of the Perfect Equilibria of Infinite Horizon Games," *Journal of Economic Theory*, 37(1), 99–125.

Harsanyi, J. C. (1973): "Games with Randomly Disturbed Payoffs: A New Rationale for Mixed-Strategy Equilibrium Points," *International Journal of Game Theory*, 2(1), 1–23.

Harsanyi, J. C., and R. Selten (1988): *A General Theory of Equilibrium Selection in Games*. MIT Press, Cambridge, MA.

Hart, S. (1985): "Nonzero-Sum Two-Person Repeated Games with Incomplete Information," *Mathematics of Operations Research*, 10(1), 117–153.

Hertel, J. (2004): "Efficient and Sustainable Risk Sharing with Adverse Selection," Princeton University.

Holmström, B. (1982): "Managerial Incentive Problems: A Dynamic Perspective," in *Essays in Economics and Management in Honour of Lars Wahlbeck*, pp. 209–230. Swedish School of Economics and Business Administration, Helsinki, published in *Review of Economic Studies* 66 (1), January 1999, 169–182.

Hörner, J. (2002): "Reputation and Competition," *American Economic Review*, 92(3), 644–663.

Hörner, J., and W. Olszewski (2005): "The Folk Theorem for Games with Private Almost-Perfect Monitoring," *Econometrica*, forthcoming.

Hrbacek, K., and T. Jech (1984): *Introduction to Set Theory*. Marcel Dekker, New York.

Israeli, E. (1999): "Sowing Doubt Optimally in Two-Person Repeated Games," *Games and Economic Behavior*, 28(2), 203–216.

Jackson, M. O., and E. Kalai (1999): "Reputation versus Social Learning," *Journal of Economic Theory*, 88(1), 40–59.

Judd, K. L., S. Yeltekin, and J. Conklin (2003): "Computing Supergame Equilibria," *Econometrica*, 71(4), 1239–1254.

Kalai, E., and E. Lehrer (1994): "Weak and Strong Merging of Opinions," *Journal of Mathematical Economics*, 23(1), 73–86.

Kalai, E., and W. Stanford (1988): "Finite Rationality and Interpersonal Complexity in Repeated Games," *Econometrica*, 56(2), 397–410.

Kandori, M. (1991a): "Cooperation in Finitely Repeated Games with Imperfect Private Information," Princeton University.

——— (1991b): "Correlated Demand Shocks and Price Wars during Booms," *Review of Economic Studies*, 58(1), 171–180.

——— (1992a): "Social Norms and Community Enforcement," *Review of Economic Studies*, 59(1), 63–80.

——— (1992b): "The Use of Information in Repeated Games with Imperfect Monitoring," *Review of Economic Studies*, 59(3), 581–594.

Kandori, M., and H. Matsushima (1998): "Private Observation, Communication and Collusion," *Econometrica*, 66(3), 627–652.

Kandori, M., and I. Obara (2003): "Less Is More: An Observability Paradox in Repeated Games," University of Tokyo and University of California at Los Angeles.

——— (2006): "Efficiency in Repeated Games Revisited: The Role of Private Strategies," *Econometrica*, 74(2), 499–519.

Kocherlakota, N. R. (1996): "Implications of Efficient Risk Sharing without Commitment," *Review of Economic Studies*, 63(3), 595–609.

Koeppl, T. V. (2003): "Differentiability of the Efficient Frontier when Commitment to Risk Sharing Is Limited," European Central Bank.

Kreps, D. M., P. R. Milgrom, J. Roberts, and R. J. Wilson (1982): "Rational Cooperation in the Finitely Repeated Prisoners' Dilemma," *Journal of Economic Theory*, 27(2), 245–252.

Kreps, D. M., and R. J. Wilson (1982a): "Reputation and Imperfect Information," *Journal of Economic Theory*, 27(2), 253–279.

——— (1982b): "Sequential Equilibrium," *Econometrica*, 50(4), 863–894.

Kydland, F. E., and E. C. Prescott (1977): "Rules Rather than Discretion: The Inconsistency of Optimal Plans," *Journal of Political Economy*, 85(3), 473–491.

Lagunoff, R., and A. Matsui (1997): "Asynchronous Choice in Repeated Coordination Games," *Econometrica*, 65(6), 1467–1477.

Lehrer, E. (1990): "Nash Equilibria of n-Player Repeated Games with Semi-Standard Information," *International Journal of Game Theory*, 19(2), 191–217.

Lehrer, E., and A. Pauzner (1999): "Repeated Games with Differential Time Preferences," *Econometrica*, 67(2), 393–412.

Lehrer, E., and S. Sorin (1998): "ε-Consistent Equilibrium in Repeated Games," *International Journal of Game Theory*, 27(2), 231–244.

Levin, J. (2003): "Relational Incentive Contracts," *American Economic Review*, 93(3), 835–857.

Levine, D. K., and W. Pesendorfer (1995): "When Are Agents Negligible?" *American Economic Review*, 85(5), 1160–1170.

Ligon, E., J. P. Thomas, and T. Worrall (2002): "Informal Insurance Arrangements with Limited Commitment: Theory and Evidence from Village Economies," *Review of Economic Studies*, 69(1), 209–244.

Ljungqvist, L., and T. J. Sargent (2004): *Recursive Macroeconomic Theory*, 2nd edition. MIT Press, Cambridge, MA.

Luciano, R., and D. Fisher (1982): *The Umpire Strikes Back*. Bantam Books, Toronto.

Mailath, G. J., S. A. Matthews, and T. Sekiguchi (2002): "Private Strategies in Finitely Repeated Games with Imperfect Public Monitoring," *Contributions to Theoretical Economics*, 2(1), article 2.

Mailath, G. J., and S. Morris (2002): "Repeated Games with Almost-Public Monitoring," *Journal of Economic Theory*, 102(1), 189–228.

——— (2005): "Coordination Failure in Repeated Games with Almost-Public Monitoring," *Theoretical Economics*, forthcoming.

Mailath, G. J., V. Nocke, and L. White (2004): "Remarks on Simple Penal Codes," University of Pennsylvania and Harvard Business School.

Mailath, G. J., I. Obara, and T. Sekiguchi (2002): "The Maximum Efficient Equilibrium Payoff in the Repeated Prisoners' Dilemma," *Games and Economic Behavior*, 40(1), 99–122.

Mailath, G. J., A. Postlewaite, and L. Samuelson (2005): "Contemporaneous Perfect Epsilon-Equilibria," *Games and Economic Behavior*, 53(1), 126–140.

Mailath, G. J., and L. Samuelson (2001): "Who Wants a Good Reputation?" *Review of Economic Studies*, 68(2), 415–441.

Mailath, G. J., L. Samuelson, and J. M. Swinkels (1994): "Normal Form Structures in Extensive Form Games," *Journal of Economic Theory*, 64(2), 325–371.

Maskin, E., and J. Tirole (1988a): "A Theory of Dynamic Oligopoly I: Overview and Quantity Competition with Large Fixed Costs," *Econometrica*, 56(3), 549–569.

——— (1988b): "A Theory of Dynamic Oligopoly II: Price Competition, Kinked Demand Curves and Fixed Costs," *Econometrica*, 56(3), 571–599.

——— (1999): "Unforeseen Contingencies and Incomplete Contracts," *Review of Economic Studies*, 66(1), 83–114.

——— (2001): "Markov Perfect Equilibrium: I. Observable Actions," *Journal of Economic Theory*, 100(2), 191–219.

Masso, J. (1996): "A Note on Reputation: More on the Chain-Store Paradox," *Games and Economic Behavior*, 15(1), 55–81.

Matsushima, H. (1989): "Efficiency in Repeated Games with Imperfect Monitoring," *Journal of Economic Theory*, 48(2), 428–442.

——— (1991): "On the Theory of Repeated Games with Private Information. Part I: Anti-Folk Theorem without Communication," *Economic Letters*, 35, 253–256.

——— (2004): "Repeated Games with Private Monitoring: Two Players," *Econometrica*, 72(3), 823–852.

Mertens, J.-F. (2002): "Stochastic Games," in *Handbook of Game Theory, Volume 3*, ed. by R. J. Aumann and S. Hart, pp. 1809–1832. North Holland, New York.

Mertens, J.-F., S. Sorin, and S. Zamir (1994): "Repeated Games," Core discussion papers 9420–9422, Universite Chatholique de Louvain.

Milgrom, P. R., and J. Roberts (1982): "Predation, Reputation and Entry Deterrence," *Journal of Economic Theory*, 27(2), 280–312.

Monderer, D., and D. Samet (1989): "Approximating Common Knowledge with Common Beliefs," *Games and Economic Behavior*, 1(2), 170–190.

Morris, S. (2001): "Political Correctness," *Journal of Political Economy*, 109(2), 231–265.

Nash, J. F. (1951): "Non-Cooperative Games," *Annals of Mathematics*, 54(1), 286–295.

Neyman, A. (1985): "Bounded Complexity Justifies Cooperation in the Finitely Repeated Prisoners' Dilemma," *Economics Letters*, 19(3), 227–229.

——— (1999): "Cooperation in Repeated Games When the Number of Stages Is Not Commonly Known," *Econometrica*, 67(1), 45–64.

Okuno-Fujiwara, M., and A. Postlewaite (1995): "Social Norms in Random Matching Games," *Games and Economic Behavior*, 9(1), 79–109.

Osborne, M. J., and A. Rubinstein (1994): *A Course in Game Theory*. MIT Press, Cambridge, MA.

Phelan, C. (2001): "Public Trust and Government Betrayal," *Journal of Economic Theory*, forthcoming.

Piccione, M. (2002): "The Repeated Prisoner's Dilemma with Imperfect Private Monitoring," *Journal of Economic Theory*, 102(1), 70–83.

Porter, R. H. (1983a): "Optimal Cartel Trigger Price Strategies," *Journal of Economic Theory*, 29(2), 313–338.

——— (1983b): "A Study of Cartel Stability: The Joint Executive Committee, 1880–1886," *Bell Journal of Economics*, 14(2), 301–314.

Radner, R. (1980): "Collusive Behaviour in Noncooperative Epsilon-Equilibria of Oligopolies with Long but Finite Lives," *Journal of Economic Theory*, 22(2), 136–154.

——— (1981): "Monitoring Cooperative Agreements in a Repeated Principal-Agent Relationship," *Econometrica*, 49(5), 1127–1148.

——— (1985): "Repeated Principal-Agent Games with Discounting," *Econometrica*, 53(5), 1173–1198.

Radner, R., R. Myerson, and E. Maskin (1986): "An Example of a Repeated Partnership Game with Discounting and with Uniformly Inefficient Equilibria," *Review of Economic Studies*, 53(1), 59–69.

Ritzberger, K. (2002): *Foundations of Non-Cooperative Game Theory*. Oxford University Press, New York.

Rockafellar, R. T. (1970): *Convex Analysis*. Princeton University Press, Princeton, NJ.

Rogerson, W. P. (1985): "The First-Order Approach to Principal-Agent Problems," *Econometrica*, 53(6), 1357–1367.

Rosenthal, R. (1981): "Games of Perfect Information, Predatory Pricing and the Chain-Store Paradox," *Journal of Economic Theory*, 25(1), 92–100.

Ross, D. A. (2005): "An Elementary Proof of Lyapunov's Theorem," *American Mathematical Monthly*, 112(7), 651–653.

Rotemberg, J. J., and G. Saloner (1986): "A Supergame-Theoretic Model of Price Wars during Booms," *American Economic Review*, 76(3), 390–407.

Royden, H. L. (1988): *Real Analysis*, 3rd edition. Prentice Hall, Englewood Cliffs, NJ.

Rubinstein, A. (1977): "Equilibrium in Supergames," Master's thesis, Hebrew Univeristy of Jerusalem, reprinted in *Essays in Game Theory in Honor of Michael Maschler*, ed. by N. Megiddo (1994), pp. 17–28, Springer-Verlag, New York.

——— (1979a): "Equilibrium in Supergames with the Overtaking Criterion," *Journal of Economic Theory*, 21(1), 1–9.

——— (1979b): "An Optimal Conviction Policy for Offenses That May Have Been Committed by Accident," in *Applied Game Theory*, ed. by S. J. Brams, A. Schotter, and G. Schwodiauer, pp. 406–413. Physical-Verlag, Würzburg.

——— (1982): "Perfect Equilibrium in a Bargaining Model," *Econometrica*, 50(1), 97–109.

——— (1986): "Finite Automata Play the Repeated Prisoners' Dilemma," *Journal of Economic Theory*, 39(1), 83–96.

Rubinstein, A., and A. Wolinksy (1995): "Remarks on Infinitely Repeated Extensive-Form Games," *Games and Economic Behavior*, 9(4), 110–115.

Rubinstein, A., and M. E. Yaari (1983): "Repeated Insurance Contracts and Moral Hazard," *Journal of Economic Theory*, 30(1), 74–97.

Sabourian, H. (1990): "Anonymous Repeated Games with a Large Number of Players and Random Outcomes," *Journal of Economic Theory*, 51(1), 92–110.

Samuelson, L. (2006): "The Economics of Relationships," in *Advances in Economics and Econometrics: Theory and Applications, Ninth World Congress*, ed. by R. Blundell, T. Persson, and W. Newey. Cambridge University Press, New York, NY.

Sannikov, Y. (2004): "Games with Imperfectly Observable Actions in Continuous Time," Stanford University.

Schmidt, K. M. (1993a): "Commitment through Incomplete Information in a Simple Repeated Bargaining Game," *Journal of Economic Theory*, 60(1), 114–139.

——— (1993b): "Reputation and Equilibrium Characterization in Repeated Games of Conflicting Interests," *Econometrica*, 61(2), 325–351.

Sekiguchi, T. (1997): "Efficiency in Repeated Prisoner's Dilemma with Private Monitoring," *Journal of Economic Theory*, 76(2), 345–361.

Selten, R. (1978): "The Chain-Store Paradox," *Theory and Decision*, 9(2), 127–159.

Shalev, J. (1994): "Nonzero-Sum Two-Person Repeated Games with Incomplete Information and Known-Own Payoffs," *Games and Economic Behavior*, 7(2), 246–259.

Shapiro, C., and J. Stiglitz (1984): "Equilibrium Unemployment as a Worker Discipline Device," *American Economic Review*, 74(3), 433–444.

Sorin, S. (1986): "On Repeated Games with Complete Information," *Mathematics of Operations Research*, 11(1), 147–160.

——— (1995): "A Note on Repeated Extensive Games," *Games and Economic Behavior*, 9(1), 116–123.

——— (1999): "Merging, Reputation, and Repeated Games with Incomplete Information," *Games and Economic Behavior*, 29(1/2), 274–308.

——— (2002): *A First Course on Zero-Sum Repeated Games*. Springer-Verlag, Berlin.

Stahl, D. O. II. (1991): "The Graph of Prisoners' Dilemma Supergame Payoffs as a Function of the Discount Factor," *Games and Economic Behavior*, 3(3), 368–384.

Stigler, G. J. (1964): "A Theory of Oligopoly," *Journal of Political Economy*, 72(1), 44–61.

Stokey, N., and R. E. Lucas Jr. (1989): *Recursive Methods in Economic Dynamics*. Harvard University Press, Cambridge, MA.

Tadelis, S. (1999): "What's in a Name? Reputation as a Tradeable Asset," *American Economic Review*, 89(3), 548–563.

——— (2002): "The Market for Reputations as an Incentive Mechanism," *Journal of Political Economy*, 110(4), 854–882.

——— (2003): "Firm Reputation with Hidden Information," *Economic Theory*, 21(2–3), 635–651.

Thomas, J., and T. Worrall (1988): "Self-Enforcing Wage Contracts," *Review of Economic Studies*, 55(4), 541–553.

——— (1990): "Income Fluctuation and Asymmetric Information: An Example of a Repeated Principal-Agent Problem," *Journal of Economic Theory*, 51(2), 367–390.

van Damme, E. (1989): "Renegotiation-Proof Equilibria in Repeated Prisoners' Dilemma," *Journal of Economic Theory*, 47(1), 206–217.

Vieille, N. (2002): "Stochastic Games: Recent Results," in *Handbook of Game Theory, Volume 3*, ed. by R. J. Aumann and S. Hart, pp. 1833–1850. North Holland, New York.

Wang, C. (1995): "Dynamic Insurance with Private Information and Balanced Budgets," *Review of Economic Studies*, 62(4), 577–595.

Watson, J. (1994): "Cooperation in the Infinitely Repeated Prisoners' Dilemma with Perturbations," *Games and Economic Behavior*, 7(2), 260–285.

——— (1999): "Starting Small and Renegotiation," *Journal of Economic Theory*, 85(1), 52–90.

——— (2002): "Starting Small and Commitment," *Games and Economic Behavior*, 38(1), 176–199.

Wen, Q. (1994): "The "Folk Theorem" for Repeated Games with Complete Information," *Econometrica*, 62(4), 949–954.

——— (1996): "On Renegotiation-Proof Equilibria in Finitely Repeated Games," *Games and Economic Behavior*, 13(2), 286–300.

——— (2002): "A Folk Theorem for Repeated Sequential Games," *Review of Economic Studies*, 69(2), 493–512.

Wiseman, T. (2005): "A Partial Folk Theorem for Games with Unknown Payoff Distributions," *Econometrica*, 73(2), 629–645.

Zamir, S. (1992): "Repeated Games of Incomplete Information: Zero-Sum," in *Handbook of Game Theory, Volume 1*, ed. by R. J. Aumann and S. Hart, pp. 109–154. North Holland, New York.

Symbols

$|\cdot|$, Euclidean distance or cardinality of a set, 131

$\|\cdot\|$, max or sup norm, 131

$\sum_{a_i} \alpha_i(a_i) \circ a_i$, mixture assigning probability $\alpha_i(a_i)$ to a_i, 17

$(\mathcal{W}, w^0, f, \tau)$, an automaton, 29

$\hat{\alpha}^i$, i's minmax profile, 17

$\hat{\alpha}^i$, i's minmax profile with short-lived players, 63

$\hat{a}^i$, i's pure-action minmax profile, 16

$\mathbf{a}$, outcome path, 20

$\mathbf{a}(\sigma)$, outcome path induced by σ, 20

a_1^*, Stackelberg action, 465

$B : \prod_{i=1}^{n} \Delta(A_i) \rightrightarrows \prod_{i=n+1}^{N} \Delta(A_i)$, restriction on short-lived players' actions, 63

$\mathbf{B} \subset \prod_{i=1}^{N} \Delta A_i$, set of feasible profiles when $i = n+1, \ldots, N$ are short-lived, 63

$\mathcal{B}(\mathcal{W})$, set of decomposed payoffs on $\mathcal{W}$, 243

$\mathcal{B}(\mathcal{W}; \delta)$, set of decomposed payoffs on $\mathcal{W}$ given δ, 243

$\mathcal{B}(\mathcal{W}; \delta, \alpha)$, set of payoffs decomposed on $\mathcal{W}$ by α, given δ, 248

$\mathcal{B}^p(\mathcal{W})$, set of payoffs decomposed on $\mathcal{W}$ using pure actions, 248

bd $\mathcal{W}$, boundary of $\mathcal{W}$, 294

$B_\varepsilon(v)$, open ball of radius ε centered at v, 99

$\overline{B}_\varepsilon(v)$, closed ball of radius ε centered at v, 247

χ_C, indicator or characteristic function of C, 151

$\overline{\mathcal{A}}$, closure of $\mathcal{A}$, 39

co$\mathcal{A}$, convex hull of $\mathcal{A}$, 15

$\Delta(A_i)$, set of probability distributions on A_i, 15

ext$\mathcal{W}$, set of extreme points of $\mathcal{W}$, 251

e_i, ith-standard basis vector, 254

$\mathcal{E}(\delta)$, set of PPE payoffs, 231

$\mathcal{E}^E(\delta)$, set of equilibrium payoffs of a repeated extensive form, 163

$\mathcal{E}^p$, set of pure-strategy PPE payoffs, 37

$\mathcal{E}^s(\delta)$, set of strict PPE payoffs, 302

$\mathcal{F}^*$, feasible and individually rational payoffs, 17

$\mathcal{F}^{\dagger p}$, feasible and pure-action individually rational payoffs, 16

$\mathcal{F}^\dagger$, feasible payoffs, 15

$\mathcal{F}^p$, pure-action individually rational payoffs generated by pure actions, 16

$f : \mathcal{W} \to \prod_i \Delta(A_i)$, the output function of an automaton, 29

$\gamma : A \to \mathbb{R}^n$, continuation values, 37

$H(\lambda, k) = \{v \in \mathbb{R}^n : \lambda \cdot v \leq k\}$, half space, 274

$H^*(\lambda) = H(\lambda, k^*(\lambda))$, maximal half space, 277

$\mathcal{H}^t \equiv A^t$, period t histories under perfect monitoring, 19

$\mathcal{H} \equiv \cup_{t=0}^{\infty} \mathcal{H}^t$, set of all histories under perfect monitoring, 19

$\tilde{\mathcal{H}}^t = (S \times A)^t \times S$, set of all ex post histories in a dynamic game, 175

$\mathcal{H} \equiv \cup_{t=0}^{\infty} Y^t$, set of all public histories under public monitoring, 226

$\mathcal{H}_i \equiv \cup_{t=0}^{\infty} (A_i \times Y)^t$, set of i's private histories under public monitoring, 227

$\mathcal{H}_i \equiv \cup_{t=0}^{\infty} (A_i \times Z_i)^t$, set of i's private histories under private monitoring, 394

int$\mathcal{A}$, interior of the set $\mathcal{A}$, 83

$\langle \cdot, \cdot \rangle$, inner product on $\mathbb{R}^m$, 254

$k^*(\alpha, \lambda)$, 276

$k^*(\lambda)$, 277

$\lambda \in \mathbb{R}^n \setminus \{\mathbf{0}\}$, direction, 273
$\mathscr{M}$, the "limit" set, 277
$\mathscr{M}^p$, the "limit" set with pure actions, 278
$M = \max_{i,a} u_i(a)$, 26
$m = \min_{i,a} u_i(a)$, 26
$\mathbb{N} \equiv \{1, 2, \ldots\}$, the natural numbers, 102
$\mathbb{N}_0 \equiv \mathbb{N} \cup \{0\}$, 409
$\mathbf{P}$, measure over outcomes, 485
π, private monitoring distribution, 394
φ_i, Bayesian belief updating function, 407, 561, 573
rank R, rank of the matrix R, 306
ρ, public monitoring distribution, 226
σ_i, repeated game strategy for i, 19
$\sigma_i|_{h^t}$, continuation strategy induced by h^t, 20
supp(α), support of distribution α, 67
$\tau : \mathscr{W} \times A \to \mathscr{W}$, transition function of an automaton, 29
$\xi(\alpha_1)$, simple commitment type committed to $\alpha_1 \in \Delta(A_1)$, 463
$\xi(\hat{\sigma}_1)$, commitment type committed to $\hat{\sigma}_1 : \mathscr{H}_1 \to \Delta(A_1)$, 463
$u(\alpha)$, vector of stage-game payoffs, 15
$u(\alpha)$, vector of long-lived players' stage-game payoffs, 242
U, payoff function in infinitely repeated game, 21, 227

$V : \mathscr{W} \to \mathbb{R}^n$, value function of automaton, 32, 232
$\bar{v}_i$, i's maximum feasible payoff with short-lived players, 67
$\underline{v}_i$, i's minmax payoff, 17
$\underline{v}_i$, i's minmax payoff with short-lived players, 63
$\underline{v}_i^p$, i's pure-action minmax payoff, 16
v_1^{**}, a reputation lower bound, 481
v, payoff profile, 15
v_1^*, pure-action Stackelberg payoff, 465
$v_1^*(a_1) \equiv \min_{a_2 \in B(a_1)} u_1(a_1, a_2)$, one-shot bound, 474
$\underline{v}_1(\xi_0, \mu, \delta)$, infimum over normal type's payoff in any Nash equilibrium, given $\mu \in \Delta(\Xi)$ and δ, 474
Ω, space of outcomes in reputation games, 470
$\mathscr{W}$, set of states of an automaton, 29
ω, a realization of the public correlating device, 18
$\omega \in \Omega$, outcome in reputation games, 470
ξ^*, (pure-action) Stackelberg type, 465
Ξ, set of types, 463
Ξ_1, set of payoff types, 463
Ξ_2, set of commitment types, 463
Y, set of public signals, 226
Z_i, set of private signals, 394

Index

Abreu, D., 30, 37, 53, 54, 59, 87, 90, 101, 134, 141, 226, 241, 245, 248, 251, 254, 326, 377, 453, 546
action-free strategy, 395
adverse selection,
    approach to reputations, 459–463
    relationships, 155–158
    repeated games, 305, 309, 354–364
Ahn, I., 152
Alaoglu's theorem, 254
Aliprantis, C. D., 252, 254, 255
almost perfect monitoring, 394, 441
almost public monitoring, 387, 415
    minimally private, 416
    infinitely repeated prisoners' dilemma, 397–404, 425–427
    rich monitoring, 432–434
    two-period example, 387–389
    uninterpretable signals, 425
almost public monitoring games
    best responses, 421
    bounded recall, 398, 415, 423–425
    connectedness, 436
    continuation values, 419–420
    coordination failure, 425–434
    equilibria in nearby games, 421–423, 437–441
    folk theorem, 441–443
    incentives to deviate, 427–428
    nearby games, 418–423
    patient players, 434–441
    patient strictness, 435–437
    payoffs, 418–419
    rich monitoring, 432–434
    separating profiles, 428–432
    unbounded recall, 425–434
Al-Najjar, N. I., 271, 570
anonymous players, 61–62, 269–271, 492–493
Aoyagi, M., 524
approximate equilibria, finitely repeated games, 118–120
APS (Abreu, Pearce, and Stacchetti) technique. *See* decomposability; enforceability; self-generation
asymmetric profiles, symmetric payoffs from, 267
asymptotic beliefs, reputations, 494–496
asymptotic equilibrium play, reputations, 497–500
asynchronous move games, 169–172
Athey, S., 228, 355, 359
Atkeson, A., 365
Aumann, R. J., 18, 73, 121, 252, 334, 544, 549
Ausubel, L. M., 181
automata, 29–31
    accessible states, 30, 31, 230
    connected, 436
    continuation profiles, 30
    grim trigger, 31
    inclusive, 446
    individual, 30
    minimal, 230
    perfect monitoring, 29
    private monitoring, 396, 445
    public correlation, 31
    public monitoring, 230
    recursive structure of public strategies, 230
    state, 29, 178
    tit-for-tat, 35
Axelrod, R. M., 1

bad reputations, 600–610
  captive consumers, 608
  good firms, 607
  incomplete information, 603
  overproviding firms, 603
  reputation failure, 603
  stage game, 600
  repeated game, 601
bad types, modeling reputations, 580–581
Bagwell, K., 204, 228, 355, 359, 389
Baker, G., 372
Banach limits, preferences, 74
bang-bang result, 251–255
  failure of, 264
bargaining, reputations and, 546–547
Barro, R. J., 459
Battigali, P., 467
battle of the sexes, 21–22
  consistent bargaining equilibrium, 141–142
  reputations with long-lived players, 523
belief-free equilibria
  definition and examples, 445–453
  further characterization and folk theorems, 455–456
  idiosyncratic consumers, 585–586
  modeling reputations, 578–579, 581
  partial folk theorem for perfect monitoring, 450–451
  partial folk theorem for private monitoring, 453
  payoffs have product structure, 446
  private monitoring, product-choice game, 410–413
  private strategies in public monitoring, 343
  repeated prisoners' dilemma with perfect monitoring, 447–451
  repeated prisoners' dilemma with private monitoring, 451–453
  strong self-generation, 453–456
beliefs, private monitoring games, 396
Benabou, R., 493
Benoit, J., 90, 112, 115, 116, 122, 133
Bernheim, B. D., 108, 110, 134, 138, 162
best response, ε-confirmed, 480
Bhaskar, V., 385, 391, 394, 404, 407, 410
Billingsley, P., 486, 502
binding moral hazard
  inefficiency due to, 261, 281

  role of short-lived players, 281–282
  review strategies, and, 381
Binmore, K., 3, 9
Blackwell, D., 249
Blackwell comparison of experiments, 249
Börgers, T., 120
Bond, E. W., 181
Border, K. C., 252, 254, 255
bounded recall
  public profiles with, 398, 423–425
  reputations with long-lived players, 544–546
  scope of, strategies, 425
bounding perfect public equilibrium (PPE) payoffs
  binding moral hazard, 281–282
  decomposing on half-spaces, 273–278
  inefficiency of strongly symmetric equilibria, 278–280
  limit set of PPE payoffs, 277, 293–298
  prisoners' dilemma, 282–292
  short-lived players, 280–282
  upper bound on payoffs, 280
building a reputation, perfect monitoring games, 470–474
buying reputations, 610–618

canonical public monitoring games, 229, 479
  temporary reputations, 496
Canzoneri, M. B., 459
Carlton, D. W., 17
Carmichael, H. L., 155
carrot-and-stick punishment, oligopoly, 56–60
Celentani, M., 478, 513, 514, 524, 534
chain store game
  extensive form, 163–165
  finitely repeated game, 550–554
  reputations with long-lived players, 519
chain store paradox, 549
Chan, J., 540
Chari, V. V., 208
Chatterjee, K., 546
"chicken," game of, 18
Chung, K. L., 503
Clemhout, S., 11
coarsening of a partition, 190
Coase conjecture, 185
Coate, S., 121

Index

coherent consistency, dynamic games, 188–190
Cole, H. L., 425, 462
commitment problem, 206
commitment types, 463–466
    action, 463
    behavioral, 463
    punishing, 531–533
    simple, 463
common consumers and reputations, 584–596
    competitive markets, 594–596
    continuity at boundary and Markov equilibria, 590–594
    Markov strategy, 586–587
    replacements, 588–590
    reputations, 587–588
common discount factors, repeated game, 21–22
common interest game, 534
common interests, 534
    reputations with long-lived players, 534–537
    bounded recall, 544–546
common resource pool, 178–181
competitive markets, common consumers, 594–596
Compte, O., 12, 397, 404
conditionally-independent monitoring, 387, 389–393
conflicting interests, 515
    chain store game, 519
    prisoners' dilemma, 518–519
    product-choice game, 519
    reputations with long-lived players, 515–521, 537–539
    strictly, 541–543
    ultimatum game, 519–520
Conklin, J., 247
Conlon, J., 550, 554, 558
consistent bargaining equilibrium, 141–143
    battle-of-the-sexes game, 141–142
    prisoners' dilemma, 142–143
consistent partitions, dynamic games, 187–188
consumption dynamics and risk sharing, perfect monitoring, 208–222
    imperfect monitoring, 365–370
continuation promises, credible, 32–36

continuation strategy, 20
    automaton representation, 30
continuations, normalized, 275
continuum action case, 16, 226
continuum action space, 16, 226
contracts, incomplete
    repeated game, 372–374
    stage game, 371–372
Cooper, R., 121
coordinate direction, 275
coordination failure, unbounded recall, 425–434
coordination game
    asynchronous moves, 170
    efficiency and risk dominance, 121
correlated equilibrium, 334
correlated price shocks, price wars, 203–204
Coslett, S. R., 348
Cournot oligopoly model. *See* oligopoly
Cramton, P., 181
Crawford, V. P., 12
credible continuation promises
    perfect monitoring, 32–36
    public correlation, 34
Cripps, M. W., 493, 494, 500, 501, 513, 521, 533, 534, 539, 541, 543, 548, 550

Dasgupta, A., 108, 110
decomposability, 37–40, 241–249. *See also* enforceability; self-generation
    dynamic games, 194
    factorization, 245
    half-space, on, 274
    hidden short-lived players, 242
    perfect public equilibrium (PPE) payoffs, 247–248
    prisoners' dilemma, 40–51
    pure-action, 37, 39
    pure strategy restriction, 248–249
    strong, 454
    subgame, 312
    symmetric incomplete information, 319
Dekel, E., 541, 543
Deneckere, R. J., 181
Diamond, D. W., 121, 158
dimension of a convex set, 83
direction
    coordinate, 275
    pairwise, 275

discount factors
  common, 21–22
  declining, 107–112
  different, 21–22
  role in perfect monitoring games, 477
dispensability, public correlation, 83, 96–101
domination
  strict, 15
  weak, 15
Dow, J., 462
durable good monopoly problem, 181–186
Dutta, P., 87, 90, 101, 196, 197, 199
Dybvig, P. H., 121
dynamic games, 174–199
  coherent consistency, 188–190
  consistent partitions, 187–188
  decomposability 194
  enforceable, 194
  equilibrium, 192–199
  folk theorem, 195–199
  foundations, 186–192
  incomplete information, 192
  Markov equilibrium, 190–192
  repeated games with random states, 194–195
  self-generation, 194
  structure of equilibria, 192–195

efficiency wages, 155
efficient, 15
  constrained strongly, 305
  Pareto, 15
  strongly, 15
Ellickson, R. C., 9
Ellison, G., 145, 151, 348, 409
Ely, J. C., 445, 447, 448, 450, 453, 455, 600, 603, 604, 607, 608
enforceability, 37, 242, 275. *See also* decomposability; self-generation
  asymmetric, 285, 291–292
  ex ante, 355
  identifiability, 305–309
  informativeness of signals, 302
  in the direction λ, 275
  interim, 355
  mixed action, 267
  orthogonal, 275
  strong, 454
  subgame, 312
  supports, 454

English, W. B., 462
ε-equilibrium, 118
ε-confirmed best response, 480
equal discount factors and reputations, 533–548
  bounded recall, 544–546
  common interests, 534–537
  conflicting interests, 537–539
  reputations and bargaining, 546–547
  strictly conflicting interests, 541–543
  strictly dominant action games, 540–541
equilibrium
  approximate, 118–120
  correlated, 334
  dynamic games, 192–199
  in symmetric incomplete information games, 318–320
  Markov, 177, 191, 587
  Markov perfect, 177
  Nash equilibrium, 22, 62–63, 466, 479
  renegotiation-proof, 123
  sequential, 147, 319, 329, 330, 395
  sequential, in private strategies, 329–330
  subgame-perfect equilibrium, 23, 63, 163, 312
  strong renegotiation-proof, 138
  weak Markov, 192, 566
  weak renegotiation-proof, 134
Evans, R., 134, 139, 531, 532, 533
extensive-form games, 162, 311
extreme point, 251

factorization, 245
Farrell, J., 134, 136, 137, 138, 139
feasible payoffs, 15
filtration, 485
finite-horizon models, 105
finitely repeated games, 112–133, 549–566
  approximate equilibria, 118–120
  chain store game, 550–554
  mutually minmaxing in two-player game, 116–118
  prisoners' dilemma, 119, 554–560
  product-choice game, 560–566
  renegotiation, 122–133
Fisher, D., 1
folk theorem, 69
  implications, 72–73
  interpretation, 72–76
  patience and incentives, 75–76

patient players, 73–75
short period length, 75, 326–328
folk theorem, perfect monitoring, 83, 90, 101, 303
  belief-free equilibrium, 450
  convexifying equilibrium payoff set without public correlation, 96–101
  dynamic games, 195–199
  self-generation, prisoners' dilemma, 41–44
  mixed-action individual rationality, 101–104
  more than two players, 80–87
  multiplayer counterexample, 80–81
  Nash reversion, 76
  nonequivalent utilities (NEU), 87–91
  long-lived and short-lived players, 91–96, 303–305
  player-specific punishments, 82–87
  pure-action, for two players, 76–80
  pure-minmax perfect monitoring, 83–87
  pure-minmax perfect monitoring, NEU, 90
  rank conditions, 303–305
folk theorem, private monitoring,
  almost public monitoring, 441–443
  belief-free equilibrium, 453, 455–456
folk theorem, public monitoring, 301
  enforceability and identifiability, 305–309
  identifiability and informativeness, 302
  Nash-threat, 309
  pairwise identifiability, 306
  product structure, games with, 311
  pure strategy PPE, 298
  rank conditions, 298–303, 305–309
  repeated extensive form, 313–315
  strict PPE, 302
  symmetric incomplete information, 320–326
Forges, F., 11
forgiving grim triggers, prisoners' dilemma, 264–267, 404, 426–427
forgiving strategies, 235–239, 398–400, 423–425
frictionless market, 153–154
Friedman, A., 11
Friedman, J. W., 76, 114
Fudenberg, D., 16, 27, 69, 80, 83, 91, 99, 120, 191, 192, 271, 273, 281, 298, 301, 302, 306, 309, 330, 454, 464, 466, 467, 478, 479, 488, 494, 513, 514, 524, 534, 549, 607, 608
full insurance allocations, risk sharing, 210–212, 366
full rank
  individual, 298
  pairwise, 291, 299
full support, 226
full-dimensionality condition, 83, 116, 303

Gale, D., 299, 300
game. *See* battle-of-the-sexes; chain-store; common interest; conflicting interests; coordination; oligopoly; prisoners' dilemma; product-choice; purchase; strictly conflicting interests; strictly dominant action; ultimatum
game state, notion of state, 178
garbling, 250
generating function, 243
Ghosh, P., 155
Gianni Schicchi, 1–2
Gibbons, R., 372
gift exchange, 155
Glicksberg, I. L., 15
good firms, modeling reputations, 607–608
good types, modeling reputations, 581–584
Gordon, D. B., 459
Gossner, O., 116
Green, E. J., 228, 271, 301, 347, 349, 350, 385
Greif, A., 9
grim trigger, 52, 233, 400
  automaton representation, 31
  forgiving, 264–267, 404, 426–427
  one-shot deviation, 24
  public correlation, 239–241
Gul, F., 546, 547

half-space
  decomposability on, 274
  maximal, 277, 293
Haltiwanger, J., 204
Hamermesh, D. S., 1
Harrington, J. E. Jr., 152, 204
Harris, C., 120
Harsanyi, J. C., 121, 393, 445
Hart, S., 502
Hausdorff metric (distance), 123, 131, 293, 294

Hertel, J., 365
hidden actions, principal-agent problems, 370–371
histories, 19
  announcement, 366
  ex ante, 175, 210, 366
  ex post, 175, 210
  private, 394, 478
  public, 226–227, 466, 478–479
Hörner, J., 445, 455, 456, 594
Holmström, B., 462, 579
horizons
  finite or infinite, 105–106
  uncertain, 106–107
Hrbacek, K., 190

identifiability, 302, 305–309
idiosyncratic consumers
  belief-free equilibria, 585–586
  reputations, 569–576
idiosyncratic signals
  belief-free equilibria, 585–586
  modeling reputations, 570–571
  small players with, 492–493
impact of increased precision on PPE, 249–251
imperfect monitoring. *See* public monitoring
inclusive strategy profile, belief-free equilibria, 446
incomplete contracts
  repeated game, 372–374
  stage game, 371–372
incomplete information games
  Markov strategy, 192
  reputations, 460
  symmetric, 316–326
inconsistent beliefs, two-period game, 338–340
independent monitoring, conditionally belief-based equilibrium for infinitely repeated prisoners' dilemma, 404–410
  two-period game, 389–392
  public correlation, 410
independent price shocks, price wars, 201–203
individual full rank, 298
individual rationality, strict, 16
inefficient payoff, 15

inefficiency
  binding moral hazard, 261, 281–282
  repeated prisoners' dilemma, 239–241, 282–289
  strongly symmetric equilibria, 239–241, 264–269, 278–280
information sets, 19
insurance
  full, allocations, 210–212, 366
  partial, 212–213, 366
intertemporal consumption smoothing, 208–222, 365–370
Israeli, E., 533

Jackson, M. O., 550
Jech, T., 190
John, A., 121
jointly controlled lotteries, public correlation, 18
Judd, K. L., 247

Kalai, E., 30, 487, 550
Kandori, M., 12, 145, 151, 152, 204, 249, 250, 291, 298, 335, 340, 346, 392, 396, 397
Kehoe, P. J., 208
Kocherlakota, N. R., 208, 209, 425
Koeppl, T. V., 208
Kreps, D. M., 91, 147, 228, 330, 395, 469, 549, 552, 554, 566
Krishna, V., 90, 112, 115, 116, 122, 133
Kydland, R.E., 206

Lagunoff, R., 169
large players, 61. *See also* anonymous players
Laroque, G., 493
Lee, L., 348
Lehrer, E., 22, 119, 228, 487
Levin, J., 372
Levine, D. K., 120, 191, 192, 271, 281, 298, 301, 302, 306, 309, 330, 464, 466, 467, 478, 479, 488, 494, 513, 514, 524, 534, 607, 608
Ligon, E., 208
likelihood ratios, 349–350, 371
limit-of-means preference, patient players, 73–75
Ljungqvist, L., 208, 209, 219, 365, 459
long-lived and short-lived players, 61–66
  anonymity assumption, 61–62

Index

belief-free equilibria, 410–413
binding moral hazard, 261, 281–282
constraints on payoffs, 66–67, 280–282
failure of bang-bang, 264
folk theorem, 91–96, 303–305
hidden short-lived players, 242
inefficiency due to binding moral hazard, 261, 281–282
minmax payoffs, 63–66
one-shot deviation principle, 63
product-choice game, 64–65, 255–264
reputations, 459–510
Stackelberg payoff, 66
upper bound on payoffs, 66–67, 280
long-lived (long-run) players. *See also* long-lived and short-lived players
belief-free equilibria, 413
failure of reputation effects, 512–515
folk theorem with, 303
reputations, 511–548
long-run players. *See* long-lived and short-lived players
lost consumers
bad reputations, 600–610
purchase game, 599–600
Loury, G. C., 121
low-effort equilibrium, modeling reputations, 574
Lucas, R. E. Jr., 365, 436
Luciano, R., 1
Lyapunov's convexity theorem, 251, 252

machine. *See* automata
MacLeod, W. B., 155
Mailath, G. J., 44, 120, 163, 172, 334, 397, 416, 425, 431, 434, 435, 443, 462, 493, 494, 500, 501, 548, 550, 593, 615
markets
competitive, 594–596
for reputations, 610–618
frictionless, 153–154
modeling reputations, 570–573
Markov equilibrium, 177
maximally coarse consistent partition, 190–192
common consumers, 587, 590–594
dynamic games, 177–178
nonexistence in incomplete information games, 566
foundations, 190–192
incomplete information games, 192
stationary, 177
weak, 192, 566
Markov perfect equilibrium, 177
Markov state, notion of state, 178
Markov strategy profile, 191, 586–587
martingale, 485
martingale convergence theorem, 485, 486
Maschler, M. B., 18
Maskin, E., 69, 80, 83, 91, 99, 134, 136, 137, 138, 139, 169, 186, 189, 190, 192, 298, 301, 302, 306, 309, 372, 381, 549
Masso, J., 549
matching game. *See* random matching
Matsui, A., 169
Matsushima, H., 12, 273, 298, 394, 396, 397, 456
Matthews, S. A., 334
measurable, 187
under ξ, 419
mechanism design approach, repeated games of adverse selection, 359
merging, 487
Mertens, J.-F., 11, 19
Milgrom, P., 9, 326, 377, 549, 554
minimal automata, public profiles, 230–231
minmax
mixed-action, 17
mutual, 77
punishments, 80
pure-action, 16
with short-lived players, 63
minmax-action reputations, 515–521
mixed-action Stackelberg types, 465–466
mixed actions, enforcing, 267–269, 284–287, 291–292
mixtures, observable, 70, 76
modeling reputations, 567
alternative model of reputations, 568–576
alternative model, comparison, 576
bad reputations, 600–610
bad types, 580–581
belief-free equilibria with idiosyncratic consumers, 585–586
common consumers, 584–594
competitive markets, 594–596
discrete choices, 596–599
good types, 581–584
idiosyncratic consumers, 568–576
lost consumers, 599–610

modeling reputations (*continued*)
  markets for reputations, 610–618
  Markov strategy, 586–587
  motivation, 567–568
  patience, role of, 570
  product-choice game, 460, 461, 568–569, 580–581, 596–597
  purchase game, 599–600
  replacements, 572, 573–576, 588–590
  replacements, interpreting, 569–570
  replacements, role of, 576–579
Monderer, D., 447
monotone likelihood ratio property, 351, 371
monotonicity of PPE payoffs, 46–49, 247–248
moral hazard, binding, 281–282
  inefficiency, 261, 281–282
moral-hazard mixing game, 281
Morris, S., 12, 397, 416, 425, 431, 434, 435, 443
multimarket interactions, 161–162
multiple short-lived players, reputations, 476–477
Murphy, K. J., 372
Myerson, R., 381

Nash, J. F., 15
Nash equilibrium, 22
  imperfect monitoring reputations, 479
  one-shot deviation principle, failure of, 28
  perfect monitoring reputations, 466
  short-lived players, 62–63
Nash reversion, 52
  subgame-perfect equilibrium, 76–77
  trigger strategies, 52
Nash-threat folk theorem, 309
NEU, 88, 311
Neyman, A., 30, 107
Nocke, V., 172
non-equivalent utilities. *See also* NEU
  folk theorem, 90
nonstationary-outcome equilibria, 194–195
  repeated games of adverse selection, 360–364
  risk sharing, 219–222
no observable deviations, 395
normalized continuations, 275

Obara, I., 44, 291, 340, 346, 391, 404, 407, 410

observable mixtures, 70, 76
Okuno-Fujiwara, M., 152
oligopoly, 4–5
  bang-bang, 348
  constructing equilibria, 54–60
  carrot-and-stick punishment, 56–60
  folk theorem, 71–72
  imperfect collusion, 352–354
  imperfect public monitoring, 347–354
  optimal collusion, 348–350
  price wars, 201–204
  repeated adverse selection, 355–364
  simple strategies, 54–60
  trigger prices, 350–352
Olszewski, W., 445, 455, 456
one-shot bound, 474
one-shot deviation, 24, 231, 396
  prisoners' dilemma examples, 24–25
  profitable, 24
one-shot deviation principle,
  automata version, 33, 232, 397
  failure for Nash, 28
  perfect monitoring, 24–28
  private monitoring, 397
  public monitoring, 231–232
  short-lived players, 63
optimal collusion, oligopoly with imperfect monitoring, 348–350
optimal penal code, 52–53
Osborne, M. J., 27, 30, 74, 105
outcome path, 20
overtaking preferences, patient players, 73–74

pairwise direction, 275
pairwise full rank, 299
pairwise identifiability, 306
Pareto efficient, 15
partial insurance, risk sharing, 212–222, 365–370
partnership game, prisoners' dilemma, 3
path of play, 20
patience
  Banach limits, 74
  limit of means, 73–74
  modeling, 73–75
  overtaking, 73–74
  role in folk theorem, 75–76
  role in reputations, 477, 570

patient
    incentives, 238–239, 435–437
    strictness, 435–437
Pauzner, A., 22
payoff-relevant histories, 191
payoff types, 463–465, 469
Pearce, D., 37, 134, 141, 226, 241, 245, 248, 251, 254, 326, 377, 453
Peleg, B., 134
penal codes, 52
    optimal, 52–53
perfect monitoring examples
    chain store, 550–554
    oligopoly, 54–60, 201–204
    price wars, 201–204
    prisoners' dilemma, 40–51, 447–451, 554–560
    product-choice game, 64, 256–260, 580–584
    risk sharing and consumption dynamics, 208–222
    short-lived players, 256–260
    three-player counterexample, 80–81
    time consistency, 204–208
perfect monitoring games. *See also* reputations under perfect monitoring
    antomaton representations of strategy profiles, 29–31
    canonical repeated game, 15–24
    credible continuation promises, 32–36
    existence of equilibrium, 23
    folk theorem, 83, 90, 101, 303
    generating equilibria, 37–51
    long-lived and short-lived players, 61–67, 91–96, 280–282, 303–305
    mixed-minmax folk theorem, 101–104
    one-shot deviation principle, 24–28
    public correlation, 17–19
    pure-minmax folk theorem NEU, 90–91
    self-generation, 37–51
    simple strategies and penal codes, 51–54
    stage game, 15–17
    subgame-perfect equilibrium, 22–24
    two-player folk theorem, 77–78
perfect public equilibrium (PPE). *See also* bounding perfect public equilibrium (PPE) payoffs
    bang-bang property, 251
    characterizing limit set of PPE payoffs, 293–298
    definition, 231
    inefficiency of strongly symmetric, 278–280
    limit set, 277
    monotonicity of PPE payoffs, 46–49, 247–248
    patiently strict, 437
    pure strategy, 248–249, 278, 298
    recursive structure, 228–232
    scope of bounded recall strategies, 425
    strict, 231
    strongly symmetric, 231, 239, 254, 278–280
    uniformly strict, 232
Perloff, J. M., 17
personal histories, 147
Pesendorfer, W., 271, 478, 513, 514, 524, 534, 541, 543
Phelan, C., 462, 584
Piccione, M., 445
player-specific punishments, 82
Porter, R. H., 228, 301, 347–350, 385
Postlewaite, A., 120, 152
Prescott, E. C., 206
price wars
    correlated price shocks, 203–204
    independent price shocks, 201–203
principal-agent problems, relational contracting, 370–374
    bang-bang, 374
    hidden actions, 370
    incomplete contracts, 371
    selling the firm, 371
principal-agent problems, review strategies, 374–381
    binding moral hazard, 381
    punishment phase, 376
    risk aversion, 374
    repeated partnership game, 381
    review phase, 376
prisoners' dilemma, 3–4
    almost public monitoring, 397–404, 425–427
    anonymity assumption, 61–62, 269–271
    asymmetry, efficient, 291–292
    belief-free equilibria, perfect monitoring, 447–451
    belief-free equilibria, private monitoring, 451–453

prisoners' dilemma (*continued*)
  bounds on efficiency: mixed actions, 284–287
  bounds on efficiency: pure actions, 282–284
  characterization with two signals, 287–289
  decomposability, perfect monitoring, 40–51
  decomposability, public monitoring, 264–269
  determinacy of equilibria, 346, 448–449
  efficiency with three signals, 289–291
  efficient asymmetry, 291–292
  enforceability, perfect monitoring, 40–51
  enforceability, public monitoring, 264–269
  enforcing mixed-action profile, 267–269
  finitely repeated game, 119, 554–560
  folk theorem, perfect monitoring, 70–71
  forgiving strategies, 235–239, 398–400
  forgiving grim trigger, 264–267, 404, 426–427
  grim trigger, 52, 233–235, 239–241, 264–267, 404, 426–427
  imperfect monitoring, 5–6, 232–241, 264–269, 282–292
  mixed-action profile, enforcing, 267–269
  mutual minmax is Nash-reversion, 80
  one-shot deviation, 24, 27–28
  partnership game, 3
  partnership game, extending, 159
  patient incentives, 238–239
  perfect monitoring, 40–51
  private monitoring, 451–453
  private strategies, 340–346
  public correlation, 49–51
  public monitoring, 5–6, 232–241, 264–269, 282–292
  random matching, 145–152
  renegotiation, 134–135
  reputations with long-lived players, 518–519, 554–560
  self-generation, perfect monitoring, 40–51
  self-generation, public monitoring, 264–269
  strongly symmetric behavior implying inefficiency, 239–241, 264–269
  symmetric payoffs from asymmetric profiles, 267
  three versions, 142–143
  tit-for-tat, 35
private beliefs, temporary reputations, 500
private monitoring,
  almost perfect, 387, 394, 441
  almost public, 387, 397, 415
  conditionally independent, 387, 404–410
  ε-perfect, 394
  full support, 400
  minimally, 416
private monitoring examples
  belief-free, 410–413, 451–453
  prisoners' dilemma, 397–410, 425–427, 451–453
  product-choice, 410–413
  two period, 385–394
private monitoring games. *See also* reputations under imperfect monitoring
  automata and beliefs, 396
  basic structure, 394–397
  behavior strategy, 394
  belief-free example, 410–413
  belief-free equilibrium, 445–453
  folk theorem, 453
  one-shot deviation principle, 397
  repeated prisoners' dilemma with, 451–453
  sequential equilibrium, 395–396
  symmetric, 451
private profile, private monitoring, 415
  induced by public profile, 419
private strategies in public monitoring games, 228, 329–346
  infinitely repeated prisoners' dilemma, 340–346
  sequential equilibrium, 329–330
  two-period examples, 334–340
private values, adverse selection, 354
product-choice game, 7–8
  belief-free equilibria, 410–413
  binding moral hazard, 261
  finitely repeated game, 560–566
  folk theorem, 71
  good and bad commitment types, 580–584
  Stackelberg payoff, 467–469, 481
  minmax payoffs, 64–65
  modeling reputations, 460, 461, 568–569, 580–581, 596–597
  perfect monitoring, 255–260
  private monitoring, 410–413

public monitoring, 260–264
  renegotiation, 137
  reputations with long-lived players, 519
  reputations with short-lived players, 460, 461, 469, 481
  short-lived players, 64, 255–264, 410–413
  temporary reputations, 499–500
product structure, games with, 309–311
public correlation, 18, 48, 239, 253, 391
  automaton representation, 31, 231
  bang-bang, 253
  canonical repeated game, 17–19
  convexifying equilibrium payoff set without, 83, 96–101
  dispensability, 83, 96–101
  generating equilibria, prisoners' dilemma, 49–51
  jointly controlled lotteries, 18
  notation, 21, 231
  self-generation, 39
public histories, 226–227, 466
  random matching, 146–147
public monitoring
  η-perfect, 441
  full support, 231, 417
public monitoring examples
  oligopoly, 347–354, 355–364
  principal-agent problem, 370–381
  prisoners' dilemma, 5–6, 232–241, 264–269, 282–292
  product-choice, 260–264
  purchase game, 482–484
  repeated adverse selection, 354–364
  risk sharing and consumption dynamics, 365–370
public monitoring games. *See also* reputations under imperfect monitoring
  anonymous players, 269–271
  bang-bang result, 251–255
  binding moral hazard, 281–282
  canonical, 229, 479
  decomposability, 241–249
  enforceability, 243, 305–309
  folk theorem, 301, 302, 309
  identifiability, 305–309
  impact of increased precision, 249–251
  public strategies and perfect public equilibria, 228–232
  rank conditions, 298–303, 305–309
  recursive structure, 228–232
  self-generation, 241–249
  small players with idiosyncratic signals, 492–493
  special cases, 227–228
  strongly symmetric behavior implying inefficiency, 239–241, 264–269, 278–280
public profile, 228, 415
  bounded recall, 398, 423–425
  connected, 436
  cyclically separating, 430
  finite, 435
  recursive structure, 228–232
  scope of bounded recall strategies, 425
  separating profiles, 428–432
punishment
  carrot-and-stick, 56
  grim, 52
  Nash reversion, 52
  player-specific, 82
punishment type, 512–514, 531–533
purchase game
  imperfect monitoring, 482–484
  lost consumers, 599–600
pure strategy restriction, 16, 226, 248

Radner, R., 119, 228, 347, 375, 381
random matching, 145–161
  adverse selection, 155–158
  building trust, 158–160
  contagion of shirking, 149–152
  frictionless matching, 153–154
  imperfect observability, 147–152
  perfect observability, 146–147
  personal histories, 147–152
  public histories, 146–147
  sequential equilibrium, 147
  starting small, 158–160
  sunk costs, 154–155
random states, repeated games with, 177, 194–195
rank conditions, 298–303
  enforceability and identifiability, 305–309
  identifiability and informativeness, 302
Ray, D., 134, 138, 155
realization equivalence
  mixed and behavior strategies, 19
  pure and public strategies, 229–230
recursive structure, public monitoring, 228–232

relational contract, 372
renegotiation
    efficiency and, 121–122, 127, 139
    finitely repeated games, 122–133
    infinitely repeated games, 134–143
    prisoners' dilemma, 134–135
    relationships, 152–153
    risk dominance, 121
    vulnerability of grim trigger, 120–121
renegotiation-proof equilibrium, 123, 373
    consistent bargaining, 141
    strong, 138
    weak, 134
repeated extensive forms
    additional subgames, 163–165
    asynchronous moves, 169–172
    folk theorem, 162, 311–315
    imperfect monitoring and, 167–168, 311
    player-specific punishments, 165–167
    simple strategies, 172–174
    weakly individual rationality, 168–169
repeated games with random states, 177, 194–195
replacements
    common consumers, 588–590
    interpreting, 569–570
    modeling reputations, 568–573
    reputation with, 573–576
    necessary for reputations, 576–579
reputation bound
    long-lived players and imperfect monitoring, 526
    long-lived players and perfect monitoring, 516, 521, 532–533
    short-lived players and imperfect monitoring, 484–485
    short-lived players and perfect monitoring, 474
    Stackelberg bound, 475–476
reputation effects, failure of, 512–514
reputations. *See also* modeling reputations
    adverse selection approach, 459–463
reputations, finitely repeated games,
    chain store game, 550–554
    prisoners' dilemma, 554–560
    product-choice game, 560–566
    temporary reputations, 550
reputations under imperfect monitoring,
    long-lived players, 524–530
    new possibilities for equilibrium payoffs, 481–482
    short-lived players, 478–493
    Stackelberg payoffs, 465–466, 480–484
    temporary, 493–510, 547–548, 550
reputations under perfect monitoring
    failure of reputation effects for long-lived players, 512–515
    long-lived players, 511–524, 531–547
    new possibilities for equilibrium payoffs, 476, 481–482
    short-lived players, 459–478
    Stackelberg payoffs, 465, 475–476, 515, 532–533
reputations with equally patient players
    bounded recall, 544–546
    common interests, 534–537, 544–546
    conflicting interests, 537–539
    reputations and bargaining, 546–547
    strictly conflicting interests, 541–543
    strictly dominant action games, 540–541
reputations with long-lived players, 511–524
    bounded recall, 544–546
    chain store game, 512–514, 519
    punishment types, 531–533
    common interests, 534–537
    conflicting interests, 515–521, 537–539
    equal discount factors, 533–547
    failure of reputation effects under perfect monitoring, 512–515
    finitely repeated prisoners' dilemma, 554–560
    payoff bounds, 526, 532–533
    public monitoring, 524–530
    minmax-action reputations, 515–521
    perfect monitoring, 515–524, 531–547
    prisoners' dilemma and conflicting interests, 518–519
    prisoners' dilemma, finitely repeated, 554–560
    product-choice game, 519
    reputations and bargaining, 546–547
    Stackelberg payoff, 515, 532–533
    strictly conflicting interests, 541–543
    strictly dominant action games, 540–541
    temporary reputations, 547–548
    two-sided incomplete information, 520–521, 523–524
    ultimatum game, 519–520

Index

reputations with short-lived players, 459–463
  asymptotic equilibrium play, 497–500
  building a reputation, 470–474
  canonical public monitoring, 479–480
  chain store, finitely repeated, 550–554
  commitment types, 463–466
  complete information payoffs, 476
  continuum short-lived player action set, 476
  diffuse prior beliefs, 476
  existence of equilibrium, 466–467
  imperfect monitoring games, 478–493
  mixed-action Stackelberg types, 465–466
  multiple short-lived players, 476–477
  new possibilities for equilibrium payoffs, 476, 481–482
  payoff or commitment types, 463–465, 469
  perfect monitoring games, 466–478
  product-choice game, 460, 461, 469, 481
  product-choice game, finitely, repeated, 560–566
  public monitoring, 479–480
  purchase game, 482–484
  reputation bound, 474, 484–485
  role of discounting, 477
  sequential equilibrium, 467
  small players with idiosyncratic signals, 492–493
  Stackelberg payoffs, 465–466, 475–476, 480–484
  Stackelberg type, 465–466
  temporary reputations, 493–510
  time consistency example, 477–478
  uniformly disappearing reputations, 496–497
resource extraction, 178–181
review strategy, 375
rich monitoring, almost public monitoring games, 432–434
Riesz representative theorem, 254
risk dominance, coordination game, 121
risk sharing and consumption dynamics, perfect monitoring, 208–222
  imperfect monitoring, 365–370
Ritzberger, K., 19
Roberts, J., 549, 554
Rockafellar, R. T., 253, 273, 277
Rogerson, W. P., 371

Rosenthal, R., 549
Ross, D. A., 252
Rotemberg, J. J., 201, 348
Royden, H. L., 254
Rubinstein, A., 27, 30, 73, 74, 75, 105, 165, 168, 228, 347, 375, 546, 547

Sabourian, H., 271
Saloner, G., 201, 348
Samet, D., 417
Samuelson, L., 10, 120, 163, 181, 462, 493, 494, 500, 501, 546, 548, 550, 593, 615
Sanchirico, C., 355, 359
Sannikov, Y., 328
Sargent, T. J., 208, 209, 219, 365, 459
Schmidt, K. M., 515, 521, 533, 539, 547
Sekiguchi, T., 44, 334, 396, 404
self-generation, 37–40, 241–249. *See also* decomposability; enforceability
  dynamic games, 194
  equilibrium behavior, 244–245
  factorization, 245
  fixed point characterization of equilibrium, 39, 246
  hidden short-lived players, 242
  incomplete information, 319
  local, 249
  multiple self-generating sets, 266
  perfect public equilibrium (PPE) payoffs, 247–248
  prisoners' dilemma, 40–51
  pure action, 37, 39
  pure strategy restriction, 248–249
  strong, 454
  subgame, 312
  symmetric incomplete information, 319
Selten, R., 121, 549
separating profile, 428–432
  cyclicly, 430–431
sequential equilibrium
  random matching, 147
  perfect monitoring games of incomplete information, 467
  private monitoring games, 395–396
  private strategies of public monitoring games, 329–330
  symmetric incomplete information games, 319
sequential rationality, 23

Shalev, J., 533
Shapiro, C., 155
Shapley, L. S., 73
short-lived (short-run) players. *See* long-lived and short-lived players; reputations with short-lived players
short period length, 75, 181, 326–328
signal interpretation, 416
simple strategies
    constructing equilibria, 51–54
    multi-player folk theorem, 84
    repeated extensive forms, 172–174
    two player folk theorem, 78
small players, 61. *See also* anonymous players
Smith, L., 87, 90, 101
Smorodinsky, R., 271
Sobel, J., 12
Sonnenschein, H., 547
Sorin, S., 11, 19, 97, 119, 167, 487, 544
Stacchetti, E., 37, 134, 141, 226, 241, 245, 248, 251, 254, 453
Stackelberg bound. *See* reputation bound
Stackelberg payoffs
    mixed action, 66, 465
    pure action, 66, 465–466
    reputations, 465–466, 475–476, 480–484
Stackelberg types,
    mixed action, 465
    pure action, 465
stage game
    Nash equilibrium, 15–16
    perfect monitoring, 15–19
    private monitoring, 394
    public monitoring, 225–226
Stahl, D. O. II., 50
Staiger, R. W., 204
Stanford, W., 30
starting small, relationships, 158–161
state, 178
    automaton, 29
    game, 174
    Markov, 178, 191
stationary Markov equilibrium, 177
stationary-outcome equilibrium, 195, 212, 357
Stearns, R. E., 18
Stigler, G. J., 385

Stiglitz, J., 155
stochastic games, 174
stochastic matrix, 250
Stokey, N., 436
strategy profiles. *See also* private strategies in public monitoring games
    automaton representations, 29–31
    continuation, 20
    inclusive, 446
    one-shot deviation, 24
    private, 228
    public, 228
    repeated game, 19–21
    review, 375
    simple, 51
strictly conflicting interests, 541–543
strictly dominant action games, 540–541
strictly dominates, 15
strong symmetry, 55, 231, 364
subgame enforceability, 312
subgame-perfect equilibrium
    perfect monitoring, 22–24
    repeated extensive forms, 163, 312
    short-lived players, 63
Suominen, M., 152
submartingale, 485
supergame, 19
supermartingale, 485
sustainable plans, 208
Swinkels, J. M., 163
symmetric behavior implying inefficiency, prisoners' dilemma, 239–241, 264–269
symmetric games, 278
    bang-bang, 254
    failure of rank condition, 301
    inefficiency of strongly symmetric equilibrium, 278–280
symmetric incomplete information games, 316–326
    decomposability, 319–320
    folk theorem, 320
    learning facilitating punishment, 317–318
    players need not learn, 316–317
    sequential equilibrium, 319
    self-generation, 319–320
symmetric payoffs from asymmetric profiles, 267
symmetry, strong, of nonstationary outcomes, 364

Index

Tadelis, S., 610
temporary reputations
  asymptotic equilibrium play, 497–500
  canonical public monitoring, 496
  changing types and, 462–463
  finitely repeated games, 550
  interpretation, 462
  long-lived players, 547–548
  payoff bounds, and, 462
  short-lived players, 493–500, 510
  uniformly disappearing reputations, 496–497
Thomas, J. P., 208, 209, 365, 513, 521, 531–534, 539
time consistency, 204–208, 477–478
Tirole, J., 16, 27, 169, 186, 189, 190, 192, 372
tit-for-tat
  Nash equilibrium, 36
  prisoners' dilemma, 35
trigger strategy profile, 52
  grim, 52
  Nash reversion, 52
two-sided incomplete information, reputations with long-lived players, 520–521, 523–524
type
  action, 463
  behavioral, 463
  changing, 462–463, 569, 572
  commitment, 463
  crazy, 464
  normal, 463
  payoff, 463

simple commitment, 463
Stackelberg, 465

ultimatum game, 519–520
unbounded recall
  failure of coordination under, 425–434
  repeated prisoners' dilemma examples, 425–427
  separating profiles, 428–432
uncertain horizons, 106–107
uniformly disappearing reputations, 496–497

Välimäki, J., 445, 447, 448, 450, 453, 600, 603, 604
van Damme, E., 135, 385, 391, 394
Vieille, N., 11

Wan, H. Y., Jr., 11
Wang, C., 365
Watson, J., 119, 158, 467
weakly dominates, 15
wedding rings, 155
Weingast, B. R., 9
Wen, Q., 91, 122, 311
Whinston, M. D., 134, 162
White, L., 172
Wilson, R. J., 147, 228, 330, 395, 469, 547, 549, 552, 554, 566
Wiseman, T., 318, 320, 325
Wolinksy, A., 165, 167
Worrall, T., 208, 209, 365

Yaari, M. E., 228, 347, 375
Yeltekin, S., 247

Zamir, S., 11, 19